AF692995

Sustainable Agricultural Development Economics and Policy

Sustainable Agricultural Development Economics and Policy

Editor

Aaron K. Hoshide

MDPI • Basel • Beijing • Wuhan • Barcelona • Belgrade • Manchester • Tokyo • Cluj • Tianjin

Editor
Aaron K. Hoshide
College of Natural Sciences,
Forestry and Agriculture
The University of Maine
Orono
United States

Editorial Office
MDPI
St. Alban-Anlage 66
4052 Basel, Switzerland

This is a reprint of articles from the Special Issue published online in the open access journal *Sustainability* (ISSN 2071-1050) (available at: https://www.mdpi.com/journal/sustainability/special_issues/Sustainable_Agricultural_Development).

For citation purposes, cite each article independently as indicated on the article page online and as indicated below:

LastName, A.A.; LastName, B.B.; LastName, C.C. Article Title. *Journal Name* **Year**, *Volume Number*, Page Range.

ISBN 978-3-0365-8178-1 (Hbk)
ISBN 978-3-0365-8179-8 (PDF)

Cover image courtesy of Gabriel Rezende Faria

Contents

About the Editor

Aaron K. Hoshide

Aaron Kinyu Hoshide, PhD, is a sustainable agriculture researcher in the College of Natural Sciences, Forestry and Agriculture at the University of Maine, USA. Dr. Hoshide conducts applied agricultural economic research on crops, ruminant livestock, and/or agro-forestry systems in Maine, USA, and Brazil, using sustainability indicators, enterprise/whole-farm budgets, risk analysis, and the integrated farm system model (IFSM). Dr. Hoshide is currently involved in estimating the economic threshold at which to rent honeybees or rely on native bees for Maine wild blueberries. He is also updating production costs for conventional and organic Maine wild blueberries for the U.S. Department of Agriculture. Dr. Hoshide is advising graduate students modeling beef production systems and conducting extension outreach to agricultural producers in Mato Grosso state, Brazil. He is excited to continue collaborating with international students and/or researchers in Brazil, Pakistan, and other countries outside of the United States.

Preface to "Sustainable Agricultural Development Economics and Policy"

Agriculture, in developing and developed nations alike, will face huge challenges over the next century in meeting human food needs and shifting preferences. Agricultural economic development, from the personal and local level to the global and industrial level, must be balanced with communal needs (e.g., food sovereignty, self-sufficiency) and address environmental challenges (e.g., climate change, ecosystem degradation). Local, national, and global policies must support sustainable agricultural economic development, while also addressing future environmental and community impacts.

This Special Issue focuses on agricultural systems and forest management in developing or developed nations in Africa, Asia, South America, and North America, and spans conversations from the personal and local level all the way up to the production and circulation of global commodities. Analytical methods from published articles are employed to focus on socio-economic surveys, field experiments, remote sensing, and public policy proposals. The research stresses that sustainable agricultural development can be economically viable while also reducing the environmental impact of agricultural activities and strengthening local communities. An improved understanding of sustainable agricultural and forestry systems can help farmers, researchers, students, and policy makers to design and implement similar systems.

From an economic perspective, sustainability can be achieved through "economies of scale". In simple terms, this involves increasing economic efficiency and agricultural productivity to spare land in the short run, reducing the need to convert natural habitats into agricultural areas. While export commodity agriculture can employ local workers, the diversified food needs of local communities may not be addressed under such systems. The agricultural development of both intensive and extensive systems may be more challenging in the future given changes in climate, agroecosystem degradation, and diminishing resource availability. Agricultural and forestry systems involving commodities may be less sustainable in the future. However, these systems can be designed to be more durable to future shocks in order to address the sustainability shortcomings of the "economies of scale" approach.

Alternatively, sustainable agricultural development can use "economies of scope", where agricultural producers diversify production and input use using systems-based approaches. While such diversification can be profitable, minimizes environmental impacts, and meets local community food needs, the use of these systems may be challenging due to the complexity of managing farms like an ecosystem, reducing input use, the need to sell directly to consumers, or a lack of available capital. Government policies can be structured to support more diversified agricultural production.

Sustainable development involves specialization and diversification. Despite the potential for global agriculture to undergo intensification, this may not be environmentally sustainable. Diversification can involve enterprise diversification and ecological intensification. Regional case studies highlighted in this Special Issue focus on diversified agricultural systems for the creation of more sustainable future food systems.

We are grateful to the efforts of all researchers who submitted manuscript submissions to this Special Issue of *Sustainability*. Your research efforts have gone a long way to improving the understanding of more sustainable agricultural and forestry systems. A special thanks to Ionut Spanu, the managing editor of this Special Issue, for his invaluable editorial and publication support over the past two years. Gabriel Rezende Faria, a journalist and public relations officer at Embrapa, Brazil, graciously provided the cover photograph for this Special Issue. Many thanks also to my family who made this work possible.

Aaron K. Hoshide
Editor

sustainability

MDPI

Editorial

Sustainable Development Agricultural Economics and Policy: Intensification versus Diversification

Aaron Kinyu Hoshide [1,2]

[1] College of Natural Sciences, Forestry and Agriculture, The University of Maine, Orono, ME 04469, USA; aaron.hoshide@maine.edu; Tel.: +1-207-659-4808

[2] AgriSciences, Universidade Federal de Mato Grosso, Sinop 78555-267, MT, Brazil

Sustainable development of agriculture in both the developed and developing world is not only dependent on economics and policy but also decisions to increase sustainability through either (1) specialization (e.g., sustainable intensification) or (2) diversification (e.g., ecological intensification), as demonstrated in the "Sustainable Development Agricultural Economics and Policy" Special Issue. Understanding the historical context of the region being evaluated is critical to selecting the most promising strategies. For example in the state of Maine USA, agricultural specialization tends to result in longer-term cycles of boom and bust, while historical diversification has been related to social movements such as the back-to-the-land movement of the 1970's and the recent local food movement over the past two decades [1]. Sustainable development can follow different pathways depending on the emphasis on either specialization or diversification.

Specialization during agricultural development is typically concentrated in specific geographic areas with optimal agricultural production compared to other production areas. However, there can be sustainability tradeoffs to such regional comparative advantages. For example California USA generates ~80% of global exports for almonds. However, there is increased global production risk due to drought in addition to the retaliatory trade tariffs [2]. Another example of tradeoffs in agricultural specialization is sugarcane production in southeastern Brazil. Brazil is the world's largest sugarcane producer but sandy soils in this major production area limit crop yields due to the lower water holding capacity of these soils [3].

Agricultural specialization can also be more dependent on external inputs, government support, and interdependence with other countries. China is a great example of this with potential for sustainable agricultural intensification limited by water availability and the need for more investments in irrigation [4]. Additionally, China's shift from more labor intensive to more capital intensive agricultural production requires substantial investments in agricultural mechanization which is influenced by economics, government policies, and environmental goals [5]. Top-down government policies such as Chinese agricultural subsidies can encourage agricultural enterprises to grow more favorably [6], which can alleviate extreme poverty [7]. Agricultural specialization and comparative advantage makes global trade more critical and this is especially the case for countries along China's "Belt and Road" [8]. However, Chinese agricultural economic growth is projected to be stagnant in the future despite substantial recent growth over the past 20 years [9].

Despite the potential for global agriculture to sustainably intensify in the future, such sustainable intensification may not be environmentally sustainable. Environmental impacts of agricultural development include land use change in Brazil's Midwest where native habitat has been converted to commodity crops (e.g., soybeans, maize, cotton) at a rapid rate over the past 25 years [10]. Agricultural row crop expansion and urban development in this region of Brazil has also increased suspended sediment in rivers [11].

Addressing the economic and environmental challenges of specialized agricultural production focuses on detailed models and field experiments to help balance yield and

Citation: Hoshide, A.K. Sustainable Development Agricultural Economics and Policy: Intensification versus Diversification. *Sustainability* **2023**, *15*, 9716. https://doi.org/10.3390/su15129716

Received: 31 May 2023
Revised: 13 June 2023
Accepted: 15 June 2023
Published: 18 June 2023

profit maximization with reducing adverse environmental impacts. For example, biophysical modeling can be used to evaluate and improve sustainability. In Brazil, use of growth-stage specific regression modeling can identify factors that limit sugarcane yield such as soil water storage during the second growth phase in sandy soils [12]. Agricultural erosion modeling using GeoWEPP for crops, pasture, and natural habitat in Brazil's Midwest can be validated and used to help minimize erosion at the micro-watershed scale [13]. In-field rainfall simulator experiments can suggest which combinations of ground cover and management practices are best in minimizing erosion as was demonstrated in Brazil's Midwest region [14].

Diversification can involve both enterprise diversification as well as ecological intensification both on-farm and around the farmscape. Enterprises diversification can include other crop enterprises such as mung beans and broad beans in China, which are economically promising due to lower labor requirements [15]. Diversification of enterprises can also include non-food crops such as growing and commercializing medicinal plants used for childhood diseases in South Africa [16]. There is consumer support for such indigenous plants in West Province, South Africa [17]. Enterprise diversification can also include activities not related to crops/livestock. For example, Nigerian youth diversifying into non-agricultural sectors can increase rural development and reduce dependency on the agricultural sector [18].

Ecological intensification can involve integration of livestock and agro-forestry with crops. For example in northeast Brazil, bio-fertilization of cactus for food/feed applications in dry climates can be accomplished with cattle manure [19]. Sustainable beef systems in Brazil such as integrated crop-livestock-forest systems can reduce de-forestation pressures as well as sequester global carbon emissions and have been recently encouraged by favorable government policies such as the Brazilian Forest Code, the Low Carbon Agriculture Plan, and the National Integrated Crop-Livestock-Forest Integration policy which have been updated and/or implemented over the past two decades [20].

Sustainable development in agricultural regions also involves agro-forestry as well as preserving native forests and supporting native pollinator populations. Sustainable forest plantations have critical sustainability implications in the Republic of Congo in Africa [21]. Preservation of native forest in China is dependent on ecological forest rangers [22]. Involving government agencies such as the New England USA Department of Transportation in planting native pollinator pastures can help stabilize pollinator populations which can benefit local farmers growing pollinator dependent crops such as cranberry, blueberry, and squash [23].

Despite the promise of maintaining the diversity of small shareholders in the developing world, challenges remain. Expanding chicken production by small shareholders in Nigeria is limited by the high costs of purchased poultry feed making it challenging to produce eggs cheaply without government subsidies [24]. This suggests encouraging more local concentrated feed production for livestock [1]. Farmer outreach and extension are critical for supporting agricultural producers and agricultural professionals in adopting more sustainable agricultural systems, especially in regions where agricultural specialization is dominant such as Brazil's Midwest region [25]. Future agricultural diversification can be inspired by diversified systems of the past such as diversifying into growing livestock feed for cattle and hogs to forage in-field during the fall, as was done in Maine during the mid-20th century [1]. Similar regional case studies can be used to inspire and implement diversified agricultural systems for more sustainable future food systems.

Funding: This research received no external funding.

Institutional Review Board Statement: Not applicable.

Informed Consent Statement: Not applicable.

Data Availability Statement: Data sharing not applicable. Published journal articles cited in this editorial article can be found in the "Sustainable Development Agricultural Economics and Policy" Special Issue (https://www.mdpi.com/journal/sustainability/special_issues/Sustainable_Agricultural_Development).

Acknowledgments: Thanks to all the co-authors of manuscripts submitted to this Special Issue for continuously improving the quality of their work and for improving our understanding of sustainable agriculture and forestry.

Conflicts of Interest: The author declares no conflict of interest. Supporting entities had no role in the design of the study; in the collection, analyses, or interpretation of data; in the writing of the manuscript, or in the decision to publish the results.

References

1. Hoshide, A.K. Back to the Future: Agricultural Booms, Busts, and Diversification in Maine, USA, 1840–2017. *Sustainability* **2022**, *14*, 15907. [CrossRef]
2. Ajibade, A.; Saghaian, S.U.S. Almond Exports and Retaliatory Trade Tariffs. *Sustainability* **2022**, *14*, 6409. [CrossRef]
3. Viana, J.L.; de Souza, J.L.M.; Auler, A.C.; de Oliveira, R.A.; Hoshide, A.K.; de Abreu, D.C.; da Silva, W.M. Water Dynamics and Hydraulic Functions in Sandy Soils: Limitations to Sugarcane Cultivation in Southern Brazil. *Sustainability* **2023**, *15*, 7456. [CrossRef]
4. Li, H.; Liu, H. Climate Change, Farm Irrigation Facilities, and Agriculture Total Factor Productivity: Evidence from China. *Sustainability* **2023**, *15*, 2889. [CrossRef]
5. Li, Z.; Zhu, M.; Huang, H.; Yi, Y.; Fu, J. Influencing Factors and Path Analysis of Sustainable Agricultural Mechanization: Econometric Evidence from Hubei, China. *Sustainability* **2022**, *14*, 4518. [CrossRef]
6. Wu, L.; Hu, K.; Lyulyov, O.; Pimonenko, T.; Hamid, I. The Impact of Government Subsidies on Technological Innovation in Agribusiness: The Case for China. *Sustainability* **2022**, *14*, 14003. [CrossRef]
7. Sun, Y.; Yao, S.-N. Sustainability Trade-Offs in Media Coverage of Poverty Alleviation: A Content-Based Spatiotemporal Analysis in China's Provinces. *Sustainability* **2022**, *14*, 10058. [CrossRef]
8. Zhou, L.; Tong, G. Structural Evolution and Sustainability of Agricultural Trade between China and Countries along the "Belt and Road". *Sustainability* **2022**, *14*, 9512. [CrossRef]
9. Guo, X.; Lung, P.; Sui, J.; Zhang, R.; Wang, C. Agricultural Support Policies and China's Cyclical Evolutionary Path of Agricultural Economic Growth. *Sustainability* **2021**, *13*, 6134. [CrossRef]
10. Kraeski, A.; de Almeida, F.T.; de Souza, A.P.; de Carvalho, T.M.; de Abreu, D.C.; Hoshide, A.K.; Zolin, C.A. Land Use Changes in the Teles Pires River Basin's Amazon and Cerrado Biomes, Brazil, 1986–2020. *Sustainability* **2023**, *15*, 4611. [CrossRef]
11. Paulista, R.S.D.; de Almeida, F.T.; de Souza, A.P.; Hoshide, A.K.; de Abreu, D.C.; da Silva Araujo, J.W.; Martim, C.C. Estimating Suspended Sediment Concentration using Remote Sensing for the Teles Pires River, Brazil. *Sustainability* **2023**, *15*, 7049. [CrossRef]
12. Viana, J.L.; de Souza, J.L.M.; Hoshide, A.K.; de Oliveira, R.A.; de Abreu, D.C.; da Silva, W.M. Estimating Sugarcane Yield in a Subtropical Climate Using Climatic Variables and Soil Water Storage. *Sustainability* **2023**, *15*, 4360. [CrossRef]
13. Magalhães, W.d.A.; Amorim, R.S.S.; Hunter, M.O.; Bocuti, E.D.; Raimo, L.A.D.L.D.; da Silva, W.M.; Hoshide, A.K.; de Abreu, D.C. Using the GeoWEPP Model to Predict Water Erosion in Micro-Watersheds in the Brazilian Cerrado. *Sustainability* **2023**, *15*, 4711. [CrossRef]
14. Alves, M.A.B.; Souza, A.P.; Almeida, F.T.; Hoshide, A.K.; Araújo, H.B.; Silva, A.F.; Carvalho, D.F. Effects of land use and cropping on soil erosion in agricultural frontier areas in the Cerrado-Amazon Ecotone, Brazil, using a rainfall simulator experiment. *Sustainability* **2023**, *15*, 4954. [CrossRef]
15. Ma, J.; Qu, J.; Khan, N.; Zhang, H. Towards Sustainable Agricultural Development for Edible Beans in China: Evidence from 848 Households. *Sustainability* **2022**, *14*, 9328. [CrossRef]
16. Ndhlovu, P.T.; Omotayo, A.O.; Otang-Mbeng, W.; Aremu, A.O. Commercialization Potential of Six Selected Medicinal Plants Commonly Used for Childhood Diseases in South Africa: A Review. *Sustainability* **2022**, *14*, 177. [CrossRef]
17. Omotayo, A.O.; Ndhlovu, P.T.; Tshwene, S.C.; Olagunju, K.O.; Aremu, A.O. Determinants of Household Income and Willingness to Pay for Indigenous Plants in North West Province, South Africa: A Two-Stage Heckman Approach. *Sustainability* **2021**, *13*, 5458. [CrossRef]
18. Rufai, A.M.; Ogunniyi, A.I.; Salman, K.K.; Salawu, M.B.; Omotayo, A.O. Rural Transformation and Labor Market Outcomes among Rural Youths in Nigeria. *Sustainability* **2021**, *13*, 13794. [CrossRef]
19. Silva, M.S.; Nóbrega, J.S.; Santos, C.C.; Costa, F.B.; de Abreu, D.C.; da Silva, W.M.; Hoshide, A.K.; Gomes, F.A.L.; Pereira, U.S.; Linné, J.A.; et al. Organic Fertilization with Biofertilizer Alters the Physical and Chemical Characteristics of Young Cladodes of *Opuntia stricta* (Haw.) Haw. *Sustainability* **2023**, *15*, 3841. [CrossRef]
20. Molossi, L.; Hoshide, A.K.; de Abreu, D.C.; de Oliveira, R.A. Agricultural Support and Public Policies Improving Sustainability in Brazil's Beef Industry. *Sustainability* **2023**, *15*, 4801. [CrossRef]
21. Koutika, L.-S.; Matondo, R.; Mabiala-Ngoma, A.; Tchichelle, V.S.; Toto, M.; Madzoumbou, J.-C.; Akana, J.A.; Gomat, H.Y.; Mankessi, F.; Mbou, A.T.; et al. Sustaining Forest Plantations for the United Nations' 2030 Agenda for Sustainable Development. *Sustainability* **2022**, *14*, 14624. [CrossRef]

22. Zhang, Y.; Wang, X.; Wan, S.; Zhu, H. Fuzzy Set Qualitative Comparative Analysis of the Factors Affecting Satisfaction with the Policy of Ecological Forest Rangers. *Sustainability* **2023**, *15*, 6743. [CrossRef]
23. Campanelli, J.; Kuzovkina, Y.A.; Kocurek, S. Current Impediments for New England DOTs to Transition to Sustainable Roadside Practices for Strengthening Pollinator Habitats and Health. *Sustainability* **2023**, *15*, 3639. [CrossRef]
24. Sowunmi, F.A.; Bello, A.A.; Ogunniyi, A.I.; Omotayo, A.O. Delving Deeper into Market Concentration of Poultry Feed and the Driving Factors for Brand Switching: Evidence from Commercial Egg Producers in Nigeria. *Sustainability* **2022**, *14*, 8030. [CrossRef]
25. dos Reis, J.A.V.; Hoshide, A.K.; Vreyens, J.R.; de Oliveira, A.S.; de Barros, V.A.M.; da Silva, W.M.; Molossi, L.; Viana, J.L.; de Abreu, D.C.; de Oliveira, R.A. Training Sources and Preferences for Agricultural Producers and Professionals in Middle-North Mato Grosso, Brazil. *Sustainability* **2023**, *15*, 4712. [CrossRef]

sustainability

MDPI

Article

Back to the Future: Agricultural Booms, Busts, and Diversification in Maine, USA, 1840–2017

Aaron Kinyu Hoshide [1,2]

[1] College of Natural Sciences, Forestry and Agriculture, The University of Maine, Orono, ME 04469, USA; aaron.hoshide@maine.edu; Tel.: +1-207-659-4808

[2] AgriSciences, Universidade Federal de Mato Grosso, Sinop 78555-267, MT, Brazil

Abstract: In temperate forested regions, historical agricultural production and value have been characterized by booms and busts. Agricultural diversification can encourage more stable agricultural development in the future. Agricultural Census and Survey data from 1840 to 2017 were used to estimate crop and livestock species' product production and value for Maine, USA. These data were also used to calculate agricultural diversity indicators over time such as species richness, relative abundance, effective number of species, species diversification index, evenness, Shannon-Weiner index, and composite entropy index. Maine's historical grass-based livestock systems included crops raised to feed livestock from the state's establishment until the 1950's. Since the 1950's, production and value of livestock commodity products (e.g., meat chicken, eggs) have busted after initial booms. Three categories where diversity indicators have become more favorable since the 1950's in Maine include livestock, livestock forage/feed, and potatoes and potato rotation crops. Mixed vegetables, fruits, nuts, and specialty crops as a category have had diversity increases during the 1970's back-to-the-land movement and over the past two decades. Floriculture, propagation, and X-Mas trees as a category have witnessed volatile diversity indicator changes over time. Past diversification strategies can inspire farmers to go "back to the future" to improve sustainability.

Keywords: agricultural development; sustainability; diversity indexes; cultivars; livestock breeds; Maine

Citation: Hoshide, A.K. Back to the Future: Agricultural Booms, Busts, and Diversification in Maine, USA, 1840–2017. *Sustainability* **2022**, *14*, 15907. https://doi.org/10.3390/su142315907

Academic Editor: Antonio Boggia

Received: 24 October 2022
Accepted: 24 November 2022
Published: 29 November 2022

1. Introduction

Forests contribute to global biodiversity of terrestrial species [1] especially when disturbances to forest ecosystems are moderate [2] and trees and understory plants are more diverse [3]. Diversity for temperate forests is typically greater immediately following clearcutting and during a forest's terminal and decay stages after 200 to 300 years [4]. Maine USA forests are currently in intermediate successional stages due to industrial logging requiring more active management to increase biodiversity [5]. Historical logging in Maine (Figure 1) cleared enough land to allow for relatively high percentages of the southern and central (~70%), western (~40%), and northern (~20%) parts of the state to be used for agriculture from 1860 to 1920 [6] with a sharp decline in agricultural farmland between 1950 and 1970 (Figure 2). Compared to estimating the economic value of forest biodiversity which has focused on whole ecosystems or species within such ecosystems [7], relatively little research has been done on measuring crop/livestock diversity and economic value of agricultural systems within temperate forested areas over longer historical time periods. Crop diversity for commodity field crops in the USA has declined [8,9], peaking around 1960 [9], and has been positively influenced by irrigation [10]. More nuanced analyses are needed evaluating diversity and value of vegetables, fruits, nuts, specialty crops, and livestock as influenced by farming booms and busts as well as national/regional specialization and diversification.

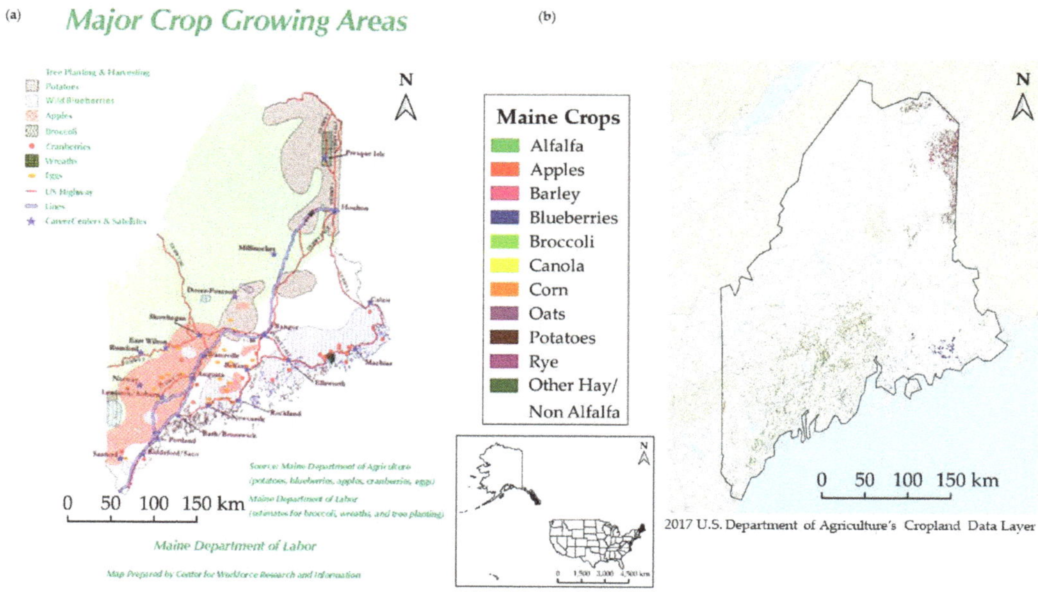

Figure 1. (**a**) Major crop growing areas and (**b**) major crops in Maine, USA. Reprinted/adapted with permission from Refs. [11,12]. 2022, Maine Department of Agriculture and Maine Department of Labor [11] and U.S. Department of Agriculture, National Agricultural Statistics Service [12].

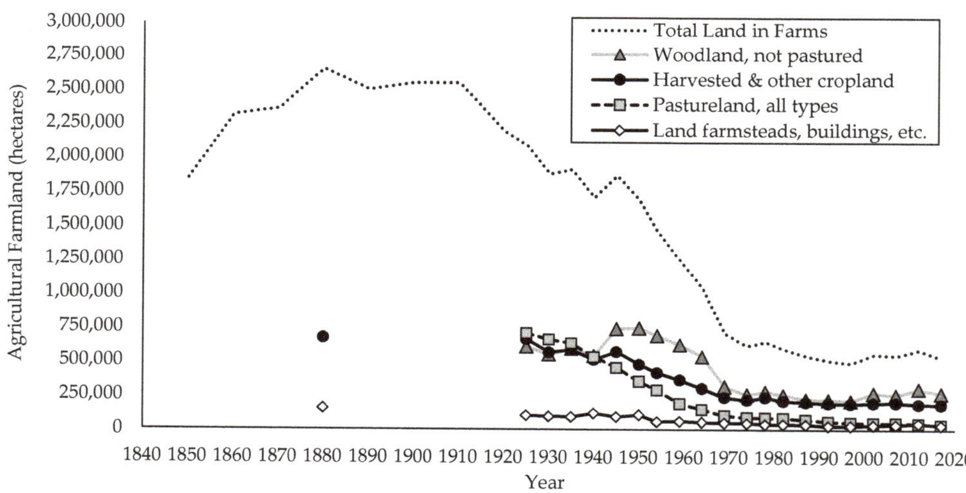

Figure 2. Agricultural farmland area (hectares) from 1840 to 2017 for Maine, USA.

Agricultural booms and busts over the past century in the USA have been driven by global trade and macroeconomics. The USA has had short-term and medium-term agricultural booms/busts from 1900 to 2015 driven by international export markets from (a) 1910–1930, (b) 1970–1990, and (c) 2000–2010 [13,14]. Economic downturns associated with these boom and bust cycles were attributed to banks aggressively lending after opening during the boom with subsequent collapse in farmland value during the bust (1910–1930) [15], the 1973–1974 oil embargo and stagflation (1970's), and the 2007–2008

Great Recession where cereals and vegetable oil were impacted but livestock was not [16]. The Midwest Corn Belt 1970–1990 boom and bust during the 1980's was triggered by a drop in export demand combined with increasing interest rates [17]. In Saskatchewan, Canada during the 1970–1990 boom and bust, the bust was delayed by crop and livestock diversification combined with expectations of temporary rather than extended down cycles. However, credit, human capital, and technical knowledge were required to diversify more into beef, pulses, and oilseeds [18].

Agricultural diversity has peaked and declined in regions and countries around the world throughout the 20th century. For example, Simpson's Index of diversity peaked in the 1950's and declined to 1992 in West Punjab, India [19]. The agricultural industry in the USA from after World War II until the mid-1970's has gotten more specialized [20] where diversity (D) measured as effective crop species weighted by the Shannon-Weiner Index has declined in the USA since the 1960's [9]. Farm-level specialization/diversification revolve around farm economics. Agricultural specialization is driven by economies of scale which maximizes production of one commodity for a specific degree of capital investment. Specialization of farms and entire agricultural industries are susceptible to agglomeration in areas of the world that provide comparative advantages of production, processing, and marketing. Areas that become less competitive lose out. Farms in dying industries must either sustainably intensify or diversify to remain viable [21]. Within-farm diversification can also be triggered by unfavorable conditions [22], such as lower market prices of agricultural products produced as well as higher inputs costs [23]. Farm characteristics that support the ability to diversify include having enough labor slack [22] and spouse/family labor [23].

The goal of this research is to estimate the production and value of commodity field crops, vegetables, fruits and nuts, specialty crops, and livestock in Maine USA from 1840 to 2017. USDA Agricultural Census and Survey data [24] was analyzed over this time frame in order to delineate Maine's boom and bust cycles of farming which have resulted in efforts to diversify its food systems. Thus the specific objectives of this study are to (1) determine production and inflation-adjusted value peaks for all agricultural crops and livestock over this 177 year period, (2) calculate the diversity of these agricultural enterprises using common ecological diversity indicators, and (3) explain recent diversification trends in Maine as responses to boom and bust of key agricultural commodities. Past and current agricultural diversification in Maine can serve as models on how to go "Back to the Future" to better diversify agricultural systems in other temperate regions.

2. Materials and Methods

2.1. Determining Historical Agricultural Production and Value

In order to identify boom and bust periods for both crops and livestock species in Maine, the production and value for each agricultural product produced from these species had to be calculated using historical data. Maine USA livestock numbers and crop area, livestock/crop farm numbers, agricultural product yields, and values were downloaded and analyzed for 29 Census of Agriculture years starting in 1840 and ending in 2017 [24]. Crops and livestock production required English to metric conversions for weights of farm-gate products produced in any given Census year. Since livestock forages and feeds had different dry matter (DM) percentages, total forage and feed production was calculated on a dry matter basis using previous assumptions for dry hay, corn silage, alfalfa hay/silage [25], sorghum silage [26], and pumpkins used for feed [27]. Root crops for livestock feed were assumed to have the same DM as forage turnips [28].

Livestock also required estimating animal live weights, carcass weights (if slaughtered), and weights of products produced (e.g., milk, meat, fiber) using appropriate conversion factors. If USDA data [24] did not provide animal product production but rather only animals sold, it was assumed this was for meat and not for breeding. Annual assumed meat production (e.g., beef, pork, poultry, rabbits, etc.) was estimated as:

$$\text{Meat production} = \text{Animals sold} \times (\text{Live weight/animal} \times \text{Dressing percentage}) = \text{Animals sold} \times \text{Carcass weight/animal} \quad (1)$$

Cattle live weights and carcass weight conversions were based on past research in Maine [25,29]. However, beef products could not be estimated since animals sold were not distinguished between feeder, slaughtered, and live breeding cattle in USDA statistics [24]. Pork livestock weight and dressing percentage were based on past work with local producers (Aaron K. Hoshide, unpublished data). Sheep and goat dressing percentage and/or live weights were from [30–33]. Horse live weight was based on [34]. Similar assumptions for poultry live weight and/or dressing percentage were used for broiler chicken [35–37], turkey [35,37], goose, duck [35], pheasant [38], guinea fowl [35], quail [39,40], pigeon [41], and emu [42]. Similar assumptions were used for bison [43], tame deer [44], and rabbit [45–47]. Conversions were used for chukar [48], partridge [49], ostrich [50], rhea [51], and chinchilla [52,53], but these animals did not have enough data to delineate boom/bust periods of production.

Non-meat animal products included milk from cows, chicken eggs, wool from sheep, and mohair from goats. Earlier Census of Agriculture years required volume to weight conversion for milk and other dairy products [54]. Live weights of laying chickens were from [55]. Egg production was available for Census years from 1880 to 1964 [24], but had to be estimated for 1969 to 2017 by multiplying the number of layers [24] by the average annual egg production per layer in Maine (1969–2007), the average in the nearby states of Massachusetts and Vermont (2012), and Vermont (2017) [56]. Sheep fleece and mohair goat fiber weights per animal were used [26].

Crop and livestock values were either available [24] or were estimated based on total production of crops/livestock multiplied by products' per unit prices. All nominal prices and values in any Agricultural Census year between 1840 and 2012 were converted to real prices and values in 2017 USD. Such adjustments for inflation were to a base year of 2017 using USA commodity specific Producer Price Indexes (PPI) when possible. Missing commodity specific PPI data from 1926 to present used composite PPI for four categories: (1) fruits and melons, fresh/dry vegetables and nuts, (2) grains, (3) hay, hayseeds, and oilseeds, and (4) slaughter livestock. Missing commodity specific PPI data from 1913–1925 used the farm products composite PPI. Crop and livestock categories without specific PPI used the all commodities PPI for 1840–2017 [57]. If the farm-gate price for a product was not available for Maine in a particular year, then a regional (e.g., New York State) or USA national price was used from USDA Agricultural Survey data [24]. The value of unthreshed oats harvested to feed livestock (1925–1950) was estimated as the sum of both grain and straw values. Oat straw prices were obtained from Andrew Plant, University of Maine Cooperative Extension in 2017.

2.2. Calculating Agricultural Diversity Indicators

USDA Agricultural Census data for crop and livestock species numbers [24] were used to determine the diversity of major categories of crops and for livestock. Seven diversity indicators were calculated from these data for four crop categories: (1) mixed vegetables, fruits, nuts, and specialty crops, (2) potatoes and annual crops rotated with potatoes, (3) floriculture, propagation, seeds, and Christmas (X-Mas) trees, and (4) livestock forage/feed crops. These seven diversity indicators were also calculated for a fifth category for all livestock species. Three of these seven indicators were related to species number (richness, effective number of species) and relative abundance. The remaining four indicators were measured on a scale of 0 (no diversity) to 1 (highest diversity) and included the Species Diversification Index, evenness, the Shannon-Weiner Index, and the Composite Entropy Index.

2.2.1. Richness, Relative Abundance, and Effective Number of Species

Agricultural diversity can be measured by the number of crop/livestock species in a particular area. For ecological systems (e.g., forests, agro-ecological agriculture), alpha diversity measures within-species diversity, beta diversity contrasts diversity between different species, and gamma diversity measures the biodiversity across an entire area, region, or biome [2,4]. For crops and livestock, richness is the number of total crop cultivars or total livestock breeds in a given time period (e.g., Agricultural Census year) for a particular area (e.g., Maine, USA).

Relative abundance for a particular crop or livestock category was calculated as the percent by area/weight relative to all other categories. So for example for crops, the relative abundance for the category livestock forage/feed is the percent of total crop area this category makes up in a particular year relative to all other crop categories. The relative abundance of eggs is its percent of total product weight relative to other livestock product categories [58].

The effective number of species (e.g., crop, livestock) or *ENS* is richness (*R*) multiplied by the natural exponent of the negative Shannon-Weiner Index (*SWI*) of diversity:

$$ENS = R \times e^{-SWI} = R \times e^{-\sum_{i=1}^{s}(p_i \times logp_i)} \tag{2}$$

where p_i is the proportion of species i within the total number (s) of species within the crop or livestock category [8–10]. So *ENS* adjusts *R* by the both the number of species and relative proportions of species within an agricultural category such as crops or livestock. For example, if there are 10 crop (livestock) species with each species making up 10% of the total category area (live weight), then e^{-SWI} equals 1, which means $ENS = R \times e^{-SWI} = R \times 1 = R$ in this particular case. However, if the number of species are <10 and/or if certain species make up a disproportionate percentage of the total category then e^{-SWI} will be less than 1 and thus $ENS < R$. If species are more evenly distributed and/or if $R > 10$, then e^{-SWI} can be greater than 1 and thus $ENS > R$.

2.2.2. Diversity Indexes

The four diversity indexes (0 to 1) evaluated were Species Diversification Index for both crop and livestock species, evenness, Shannon-Weiner Index, and the Composite Entropy Index. Species Diversification Index (*SDI*) equals one minus the Simpson's Index (*SI*) in agro-ecology (0 to 1) or one minus the sum of squared proportions of crop/livestock species:

$$SDI = 1 - SI = 1 - \sum_{i=1}^{s} p_i \tag{3}$$

where p_i is the proportion of species i within the total number (s) of species within the crop or livestock category [59]. *SI* is identical to the economic Herfindahl Index (HI). HI measures the degree of market concentration for businesses within a particular industry as the sum of squared market shares. When using market share proportions (versus percentages), HI ranges from 0 to 1 (1 to 10,000 when using percentages). For a monopoly dominating an entire industry (1 = 100%), $HI = 1^2 = 1 \times 1 = 1$. For a perfectly competitive industry with a large number of equally sized firms, HI approaches 0.

Evenness (*E*) is how balanced crop or livestock species are in a particular category. So *E* is lower if fewer species make up a disproportionally large percent of the total category [58]. *E* is calculated as *SI* divided by the natural log of the richness of species and ranges from 0 to 1:

$$E = \frac{SI}{lnR} \tag{4}$$

As specified as part of Equation (2), the Shannon-Weiner Index (*SWI*) is the negative value of the sum of squared proportions times the log of proportions:

$$SWI = -\sum_{i=1}^{s}(p_i \times logp_i) \tag{5}$$

with values ranging from 0 to positive 1. The Composite Entropy Index (*CEI*) weights the *SWI* by $1 - (1/N)$ where N equals the total number of crops or livestock species in particular agricultural category. *CEI* is defined as:

$$CEI = -[\sum_{i=1}^{S} (p_i \times logp_i)] \times [1 - \frac{1}{N}] \quad (6)$$

where N equals the total number of species of crops or livestock. So if there is only one species, then $1 - (1/N) = 1 - (1/1) = 1 - 1 = 0$ so *CEI* will equal 0 (no diversity). If there is a very large number of species, then *CEI* will be much closer in value to *SWI* [60,61].

3. Results

3.1. Agricultural Booms and Busts

Major crop production categories of potatoes, grains and oilseeds, and dry matter of livestock forages/feeds went through different boom and bust periods with production peaking in different years. Potato production in Maine USA peaked at 2,176,798 metric tons (t) in 1950 (Figure 3a) on 58,028 hectares (ha) (Figure 3b). For grains and oilseeds, production peaked in 1860 at 140,059 t harvested from 91,118 ha with a more recent rebound since 1970 to 95,885 t grown on 19,278 ha in 2017 (Figure 4a). Grain/oilseed production in 2017 was only 68.5% of production and 21.2% of crop area compared to the historical production peak in 1860. Agricultural Census data for Maine [24] did not have comprehensive production and value data for mixed vegetables/fruits (Table 1) except for the year 1900. Historical trends in other crops included declines in orchard fruit and dry beans, increase in berries (Figure 4b), and a brief boom/bust period for sugar beets around 1969 (Figure 5).

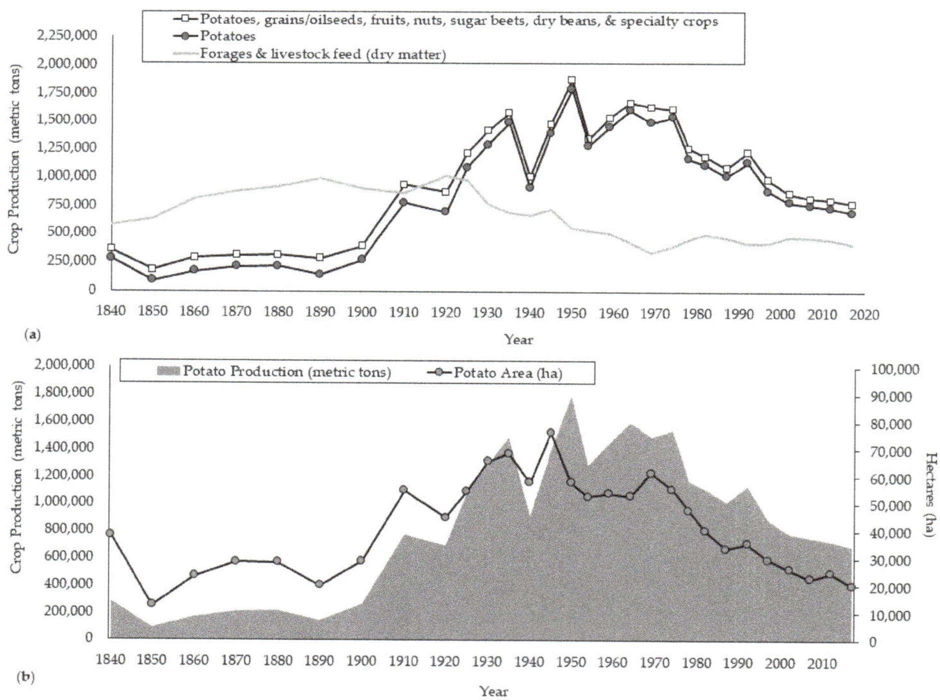

Figure 3. (**a**) Crop category weights in metric tons (t) and (**b**) potato production in metric tons (t) and area (hectares) from 1840 to 2017 for Maine, USA.

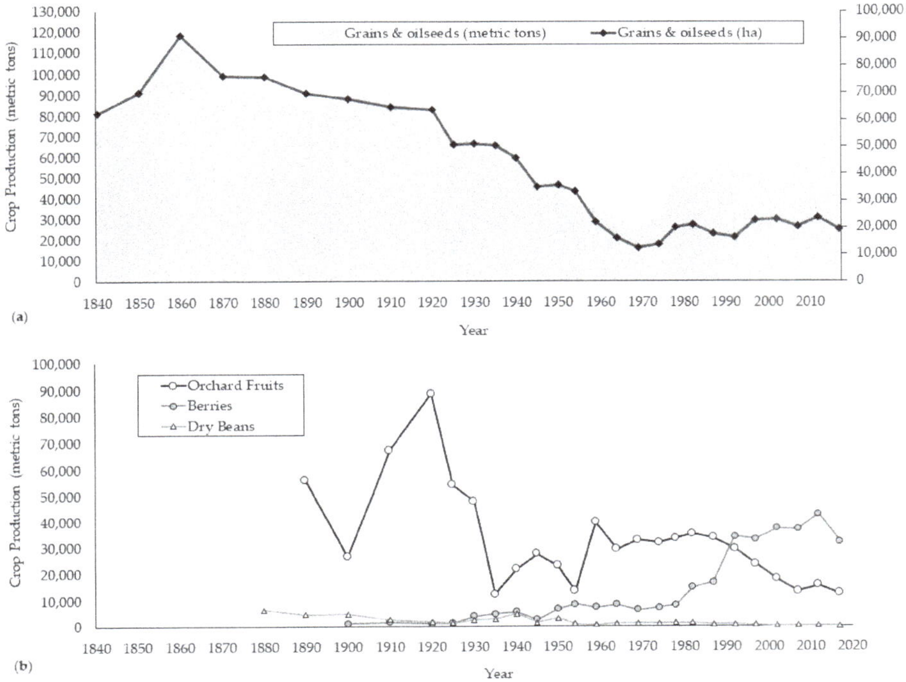

Figure 4. (**a**) Grain/oilseed production in metric tons (t) and area (hectares) and (**b**) orchard fruit, berry, and dry bean production in metric tons (t) from 1840 to 2017 for Maine, USA.

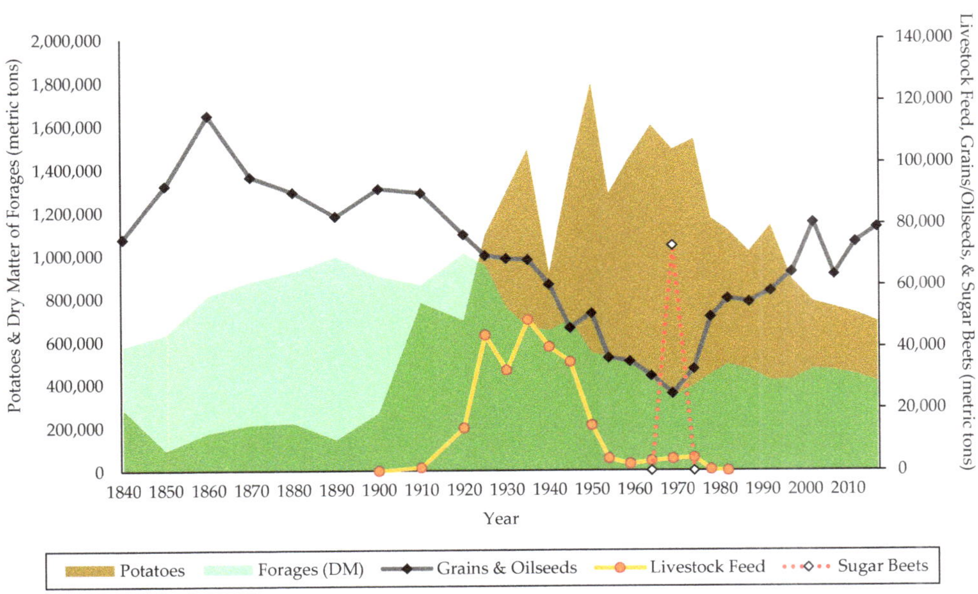

Figure 5. Potato, forage dry matter, livestock feed, and sugar beets in metric tons (t) from 1840 to 2017 for Maine, USA.

Table 1. Earlier agricultural booms and busts for livestock and crop products in Maine, USA, calculated or summarized from publicly available USDA-NASS statistics. Reprinted/adapted with permission from Ref. [24]. 2022, U.S. Department of Agriculture.

Agricultural Category and Enterprise	Latin Name (*Genus Species Subspecies*)	Boom/Bust Years (Start–*Peak*–End)	----Peak Production Year Estimate----				2017 Percent of Peak Prod. Year	
			Farms	Area (ha)	Product [1] (metric t)	Real Value (2017 USD)	Product	Value
LIVESTOCK								
Hogs	*Sus domesticus*	1840–***1840***–1959	35,101	-	10,721	3,423,882	9.9%	55.3%
Sheep (wool)	*Ovis aries*	1840–***1880***–1900	36,396	-	1259	11,931,087	3.5%	0.3%
Geese	*Anser* spp. *domesticus*	1880–***1890***–1910	1911	-	13	71,179	3%	1.7%
Horses [2]	*Equus ferus caballus*	1840–***1890***–1945	47,420	-	-	-	-	-
Pigeons	*Columba livia domestica*	1910–***1910***–1910	287	-	0.40	6937	43.3%	75.7%
Guinea Fowl	*Numida meleagris*	1910–***1910***–1910	1073	-	3.64	27,828	10.6%	2.3%
Angora Goat	*Capra hircus aegagrus*	1900–***1910***–1920	39 [3]	-	0.29 [3]	3152 [3]	85.4% [3]	22.8% [3]
FORAGES/ FEED								
Seed	*Fabaceae/Poaceae* family	1850–***1860***–1890	9818	15,227	1460	6,223,061	1.1%	1.8%
Forage hay	*Poa* spp. & others	1840–***1920***–1959	46,790	496,582	1,094,358	212,351,008	30.2%	22.8%
Root crops	Not specified	1910–***1920***–1950	1872	633	8411	1,551,150	0%	0%
Oats unthrsh.	*Avena sativa*	1925–***1925***–1950	6272	6582	20,724	3,653,218	0%	0%
Corn hogged	*Zea mays*	1920–***1935***–1978	2273	1512	34,336	4,028,874	0%	0%
Pumpkins	*Cucurbita pepo*	1940–***1940***–1945	31	8	58	8850	0%	0%
GRAIN/ OILSEED								
Rye	*Secale cereale*	1840–***1840***–1870	5335	5409	3504	1,171,145	59.1%	32.2%
Wheat	*Triticum aestivum*	1840–***1840***–1935	23,487	22,598	23,083	13,002,123	1.6%	0.7%
Flaxseed	*Linum usitatissimum*	1850–***1850***–1910	98	124	15	7724	0%	0%
Corn	*Zea mays*	1840–***1850***–1920	36,109	22,852	44,453	7,052,906	68.2%	68.9%
Barley	*Hordeum vulgare*	1840–***1860***–1900	14,369	13,059	17,464	4,674,916	144%	76.7%
Buckwheat	*Fagopyrum esculentum*	1840–***1900***–1954	9727	10,235	11,046	3,725,345	0.4%	1.0%
Small Grains	Not specified	1925–***1950***–1964	350	2062	n/a	n/a	0%	n/a
VEGETABLES								
Dry peas	*Pisum sativum*	1880–***1880***–1920	8531	1784	1495	1,736,056	0%	0%
Dry beans	*Fabaceae* Family spp.	1880–***1880***–1997	30,005	6852	4945	6,229,964	1.3%	0.6%
Sweet potato	*Ipomoea batatas*	1890–***1890***–1890	6	1.6	6	4309	142%	126.3%
Sweet corn	*Zea mays*	1900–***1930***–1950	7153	6654	n/a	n/a	n/a	n/a
Cabbage	*Brassica oleracea capitata*	1900–***1935***–1969	1273	242	n/a	n/a	n/a	n/a
Squash	*Cucurbita* spp.	1930–***1940***–2017	494	478	n/a	n/a	n/a	n/a
Green beans	*Phaseolus vulgaris*	1920–***1945***–1978	2397	1616	n/a	n/a	n/a	n/a
FRUITS								
Grapes	*Vitis vinifera*	1900–***1900***–1950	4350	20	125	74,441	118%	192.5%
Apricots	*Prunus armeniaca*	1910–***1910***–1910	48	7	0.57	605	0%	0%
Currants	*Ribes* spp.	1900–***1910***–1920	1076	32	35	111,189	1.1%	4.6%
Cherries	*Prunus avium/P. cerasus*	1890–***1910***–1930	3165	19	61	114,022	4.8%	6.9%
Pears	*Pyrus* spp.	1890–***1910***–1950	10,857	241	1016	234,295	3.9%	16.4%

Table 1. *Cont.*

Agricultural Category and Enterprise	Latin Name (*Genus Species Subspecies*)	Boom/Bust Years (Start–*Peak*–End)	- - - -Peak Production Year Estimate- - - -				2017 Percent of Peak Prod. Year	
			Farms	Area (ha)	Product [1] (metric t)	Real Value (2017 USD)	Product	Value
Blackberry	*Rubus* spp.	1900–***1920***–1945	2198	114	127	388,062	6%	7.7%
Plum/Prune	*Prunus domestica*	1900–***1920***–1950	6792	54	264	379,093	2.2%	1.2%
Apples	*Malus domestica*	1890–***1920***–1978	34,600	24,396	87,622	47,753,831	14%	24.8%
Strawberry	*Fragaria × ananassa*	1900–***1940***–1940	2168	293	1518	1,946,278	31%	121.9%
Raspberry	*Rubus idaeus*	1900–***1940***–1945	1783	204	179	498,673	26.4%	61.7%
Peaches	*Prunus persica*	1900–***1950***–1964	1097	19	139	216,309	60.8%	57.9%
TEXTILE/ OTHER								
Textiles	Silk, flax, & hemp	1840–***1840***–1910	76	459	35	n/a	0%	n/a
Hops	*Humulus lupulus*	1840–***1870***–1900	n/a	378	135	560,045	1.4%	4.3%

[1] Estimated livestock product/carcass, crop harvest measured in metric tons (t). [2] Horses numbered 109,156 in 1890. [3] Angora goat production as mohair for fiber (not carcass weight) from 168 goats in 1910.

Hay production peaked around 1920 at 1,094,385 t as harvested on 496,582 ha (Table 1). Dry matter (DM) production of total livestock forages and feeds plateaued from 1880 to 1925 (Figures 3a and 5) ranging from 903,100 to 1,014,792 t harvested annually with 2017 production (410,494 t) only 40.5% of 1920's peak production. Since the 1964 Agricultural Census, higher energy corn and sorghum silages, higher protein alfalfa silage, and grass silage (i.e., haylage) have replaced more traditional dry hay such as grass, alfalfa, and small grain hays (Figure 6). Peaks for crops directly fed to livestock on-farm (Table 1, Figure 5) from 1910 to 1950 included (1) 8411 t (920 t DM) of root crops in 1920, (2) 20,724 t (18,444 t DM) of unthreshed feed oats in 1925, (3) 34,336 t (10,301 t DM) for corn hogged, grazed, or cut for fodder in 1935, and (4) 58 t (4.51 t DM) of pumpkins in 1940. Forage seed production peaked around 1860 at 1460 t (Table 1).

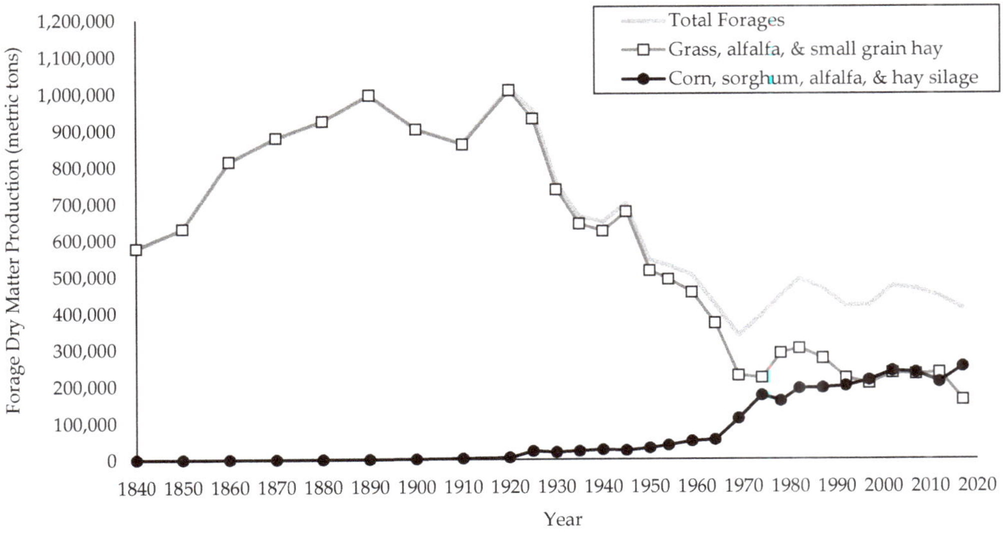

Figure 6. Forage dry matter production in metric tons (t) from 1840 to 2017 for Maine, USA.

Traditional livestock peak live weight and production could not be determined for hogs due to a lack of Agricultural Census data prior to 1840 for specific livestock and crops (Table 2). Cattle (beef and dairy) live weight was at its zenith around 1860 at 172,883 t, while 1840 live weights for both sheep (51,043 t) and hogs/pigs (16,506 t) could not be confirmed as peaks (Figure 7a) due to a lack of Agricultural Census data prior to 1840. Horse live weight peaked around 1890 at 46,551 t (Table 2, Figure 7a). Total poultry (broiler and layer chickens, turkeys, ducks, etc.) live weight reached its maximum around 1978 at 184,729 t (Figure 7a).

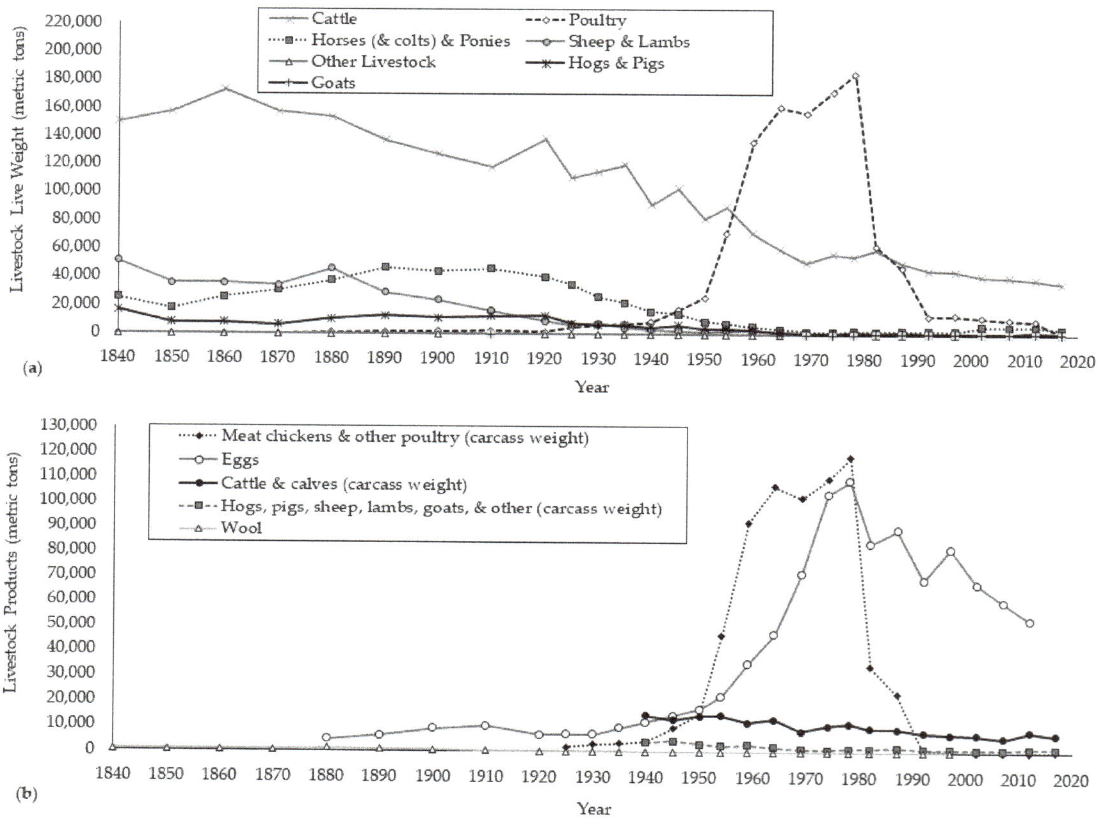

Figure 7. (**a**) Livestock live weight and (**b**) livestock product production as carcass weight or product weight in metric tons (t) from 1840 to 2017 for Maine, USA.

Table 2. Agricultural booms & busts for more recent specialized/niche systems in Maine, USA, calculated or summarized from publicly available USDA-NASS statistics. Reprinted/adapted with permission from Ref. [24]. 2022, U.S. Department of Agriculture.

Agricultural Category and Enterprise	Latin Name (*Genus Species Subspecies*)	Boom/Bust Years (Start–*Peak*–End)	----Peak Production Year Estimate----				2017 Percent of Peak Prod. Year	
			Farms	Area (ha)	Product [1] (metric t)	Real Value (2017 USD)	Product	Value
LIVESTOCK								
Cattle (dairy)	*Bos taurus*	1850–***1900***–2017	59,299	-	409,282	482,207,450 [1]	69.8%	27.9%
Turkeys	*Meleagris gallopavo domestica*	1930–***1954***–1969	567	-	1970	5,336,560	5.0%	2.6%
Broilers	*Gallus gallus domesticus*	1945–***1978***–1987	345	-	117,718	163,162,608	0.3%	0.03%
Eggs (dozen)	*Gallus gallus domesticus*	1950–***1978***–2012	822	-	108,958	162,158,469	47.8% [2]	40.7% [2]
Ducks	*Anas platrhynchos domesticus*	1974–***1978***–1982	74	-	689	654,121	6.5%	6.5%
Rabbits	*Oryctolagus cuniculus*	1978–***1992***–2002	59	-	67	414,407	31.1%	8.9%
Quail	*Coturnix coturnix*	1992–***1997***–2017	4	-	0.92	5511	73%	125.8%
Emus	*Dromaius novaehollandiae*	2002–***2002***–2012	6	-	0.94	10,076	33.3% [2]	62.8% [2]
Angora Goat	*Capra hircus aegagrus*	2002–***2007***–2017	37 [3]	-	0.81 [3]	8190 [3]	30.6% [3]	8.8% [3]
Goats	*Capra hircus*	1997–***2007***–2017	116	-	39	107,165	66.8%	141%
Guinea Fowl	*Numida meleagris*	2012–***2012***–2017	122	-	0.44	737	87.0%	87.0%
FORAGES								
Sorghum	*Sorghum bicolor*	1969–***1974***–1987	60	667	17,328	202,036	2.9%	11.9%
Corn silage	*Zea mays*	1964–***1974***–2017	805	15,122	466,462	8,226,001	83.5%	227.5%
Alfalfa (hay)	*Medicago sativa*	1954–***1992***–2017	774	13,995	52,437	13,815,513	37.9%	30.7%
GRAIN/ OILSEED								
Oats	*Avena sativa*	1840–***1910***–2017	22,029	48,963	61,432	17,683,787	32.0%	21.5%
Sorghum	*Sorghum bicolor*	1969–***1969***–1978	4	62	142	14,953	12.9%	60.2%
Wheat	*Triticum aestivum*	1974–***1974***–1982 2007–***2012***–2017	59 19	777 968	2073 2576	300,211 471,630	17.3% 13.9%	30.3% 19.3%
Canola	*Brassica napus*	1997–***2002***–2017	20	621	1131	510,770	5.1%	4.3%
Barley	*Hordeum vulgare*	1992–***2002***–2017	112	10,464	39,741	4,425,358	63.3%	81.0%
Spelt/Emmer	*Triticum spelta/T. turgidum*	2007–***2007***–2017	6	18	55	17,473	31.3%	86.5%
Triticale	×*Triticosecale* spp.	2007–***2007***–2017	4	13	38	16,813	25.8%	38.7%
Rye	*Secale cereale*	2012–***2012***–2017	23	1687	6556	968,960	31.6%	38.9%
POTATO + CROPS								
Potatoes	*Solanum tuberosum*	1910–***1950***–2017	14,904	58,028	1,791,470	313,564,930	38.5%	49.4%
Green peas	*Pisum sativum*	1935–***1964***–1997	435	4373	n/a	n/a	-	-
Sugar beets	*Beta vulgaris vulgaris*	1964–***1969***–1974	141	3840	73,064	5,559,566	0%	0%
Broccoli	*Brassica oleracea* var. *italica*	1987–***2012***–2017	145	2555	n/a	n/a	-	-
VEGETABLES								
Cucumbers	*Cucumis sativus*	1900–***1950***–1964	982	328	n/a	n/a	-	-
Lettuce	*Lactuca sativa*	1930–***1954***–1969	168	308	n/a	n/a	-	-
Carrots	*Daucus carota sativus*	1935–***1954***–1978	302	172	n/a	n/a	-	-
Dry peas	*Pisum sativum*	1969–***1969***–1978	13	250	647	548,841	0%	0%
Dry beans	*Fabaceae* Family spp.	1969–***1982***–1997	115	1955	1122	2,124,836	5.7%	1.8%
FRUITS								
Grapes	*Vitis vinifera*	1974–***1982***–2017	60	15	150	99,865	98.5%	249.2%
Strawberry	*Fragaria* × *ananassa*	1982–***1987***–2017	163	238	947	1,725,185	49.6%	137.5%
Blueberry								
Highbush	*Vaccinium corymbosum*	1940–***1987***–2017	110	982	1118	5,288,894	30.8%	31.4%
Lowbush	*Vaccinium angustifolium*	1982–***2014***–2020	500	9225	47,355	59,039,649	45.4% [2]	20.9% [2]
Cranberry	*Vaccinium oxycoccos*	1997–***2007***–2017	40	121	841	1,457,458	30.4%	11.7%

[1] Estimated livestock product/carcass, crop harvest (t) and 173,592 dairy cattle in 1900. [2] Egg and emus (2012), lowbush blueberry (2020) used to calculate percent of peak year. [3] Angora goat production as mohair.

Estimated carcass weight for hogs/pigs in 1840 was 10,721 t. Wool production peaked around 1880 at 1259 t (Table 1, Figure 7b) followed by dairy products (milk, cream, cheese, butter) around 1900 at 409,282 t (Table 2). Chicken products exceeded those of other livestock, peaking in 1978 for estimated broiler chicken carcass weight (117,718 t) and eggs (108,958 t) with the decline in egg production more gradual than for broilers since 1978 (Table 2, Figure 7b). Turkey and duck went through shorter boom and bust cycles compared to chicken topping off at estimated carcass weights of 1970 t and 689 t, respectively (Table 2). Production peaks from other specialty livestock ranged from 0.29 t (mohair in 1910) to 67 t (rabbit in 1992) (Tables 1 and 2).

Grains such as wheat, corn, buckwheat, and oats were at their highest and/or peaked between 1840 and 1920 (Tables 1 and 2). While barley and rye were grown during this 19th century time period, more recently their outputs have peaked at 39,741 t in 2002 for barley and 6556 t in 2012 for rye. Wheat has had two minor resurgences, one around 1974 and another around 2012 (Table 2). However, 2017 wheat production was only 1.6% of its historical high in 1840 (Table 1). More recent production cycles have included canola (1131 t in 2002) and non-traditional grains such as spelt/emmer and triticale (Table 2). Dry bean and dry pea production was highest in 1880 (Table 1). Most vegetables and fruits peaked production and/or growing area during the first half of the 20th century (Tables 1 and 2) with potatoes dominating at 1,791,470 t harvested on 58,028 ha in 1950 (Table 2). There were minor rebounds for dry peas (1969) and dry beans (1982). Recent peaks of vegetables/fruits include green peas (1964), grapes (1982), strawberries and highbush blueberries (1987), cranberries (2007), broccoli (2012), and wild blueberries (2014) in Maine (Table 2).

3.2. *Crop and Livestock Values*

Historical peak crop and livestock value of production adjusted for inflation to 2017 U.S. dollars (USD) were highest for dairy products in 1900 (USD 482,207,450), 1950 potatoes (USD 313,564,930), 1920 forage hay (USD 212,351,008), and chicken broilers (USD 163,162,608) and chicken eggs (USD 162,158,469) in 1978. This was followed by 2014 lowbush blueberry (USD 59,039,649), 1920 apples (USD 47,753,831), 1920 oats (USD 17,683,787), 1992 alfalfa hay (USD 13,815,513), 1840 wheat (USD 13,002,123), and 1880 wool (USD 11,931,087). All other crop and livestock products were below USD 10 million. Despite lower production compared to historical peaks, agricultural products with higher values in 2017 were corn silage, quail, meat goats, grapes, and strawberries (Tables 1 and 2). Certain livestock as suggested by their 2017 value have the potential to be future niche species such as tame deer (USD 1,397,000), bison (USD 101,000), pigeons (USD 3547), and peafowl feathers (USD 721). The 2017 crop value of corn harvested as grain is USD 4,859,275 which is 68.2% of the 1850 maximum. Other 2017 values that were historical highs were for alfalfa haylage (USD 4,268,000), peaches (USD 303,117), sweet potatoes (USD 35,084), sunflower seed (USD 3560), and sorghum grain (USD 2187) (Table 3).

3.3. *Agricultural Diversity Indicators*

3.3.1. Richness, Effective Number of Species, and Relative Abundance

Crop/livestock category species richness (*R* = number of species) increased very modestly for livestock forage/feed, floriculture/propagation/seeds/X-Mas trees, and potatoes and annual crops rotated with potatoes. For mixed vegetables, fruits, nuts, and specialty crops, *R* increased from 33 to 60 (1900 to 1950), bottomed out to 31 to 34 (1954 to 1969), spiked from 42 to 61 (1974 to 1978), and then bottomed out again to 36 (1982 to 1987), and then increased substantially from 46 to 82 (1992 to 2017). Livestock *R* increased from 5 to 26 from 1850 to 2012 (Figure 8a).

Table 3. Agricultural systems with increasing production potential in Maine, USA, calculated or summarized from publicly available USDA-NASS statistics. Reprinted/adapted with permission from Ref. [24]. 2022, U.S. Department of Agriculture.

Agricultural Category and Enterprise	Latin Name (*Genus Species Subspecies*)	Growth Years (*Peak*–Start–Recent)	------Recent Year Estimate------				2017 Percent of Peak Prod. Year	
			Farms	Area (ha)	Product [1] (Metric Tons)	Real Val. (2017 USD)	Product	Value
LIVESTOCK								
Pheasants	*Phasianus colchicus*	***None***–1992–2012	6	-	7	n/a	n/a	n/a
Bison	*Bison* spp.	***None***–2002–2017	12	-	13	101,000	100%	100%
Deer (tame)	*Odocoileus virginianus*	***None***–2002–2017	37	-	73	1,397,000	100%	100%
Pigeons	*Columba livia domestica*	***1910***–1992–2017	15	-	0.17	3547	43.3%	43.3%
Peafowl [2]	*Pavo* & *Afropavo* spp.	***None***–1910; 2012–2017	25	-	n/a	721 [2]	-	-
FORAGES								
Alfalfa (haylage)	*Medicago sativa*	***None***–2002–2017	139	5750	87,400	4,268,000	100%	100%
GRAIN/OILSEED								
Sorghum	*Sorghum bicolor*	***None***–2012–2017	3	8	142	2187	12.9%	12.9%
Sunflower seed	*Helianthus annuus*	***None***–1997–2017	1	9	9	3560	100%	100%
Corn	*Zea mays*	***1850***–1974–2017	82	2929	30,327	4,859,275	68.2%	68.2%
VEGETABLES								
Sweet potato	*Ipomoea batatas*	***None***–2002–2017	34	4	9	35,084	100%	100%
FRUITS								
Peaches	*Prunus persica*	***1950***–1982–2017	118	18	84	303,117	60.8%	60.8%

[1] Estimated livestock product/carcass, crop harvest (t). [2] Peafowl numbered 6 in 2017 valued at USD 115.64 per fowl so cannot distinguish if feathers or live animals sold.

Effective number of species (*ENS*) was similar to *R* for potatoes and annual crops rotated with potatoes as well as livestock forage/feed (Figure 8b). The *ENS* for floriculture, propagation, seeds, and X-Mas trees declined abruptly after 1964 (Figure 8b) whereas *R* fluctuated over time with a more recent increase (Figure 8a). *ENS* for the mixed vegetables, fruits, nuts, and specialty crops category had much less pronounced valleys and peak after 1950 compared to *R*. Unlike *R*, *ENS* for mixed vegetables, fruits, nuts, and specialty crops in 2017 did not exceed the 1950 peak (Figure 8).

Area of livestock forage/feed and potato rotation systems have been more relatively abundant compared to the other two crop categories of mixed vegetables, fruits, nuts, and specialty crops as well as floriculture, propagation, seeds, and X-Mas trees. The relative abundance of the livestock forage/feed category has increased, while the relative abundance of the potatoes and crops rotated with potatoes category has increased since 1969 (Figure 9a). Relative abundance of livestock product weight was dominated by broiler chickens and eggs with more recent balance relative to beef and pork (Figure 9b).

3.3.2. Diversity Indexes

Crop and livestock category diversity indexes were consistent with calculated effective number of species. For mixed vegetable, fruit, nut, and specialty crops, the Species Diversity Index (*SDI*), Shannon-Weiner Index (*SWI*), and Composite Entropy Index (*CEI*) were similar, while Evenness (*E*) gradually increased over time (Figure 10). For livestock, *SDI*, *SWI*, and *CEI* increased over time with volatility between the years 1930 to 2000, while livestock *E* declined (Figure 11a). There were increases in evenness and diversity in potato systems since the 1970's (Figure 11b) and livestock forage/feed from 1954 to 2007 (Figure 12a). Diversity indexes and *E* for floriculture, propagation, seeds, and X-Mas trees have been more volatile trending upward and downward, respectively (Figure 12b).

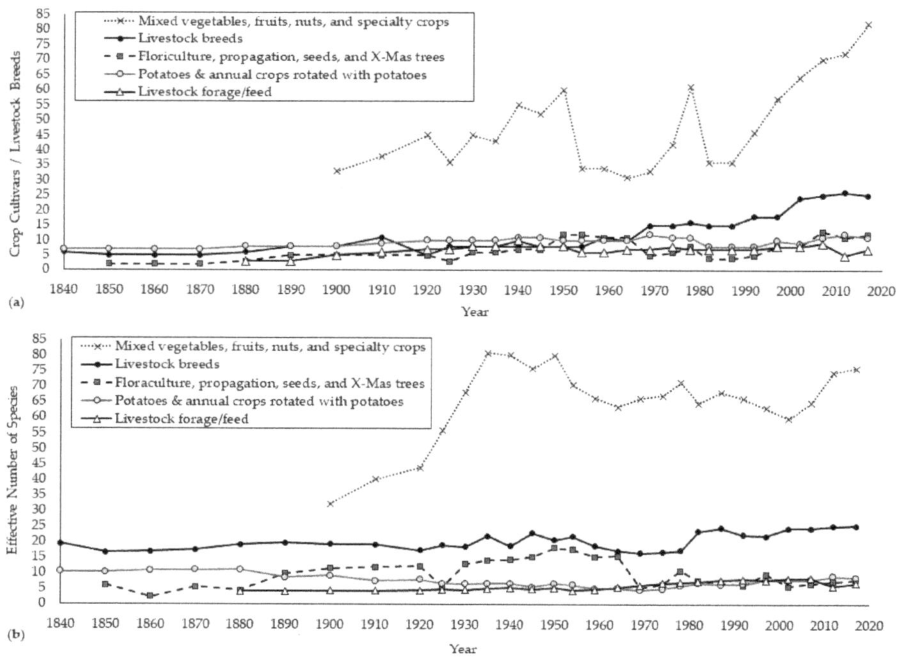

Figure 8. Crop cultivar and livestock breed (**a**) richness and (**b**) effective number of species in general categories from 1840 to 2017 for Maine, USA.

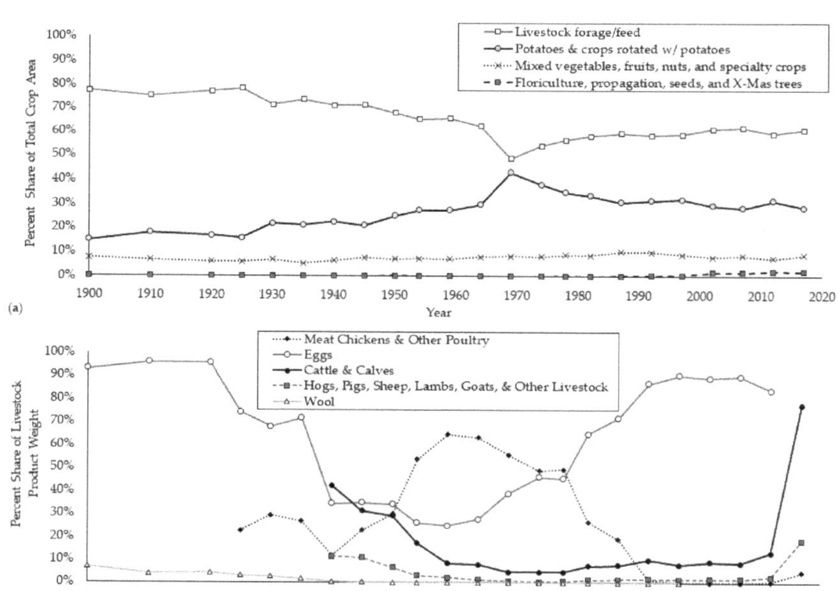

Figure 9. Relative abundance of (**a**) crop and (**b**) livestock product categories' percent share of total crop area from 1900 to 2017 for Maine, USA.

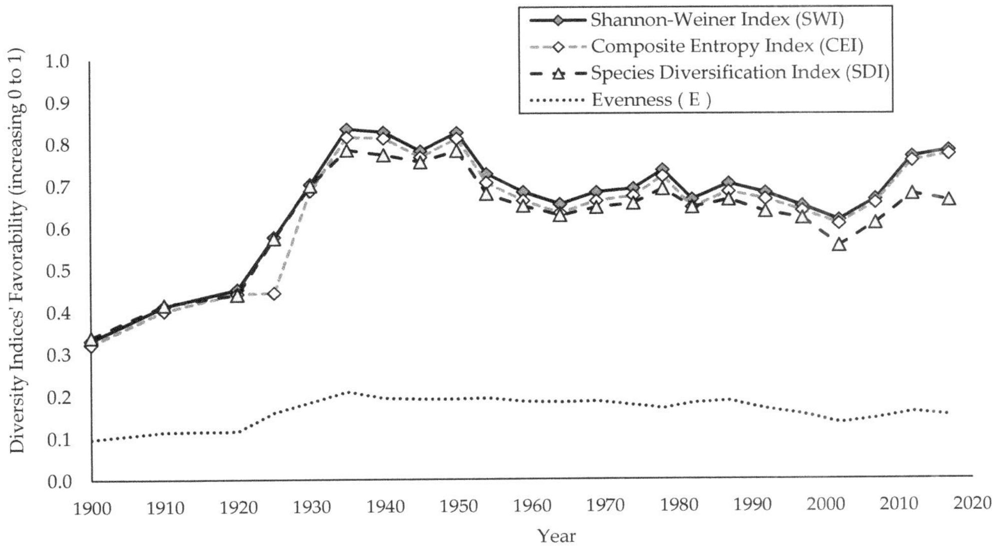

Figure 10. Diversity indexes for mixed vegetable, fruit, nut, and specialty crops from 1900 to 2017 for Maine, USA.

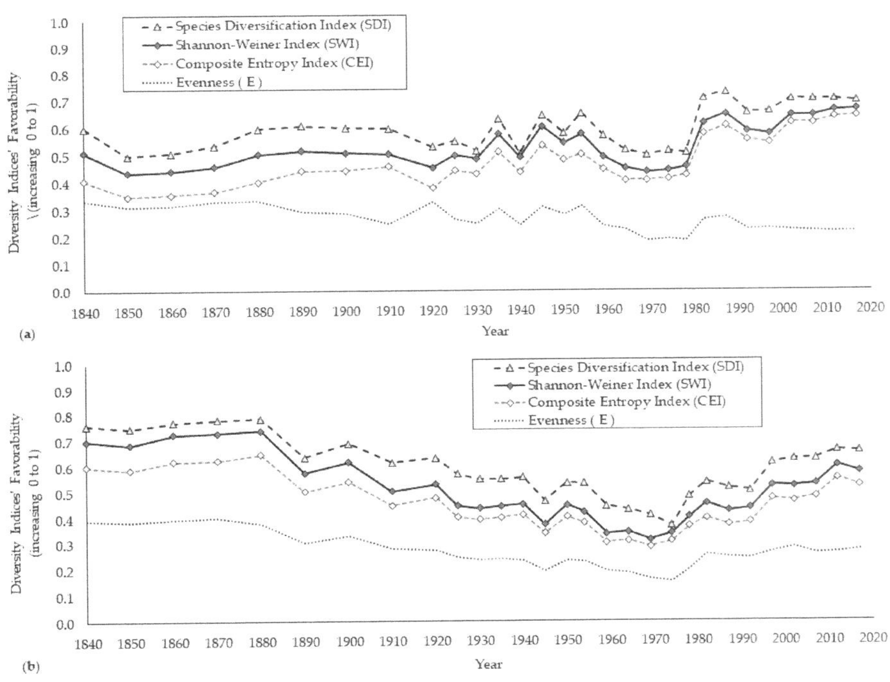

Figure 11. Diversity indexes for (**a**) livestock and (**b**) for potatoes and crops rotated with potatoes from 1840 to 2017 for Maine, USA.

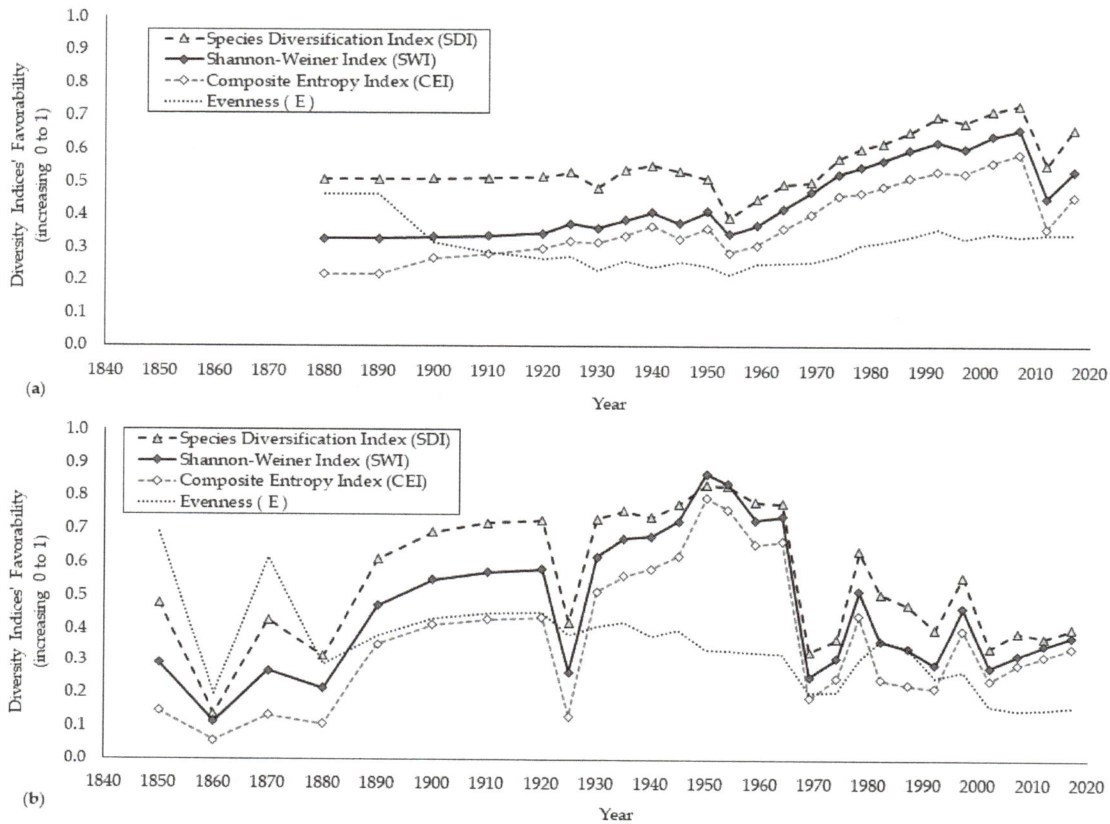

Figure 12. Diversity indexes for (**a**) livestock forage/feed and (**b**) floriculture, propagation, seeds, and X-Mas trees from 1880 to 2017 for Maine, USA.

4. Discussion and Conclusions

4.1. Comparisons and Contrasts to Prior Studies

Compared to past research measuring effective number of species of agricultural crops across the USA, results for Maine (1840 to 2017) were both consistent and different. Maine crops rotated with potatoes (Figure 13a) which are predominantly small grains such as oats (Figure 13b) had crop area following similar trends compared to a state-level USA study from 1870 to 2012 for 22 major field crops [9]. The effective number of species (*ENS*) for these 22 crops ranged between 1 and 7 and peaked during the 1940's and 1960's [9]. For Maine, *ENS* decreased from 10.1 to 3.7 from 1880 to 1969 and then rebounded to 8 in 2012 for crops rotated with potatoes (minus potatoes) for Maine (Figure 8b). A study using USA county-level data for all crop species from 1978 to 2012 found that average *ENS* increased from 5.85 in 1978 to 6.6 in 1997 and then decreased to 5.49 in 2012 for the Northern Crescent region (Great Lakes states, New York, and New England) in the USA [8]. A more recent Geographic Information Systems study analyzing USDA's Cropland Data Layer (CDL at 30 m resolution) of crop categories from 2008 to 2018 found increasing *ENS* in potato producing regions (e.g., northern Aroostook County) and decreasing *ENS* in other areas of Maine [10]. This is consistent with results for Maine's recent *ENS* trends for livestock forage/feed and small grain crops in rotation with potatoes (Figure 8b) as previously discussed.

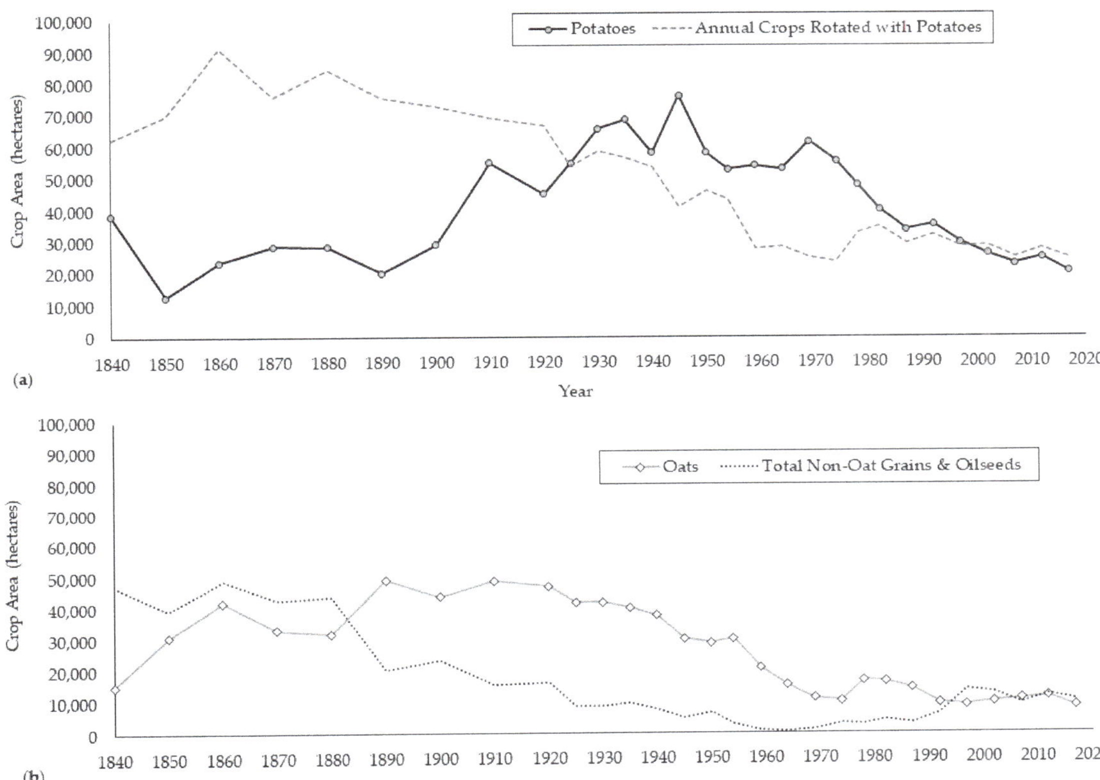

Figure 13. (**a**) Potatoes and annual crops rotated with potatoes (hectares) and (**b**) oats and non-oat grains and oilseeds (hectares) from 1840 to 2017 for Maine, USA.

Unlike other states in this region, *ENS* in Maine has increased and not decreased since 2002 for crop species with the exception of livestock forage/feed (Figure 8b) which had a pattern similar to the Northern Crescent region [8]. Maine's recent increase in crop and livestock diversification over the past 20 years has involved mixed vegetables and specialty crops as well as non-traditional livestock (Figures 8, 10 and 11a; Table 3), which have been more difficult to measure by past research on regional trends [8,9]. Additionally, the Geographic Information Systems (GIS) Cropland Data Layer (CDL) data used in [10] is too coarse to distinguish smaller diversified mixed vegetable farms and crop area which have increased in Maine over this time. From 2007 to 2017, the average number of farms growing any one of 55 categories of mixed vegetables increased from 69 to 169 corresponding to an increase from 25.5 to 36.25 average hectares per category over this time [24]. Thus Maine's *ENS* for mixed vegetables, fruits, nuts, and specialty crops increased 27% from 2002 (59.7) to 2017 (75.8) and did not decrease (Figure 8b).

On diversified produce farms in Maine, blocks or beds of mixed vegetable species can be as small as ~10 square meters (m^2) with the average sized vegetable farm in Maine being 1.42 hectares (ha) [62]. USDA's Cropland Data Layer GIS pixel size is 30 m × 30 m = 900 m^2 = 0.09 ha which could distinguish larger contiguous planting of vegetable species. However, past GIS diversification analysis aggregated grids to a 4 km × 4 km = 16 km^2 = 1600 ha [10] which distinguished consistent patterns for regions within Maine, but not at a farm-level or diversified crop-specific scale. Clearly there is a need for finer scale evaluations of crop/livestock species diversity from analyses of USDA Agricultural Census and Survey data or similar types of national statistics to complement coarser scale analyses.

4.2. Historical Determinants of Agricultural Specialization in Maine

Maine has witnessed boom and bust in both ruminant livestock and horses (Figure 4) as well as the hay, forages, and other livestock feeds (Figure 5) fed to these livestock. During the early 1800's in Maine, livestock were fed manually harvested and cured hay, highly subject to reductions in yield and quality from adverse weather conditions throughout the year [63]. Land in Maine from 1850 to 1910 used to be ~30–70% cleared for farmland as farms were dependent on horses in this pre-tractor era [6]. Maine's sheep industry peak (Figure 7) was driven by the doubling of wool prices during the Civil War [64]. Declines in livestock (Figure 7) and crops (Figure 5) during the mid- to late-1800's can also be explained by the end of the family farming era in Maine as many younger farmers moved to take advantage of Ohio's cheap yet more productive agricultural land [65].

While Maine's boom and bust for ruminant livestock and horses characterized the 19th century, the boom and bust of the potato industry in northern Aroostook County, Maine, dominated the 20th century. Aroostook County farmers were originally diversified and dependent on logging as these two industries were tied together until the 1870's to 1890's. Farmers had to work in the woods to make ends meet. It was not until railroad access was finished in the 1870's that farmers were able to specialize into potatoes by securing more reliable out-of-state markets such as that for potato starch [66,67]. Self-sufficiency in farming in the mid-1800's gave way to industrial agriculture. Additionally, contributing to the specialization boom for Maine potatoes were the establishment of the Maine Agricultural Experiment Station in 1915, local agricultural societies, fairs, and clubs as well as the Grange [67]. This followed an ivory silo issue in 1870's and 1880's where farmers were very distrustful of universities and education was biased against teaching the practice of farming instead focusing on the theory and current research of the academic disciplines related to farming [68].

A common theme behind the decline in Aroostook County's specialized potato production is The County's distance and isolation from both job and product markets. Agriculture in Maine since the early 1900's has been far away from East Coast, USA, markets compared to the rest of New England. There has also been historical brain drain of younger people seeking careers out-of-state which has made economic let alone agricultural stability more challenging. A key difference is that in 1930, more people were involved in farming and self-sufficient food production and procurement. Since the 1980's, producers have been responsible for feeding more people per farm so there is more pressure to maintain farm solvency through farm specialization versus being able to rely on other enterprises and activities to maintain the viability of the farm household [69].

Maine's potato rotations were longer prior to the early 1900's with potato production during the boom intensifying so that potato area exceeded commodity crops commonly rotated with potatoes from 1925 to 1997 (Figure 13a). The dominant crop rotated with potatoes was oats from 1890 to 1992 with more balance of potato rotation crops bookended before and after this one hundred year period (Figure 13b). Specialized agriculture capitalizing on economies of scale can result in reduced farm resilience over time especially if there is a lack of strong centralized marketing [21]. In Maine, this was exemplified by the failure to establish sugar beets as a complementary commodity rotation crop in potato rotations during the late 1960's and early 1970's (Figure 5). Agricultural industry specialization combined with Aroostook County's isolation has made recent adoption of sustainable systems involving crop-livestock integration (CLI) more challenging [70], even though there are mutual economic benefits for specialized potato and dairy farmers to integrate their cropping systems [71]. There is also a lack of regional CLI infrastructure [72] that could facilitate cost-effective movement of excess manure from larger livestock farms in central/southern Maine to northern Maine's non-integrated potato farms.

Crop boom and busts for vegetables primary grown for canning included peaks of 6654 hectares (ha) of sweet corn in 1930, 1616 ha of green beans in 1945 (Table 1), and 4373 ha of peas in 1964 (Table 2) [24]. During the latter half of the 20th century (1940 to 1985), Maine agriculture was characterized by the Green Revolution treadmill of getting bigger or getting out juxtaposed against a focus on diversification into activities not widespread at the time such as sheep production in addition to direct marketing. Specialized commodities in Maine included potatoes, dairy products, broilers, eggs, apples, wild blueberries, and cattle which made up 84% of the value of farm production in 1974 [73].

Agricultural commodity specialization, booms, and busts can be driven by global and regional factors such as trade and/or competition between nations or regions [74]. Commodity production commonly clusters around adequate agricultural support industries as well as greater availability of key agricultural inputs such as labor [75]. Legal factors (e.g., environmental regulations) can also shift entire agricultural industries. For example, the collapse of Maine's broiler chicken industry by 1992 (Figure 5) and the rapid, subsequent shift in broiler production from Maine to the Southeast USA illustrates hard to counteract national/regional forces and policies. Lower labor costs and less stringent regulations in the Southeast USA relative to Maine shifted the broiler chicken industry to this region by the late 1980's and early 1990's [18,73].

4.3. Recent and Future Diversification Directions

Maine's recent crop/livestock diversifications since the 1970's has flourished in the wake of the collapse of the traditional and conventional agricultural systems previously discussed. Rather than crop land being ~30–70% of Maine's land area from 1850 to 1910, by 1970, Maine's farmland had declined to only ~10% of total land area [6]. Despite a more limited agricultural land base compared to historical periods (Figure 2), Maine's agriculture has shifted to more diversified, smaller farms [76]. There are three areas where Maine can continue to diversify its agricultural systems: (1) growing more livestock feed in-state rather than importing this from Canada and/or the Midwest USA, (2) mixed vegetables, fruits, nuts, and specialty crops, and (3) crops rotated with commodity potatoes.

The early to mid-1900's showcased diverse livestock feed (Figure 5) such as unthreshed oats bound for feeding, corn hogged in field, as well as forage turnips and pumpkins [24]. Given increasing effective number of species (Figure 8b) and livestock diversification (Figure 11a) in Maine over the past couple of decades, re-adopting both harvested and in-field supplemental livestock feed presents a tremendous growth opportunity and a future pathway to livestock feed self-sufficiency using cover crops [77]. Livestock forage diversification research has evaluated integrating non-traditional forages into Maine's organic dairy farm systems such as triticale (*Triticosecale rimpaui* Wittm.) and brown midrib sorghum-sudangrass (*Sorghum sudanense* (Piper) Stapf) [25]. Other recent initiatives to diversify organic dairy farm crop rotations have included integrating wheat (*Triticum aestivum*), soybeans (*Glycine max* L.) [78], and sunflower [24] meal as a by-product of sunflower oil production (personal correspondence, Richard Kersbergen, University of Maine). There has been a lack of focus on forage crops consumed in-field as was prevalent in the early to mid-1900's. This presents an opportunity to go "Back to the Future" to diversify crop rotations on farms with non-confined livestock such as hogs, beef cattle, and organic dairy herds, while simultaneously reducing reliance on imported feeds.

Diversification indicators for Maine's mixed vegetable, fruits, nuts, and specialty crops category from 1840 to 1970 support the theory that farms initially become more diverse as market size increases, but then once a critical threshold is reached, farms become increasingly less diverse and more specialized where diversification indicators follow a reverse U-shape over time [79]. Unlike other regions in the USA, Maine mixed vegetables and specialty crops have recently become more diverse since 1970, not increasingly less diverse. Similar to West Bengal, India from 1970 to 2005, demonstrating diversification is influenced by smaller farms and growth of infrastructure networks [80], the Maine Organic Farmers and Gardeners Association has been instrumental in supporting smaller organic farms since 1971 [81]. This recent shift to local and regional food systems presents an opportunity to diversify farming in Aroostook County, which has the greatest potential in New England, USA, for produce distribution [82].

Maine, USA, has had two periods of increased mixed vegetable, fruit, nuts, and specialty/other crop diversification since the 1950's, (1) the back-to-the-land movement of the mid-1970's [83] and (2) the local food movement over the past 15 years. Diversified producers in Maine have focused on economies of scope, retaining a greater share of consumer expenditures [84]. One way to do this is to produce and direct market higher value crops such as sweet potatoes (Table 1), strawberries, and grapes (Tables 1 and 2). In Maine, the 2017 value for these three crops was proportionally greater relative to their inflation-adjusted historical peak values. Non-traditional crops can capitalize on early entry into the market but many of these crops require season extension (e.g., sweet potatoes, ginger, etc.). Profits could also be eroded by competitive entry from other farmers as the market for these non-traditional crops becomes more saturated or by increases in home gardening. For example, 7 out of 54 mixed vegetable/field fruit crops had production declines from 2012 to 2017 (tomatoes, peppers, turnip greens, sweet corn, cucumbers, green beans, and broccoli) with >20% drops in area for tomatoes (−54.4%), peppers (−38.2%), and turnip greens (−26.5%) [24].

Maine's potato rotations were historically more integrated with livestock forages [85]. Although there has been increased diversification in potatoes and potato rotation crops (Figure 11b) since the 1969 boom and bust in sugar beets (Figure 5), recent potato rotation crops have been dominated by commodities such as barley grown for malting with limited area devoted to higher value small grains such as wheat [24]. Barley recently peaked at 10,464 ha in 2002 compared to only a 2012 peak of 968 ha for wheat (Table 2) despite more recent interest and research on expanding organic wheat production in Maine and Vermont, USA [86]. With almost 90% of Maine's potato production concentrated in Aroostook County in the northeast corner of the state [85], there have been limited options for higher value, more profitable potato rotation crops such as broccoli [87]. Recent diversification into broccoli production in Aroostook County has peaked at 2555 ha in 2012 (Table 2). Future efforts could focus on other higher value commodity vegetables for produce or processing that can be rotated with potatoes.

Future research could also statistically test potential drivers of recent diversification trends in Maine. These potential drivers include socio-demographic characteristics of diversified farmers including off-farm income stability [59,88], species selection and input use [61], agricultural technologies such as irrigation and equipment [10,61,88], regional infrastructure [88], population density [59], and access to Extension, market information, and rural credit [88]. Future studies could also evaluate potential food security benefits of crop-livestock integration [89] in addition to better quantifying diversification and benefits from inter-cropping [90]. Aggregate crop data and agricultural statistical surveys do not measure if crops are inter-cropped so cultivar/species richness may underestimate positive synergistic impacts.

Funding: This research received no external funding.

Institutional Review Board Statement: Not applicable.

Informed Consent Statement: Not applicable.

Data Availability Statement: Publicly available datasets were analyzed in this study. These data can be found here: U.S. Department of Agriculture (USDA) Cropland Data Layer [https://nassgeodata.gmu.edu/CropScape/] accessed on 21 October 2022 and USDA Agricultural Census and Survey [https://quickstats.nass.usda.gov/] accessed on 21 October 2022.

Acknowledgments: The author would like to thank MDPI Sustainability for the opportunity to serve as guest editor for the Special Issue "Sustainable Agricultural Development Economics and Policy". Thanks also to all the researchers who have taken the time to continuously improve their manuscripts for this Special Issue. Many thanks to Ronaldo A. de Oliveira at AgriSciences, Universidade Federal de Mato Grosso, Sinop, Mato Grosso state, Brazil, for creating the Maine crops map using U.S. Department of Agriculture's (USDA's) Cropland Data Layer as recommended by Tara King at USDA Natural Resource Conservation Service in Bangor, Maine. Nancy McBrady and Willie Sawyer Grenier at Maine Department of Agriculture (MDA) authorized permission to use MDA's map of major crop growing areas in Maine. Thanks also to John Harker, Richard Judd at the University of Maine, and David Vail at Bowdoin College for advising on map research. Also, thanks to Jim Barrett at USDA National Agricultural Statistics Service. Thanks to Tim Schermerhorn at U.S. Bureau of Labor Statistics for providing historical Producer Price Index data. The organization and writing of this research was substantially improved with edits and feedback from three anonymous reviewers.

Conflicts of Interest: The author declares no conflict of interest. Supporting entities had no role in the design of the study; in the collection, analyses, or interpretation of data; in the writing of the manuscript, or in the decision to publish the results.

References

1. Hill, S.L.L.; Arnell, A.; Maney, C.; Butchart, S.H.M.; Hilton-Taylor, C.; Ciciarelli, C.; Davis, C.; Dinerstein, E.; Purvis, A.; Burgess, N.D. Measuring Forest Biodiversity Status and Changes Globally. *Front. For. Glob. Chang.* **2019**, *2*, 70. [CrossRef]
2. Viljur, M.-L.; Abella, S.R.; Adámek, M.; Alencar, J.B.R.; Barber, N.A.; Beudert, B.; Burkle, L.A.; Cagnolo, L.; Campos, B.R.; Chao, A.; et al. The effect of natural disturbances on forest biodiversity: An ecological synthesis. *Biol. Rev.* **2022**, *97*, 1930–1947. [CrossRef] [PubMed]
3. Tinya, F.; Kovács, B.; Bidló, A.; Dima, B.; Király, I.; Kutszegi, G.; Lakatos, F.; Mag, Z.; Márialigeti, A.; Nascimbene, J.; et al. Environmental drivers of forest biodiversity in temperate mixed forests—A multi-taxon approach. *Sci. Total Environ.* **2021**, *795*, 148720. [CrossRef] [PubMed]
4. Hilmers, T.; Friess, N.; Bässler, C.; Heurich, M.; Brandl, R.; Pretzsch, H.; Seidl, R.; Müller, J. Biodiversity along temperate forest succession. *J. App. Ecol.* **2018**, *55*, 2756–2766. [CrossRef]
5. Flatebo, G.; Foss, C.R.; Pelletier, S.K. *Biodiversity in the Forests of Maine: Guidelines for Land Management*; Bulletin #7147; University of Maine Cooperative Extension: Bangor, ME, USA, 1999; 167p, Available online: http://www.upperstjohnriver.com/BFM.pdf (accessed on 23 August 2022).
6. Ahn, S.; Krohn, W.B.; Plantinga, A.J.; Dalton, T.J.; Hepinstall, J.A. *Agricultural Land Changes in Maine: A Compilation and Brief Analysis of Census of Agriculture Data, 1850–1997*; Technical Bulletin 182; Maine Agricultural and Forest Experiment Station, The University of Maine: Orono, ME, USA, 2002; 31p, Available online: https://digitalcommons.library.umaine.edu/cgi/viewcontent.cgi?article=1026&context=aes_techbulletin (accessed on 3 October 2022).
7. Czajkowski, M.; Buszko-Briggs, M.; Hanley, N. Valuing changes in forest biodiversity. *Ecol. Econ.* **2009**, *68*, 2910–2917. [CrossRef]
8. Aguilar, J.; Gramig, G.G.; Hendrickson, J.R.; Archer, D.W.; Forcella, F.; Liebig, M.A. Crop Species Diversity Changes in the United States: 1978–2012. *PLoS ONE* **2015**, *10*, e0136580. [CrossRef]
9. Hijmans, R.J.; Choe, H.; Perlman, J. Spatiotemporal Patterns of Field Crop Diversity in the United States. *Agric. Environ. Lett.* **2016**, *1*, 160022. [CrossRef]
10. Goslee, S.C. Drivers of Agricultural Diversity in the Contiguous United States. *Front. Sustain. Food Sys.* **2020**, *4*, 75. [CrossRef]
11. Maine Department of Agriculture and Maine Department of Labor. Major Crop Growing Areas. Center for Workforce Research and Information, Division of Migrant & Immigrant Services. Available online: https://www.maine.gov/labor/labor_laws/crop_map.html (accessed on 18 November 2022).
12. *USDA National Agricultural Statistics Service Cropland Data Layer*; Published Crop-Specific Data Layer [Online]; USDA-NASS: Washington, DC, USA, 2017; Available online: https://nassgeodata.gmu.edu/CropScape/ (accessed on 15 November 2022).
13. Jacks, D.S. From boom to bust: A typology of real commodity prices in the long run. *Cliometrica* **2019**, *13*, 201–220. [CrossRef]
14. Henderson, J.; Gloy, B.; Boehlje, M. Agriculture's Boom-Bust Cycles: Is This Time Different? Federal Reserve Bank of Kansas City. *Econ. Rev.* **2011**, *96*, 81–103. Available online: https://www.kansascityfed.org/documents/939/2011-Agriculture\T1\textquoterights%20Boom-Bust%20Cycles:%20Is%20This%20Time%20Different%3F.pdf (accessed on 3 August 2021).
15. Jaremski, M.; Wheelock, D.C. Banking on the Boom, Tripped by the Bust: Banks and the World War I Agricultural Price Shock. *J. Money Crédit Bank.* **2020**, *52*, 1719–1754. [CrossRef]
16. Carter, C.A.; Rausser, G.C.; Smith, A. Commodity Booms and Busts. *Ann. Rev. Resour. Econ.* **2011**, *3*, 87–118. [CrossRef]
17. Heffernan, P. Nothing is forever: Boom and bust in Midwest farming. *Econ. Perspect.* **1987**, *11*, 27–31. Available online: https://www.chicagofed.org/publications/economic-perspectives/1987/17sepoct1987-part4-heffernan (accessed on 3 August 2021).
18. Schmitz, A. Boom/Bust Cycles and Ricardian Rent. *Amer. J. Agric. Econ.* **1995**, *77*, 1110–1125. Available online: https://www.jstor.org/stable/1243332 (accessed on 3 August 2021). [CrossRef]

19. Kurosaki, T. Specialization and Diversification in Agricultural Transformation: The Case of West Punjab, 1903–1992. *Am. J. Agric. Econ.* **2003**, *85*, 372–386. [CrossRef]
20. Winsberg, M.D. Agricultural Specialization in the United States since World War II. *Agric. Hist.* **1982**, *56*, 692–701. Available online: https://www.jstor.org/stable/3742350 (accessed on 2 June 2021).
21. De Roest, K.; Ferrari, P.; Knickel, K. Specialization and economies of scale or diversification and economies of scope? Assessing different agricultural development pathways. *J. Rural Stud.* **2018**, *59*, 222–231. [CrossRef]
22. Hansson, H.; Ferguson, R.; Olofsson, C. Understanding the diversification and specialization of farm businesses. *Agric. Food Sci.* **2010**, *19*, 269–283. [CrossRef]
23. Barbieri, C.; Mahoney, E.; Butler, L. Understanding the Nature and Extent of Farm and Ranch Diversification in North America. *Rural Soc.* **2008**, *73*, 205–229. [CrossRef]
24. U.S. Department of Agriculture, National Agricultural Statistics Service (NASS). 1840–2017. Census of Agriculture and Crop Production Survey. Available online: https://quickstats.nass.usda.gov/ (accessed on 21 April 2021).
25. Hoshide, A.K.; Halloran, J.M.; Kersbergen, R.J.; Griffin, T.S.; DeFauw, S.L.; Lagasse, B.J.; Jain, S. Effects of stored feed cropping systems and farm size on the profitability of Maine organic dairy farm simulations. *J. Dairy Sci.* **2011**, *94*, 5710–5723. [CrossRef] [PubMed]
26. California Certified Organic Farmers (CCOF). Average Dry Matter Percentages for Various Livestock Feeds. CCOF Foundation, 2155 Delaware Avenue, Suite 150, Santa Cruz, CA 95060. Available online: https://www.ccof.org/sites/default/files/Feed%20Type%20DMI%20Table%20Final.pdf (accessed on 9 September 2022).
27. Khatib, S.E.; Muhieddine, M. Nutritional Profile and Medicinal Properties of Pumpkin Fruit Pulp. In *The Health Benefits of Foods—Current Knowledge and Further Development*, 1st ed.; Salanta, L.C., Ed.; IntechOpen: London, UK, 2019; p. 19. [CrossRef]
28. Best Forage. Barkant Forage Turnips. 11570 E 275 S, LaGrange, IN 46761. Available online: https://www.bestforage.com/wp-content/uploads/2009/07/file_2_25.pdf (accessed on 9 September 2022).
29. Smith, S.N.; Plant, A.; Potter, D.; Hoshide, A.K. The Economics of Producing Grass-Finished and Natural Beef in Maine. Maine Agricultural Center Integrated Research and Extension Projects: 2010–2011, MAC121; Maine Agricultural Center, University of Maine, 2011; 23p, Available online: https://umaine.edu/mainefoodandagcenter/wp-content/uploads/sites/403/2016/04/MAC-projects-2010-2011.pdf (accessed on 10 August 2022).
30. Oklahoma State University, Department of Animal Science. Breeds of Livestock. Available online: http://afs.okstate.edu/breeds/sheep/corriedale/#content (accessed on 15 May 2021).
31. University of Maine Cooperative Extension. Dairy and Livestock Management Fact Sheet: Are You Thinking of Raising Sheep. Bulletin #2186; 6p, Available online: https://extension.umaine.edu/publications/wp-content/uploads/sites/52/2015/04/2186.pdf (accessed on 23 February 2021).
32. Blazavic, L.; Chessen, M.; Feal-Staub, C.; Gyauch-Lewis, D.; McKenzie, H.; Pearson, N.; Ratsimbazafy, M.; Ros, C.; Sperduto, T. Goat Production in Maine: A Report to the Somali Bantu Mutual Assistance Association of Lewiston-Auburn. Bates College, 2019; 35p, Available online: https://www.bates.edu/faculty-expertise/files/2015/08/ANTHRO339-Goat-Project-Report-December-2019.pdf (accessed on 1 March 2021).
33. Delaney, C. A Guide to Starting a Commercial Goat Dairy. University of Vermont Center for Sustainable Agriculture, 2012; 157p, Available online: https://www.uvm.edu/sites/default/files/media/goatguide.pdf (accessed on 1 March 2021).
34. Equi-Analytical Laboratory Services, 730 Warren Road, Ithaca, NY, 14850, USA. Available online: https://equi-analytical.com/resources/typical-body-weights/ (accessed on 2 June 2021).
35. Knight, C.W. Understanding Poultry Yields. University of Maine Cooperative Extension, Bulletin #2223; 2020; 4p, Available online: https://extension.umaine.edu/publications/2223e/ (accessed on 3 June 2021).
36. Fletcher, D.L.; Carpenter, J.A. Breast Meat and Parts Yields from Four Retail Brands of Broiler Chickens Obtained in the Northeast Georgia Area. *Poult. Sci.* **1993**, *72*, 2347–2352. [CrossRef]
37. Carlson, C.W.; Marion, W.W.; Miller, B.F.; Goodwin, T.L. Factors Affecting Poultry Meat Yields. Agricultural Experiment Station, South Dakota State University; Bulletins, Paper 635; 1975; 18p, Available online: https://openprairie.sdstate.edu/cgi/viewcontent.cgi?article=1634&context=agexperimentsta_bulletins (accessed on 28 February 2021).
38. Kokoszyński, D.; Bernacki, Z.; Cisowska, A. Growth and development of young game pheasants (*Phasianus colchicus*). *Archiv Tierz.* **2011**, *54*, 83–92. [CrossRef]
39. Wilkanowska, A.; Kokoszyński, D. Comparison of slaughter value in Pharaoh quail of different ages. *J. Cent. Eur. Agric.* **2011**, *12*, 145–154. [CrossRef]
40. Krishnan, L.M. A comparative study on carcass yield in male and female Japanese quail (*Coturnix coturnix japonica*). *Plant Arch.* **2019**, *19*, 2099–2100. Available online: http://plantarchives.org/19-2/2099-2100%20(5409).pdf (accessed on 19 May 2021).
41. Gao, C.Q.; Yang, J.X.; Chen, M.X.; Yan, H.C.; Wang, X.Q. Growth curves and age-related changes in carcass characteristics, organs, serum parameters, and intestinal transporter gene expression in domestic pigeon (*Columba livia*). *Poul. Sci.* **2016**, *95*, 867–877. [CrossRef]
42. Daniel, D.R. Carcass Characteristics, Composition, and Palatability of Emu Meat. Masters Thesis, Texas Tech University, Lubbock, TX, December, 1995; 72p. Available online: https://ttu-ir.tdl.org/bitstream/handle/2346/11436/31295008888272.pdf?sequence=1 (accessed on 17 June 2021).

43. Hauer, Gerald. Expected Meat Yield from a Bison Bull Carcass. Canadian Bison. Available online: https://www.canadianbison.ca/application/files/2414/8778/3203/ExpectedMeatYieldfromaBisonBullCarcass.pdf (accessed on 19 May 2021).
44. Butcher & Packer Supply Company (B&P), 1780 E. 14 Mile Rd., Madison Heights, MI, USA. Available online: https://www.butcher-packer.com (accessed on 22 May 2021).
45. Bioanospino, E.; Wechsler, F.S.; Moura, A.S.A.M.T.; Fernandes, S. Growth Traits and Dressing Percentage of Straightbred and Crossbred Rabbits. In Proceedings of the 8th Word Rabbit Conference, Puebla, Mexico, 7–10 September 2004; World Rabbit Science Association: Valencia, Spain, 2004; pp. 1360–1365. Available online: http://world-rabbit-science.com/WRSA-Proceedings/Congress-2004-Puebla/Papers/Meat%20Quality/Q-Bianospono-2.pdf (accessed on 20 May 2021).
46. Schaeffer, R.; Kime, L.F.; Harper, J.K. Rabbit Production. Agricultural Alternatives; College of Agricultural Sciences, Agricultural Research and Cooperative Extension, Penn State University, 2008; 6p, Available online: https://extension.psu.edu/rabbit-production (accessed on 20 May 2021).
47. Mississippi State University, Extension Service. Slaughtering and Dressing Rabbits. Available online: http://extension.msstate.edu/content/slaughtering-and-dressing-rabbits (accessed on 20 May 2021).
48. Jůzl, R.; Suchý, P.; Straková, E.; Rusníková, L.; Macháček, M.; Marada, P. Nutritional value of breast and thigh muscle of chukar partridge (*Alectoris chukar*) under intensive fattening conditions. *Acta Vet. Brno* **2012**, *81*, 251–255. [CrossRef]
49. Kokoszyński, D.; Bernacki, Z.; Korytkowska, H.; Wilkanowska, A.; Frieske, A. Carcass composition and meat quality of grey partridge (*Perdix perdix* L.). *J. Cent. Eur. Agric.* **2013**, *14*, 378–387. [CrossRef]
50. Hoffman, L.C.; Brand, M.M.; Muller, M.; Cloete, S.W.P. Carcass and muscle yields of ostriches as influenced by genotype. *S. Afr. J. Animal Sci.* **2007**, *37*, 256–260. [CrossRef]
51. Horbańczuk, O.K.; Wierzbicka, A. Technological and nutritional properties of ostrich, emu, and rhea meat quality. *J. Vet. Res.* **2016**, *60*, 279–286. [CrossRef]
52. Ramírez, R.M.; Mota-Rojas, D.; Reyes, A.D.L.; Becerril-Herrera, M.; Flores-Pintado, S.; Alonso-Spilsbury, M.D.L.; Cardona, L.A.; Ramírez-Necoechea, R.; Lemus-Flores, C. Slaughtering Process, Carcass Yield and Cutting Process in California and Chinchilla Rabbit Breeds. *J. Food Technol.* **2006**, *4*, 86–89. Available online: https://medwelljournals.com/abstract/?doi=jftech.2006.86.89 (accessed on 20 May 2021).
53. Kumar, S.D.; Singh, D.A.P.; Natarajan, A.; Sivakumar, K. Carcass Characteristics of Soviet Chinchilla Rabbits Supplemented with Vitamin C, E and Selenium during the Period of Heat Stress. *Int. J. Curr. Microbiol. Appl. Sci.* **2018**, *7*, 1962–1969. [CrossRef]
54. Hilker, L.D.; Caldwell, W.R. A method for calculating the weight per gallon of fluid dairy products. *J. Dairy Sci.* **1961**, *44*, 183–188. [CrossRef]
55. Kempster, H.L. The Normal Growth of Chickens. Agricultural Experiment Station, College of Agriculture, University of Missouri; Bulletin 423; 1941; 20p, Available online: https://mospace.umsystem.edu/xmlui/handle/10355/54554 (accessed on 18 May 2021).
56. U.S. Department of Agriculture, National Agricultural Statistics Service. Chickens and Eggs (1965 to 2019). Available online: https://usda.library.cornell.edu/concern/publications/8623hx75j (accessed on 14 April 2021).
57. Bureau of Labor Statistics (BLS). Producer Price Indexes (PPI), 1720–1912 (archived) & 1913–2020. 1913-Present. Available online: https://www.bls.gov/ppi/data-retrieval-guide/home.htm (accessed on 9 September 2022).
58. Eaton, D.; Windig, J.; Hiemstra, S.J.; van Veller, M.; Trach, N.X.; Hao, P.X.; Doan, B.H.; Hu, R. Indicators for Livestock and Crop Biodiversity. Center for Genetic Resources: Wageningen, The Netherlands, 2006; 56p, Available online: https://www.researchgate.net/publication/40106466_Indicators_for_livestock_and_crop_biodiversity (accessed on 11 May 2021).
59. Dessie, A.B.; Abate, T.M.; Mekie, T.M.; Liyew, Y.M. Crop diversification analysis on red pepper dominated smallholder farming system: Evidence from northwest Ethiopia. *Ecol. Process.* **2019**, *8*, 11. [CrossRef]
60. Jha, B.; Kumar, N.; Mohanty, B. Pattern of Agricultural Diversification in India. Institute of Economic Growth, University of Delhi Enclave, North Campus, Delhi, India, Working Paper Series No. E/302; 2009; 55p, Available online: https://ideas.repec.org/p/eab/develo/22923.html (accessed on 8 June 2021).
61. Ahmad, N.; Sinha, D.K.; Singh, K.M.; Kumar, A. Determinants of Crop Diversification in Bihar Agriculture—An Economic Analysis. *Environ. Ecol.* **2017**, *35*, 3683–3687. [CrossRef]
62. Brown, B.; Hoshide, A.K.; Gallandt, E.R. An economic comparison of weed management systems used in small-scale organic vegetable production. *Org. Agric.* **2018**, 11. [CrossRef]
63. Baron, W.R.; Bridges, A.E. Making Hay in Northern New England: Maine as a Case Study, 1800–1850. *Agric. Hist.* **1983**, *57*, 165–180. Available online: https://www.jstor.org/stable/3743154 (accessed on 10 August 2021).
64. Tuthill, D.F. An economic study of sheep production in Maine. Maine Agricultural Experiment Station, The University of Maine; Bulletin 619; 1964; 45p, Available online: https://digitalcommons.library.umaine.edu/cgi/viewcontent.cgi?article=1076&context=aes_bulletin (accessed on 23 February 2021).
65. Moore, E.H.; Witham, J.W. From Forest to Farm and Back Again: Land Use History as a Dimension of Ecological Research in Coastal Maine. *Environ. Hist.* **1996**, *1*, 50–69. Available online: https://www.jstor.org/stable/3985156 (accessed on 10 August 2021). [CrossRef]
66. Judd, R.W. Lumbering and the Farming Frontier in Aroostook County, Maine, 1840–1880. *J. Forest Hist.* **1984**, *28*, 56–67. Available online: https://www.jstor.org/stable/4004772 (accessed on 10 August 2021). [CrossRef]

67. Reznick, T.J.H.; Science, Technology and the Garden of Maine: Industrial Farming in Aroostook County, 1850–1900. Senior Scholars Papers in Computer Science, Colby College, Paper 2, 2007; 71p. Available online: https://digitalcommons.colby.edu/cgi/viewcontent.cgi?article=1001&context=cs_senior (accessed on 27 March 2021).
68. Marcus, A.I. The Ivory Silo: Farmer-Agricultural College Tensions in the 1870s and 1880s. *Agric. Hist.* **1986**, *60*, 22–36. Available online: https://www.jstor.org/stable/3743428 (accessed on 10 August 2021).
69. Condon, R.H. Living in Two Worlds: Rural Maine in 1930. *Maine Hist. J. Rur. Maine* **1985**, *25*, 58–87. Available online: https://digitalcommons.library.umaine.edu/cgi/viewcontent.cgi?article=1517&context=mainehistoryjournal (accessed on 10 August 2021).
70. Hoshide, A.K.; Smith, S.N. Managing Risks by Integrating Cropping and Livestock Enterprises in Aroostook County, Maine. Final Grant Results, Northeast Center for Risk Management Education, 2006 Unrestricted Risk Management Education, RME-DJ202049; 2007; 27p, Available online: http://www.agrisk.umn.edu/ProjectSearch/SearchProjects.aspx (accessed on 2 September 2022).
71. Hoshide, A.K.; Dalton, T.J.; Smith, S.N. Profitability of coupled potato and dairy farms in Maine. *Renew. Agric. Food Sys.* **2006**, *21*, 261–272. [CrossRef]
72. Asai, M.; Moraine, M.; Ryschawy, J.; de Wit, J.; Hoshide, A.K.; Martin, G. Critical factors to crop-livestock integration beyond the farm level: A cross-analysis of worldwide case studies. *Land Use Pol.* **2018**, *72*, 184–194. [CrossRef]
73. Wescott, R.; Vail, D. The Transformation of Farming in Maine, 1940–1985. *Maine Hist. J. Character N. Engl.* **1988**, *28*, 66–84. Available online: https://digitalcommons.library.umaine.edu/cgi/viewcontent.cgi?article=1422&context=mainehistoryjournal (accessed on 27 March 2021).
74. Gurgel, A.C.; Reilly, J.; Blanc, E. Agriculture and forest land use change in the continental United States: Are there tipping points? *iScience* **2021**, *24*, 102772. [CrossRef] [PubMed]
75. Carpenter, C.W.; Van Sandt, A.; Loveridge, S. Food and agricultural industry locational determinants research: Aggregation bias and size measurement in the agricultural support industry. *Agric. Res. Econ. Rev.* **2022**, *51*, 1–21. [CrossRef]
76. Smith, S.N. Maine Agriculture: Adapting to a More Complex Setting. In *Changing Maine 1960–2010*, 1st ed.; Barringer, R., Ed.; Tilbury House, Publishers: Gardiner, ME, USA, 2004; Chapter 20; pp. 388–408.
77. Chatterjee, R. Cover Crops for Forages and Livestock Grazing. In *Cover Crops and Sustainable Agriculture*, 1st ed.; Islam, R., Sherman, B., Eds.; CRC Press, Taylor & Francis: New York, NY, USA, 2021; Chapter 8; pp. 83–107.
78. Abreu, D.C.; Hoshide, A.K.; Mallory, E.B.; Roche, E.H.; Oliveira, A.S.; Kersbergen, R.J.; Lana, R.P.; Fonseca, M.A. Economic and environmental implications of wheat crop sequences on organic dairy farm simulations. *Crop Pasture Sci.* **2016**, *67*, 1127–1138. [CrossRef]
79. Emran, M.S.; Shilpi, F. The extent of the market and stages of agricultural specialization. *Can. J. Econ.* **2012**, *45*, 1125–1153. [CrossRef]
80. De, U.K.; Chattopadhyay, M. Crop diversification by poor peasants and role of infrastructure: Evidence from West Bengal. *J. Dev. Agric. Econ.* **2010**, *2*, 340–350. Available online: https://academicjournals.org/article/article1411401233_De%20and%20Chattopadhyay.pdf (accessed on 22 April 2021).
81. Maine Organic Farmers and Gardeners (MOFGA). 294 Crosby Brook Road, Unity, ME 04988. Available online: https://www.mofga.org (accessed on 20 September 2022).
82. Ge, H.; Canning, P.; Goetz, S.; Perez, A.; Li, J. Embedding Economies of Scale Concepts in the Model of Optimal Locations of Fresh Produce Aggregation Hubs. *Agric. Resour. Econ. Rev.* **2019**, *48*, 365–387. [CrossRef]
83. Britt, C.; Rozzyne, M.; Roberts, T.; Vail, D. *The Past, the Present, and the Future Competitiveness of Maine Agriculture*, 1st ed.; The Maine Consortium for Food Self-Reliance: Freeport, ME, USA, 1980; pp. 1–73.
84. Hoshide, A.K. Impacts of Technology Adoption: Comparing Returns to the Farming Sector in Maine under Alternative Technology Regimes. Master's Thesis, University of Maine, Orono, ME, USA, 21 August 2002. DigitalCommons@UMaine. Available online: https://digitalcommons.library.umaine.edu/cgi/viewcontent.cgi?article=1563&context=etd (accessed on 20 October 2022).
85. Hoshide, A.K.; Dalton, T.J.; Smith, S.N. *Representative Farm Budgets and Performance Indicators for Integrated Farming Practices in Maine*; Bulletin 850; Maine Agricultural and Forest Experiment Station, University of Maine at Orono: Orono, ME, USA, 2004; 75p, Available online: http://www.umaine.edu/mafes/elec_pubs/bulletins/B850.pdf (accessed on 19 November 2022).
86. Mallory, E.B.; Darby, H. In-Season Nitrogen Effects on Organic Hard Red Winter Wheat Yield and Quality. *Agron. J.* **2013**, *105*, 1167–1175. [CrossRef]
87. DeFauw, S.L.; English, P.J.; Larkin, R.P.; Halloran, J.M.; Hoshide, A.K. Geospatial evaluations of potato production systems in Maine. *Am. J. Potato Res.* **2012**, *89*, 471–488. [CrossRef]
88. Asante, B.O.; Villano, R.A.; Patrick, I.W.; Battese, G.E. Determinants of farm diversification in integrated crop-livestock farming systems. *Renew. Agric. Food Syst.* **2018**, *33*, 131–149. [CrossRef]
89. Mulwa, C.K.; Visser, M. Farm diversification as an adaptation strategy to climatic shocks and implications for food security in northern Namibia. *World Dev.* **2020**, *129*, 104906. [CrossRef]
90. Paut, R.; Sabatier, R.; Tchamitchian, M. Modelling crop diversification and association effects in agricultural systems. *Agric. Ecosys. Environ.* **2020**, *288*, 106711. [CrossRef]

sustainability MDPI

Article

U.S. Almond Exports and Retaliatory Trade Tariffs

Abraham Ajibade and Sayed Saghaian *

Department of Agricultural Economics, University of Kentucky, Barnhart Building, Lexington, KY 40546-0276, USA; abraham.ajibade@uky.edu
* Correspondence: ssaghaian@uky.edu; Tel.: +1-859-619-1818

Abstract: The U.S. is the top producer, exporter, and consumer of tree nuts in the world. Tree nuts are a significant part of U.S. agricultural exports to the world. In 2019, the U.S. exported about USD 9.1 billion worth of tree nuts, just behind soybean exports at USD 18.7 billion. Tree nuts, such as almonds and pistachios, are mostly produced in the state of California. California produces 100% of U.S. commercial almonds. Globally, almonds are the leading U.S. tree nut export in both value and volume. Almonds are shipped to over 90 countries annually. This study aimed to investigate factors affecting the export demand function for U.S. almonds in major destination countries and evaluate the impact of the retaliatory trade tariffs policy by some of the importing countries on the U.S. almond exports. The currently available literature does not fully address these issues. We identified the top five almond export destinations, which were in Europe and Asia, namely, China/Hong Kong, Germany, India, Japan, and Spain, which account for more than 50% of U.S. almond imports. We used a double-log export demand equation that is well referenced in the literature and economic theory to identify the significant explanatory variables affecting the U.S. almonds export demand function. We also tried to estimate the impact of retaliatory tariffs on almond exports imposed by the major importing countries. Our results showed that U.S. almond and pistachio prices, real exchange rates, and gross domestic products of importing countries were significant factors that affected U.S. almond exports. The results showed that the imposed retaliatory tariffs had no negative effect on U.S. almond exports. This could have been because the study ended in 2019 and did not involve enough data to fully evaluate the impact of the retaliatory trade tariffs policy. U.S. almond exports have market concentration and strong market power in international markets. The efforts toward more sustainable production of almonds to solidify an already established market share in the world almond markets and against substitutes, such as pistachios, seem to be a sound strategy and focus of the U.S. almond agribusinesses and exporters.

Keywords: almond nuts; export demand; retaliatory tariff policy

Citation: Ajibade, A.; Saghaian, S. U.S. Almond Exports and Retaliatory Trade Tariffs. *Sustainability* **2022**, *14*, 6409. https://doi.org/10.3390/su14116409

Academic Editor: Aaron K. Hoshide

Received: 24 April 2022
Accepted: 23 May 2022
Published: 24 May 2022

1. Introduction

According to the United States Department of Agriculture, the United States (U.S.) dominates the world's almond market as a top producer, consumer, and exporter [1]. The U.S. has a competitive advantage in tree nut production and exports and is well-positioned to maintain its global dominance over time. Since the 1980s, the U.S. is the number one producer of almonds globally (Figure 1). Other notable producers are Australia, Chile, Spain, and Italy [1]. Almond production has rapidly outpaced other U.S. tree nuts, such as walnuts, pistachios, pecans, and hazelnuts.

When it comes to consumption and exports, the U.S. has maintained a one-third to two-thirds ratio, where one-third of produced almonds are consumed locally, while two-thirds are exported to other countries (Figure 2). In 2019, the U.S. consumed 377,717 metric tons of almonds, or 33% of its total almonds production locally, and exported 731,177 metric tons, or 67% of the total to other countries. Local consumption grew in the last decade, just as exports did. To put this into context, the growth of almond consumption increased from 0.42 pounds per person in 1980 to 2.36 pounds in 2019 [2].

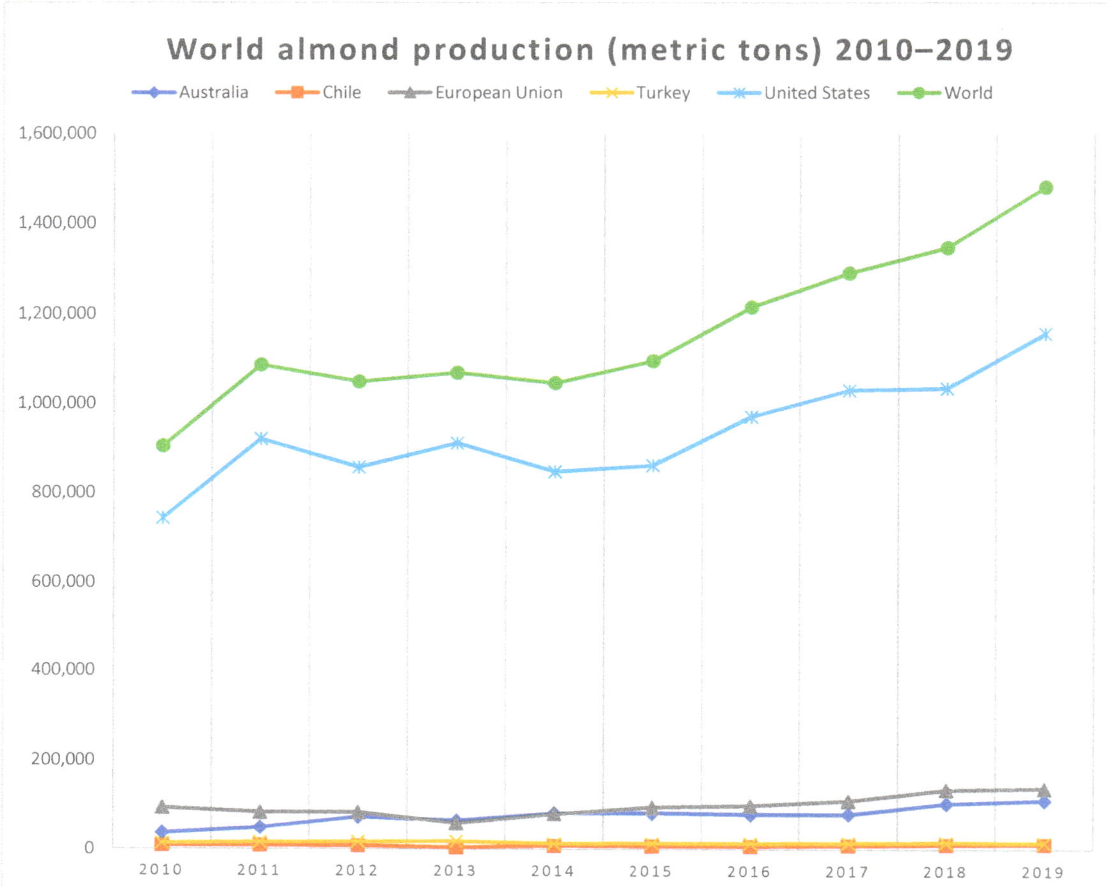

Figure 1. World almond production (in metric tons) 2010–2019. Source: USDA, Foreign Agricultural Service.

Almonds, the leading U.S. tree nut export, both in value and volume, are shipped to over 90 countries annually, with about 70% of exports going to the top 10 export destinations: China/Hong Kong, Germany, India, Italy, Japan, the Netherlands, Spain, South Korea, Turkey, and the United Arab Emirates (U.A.E) [2]. In 2019, U.S. tree nut exports were made up of 54% almonds valued at USD 4.9 billion, 22% pistachios valued at USD 2.0 billion, 14% walnuts valued at USD 1.3 billion, 5% pecans valued at USD 475 million, 4% 'mixed and other nuts' valued at USD 350 million, and 1% hazelnuts valued at USD 90 million [1].

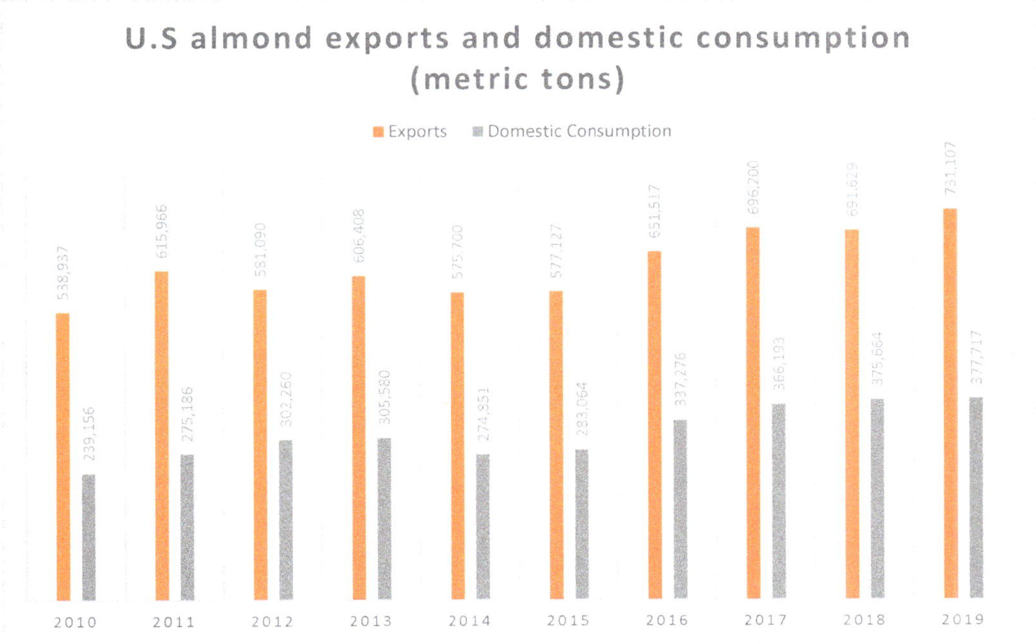

Figure 2. U.S. almond domestic consumption and exports. Source: USDA Foreign Agricultural Service.

In March 2018, the Trump administration raised tariffs on imports from key U.S. trading partners. In response, retaliatory tariffs were imposed on U.S. agricultural products. For the tree nuts industry, retaliatory tariffs were imposed on the '0802-tariff' line (both in-shell and shelled nuts), causing higher tariffs on almond exports. Tariffs on almonds were raised from 10% to 55% [3,4]. The U.S. almond industry experts suggested that these tariffs would cause harm to the U.S. almond industry and the tree nuts industry in its entirety, which would, in the long run, have a negative impact on the tree-nuts-dependent economy of California. California's economy is highly export-dependent, where 70% of almonds produced are exported to other countries [5].

This research aimed to investigate the determinants of export demand for the U.S. almond industry and assess the impact of the retaliatory tariffs policy imposed by key almond-importing nations on U.S. almond exports. Hence, the motivation for this study was twofold: (1) understanding the factors affecting almond exports in major importing countries and (2) evaluating the impact of recent retaliatory trade tariffs that some importing countries imposed on the U.S. exports in reaction to President Trump's policies of increasing tariffs on some of the imports to the U.S. from those countries. This study is the first of its kind investigating the impact of this specific retaliatory tariffs policy. However, since the study ended in 2019, it did not involve enough data to fully evaluate the impact of this retaliatory trade tariffs policy.

2. Background

2.1. Changing Trends in World Almond Exports

A significant part of U.S. agricultural exports to the world are tree nuts. In 2019, the U.S. exported about USD 9.1 billion worth of tree nuts, just behind soybean exports at USD 18.7 billion. The U.S. tree nut exports include almonds, pistachios, walnuts, pecans, and hazelnuts. The U.S. is the leading exporter of almonds in the world. The trends of

U.S. almond exports have changed over the last 20 years (Figure 3). In the late 1990s and early 2000s, most U.S. almond exports went to the European Union (E.U) and Asia, with Germany being the major importer of U.S. almonds in the E.U., while Japan was the major importer in Asia alongside China/Hong Kong, South Korea, and Taiwan. Emerging markets, such as India and the United Arab Emirates (U.A.E), were relatively untapped at that time [6,7].

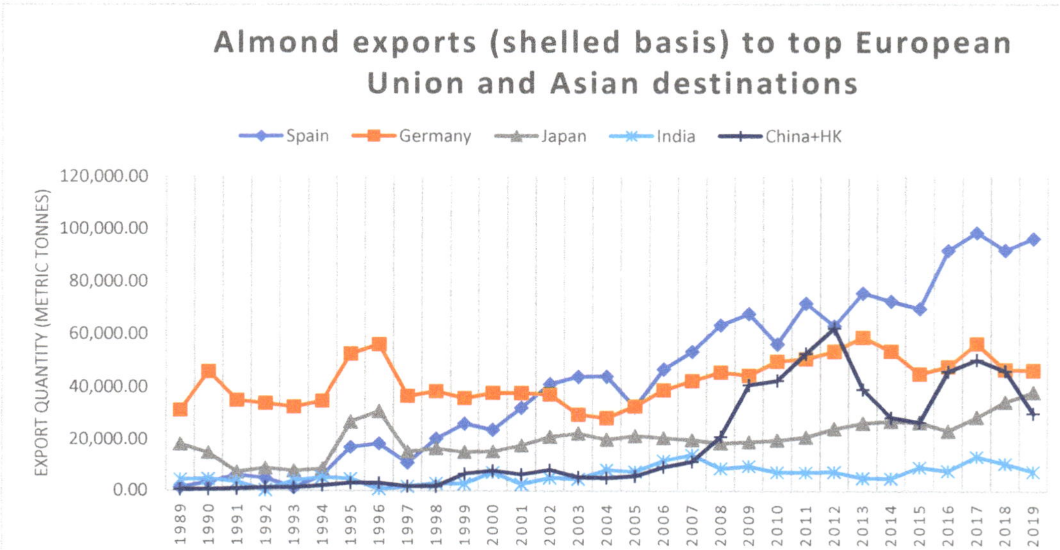

Figure 3. U.S. almond export trends to the top European Union and Asian countries. Source: USDA Foreign Agricultural Service.

Today, the E.U. remains the largest market for U.S. almond exports, importing almost 40%, and Spain now accounts for most U.S. almond imports, with Germany close behind as the second-largest E.U. importer. The Asian market, however, has seen a much more drastic change, as new export destinations, such as India, China/Hong Kong, and the U.A.E, are beginning to catch up with Japan for the share of U.S. almond exports to Asia [2]. These changing trends are the results of several years of nutrition research and global market development programs to increase U.S. almond exports. These programs are funded by both the public (the federal government) and private partners in the U.S. almond industry. The Almond Board of California (A.B.C) spent about 61% of its global budget on global market development programs in the fiscal year 2019 to boost exports to those markets [2].

2.2. Recent Trade Wars and Retaliatory Tariffs on U.S. Agricultural Exports

The dominance of U.S. tree nut exports has faced several challenges over the years, with food safety and aflatoxin concerns being among these challenges [8]. In recent years, another concern has been the retaliatory tariffs imposed on U.S. agricultural exports. After spending just over a year in office, the Trump administration announced two major tariffs against products from key trading partners: Sections 232 and 301 tariffs. Section 232 of the Trade Expansion Act of 1962 allows the President to adjust imports if the Department of Commerce finds certain products are imported in certain quantities or under certain circumstances that threaten U.S. national security, while Section 301 of the Trade Act of 1974 allows the U.S. Trade Representative (USTR) to suspend trade agreement concessions or impose new import restrictions if it finds a U.S. trading partner violates trade agreement commitments or engages in discriminatory or unreasonable practices that burden or restrict U.S. commerce.

U.S. Section 232 tariffs were imposed on steel and aluminum imports from the European Union, as well as countries, such as China, Canada, Mexico, and Turkey. U.S. Section 301 tariffs were levied on imports from China [3,4]. These tariffs caused retaliatory tariffs imposed on major U.S. agricultural exports, such as meats, grains, dairy, and horticultural crops. Two of the export destinations considered in this study, namely, China and India, both imposed retaliatory tariffs on U.S. tree nut exports starting in 2018 and 2019, respectively. In April 2018, U.S. tree nuts, including almonds, pistachios, walnuts, and pecans, were slapped with retaliatory tariffs, and China and India imposed retaliatory tariffs on almonds and walnuts a year later in June 2019 [7,9].

Given the importance and contributions of almond production and exports to the California economy, the harm caused to the almond industry was transferred as an adverse effect on the Californian economy [9,10]. According to experts in this area, the tree nuts industry is export-oriented. Thus, it is adversely affected in the long run as shifting markets is expensive and time-consuming, and export promotion rewards are not gained quickly in new markets.

2.3. *The Important Role of California in the U.S. Almond Crop Production*

California is the only commercial producer of almonds in the U.S. and the leading supplier and exporter of almonds worldwide. California produces 100% of U.S. commercial almonds and U.S. imports remain negligible. The earliest varieties of almonds were native to western Asia and were introduced to California by Spanish explorers in the 1700s. The Spanish are generally referred to as 'the originators' because Franciscan padres from Spain originally introduced the almond trees to California [11]. Today, almonds are California's top agricultural export and largest tree nut crop in total dollar value and acreage. Almonds also rank as the largest U.S. specialty crop export, generating about USD 4.9 billion in 2019 [11]. Almond production in California is carried out within a well-defined community of farmers. There are about 7600 almond farms in California. Most of these farms (91%) are family-owned and run by third- and fourth-generation farmers carrying on the family legacy in almond production. Approximately one-third of California almond farms are 100 acres or more in size.

The almond industry community includes almond handlers who move almonds from farm gates to trade points as local or export shipments. The handlers may also carry out processes such as cleaning, sizing, sorting, and bulk packaging. The handling end of the supply chain is like the production end, as most handlers of almonds are also family-owned entities [2,12].

The single most important factor determining a good almond yield is pollination during the bloom period. California almond varieties are self-incompatible, i.e., they require cross-pollination with other varieties to produce the crop. In cases where the varieties are self-compatible, they still require the transfer of pollen within the flower [13]. During the almond-growing season between February and March, almond tree buds bloom in preparation for pollination by managed (mostly imported) honeybee colonies. As blooming occurs, honeybees in search of pollen and nectar go into the almond orchards and as they move around the orchards, they pollinate the blooming flowers, allowing for the fertilization process to occur, and eventually, fertilized flowers grow into almond nuts [11].

2.4. *Almond Industry Sustainability Crisis*

The almond crop has high water demands for growth and ranks highest among all tree nuts in terms of its water footprint on the environment. Almonds rank higher in water use compared with both pistachios and walnuts. Thus, there are questions surrounding the sustainability of water resources and the economic cost of water in the long run if almonds are to be produced in an environmentally sound and cost-friendly manner [14].

From an economic standpoint, having over 80% of the world's almond exports coming from California could be unsustainable in the long run. Finding other locations across the world where almonds could also be grown (such as Spain and Australia) could help to sustain the world supply of almonds [15]. High costs and supply uncertainty of water due to lengthy California droughts could cause long-run economic problems for almond producers, as they would have to pay more money for water due to water scarcity and the high demand for water by the Californian agricultural industry [16,17].

In 2018, the California almond industry used up approximately 70% of honeybee colonies in the U.S. during its pollination season (between February and March). Each almond kernel must be individually pollinated for fruit setting to occur since almond trees are self-incompatible, requiring cross-pollination to produce nuts [18–20]. Commercial bee pollinators flock to California due to the high pollination prices offered by the almond producers. Earning income from almond pollination services is very important to commercial bee pollinators [20]. However, the long-term sustainability of the commercial pollination industry has also been put into question, as the mass transshipment of bee colonies has led to declines in managed pollinator populations (colony collapse disorder). Cross-country transport of migratory, managed honeybee colonies can be stressful to honeybees, as the trip to pollinate California almonds for over a month (4 to 6 weeks) demonstrates [21].

3. Literature Review

Specification of export demand functions is a widely studied research area in international trade literature and remains an important source of information for industry experts. Much of the previous literature focused on how the importing countries' income and exchange rates affect the export demand function. Aggregate export demand forecasts and estimates serve as a tool for long-term international trade planning and policy formulation [22]. Many empirical studies used the export demand function in the past. Research was conducted on U.S. export demand for different commodities in the agricultural sector (e.g., [7,8,23–40]).

In this study, we focused on the export demand model to estimate the determinants and factors that affected export demand for U.S. almonds. In addition, this study aimed to assess the impact of the retaliatory tariffs imposed by some importing countries on U.S. almond exports. The retaliatory tariffs were used in the model as percentage increments.

Previous research estimated the major factors affecting U.S. almond export demand in Asia and the European Union (E.U.), with a focus on the impact of federal promotion programs on the export demand [7]. Results showed that own-price elasticities for almond exports were negative, and the cross-price elasticities with respect to walnuts were positive, indicating walnuts to be substitutes for almonds. Interestingly, they found mixed results for the income elasticity, with a negative income elasticity for Asia, suggesting almonds here are an inferior good. That is, with an increase in income, almond consumption decreased. However, income elasticity for the E.U. was positive and highly elastic, suggesting almonds to be a luxury good.

Past research [25] showed the relation between U.S. export demand and exchange rates. This study focused on estimating the export demand function for U.S. wheat. These results suggested exchange rate changes had a significant impact on U.S. wheat exports. Another study [26] demonstrated the determinants of trade flows in international markets. The reasons for the decline in export demand in agriculture markets back in 1986 were old and wrong policy implications, lower-than-normal levels of stocks, and government intervention in agricultural trade, which caused confusion in the markets, leading to an increase in demand and a decrease in supply, which, in return, increased prices in the markets [27].

Previous studies analyzed export demand for specific countries and the factors impacting such export demand. These include studies on export demand for U.S. cotton [28], orange juice [29], corn and soybean [30], and beef [31]. In an analysis of export demand elasticities for 53 developed and developing countries, the trading country's income and relative commodity prices were found to be statistically significant in impacting export demand [32]. Turkish aggregated export demand was found to be inelastic (not responsive) with respect to the real exchange rate, but elastic (responsive) with respect to foreign income [33].

Past research has also demonstrated the benefits of marketing orders and the impacts of the conditions and costs of global trade, including tariffs. The benefits of the federal marketing order over the past 50 years for California pistachios were found to greatly exceed costs [34]. U.S. peanut exports were found to be influenced by both the price of Chinese peanut exports, as well as the real gross domestic product (*GDP*) of China [35]. The export demand function was modeled in Indonesia [36] and for U.S. corn seed export to 48 countries [37], which concluded that trade costs matter, mostly as tariffs, and that all such costs of global trade have a negative impact on exports.

The export demands for pistachios in 21 major export destinations were analyzed using a single framework logarithmic model [8]. The results showed that variation in export demand for U.S. pistachios in these markets was significantly affected by export prices of U.S. pistachios, export prices of other U.S. tree nuts (pecans, almonds, and walnuts), and food safety concerns for pistachios produced in the U.S. and its main competitor, namely, Iran. It was argued that the U.S. producers could expand the export demand for U.S. pistachios by taking advantage of the advanced production technologies they employ to improve food safety and quality and differentiate their products in international markets.

The export demand function for U.S. raisins was also investigated [38]. Other studies regarding the raisin situation were more focused on consumer marketing issues [39] and consumer demand [40].

Overall, these studies estimated the determinants of export demand and usually showed factors such as product own-price, cross-prices (product substitute/complement prices), exchange rates, and importing countries' *GDP* to be significant factors affecting the export demand function. Our research adds to the literature in this area by investigating the export demand function for U.S. almonds, which is the leading U.S. tree nut export both in value and volume, when retaliatory tariffs were imposed on U.S. agricultural exports in reaction to the U.S. government policy announcing major tariffs against products from other trading countries.

4. Analytical Framework and Data Description

4.1. The Theoretical Model

Economic theory postulates that the quantity of a good demanded is a function of its own price, cross prices (prices of substitutes and complements), income, and other variables, such as tastes and preferences (e.g., advertising) [41]. The export demand function is tailored similarly, including other factors such as bilateral exchange rates, prices charged by foreign competitors, bilateral trade agreements, tariffs, quotas, and income (gross domestic product of the importing country), which affect the quantity of goods exported. Other factors that could influence the export demand of a product are product quality and safety (food safety), reliability of goods, delivery time, and export promotion programs [42,43].

Following the literature in this area, we employed the general export demand function specified in economic theory as the export demand equation for U.S. almonds:

$$(EXQ) = f\ (EXPa, EXPcom, EXPnuts, GDP, RER) \tag{1}$$

where *EXQ* is the quantity of almond exports; *EXPa* is the export price of almonds; *EXPcom* is the export price of almonds produced by competitors; and *EXPnuts* is the export prices of substitutes, such as walnuts and pistachios. Previous studies [7,8] showed that almonds, pistachios, and walnuts are substituted for each other. *GDP* is the gross domestic product of the importing country (used as a proxy for income), and *RER* is the real bilateral exchange rate, which is the ratio of importing country's currency to U.S. dollars. While the international trade literature postulates and supports the functional form stated in equation

1 [8,15–17], the retaliatory tariffs are also included in the model as percentage increments to investigate their impacts. This is because the countries in this study use both ad valorem and specific tariffs on almond imports.

4.2. The Empirical Model

Built on previous studies (e.g., [7,8]), this study used the double-logarithm equation model to estimate the export demand for U.S. almonds exported to the five major destinations. There are three main variables used in the empirical export demand model [22,23]. The first is the product price, which is the main explanatory variable; the second is foreign income (*GDP*), which represents the economic activity and purchasing power of the trading country; and the third is the exchange rate, which is a relative price that is crucial in affecting imports.

The explanatory variables included in the model include the export price of U.S. almonds (both shelled and in-shell), export price of U.S. pistachios (both shelled and in-shell), and export price of U.S. walnuts (both shelled and in-shell). Pistachios and walnuts are alternative nuts that could be substituted for almonds in foreign markets. The model utilized a logarithmic functional form to allow for more flexibility in the interpretation of estimated coefficients. The log-log form provides an advantage because the coefficients are elasticities [8]. In addition, the variables in a logarithm format have reduced outlier effects.

In addition, free trade agreement (*FTA*) status, as well as gross domestic product and real exchange rates, were the explanatory variables included in the model. China and Hong Kong were separated in this analysis, given that they have different gross domestic products and real exchange rates, and individual export data exists for both countries. Hence, the data had six cross-sections.

Furthermore, the retaliatory tariff changes were included as percentage increments since the mix of countries in the study (i.e., China/Hong Kong, Germany, India, Japan, and Spain) use both ad valorem and specific tariffs on almond imports. That is, a 20% increase in tariffs would be represented as 1.2 compared with prior to retaliation, which is represented as 1. The free trade agreement (*FTA*) is also used in the study as a dummy (0 for non-*FTA* nations and 1 for *FTA* nations). The dummy is used to discern between the importing nations with a free trade agreement with the U.S., which allows for zero tariffs on the '0802 line' (shelled and in-shell tree nuts), as this would help see the overall effect of the policy on U.S. almond exports. The export price of almonds from other countries was excluded from the model because the U.S. has a tight grip on the world almond export markets and the effect of competitors is negligible [7].

Hence, the empirical export demand function for U.S. almonds was specified as:

$$\ln(Q_{it}) = \beta_0 + \beta_1 \times \ln(Pas_{it}) + \beta_2 \times \ln(Pans_{it}) + \beta_3 \times \ln(Pps_{it}) + \beta_4 \times \ln(Ppns_{it}) + \beta_5 \times \ln(Pws_{it}) + \beta_6 \times \ln(Pwns_{it}) + \beta_7 \times \ln(GDP_{it}) + \beta_8 \times \ln(RER_{it}) + \beta_9 \times (TARIFF_{it}) + \beta_{10} \times (FTA_{it}) + \varepsilon \quad (2)$$

where (Q_{it}) is the quantity of shelled/in-shell U.S. almonds exported to country *i* for quarter *t*; (Pas_{it}) and $(Pans_{it})$ are the export prices of shelled and in-shell almonds, respectively, to country *i* for quarter *t*; (Pps_{it}) and $(Ppns_{it})$ are the export prices of shelled and in-shell pistachios, respectively, to country *i* for quarter *t*; (Pws_{it}) and $(Pwns_{it})$ are the export prices of shelled and in-shell walnuts, respectively, to country *i* for quarter *t*; (GDP_{it}) is the gross domestic product of a country *i* for quarter *t*; (RER_{it}) is the real exchange rate of local currency per U.S. dollar for country *i* for quarter *t*; (FTA_{it}) is the free trade agreement status of the importing nation (1 for an *FTA* nation and 0 for a non-*FTA* nation); $(TARIFF_{it})$ is the tariff increments as a percentage on U.S. almonds (base = 1 for pre-retaliatory tariff months). Table 1 summarizes the model variables and their expected signs.

Table 1. Variable description and expected signs.

Variable Description	Crop	Crop Type	Variable Name	Unit (t—Metric Ton)	Expected Sign
Export price to country *i* for quarter *t*	Almonds	Shelled	Pas_{it}	USD/t	−
		In-shell	$Pans_{it}$	USD/t	−
	Pistachios	Shelled	Pps_{it}	USD/t	+/−
		In-Shell	$Ppns_{it}$	USD/t	+/−
	Walnuts	Shelled	Pws_{it}	USD/t	+/−
		In-shell	$Pwns_{it}$	USD/t	+/−
Gross domestic product (*GDP*) of country *i* for quarter *t*	n/a	n/a	GDP_{it}	USD	+
Real exchange rate to USD for local currency of country *i* for quarter *t*	n/a	n/a	RER_{it}	Local currency/USD	−
Free trade agreement status	n/a	n/a	FTA_{it}	Dummy variable	+
Tariff increase	n/a	n/a	$TARIFF_{it}$	Percentage increment	TBD

4.3. Data Description

We selected the following top five export destinations for U.S. almonds for this study: China/Hong Kong, Germany, India, Japan, and Spain. These five destinations make up about 50% of the total U.S. almond exports annually. China and Hong Kong were treated separately for the reasons stated previously. The data for this study was 240 quarterly observations from 2010 to 2019 (i.e., 6 cross-sections times 40 time-series data). The data for each nut type (shelled and in-shell) for the export unit values and quantities were from the USDA General Agreement on Trade and Services (GATS) database. Data for the real exchange rates and GDPs were from the USDA Economic Research Service (ERS) database. The timeline for the tariff increments was determined using yearly USDA FAS *Tree Nuts: World Markets and Trade* publications, and congressional service reports (CSR). Google search provided the timeline for the implementation of the retaliatory tariffs on U.S. almond exports. All the variables were in real terms 2015 U.S. dollars to allow for uniformity of the values. Table 2 provides summary statistics for the dataset.

Table 2. Summary statistics for dependent and independent variable(s).

Variable	Mean	Standard Deviation	Minimum	Maximum
Q_{it} (shelled)	8629.313	6990.599	621.2	28,927.4
Q_{it} (in-shell)	5984.027	9593.538	14.6	49,544.6
Pas_{it}	6228.745	1518.465	3650.5	10,579.5
$Pans_{it}$	4930.853	1330.011	2347.4	10,671.1
Pps_{it}	7915.2	3425.189	3268.8	19,781.3
$Ppns_{it}$	7793.552	1470.735	4502.8	11,577.2
Pws_{it}	7647.324	1994.769	2655.3	12,117.9
$Pwns_{it}$	3974.492	1421.821	1318.2	11,055.5
GDP_{it}	3685.353	3571.144	267.5503	14,317.78
RER_{it}	30.57722	39.83349	0.6815718	124.3476
FTA_{it}	0.1666667	0.3734568	0	1
$TARIFF_{it}$	1.100417	0.6014277	1	5.5

5. Results

We began the analysis with panel-robust checks to adjust the standard errors of the results for general forms of heteroscedasticity and autocorrelation that could exist in the dataset [7,41]. The first difference was also applied to the explanatory variables to make sure they were stationary to obtain meaningful estimations. We tested for stationarity before conducting the analysis, as a non-stationary variable could cause model misspecifications (OLS estimates will no longer be the best linear unbiased estimates (BLUEs) if the data are non-stationary). Due to the panel nature of the data, two possible models were feasible to arrive at the results of the analysis, the fixed-effects model and the random-effects model.

A Hausman test was carried out to determine which model was best suited to analyze the data [44]. The results of the Hausman test suggested that the random-effects model was more appropriate to analyze the shelled export demand, while the fixed-effects model was used to analyze the in-shell export demand.

The results for the U.S. shelled almond prices indicated that the own-price elasticity of almond prices was statistically significant and negative, which is consistent with economic theory. The cross-price elasticity with respect to in-shell U.S. pistachio prices was statistically significant and positive, indicating in-shell pistachios to be substitutes for almonds.

Interestingly, we found the income elasticity (gross domestic product of importing countries) to be significant with a negative sign. That is, with an increase in income, a lower quantity of almonds was demanded, indicating almonds to be an inferior good, with consumers switching to higher-priced luxury nuts, such as pistachios. This was also observed by the substitution effect of in-shell U.S. pistachio prices with both shelled and in-shell U.S. almonds. These results should be a concern to almond exporters as the value and volume of U.S. pistachio exports might be taking away market share from almonds in import destinations. The real exchange rate variable was also statistically significant with a negative sign, which is consistent with economic theory. As the value of the USD appreciated, almonds became more expensive and consumers decreased the quantity demanded. The retaliatory tariffs and free trade agreements were both statistically significant variables, explaining the shelled almond export demand function.

Overall, the estimated results of the export demand function for shelled almonds showed factors such as product own-price, cross-prices (product substitute), exchange rates, and importing countries' *GDP*, as well as tariffs and *FTA* variables, to be significant factors that affected the export demand function. Our results showed that the variation in export demand for in-shell U.S. almonds was significantly affected by own-prices, in-shell U.S. pistachio prices (substitutes), and tariffs. The statistically significant results support previous studies and economic theory in general, except for the retaliatory tariff variable. Table 3 summarizes our export demand model estimation results.

Our results showed that the retaliatory tariff coefficient was statistically significant with a positive sign. That is, more tariffs mean more exports, which is normally contrary to economic theory. This result could be due to several factors. First, the tariff coefficient represents the aggregate effect of the explanatory variable for all the countries in the sample, while in that sample, China was the only country that imposed retaliatory tariffs on the U.S. tree nut industry effectively. Second, the tariffs were imposed in 2018, while the weekly dataset covered up to 2019, which is, relatively speaking, a very short time for the retaliatory tariffs to take effect. In the short run, the tariffs translated into Chinese consumers paying higher prices. In the long run, the tree nuts industry, being export-oriented, is expected to be affected adversely by the tariffs. Third, another important factor is the dominant position of U.S. almonds in international markets. U.S. almonds have a huge concentration and strong market power in international markets. U.S. almonds have very little competition in international markets according to industry experts who analyzed the impact of the U.S.–China Trade war on California agriculture. With retaliatory tariffs in place, Chinese consumers paid higher prices for almonds and pistachios, allowing the imports of almonds and pistachios from the U.S. to remain constant due to the dominant position of the U.S. in these tree nuts. However, walnuts suffered because of the trade war since China is overall a net exporter of walnuts. Here, the losses were offset by diverting exports to other countries [10].

Table 3. The export demand model estimation results.

Variables	Shelled U.S. Almonds Demand	In-Shell U.S. Almonds Demand
Export price of shelled U.S. almonds	**−0.8121583** **	−0.2062182
Export price of in-shell U.S. almonds	−0.2002172	**−1.262263** **
Export price of shelled U.S. pistachios	0.0980747	0.317252
Export price of in-shell U.S. pistachios	**1.531595** **	**1.173499** **
Export price of shelled U.S. walnuts	−0.1754965	0.5463877
Export price of in-shell U.S. walnuts	−0.0132256	0.1278927
Gross domestic product	**−0.3146826** **	−0.0972651
Real exchange rates	**−0.5453288** **	−1.872415
Free trade agreement status	**2.185382** **	-
Tariff increase	**0.3742759** **	**0.6287851** **
R-Squared	0.7328	0.3420

** Significant at 5%.

6. Discussion

6.1. Linkages to Previous Studies

The own-price elasticity results for both almond types were in line with economic theory (quantity demanded dropped with increased own-prices) and previous studies [7,8] also found that almonds and pistachios had negative own-price elasticity, respectively.

The cross-price elasticity results showed that walnut prices did not significantly affect shelled and in-shell almond demand functions, but in-shell U.S. pistachio prices affected the demand for both shelled and in-shell almonds and were a substitute with both almond types. In a previous study [7], the authors found walnuts to be substitutes for almonds. It is worthy to note that in the late 1990s, U.S pistachios rallied in the world market due to food safety issues (due to the Iranian aflatoxin incident) over other tree nut markets and gained market share over other nuts, such as walnuts and pecans, where pecans are second only to almonds [8].

The gross domestic product results showed that almonds had a negative income elasticity. This is supported by a study [7] that found tree nuts to have both positive and negative income elasticities. A previous study [7] estimated the major factors affecting U.S. almond export demand in Asia and the European Union (E.U.) using the export demand function. Interestingly, the income elasticity results were mixed, with a negative income elasticity for Asia, indicating almonds to be an inferior good. That is, with an increase in income, almond consumption decreased, just like the results in this study. However, income elasticity for the European Union (E.U.) was positive and highly elastic, indicating almonds to be a luxury good.

In addition, coupled with the substitution relationship with shelled U.S. pistachios prices, negative income elasticity is not far-fetched, as increased income/prosperity could make consumers move to the more expensive and more environmentally friendly pistachio nuts.

The real exchange rate results show that almonds have a negative exchange rate. This is consistent with prior literature [8], which found U.S. pistachios to have a negative real exchange rate elasticity. Unlike our results, pistachios in this study were found to be elastic with respect to real exchange rates. We found almonds to be inelastic with respect to real exchange rates. Another study on U.S wheat also found that exchange rates have a significant effect on crop export demand [25].

The free trade agreement results show that almonds have higher export demand in situations where free trade agreements exist, i.e., where tariffs are eliminated or are substantially lower. Previous studies (e.g., [45]) found that free trade agreements do increase the volume of trade between trading nations.

The results for the effect of the retaliatory trade tariff policy are novel and indicate that U.S almond exports have not been negatively affected by the retaliatory tariffs imposed by some importing nations (two of which are included in this study). Trade deflections were shown to keep export levels unaffected [45], while export promotion by the Almond Board of California caused increased demand from other markets, such as the U.A.E [2].

It is worthy to note that the timeline of the retaliatory tariffs was best cut off at the end of 2019. U.S. government officials met with both India and China in early 2019 to renegotiate trade terms, which kicked in at the start of 2020. Furthermore, purchase promises were made, tariffs were suspended for crops based on the 'Phase One Agreement,' and this rendered post-2019 figures somewhat biased and were consequently ignored in this study [46].

6.2. Implications and Major Concerns

From an economic standpoint, the dominance of the U.S. in the world's almond exports is of concern to competitors in the global markets. An extreme weather event, such as the extended drought in California, is also a major concern that could render U.S. almond producers helpless in the face of almond production challenges. Diversifying almond supply sources would be a good way to ensure that the world almond market supply keeps up with demand, but traditional almond producers and exporters, such as Spain, Iran, Morocco, Syria, Turkey, and Italy, have aging, low-yielding trees that show little potential for expanded production. Currently, those countries mainly serve their domestic market demand [47–49].

Another major producer is Australia, where the almond acreage and production are steadily expanding. However, Australian almond producers face a similar problem to the California almond producers, that is, water shortages. Australia has most of its agricultural production in the Murray–Darling Basin and just like the Central Valley in California, that area also faces many challenges due to a lack of water resources. Even though Australia could represent a major threat to California's dominance in supplying almonds to the world, this threat is muted by water issues in the Murray–Darling Basin [50].

From an agro-ecological perspective, consumers could become more concerned about the environmental impact of consuming tree nuts. Almonds and other tree nuts, such as walnuts and pistachios, have similar water footprints, but the massive use of commercial pollination by the almond industry is another concern. Non-native bee colonies are shipped to California during the pollination season to help the almond fruit-setting process and this could have detrimental environmental impacts. The loss of managed honeybee colonies from colony collapse disorder is a major concern due to the long distances of migratory honeybee routes and a lack of dietary diversity during the pollination season. Pistachios do not require any invertebrate (e.g., insect)-mediated pollination since pistachios are wind-pollinated. Pistachios require less water than almonds. Therefore, shifting from almonds to pistachios could reduce not only the water use but also dependence on increasingly scant migratory honeybee colonies.

Contemporary consumers are much more environmentally conscious and could factor in the environmental effects of growing almonds and other tree nuts in their consumption decisions. Our results indicated that shelled U.S. pistachios are quietly taking away market shares from U.S. almonds. Therefore, an argument could be made that such a substitution of almonds with pistachios by consumers has potential environmental benefits, albeit at the expense of the U.S. almond industry.

7. Conclusions

This study estimated the factors that affect export demand for U.S. almonds and analyzed the effects of recent retaliatory tariffs on almond export demand by some importing countries. A double-log equation of the export demand function was employed to estimate the affecting factors for the top five export markets, namely, China/Hong Kong, Germany, India, Japan, and Spain. The results for shelled U.S. almonds indicated that shelled U.S. almond prices, in-shell U.S. pistachio prices, gross domestic product, real exchange rates, tariffs, and free trade agreements were the major determinants of the export demand. The results for in-shell U.S. almonds indicated that in-shell U.S. almond prices, in-shell U.S. pistachio prices, and tariffs determined the export demand for almonds.

These results provided some key insights for the U.S. almond exporters about the changing export trends. First, the retaliatory tariff increases have had no serious impact on the U.S. almond exports in the short run and there was no immediate cause for alarm for the almond export industry. Only a few major importing countries imposed those tariffs, but that did not have a serious negative impact on almond exports, even though those importing countries that imposed the tariffs were major importers of U.S. almonds. On the contrary, almond exports even increased in face of the retaliatory tariffs. That is, the downward trend in demand for U.S. almonds by the countries imposing the retaliatory tariffs was made up by trade deflections and redirections to other countries. Almond exports even increased due to export-promotion programs.

Second, U.S. pistachios are fast gaining ground on almonds in export destinations, and in due time, market shares would have to be sacrificed if almond exporters do not proactively seek ways to promote almonds to protect their high market shares. U.S. exporters could increase the budget for almond promotion and marketing expenditures and employ more innovative export promotion programs. The immediate strategic response could be increasing the frequency of food science/educational workshops and symposiums, focusing on providing information to consumers about the benefits and advantages of almond consumption to promote almond consumption and keep the world consumers from turning to alternative tree nuts.

Another strategic response would be the provision of food safety assurances as an indication of higher quality and product differentiation, coupled with the existing programs that were rolled out by the Almond Board of California over recent years. Resolving trade restrictions, avoiding trade wars, and facilitating free trade agreements with key importing countries is always an appropriate strategic response to increasing the quantity of almonds exported to the global community. The absence of tariffs and other barriers to trade would help stimulate greater exports worldwide. Among the top five export countries in our sample, only Japan currently has a free trade agreement in place with the U.S.

The limitations of this study are several: We were unable to fully observe the expected effect of retaliatory tariffs using this study because the study ended in 2019 and did not involve enough data to fully evaluate the impact of the retaliatory trade tariffs policy. In addition, the results of this study are unique to almonds exported to the top five U.S. almond export destinations. More research is required to include other almond-importing countries to investigate the impact of retaliatory trade tariffs and trade barriers on the U.S. tree nut industry and their products. Future studies in this area could focus on how to diminish the overall effects of tariffs on U.S. farmers and exporters.

Furthermore, the countries included in the sample for each group had different economic structures regarding their agricultural sectors and food production. The countries also had different policies on food prices, such as production and consumption subsidies and price controls. In a more general picture, the sample countries had different institutional and political systems that affected how they responded to changing market conditions. Fixed effect estimation may control for some of these time-invariant specific destination-country factors to some extent, but more research is required to address these issues.

Author Contributions: Conceptualization, A.A. and S.S.; methodology, A.A. and S.S.; software, A.A.; validation, A.A. and S.S.; formal analysis, A.A. and S.S.; investigation, A.A. and S.S.; resources, A.A. and S.S.; data curation, A.A.; writing—original draft preparation, A.A. and S.S.; writing—review and editing, A.A. and S.S.; visualization, A.A. and S.S.; supervision, S.S.; project administration, A.A. and S.S. All authors have read and agreed to the published version of the manuscript..

Funding: This research received no external funding.

Institutional Review Board Statement: Not applicable.

Informed Consent Statement: Not applicable.

Data Availability Statement: The data are available upon request.

Acknowledgments: The authors would like to thank the editors and the reviewers of Sustainability for all their efforts, especially Academic Editor: Aaron K. Hoshide. Sayed Saghaian acknowledges the support from the United States Department of Agriculture, National Institute of Food and Agriculture, Hatch project No. KY004052, under accession number 1012994.

Conflicts of Interest: The authors declare no conflict of interest.

References

1. United States Department of Agriculture; Foreign Agricultural Service. *United States Agricultural Export Yearbook*; U.S. Department of Agriculture, Foreign Agricultural Service: Washington, DC, USA, 2019.
2. Almond Board of California. *Almond Almanac*; Almond Board of California: Modesto, CA, USA, 2020.
3. Regmi, A. *China's Retaliatory Tariffs on U.S. Agriculture: In Brief*; Congressional Research Service: Washington, DC, USA, 2019.
4. Williams, B.R. *Trump Administration Tariff Actions: Frequently Asked Questions*; Congressional Research Service: Washington, DC, USA, 2020.
5. Konduru, S.; Asci, S. A Study of the Chinese Retaliatory Tariffs on Tree Nuts Industry of California. *Theor. Econ. Lett.* **2019**, *9*, 2747–2755. [CrossRef]
6. Johnson, D.C. United States is World Leader in Tree Nut Production and Trade. In *USDA-ERS Fruit and Tree Nuts Situation and Outlook. FTS-280*; U.S. Department of Agriculture—Economic Research Service: Washington, DC, USA, 1997.
7. Onunkwo, I.M.; Epperson, J.E. Export Demand for U.S. Almonds: Impacts of U.S. Export Promotion Programs. *J. Food Distrib. Res.* **2001**, *32*, 140–151.
8. Zheng, Z.; Saghaian, S.H.; Reed, M.R. Factors Affecting the Export Demand for U.S. Pistachios. *Int. Food Agribus. Manag. Rev.* **2012**, *15*, 139–154.
9. Sumner, D.A.; Matthews, W.A.; Medellín-Azuara, J.; Bradley, A. *The Economic Impacts of the California Almond Industry*; University of California Agricultural Issues Center: Berkeley, CA, USA, 2014.
10. Carter, C.A.; Steinbach, S. Impact of the U.S-China Trade War on California Agriculture. *ARE Update* **2019**, *23*, 9–11.
11. Almond Board of California. *Almond Almanac*; Almond Board of California: Modesto, CA, USA, 2019.
12. Sumner, D.A.; Hanon, T.; Matthews, W.A. Implication of Trade Policy Turmoil for Perennial Crops. *Choices* **2019**, *34*, 1–9.
13. Almond Board of California. *California Almond Industry Facts*; Almond Board of California: Modesto, CA, USA, 2016.
14. Fulton, J.; Norton, M.; Shilling, F. Water-indexed Benefits and Impacts of California almonds. *Ecol. Indic.* **2019**, *96*, 711–717. [CrossRef]
15. Marston, L.; Konar, M. Drought Impacts to Water Footprints and Virtual Water Transfers of the Central Valley of California. *Water Resour. Res.* **2017**, *53*, 5756–5773. [CrossRef]
16. Howitt, R.; Medellín-Azuara, J.; MacEwan, D.; Lund, J.R.; Sumner, D. *Economic Analysis of the 2014 Drought for California Agriculture*; Center for Watershed Sciences, University of California: Davis, CA, USA, 2014.
17. Sumner, D.A.; Hanak, E.; Mount, J.; Medellín-Azuara, J.; Lund, J.R.; Howitt, R.E.; MacEwan, D. The Economics of the Drought for California Food and Agriculture. *Agric. Resour. Econ. Update* **2015**, *18*, 1–6.
18. Sumner, D.A.; Boriss, H. Bee-conomics and the Leap in Pollination Fees. *Agric. Resour. Econ. Update* **2006**, *9*, 9–11.
19. United States Department of Agriculture. *Honeybee Colonies*; USDA National Agricultural Statistics Service: Washington, DC, USA, 2018.
20. Lee, H.; Sumner, D.A.; Champetier, A. Pollination Markets and the Coupled Futures of Almonds and Honeybees: Simulating Impacts of Shifts in Demands and Costs. *Am. J. Agric. Econ.* **2019**, *101*, 230–249. [CrossRef]
21. Aebi, A.; Neumann, P. Endosymbionts and Honeybee Colony Losses? *Trends Ecol. Evol.* **2011**, *26*, 494. [CrossRef] [PubMed]
22. Arize, A.C. Traditional Export Demand Relation and Parameter Instability: An Empirical Investigation. *J. Econ. Stud.* **2001**, *28*, 378–396. [CrossRef]
23. Guci, L. *Exchange Rates and The Export Demand for US Grapefruit Juice*; Research Papers 36816; Florida Department of Citrus: Bartow, CA, USA, 2008.
24. Hooy, C.W.; Choong, C.K. Export Demand within SAARC Members: Does Exchange Rate Volatility Matter. *Int. J. Econ. Manag.* **2010**, *4*, 373–390.
25. Konandreas, P.; Bushnell, P.; Green, R. Estimation of Export Demand Functions for U.S. Wheat. *J. West. Econ.* **1978**, *3*, 39–49.
26. Bahmani-Oskooee, M. Determination of International Trade Flows. *J. Dev. Econ.* **1986**, *20*, 107–123. [CrossRef]
27. Haniotis, T.; Baffes, J.; Ames, G.W. The Demand and Supply of U.S. Agricultural Exports: The Case of Wheat, Corn and Soybean. *South. J. Agric. Econ.* **1988**, *20*, 45–56. [CrossRef]
28. Duffy, P.A.; Wohlgenant, M.K.; Richardson, J.W. The Elasticity of Export Demand for U.S. Cotton. *Am. J. Agric. Econ.* **1990**, *72*, 468–474. [CrossRef]
29. Armah, B.K., Jr.; Epperson, J.E. Export Demand for U.S. Orange Juice: Impacts of U.S. Export Promotion Programs. *Agribusiness* **1997**, *13*, 1–10. [CrossRef]
30. Saghaian, Y.; Reed, M.; Saghaian, S. Export Demand Estimation for U.S. Corn and Soybeans to Major Destinations. In Proceedings of the 2014 Southern Agricultural Economics Association (SAEA) Annual Meeting in Dallas, Dallas, TX, USA, 1–4 February 2014.

31. Eenoo, E.V.; Peterson, E.; Purcell, W. *Impact of Exports on the U.S. Beef Industry*; Staff Papers 232380; Department of Agricultural and Applied Economics, Virginia Polytechnic Institute and State University: Blacksburg, VI, USA, 2000.
32. Senhadji, A.S.; Montenegro, C.E. Time Series Analysis of Export Demand Equations: A Cross-Country Analysis. *IMF Staff. Pap.* **1999**, *46*, 259–273. [CrossRef]
33. Cosar, E.E. Price and Income Elasticities of Turkish Export Demand: A Panel Data Application. *Cent. Bank Rev.* **2002**, *2*, 19–53.
34. Gray, R.S.; Sumner, D.A.; Alston, J.M.; Brunke, H.; Acquaye, A.K. *Economic Consequences of Mandated Grading and Food Safety Assurance: Ex ante Analysis of the Federal Marketing Order for California Pistachios*; University of California: Davis, CA, USA, 2005.
35. Boonsaeng, T.; Fletcher, S.M. The Impact of U.S. Non-price Export Promotion Program on Export Demand for U.S. Peanuts in North America. *J. Peanut Sci.* **2010**, *37*, 70–77. [CrossRef]
36. Hussein, A. Structural Change in the Export Demand Function for Indonesia: Estimation, Analysis and Policy Implications. *J. Policy Mark.* **2009**, *31*, 260–271. [CrossRef]
37. Jayasinghe, S.; Beghin, J.C.; Moschini, G. Determinants of World Demand for U.S. Corn Seeds: The Role of Trade Costs. *Am. J. Agric. Econ.* **2010**, *92*, 999–1010. [CrossRef]
38. Soltani, M.; Saghaian, S. Export Demand Function Estimation for U.S. Raisins. Presented at the 2012 Southern Agricultural Economics Association (SAEA) Annual Meeting in Birmingham, Birmingham, AL, USA, 4–7 February 2012.
39. Keeling, J.J.; Andersen, M.A. Welfare Analysis and Policy Recommendations for the California Raisin Marketing Order. Presented at the American Agricultural Economics Association Annual Meeting in Denver, Denver, CO, USA, 1–4 August 2004.
40. Brant, M.; Marsh, T.L.; Featherstone, A.M.; Crespi, J.M. Multivariate AIM Consumer Demand Model Applied to Dried Fruit, Raisins, and Dried Plums (No. 378-2016-21219). In Proceedings of the American Agricultural Economics Association Annual Meeting, Providence, RI, USA, 24–27 July 2005.
41. Boriss, H.; Brunke, H. *Commodity Profile: Almonds*; Agricultural Issues Center, University of California: Davis, CA, USA, 2005.
42. Thirlwall, A.P. *Balance of Payments Theory and the United Kingdom Experience*; Springer: London, UK, 1986.
43. Verbeek, M. *A Guide to Modern Econometrics*, 5th ed.; John Wiley & Sons, Inc.: Hoboken, NJ, USA, 2017.
44. Hausman, J.A. Specification Tests in Econometrics. *Econometrics* **1978**, *46*, 1251–1271. [CrossRef]
45. Baier, S.L.; Bergstrand, J.H. Do Free Trade Agreements Actually Increase Members' International Trade? *J. Int. Econ.* **2007**, *71*, 72–95. [CrossRef]
46. Williams, B.; Hammond, K.; Morrison, W.M. *Escalating US Tariffs: Timeline*; Congressional Research Service: Washington, DC, USA, 2020.
47. U.S. Department of Agriculture, National Agricultural Statistics Services. 2012 Census of Agriculture. 2014. Available online: http://www.agcensus.usda.gov/Publications/2012/ (accessed on 12 May 2022).
48. Bruno, E.M.; Goodrich, B.; Sexton, R.J. The Outlook for California's Almond Market. *Calif. Almond Acreage Rep.* **2021**, *24*, 9–11.
49. U.S. Department of Agriculture, National Agricultural Statistics Service 2020. California Almond Acreage Report. 2021. Available online: https://bit.ly/3wUXK0x (accessed on 12 May 2022).
50. Davies, A. Tough Nut to Crack: The Almond Boom and Its Drain on the Murray-Darling. The Guardian. 2019. Available online: https://bit.ly/3hWPdpm (accessed on 12 May 2022).

sustainability

MDPI

Article

Water Dynamics and Hydraulic Functions in Sandy Soils: Limitations to Sugarcane Cultivation in Southern Brazil

Jessica Lima Viana [1,*], Jorge Luiz Moretti de Souza [2], André Carlos Auler [2], Ricardo Augusto de Oliveira [3], Renã Moreira Araújo [3], Aaron Kinyu Hoshide [1,4], Daniel Carneiro de Abreu [1,5] and Wininton Mendes da Silva [6]

1 AgriSciences, Universidade Federal de Mato Grosso, Caixa Postal 729, Sinop 78550-970, MT, Brazil
2 Departamento de Solos e Engenharia Agrícola (DSEA), Universidade Federal do Paraná, Campus Universitário de Curitiba, Rua dos Funcionários 1540, Curitiba 80035-050, PR, Brazil
3 Departamento de Fitotecnia e Fitossanitarismo (DFF), Universidade Federal do Paraná, Campus Universitário de Curitiba, Rua dos Funcionários 1540, Curitiba 80035-050, PR, Brazil
4 College of Natural Sciences, Forestry and Agriculture, The University of Maine, Orono, ME 04469, USA
5 Instituto de Ciências Agrárias e Ambientais (ICAA), Universidade Federal de Mato Grosso, Campus Universitário de Sinop, Avenida Alexandre Ferronato 1200, Sinop 78550-728, MT, Brazil
6 Empresa Mato-Grossense de Pesquisa, Assistência e Extensão Rural (EMPAER-MT), Centro Político Administrativo, Cuiabá 78049-903, MT, Brazil
* Correspondence: jessica.limavp@gmail.com; Tel.: +55-66-9966-6763

Abstract: Crop cultivation on sandy soils is susceptible to water stress. Therefore, we determined the physical-hydric attributes of a Latossolo Vermelho distrófico (Oxisol) in northwestern Paraná state, Brazil. Soil samples were collected at depth ranges of 0 to 0.2 m, 0.2 to 0.4 m, and 0.4 to 0.6 m. We measured clay, silt, sand, fine and coarse sand contents, soil particle density, soil bulk density, total porosity, microporosity, and macroporosity. We also measured soil characteristics such as saturated and unsaturated soil hydraulic conductivities, pore distribution, water retention, available water capacity, and easily available water. We also estimated soil moisture, matric potential at field capacity, and time at field capacity. Validation of associations among these soil physical-hydric attributes was performed using principal component analysis. For the sandy soils analyzed, the distributions of coarse and fine sand fractions were measured for better evaluation of the soil's physical and hydric attributes. Higher coarse sand contents increased soil hydraulic conductivities, maximum pore diameter, and macroporosity while reducing microporosity. Fine sand content reduced conductivity and increased soil water retention in subsurface layers. Simulated sugarcane yield increased with soil water storage. These results support improving crop simulation modeling of sugarcane to support sustainable intensification in regions with sandy soils.

Keywords: available water capacity; hydraulic conductivity; pore distribution curve; retention curve; soil texture

Citation: Viana, J.L.; de Souza, J.L.M.; Auler, A.C.; de Oliveira, R.A.; Araújo, R.M.; Hoshide, A.K.; de Abreu, D.C.; da Silva, W.M. Water Dynamics and Hydraulic Functions in Sandy Soils: Limitations to Sugarcane Cultivation in Southern Brazil. *Sustainability* **2023**, *15*, 7456. https://doi.org/10.3390/su15097456

Academic Editor: Jan Hopmans

Received: 31 January 2023
Revised: 22 April 2023
Accepted: 25 April 2023
Published: 1 May 2023

1. Introduction

Physical-hydric attributes are used to understand and evaluate soil variability and quality, aeration, hydraulic conductivity, water redistribution, storage capacity, water availability to plants, and root growth [1–3]. Soil density, texture, total porosity, and saturated soil hydraulic conductivity are physical-hydric attributes used as primary indicators to monitor soil quality [4]. Sandy soils are associated with high hydraulic conductivity, and higher proportions of the coarse sand fraction are associated with reduced water availability [5,6].

In addition to intrinsic factors, soil use and management interfere with soil water retention [7] and consequently water availability for plants. A previous study on soil developed from sandstone in the 0 to 0.4 m layer, found 48.6 mm and 58.7 mm of available water capacity (AWC) in soil under crop-livestock integration and pineapple, respectively [6]. This difference of 10.1 mm was influenced by land use and management. Due to the intense management of the sugarcane crop, knowledge of the physical-hydric attributes of the soil is important, especially for crops established on sandy soils. The effect of sugarcane cultivation can increase soil bulk density, reaching critical values of 1760 and 1770 kg/m^3 in layers 0 to 0.2 m and 0.2 to 0.4 m, respectively, in medium texture Oxisol [8].

In sandy soils, there are differences in soil water retention due to the particle size composition of the sand fraction [9]. High proportions of medium and fine sand in sandy soils favor the formation of a capillary distribution network with smaller diameter pores, allowing water retention between soil particles. This matrix provides less soil moisture loss from percolation and, consequently, greater water storage due to the high percentages of finer particles (between 40% and 75% of total sand), combined with low clay content (i.e., particle diameters < 0.02 mm). Although typically less of a contributing factor, clay content can also contribute to increased water retention in sandy soils [9].

The Experimental Station of the Sugarcane Genetic Improvement at the Federal University of Paraná, in collaboration with the Inter-University Network for the Development of Sugarcane Energy, annually conducts numerous studies aimed at genetic improvement and productivity of sugarcane [10,11]. The results obtained in the experiments developed at this Experimental Station are a reference for numerous agricultural planning activities in the northwest region of Paraná state in southern Brazil. However, no detailed study of the physical-hydric attributes has yet been conducted for soils on the Experimental Station that represent a large number of Oxisols in the region. Obtaining information about the dynamics of water in the soil, especially because it is a region dominated by soils developed in the Caiuá Sandstone, may help in decision making regarding the selection of sugarcane varieties better adapted to water stress conditions. The survey of physical and hydric attributes can also subsidize water management in irrigated areas in sandy soils in northwestern Paraná. In this context, the goals and objectives of our research were to (1) determine the main physical-hydric attributes of the Latossolo Vermelho distrófico (Oxisol) cultivated with sugarcane in the northwestern region of Paraná state in Brazil, and (2) evaluate this soil's physical qualities for supporting the sugarcane crop.

2. Materials and Methods

2.1. Study Site Characteristics

This work was developed in the Laboratory of Modeling of Agricultural Systems (LAMOSA) at the Setor de Ciências Agrárias (SCA) at the Federal University of Paraná (UFPR). The analyses were performed using soil samples taken from the Experimental Station of the Sugarcane Genetic Improvement Program (PMGCA), belonging to UFPR and to the Inter-University Network for the Development of Sugarcane Energy (RIDESA). The soil of the Paranavaí Experiment Station is classified as Latossolo Vermelho distrófico (Oxisol) [12]. The PMGCA is located at 22°58′ South latitude, 52°28′ West longitude and is at 470 m of average altitude in the municipality of Paranavaí, located in the northwestern region of the Paraná state, Brazil. The climate of the region, according to the Köppen classification is Cfa (subtropical climate) and has an average annual air temperature between 22.1 and 23 °C, with average annual precipitation between 1400 and 1600 mm [13].

The geomorphology of northwestern Paraná is characterized by the Caiuá sandstone, with an area of 3.2 million hectares, representing 16% of the state area [14]. The Caiuá formation consists of deposits of eolic and fluvial environments, represented by fine to medium sandstones that are purplish with large cross-stratification [15]. According to Costa et al., 2020, the low natural fertility is due to the high sand content associated with the predominance of minerals of low activity in the clay fraction [16]. Latossolo soils associated with the Caiuá sandstone can also be highly susceptible to erosion as well as vertical subsurface soil water movement [17]. According to Pereira 2016, the Caiuá Aquifer is an important water source for the northwest region of the state of Paraná, with an outcrop area of approximately 30,000 km^2. This aquifer is characterized by being free and porous, with bicarbonated calcic to calcium-magnesian waters. The aquifer is composed of aeolian sandstones from the Caiuá Group, which were deposited during the Upper Cretaceous over a paleo-depression formed by the basaltic flows of the Serra Geral Formation [18].

2.2. Soil Attributes

In order to determine soil physical-hydric attributes, we collected soil samples at ten representative points in the experimental area during July 2019. Interval soil sampling was conducted for the soil layers at depths of 0 to 0.2, 0.2 to 0.4, and 0.4 to 0.6 m. Soil analyses were performed using the Soil Physics Laboratories of the Soil and Agricultural Engineering Department at UFPR and in the Rural Development Institute (IDR) in Londrina, Paraná state in Brazil.

The deformed samples were collected with a flat bottom scoop, dried in an oven at 40 °C, sieved through a 2 mm mesh, and stored in plastic bags. After that, the clay, silt, and sand contents were gravimetrically determined using the pipette method [19]. Separation of the total sand fraction into coarse sand (0.2 to 2 mm) and fine sand (0.05 to 0.2 mm) was also performed. These grain fractions were then weighed (g/kg), according to the methodology developed by Embrapa [19]. Soil particle density (ρ_P; kg/m^3) was determined by the modified volumetric balloon method [20] using the formula:

$$\rho_P = \frac{M_{VS} - M_V}{50 - \left(\frac{M_{VSa} - M_{VS}}{\rho_a}\right)} \quad (1)$$

where ρ_P is the particle density of soil (kg/m^3), M_{vs} is the mass of the volumetric flask containing soil (kg), M_v is the mass of the volumetric flask (kg), M_{vsa} is the mass of the volumetric flask containing soil plus alcohol (kg), ρ_a is the density of the alcohol (kg/m^3), determined by the relationship between M_a and V_v, where M_a is the mass of alcohol (kg), and V_v is the volume of the flask (m^3).

2.3. Saturated Soil Hydraulic Conductivity

The un-deformed soil samples were collected in duplicate at each sampling point with the aid of volumetric rings (4.69 cm internal diameter and 3.51 cm height) and an Uhland sampler. After adequate preparation, the samples were saturated using capillary rise. After saturation, the saturated soil hydraulic conductivity (K_S measured as cm/hour) was determined by the Constant Load Permeameter method [21]:

$$K_S = \frac{V \cdot H}{(H + h) \cdot A \cdot t} \quad (2)$$

where K_S is the saturated hydraulic conductivity (cm/hour), V is the volume of percolated water (cm^3), H is the height of the cylindrical ring with soil (cm), h is the water sheet over the sample (cm), A is the cross-sectional area of the sample (cm^2), and t is the percolation time (hours).

Total porosity (TP), measured in cubic meters (m^3) of air space per volume of soil (m^3), was considered equal to the volumetric soil moisture at saturation (θs), which was also measured in m^3/m^3 of soil. The microporosity measured in m^3/m^3 of soil corresponded to the volumetric moisture of the soil sample submitted to a tension of −6 kilopascals (kPa). The macroporosity also measured in m^3/m^3 of soil was obtained by taking the difference between the value for TP and the value for microporosity [19].

2.4. Soil Water Retention

The soil samples were submitted to tensions of −6 and −10 kPa in a tension table and −33, −100, −500, and −1500 kPa in a Richards chamber. After each applied matric potential, the samples were dried at 105°C for 48 h for the determination of soil bulk density and volumetric moisture (θ), according to the methodologies developed at Embrapa [19]. The "θ vs. ψ_m" points and the moisture at saturation were used to determine the soil water retention curve (SWRC), fitted with the model of van Genuchten 1980 [22] using the Burdine

1953 restriction [23]. The adjustment of the parameters of the van Genuchten model and estimation of the volumetric soil moisture ($\hat{\theta}$) was performed with the *nls* function of the open-course statistical software program R version 4.0.0 [24]:

$$\Theta = \frac{\theta(\psi_m) - \theta r}{\theta s - \theta r} = \frac{1}{[1 + |\alpha \cdot \psi_m|^n]^m} \tag{3}$$

where Θ is the effective saturation (dimensionless), $\theta(\psi_m)$ is the volumetric soil moisture (m^3/m^3), θr is the residual volumetric soil moisture (m^3/m^3), θs is the volumetric soil moisture at saturation (m^3/m^3), ψ_m is the matric potential of water in the soil (kPa), α the value of air input per kPa, n is the empirical parameter of the fit (dimensionless), and m the Burdine restriction where $m = 1 - (2/n)$. The volumetric soil moisture (θ_{hco}) measured in m^3/m^3 at the pore water suction point in the hydraulic cut-off (ψ_{hco} measured in kPa) was calculated according to Czyz and Dexter 2013 [25] considering $\psi_{hco} = -1030$ kPa.

2.5. Unsaturated Soil Hydraulic Conductivity and Soil Water Redistribution

The unsaturated soil hydraulic conductivity ($K(\theta)$), measured in cm/hour, was calculated using the van Genuchten–Burdine equation:

$$K(\theta) = K_S \cdot \Theta^2 \cdot \left[1 - \left(1 - \Theta^{\frac{1}{m}}\right)^m\right] \tag{4}$$

The volumetric soil moisture (θ) at potentials -1 to -6 kPa was calculated (for matric potential or ψ less than -6 kPa the water is retained) to estimate $K(\theta)$. Subsequently, the $K(\psi)$ plot was prepared for soil layers at 0 to 0.2 m, 0.2 to 0.4 m, and 0.4 to 0.6 m depths.

In order to estimate the effective saturation (Θ), which was calculated using the actual volumetric moisture at field capacity ($\hat{\theta}_{CC}$):

$$\Theta = \frac{\hat{\theta}_{CC} - \theta r}{\theta s - \theta r} = \frac{1}{(1 + |\alpha \cdot \psi_{CC}|^n)^m} \tag{5}$$

and in order to estimate the matric potential at field capacity (ψ_{CC}):

$$\psi_{CC} = \frac{\left[\left(\frac{\theta s - \theta r}{\hat{\theta}_{CC} - \theta r}\right)^{\frac{1}{m}} - 1\right]^{\frac{1}{n}}}{\alpha} \tag{6}$$

It was assumed that this condition was reached when the drainage rate (τ) was defined as a percentage (p) of the saturated hydraulic conductivity (K_S) based on Prevedello 1999 [26] as restricted by Burdine 1953 [23]. The equations involving this assumption are as follows:

$$\tau = K(\theta) \tag{7}$$

$$\tau = K_S \cdot \Theta^2 \cdot \left[1 - \left(1 - \Theta^{\frac{1}{m}}\right)^m\right] \tag{8}$$

$$p = \frac{\tau}{K_S} = \Theta^2 \cdot \left[1 - \left(1 - \Theta^{\frac{1}{m}}\right)^m\right] \tag{9}$$

Based on Sisson et al., 1980 (Equation (10)) [27] and Prevedello and Armindo 2015 (Equation (11)) [28], an analytical model of soil water redistribution was obtained. For this, Equation (4) was derived with respect to θ and equalized to z/t:

$$\frac{z}{t} = \frac{dK}{d\theta} \tag{10}$$

$$\frac{dK}{d\theta} = \frac{dK}{d\Theta} \cdot \frac{d\Theta}{d\theta} \tag{11}$$

Then, the time corresponding to the field capacity in each soil layer was obtained:

$$\frac{z}{t_{CC}} = \left[\left(K_S \cdot (2 \cdot \Theta) \cdot \left(1 - \left(1 - (\Theta)^{\frac{1}{m}}\right)^{m}\right) + K_S \cdot (\Theta)^2 \cdot \left(\left(1 - (\Theta)^{\frac{1}{m}}\right)^{m-1} \cdot \left(m \cdot \left((\Theta)^{\frac{1}{m}-1} \cdot \frac{1}{m}\right)\right)\right)\right) \cdot \left(\frac{1}{(\theta s - \theta r)}\right)\right] \quad (12)$$

where p is the percentage of K_S (where we considered the values: 0.005; 0.010; 0.015; 0.020; 0.025; 0.030; 0.040; and 0.050), Θ is the effective saturation (dimensionless), m is the parameter from the Burdine (1953) restriction (dimensionless) [23]; t_{CC} is the time corresponding to the field capacity (h), z is the soil layer (m), and K_S is the saturated hydraulic conductivity (m/h). Estimation of the Θ function was performed in a routine developed in the R statistical program [24]. The moisture at the inflection point ($\hat{\theta}_{IP}$ measured in m^3/m^3) of the soil water retention curve was determined according to Dexter and Bird (2001) [29], and the respective matric potential at the inflection point (ψ_{IP} measured in kPa) was estimated using Equation (9) as:

$$\hat{\theta}_{IP} = (\theta s - \theta r) \cdot \left(1 + \frac{1}{m}\right)^{-m} + \theta r \quad (13)$$

2.6. Soil Water Capacity Function

The soil pore size distribution, represented by the water capacity function, was estimated using the retention curve model and capillarity theory [3,30]. The van Genuchten (1980) model was modified by replacing the matric potential ψ_m with the pore radius (r), using the capillarity equation [3]:

$$\theta = \theta r + \frac{(\theta s - \theta r)}{\left[1 + \left(\frac{A}{r}\right)^{n}\right]^{m}} \quad (14)$$

$$A = 2 \cdot \sigma \cdot \alpha \cdot 10^3 \quad (15)$$

With the derivative of the modified model, one obtains the following expressions for r_{max} (maximum pore radius) and d_{max} (maximum pore diameter):

$$\frac{d\theta}{d\log r} = (\theta s - \theta r) \cdot m \cdot n \cdot A^n \cdot r^{-n} \cdot (1 + A^n \cdot r^{-n})^{-m-1} \quad (16)$$

Deriving and equating to zero, one has:

$$r_{max} = A \cdot \left(\frac{1}{m}\right)^{-\frac{1}{n}} \quad (17)$$

$$d_{max} = 2 \cdot r_{max} \quad (18)$$

where σ is the surface tension of water ($N\ m^{-1}$); r is the pore radius measured in micrometers (μm).

2.7. Theoretical and Actual Soil Water Storage and Availability

The available soil water capacity (AWC) was determined from soil moisture measured at −6 kPa (AWC_1) and estimated from matric potential at field capacity (AWC_2), relative to the −1500 kPa matric potential:

$$AWC_1 = \sum_{i=1}^{n} (\theta_{CCi} - \theta_{PMPi}) \cdot z_i \quad (19)$$

$$AWC_2 = \sum_{i=1}^{n} (\hat{\theta}_{CCi} - \theta_{PMPi}) \cdot z_i \ (20) \quad (20)$$

where AWC is measured in millimeters; θ_{CCi} the volumetric soil moisture at field capacity in the *i*th layer (m^3/m^3) at the matric potential of −6 kPa; θ_{PMPi} the volumetric soil moisture at the permanent wilting point in the *i*th layer (m^3/m^3) at the matric potential of −1500 kPa, $\hat{\theta}_{CC}$ the estimated soil volumetric moisture at field capacity in the *i*th layer (m^3/m^3) at the matric potential ψ_{CC}; z_i is the thickness of the *i*th soil layer (mm); and n the number of layers considered, which is dimensionless.

Considering the results obtained from the soil water redistribution and point of hydraulic cut-off, the easily available water (EAW) at matric potential −6 kPa (EAW_1) and ψ_{CC} (EAW_2) was determined in relation to the volumetric moisture at 1030 kPa −1030 kPa (ψ_{hco}):

$$EAW_1 = \sum_{i=1}^{n} (\theta_{CCi} - \theta_{hco}) \cdot z_i \quad (21)$$

$$EAW_2 = \sum_{i=1}^{n} (\hat{\theta}_{CCi} - \theta_{hco}) \cdot z_i \quad (22)$$

where EAW in the soil is measured in mm, θ_{hco} is the volumetric soil moisture at the matric potential of −1030 kPa in the *i*th layer (m^3/m^3), z_i is the thickness of the *i*th soil layer (mm), and n is the number of layers considered and is dimensionless. Available soil water capacity (AWC) and easily available water (EAW) were calculated for different soil layers at four percentages of AWC and EAW: 100%, 80%, 60%, and 40%. These values were then summarized using R software's *ggplot2* function [24].

2.8. Statistical Analyses

Descriptive statistics were used to analyze our results with the help of the boxplot graph, and the discriminant values (outliers) were removed from the subsequent analyses with 6 repetitions. The assumptions of normality and homogeneity of variances were verified with the Shapiro–Wilk and Bartlett tests, respectively. Once these assumptions were met, the data were submitted for analysis of variance and the averages were compared using the Tukey test at the 5% significance level. The t_{CC} data were submitted to the Kruskal–Wallis test. Statistical errors and index between observed (Y) and estimated ($\hat{Y}$) values were quantified with the following expressions for root mean square error (RMSE), the root mean square error normalized by the mean (NRMSE), the ratio of RMSE to the standard deviation (RSR), and Willmott's agreement index (d) [31,32]:

$$RMSE = \sqrt{\frac{\sum_{i=1}^{n} (\hat{Y}_i - Y_i)^2}{n}} \quad (23)$$

$$NRMSE = \frac{\sqrt{\frac{\sum_{i=1}^{n} (\hat{Y}_i - Y_i)^2}{n}}}{\overline{Y}} \quad (24)$$

$$RSR = \frac{RMSE}{Dp_Y} \quad (25)$$

$$d = 1 - \left[\frac{\sum_{i=1}^{n} (\hat{Y}_i - Y_i)^2}{\sum_{i=1}^{n} (|\hat{Y}_i - \overline{Y}| + |Y_i - \overline{Y}|)^2} \right] \quad (26)$$

with measurement units specified for RMSE (m^3/m^3), NRMSE (%), RSR (dimensionless), and d (dimensionless). For RMSE, NRMSE, and d, $\hat{Y}_i$ is the *i*th value of the estimated variable (m^3/m^3), Y_i is the *i*th value of the observed variable (m^3/m^3), and n is the amount of data for the observed variable (units of variable). When calculating NRMSE and d, $\overline{Y}$ is the mean of the values of the observed variable (m^3/m^3). For RSR calculation, Dp_Y is the standard deviation of the observed data (m^3/m^3).

Validation of the associations among soil physical-hydric attributes was performed using principal component analysis (PCA). The correlation matrix of the standardized variables (unit variances and zero means) of each soil layer was used in PCA. The *prcomp* function was used to perform PCA and the *ggfortify* package [33] of the R software [24] was used for graphing.

2.9. Sugarcane Yield Modeling

The 12 years at the research site where sugarcane cultivars were evaluated were 1998–2006, 2008, and 2018–2019. Over these years, there were four sugarcane cultivars that were tested. RB72454 was planted from 1998 to 2006 and in 2008. RB867515 was trialed from 2004 to 2006 and from 2018 to 2019. RB966928 was evaluated from 2004 to 2005 and from 2018 to 2019. Finally, RB036066 was planted in 2018 and 2019. Sugarcane was grown in all years without irrigation.

Simulated sugarcane yields were estimated using soil water storage values (AWC and EAW, considered at 100%) from this study entered into a model validated from 12 years of field data from UFPR and RIDESA by Viana et al., 2023 [34]:

$$\mathrm{TCH} = -986.016 - 0.288 \times \mathrm{ADD_I} + 165.645 \times \mathrm{SWS_{II}} \quad (27)$$

In the simulation, the logarithmic transformation of the multiple linear regression model was performed:

$$\mathrm{TCH} = e^{-11.3374} \times \mathrm{ADD_I}^{-0.8224} \times \mathrm{SWS_{II}}^{10.5272} \quad (28)$$

where TCH is metric tons of sugarcane stalk per hectare, $\mathrm{ADD_I}$ is growing degree days (°C) during the first stage of sugarcane crop development (July through October), and $\mathrm{SWS_{II}}$ is soil water storage during the second development (October through March) for a 12-month production period. The third stage of development runs from March through July where the crop accumulates sucrose [34]. Simulated yields (y-axis) were then graphed three-dimensionally against $\mathrm{SWS_{II}}$ (AWC and EAW) measured in cm (x-axis) and $\mathrm{ADD_I}$ measured in °C (z-axis) for all data points. Graphing was performed using R's *scatter3d plotly r* function [24].

3. Results

3.1. Texture and Sandy Soil Structure

Clay content was significantly higher in the 0.2 to 0.4 and 0.4 to 0.6 m layers of soil compared to the topsoil layer at 0 to 0.2 m depth (Table 1). There was an inverse relationship between coarse and fine sand fractions in the soil profile (Table 1). The soil bulk density (ρ_S) and particle density (ρ_P) did not show significant differences among the soil layers (Table 1). The magnitude of the values we found is consistent with what is expected from sandy texture soils [1]. The soil bulk densities we measured (Table 1) ranged from 1582 to 1690 kg/m^3 which were consistent with an expected range of 1300 to 1800 kg/m^3 [30]. The average value of total porosity (TP = 0.33 m^3/m^3) was low, as commonly observed in sandy soils with fine sand contents [2,35]. As macroporosity decreased, microporosity increased, but ρ_S did not increase and TP decreased ($p < 0.05$) in the subsurface layers (Table 1). There was an increase in microporosity and clay contents at depth.

Table 1. Physical-hydric attributes of *Latossolo Vermelho distrófico* (Oxisol) soil layers at the Sugarcane Genetic Improvement Program Experimental Station cultivated with sugarcane in Paranavaí, Paraná state, Brazil.

Soil Layer	Sand Classification [1]			Silt	Clay	Soil Density [1]		Soil Porosity [1]		
	Total	Coarse	Fine			Particle	Bulk	Total	Macro	Micro
m			g/kg			kg/m^3			m^3/m^3	
0.00–0.20	838 a	770 a	121 b	67.6 a	94.8 b	2710 a	1580 a	0.331 a	0.158 a	0.173 b
0.20–0.40	793 b	685 b	135 b	64.3 b	143 a	2722 a	1690 a	0.316 a	0.0906 b	0.225 a
0.40–0.60	785 b	627 b	167 a	65.4 b	150 a	2703 a	1650 a	0.329 a	0.0883 b	0.241 a
Average	805	694	140	65.8	129	2712	1640	0.325	0.112	0.213
Var [2]	571	3620	402	1.99	633	2216.8	10,600	<0.001	0.00167	0.00107
Stdev [3]	23.9	60.2	20.1	1.41	25.2	47.08	103	0.0205	0.0409	0.0326
CV [4]	2.97%	8.67%	14.2%	2.15%	19.5%	1.74%	6.27%	6.31%	36.4%	15.3%

[1] Averages followed by equal letters in the column do not differ by the Tukey test at 5% probability. [2] Var is the variance (variable unit). [3] Stdev is the standard deviation (variable unit). [4] CV is the coefficient of variation (%).

3.2. Water Movement in Sandy Soil

The saturated hydraulic conductivity (K_S) was higher in the 0 to 0.2 m layer and reduced abruptly at depth (Figure 1a) due to macroporosity facilitating water drainage in saturated soil [35]. Like Ks, unsaturated hydraulic conductivity K(θ) was also higher in the uppermost soil layer. However, the decrease in K(θ) with the reduction in soil moisture was more significant in the 0 to 0.2 m layer, compared to the others (Figure 1b). The K(ψ) in the surface layer showed greater decay (Figure 1c) with increasing matric potential due to the smaller number of capillary pores [35].

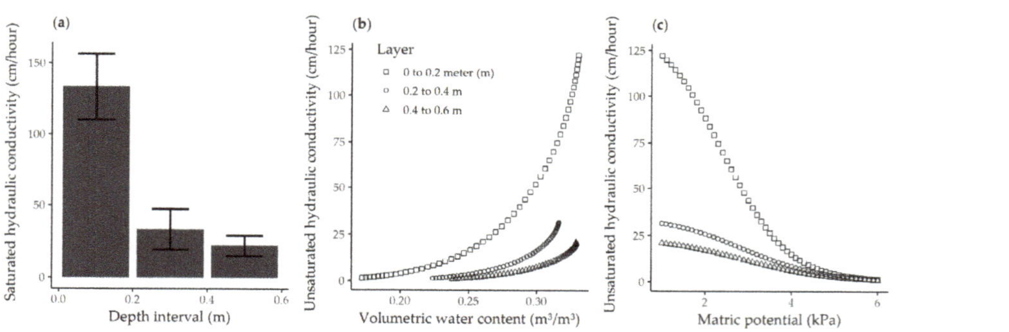

Figure 1. (**a**) Saturated (K_S) conductivity with error bars in cm/hour; (**b**) unsaturated (K(θ)) hydraulic conductivity versus volumetric water content; and (**c**) versus matric potential in the soil layers 0 to 0.20 m (m), 0.2 to 0.4 m, and 0.4 to 0.6 m for the Latossolo Vermelho distrófico (Oxisol) at the Sugarcane Genetic Improvement Program Experimental Station cultivated with sugarcane in Paranavaí, Paraná state, Brazil.

3.3. Water Retention and Pore Distribution

The soil water retention curves showed no evidence of compaction effects (Figure 2), as the soil bulk density (ρ_S) values are below the critical value of $\rho_S > 1.70$ g/cm^3 [2]. The parameters of the van Genuchten model (α and n) were significant ($p < 0.05$) to represent the soil water retention curve for all soil layers (Table 2). Statistical errors and indices between observed (θ) and estimated ($\hat{\theta}$) soil volumetric moisture indicated that fits with the van Genuchten–Burdine model were satisfactory for the analyzed layers (Table 3). Among the errors, the NRMSE was higher due to the use of the six repetitions at each matric potential to perform the adjustment of the soil water retention curve, which improved the significance of the adjustment.

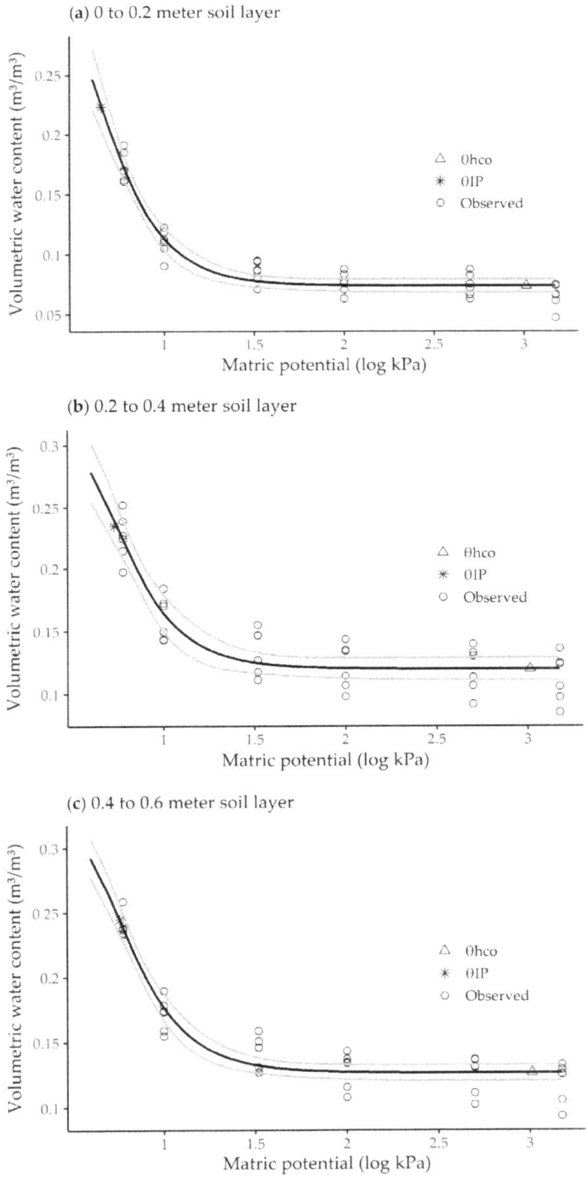

Figure 2. The water retention curve of the Latossolo Vermelho distrófico (Oxisol) in the layers (**a**) 0 to 0.2 m; (**b**) 0.2 to 0.4 m; and (**c**) 0.4 to 0.6 m. The fitted values are accompanied by the confidence band for the predicted value.

The RSR is one of the quantitative statistics recommended by Moriasi et al. (2007) [36] to evaluate models. The smallest magnitudes of RSR were observed at different p in the soil layers due to the physical-hydric attributes of the soil (Table 4). Lower matric potential ψ_{IP} was found in the 0 to 0.2 m (m) layer due to the soil texture (Table 5). The pore distribution curve (Figure 3) showed a maximum pore diameter (d_{max}) of 64.19 μm (μm) at a soil depth of 0 to 0.2 m, 51.96 μm at 0.2 to 0.4 m, and 50.6 μm at 0.4 to 0.6 m. The maximum point of the pore diameter frequency curve dθ/dlog(r) corresponded to 0.192 at a depth of 0 to 0.2 m, 0.1458 at 0.2 to 0.4 m, and 0.1463 at 0.4 to 0.6 m in the analyzed soil layers.

Table 2. Fitted parameters for the van Genuchten model and their confidence intervals.

Soil Layer Depth (Meter)	Parameter [1]	Estimated Value	Error	t-Statistic	Significance (*p*-Value)	95% Confidence Interval ($CI_{95\%}$)	Goodness of Fit (R^2)
0 to 0.2	θs (m^3/m^3)	0.331	0.00517	63.9	<0.001	[0.321; 0.341]	0.982
	θr (m^3/m^3)	0.0737	0.00283	26.0	<0.001	[0.0681; 0.0792]	
	α (1/kPa)	0.265	0.0275	9.63	<0.001	[0.211; 0.318]	
	n (ad.)	3.92	0.331	11.8	<0.001	[3.27; 4.57]	
0.2 to 0.4	θs (m^3/m^3)	0.316	0.008	39.1	<0.001	[0.300; 0.332]	0.932
	θr (m^3/m^3)	0.120	0.004	26.8	<0.001	[0.111; 0.128]	
	α (1/kPa)	0.215	0.0282	7.6	<0.001	[0.159; 0.270]	
	n (ad.)	3.90	0.507	7.68	<0.001	[2.90; 4.89]	
0.4 to 0.6	θs (m^3/m^3)	0.329	0.00558	59.0	<0.001	[0.318; 0.340]	0.969
	θr (m^3/m^3)	0.126	0.00311	40.4	<0.001	[0.120; 0.132]	
	α (1/kPa)	0.211	0.0183	11.6	<0.001	[0.175; 0.247]	
	n (ad.)	3.82	0.318	12.0	<0.001	[3.20; 4.45]	

[1] θs and θr are the soil moisture at saturation and residual, respectively, while α and n are the parameters of the van Genuchten model.

Table 3. Errors and statistical index between observed (θ) and estimated ($\hat{\theta}$) soil volumetric moisture with van Genuchten model using Burdine restriction.

Soil Layer Depth	Root Mean Square Error (RMSE)	Normalized RMSE	RMSE/Stdev (RSR)	Willmott's Agreement Index (d)
(m)	**(m^3/m^3)**	**(%)**	**(ad.)**	**(ad.)**
0 to 0.2	0.0121	13.4	0.134	0.995
0.2 to 0.4	0.0188	25.8	0.258	0.982
0.4 to 0.6	0.0130	17.5	0.175	0.992

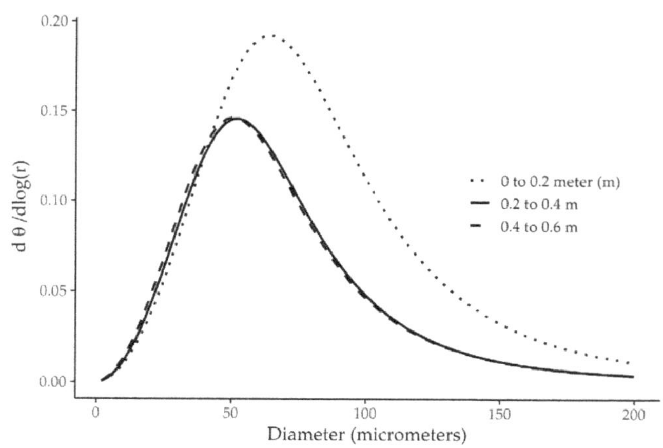

Figure 3. Pore distribution curve for soil layers at depths of 0 to 0.2 m, 0.2 to 0.4 m, and 0.4 to 0.6 m of Latossolo Vermelho distrófico (Oxisol) used for sugarcane cultivation at the Experimental Station of the Sugarcane Genetic Improvement Program at the Federal University of Paraná and the Inter-University Network for the Development of Sugarcane Energy.

Table 4. Estimated actual moisture ($\hat{\theta}_{CC}$), matric potential (ψ_{CC}), and time (t_{CC}) at field capacity and statistical error between observed (θ_{CC}) at –6 kPa and estimated ($\hat{\theta}_{CC}$) volumetric moisture at field capacity.

Soil Layer	**Percentage Ratio (p) of Saturated Hydraulic Conductivity (K_S)**							
Depth (m)	**0.005**	**0.010**	**0.015**	**0.020**	**0.025**	**0.030**	**0.040**	**0.050**
	Estimated actual moisture at field capacity $\hat{\theta}_{CC}$ (m^3/m^3) [1]							
0 to 0.2	0.157 c	0.172 c	0.182 b	0.190 b	0.196 b	0.202 b	0.211 b	0.218 a
0.2 to 0.4	0.184 ab	0.195 ab	0.203 ab	0.209 ab	0.214 ab	0.218 ab	0.225 ab	0.230 a
0.4 to 0.6	0.192 a	0.204 a	0.212 a	0.218 a	0.223 a	0.228 a	0.235 a	0.240 a
Average	0.178	0.191	0.199	0.206	0.211	0.216	0.223	0.230
Var	0.000404	0.000366	0.000346	0.000332	0.000323	0.000315	0.000306	0.000300
Stdev	0.0201	0.0191	0.0186	0.0182	0.0180	0.0178	0.0175	0.0173
CV	11.3	10.1	9.34	8.87	8.51	8.24	7.83	7.54
	Root mean square error/standard deviation (RSR) (dimensionless)							
0 to 0.2	1.96	1.57	1.78	2.15	2.52	2.87	3.49	4.02
0.2 to 0.4	2.41	1.88	1.57	1.37	1.23	1.15	1.10	1.15
0.4 to 0.6	5.20	4.00	3.23	2.66	2.23	1.89	1.47	1.35
	Matric potential at field capacity (ψ_{CC} in kPa) [1]							
0 to 0.2	6.55 c	5.93 c	5.59 c	5.35 c	5.16 b	5.01 b	4.78 b	4.60 b
0.2 to 0.4	8.16 ab	7.38 ab	6.95 ab	6.64 ab	6.41 ab	6.23 ab	5.94 ab	5.71 ab
0.4 to 0.6	8.56 a	7.73 a	7.26 a	6.94 a	6.69 a	6.49 a	6.18 a	5.94 a
Average	7.76	7.01	6.60	6.31	6.09	5.91	5.63	5.42
Var	1.48	1.30	1.21	1.15	1.10	1.07	1.01	0.969
Stdev	1.22	1.14	1.10	1.07	1.05	1.03	1.01	0.985
CV	15.7	16.3	16.7	17.0	17.3	17.5	17.9	18.2
	Time at field capacity (t_{CC} in hours) [1]							
0 to 0.2	0.604 c	0.355 c	0.259 c	0.207 c	0.174 c	0.151 c	0.120 c	0.101 c
0.2 to 0.4	1.85 b	1.08 b	0.791 b	0.633 b	0.532 b	0.461 b	0.367 b	0.307 b
0.4 to 0.6	2.84 a	1.67 a	1.22 a	0.975 a	0.819 a	0.710 a	0.566 a	0.473 a
Average	1.76	1.04	0.757	0.605	0.508	0.441	0.351	0.294
Var	0.953	0.328	0.175	0.112	0.079	0.060	0.038	0.027
Stdev	0.976	0.573	0.419	0.335	0.281	0.244	0.194	0.163
CV	55.3	55.3	55.3	55.3	55.4	55.4	55.4	55.4

[1] Averages followed by equal letters in the column do not differ by Tukey's test at 5% probability ($\hat{\theta}_{CC}$ and ψ_{CC}) and t_{CC} by Kruskal–Wallis test.

Table 5. Estimated moisture ($\hat{\theta}_{IP}$) and matric potential (ψ_{IP}) at the inflection point, and statistical error between observed (θ_{CC}) at –6 kPa and estimated ($\hat{\theta}_{IP}$) volumetric moisture at field capacity.

Soil Layer Depth (Meter)	**Estimated Moisture $\hat{\theta}_{IP}$ (m^3/m^3) [1]**	**Matric Potential ψ_{IP} (kPa) [1]**	**RSR (Dimensionless) [5]**
0 to 0.2	0.223 b	4.48 b	4.29
0.2 to 0.4	0.234 ab	5.37 ab	1.08
0.4 to 0.6	0.244 a	5.79 a	1.37
Average	0.234	5.21	-
Var [2]	0.000229	0.879	-
Stdev [3]	0.0151	0.938	-
CV [4]	6.48	18.0	-

[1] Averages followed by equal letters in the column do not differ by the Tukey test at 5% probability. [2] Var is the variance (variable unit). [3] Stdev is the standard deviation (variable unit). [4] CV is the coefficient of variation (%). [5] RSR—Root mean square error/Standard deviation (dimensionless).

3.4. Water Availability

In general, the soil of the experimental area cultivated with sugarcane showed low water availability for plants in the layers analyzed (Table 6), which is characteristic of soils with low clay contents [5,37]. There was no statistical difference between the layers

($p > 0.05$), but the highest AWC_1 and EAW_1 occur in the soil layers at 0.2 to 0.4 m and 0.4 to 0.6 m depth due to the higher amount of clay and microporosity [5]. The highest AWC_2 and EAW_2 occurred in the surface layer ($p < 0.05$) due to the higher organic carbon content that increases water retention and availability for plants [38]. This surface layer corresponds to the soil layer with the highest volume and density of root length for sugarcane [34]. Figure 4 shows the variation in the percentage of water availability for the soil in the experimental area at the Experimental Station of the Sugarcane Genetic Improvement Program at the Federal University of Paraná and the Inter-University Network for the Development of Sugarcane Energy.

Table 6. Hydric attributes of the *Latossolo Vermelho distrófico* (Oxisol), cultivated with sugarcane at the Experimental Station of the Sugarcane Genetic Improvement Program at the Federal University of Paraná and the Inter-University Network for the Development of Sugarcane Energy.

Soil Layer	θ_{CC} [1]	$\hat{\theta}_{CC}$ [1]	θ_{PMP} [1]	θ_{hco} [1]	AWC_1 [1]	AWC_2 [1]	EAW_1 [1]	EAW_2 [1]
Depth (m)	m^3/m^3				mm			
0.00–0.20	0.173 b	0.218 a	0.0650 b	0.0735 b	21.6 a	30.6 a	19.9 a	28.9 a
0.20–0.40	0.225 a	0.230 a	0.112 a	0.119 a	22.7 a	23.6 b	21.2 a	22.1 b
0.40–0.60	0.241 a	0.240 a	0.118 a	0.126 a	24.5 a	24.4 b	22.9 a	22.9 b
Average	0.213	0.229	0.0983	0.106	22.9	26.2	21.3	24.6
Var [2]	0.00107	0.000300	0.000804	0.000753	9.34	21.2	7.91	20.7
Stdev [3]	0.0326	0.0173	0.0283	0.0274	3.06	4.61	2.81	4.55
CV [4]	15.3	7.54	28.8	25.8	13.3	17.6	13.2	18.4

[1] Averages followed by equal letters in the column do not differ by the Tukey test at 5% probability. [2] Var is the variance (variable unit). [3] Stdev is the standard deviation (variable unit). [4] CV is the coefficient of variation (%).

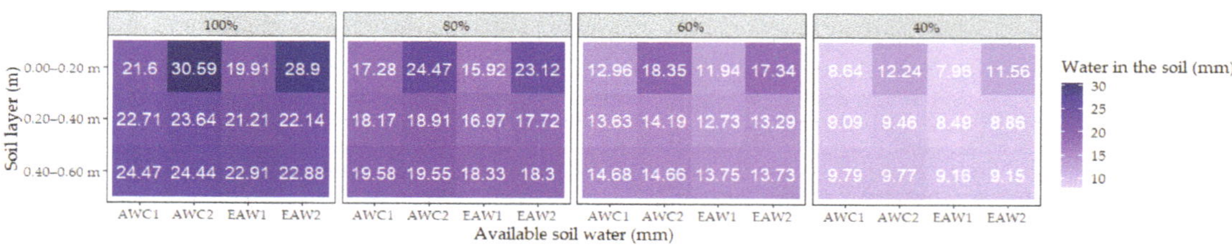

Figure 4. Water in the soil (mm) for different percentages of available soil water capacity (AWC_1 and AWC_2) and easily available water (EAW_1 and EAW_2) and for different soil layers (m).

3.5. Multivariate Relationships among Soil Physical-Hydric Attributes

Principal component analysis (PCA) was used to validate the previously established associations between the structural and hydric physical attributes. We observed that the first principal component (PC1) explained 60.6% of the total variance. PC1 (eigenvalue of 3.114) and PC2 (eigenvalue of 1.495) together explained 74.6% of the total variance (Figure 5).

PCA was efficient in distinguishing between soil layers. The 0 to 0.2 m layer was characterized and influenced mainly by the physical-hydric attributes of macroporosity, silt, saturated hydraulic conductivity, unsaturated hydraulic conductivity, sand, coarse sand, and maximum pore diameter. The variables fine sand, clay, microporosity, volumetric soil moisture at the permanent wilting point, volumetric soil moisture at the inflection point, soil density, and available soil water capacity determined at –6 kPa were higher and more expressive in the physical-hydric attributes of the soil layers at depths of 0.2 to 0.4 m and 0.4 to 0.6 m (Figure 5). It was found that saturated hydraulic conductivity showed a direct relationship with sand, coarse sand, silt, macroporosity, and maximum pore diameter, and inverse with microporosity, clay, volumetric soil moisture at the permanent wilting point, and fine sand. The soil density showed an inverse relationship with macroporosity.

Microporosity was directly related to clay, fine sand, and volumetric soil moisture at the permanent wilting point, and inversely with saturated hydraulic conductivity, sand, coarse sand, silt, macroporosity, and maximum pore diameter.

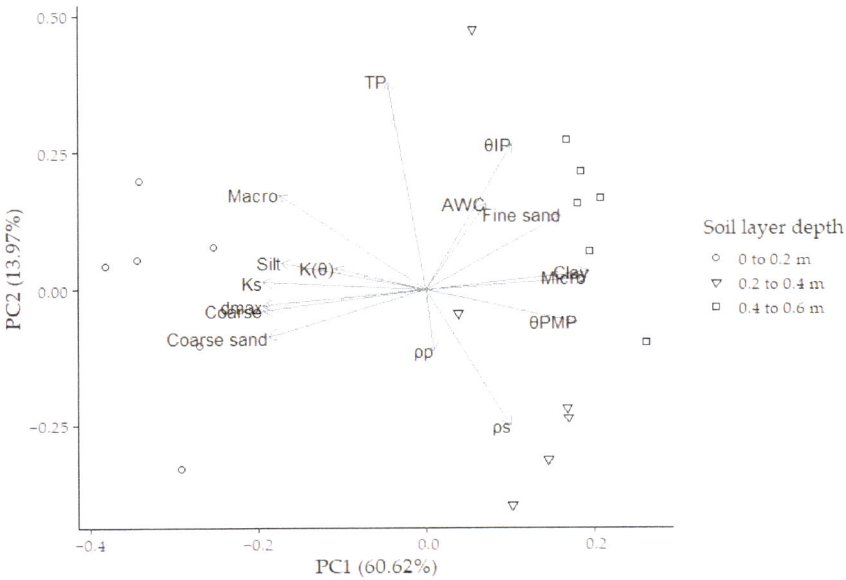

Figure 5. Principal component analysis (PCA) of the physical-hydric attributes of the Latossolo Vermelho distrófico (Oxisol), cultivated with sugarcane in the Experimental Station of the Sugarcane Genetic Improvement Program at the Federal University of Paraná and the Inter-University Network for the Development of Sugarcane Energy. Labels for (1) "dmax" and "Coarse" and (2) "Clay" and "Micro" slightly overlap with each other due to close proximity of these two pairs of vectors.

3.6. Sugarcane Yield Simulations

For 12 years at the research site, sugarcane stalk yield (TCH) averaged 114.68 metric tons (t)/hectare (ha). TCH was at a minimum of 65.997 t/ha in 2005 and a maximum of 164.190 t/ha in 2006 (Figure 6). Fracaro et al., 2014 [39] found sugarcane TCH of 170 t/ha for RB036152 and an average TCH of 110 t/ha across 26 genotypes from RIDESA in the 2013–2014 crop season in the municipality of Viamão, Rio Grande do Sul state, Brazil. The estimated Brazilian average national productivity of sugarcane for the 2022–2023 crop season, according to CONAB [40], is 72 t/ha. The 2022–2023 crop season was characterized by reduced rainfall and low temperatures in the Central-South Region, which accounts for around 90% of Brazil's total sugarcane production [40].

In the sugarcane model simulation with AWC and EAW set at a value of 100% (Figure 7), TCH for: (1) EAW_1 averaged 33.94 t/ha with a minimum of 26.84 t/ha and a maximum of 42.27 t/ha; (2) AWC_1 averaged 72.10 t/ha with a minimum of 57.00 t/ha and a maximum of 89.78 t/ha; (3) EAW_2 averaged 153.96 t/ha with a minimum of 121.73 t/ha and a maximum of 191.73 t/ha; and (4) AWC_2 averaged 296.58 t/ha with a minimum of 234.50 t/ha and a maximum of 369.34 t/ha. Under the 100% water availability condition in Phase II, EAW_1 underestimates and AWC_2 overestimates sugarcane yield. Analyzing the simulated values with EAW_2 (set at 100%), the potential yield of sugarcane can be achieved with sustainable management practices, genetic improvement, and favorable climate conditions.

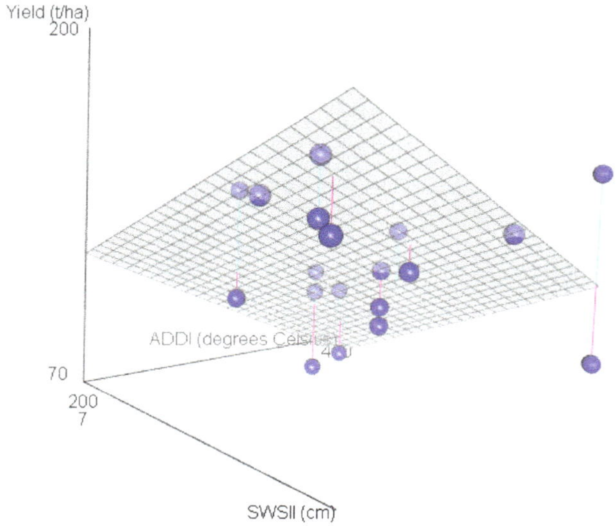

Figure 6. Sugarcane yield model validated from 12 years of field data from UFPR and RIDESA by Viana et al., 2023 [34].

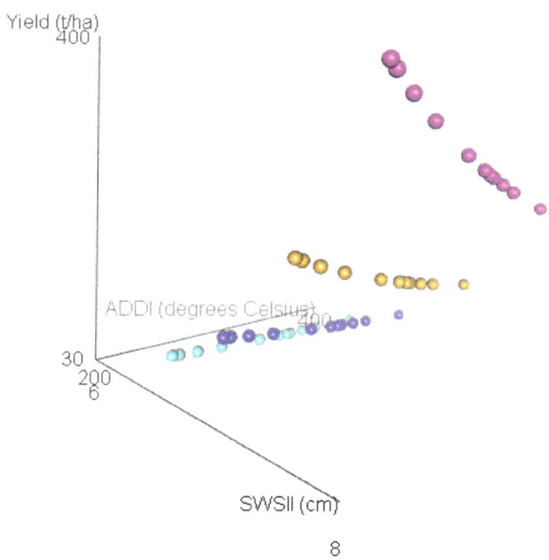

Figure 7. Simulated sugarcane stalk yield (TCH = t/ha) as a function of growing degree days during development phase I (ADD_I) and soil water storage during development phase II (SWS_{II}) using both AWC and EAW (100%), with the logarithmic transformation of the validated model by Viana et al., 2023 [34].

4. Discussion

4.1. Comparisons to Previous Research

4.1.1. Soil Characteristics

Although there was an increase in clay content in the subsurface layers, this did not result in abrupt textural changes in the soil. Greater clay content in lower soil surface layers can result in increased water retention for sandy soils. As a consequence, the sand and silt contents were lower in the 0.2 to 0.4 m and 0.4 to 0.6 m soil layers. Thus, the analyzed soil is in the textural classes of sand in the soil layer at the depth range of 0 to 0.20 m, and loamy sand in soil layers 0.2 to 0.4 m and 0.4 to 0.6 m [41], characteristic of the Paranavaí Sandstone Formation in the northwestern region of Paraná [6]. The higher coarse sand content of the sand fraction is characteristic of soils of the Paranavaí region, confirmed by the granulometric characterization in areas of crop-livestock and pineapple integration in the Caiuá and Paranavaí Sandstone Formations [6]. Fidalski 2017 [42] considered that the granulometric characterization is a criterion to guide the use and management of soils of the Caiuá Sandstone in the northwestern region of Paraná. This is due to the inverse relationship of the sand contents (total or coarse) with the available water in the soil.

Souza et al., 2015 [37] found an average soil bulk density of 1620 kg/m^3 at the RIDESA. This was close to the average ρ_S obtained in this study of 1640 kg/m^3. In the literature, there are reports of root growth restriction in sugarcane for ρ_S = 1700 kg/m^3 [2]. Thus, the bulk density values obtained in this study between 1600 and 1700 kg/m^3 are at the threshold of possibly limiting crop root growth. Therefore, management practices should be adopted to minimize soil compaction.

Ortiz et al., 2017 [43] found TP between 0.29 and 0.38 m^3/m^3 in sandy soil grown in the first, third, and fifth sugarcane cycles in 44, 40, and 15 years of production, respectively. Low macroporosity pore values (0.09 m^3/m^3) were obtained in the soil layers at 0.2 to 0.4 m and 0.4 to 0.6 m (Table 1). This may be a problem since values lower than 0.1 m^3/m^3 are limiting to soil aeration and harmful to agricultural production [1,44,45]. These results support recommendations for better managing sandy soils, including reducing soil bulk density and increasing total porosity and macroporosity, especially in deeper soil layers [44].

4.1.2. Saturated Soil Hydraulic Conductivity

Our results indicate that the water dynamics in the soils that were analyzed can be modified between their layers since the microporosity is responsible for the retention and conduction of water in unsaturated soil, and macroporosity is responsible for the drainage and aeration of the soil [1,35]. High K_S values in more superficial layers are characteristic of sandy textured and well-drained soils [5]. However, especially in the 0.4 to 0.6 m layer, the abrupt change in K_S can make the soil susceptible to shear which may lead to erosion. The combination of high K_S and sand content account for more than 80% of the variability in soil loss [46]. Our results reinforce the need to increase macroporosity at depth to better manage sugarcane.

Even though the K_S in the surface layer was higher than the others (Figure 1a), the values were lower than in soils of other textures also cultivated with sugarcane [47,48]. Due to the effects of soil preparation recently performed for sugarcane re-planting such as chiseling and disking, Cherubin et al., 2016 [47] found K_S (measured in the field) of 250 cm/h in soil with medium clay texture, in the surface layer. The authors found that K_S showed a direct correlation with total porosity and macroporosity, and an inverse relationship with ρ_S and the degree of soil compaction, both of which may be directly influenced by soil management. Silva et al., 2018 [48] observed in sandy soils that K_S decreased with increasing depth in sandy soils, associated with increasing bulk density and microporosity. Silva et al., 2018 [48] found a direct correlation between K_S with coarse sand and medium sand fractions and an inverse correlation with the available water capacity. In addition, it was found that the available water capacity showed an inverse linear correlation with the coarse sand and medium sand grain size fractions, and direct linear correlation with the micropore volume.

The $K(\psi)$ between soil layers was similar at matric potentials greater than -5 kPa, suggesting that the field capacity of the sandy soil was reached before -6 kPa [26]. The dependent relationship of unsaturated hydraulic conductivity $K(\theta)$ with the matric potential indicated that there are modifications in the soil water retention as a consequence of some variation in the pore distribution curve. Thus, the results of $K(\psi)$ of the analyzed sandy soils demonstrated that the evaluation of soil quality cannot be limited to the separation of TP into macroporosity and microporosity, but a detailed determination of the pore distribution curve is necessary [9].

4.1.3. Soil Water Retention

Higher soil water retention was observed in the subsurface layers (Table 2; Figure 2) due to the increase in microporosity, clay and silt contents, and reduction in coarse sand content at depth (Table 1). The relative error corresponded to 33.5% and 37.6% in the 0.2 to 0.4 m and 0.4 to 0.6 m layers relative to the subsurface layer, respectively. At potentials of −33 kPa, −100 kPa, and −500 kPa, the average relative errors were 35.6% and 39.5%. The largest relative errors (40.3% and 43.9%) occurred at the −1500 kPa matric potential.

Soil retains water due to capillary and adsorption matrix forces, which are responsible for water retention in the capillary pores of aggregates and on the surfaces of soil particles, respectively [49]. The surfaces of clays are negatively charged and adsorb a large number of water molecules on the soil surface due to their high specific surface area. For example, kaolinite can absorb 70,000 to 300,000 cm^2 of water/gram (g) per gram of kaolinite [3,50]. Thus, soil layers with higher clay contents retain more water in the soil, reinforcing the results obtained in this study.

The soils of the Caiuá Sandstone Formation have low clay content with a predominance of kaolinite (>80%), the presence of illite, a ratio of 2:1 clay to minerals, and small amounts of iron and aluminum oxides [16]. The Caiuá Sandstone Formation soils also have a high sand content, with a predominance of large pore diameters that are greater than 30 μm (Figure 3), which is drained at low matric potentials [9,51], reducing the water holding capacity of the soil.

The soil layer with higher fine sand content retained more soil water, especially at low matric potentials (less than −33 kPa; Figure 2). The obtained result reinforces the importance of the fractionation of total sand in sandy soils due to the higher specific surface area of fine sand at 318 cm^2/g compared to coarse sand at 79 cm^2/g [6,50]. The distribution between fine and coarse sand fractions is an inherent characteristic of soil formation and is not altered by management. Thus, it is worth emphasizing the importance that the accumulation of organic matter has in sandy soils because, besides increasing aggregation, it improves water retention and availability in soils [1,16,38,52,53].

Among the results obtained in the analyses of the soil water retention curves, it was verified that (1) the residual moisture (θr) was similar to the soil volumetric moisture at the hydraulic cut-off point (θ_{hco}) in the soil layers at depth ranges of 0 to 0.2 m (0.074 m^3/m^3), 0.2 to 0.4 m (0.120 m^3/m^3), and 0.4 to 0.6 m (0.126 m^3/m^3) (Table 2). Moreover, (2) the residual moisture (θr) was greater than the observed volumetric moisture at –1500 kPa. Finally, (3) the average relative difference in volumetric moisture at –500 and –1500 kPa was 12.92%, 6.03%, and 5.09% in the 0 to 0.2 m, 0.2 to 0.4 m, and 0.4 to 0.6 m layers, respectively. There are indications that the matric potential at the permanent wilting point occurs between −500 kPa and −1030 kPa at the hydraulic cut-off point.

Obtaining the true value of matric potential for improved agricultural crops has been a recurring discussion [54,55]. Torres et al., 2021 [55] found values for matric potential $\psi_{PMP} < -1500$ kPa and found that sunflower, corn, and soybean crops wilted at potentials lower than the estimated matric potentials in the hydraulic cut-off. When plants wilt at water suction lower than the hydraulic cut-off, the permanent wilting point is plant-limited, and when plants wilt at suctions greater than the hydraulic cut-off the limitation is from the soil [25].

Therefore, in sandy soils, obtaining the actual permanent wilting point should receive more attention. In the literature, in sandy texture soil (803 g/kg), there are indications that the average matric potential at the physiological permanent wilting point was –163.7 kPa (Mestre) and –241.7 kPa (Noble) for two modern wheat cultivars, and –90.9 kPa (ANAG 01) and –45.9 kPa (BRS Brau) for two barley cultivars [54].

The relative difference in volumetric moisture at –200 and –1030 kPa was 0.164%, 0.125% and 0.166% in soil layers 0 to 0.2 m, 0.2 to 0.4 m, and 0.4 to 0.6 m, respectively. However, it is critical to emphasize that no studies were found in the literature reporting the estimated matric potential at the physiological permanent wilting point for modern sugarcane cultivars from breeding programs. The hydraulic cut-off suction (–1030 kPa) lower than –1500 kPa reinforces the need for further studies for the sandy soil type analyzed.

Prevedello 1999 [26] found, in homogeneous and heterogeneous soil, that the drainage rate (τ) corresponded to $p = 0.01$ of the saturated soil hydraulic conductivity ($\tau = 0.01$, K_S). Andrade and Stone 2011 [56] found matric potential at field capacity $\psi_{CC} = -6.5$ kPa, corresponding to $\hat{\theta}_{CC}$, for τ of 0.01 of K_S, in a sandy texture soil in the Cerrado. The moisture at the inflection point ($\hat{\theta}_{IP}$) is considered optimal for soil preparation [29] and correlates well with moisture at field capacity [56]. At the inflection point of the retention curve the first derivative changes from negative (convexity) to positive (concavity), and the second derivative is zero [57], which is relevant to determine the estimation of $\hat{\theta}_{CC}$.

An analysis of the moisture at the inflection point of the soil water retention curve (Table 5; Figure 2) indicated that the drainage rate of the soil corresponded to $p = 0.050$ of the K_S. The higher magnitude of the drainage rate of the analyzed soil was due to the texture, predominantly of the coarse sand fraction. For $p = 0.050$, there was ψ_{CC} of -4.602 kPa (0 to 0.2 m), -5.711 kPa (0.2 to 0.4 m), and -5.942 kPa (0.4 to 0.6 m). Regarding the time for field capacity to occur (t_{CC} measured in hours), the soil showed rapid vertical drainage due to the magnitude of the saturated hydraulic conductivity (K_S). In this case, the high magnitude of K_S influenced the low values of t_{CC}. The sand fraction of the soil also interferes with the matric potential corresponding to the permanent wilting point.

A higher magnitude of pore diameter was observed in the topsoil layer due to the higher coarse sand content (Table 1). The influence of the coarse sand fraction on the larger pore diameter and, consequently, lower water availability in predominantly sandy soils was also found by Fidalski et al., 2013 [6]. In soils with a predominance of sand, especially coarse sand, soil aggregation is reduced, not allowing the development of a more diverse pore network, especially in the smaller diameter pores, predominantly formed by organo-mineral interactions in the clay fraction [9,16]. Importantly, larger pore diameters require lower energy for soil water removal [9], consistent with results obtained for saturated hydraulic conductivity Ks and unsaturated hydraulic conductivity $K(\theta)$ (Figure 1).

4.1.4. Water Availability to Plants

Fidalski et al., 2013 [6] studied soils of the Paranavaí Sandstone Formation, in the northwestern region of Paraná and found that soil use and management interfered with water availability to plants. Fidalski et al., 2013 [6] also found available water capacity (AWC) in the layer between 0 and 0.4 m of 48.60 mm and 58.70 mm in soil under crop-livestock integration and pineapple, respectively. Helbel Junior and Fidalski 2017 [58] found an average water storage capacity of 1 mm/cm in the 0 to 0.6 m layer, in *Latossolo Vermelho distrófico* (Oxisol), medium texture, in the northwestern region of Paraná.

In the 0 to 0.6 m soil layer, the following were verified: $AWC_1 = 68.78$ mm; $AWC_2 = 78.67$ mm; $EAW_1 = 64.03$ mm and $EAW_2 = 73.92$ mm. AWC_2 overestimates soil water availability because the actual volumetric moisture at the permanent wilting point of the analyzed soil occurs before –1500 kPa (θ_{PMP}). EAW_1 underestimates soil water availability because the actual volumetric moisture at field capacity occurs at matric potentials lower than –6 kPa (θ_{CC}), highlighting the importance of the actual estimation of volumetric moisture at field capacity [56] and the hydraulic cut-off point [25].

Low water availability to plants in sandy soils can provide the occurrence of water deficit during the sugarcane cycle, as evidenced by Gurski et al., 2020 [59] and Araújo 2019 [60]. High soil hydraulic conductivity ($K(\theta)$) and low time to reach moisture at field capacity (t_{CC}) intensify the occurrence of water deficit. Gurski et al., 2020 [59] verified that water deficiency and surplus were concentrated mainly in development phase II (vegetative growth), in the 1997-98 to 2008-09 crops, in Paranavaí, Paraná, requiring rescue irrigation.

Thus, the physical-hydric attributes determined in this study should be considered in the management of irrigation and the plant in each phase of its development.

The effect of sand contents on the water retention of the Caiuá Sandstone, from the northwestern Paraná region, was confirmed with the inverse correlation of coarse sand contents with microporosity, θ_{PMP}, clay, fine sand, and θ_{IP}, agreeing with [42]. The results again indicated the importance of a more detailed evaluation of the physical-hydric attributes of sandy soils, especially for monitoring the physical quality and water management in the analyzed soil.

4.2. Sustainability Implications

Soil type and properties at different depths influence the development of sugarcane [61]. While a shallow groundwater table can impact soil moisture dynamics [62], this research was conducted in an area that has a water table that is from 5 to 50 m deep [18], which is associated with vertical, subsurface soil water movement [17]. Yield modeling of sugarcane involves a detailed understanding of the impact of soil water storage dynamics on crop yield in Brazil. This is especially important for soils with low water holding capacity across water deficit gradients which have been modeled in northwestern São Paulo state using a crop water balance model combined with GIS [63]. In Köppen classification Cfa climates like in this study, the average yield response to full irrigation was 22 t/ha using the FAO-AZM simulation model [64]. Greater yield responses were in sandy soils with lower soil water holding capacities, where moisture from rainfall concentrated at the start of the dry season is not retained for as long as other soil types [64]. This research improves understanding of water dynamics for sugarcane grown in such sandy soils and how this relates to crop yield.

More accurate sugarcane models for Brazil's sandy soils are essential to optimize water management decisions to (1) invest in irrigation in order to increase crop productivity and to (2) withhold water, especially in irrigated systems prior to harvest to maximize sugar content. Yield gaps for sugarcane can be closed by using crop breeding for drought-tolerant cultivars [65] and irrigation [65,66]. If future climate changes increase sugarcane yield, payback for irrigation investments could decrease from 14 years to only 3 years, making such investments more feasible [67]. However, irrigation investments may be challenging since medium-sized sugarcane farms tend to be the most profitable [68], with potentially less capital available for such upfront purchases. Dias and Sentelhas 2018 [69] simulated sugarcane with APSIM-Sugar, showing stalk biomass reductions of 4% are ideal for maximizing sucrose yields. This can be accomplished by withholding water for drying off prior to harvest for 17 to 31 days for sandy soils with low water holding capacity [69].

The sustainable intensification of sugarcane, which includes increasing yield, has potential environmental benefits in addition to increased crop productivity. Sugarcane has been used recently to rehabilitate degraded pastures in Brazil. Luz et al., 2020 suggest that aside from compaction in the 10 to 20 cm soil layer, conversion from degraded pasture to sugarcane did not result in more soil degradation, whereas conversion from native habitat to extensive pasture resulted in severe soil degradation over time [70]. However, Dias et al., 2021 used APSIM-Sugar to show that climate change from RCP8.5 and RCP4.5 over the next century could result in sugarcane yield reductions in Brazil primarily due to water stress even with irrigation [64]. Future research is needed to improve the accuracy of sugarcane simulation models in sandy soils by validating models developed from experimental results from long-term sugarcane experiments with process-based sugarcane simulation models such as DSSAT-CANEGRO and APSIM-Sugar [71].

5. Conclusions

In this study, the main physical and hydric attributes of sandy soil were investigated for sugarcane cultivation in southern Brazil. Our results can guide future breeding programs to select sugarcane cultivars more tolerant to water deficit. In sandy soils with a total sand content close to 800 g/kg, the distribution of both coarse sand and fine sand fractions should be measured to evaluate soil physical and hydric attributes, not just total sand content. The particle size of coarse sand ranges from 2 mm down to 0.2 mm while fine sand ranges from 0.2 to 0.05 mm. Higher coarse sand content increases saturated and unsaturated soil hydraulic conductivity, maximum pore diameter, and macroporosity, as well as reducing microporosity. A higher proportion of fine sand reduces conductivity and increases water retention, especially in deeper soil layers 0.4 to 0.6 m below the soil surface. The actual matric potential at field capacity was −4.60 kPa which is below the matric potential of −6 kPa commonly found in sandy soils in the topmost soil layer. The soil hydraulic cut-off point indicated that the matric potential of the permanent wilting point for sugarcane is reached at −1030 kPa, not −1500 kPa.

Although soil water retention increases with depth, its availability is greatest in the 0 to 0.2 m soil layer, with changes in the actual moisture at field capacity and permanent wilting point. The drainage rate of water at the soil surface is high. Therefore, there is a need for a continuous inflow of water into the soil, whether by precipitation or irrigation in order to ensure adequate soil moisture for the sugarcane crop. Sugarcane yield simulation using a locally validated model demonstrated that adequate crop yields are possible if there is adequate soil water storage for the sandy soils evaluated in this study. Future research can use actual moisture at field capacity and the permanent wilting point to develop a model for estimating sugarcane yield, considering the variation in soil water availability during the growth phase of sugarcane (phase II). This can be used to develop maps of soil water availability and yield estimates in the main sugarcane-producing regions for sandy soils.

Author Contributions: Conceptualization, J.L.V. and J.L.M.d.S.; methodology, J.L.V. and A.C.A.; software, J.L.V.; validation, J.L.V.; formal analysis, J.L.V.; investigation, J.L.V.; data curation, J.L.V.; writing—original draft preparation, J.L.V. and J.L.M.d.S.; writing—review and editing, J.L.V.; J.L.M.d.S., A.C.A., R.A.d.O., A.K.H., D.C.d.A., W.M.d.S. and R.M.A.; supervision, J.L.M.d.S.; project administration, R.A.d.O.; resources, A.K.H. All authors have read and agreed to the published version of the manuscript.

Funding: This research received funding support from Capes for the Doctoral Program of the lead author.

Institutional Review Board Statement: Not applicable.

Informed Consent Statement: Not applicable.

Data Availability Statement: Study data can be obtained by request to the corresponding author or the first author, via e-mail. It is not available on the website as the research project is still under development.

Acknowledgments: We would like to thank RIDESA for the availability of the experimental area and contribution to the research. We would also like to thank the Post-Graduate Program in Soil Science (UFPR), the Modeling of Agricultural Systems Laboratory (LAMOSA)/SCA/UFPR, and C. F. de Araújo-Júnior from the Rural Development Institute (IDR), Londrina, Paraná State. We are also grateful for the comments and edits from three anonymous reviewers which improved the quality of this work.

Conflicts of Interest: The authors declare no conflict of interest. Supporting entities had no role in the design of the study; in the collection, analyses, or interpretation of data; in the writing of the manuscript, or in the decision to publish the results.

References

1. Ferreira, M. Caracterização física do solo. In *Física do Solo*, 1st ed.; de Jong van Lier, Q., Ed.; Sociedade Brasileira de Ciência do Solo: Viçosa, Brasil, 2010; pp. 1–27.
2. Barbosa, L.C. Atributos Físicos do Solo e Desenvolvimento Radicular à Cana-Planta em Diferentes Sistemas De manejo. Master's Thesis, Universidade Estadual de Campinas, Faculdade de Engenharia Agrícola, Campinas, Brazil, 2015.
3. Silva, M.L.N.; Libardi, P.L.; Gimenes, F.H.S. Soil water retention curve as affected by sample height. *Rev. Bras. Ciênc. Solo* **2018**, *42*, 1–13. [CrossRef]

4. Stefanoski, D.C.; Santos, G.G.; Marchão, R.L.; Petter, F.A.; Pacheco, L.P. Uso e manejo do solo e seus impactos sobre a qualidade física. *Rev. Bras. Eng. Agríc. Ambient.* **2013**, *17*, 1301–1309. [CrossRef]
5. Souza, J.L.M.; Fezer, K.F.; Gurski, B.C.; Jerszurki, D.; Pachechenik, P.E.; Evangelista, A.W.P. Atributos físicos e balanço hídrico do solo com Floresta Ombrófila Mista, em Latossolo Vermelho-Amarelo, em Telêmaco Borba—PR. *Ciên. Florest.* **2018**, *28*, 90–101. [CrossRef]
6. Fidalski, J.; Tormena, C.A.; Alves, S.J.; Auler, P.A.M. Influência das frações de areia na retenção e disponibilidade de água em solos das Formações Caiuá e Paranavaí. *Rev. Bras. Ciên. Solo* **2013**, *37*, 613–621. [CrossRef]
7. Machado, J.L.; Tormena, C.A.; Fidalski, J.; Scapim, C.A. Inter-relações entre as propriedades físicas e os coeficientes da curva de retenção de água de um Latossolo sob diferentes sistemas de uso. *Rev. Bras. Ciênc. Solo* **2008**, *32*, 495–502. [CrossRef]
8. Cavalieri, K.M.V.; Carvalho, L.A.; Silva, A.P.; Libardi, P.L.; Tormena, C.A. Qualidade física de três solos sob colheita mecanizada de cana-de-açúcar. *Rev. Bras. Ciênc. Solo* **2011**, *35*, 1541–1549. [CrossRef]
9. Parahyba, R.B.V.; Araújo, M.S.B.; Almeida, B.G.; Rolim Neto, F.C.; Sampaio, E.V.S.B.; Caldas, A.M. Water retention capacity in Arenosols and Ferralsols in a semiarid area in the state of Bahia, Brazil. *An. Acad. Bras. Ciênc.* **2019**, *91*, 1–20. [CrossRef]
10. Daros, E.; Oliveria, R.A.; Zambon, J.L.C.; Bespalhok Filho, J.C.; Brasileiro, B.P.; Ido, O.T.; Ruaro, L.; Weber, H. RB036066—A sugarcane cultivar with high adaptability and yield stability to Brazilian South-Central region. *Crop Breed. Appl. Biotechnol.* **2018**, *18*, 325–329. [CrossRef]
11. Berton, G.S.; de Daros, O.R.A.E.; Zambon, J.L.C.; Bespalhok Filho, J.C.; Brasileiro, B.P.; Ido, O.T.; Ruaro, L.; Weber, H. RB036091—An early-maturing sugarcane cultivar for the Central South of Brazil. *Crop Breed. Appl. Biotechnol.* **2020**, *20*, 1–5. [CrossRef]
12. Santos, H.G.; Jacomine, P.K.T.; Anjos, L.H.C.; Oliveira, V.A.; Lumbreras, J.F.; Coelho, M.R.; Almeida, J.A.; Araújo Filho, J.C.; Oliverira, J.B.; Cunha, T.J.F. *Sistema Brasileiro de Classificação de Solos*, 5th ed.; Embrapa: Brasília, Brazil, 2018.
13. Nitsche, P.R.; Caramori, P.H.; Ricce, W.S.; Pinto, L.F.D. *Atlas Climático do Estado do Paraná*; Instituto de Desenvolvimento Rural do Paraná—IAPAR-EMATER: Londrina, Brazil, 2019.
14. FranchinI, J.C.; Veliini, C.L.; Balbinot Junior, A.A.; Debiasi, H.; Watanabe, R.H. *Integração Lavoura-Pecuária em Solo Arenoso e Clima Quente: Duas décadas de Experiência*; Circular Técnica 118; Embrapa Soja: Londrina, Brazil, 2016; pp. 1–11, ISSN 2176-2864. Available online: https://www.embrapa.br/en/busca-de-publicacoes/-/publicacao/1047856/integracao-lavoura-pecuaria-em-solo-arenoso-e-clima-quente-duas-decadas-de-experiencia (accessed on 24 April 2023).
15. MINOPAR—Minerais do Paraná S/A. Atlas geológico do Estado do Paraná. 2001. Available online: https://www.mineropar.gov.pr.br (accessed on 22 March 2023).
16. Costa, A.C.S.; Souza Junior, I.G.; Canton, L.C.; Gil, L.G.; Figueiredo, R. Contribution of the chemical and mineralogical properties of sandy-loam tropical soils to the cation exchange capacity. *Rev. Bras. Ciênc. Solo* **2020**, *44*, 1–18. [CrossRef]
17. Thomaz, E.L.; Fidalski, J. Interill erodibility of different sandy soils increases along a catena in the Caiuá Sandstone Formation. *Rev. Bras. Ciênc. Solo* **2020**, *44*, e0190064. [CrossRef]
18. Pereira, T.Y.Z. Avaliação da Vulnerabilidade e Fontes de Contaminações Difusas e Pontuais do Aquífero Caiuá. In *Monograph Course Conclusion Work (GC-119)*; Setor de Ciências da Terra, Curso de Geologia, Unversidade Federal do Paraná: Curitiba, Paraná, Brazil, 2016; pp. 1–32. Available online: https://www.geologia.ufpr.br/portal/wp-content/uploads/2018/10/Thomaz-Zulpo-TCC.pdf (accessed on 31 March 2023).
19. Empresa Brasileira de Pesquisa Agropecuária (EMBRAPA). *Manual de Métodos de Análise de Solo*, 2nd ed.; Embrapa Solos: Rio de Janeiro, Brazil, 2011; pp. 1–230.
20. Gubiani, P.I.; Reinert, D.J.; Reichert, J.M. Método alternativo para a determinação da densidade de partículas do solo—Exatidão, precisão e tempo de processamento. *Ciênc. Rural* **2006**, *36*, 664–668. [CrossRef]
21. Youngs, E.G. Hydraulic conductivity of saturated soils. In *Soil Analysis: Physical Methods*; Smith, K.A., Mullins, C.E., Eds.; Marcel Dekker: New York, NY, USA, 1991; Chapter 4; pp. 161–207.
22. Van Genuchten, M.T. A closed-form equation for predicting the hydraulic conductivity of unsaturated soils. *Soil Sci. Soc. Am. J.* **1980**, *44*, 892–898. [CrossRef]
23. Burdine, N.T. Relative permeability calculations from pore size distribution data. Petroleum Transactions. *Amer. Instit. Min. Metall. Eng.* **1953**, *198*, 71–78. [CrossRef]
24. R Core Team. *R: A Language and Environment for Statistical Computing*; R Foundation for Statistical Computing: Vienna, Austria, 2020; Available online: https://www.R-project.org/ (accessed on 3 January 2021).
25. Czyz, E.A.; Dexter, A.R. Influence of soil type on the wilting of plants. *Int. Agrophys.* **2013**, *27*, 385–390. [CrossRef]
26. Prevedello, C.L. Novo método para estimativa da umidade do solo na condição de capacidade de campo. *Rev. Bras. Recur. Hídr.* **1999**, *4*, 23–28. [CrossRef]
27. Sisson, J.B.; Fergusson, A.H.; van Genuchten, M.T. Simple method for predicting drainage from field plots. *Soil Sci. Soc. Am. J.* **1980**, *44*, 1147–1152. [CrossRef]
28. Prevedello, C.L.; Armindo, R.A. *Física do Solo com Problemas Resolvidos*, 2nd ed.; Celso Luiz Prevedello: Curitiba, Brazil, 2015; pp. 1–474, ISBN 978-8590757429.
29. Dexter, A.R.; Bird, N.R.A. Methods for predicting the optimum and the range of soil water contents for tillage based on the water retention curve. *Soil Tillage Res.* **2001**, *57*, 203–212. [CrossRef]
30. Libardi, P.L. *Dinâmica da Água no Sol*, 2nd ed.; Editora da Universidade de São Paulo: São Paulo, Brazil, 2012; ISBN 978-8531417054.
31. Willmott, C.J. Some comments on the evaluation of model performance. *Bull. Am. Meteorol. Soc.* **1982**, *63*, 1309–1313. [CrossRef]

32. Mangiafico, S.S. Summary and Analysis of Extension Program Evaluation in R, Version 1.18.1. 2016. Available online: https://rcompanion.org/handbook (accessed on 10 January 2021).
33. Horikoshi, M.; Tang, Y. ggfortify: Data Visualization Tools for Statistical Analysis Results. 2016. Available online: https://CRAN.R-project.org/package=ggfortify (accessed on 28 January 2021).
34. Viana, J.L.; de Souza, J.L.M.; Hoshide, A.K.; de Oliveira, R.A.; de Abreu, D.C.; da Silva, W.M. Estimating Sugarcane Yield in a Subtropical Climate Using Climatic Variables and Soil Water Storage. *Sustainability* **2023**, *15*, 4360. [CrossRef]
35. Parahyba, R.B.V.; Almeida, B.G.; Rolim Neto, F.C.; Araújo, M.S.B. Condutividade hidráulica dos solos arenosos da região semiárida da Bacia sedimentar do Tucano no município de Glória, Bahia, Brasil. In Proceedings of the III Reunião Nordestina de Ciência do Solo: Integração e Uso do Conhecimento para Uma Agricultura Sustentável no Nordeste, Aracajú, Brazil, 12–15 September 2016; Available online: https://www.embrapa.br/en/busca-de-publicacoes/-/publicacao/1058873/condutividade-hidraulica-dos-solos-arenosos-da-regiao-semiarida-da-bacia-sedimentar-do-tucano-no-municipio-de-gloria-bahia-brasil (accessed on 24 April 2023).
36. Moriasi, D.N.; Arnold, J.G.; van Liew, M.W.; Bingner, R.L.; Harmel, R.D.; Veith, T.L. Model evaluation guidelines for systematic quantification of accuracy in watershed simulations. *Trans. ASABE* **2007**, *50*, 885–900. [CrossRef]
37. Souza, J.L.M.; Gerstemberger, E.; Gurski, B.C.; Oliveira, R.A. Adjustment of water-crop production models for ratoon sugarcane. *Pesqui. Agropecu. Trop.* **2015**, *45*, 426–433. [CrossRef]
38. Rawls, W.J.; Pachepsky, Y.A.; Ritchie, J.C.; Sobecki, T.M.; Bloodworth, H. Effect of soil organic carbon on soil water retention. *Geoderma* **2003**, *116*, 61–76. [CrossRef]
39. Soil Science Division Staff. *Soil Survey Manual, USDA Handbook 18*; Ditzler, C., Scheffe, K., Monger, H.C., Eds.; Government Printing Office: Washington, DC, USA, 2017; pp. 1–603.
40. Fracaro, F.; Lamb, C.R.C.; Straaburge, A.S.; Lattuada, D.S.; Buttow, M.V.; Monttero, C.R.S.; Silva, S.D.A. Avaliação agronômica de genótipos de cana-de-açúcar em Via-mão/RS. In *Simpósio Estadual de Aroenergia; Reunião Técnica de Agroenergia—RS, 5*; Encontro de Energias Renováveis na Agricultura Familiar, 2. Pelotas; Embrapa Clima Temperado: Pelotas, Brazil, 2014; pp. 1–4. Available online: https://www.embrapa.br/en/busca-de-publicacoes/-/publicacao/1005818/avaliacao-agronomica-de-genotipos-de-cana-de-acucar-em-viamaors (accessed on 4 April 2023).
41. CONAB (Companhia Nacional de Abastecimento). Acompanhamento da Safra Bra-Sileira de Cana-de-Açúcar, Safra 2022/23–Terceiro Levantamento, Brasília, Distrito Federal, Brazil. Available online: https://www.conab.gov.br/info-agro/safras/cana (accessed on 4 April 2023).
42. Fidalski, J. Disponibilidade de água e comportamento físico dos solos da Formação Arenito Caiuá. In *V Reunião Paranaense de Ciência do Solo e II Simpósio Brasileiro de Solos Arenosos*; Maringá, Livro Eletrônico; SBCS/NEPAR: Curitiba, Brazil; IAPAR: Londrina, Brazil, 2017.
43. Ortiz, P.F.S.; Rolim, M.M.; Lima, J.L.P.; Pedrosa, E.M.R.; Dantas, M.S.M.; Tavares, U.E. Physical qualities of an Ultisol under sugarcane and Atlantic forest in Brazil. *Geoderma Reg.* **2017**, *11*, 62–70. [CrossRef]
44. Cabral, M.C.M.; Carvalho, L.A.; Novak, E.; Schicarelli, C.D.S. Sistema de preparo de solo em área de reforma de canavial e as alterações físicas do solo. *Rev. Agrar.* **2015**, *8*, 376–386.
45. Arcoverde, S.N.S.; Souza, C.M.A.; Suarez, A.H.T.; Colman, B.A.; Nagahama, H.J. Atributos físicos do solo cultivado com cana-de-açúcar em função do preparo e época de amostragem. *Rev. Agric. Neotrop.* **2019**, *6*, 41–47. [CrossRef]
46. Jadczyszyn, J.; Niedzwiecki, J. Relation of Saturated Hydraulic Conductivity to Soil Losses. *Pol. J. Environ. Stud.* **2005**, *14*, 431–435.
47. Cherubin, M.R.; Karlen, D.L.; Franco, A.L.; Tormena, C.A.; Cerri, C.E.; Davies, C.A.; Cerri, C.C. Soil physical quality response to sugarcane expansion in Brazil. *Geoderma* **2016**, *267*, 156–168. [CrossRef]
48. Silva, P.L.F.; Oliveira, F.P.; Borba, J.O.M.; Tavares, D.D.; do Amaral, A.J.; Martins, A.F. Solos arenosos para Sistemas de integração lavoura-pecuária-floresta em Arez, Rio Grande do Norte. Rev. *Verde de Agroecol. Desenvolv. Sust.* **2018**, *13*, 581–589. [CrossRef]
49. Libardi, P.L. Água no solo. In *Física do Solo*, 1st ed.; de Jong van Lier, Q., Ed.; Sociedade Brasileira de Ciência do Solo: Viçosa, Brazil, 2010; pp. 101–152.
50. Giasson, E. Introdução ao estudo dos solos. In *Fundamentos de Química do Solo*, 6th ed.; Meurer, E.J., Ed.; Evangraf Publishing, Ltd.: Porto Alegre, Brazil, 2017; pp. 11–28.
51. Costa, A.; Albuquerque, J.A.; Costa, A.; Pértile, P.; Silva, F.R. Water retention and availability in soils of the state of Santa Catarina-Brazil: Effect of textural classes, soil classes and lithology. *Rev. Bras. Ciênc. Solo* **2013**, *37*, 1535–1548. [CrossRef]
52. Lopes, I.M.; Ziviani, M.M.; Pinto, L.A.S.; Pinheiro, E.F.M.; Weber, H.; Pereira, M.G.; Lima, E.; Campos, D.V.B. Agregação e distribuição do carbono nos agregados de Latossolo Vermelho sob diferentes níveis de palhada da cana-de-açúcar em Paranavaí (PR). *Rev. Virtual Quím.* **2017**, *9*, 1985–1995. [CrossRef]
53. Auler, A.C.; Romaniw, J.; Sá, J.C.M.; Pires, L.F.; Hartman, D.C.; Inagaki, T.M.; Rosa, J.A. Improvement on soil structure and water retention after application of industrial organic waste as a crop fertilizer. *J. Soils Sedim.* **2020**, *20*, 2771–2783. [CrossRef]
54. Wiecheteck, L.H.; Giarola, N.F.B.; Lima, R.P.; Tormena, C.A.; Torres, L.C.; Paula, A.L. Comparing the classical permanent wilting point concept of soil (−15,000 hPa) to biological wilting of wheat and barley plants under contrasting soil textures. *Agric. Water Manag.* **2020**, *230*, 105965. [CrossRef]
55. Torres, L.C.; Keller, T.; Lima, R.P.; Tormena, C.A.; Lima, H.V.; Giarola, N.F.B. Impacts of soil type and crop species on permanent wilting of plants. *Geoderma* **2020**, *384*, 114798. [CrossRef]

56. Andrade, R.S.; Stone, L.F. Estimativa da umidade na capacidade de campo em solos sob Cerrado. *Rev. Bras. Eng. Agríc. Ambient.* **2021**, *15*, 111–116. [CrossRef]
57. Silva, A.P.; Tormena, C.A.; de Dias Júnior, M.S.; Imhoff, S.; Klein, V.A. Indicadores da qualidade física do solo. In *Física do Solo*, 1st ed.; de Jong van Lier, Q., Ed.; Sociedade Brasileira de Ciência do Solo: Viçosa, Brazil, 2010; pp. 241–281.
58. Helbel Junior, C.; Fidalski, J. Parâmetros técnicos sobre o armazenamento de água no solo na Região Noroeste do Paraná. In *V Reunião Paranaense de Ciência do Solo e II Simpósio Brasileiro de Solos Arenosos. Livro Eletrônico*; SBCS/NEPAR: Curitiba, Brazil; IAPAR: Londrina, Brazil, 2017.
59. Gurski, B.C.; Souza, J.L.M.; Gerstemberg, E.; Oliveira, R.A. Water requirements and restrictions to sugarcane in cane plants and ratoon cane cycles in Southern Brazil. *Acta Agron.* **2020**, *69*, 136–144. [CrossRef]
60. Araújo, R.M. Influência das Condições Climáticas no Crescimento e Desenvolvimento da Cana-de-Açúcar na Região sul do Brasil. Ph.D. Thesis, Programa de Pós-Graduação em Agronomia (Produção Vegetal), Setor de Ciências Agrárias, Universidade Federal do Paraná, Curitiba, Brazil, 2019.
61. Amorim, M.T.A.; Silverio, N.E.Q.; Bellinaso, H.; Gómez, A.M.R.; Greschuk, L.T.; Campos, L.R.; Demattê, J.A.M. Impact of soil types on sugarcane development monitored over time by remote sensing. *Precis. Agric.* **2022**, *23*, 1532–1552. [CrossRef]
62. Liu, Z.L.; Wang, C.; Zhang, L.; Wang, X.; Huang, G.; Xu, X.; Steenhuis, T.S. A field-validated surrogate crop model for predicting root-zone moisture and salt content in regions with shallow groundwater. *Hydrol. Earth Syst. Sci.* **2020**, *24*, 4213–4237. [CrossRef]
63. Perin, V.; Sentelhas, P.C.; Dias, H.B.; Santos, E.A. Sugarcane irrigation potential in Northwestern São Paulo, Brazil, by integrating Agrometeorological and GIS tools. *Agric. Water Manag.* **2019**, *220*, 50–58. [CrossRef]
64. Dias, H.B.; Sentelhas, P.C.; Inman-Bamber, G.; Everingham, Y. Sugarcane yield future scenarios in Brazil as projected by the APSIM-Sugar model. *Ind. Crops Prod.* **2021**, *171*, 113918. [CrossRef]
65. Xu, C.; Mao, J.; Li, X.; Burner, D.M.; Li, C.; Hussin, S.H.H.; Lin, X.; Liu, H.; Zhao, P.; Lu, X.; et al. Evaluating of drought tolerance in sugarcane genotypes using the membership function value of drought tolerance (MFVD). *Euphytica* **2023**, *219*, 37. [CrossRef]
66. Dias, H.B.; Sentelhas, P.C. Sugarcane yield gap analysis in Brazil—A multi-model approach for determining magnitudes and causes. *Sci. Total Environ.* **2018**, *637–638*, 1127–1136. [CrossRef] [PubMed]
67. de Castro Mattos, E.; Tomé, T.C.H.; Menillo, R.B.; Marin, F.R. Simulating of Future Climate Conditions on the Effect of Irrigation on Sugarcane Yield in Southern Brazil. *Braz. J. Agric.* **2021**, *96*, 446–470. [CrossRef]
68. Patino, M.T.O.; de Amorim, F.R.; de Andrade, A.G.; Alam, M.J.; Solfa, F.D.G. Costs of Agronomic Practices: Profitability at Different Scales of Sugarcane Production in Brazil. *Int. J. Bus. Adm.* **2022**, *13*, 32–43. [CrossRef]
69. Dias, H.B.; Sentelhas, P.C. Drying-off Periods for Irrigated Sugarcane to Maximize Sucrose Yields under Brazilian Conditions. *Irrig. Drain.* **2018**, *67*, 527–537. [CrossRef]
70. da Luz, F.B.; Carvalho, M.L.; de Borba, D.A.; Schiebelbein, B.E.; de Lima, R.P.; Cherubin, M.R. Linking Soil Water Changes to Soil Physical Quality in Sugarcane Expansion Areas in Brazil. *Water* **2020**, *12*, 3156. [CrossRef]
71. Dias, H.B.; Inman-Bamber, G. Sugarcane: Contribution of Process-Based Models for Understanding and Mitigating Impacts of Climate Variability and Change on Production. In *Systems Modeling*; Springer Nature: Singapore, 2020; Chapter 8; pp. 217–260. [CrossRef]

sustainability MDPI

Article

Climate Change, Farm Irrigation Facilities, and Agriculture Total Factor Productivity: Evidence from China

Hai Li and Hui Liu *

School of Economics, Hunan Agricultural University, Changsha 410128, China
* Correspondence: liuh1220@163.com

Abstract: Due to the trend of global warming, individuals from all walks of life have paid close attention to how climate change affects food security. China is a sizable nation with a rich climate and a diverse range of food crops that are of interest to researchers. Additionally, there is little mention of agricultural technology and farm irrigation facilities in academic research on climate change and agricultural economic growth in China. As a result, this study uses the SBM model, panel fixed effect model, and SYS-GMM model to examine the development trend of climate change and food security based on the panel data of Chinese provinces from 2000 to 2020. The study found that China has maintained an average annual growth rate of 4.3% in agricultural total factor productivity (TFP) in recent years, despite the impact of extreme weather. The average annual precipitation has a depressing influence on the TFP in agriculture, while the average annual temperature has the opposite effect. The farm irrigation facilities and agricultural technology's moderating impact is mostly shown in how well they attenuate the impact of climate change on the TFP in agriculture. Food crops have thereby improved their ability to survive natural risks and attain higher yields as a result of advancements in agricultural technology and increasing investment in contemporary farm irrigation facilities. The study's conclusions are used in the article to make the suggestion that strengthening climate change adaptation is necessary to ensure food security. The strategic policy of "storing grain in technology and storing grain in the soil" and the advancement of contemporary agricultural technology must be put into reality while the management system for grain reserves is being improved.

Keywords: climate change; farm irrigation facilities; agriculture total factor productivity (TFP); technical advancement

Citation: Li, H.; Liu, H. Climate Change, Farm Irrigation Facilities, and Agriculture Total Factor Productivity: Evidence from China. *Sustainability* **2023**, *15*, 2889. https://doi.org/10.3390/su15042889

Academic Editor: Aaron K. Hoshide

Received: 16 December 2022
Revised: 31 January 2023
Accepted: 2 February 2023
Published: 6 February 2023

1. Introduction

The sustainability of agricultural development, which is essential for human survival, has been severely threatened by climate change [1]. The average annual surface temperature in China is increasing due to global warming at a rate of 0.23 °C every ten years. The distribution of temperature and precipitation over space and time will continue to vary due to climate change, which will also increase the frequency and severity of extreme events, including torrential rainstorms, floods, droughts, and insect outbreaks [2]. China has been able to feed 22% of the world's people on only 8% of the planet's territory while using a high-input, high-pollution agricultural growth model, but at a significant cost to resources and the environment. China's agriculture will need to adapt to a more effective, resource-efficient, and environmentally friendly development model in the future against the backdrop of high-quality and sustainable agricultural development as agricultural modernization progresses. Increasing agriculture's total factor productivity (TFP), which is often calculated as the ratio of the total agricultural output to total factor input, is the key to maintaining agricultural economic growth [3]. However, because of how it affects agricultural production and input levels, climate change, particularly extreme weather, has raised a great deal of uncertainty regarding the improvement of the agricultural TFP, making it necessary to find a solution [4]. Smallholder farmers can adopt a variety of adaptive strategies in response to climate change, including crop restructuring, more irrigation, and variety selection. However, it is insufficient to rely solely on farmers to take adaptive action [5]. Therefore, it is only possible to make up for the shortcomings of smallholder farmers in adapting to climate change by properly understanding the policy perspective.

Academics have conducted a great deal of research on agricultural TFP, farm irrigation systems, and climate change. First off, there are numerous studies on the effects of climate change on agricultural output in the body of existing literature, and this literature holds a dominant position [6–9]. Additionally, the effects of climate change on agricultural output are mainly detrimental [10]. There is still no agreement among academics regarding the effects of climate change on agricultural productivity due to the significant geographical differences in these effects. In addition, the model that uses unit yield as the explanatory variable allows researchers to examine how climate change affects various crop yields, but it falls short in its understanding of input-output efficiency [11,12].

Based on the difference in climate change indicators, Villavicencio et al. [13] examined the impacts of climate change on the TFP in U.S. agriculture from two perspectives: temperature and precipitation. The results show that annual precipitation had a significant positive effect on the TFP, but the precipitation density had a significant negative effect on the TFP, and temperature change did not have a significant impact on the TFP in most regions. Liang et al. [14] analyzed the effects of climate change on the TFP in agriculture at the seasonal and regional levels, showing that temperature and precipitation in different agricultural regions and seasons accounted for 70% of the TFP changes in U.S. agriculture from 1981 to 2010.

Based on the agricultural TFP of different industries, crops, or cash crops, further research on the effects of climate change on rice yield per plant (TFP) in Japan by Kunimitsu et al. [15] revealed that climate change has different influences on the TFP in different regions. According to Chen and Gong [16], in the short term, extremely high temperatures will negatively affect China's planting industry's TFP, leading to a greater loss in land output. The usual growth cycle of food crops is shortened by a rise in temperature, which also concerns the food supply and consumption due to aberrant precipitation and incalculable harm to grain production per unit area [17,18]. Due to changes in rainfall, evaporation, runoff, and other water-related processes, as well as the subsequent redistribution of water resources over time and geography and subsequent changes in soil moisture, food production is impacted by inadequate water supplies [19,20].

Second, many techniques are now available to measure the change in the agricultural TFP at various phases. The measurement results differ slightly as a result of the different input–output indicators and periods chosen, although the TFP is infrequently considered in studies of farm irrigation facilities. Numerous studies have been conducted on farm irrigation facilities, the majority of which are performance-focused and use the economic growth model as their theoretical framework. These studies examine the contribution of farm irrigation facilities to agricultural output with a focus on the associations between these facilities and economic growth, grain production, farmers' income, and the environment [21]. Scholars build performance evaluation index systems of governance and assess their direct performance, indirect performance, and total performance using the DEA model and network analysis following pertinent evaluation criteria such as the 3E, 4E, and IOO models [22]. The DEA Tobit two-step approach, S-SBM model, Malmquist–Luenberger index, three-stage DEA model, and UHSBM model were derived to provide an empirical analysis of infrastructure supply and investment performance from both static and dynamic perspectives, respectively. These models address the shortcomings of unintended outputs, environmental factors, and random factors.

Additionally, the impacts of farm irrigation infrastructure on agricultural production are mostly seen in four areas: agricultural growth, lowering production costs, fostering structural adjustment in the agricultural business, and fostering agricultural technology advancement [23,24]. The Cobb–Douglas production function model typically includes variables representing infrastructure investment for assessing the marginal effects of farm irrigation facilities on agricultural development [25]. Some academics have recently concentrated on the ecological and production value of agricultural production-related farm irrigation facilities [26]. We can assess the value of ecosystem services in terms of controlling greenhouse gas emissions and controlling the climate as well as conserving water, soil, biodiversity, and the environment to further investigate the effects of farm irrigation facilities on agricultural environmental efficiency [27–29].

The results of the literature search revealed that the majority of academics have also talked about how climate extremes affect food production. Furthermore, it has been established that boosting the building of farm irrigation facilities can significantly raise the quantity and quality of food produced. The relationship between climate change, farm irrigation facilities, and TFP in agriculture, however, has not received much attention from researchers. In addition, many academics have overlooked the impacts of both undesirable outputs and climate change when measuring the TFP in Chinese agriculture. Few academics have concentrated on the effect of farm irrigation facilities on the TFP in agriculture, notably the significance of agricultural technology, when researching the growth processes of China's agricultural economy.

The input structure varies greatly across China's huge territory and wealth of resources, climatic change, and farm irrigation facilities. Additionally, "living off the sky" continues to be the norm in most places. It is clear that two factors need to be prioritized to improve the agricultural TFP. To start, there are differences in the initial factor input for farm irrigation facilities, the structure of endowments, and the rate at which various regions develop agricultural technology. The second is the fluctuation of climatic conditions. The production structure of agricultural economic development changes with the environment. Thus, we need to precisely study how investments in farm irrigation facilities and climate change affect the trajectory of the agricultural TFP.

In light of this, using inter-provincial panel data and a theoretical model of economic growth, this study undertakes an empirical test. The SBM model is utilized in this work to assess the influence of undesirable production on China's agricultural TFP. In addition, the panel fixed-effect model and SYS-GMM model are suggested to precisely investigate the relationship between climate change, farm irrigation facilities, and China's agricultural TFP. On the one hand, we examine the crucial part that farm irrigation facilities play in the process of how harsh climate impacts food production. On the other hand, in order to supplement the already-present data on the variables impacting TFP in agriculture, the crucial component of technological advancement is introduced. It offers practical policy recommendations for managing the shocks brought on by climate change and guaranteeing food security.

This study looks at the effects of climate change, agricultural technology, and farm irrigation facilities on agricultural TFP and aims to innovate in the following areas:

(1) The impact of extreme weather on agricultural production. With global warming, extreme meteorological disaster events such as heavy rainfall, floods as well as droughts are frequent. The annual average surface temperature in China is rising at a rate of 0.26 °C/decade, bringing unpredictable changes to the spatial and temporal distribution of temperature and precipitation and increasing challenges to sustainable agricultural development. Therefore, this paper chooses the average annual precipitation and average temperature as the indicators of climate change, which is different from some scholars.

(2) The impact of non-desired output, i.e., agricultural carbon emissions, is taken into account when measuring TFP in agriculture. In addition, this method is different from other studies, such as DEA.

(3) Introducing two intermediary variables—farm irrigation facilities and agricultural technology—to broaden the scope of the research.

(4) Using the most recent years of data from 2000 to 2020 for the agricultural TFP measurement. Furthermore, this paper provides a scientific foundation and policy suggestions for future food security.

To this end, this paper is organized as follows: Section 2 illustrates the study area, methodology, models, indicator selection, and data sources used in this study. Section 3 discusses theoretical analysis and analyzes the impact of climate change and farm irrigation facilities on the total factor productivity growth in agriculture using an econometric model. Section 4 contains research conclusions and policy implications.

2. Materials and Methods

2.1. Study Area

The study area of the article is located in the eastern part of Asia and the west coast of the Pacific Ocean. At present, there are 34 provincial administrative regions in China, including 23 provinces, 5 autonomous regions, 4 municipalities directly under the central government, and 2 special administrative regions. Due to the limitation of data availability, this part of the empirical sample data mainly comes from relevant statistics for 31 provinces in China. The study area is shown in Figure 1.

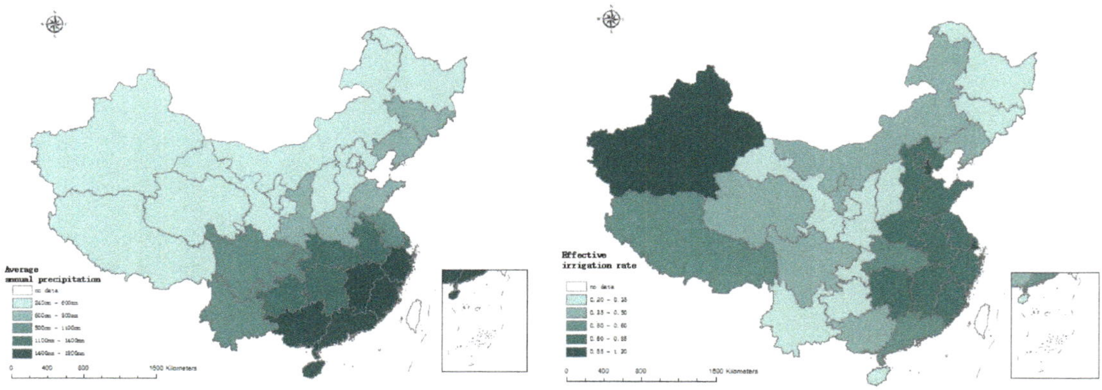

(a) (b)

Figure 1. Digital elevation model of the study area. (**a**) Average annual precipitation in China from 2000 to 2020. (**b**) Prevalence of irrigation in China from 2000 to 2020; the ratio of the effective irrigated area to the cultivated area is known as the prevalence of irrigation. The figure includes the provincial boundaries (gray line). The files used to create the map are licensed under GS (2019) 1822 available at: http://www.gov.cn (accessed on 1 December 2022). The data on annual precipitation are taken from the National Weather Data Network (http://data.cma.cn, accessed on 1 December 2022). The data on the effective irrigated area to the cultivated area are taken from the EPS database (https://www.epsnet.com.cn, accessed on 1 December 2022), the *China Statistical Yearbook*, and the *China Rural Statistical Yearbook* (http://www.stats.gov.cn, accessed on 1 December 2022).

There are five different types of terrain in China's general topography: mountains, plateaus, basins, plains, and hills. The country is high in the west and low in the east. This offers a range of possibilities and settings for the growth of China's industry and agriculture. In China, the north and south experience significantly different wintertime temperatures, and summertime highs are typical across the board. Furthermore, the distribution of yearly precipitation is more southerly than northern, with more rain falling in the summer and autumn and less in the winter and spring. Agricultural productivity and other things are intimately tied to rainfall conditions.

Due to its size and the huge variations in its temperature and precipitation patterns, China has a complex and varied climate. This makes China a good location for growing the majority of the world's crops. China's climate has several elements that are helpful for the growth of agricultural production, but it also has some disadvantages. Droughts, floods, cold waves, and typhoons are the most common severe weather phenomena that have a significant influence on China. These disasters frequently have a negative impact on agricultural productivity and farmers' lives.

2.2. Methods

The impact of undesirable outputs is not taken into account when measuring efficiency using the conventional data envelopment analysis (DEA), which only considers economic benefits. This ignores the issue of input–output slackness and is at odds with the actual agricultural production process. This study employs an over-efficient SBM model with non-desired outcomes, which explicitly incorporates the slack variables of each input and output into the objective function. The impact of the slack variables on the measured values is discussed as the total factor productivity (TFP) of each province in the nation is measured using the Max DEA program from 2000 to 2020. The model expressions are as follows:

$$\rho^* = min\frac{1-\frac{1}{m}\sum_{i=1}^{m}\frac{S_i^-}{x_{i0}}}{1+\frac{1}{S_1+S_2}\left(\sum_{j=1}^{S_1}\frac{S_j^g}{y_{j0}^g}+\sum_{k=1}^{S_2}\frac{S_k^b}{z_{k0}^b}\right)}$$
$$\text{s.t.}\begin{cases} x_0 = X\lambda + S^-, y_0^g = Y^g\lambda - S^g, z_0^b = Z^b\lambda + S^b \\ S^- \geq 0, S^g \geq 0, S^b \geq 0, \lambda \geq 0 \end{cases} \tag{1}$$

where, in the formula, ρ^* is the agricultural TFP, where $0 < \rho^* \leq 1$; S^-, S^g, and S^b are the slack vectors of inputs, desired outputs, and undesired outputs, respectively. x_i, y_j^g, and z_k^b are the input of i, the desired output of j, and the non-desired output vector of k, respectively. "0" is the evaluated unit. m, S_1, and S_2 are the number of input, desired output, and non-desired output elements. X, Y^g, and Z^b are matrices consisting of inputs, desired outputs, and undesired outputs. λ is the weight vector. When $S^- = S^g = S^b = 0$, $\rho^* = 1$, signifying that the decision unit is totally legitimate; otherwise, it denotes a loss and necessitates adjusting the input and output quantities.

Malmquist's proposal for the Malmquist Index was merged with the DEA theory to create the intertemporally variable TFP in 1953, and the equation is as follows:

$$TFP = \left[\frac{D_0^{t+1}(x_{t+1}, y_{t+1})}{D_0^{t+1}(x_t, y_t)} \times \frac{D_0^t(x_{t+1}, y_{t+1})}{D_0^t(x_t, y_t)}\right]^{1/2} = EC \times TC \tag{2}$$

where the technical efficiency change index and the technical progress change index, respectively, are denoted by the letters EC and TC. The expressions for each of these two are as follows:

$$EC = SEC \times PEC = \frac{S_0^t(x_t, y_t)}{S_0^t(x_{t+1}, y_{t+1})} \times \frac{D_0^t(x_{t+1}, y_{t+1})}{D_0^t(x_t, y_t)}$$
$$TC = \left[\frac{D_0^t(x_{t+1}, y_{t+1})}{D_0^{t+1}(x_{t+1}, y_{t+1})} \times \frac{D_0^t(x_t, y_t)}{D_0^{t+1}(x_t, y_t)}\right]^{1/2} \tag{3}$$

where, in the formula, SEC and PEC are the scale efficiency change index and pure technical efficiency change index, respectively. The criteria for determining EC, TC, SEC, and PEC are the same. TFP > 1 denotes that the total factor productivity increases, TFP < 1 denotes that the total factor productivity decreases, and TFP = 1 denotes that the total factor productivity does not cause changes.

2.3. Model Specification

The article analyzes the data by constructing an adjustment model and mediation model using the panel fixed effect method and the SYS-GMM method. The text that follows displays the model's precise form.

2.3.1. Adjustment Model

The agricultural TFP is influenced by climate change and farm irrigation facilities. In order to study the adjustment effect of agricultural technology, the interaction term was set separately for precipitation and temperature using it. And, build the following model.

$$\begin{aligned} TFP_{i,t} &= \alpha_0 + \alpha_1 pre_{i,t} + \alpha_2 tem_{i,t} + \alpha_3 fru_{i,t} + \alpha_4 x_{i,t} + \mu_i + \varepsilon_{i,t} \\ TFP_{i,t} &= \alpha_0 + \alpha_1 pre_{i,t} + \alpha_2 tem_{i,t} + \alpha_3 fru_{i,t} + \alpha_4 tc_{i,t} + \alpha_5 tc_{i,t} * pre_{i,t} \\ &\quad + \alpha_6 tc_{i,t} * tem_{i,t} + \alpha_7 x_{i,t} + \mu_i + \varepsilon_{i,t}, \end{aligned} \tag{4}$$

2.3.2. Mediation Model

A dynamic process affecting the agricultural TFP is influenced by beginning conditions as well as current factors affecting the agricultural TFP, among other things. In order to determine how climate change and farm irrigation facilities affect the agricultural TFP, this study attempts to control the initial conditions, incorporate the lagging term of the agricultural TFP ($TFP_{i,t-1}$) into the regression model, and build the following model.

$$\begin{aligned} TFP_{i,t} &= \alpha_0 + \alpha_1 pre_{i,t} + \alpha_2 tem_{i,t} + \alpha_3 fru_{i,t} + \alpha_4 x_{i,t} + \mu_i + \varepsilon_{i,t} \\ TC_{i,t} &= \alpha_0 + \alpha_1 pre_{i,t} + \alpha_2 tem_{i,t} + \alpha_3 fru_{i,t} + \alpha_4 x_{i,t} + \mu_i + \varepsilon_{i,t} \\ TFP_{i,t} &= \alpha_0 + \alpha_1 pre_{i,t} + \alpha_2 tem_{i,t} + \alpha_3 fru_{i,t} + \alpha_4 tc_{i,t} + \alpha_5 x_{i,t} + \mu_i + \varepsilon_{i,t} \\ TFP_{i,t} &= \alpha_0 + \alpha_1 TFP_{i,t-1} + \alpha_2 cli_{i,t} + \alpha_3 tem_{i,t} + \alpha_4 fru_{i,t} + \alpha_5 tc_{i,t} + \alpha_6 x_{i,t} + \mu_i + \varepsilon_{i,t} \end{aligned} \tag{5}$$

In Formulas (4) and (5), TFP_{it} is the agricultural TFP of province i in year t; pre_{it} represents the climate variable of province i in year t, that is, the yearly precipitation; tem_{it} is the average temperature; fru_{it} is the input of farm irrigation facilities in year t in province i, that is, the effective irrigation area per capita; tc_{it} is the input variable for the Malmquist–Luenberger index in year t in province i; x_{it} is a control variable, including the rural road density, agricultural structure, population density, and fiscal decentralization; α is the parameter to be estimated; μ_i represents the fixed effect of each province; and $\varepsilon_{i,t}$ is the random error term. All variables are logged to avoid heteroskedasticity in the model.

2.4. Variables

2.4.1. The Agricultural TFP

Output variables: Including the expected output and undesired output, the expected output makes use of the grain output, whereas undesired output typically refers to non-point-source pollutants such as chemical oxygen demand (COD), nitrogen (N), or phosphorus (P), as well as other pollutants or agricultural carbon emissions. This paper intends to use agricultural carbon emissions as the undesired output.

The sources of agricultural carbon emissions have diverse characteristics. The sources of agricultural carbon emissions have been comprehensively identified as chemical fertilizers, pesticides, agricultural diesel, agricultural film, land plowing, and irrigation power consumption [30,31]. A method for calculating agricultural carbon emissions is built based on the sources of carbon in agriculture that have been recognized:

$$E = \sum E_i = \sum T_i \times \delta_i \tag{6}$$

where, in the formula, E is the total amount of agricultural carbon emissions, i is the type of agricultural carbon sources, T_i is the consumption of each carbon source, and δ_i is the carbon emission coefficient of each carbon source. The coefficient is 0.8956 kgC/kg, the carbon emission coefficient of pesticides is 4.9341 kgC/kg, the carbon emission coefficient of the agricultural film is 5.18 kgC/kg, the carbon emission coefficient of diesel is 0.5972 kgC/kg, the carbon emission coefficient of tillage is 312.6 kgC/hm^2, and the carbon emission coefficient of agricultural irrigation should be 25 kgC/hm^2 [32,33].

Input variables: This paper only chooses seven indicators as input variables—labor, machinery, fertilizer, pesticide, agricultural film, diesel oil, and land—in order to comply with the empirical rule of the decision-making unit (DMU) and the number of input variables in DEA analysis [34–36]. The mechanical power input, for example, refers to the total power index of various agricultural machines, including tractors, balers, and seeders used in agricultural output, etc. [37]. The labor force input is based on the number of rural residents per unit area in each region [38]. The amount of agricultural chemical fertilizers applied in each region (in pure volume), pesticide input (the amount of pesticides per unit area in each region), agricultural film input (the amount of agricultural plastic film per unit area in each region), diesel input (the amount of agricultural diesel per unit area in each region), and land input (the effective irrigated area at the end of each region) are used because considering that the article measures the TFP of agriculture, combined with the common phenomenon of agricultural replanting, fallow, and abandonment in China, the ratio of effective irrigated area to cultivated land area and crop sown area are used instead. Agricultural land input is more accurate.

2.4.2. Climate Variables

The influence of climate change on agricultural production is mostly evident in temperature and precipitation [39]. Since this paper focuses on the agricultural TFP of food crop production, annual precipitation is used to measure the impact of climate change on agriculture. The accumulated temperature variable, which is frequently used in agronomy, reflects the impact of temperature on the growth and development of food crops from two aspects: temperature and time [40]. The TFP's impact is comparatively more consistent. In order to quantify the impact of climate change on the agricultural TFP, this research constructs the relationship between annual precipitation and the average temperature and TFP [41].

2.4.3. Farm Irrigation Facilities

At present, scholars usually utilize two types of indicators, monetary and physical, for measurement. However, monetary indicators tend to depart from the true worth of infrastructure since they are usually direct sums of investment volumes, while physical indicators are similarly incorrect due to variances in units of measurement. Usually, the more accurate way is to utilize the perpetual inventory method to estimate agricultural irrigation facilities, but the selection of the depreciation rate and initial capital stock has a significant impact on the inventory results [42,43]. Therefore, experts have varying estimations of farm irrigation facilities. Therefore, based on synthesizing the available research literature, this work selects the effective irrigated area as an index to measure the input of farm irrigation infrastructure.

2.4.4. Control Variables

Several factors affect grain output, farmers' income, and economic development level. In order to objectively evaluate the impact of climate change and farm irrigation facilities on the agricultural TFP, combined with previous research and actual conditions, the model also selected the agricultural structure, and seven disaster factors that have a significant impact on the agricultural TFP are used as control variables (see Table 1).

Table 1. Index system for evaluation of agricultural infrastructure governance efficiency.

First-Level Indicator	Secondary Indicators	Specific Description
Explained variable	*tfp*	Agricultural total factor productivity
Explanatory variables	*pre*	Average annual precipitation
	tem	Average temperature
Moderator	*fru*	Effective irrigation area per capita
	tc	Malmquist index
Control variable	*aff*	Affected area/crop sown area
	cap	Water conservancy investment
	eng	Food consumption expenditure/rural household consumption expenditure
	fer	Fertilizer application per unit area by region
	fil	Usage of agricultural plastic film per unit area by region
	inv	Water conservancy investment
	wat	Waterlogging area per capita

Note: The table is organized by the author.

2.5. Data Sources and Statistical Analysis

The time range of the above variables is from 2000 to 2020. The source data are primarily taken from the EPS database (https://www.epsnet.com.cn, accessed on 1 December 2022), the *China Statistical Yearbook*, and the *China Rural Statistical Yearbook* (http://www.stats.gov.cn, accessed on 1 December 2022).

The economic data of agricultural inputs and outputs as well as climate data for each province from 2000 to 2020 were compiled into a table in accordance with pertinent databases and statistical yearbooks. A descriptive statistical analysis was then carried out for each indicator's data, and the results are shown in Table 2.

Table 2. Descriptive statistics of different variables in China from 2000 to 2020.

Variables	Mean	Standard Deviation	Minimum Value	Maximum Value
tfp	1.01682	0. 0936045	0. 7184745	1.30906
pre	1384.856	2262.111	167	15,574.63
tem	13.84364	5.379699	2.998849	24.80421
fru	1062.781	775.7506	217.1962	4066.159
tc	1.012202	0.1122717	0.6795132	1.363629
aff	0.2085708	0.145902	0	0.6460758
cap	407.8922	308.2451	38.0959	1598.9
eng	0.3802011	0.0760678	0.259	0.56
fer	579.413	264.1651	170.0166	1396.03
fil	31.38759	30.8204	4.710912	148.1515
inv	1218.027	1278.347	52.809	5638.867
wat	3644.95	4690.888	24.46483	21,561.78

Note: Data compiled by the authors.

3. Results

3.1. Trend and Regional Analysis of Agricultural Total Factor Productivity (TFP)

The agricultural TFP for the entire country is calculated in this research using the Max DEA software and the super-efficiency SBM model, incorporating unexpected production. The dynamic trajectory of the agricultural TFP in various grain-producing regions is considerably diverse from 2000 to 2020, as illustrated in Figure 2. The primary regions for producing grain have an upward tendency overall. The agricultural TFP in this region has been in a good and steady state for a long time, and the agricultural output has resulted in some economic gains, except in 2003, when the governance efficiency was lower than 0.5; the agricultural TFP in the main grain-selling locations varies substantially.

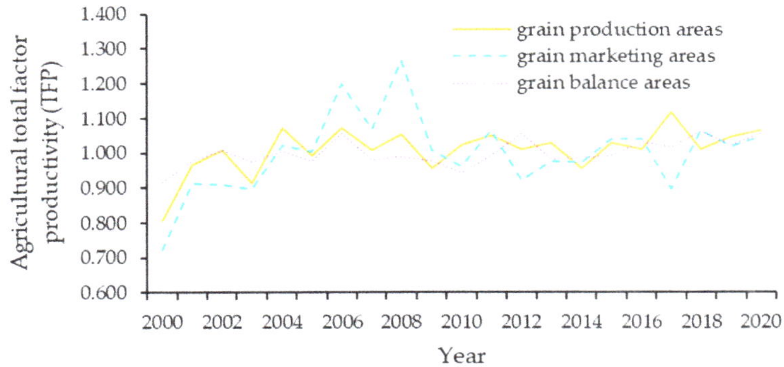

Figure 2. Internal mechanism of farm irrigation facilities affecting the agricultural TFP operation diagram. (China can be divided into three categories, including grain production areas, grain marketing areas, and grain balance areas. Of these, there are 13 grain production areas, including Heilongjiang, Jilin, Liaoning, Inner Mongolia, Hebei, Henan, Shandong, Jiangsu, Anhui, Jiangxi, Hubei, Hunan, and Sichuan; 7 grain marketing areas, including Beijing, Tianjin, Shanghai, Zhejiang, Fujian, Guangdong, and Hainan; and 11 grain balance areas, including Shanxi, Ningxia, Qinghai, Gansu, Tibet, Yunnan, Guizhou, Chongqing, Guangxi, Shaanxi, and Xinjiang).

The other 25 Chinese provinces increased positively between 2000 and 2020, as indicated in Table 3, except for Guangdong, Hainan, Guizhou, and Shaanxi, where the TFP was less than 1. The fact that it was larger than 0.970 and that Jilin Province had the highest score and Guizhou Province had the lowest index among them showed that TFP in China was still progressing well. Additionally, the average annual growth rate of the agricultural TFP in the 29 provinces was 4.3%, slightly higher than the findings of the other scholar. The ability to tolerate natural risks has increased grain production in recent years, despite the impact of harsh weather. This is because investments in modern irrigation and water conservation facilities have increased due to technological advancement. However, the precise causes demand further investigation.

Ningxia and Shanghai had the lowest indexes, both of which were not high, showing that further advancement in agricultural technology is needed to accelerate the development of modern agriculture. Less than half of the provinces reached above 1 on the technological progress and change index (TC) from 2000 to 2020, with the lowest indexes being low values in both cases. Only four provinces had an index of technical efficiency change (EC) below 1, which was consistent with the TFP comparison result and showed that China's agricultural technological level was in good shape. There are 11 provinces in which all three major indexes are larger than 1, and EC is primarily responsible for the rise in TFP, according to the three major indexes. It is clear that in order to raise the local TFP, each province in the nation should concentrate on raising its technological level and increasing its investment in agricultural production.

The national agricultural total factor productivity (TFP) value, which accounts for inputs from farmland and water facilities, can be seen in Figure 3. It fluctuated from 0.821 in 2000 to 1.040 in 2004 and then from 2005 to 2020, except in 2005, 2009, 2010, and 2014, wherein the TFP remained above 1. The average TFP showed an upward trend from 2000 to 2004 and an upward trend from 2005 to 2020, with the smaller fluctuations being brought on by the varying trend of the technological progress change index (TC). The technical efficiency index (EC) is influenced by both the scale efficiency change index (SEC) and the pure technical efficiency change index (PEC), with the mean value of the pure technical efficiency change index (PEC) trending more similarly while the scale efficiency change index has a different trend. Additionally, the mean value of total factor productivity (TFP) and mean value of the technical efficiency index (EC) are relatively similar (SEC). This suggests that the technical efficiency index (EC) and the pure technical efficiency index of change have a significant impact on the total factor productivity (TFP) in agriculture (PEC).

Table 3. TFP of agriculture in China from 2000 to 2020.

Province	TFP	EC	TC
Hebei Province	1.028	1.030	0.997
The Nei Monggol Autonomous Region	1.012	1.004	1.007
Liaoning Province	1.056	1.063	0.993
Jilin Province	1.328	1.326	1.001
Heilongjiang Province	1.014	1.016	0.998
Jiangsu Province	1.230	1.225	1.004
Anhui Province	1.019	1.021	0.999
Jiangxi Province	1.008	1.003	1.005
Shandong Province	1.064	1.067	0.998
Henan Province	1.034	1.033	1.001
Hubei province	1.025	1.024	1.001
Hunan Province	1.026	1.025	1.001
Sichuan Province	1.025	1.023	1.002
Beijing	1.071	1.046	1.024
Tianjin	1.116	1.127	0.990
Shanghai	1.006	1.023	0.984
Zhejiang Province	1.112	1.108	1.003
Fujian Province	1.000	0.990	1.010
Guangdong Province	0.998	1.012	0.987
Hainan	0.980	0.962	1.019
Shanxi Province	1.041	1.048	0.994
The Guangxi Zhuang Autonomous Region	1.015	1.017	0.998
Guizhou Province	0.978	0.991	0.987
Yunnan Province	1.017	1.023	0.994
Shaanxi Province	0.998	0.932	1.071
Gansu Province	1.010	1.018	0.992
Qinghai Province	1.028	1.033	0.995
The Ningxia Hui Autonomous Region	1.023	1.040	0.984
The Xinjiang Uygur Autonomous Region	1.009	1.002	1.007

Note: Data are calculated by Max DEA software. TFP = EC × TC = PEC × SEC × TC.

3.2. *Analysis of the Impact of Climate Change on the Agricultural TFP*

The Hausman test findings for model (1) in Table 4 indicate that the panel fixed effect model is preferable to the mixed regression and random effect models. In terms of the regression results of the explanatory variables, while precipitation has played a negative, but not significant, role on the agricultural TFP, temperature has a significant positive effect on the agricultural TFP. The reason is that precipitation irregularity increases the likelihood of geological disasters such as floods and damage to food production, transportation, and other links, endangering food supply and utilization. The annual precipitation varies greatly across the entire nation and exhibits an upward trend, with four peaks occurring in 2010, 2012, 2016, and 2020, so it is not significant. In comparison to other years, 2016 had much more precipitation, whereas the highest amount of precipitation in 2011 was very low (as shown in Figure 4). Data analysis demonstrates that China's average annual accumulated temperature of 13 to 15 °C in the range of floating, in general, is rising. In 2007, 2017, and 2019, there were three relatively obvious annual accumulated temperature peaks, and in 2012, there were the most resources; climate change is the most direct characteristic of climate warming. For crops well north, the winter cold brings beneficial effects, resulting in increased grain production [44].

Figure 3. Trend chart of the agricultural total factor productivity (TFP) and its decomposition indicators from 2000 to 2020. (TFP = EC × TC = PEC × SEC × TC, TFP is agricultural total factor productivity, EC is the technical efficiency index, TC is the technological progress change index, PEC is the pure technical efficiency change index, and SEC is the scale efficiency change index).

The moderating factors were decentralized to prevent multicollinearity, and the moderating effect was examined using the panel fixed-effect model. The panel fixed-effect model outperformed the random-effect model, according to the Hausman test. According to model (2)'s findings, the temperature had a positive main effect coefficient on the agricultural TFP, whereas precipitation had a negative main effect coefficient. Agricultural technology served as a moderating factor that attenuated some of the negative effects of rising precipitation on the agricultural TFP. The coefficient of the interaction term between agricultural technology and precipitation was positive but insignificant. With the advancement of science and technology, factors such as breeding technology, greenhouses, and the widespread use of agricultural technology have significantly reduced the impact of natural disasters on food production and improved the security of grain production [45–47]. These factors, along with climate change brought on by a lack of resources such as water, heat, and other resources, have contributed to global warming [48]. The negligible coefficient of the interaction term between agricultural technology and precipitation, however, may have two causes. On the one hand, there is an irregular tendency in the current precipitation situation in China. According to research, a 20% decrease in rainfall will render useless any technology used to boost the productivity of food crops. Agricultural technology has a limited ability to reduce precipitation. On the other hand, the regulation effect of irrigation technology on precipitation instability may be weakened due to the early construction and heavy damage of farm irrigation facilities and the serious abandonment of mechanical wells caused by the beginning of the groundwater protection plan, which is consistent with the small regression coefficient of the interaction term between climate change and the input of farm irrigation facilities mentioned above [49].

Table 4. Estimation results of climate change on the agricultural TFP.

Variables	(1)	(2)
lnpre	−0.0160 *	−0.00997
	(−1.80)	(−1.17)
tem	0.0264 **	0.0289 **
	(2.16)	(2.52)
lnfru	0.108 ***	0.0818 **
	(3.05)	(2.47)
lncap	0.0331	0.0452 *
	(1.33)	(1.93)
lnaff	−0.370 ***	−0.266 ***
	(−6.47)	(−4.74)
lneng	0.0563	0.0275
	(1.10)	(0.57)
lnfer	−0.0756 **	−0.0661 *
	(−2.06)	(−1.92)
lninv	−0.0177 *	−0.0124
	(−1.68)	(−1.25)
lnfil	−0.0115	−0.0143
	(−1.10)	(−1.44)
lnwat	0.0120	0.0150
	(0.39)	(0.52)
tc		0.324 ***
		(7.80)
c_lncli_tc		0.0315
		(0.55)
c_tem_tc		−0.0401 ***
		(−3.95)
_cons	−0.504	−0.954
	(−1.20)	(−2.38)
N	522	522
adj. R2	0.0658	0.1825

Note: ***, **, and * denote significance levels of 1%, 5%, and 10%, respectively, with standard errors in parentheses; the same below.

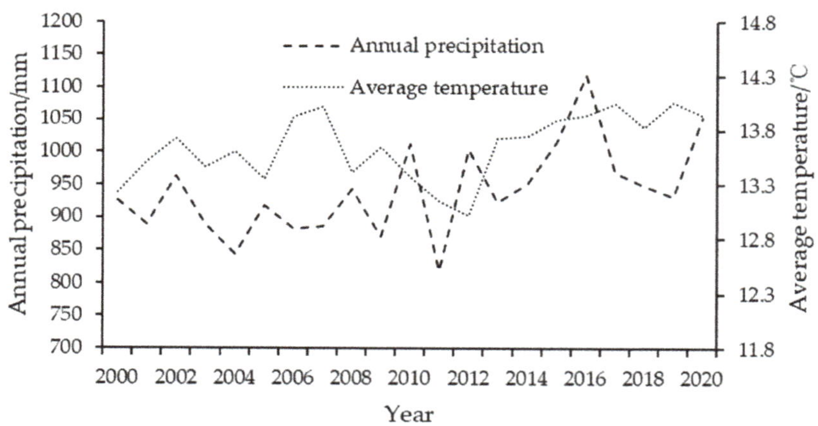

Figure 4. Trend chart of the average annual precipitation and the average temperature in China from 2000 to 2020 (http://data.cma.cn, accessed on 1 December 2022).

Agricultural technology, as a moderator variable, significantly reduces the temperature rise's positive influence on the agricultural TFP due to the sharp rise in temperature and aggravation of drought, plant diseases, and insect pests; it also affects the crop quality, causes an increase in extreme weather, and poses a threat to agricultural production [50]. Agricultural technology and a temperature interaction coefficient under the 1% level are significantly negative.

The paper offers a simple theoretical model to explain the influence of climate change on the agricultural TFP and the function of farm irrigation facilities based on the aforementioned literature review. Agriculture is considered to conform to the standard C-D production function form:

$$Y = AF(K, L) \tag{7}$$

where Y stands for the agricultural output; A stands for the agricultural TFP; and K and L stand for the capital and labor inputs for agricultural production, respectively.

In addition, A = θT, that is, A is a function of factor allocation (θ) and agricultural technological progress (T). Farm irrigation facilities do, however, have a ceiling effect ($\theta \leq 1$) that reduces the efficiency with which agricultural output elements are distributed. Therefore, $T'(f) = 0$, $\theta'(f) \geq 0$, $\theta''(f) \leq 0$, and f stands for farm irrigation facilities.

This paper analyses the impact of climate factors on the agricultural TFP, which is denoted by *cli*. The formula is as follows:

$$\frac{dA}{dcli} = \frac{\partial A}{\partial cli} + \frac{\partial A}{\partial \theta(f)T(f)} \cdot \frac{\partial \theta(f)}{\partial cli} \cdot \frac{\partial T(f)}{\partial cli} = \frac{\partial A}{\partial cli} + \frac{\partial A}{\partial f} \cdot \frac{\partial f}{\partial cli}, \tag{8}$$

The influence of climate change on the agricultural TFP is dependent on two factors: the direct effect of climate on the TFP and the indirect effect of climate change on the TFP, which is the input of farm irrigation facilities on the effectiveness of factor allocation θ and technical advancement.

Conclusion 1: Technological progress plays a regulatory role.

3.3. Analysis of the Impact of Farm Irrigation Facilities on the Agricultural TFP

The panel fixed effect model is used in this study to investigate the relationship between farm irrigation facilities, technological advancement, and the agricultural TFP based on the intermediary effect. The first-order lagged TFP is used as the instrumental variable in the systematic GMM estimate approach to show how climate change affects TFP, and the data passed the robustness test. Table 5 displays the findings, which demonstrate that model (6) passed the test for instrumental variables. The residual terms only possessed a first-order serial correlation, according to the findings of the AR (1) and AR (2) tests, and there was no second-order autocorrelation. All of the Hansen statistics' P-values were higher than 0.1, proving the validity of the instrumental variables.

According to the results of model (5)'s estimation, the agricultural TFP with one lag period passed the positive significance test at the 10% level in all models, suggesting that capital accumulation in the early stages may not have a positive impact on agricultural economic growth in the later stages and that there may be a phenomenon known as diminishing marginal utility. It has been demonstrated, however, that TFP does exhibit "inertia" in time series. A continuous accumulation adjustment process is being used to improve the agricultural TFP.

The coefficient of farm irrigation facilities in model (3) is significantly positive based on the regression results, indicating that these facilities have a positive spillover effect on grain production growth and that accelerating infrastructure investment is a key strategy for enhancing the agricultural TFP. The accumulation of farm irrigation facilities is the primary internal component of technical advancement, which is consistent with model (4)'s finding that the coefficient of farm irrigation facilities is significantly positive. Model (5) shows that agricultural technology and farm irrigation facilities have a significant positive impact on the agricultural TFP, suggesting that technological advancement has a mediating effect. Model (6) shows that accelerated technical innovation and progress can boost the improvement of the agricultural TFP. These results are consistent with the estimated results of the fixed effects.

Table 5. Estimation results of farm irrigation facilities on the agricultural TFP.

Variables	(3) *lntfp*	(4) *tc*	(5) *lntfp*	(6) *lntfp*
lnpre	−0.0160 * (−1.80)	−0.00108 (−0.12)	−0.0157 * (−1.85)	−0.0189 (−1.47)
tem	0.0264 ** (2.16)	−0.0000847 (−0.01)	0.0264 ** (2.27)	0.0116 ** (2.20)
lnfru	0.108 *** (3.05)	0.0879 * (2.39)	0.0818 ** (2.42)	0.0884 (1.52)
lncap	0.0331 (1.33)	−0.0171 (−0.66)	0.0381 (1.60)	0.0405 (1.44)
lnaff	−0.370 *** (−6.47)	−0.150 * (−2.52)	−0.326 *** (−5.94)	−0.346 ** (−2.73)
lneng	0.0563 (1.10)	0.110 * (2.06)	0.0239 (0.49)	−0.122 * (−1.77)
lnfer	−0.0756 ** (−2.06)	0.00322 (0.08)	−0.0765 ** (−2.19)	−0.103 * (−2.74)
lninv	−0.0177 * (−1.68)	−0.00704 (−0.64)	−0.0156 (−1.56)	−0.0390 ** (−2.12)
lnfil	−0.0115 (−1.10)	0.0266 * (2.44)	−0.0193 * (−1.92)	0.0000478 (0.00)
lnwat	0.0120 (0.39)	0.0338 (1.05)	0.00206 (0.07)	−0.0343 (−1.20)
tc			0.293 *** (7.06)	0.248 * (1.85)
L.lntfp				−0.153 * (−2.02)
_cons	−0.504 (−1.20)	0.355 (0.81)	−0.608 (−1.52)	
N	522	522	522	493
adj. R^2	0.0658	0.0290	0.1516	
AR(1)				−2.84
AR(1) p-value				0.005
AR(2)				−0.14
AR(2) p-value				0.888
Sargan-test				23.03
Sargan-test p-value				0.113

Note: L.TFP is the agricultural TFP of the lag period. ***, **, and * denote significance levels of 1%, 5%, and 10%, respectively, with standard errors in parentheses.

This research also investigates the underlying mechanisms through which the investment in farm irrigation facilities impacts the agricultural TFP. To find the solution to this question, we can simultaneously derive the input of agricultural irrigation facilities on both sides of Equation (8):

$$\frac{d^2A}{dclidf} = \frac{\partial^2 A}{\partial cli \partial f} + \frac{\partial^2 A}{\partial^2 f} \cdot \frac{\partial f}{\partial cli}, \quad (9)$$

The left side of Equation (9), which can be broken down into direct impacts and indirect effects of farm irrigation facility investment, shows how farm irrigation facility investments are used to deal with the consequences of climate change on the agricultural TFP.

The technical effects of farm irrigation facilities primarily consist of two aspects, as depicted in the figure: (1) horizontal effect: assuming that the agricultural technical conditions of T1 do not change, the increase in the stock level of farmland water conservancy facilities (as shown in Figure 5, from f1 to f2) will increase the allocation efficiency of agricultural production factors, and the agricultural TFP will also gradually rise; (2) growth effect: based on the assumption that exogenous agricultural technical circumstances exist, the agricultural TFP will steadily rise from T1 to T2, while the marginal utility will fall. This means that a higher level of farmland and farm irrigation facilities will correlate to a lower θ' and A$'$.

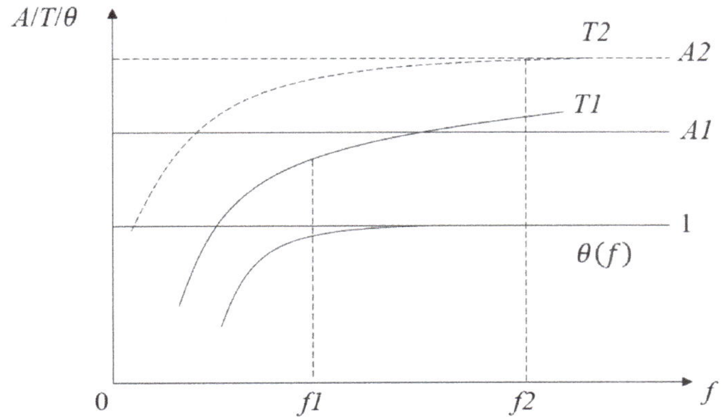

Figure 5. Diagram of internal mechanisms of farm irrigation facilities affecting agricultural TFP operations.

Conclusion 2: Technological progress plays an intermediary role in the process of TFP affected by farm irrigation facilities.

3.4. Limitations and Implications

Climate change is one of the shifts the world is going through. According to this paper's analysis of climate change in China from 2000 to 2020, which is also in line with the conclusions of Li et al. [51], China is currently undergoing extreme warming. Alexander et al. [52] conducted research on extreme climate change, and the results showed that 70% of the world's land area exhibits a growing trend toward a continuous decline in the number of cold night days and a continuous increase in the number of warm night days. As a result, the Chinese region fits with the trend of global climate change. The results of earlier investigations are more compatible with the finding that the intensity of precipitation increases with some variations in precipitation variability [53].

Precipitation was shown to be the principal growth factor impacting grain output and to have a suppressive effect on grain production when the effects of temperature and precipitation on the TFP in agriculture were compared, which was in line with the findings of Yang et al. [54]. The data analysis results revealed that, in contrast to the findings of other studies, the rise in temperature was accompanied by an increase in the agricultural TFP. Combining the research findings of Liu et al. and Haider et al. [55,56] on the impact of temperature change on grain production in a particular country, it is possible to conclude that, despite China's ongoing warming trend, the country's warming climate has advanced the start of the warmer season in spring and postponed the start of the cooler season in autumn. Crops have more time to grow, absorb sunlight, photosynthesize, and change matter as a result. This theoretically increases their production capacity and may, in part, result in larger grain yields.

To sum up, the analysis of this work still has numerous limitations.

Firstly, there are other elements besides climate change that have an impact on the TFP in agriculture. Human factors are also crucial. In theory, investing in farm irrigation facilities can increase the TFP [57]. The policymakers of farm irrigation facility investment are unable to determine the degree of farm irrigation facility investment that is most appropriate for all types of climate change challenges in all places due to factors such as food security [58,59]. How to open the "last mile" of agricultural technology promotion, which is related to the development of digital platforms for agricultural technology services, as well as how to allow the most cutting-edge and advanced agricultural science and technology achievements to permeate the village, home, and field, is critical. It is essential to obtain local government spending at this time. As a result, future research will continue to focus more on the processes of harsh climate and the effects of anthropogenic variables on the food supply.

Second, this study measures the effect of climatic extremes on agricultural production using average temperature and precipitation density rather than measuring climate extremes directly. According to preliminary findings, temperature change has a large positive impact on the agricultural output increase, while extreme precipitation weather has a negative impact. The measurement and evaluation of extreme weather events as well as the processes by which they affect the productivity of all factors in agriculture, however, require more in-depth study.

Third, due to the size of the research region, there are also significant variations in hydrothermal conditions, particularly for the many food crops that have distinct spatial distribution patterns. Regions or types of changes affected directly by climate change vary greatly [60]. For instance, whereas precipitation increases in South China may increase the likelihood of agricultural disasters, precipitation increases in Northwest China will greatly enhance agricultural output [61,62]. Therefore, future research will concentrate on how extreme climatic change affects the agricultural TFP in various regions of China and for various food crops.

4. Conclusions

In this study, we examined the mechanism underlying the impact of climate change on the agricultural total factor productivity (TFP), developed a model of the moderating effect, examined the moderating impact of farm irrigation facilities and agricultural technology, and empirically tested the model using panel data from all provinces. The result is the same as that used by Li et al., Alexander et al., and Yang et al. The present paper confirms their findings using the most recent data, different methods, and more scientific indicators of agricultural TFP measurement. The main conclusions are as follows:

(1) Similar to the findings of other studies, this paper finds that climate extremes and farm water facilities have an impact on food output. Agriculture's technological advancements also have a moderating and mediating effect;

(2) In contrast to the findings of other studies, the paper discovers that China's average agricultural TFP between 2000 and 2020 is 4.3%, with a TFP above 1 in 25 provinces, at a somewhat faster rate than other research. Furthermore, this paper makes the case that the average annual cumulative temperature in China varies between 13 and 15 degrees with a general upward trend, and the warming is beneficial for crop growth and preventing freezing calamities, which has a large and positive impact on the agricultural TFP.

The article does, however, have certain shortcomings. Instead of relying solely on data that are available to the public, we must do additional field research for micro-subjects in future studies. Additionally, we must broaden our choice of indicators, particularly for extreme climate change.

The aforementioned findings have significant policy ramifications for combating climate change and guaranteeing domestic food security: (1) in order to handle the food crisis in the short term while strengthening the ability to foresee meteorological disasters, the method for managing grain reserves should be strengthened in light of the constricting effect of climate change on the agricultural TFP; (2) we must aim at the moderating influence of agricultural technology and farm irrigation systems in the process of climate change on the agricultural TFP. On the one hand, we must firmly implement the "storing grain in technology, storing grain in the land" strategy, "short board" swallow the infrastructure, adjust measures to local conditions to adjust the structure of grain cropping systems and planting, and achieve cultivation optimization of new varieties of crops, taking the comprehensive technology to improve the ability to withstand natural disasters and improve crops' adaptability to environmental changes. On the other hand, although agricultural technology has controlled the precipitation process, its influence has not been substantial, indicating that China's response to climate change and the overall rate of the agricultural technology extension mechanism transformation still need to be improved.

Author Contributions: H.L. (Hai Li) generated the conceptualization, methodology, software, and validation; carried out formal analysis, investigation, resources, and data curation; and wrote the original draft, including preparation, review, visualization, supervision, and editing. H.L. (Hui Liu) provided statistical assistance and read, revised, and shaped the manuscript to the present form. H.L. (Hai Li) and H.L. (Hui Liu) co-provided project administration and funding acquisition. All authors have read and agreed to the published version of the manuscript.

Funding: This research was funded by the Key Science Fund Project of Hunan Provincial Department of Education, grant number 18A085; Hunan Provincial Philosophy and Social Science Fund Project, grant number 21YBA079; and Hunan Postgraduate Scientific Research Innovation Project, grant number CX2018B403.

Institutional Review Board Statement: Not applicable.

Informed Consent Statement: Not applicable.

Data Availability Statement: This is not applicable to this article since no datasets were generated.

Conflicts of Interest: The authors declare that they have no conflict of interest.

References

1. Ur Rahim, H.; Qaswar, M.; Uddin, M.; Giannini, C.; Herrera, M.L.; Rea, G. Nano-Enable Materials Promoting Sustainability and Resilience in Modern Agriculture. *Nanomaterials* **2021**, *11*, 2068. [CrossRef] [PubMed]
2. Farooq, M.S.; Uzaiir, M.; Raza, A.; Habib, M.; Xu, Y.; Yousuf, M.; Yang, S.H.; Ramzan Khan, M. Uncovering the Research Gaps to Alleviate the Negative Impacts of Climate Change on Food Security: A Review. *Front. Plant. Sci.* **2022**, *13*, 2334. [CrossRef] [PubMed]
3. Zhong, S.; Li, Y.; Li, J.; Yang, H. Measurement of total factor productivity of green agriculture in China: Analysis of the regional differences based on China. *PLoS ONE* **2021**, *16*, e0257239. [CrossRef] [PubMed]
4. Zawalińska, K.; Wąs, A.; Kobus, P.; Bańkowska, K. A framework linking farming resilience with productivity: Empirical validation from Poland in times of crises. *Sustain. Sci.* **2022**, *17*, 81–103. [CrossRef]
5. Jalal, M.J.E.; Khan, M.A.; Hossain, M.E.; Yedla, S.; Alam, G.M. Does climate change stimulate household vulnerability and income diversity? Evidence from southern coastal region of Bangladesh. *Heliyon* **2021**, *7*, e07990. [CrossRef]
6. Deaton, B.J.; Deaton, B.J. Food security and Canada's agricultural system challenged by COVID-19. *Can. J. Agr. Econ.* **2020**, *68*, 143–149. [CrossRef]
7. Kogo, B.K.; Kumar, L.; Koech, R. Climate change and variability in Kenya: A review of impacts on agriculture and food security. *Environ. Dev. Sustain.* **2020**, *23*, 23–43. [CrossRef]
8. Ainehvand, S.; Raeissi, P.; Ravaghi, H.; Maleki, M. Natural disasters and challenges toward achieving food security response in Iran. *J. Educ. Health Promot.* **2019**, *8*, 51.
9. Xie, W.; Huang, J.; Wang, J.; Cui, Q.; Robertson, R.; Chen, K. Climate change impacts on China's agriculture: The responses from market and trade. *China Econ. Rev.* **2020**, *62*, 101256. [CrossRef]
10. Renner, T.; Meijerink, S. Policy entrepreneurs in international river basins-getting climate adaptation on the cross-border water policy agenda. *Reg. Environ. Chang.* **2018**, *18*, 1287–1298. [CrossRef]
11. Mkuhlani, S.; Zinyengere, N.; Kumi, N.; Crespo, O. Lessons from integrated seasonal forecast-crop modelling in Africa: A systematic review. *Open Life Sci.* **2022**, *17*, 1398–1417. [CrossRef]
12. Ortiz-Bobea, A.; Knippenberg, E.; Chambers, R.G. Growing climatic sensitivity of U.S. agriculture linked to technological change and regional specialization. *Sci. Adv.* **2018**, *4*, eaat4343. [CrossRef]
13. Villavicencio, X.; McCarl, B.A.; Wu, X.; Huffman, W.E. Climate change influences on agricultural research productivity. *Clim. Chang.* **2013**, *119*, 815–824. [CrossRef]
14. Liang, X.Z.; Wu, Y.; Chambers, R.G.; Schmoldt, D.L.; Gao, W.; Liu, C.; Liu, Y.A.; Sun, C.; Kennedy, J.A. Determining climate effects on US total agricultural productivity. *Proc. Natl. Acad. Sci. USA* **2017**, *114*, E2285–E2292. [CrossRef]
15. Kunimitsu, Y.; Kudo, R.; Iizumi, T.; Yokozawa, M. Technological spillover in Japanese rice productivity under long-term climate change: Evidence from the spatial econometric model. *Paddy Water Environ.* **2015**, *14*, 131–144. [CrossRef]
16. Chen, S.; Gong, B. Response and adaptation of agriculture to climate change: Evidence from China. *J. Devel. Econ.* **2021**, *148*, 102557. [CrossRef]

17. Kusainova, A.A.; Mezentseva, O.V.; Tusupbekov, Z.A. Influence of precipitation variability and temperature conditions on the yield of grain crops in Northern Kazakhstan. *IOP Conf. Ser. Earth Environ. Sci.* **2020**, *548*, 042026. [CrossRef]
18. Waller, D.M.; Meyer, A.G.; Raff, Z.; Apfelbaum, S.I. Shifts in precipitation and agricultural intensity increase phosphorus concentrations and loads in an agricultural watershed. *J. Environ. Manag.* **2021**, *284*, 112019. [CrossRef]
19. Holtermann, L. Precipitation anomalies, economic production, and the role of "first-nature" and "second-nature" geographies: A disaggregated analysis in high-income countries. *Glob. Environ. Chang.* **2020**, *65*, 102167. [CrossRef]
20. Zhang, D.; Shen, J.; Liu, P.; Zhang, Q.; Sun, F. Use of Fuzzy Analytic Hierarchy Process and Environmental Gini Coefficient for Allocation of Regional Flood Drainage Rights. *Int. J. Environ. Res. Public Health* **2020**, *17*, 2063. [CrossRef]
21. An, M.; Fan, L.; Huang, J.; Yang, W.; Wu, H.; Wang, X.; Khanal, R. The gap of water supply-Demand and its driving factors: From water footprint view in Huaihe River Basin. *PLoS ONE* **2021**, *16*, e0247604. [CrossRef] [PubMed]
22. Liu, Z.; Qiu, C.; Sun, M.; Zhang, D. Environmental Performance Evaluation of Key Polluting Industries in China-Taking the Power Industry as an Example. *Int. J. Environ. Res. Public Health* **2022**, *19*, 7295. [CrossRef] [PubMed]
23. Cao, L.; Niu, H.; Wang, Y. Utility analysis of digital villages to empower balanced urban-rural development based on the three-stage DEA-Malmquist model. *PLoS ONE* **2022**, *17*, e0270952. [CrossRef] [PubMed]
24. Azrour, M.; Mabrouki, J.; Fattah, G.; Guezzaz, A.; Aziz, F. Machine learning algorithms for efficient water quality prediction. *Model. Earth Syst. Environ.* **2021**, *8*, 2793–2801. [CrossRef]
25. Shen, Z.; Liu, N.; Li, X.; Kang, Z.; Cao, N. A Method for Evaluating the Green Economic Efficiency of Resource-Based Cities Based on Neural Network Improved DEA Model. *Comput. Intell. Neurosci.* **2022**, *2022*, 9521107. [CrossRef] [PubMed]
26. Li, M.; Zhong, R.; Zhu, S.; Ramsay, L.C.; Li, F.; Coyte, P.C. Access to Community Living Infrastructure and Its Impact on the Establishment of Community-Based Day Care Centres for Seniors in Rural China. *Int. J. Environ. Res. Public Health* **2018**, *15*, 1184. [CrossRef]
27. Li, X.; Wang, X.; Song, X. Impacts of Agricultural Capitalization on Regional Paddy Field Change: A Production-Factor Substitution Perspective. *Int. J. Environ. Res. Public Health* **2021**, *18*, 1729. [CrossRef]
28. Liu, C.; Qi, Y.; Wang, Z.; Yu, J.; Li, S.; Yao, H.; Ni, T. Deep learning: To better understand how human activities affect the value of ecosystem services-A case study of Nanjing. *PLoS ONE* **2020**, *15*, e0238789. [CrossRef]
29. Mabrouki, J.; Azrour, M.; Hajjaji, S.E. Use of internet of things for monitoring and evaluating water's quality: A comparative study. *IJCC* **2021**, *10*, 633–644. [CrossRef]
30. Tang, Y.; Chen, M. The Impact Mechanism and Spillover Effect of Digital Rural Construction on the Efficiency of Green Transformation for Cultivated Land Use in China. *Int. J. Environ. Res. Public Health* **2022**, *19*, 16159. [CrossRef]
31. Shan, T.; Xia, Y.; Hu, C.; Zhang, S.; Zhang, J.; Xiao, Y.; Dan, F. Analysis of regional agricultural carbon emission efficiency and influencing factors: Case study of Hubei Province in China. *PLoS ONE* **2022**, *17*, e0266172. [CrossRef]
32. Yang, H.; Zheng, H.; Liu, H.; Wu, Q. NonLinear Effects of Environmental Regulation on Eco-Efficiency under the Constraint of Land Use Carbon Emissions: Evidence Based on a Bootstrapping Approach and Panel Threshold Model. *Int. J. Environ. Res. Public Health* **2019**, *16*, 1679. [CrossRef]
33. Zhong, R.; He, Q.; Qi, Y. Digital Economy, Agricultural Technological Progress, and Agricultural Carbon Intensity: Evidence from China. *Int. J. Environ. Res. Public Health* **2022**, *19*, 6488. [CrossRef]
34. Yang, X.; Shang, G. Smallholders' Agricultural Production Efficiency of Conservation Tillage in Jianghan Plain, China-Based on a Three-Stage DEA Model. *Int. J. Environ. Res. Public Health* **2020**, *17*, 7470. [CrossRef]
35. Zhang, H.; Guo, S.; Qian, Y.; Liu, Y.; Lu, C. Dynamic analysis of agricultural carbon emissions efficiency in Chinese provinces along the Belt and Road. *PLoS ONE* **2020**, *15*, e0228223. [CrossRef]
36. Li, J.; Lin, Q. Can the Adjustment of China's Grain Purchase and Storage Policy Improve Its Green Productivity? *Int. J. Environ. Res. Public Health* **2022**, *19*, 6310. [CrossRef]
37. Hou, M.; Xi, Z.; Zhao, S. Evaluating the Heterogeneity Effect of Fertilizer Use Intensity on Agricultural Eco-Efficiency in China: Evidence from a Panel Quantile Regression Model. *Int. J. Environ. Res. Public Health* **2022**, *19*, 6612. [CrossRef]
38. Chen, K.; Wang, Z. Evaluation of Financial Subsidy for Agriculture Based on Combined Algorithm. *Comput. Intell. Neurosci.* **2022**, *2022*, 6587460. [CrossRef]
39. Randell, H.; Gray, C. Climate variability and educational attainment: Evidence from rural Ethiopia. *Glob. Environ. Chang.* **2016**, *41*, 111–123. [CrossRef]
40. Yi, F.J.; Feng, J.A.; Wang, Y.J.; Jiang, F. Influence of surface ozone on crop yield of maize in China. *J. Integr. Agric.* **2020**, *19*, 578–589. [CrossRef]
41. Chen, X.; Yang, L. Temperature and industrial output: Firm-level evidence from China. *J. Environ. Econ. Manag.* **2019**, *95*, 257–274. [CrossRef]
42. Wang, Y.; Xiong, J.; Li, W.; Na, M.; Yao, M. The Effect of Social Capital on Environmental Pollution in China-Suppression or Promotion? *Int. J. Environ. Res. Public Health* **2020**, *17*, 9459. [CrossRef] [PubMed]
43. Zhao, Q.; Yuan, C.H. Can China's industrial policies enhance the green competitiveness of the manufacturing industry? *PLoS ONE* **2021**, *16*, e0253774. [CrossRef]
44. Volsi, B.; Higashi, G.E.; Bordin, I.; Telles, T.S. The diversification of species in crop rotation increases the profitability of grain production systems. *Sci. Rep.* **2022**, *12*, 19849. [CrossRef] [PubMed]

45. Chou, J.; Xu, Y.; Dong, W.; Xian, T.; Xu, H.; Wang, Z. Comprehensive climate factor characteristics and quantitative analysis of their impacts on grain yields in China's grain-producing areas. *Heliyon* **2019**, *5*, e02846. [CrossRef]
46. Mabrouki, J.; Azoulay, K.; Elfanssi, S.; Bouhachlaf, L.; Mousli, F.; Azrour, M.; El Hajjaji, S. Smart System for Monitoring and Controlling of Agricultural Production by the IoT. In *IoT and Smart Devices for Sustainable Environment*; Azrour, M., Irshad, A., Chaganti, R., Eds.; Springer International Publishing: Cham, Switzerland, 2022; pp. 103–115.
47. Mabrouki, J.; Fattah, G.; Kherraf, S.; Abrouki, Y.; Azrour, M.; El Hajjaji, S. Artificial intelligence system for intelligent monitoring and management of water treatment plants. In *Emerging Real-World Applications of Internet of Things*; Verma, A., Verma, A., Farhaoui, Y., Lv, Z., Eds.; CRC Press: Boca Raton, FL, USA, 2022; pp. 69–87.
48. Guo, H.; Huang, L.; Liang, D. Further promotion of sustainable development goals using science, technology, and innovation. *Innovation* **2022**, *3*, 100325. [CrossRef]
49. Zhang, D.; Wang, H.; Lou, S. Research on grain production efficiency in China's main grain producing areas from the perspective of financial support. *PLoS ONE* **2021**, *16*, e0247610. [CrossRef]
50. Ovalle-Rivera, O.; Läderach, P.; Bunn, C.; Obersteiner, M.; Schroth, G. Projected shifts in Coffea arabica suitability among major global producing regions due to climate change. *PLoS ONE* **2015**, *10*, e0124155. [CrossRef]
51. Li, C.; Wang, J.; Hu, R.; Yin, S.; Bao, Y.; Ayal, D.Y. Relationship between vegetation change and extreme climate indices on the Inner Mongolia Plateau, China, from 1982 to 2013. *Ecol. Indic.* **2018**, *89*, 101–109. [CrossRef]
52. Alexander, L.V.; Zhang, X.; Peterson, T.C.; Caesar, J.; Gleason, B.; Klein Tank, A.M.G.; Haylock, M.; Collins, D.; Trewin, B.; Rahimzadeh, F.; et al. Global observed changes in daily climate extremes of temperature and precipitation. *J. Geophys. Res.* **2006**, *111*, D05109. [CrossRef]
53. Gray, C.; Wise, E. Country-Specific Effects of Climate Variability on Human Migration. *Clim. Chang.* **2016**, *135*, 555–568. [CrossRef]
54. Yang, Z.; Li, B.; Wu, H.; Li, M.; Fan, J.; Chen, M.; Long, J. Water consumption prediction and influencing factor analysis based on PCA-BP neural network in karst regions: A case study of Guizhou Province. *Environ. Sci. Pollut. Res. Int.* **2022**, *8*, 1–12. [CrossRef]
55. Liu, T.; Zeng, W.; Lin, H.; Rutherford, S.; Xiao, J.; Li, X.; Li, Z.; Qian, Z.; Feng, B.; Ma, W. Tempo-Spatial Variations of Ambient Ozone-Mortality Associations in the USA: Results from the NMMAPS Data. *Int. J. Environ. Res. Public Health* **2016**, *13*, 851. [CrossRef]
56. Haider, N.; Kirkeby, C.; Kristensen, B.; Kjær, L.J.; Sørensen, J.H.; Bødker, R. Microclimatic temperatures increase the potential for vector-borne disease transmission in the Scandinavian climate. *Sci. Rep.* **2017**, *7*, 8175. [CrossRef]
57. Zhang, Q.; Shi, F.; Abdullahi, N.M.; Shao, L.; Huo, X. An empirical study on spatial-temporal dynamics and influencing factors of apple production in China. *PLoS ONE* **2020**, *15*, e0240140. [CrossRef]
58. Bolang, P.D.; Osumanu, I.K. Formal sector workers' participation in urban agriculture in Ghana: Perspectives from the Wa Municipality. *Heliyon* **2019**, *5*, e02230. [CrossRef]
59. Baca, M.; Läderach, P.; Haggar, J.; Schroth, G.; Ovalle, O. An integrated framework for assessing vulnerability to climate change and developing adaptation strategies for coffee growing families in Mesoamerica. *PLoS ONE* **2014**, *9*, e88463. [CrossRef]
60. Zhao, L.; Chen, Y.; Wang, X.; Su, M.; Xu, H.; Zhang, P. Spatiotemporal changes in net primary productivity before and after the development of unused land in the hilly areas of Hebei, China. *PLoS ONE* **2022**, *17*, e0270010. [CrossRef]
61. Liu, M.; Zhang, Q.; Luo, Y.; Mason, R.P.; Ge, S.; He, Y.; Yu, C.; Sa, R.; Cao, H.; Wang, X.; et al. Impact of Water-Induced Soil Erosion on the Terrestrial Transport and Atmospheric Emission of Mercury in China. *Environ. Sci. Technol.* **2018**, *52*, 6945–6956. [CrossRef]
62. Liu, D.; Li, Y.; Wang, P.; Zhong, H.; Wang, P. Sustainable Agriculture Development in Northwest China Under the Impacts of Global Climate Change. *Front. Nutr.* **2021**, *8*, 706552. [CrossRef]

sustainability

MDPI

Article

Influencing Factors and Path Analysis of Sustainable Agricultural Mechanization: Econometric Evidence from Hubei, China

Zhi Li [1,2], Ming Zhu [1], Huang Huang [1], Yu Yi [3,*] and Jingyi Fu [4]

1 College of Engineering, Huazhong Agricultural University, Wuhan 430070, China; hzau_lizhi@mail.hzau.edu.cn (Z.L.); 13801392760@163.com (M.Z.); huanghuang@mail.hzau.edu.cn (H.H.)
2 College of Science, Huazhong Agricultural University, Wuhan 430070, China
3 Law School, Zhongnan University of Economics and Law, Wuhan 430073, China
4 Zhengzhou Central Sub Branch, People's Bank of China, Zhengzhou 450018, China; kaelynn@163.com
* Correspondence: yiyulucky@zuel.edu.cn

Abstract: The importance of supporting agricultural mechanization in agri-food supply chains to achieve agricultural and rural development has been comprehensively recognized. There has been a surge in the attention given to Sustainable Agricultural Mechanization (SAM) in the context of developing countries. However, it is important to address the major challenge of studying the important factors and the influencing path of SAM. As a representative province of China's agricultural development, Hubei has developed significantly in terms of agricultural mechanization in the past 20 years. Therefore, using a literature review, representative field survey data, and statistical analytical approaches, 28 relevant factors related to SAM were extracted, and the main influencing factors of SAM were determined by building an integrative conceptual framework and using the corresponding structural equation model based on partial least squares (PLS-SEM). The relationships and influencing paths between the factors were analyzed, and a confirmatory measurement model and a structural model of the effects on sustainable agricultural mechanization were constructed. The results show that (1) the PLS-SEM model fits the experimental data well and can effectively reflect the relationships among factors in this complex system; (2) within the factors influencing the development level of SAM in Hubei, China, the economic factors have the greatest weight, whereas government policy factors are the core elements promoting development, and environmental factors are the most noteworthy outcome factors; and (3) economic and policy factors play a very obvious role in promoting SAM through the influencing paths of agricultural production and agricultural machinery production and sales. Ultimately, corresponding suggestions have been put forward for decisions regarding the implementation of SAM for similar countries and regions.

Keywords: sustainable; agricultural mechanization; structural equation model (SEM); partial least square (PLS); affecting factors; agri-food supply chain

Citation: Li, Z.; Zhu, M.; Huang, H.; Yi, Y.; Fu, J. Influencing Factors and Path Analysis of Sustainable Agricultural Mechanization: Econometric Evidence from Hubei, China. *Sustainability* **2022**, *14*, 4518. https://doi.org/10.3390/su14084518

Academic Editor: Aaron K. Hoshide

Received: 7 March 2022
Accepted: 8 April 2022
Published: 11 April 2022

1. Introduction

Mechanization is a crucial input for agricultural crop production and one that historically has been neglected in the context of developing countries [1], especially in sub-Saharan Africa, Southeast Asia, South Asia, and Latin America. Mechanization contributes significantly to the development of food supply chains through improved agricultural practices for increased production and enhanced food security. It eases and reduces hard labor, relieves labor shortage, and improves the productivity and timeliness of agricultural operations [2,3].

The issue of Sustainable Agricultural Mechanization (SAM) has received considerable critical attention in recent years. The research of SAM continues the typical paradigm of sustainable agriculture [4], wherein SAM can be described as mechanization that is economically viable, environmentally sensitive, and socially acceptable [3]. The United Nations (UN) Food and Agriculture Organization (FAO) SAM website noted that sustainable mechanization is important, as farmers who have access to improved agricultural tools and powered technologies can shift from subsistence farming to more market-oriented farming, making the agricultural sector more attractive to rural youth [2]. SAM can improve the efficient use of resources, enhance market access, and contribute to mitigating climate-related hazards, as it has the potential to render producing, processing, and marketing activities and functions more efficient, economically feasible, socially acceptable, and environmentally friendly [5]. As the effects of climate change and natural resource depletion become more visible, sustainable mechanization has adopted Conservation Agriculture principles, and the "Save and Grow" paradigm-which aims to protect the soil, use less energy, and encourage more efficient and precise use of inputs-will be essential to maintain and sustainably improve food production and distribution. Analyses of SAM have to account for not just the technical, economical, and engineering aspects, but also the linkages and inter-dependencies with other sectors, such as social, environmental, cultural, and policy aspects, and consider their role when contributing to the sustainable development of the food and agriculture sector. Overcoming the environmental and social challenges of today is not an isolated action but is part of a comprehensive view of agriculture that considers efficiency and ecology [2].

As a representative of developing countries, the agricultural mechanization of China has made remarkable achievements in the past 20 years. The comprehensive mechanization rate of crop cultivation and harvesting in China has risen from 45.8% in 2008 to 71.3% in 2020, an average annual increase of about 2%. However, with the slowing of economic growth and the "The New Normal" of agriculture, the development of agricultural mechanization has recently faced many unsustainable problems [6]. The challenges to agricultural machinery and equipment, production technology, and professional and technical personnel are structural shortages; issues with public service funding and an insufficient effective supply of social service systems and policy support increase pressure on agricultural resources, the ecological environment, and the cost of agricultural machinery [7]. Therefore, this present research focuses on the role of the influencing factors, the development paths, and the development mode of SAM.

The purpose of this study was to take Hubei, a typical region of China, as an example for empirical analysis to estimate the effect of various factors and the development paths of SAM in an integrated analytic framework. The results will enable us to understand the mutual influence of SAM on agriculture, society, the economy, and the environment and can be used to help policymakers and project implementers of agricultural machinery purchase subsidy policies and further formulate and implement their policies' strategy and development path, thus promoting steady and efficient improvements in SAM.

2. Literature Review and Conceptual Framework

2.1. Literature Review

In line with new efforts and opportunities to promote mechanization, there is a growing body of empirical research on the topic of SAM. Research on adaptation to SAM is diverse but mainly focuses on two aspects: (1) the relationship between SAM and economic, environmental, and social sustainability and policy factors; (2) the influencing factors of mechanization development and effective implementation. These two aspects complement each other.

As a sub-element of sustainable development of agriculture, SAM is bound to interact with multiple systems. Some scholars have tried to explore the agronomic, environmental, and socioeconomic effects of mechanization, thereby revealing linkages and trade-offs. For example, the economy has a driving effect on mechanization, which is a direct requirement for improving agricultural output; mechanization is bound to have an impact on the environment, and the machinery industry can promote mechanization [8–12]. Some research has given a voice to the rural population in Africa regarding mechanization and allowed researchers to identify causal impact chains [13]. Other scholars have researched and analyzed the effects of policy formulation. Governments must create an enabling environment to allow the multiple dimensions of SAM to develop. This policy environment includes mechanization policy instruments, including appropriate short-term subsidies for purchasing and leasing equipment [14,15], and law [16]. Sustainability requires the mechanization pathways promoted through policies to be thought through carefully. Formulating adaptation strategies or frameworks are the most common means used by governments to carry out SAM actions, which can guide countries or regions [17]. According to different national conditions, some countries have issued national promotion policies or laws to guide practice, while others have issued action plans that match the strategies. Some studies have also empirically analyzed the relationships of agricultural mechanization with agricultural carbon emissions [18–20], green agricultural transformation [21–23], a low-carbon economy, and food safety [24]. Table 1 lists various agricultural sustainability- and SAM-related policies introduced by developing countries in the past two decades.

Table 1. Agricultural sustainability/SAM related policies.

Country/Region/ International Organization	Policy	Year	Literature Source
China	China's Agricultural Mechanization Promotion Law	2004	[14]
Tanzania	Tanzania Agricultural Mechanization Strategy	2006	[16]
Mexico	MasAgro program of a government public policy framework	2009	[15]
Bangladesh	Cereal Systems Initiative for South Asia—Mechanization and Irrigation (CSISA-MI) project	2013	[15]
India	National Agricultural Extension and Technology Mission (NMAET), Sub Mission on Agricultural Mechanization (SMAM)	2014	[3]
Nepal	Agricultural Mechanization Promotion Policy (AMPP)	2014	[17]
Ethiopia	Ethiopia National Agricultural Mechanization Strategy	2014	[16]
Kenya	National Agricultural Mechanization Policy	2016	[16]
FAO and AUC	Sustainable Agricultural Mechanization for Africa	2018	[16]

With the continuous deepening of SAM research, scholars have begun to pay attention to the influencing factors of SAM. Few previous studies have looked at the potential effects of mechanization empirically but rather have mostly focused on yields and labor alone [13]. However, the factors involved in SAM are likely to be more complex. However, because of the differences in the research objects, research perspectives, or sample selection, the conclusions of the different studies are different. In China, the research related to SAM can be roughly divided into three categories: (1) qualitative policy analysis [7]; (2) mechanization as a sub-element of agricultural sustainability [10,25]; and (3) a discussion of factors related to SAM, including the environment [11,21,22], agricultural carbon emissions [19,20,23], mechanization level [14,26,27], agricultural machinery industry [18], etc. However, there is a lack of quantitative and systematic research on SAM in China.

This study focused on the interactions among SAM factors and undertook an overall and systematic quantitative empirical study to make up for the shortcomings in the existing literature. At the same time, an analysis system covering education and training, science and technology, and other influencing factors was constructed, which expanded the scope of influencing factors and the path of research by including the ecological environment in the influencing factors of agricultural mechanization. This part of the research is an important complement to the existing literature on SAM. These two aspects are the important innovation points of this study, which are different from those in previous studies.

2.2. Analysis of the Influencing Factors of SAM

There is a wide range of factors affecting SAM. Each country has different land conditions, planting bases, and climate backgrounds, and there are great differences in the mechanization process. Therefore, research on the development mode mechanization and appropriate strategies should ensure the application of mechanization theory at the decomposition level. The types of strategies needed to promote the development of SAM must account for the conditions of specific sites, each of the factors and the mechanisms, and the extent to which these influence SAM will vary from country to country, potentially even within countries. According to investigation and research, literature reviews, and policy analyses, combined with the actual situation in different regions, it can be concluded that the factors affecting the SAM include those summarized in Table 2. Of course, one should not ignore that there are potentially several adverse propositions that have emerged from using agricultural mechanization, such as "mechanization leads to unemployment" or "smallholders cannot benefit from mechanization" (particularly in developing countries) [3,28]. Such topics also can affect policies and programs regarding mechanization.

Table 2. Influencing factors of SAM.

Country/Area, Continent	Influencing Factors	Literature Source
Benin, Kenya, Nigeria, and Mali, Africa	Soil, terrain and rainfall, institutional environments Social objectives of societies, labor shortages, timeliness Land preparation, higher yields, soil fertility, deforestation	[13,28]
Eleven countries, Africa	Size of the household, gender of the household, participation in off-farm economic activities, farm size, land tenure, distance to the input and output markets, type of farming system, access to extension services, use of fertilizer and pesticides	[29]
Ghana, Africa	Population pressure, better market access	[30]
Ethiopia, Africa	Rising rural wages, working animal costs	[31]
Africa	Education level, area cropped, access to land, access to credit and agroecological zone	[32]
Asia	Household assets, credit availability, electrification, road density, substantial capital investment, purchases and rental services	[33]
Nepal, Asia	Land consolidation, business mergers, more intensive cropping, labor displacement	[17]
India, Asia	Irrigation, access to institutional credit, size of land holdings, age-old customs	[34]
Bangladesh/South Asia, Asia	Male headship, access to credit and extension services, economic status, training positively, rental services, educational level	[35,36]
Myanmar, Asia	Structural transformation, timeliness, speed, drudgery, risk, yields, financing and machinery prices, policies and interventions	[37]
China, Asia	Scale of farmland management, agricultural labor transfer, farmers' income level, the development level of the agricultural machinery industry, the cost of using agricultural machinery products Agricultural equipment level, regional economic development, land resources, policy, environment Economic development, scale of farmland, agricultural planting structure	[7,14,21,25–27]

From the analysis above, it can be seen that the factors affecting the sustainable development of agricultural mechanization mainly include economics, society, the population and labor force, agricultural production, land resources, industrial technology development, education, the energy environment, ecology, and policies and regulations. There are many corresponding component indicators with each aspect. The relationships are also more complicated, and the mutually influencing relationship paths are often not clear, so traditional methods of research are more difficult. Therefore, this article focused on the use of structured statistical research methods to comprehensively and quantitatively analyze the relationships among the influencing factors of agricultural mechanization and the path and intensity, as well as to quantitatively verify the conclusions of the qualitative analysis. According to the analysis above and index-selection principles, 28 representative indicators were finally selected from the different categories (socioeconomic, environmental, production and land resource, agricultural machinery industry and technology, agricultural mechanization status and policy support, etc.).

3. Materials and Methods

3.1. Research Area and Data Sources

The regional area of Hubei Province (185,900 km^2) is equivalent to that of a medium-sized developing country, such as Uganda, Ghana, or Cambodia. The terrain includes plains, hills, mountains, and lakes. There are various agricultural planting operations, and they have been dominated by small farmers and small business owners for a long time. The development strategy of SAM is highly typical of quite a few developing countries. The original data of Hubei Province collected in this article came from China Statistical Yearbook, China Agricultural Machinery Industry Yearbook, Hubei Statistical Yearbook, Hubei Rural Statistical Yearbook, and some field investigations. For some of the missing data and unreasonable data, we estimated the missing values through mean replacement and regression interpolation, then completed data preprocessing and finally obtained 392 valid data for the 28 measurement indicators used in this article. The descriptive statistical results of indicators data are shown in Table A1 (see in Appendix A).

To eliminate the effects of the different orders of magnitude and dimensions of different variables, the data of all variables were standardized. The method used for standardization of the variables was the Min-Max standardization method [14]. That is, all variables were transformed linearly. If MinX and MaxX are the minimum and maximum values of variable X, after standardization, $X' = (X - MinX)/(MaxX - MinX)$. It is also difficult to deal with the complexity of SAM via traditional methods. Furthermore, in this study, there were several latent influencing variables (latent variables) of practical significance for agricultural mechanization, and there were also several different observation variables or manifest variables for each latent variable, which may have also affected other latent variables. These can be influenced by the internal and external relationships of SAM within the model, and it was necessary to evaluate the influencing relationships and size from different aspects. The six aspects of the influencing factors can be regarded as latent variables, and the influencing factors themselves can be regarded as manifest variables. This article established the latent variables as economic and population factors (EP), agricultural production (AP), the agricultural mechanization development level (AMDL), the agricultural machinery industry and agricultural technology (AMIAT), policies (P), and the environment (E). The final results are shown in Table 3.

Table 3. Impact factors of SAM.

Latent Variable	Manifest Variable	Variable Codes	Latent Variable	Manifest Variable	Variable Codes
EP	Gross Domestic Product of the region	s1	AMDL	Machine farming area	s15
EP	Agricultural investment in fixed assets	s2	AMDL	Machine collecting area	s16
EP	Number of employees in agriculture, forestry, animal husbandry, and fishery	s3	AMDL	Total power of agricultural machinery	s17
EP	The proportion of rural labor force with junior high school education or above	s4	AMDL	Comprehensive operation rate of main crops cultivation and harvesting	s18
AP	The sown area of food crops	s5	AMDL	Number of farm machinery households	s19
AMIAT	The sales output value of agricultural machinery industry	s6	EP	The original value of agricultural machinery	s20
AMIAT	The contribution rate of agricultural science and technology progress	s7	EP	Total profit of agricultural machinery	s21
AMIAT	Informationization level	s8	EP	Per capita net income of farmers	s22
EP	Total investment in agricultural mechanization	s9	AP	Per capita food production	s23
EP	Farm machinery purchase cost	s10	AP	Agricultural output value	s24
EP	Fixed base price index of mechanized farm tools	s11	E	Agricultural diesel consumption	s25
AMDL	Number of service organizations of agricultural mechanization	s12	E	Agricultural carbon emissions	s26
AMDL	Number of trainees in agricultural mechanization	s13	P	Government Agricultural Machinery Policy Subsidies—National	s27
AMDL	Machine sowing area	s14	AMDL	Number of agricultural mechanization technology promotion agencies	s28

3.2. Basic Hypotheses

Our research set the first-level indicators, divided their corresponding explicit variables, and established the causal relationships among latent variables. Since the assignment of indicators and the setting process of causality are subjective and referential, the set construction and adjustment process relied on the overall assumptions of the model described in the following hypotheses:

Hypothesis 1. *The correlation between the latent variable and its corresponding explicit variable can be expressed by linear equations; the latent variables do not cross each other in the theoretical sense.*

Hypothesis 2. *According to the actual meaning of the selected indicators, the selected latent variables are directly related to each other, and they may have indirect secondary path effects through other latent variables.*

According to the influencing factors and the related relationships analyzed in the literature review, the following assumptions were put forward: The impact of agricultural mechanization and agricultural economic development has a strong two-way positive effect. Conversely, to promote the development of agricultural mechanization, capital investment is indispensable. At the same time, the development of the agricultural machinery industry is an important basic guarantee for the sustainable development of agricultural mechanization. The development of agricultural mechanization and the agricultural machinery industry complement each other. In addition, making the input of agricultural machinery produce real profits and increasing wealth by using production machinery is the only way to encourage agricultural machinery users to further invest and expand production [38]. Thus, based on the above view, we hypothesize:

Hypothesis 2a. *The agricultural mechanization development level (AMDL) is affected by economic factors, agricultural production, policy, and the agricultural machinery industry and agricultural technology (AMIAT) factor.*

The healthy development of agricultural mechanization can directly increase the output and efficiency of agricultural workers, directly increase labor income, and stimulate the overall development of the agricultural economy [9]. Many studies have also discussed the impact of policies and the agricultural machinery industry on the growth of agricultural products [6–8,12]. Thus, based on the above view, we hypothesize:

Hypothesis 2b. *Economic factors, policy factors, the AMDL, and AMIAT can promote agricultural production.*

With their rapid development, modern science and technology have become widely used in agricultural production, including high-tech informatization and intelligent agricultural machinery used in innovative crop production methods. Improved machinery operation capabilities are used to implement precision production operations, saving labor while improving efficiency. It is no doubt that science and technology play a key role in the development of modern agricultural mechanization. At the same time, the agricultural machinery industry must strive to improve its innovation and investment, which is an important new growth point to realize the development of SAM for developing countries [39]. Moreover, the implementation of China's Agricultural Mechanization Promotion Law in 2004 and the subsidy policy for the purchase of agricultural machinery in 1998 played significant roles in improving the agricultural machinery industry and agricultural mechanization [14]. Thus, based on the above view, we hypothesize:

Hypothesis 2c. *AMIAT is positively correlated with economy and policy.*

Physical limits to land and water availability within ecosystems are often worsened by climate change. By including SAM in its projects, FAO promotes conservation agriculture practices that contribute to soil conservation and water use efficiency [2]. The development of SAM must be organically combined with energy-conservation technology, emission control, and ecological protection, and must strive to achieve harmonious coexistence between human activities and nature. To advocate for a green economy [11,19,20], there needs to be active promotion by the government. We are sure that the change in the environment must be the product of a comprehensive effect [16]. Thus, based on the above view, we hypothesize:

Hypothesis 2d. *Environmental factors are affected by economic factors, agricultural production, AMIAT, AMDL, and policy factors at the same time.*

Relevant national policies and regulations can ensure that capital investment and subsidy policies can effectively reduce purchasing costs so that they can effectively promote the sound and rapid development of agricultural mechanization and the agricultural machinery industry, which is one of the main ways to effectively promote the popularization and extension of agricultural machinery [7]. Meanwhile, policies and regulations can also manage and coordinate various development goals and promote the balanced development of society. Therefore, national policies provide strong support and guarantee the sustainable development of agricultural mechanization. Thus, based on the above view, we hypothesize:

Hypothesis 2e. *Economic factors are affected by policy factors.*

Based on these assumptions, this research first established an initial path graph structure in a fully connected form and then continuously made corrections based on the analysis results to create the final improved model. In the establishment of the measurement model, the corresponding relationships and influencing paths between the observed variables and latent variables were set according to the actual meaning of the indicators. All the indicators were then matched to the latent variables to achieve a causal equilibrium. The initial hypothesis structure is shown in Figure 1.

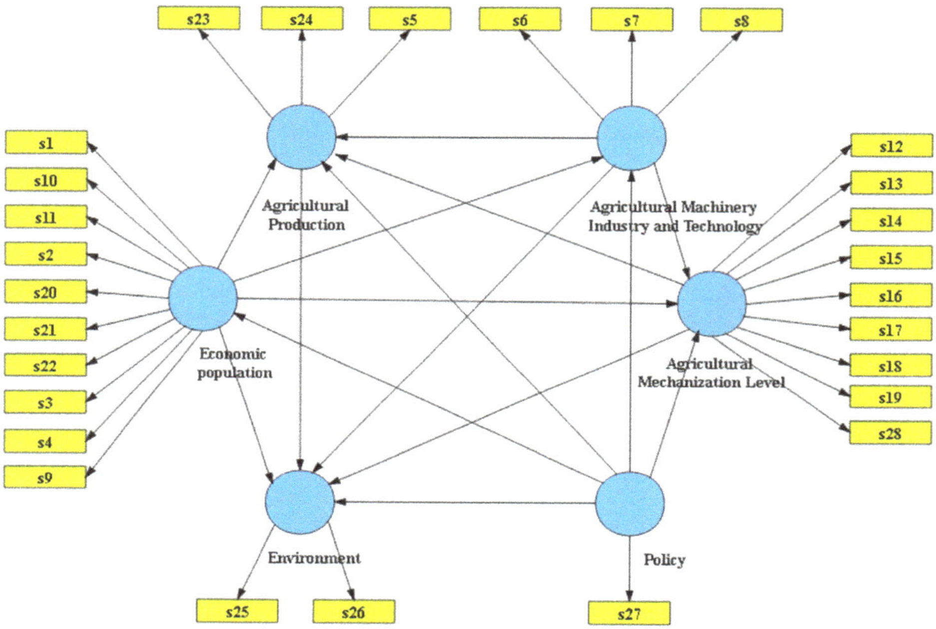

Figure 1. Hypothetical structural equation framework of SAM.

3.3. Statistical Modeling Methods

Traditional statistical analysis methods, such as linear regression and principal component analysis, cannot effectively deal with these latent variables, but they can be studied with the help of structural equations. Structural equation modeling (SEM) is a systematic analysis method that integrates factor analysis and path analysis. SEM has the advantages of simultaneously processing multiple dependent variables, allowing independent variables and dependent variables to contain measurement errors; estimating factor structures and factor relationships; and estimating the fitting degree of the whole model [14]. It uses a structure of linear equations to deal with the relationships between manifest variables and latent variables and the relationships between latent variables.

SEM can be divided into two types according to the nature and relationship characteristics of the variables. One is the measurement model, and the other is the structural model, which uses a similar path-analysis method to establish the structural relationships between latent variables. The following equations show the specific forms of the measurement model and the structural model. The measurement model is [40]:

$$x = \Lambda_x \xi + \delta \quad (1)$$

$$y = \Lambda_y \eta + \varepsilon \quad (2)$$

In Formula (1), x is a $p \times 1$ dimensional vector formed by p exogenous manifest variables, ξ is an $m \times 1$ dimensional vector formed by m exogenous latent variables, Λ_x is a $p \times m$ dimensional load matrix, and δ is a $p \times 1$ dimensional vector composed of p measurement errors. In Formula (2), y is a $q \times 1$ dimensional vector formed by q exogenous manifest variables, η is an $n \times 1$ dimensional vector formed by n exogenous latent variables, Λ_y is a $q \times n$ dimensional load matrix, and ε is a $q \times 1$ dimensional vector composed of q measurement errors.

The structural model is:

$$\eta = B\eta + \Gamma\xi + \zeta \quad (3)$$

where B is an $n \times n$ dimensional correlation coefficient matrix, which is used to reflect the relationships among the various endogenous latent variables; Γ is an $n \times m$ dimensional correlation coefficient matrix, which is used to reflect the relationship between the exogenous latent variable ξ and the endogenous latent variable η; ζ is an $n \times 1$ dimensional vector composed of interpretation errors. The correlation coefficient is a standardized path coefficient. This path coefficient is used to measure the degree of correlation between two variables and is generally used to indicate reliability, under the premise that when the significance of the path coefficient is larger, the indicator has a greater impact [41].

In a realistic structural model, the variables may be both dependent variables and independent variables, and there may be not only direct but also indirect relationships among the variables. Some causal variables will affect the result variable through one or more intermediate variables, which are called indirect effects. The path coefficient of the indirect path is the product of the direct path coefficient involved in each path. The meaning of the total effect is the sum of the result variables affected by the causal variables, which is expressed as the sum of the direct effects and indirect effects, which can be used to verify the rationality of the hypothetical effect through path analysis.

Due to the number of samples collected, the maximum likelihood estimation method was not used, but the partial least squares (PLSs) method based on nonparametric estimation was used, as it has no strict assumptions about the sample size and sample distribution. Using the PLSs method to solve the SEM can avoid the situation where the model cannot be recognized because of a non-positive definite covariance matrix, and the method is more extensive. In summary, the study finally conducted an empirical analysis of the factors affecting the SAM by using the PLS-SEM method.

4. Results of the Case Study

In this study, Smart-PLS3 [42] was used to estimate Equations (1)–(3) based on the standardized data. Model evaluation and testing used multiple test statistical indicators to carry out correlation reliability and validity tests, such as Cronbach's alpha coefficient (Cronbach's α) and composite reliability. Hair et al. stated that it is acceptable for Cronbach's α to be greater than 0.7 for verification purposes [43].

4.1. Model Specification Tests

The test results of the reliability and the goodness of fit of the model are shown in Table 4. It can be seen from Table 4 that the model passed the reliability and validity test. The outer loading of the measurement model and the path coefficient of the structural model were calculated by the PLSs method. Then, the bootstrapping method was used to test and evaluate the estimated results of the two coefficients, as shown in Tables A2 and A3.

Table 4. Test result of reliability and goodness of model.

Latent Variables	Average Variance Extracted	Composite Reliability	R Square	Cronbach's Alpha
AP	0.9611	0.9867	0.9796	0.9797
AMDL	0.8220	0.9442	0.9821	0.7682
AMIAT	0.9527	0.9837	0.9745	0.9751
P	1.0000	1.0000		1. 0000
E	0.9771	0.9884	0.9941	0.9766
EP	0.8442	0.9557	0.9642	0.7932

It can be seen from Table A2 that the loading of each path in the measurement model passed the significance test. The general test passing standard is that the significance level is 0.05, and the t-statistic is greater than 1.96. The parameter estimation results of the structural model are shown in Table A3. In Table A3, the estimated values of the direct path coefficients of four paths did not pass the significance test. As these direct relationships were not supported by the test results, these four paths were excluded from the model. Generally speaking, the path coefficient relates to the number, type, and nature of the observed variables corresponding to the latent variables. An insignificant path coefficient in the internal model does not prove that there is no causal relationship between these latent variables. The current variables and model settings were not enough to prove their relationship, and other latent variables can be used as intermediaries to supplement the path. Since the model passed the best test, the significant initial variables used when setting the measurement model did not change, but the causal relationships between the latent variables in the structural model were adjusted, and several insignificant paths were removed. The model was then tested again. The result was that in the revised structural model, all of the path coefficients also passed the significance test, and therefore, the model is desirable. The structural equation model obtained after the final adjustment is referred to as Model B, as shown in Figure 2.

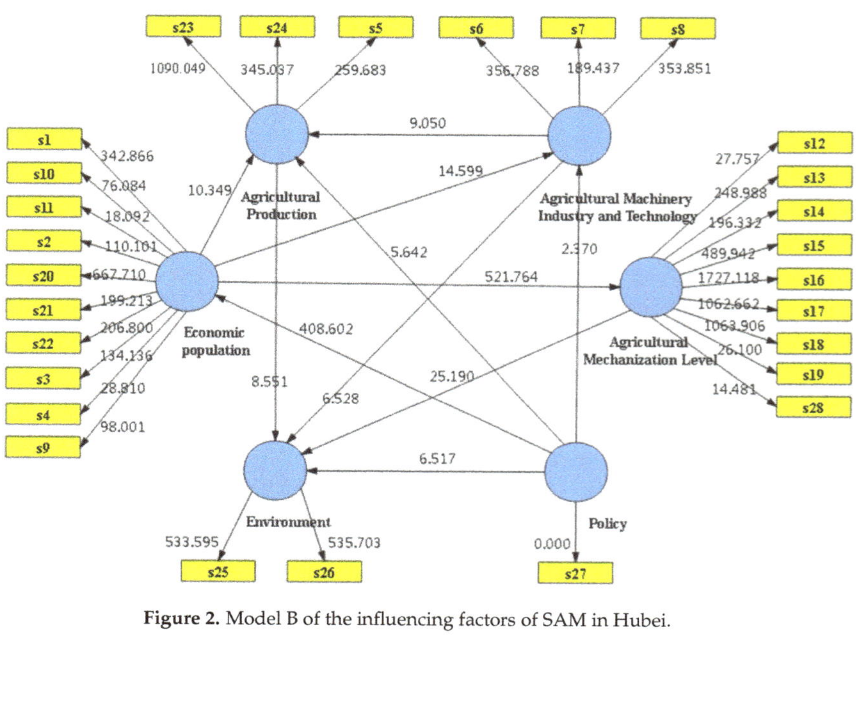

Figure 2. Model B of the influencing factors of SAM in Hubei.

4.2. Results of the First-Order PLS-SEM Model

The calculation results of measurement Model B are shown in Table A4. The factor loadings in Table A4 indicate that most of the indicators have a higher explanatory degree, reflecting that the selection of indicators is more representative and indicating that the measurement model's interpretation ability is good. The negative numbers reflect negative correlations between the indicator and Sustainable Agricultural Mechanization (SAM).

According to measurement Model B, the latent variables of the original value of agricultural machinery, GDP, and agricultural machinery profit were the three most important economic factors, indicating that the overall economic environment and the economic conditions of the agricultural machinery industry are important economic factors of SAM. The agricultural machinery price index is negatively related to the economic factors, indicating that the higher the price of agricultural machinery, the lower the market for agricultural machinery, in line with the actual situation. The relationship between the SAM and the number of people in the labor force is also negatively related, consistent with theoretical analysis. It is also worth noting that the degree of education (the proportion of the population educated in junior high school) is about 0.78. Mechanization is related to the quality of workers, but with the popularization of education, the degree of relevance of the impact of this indicator is not particularly sensitive. The correlations of the three indicators of agricultural production are high, which indicates that the benefits of agricultural mechanization are obvious from the statistical point of view. It is noted that the correlations between several indicators of the agricultural machinery industry and scientific and technological factors are also strong, indicating that the contribution of scientific and technological input to SAM is increasing, and the previous qualitative analysis is verified.

In the indicator corresponding to the latent variable "agricultural level", most of the indicators' factor loadings are relatively large. The correlation coefficient of the number of households with agricultural machinery is about 0.76, which does not show a high correlation, indicating that SAM has slowly reflected the trend of increasing through quality and intensive development, rather than a simple absolute increase in quantity, which may also be reflected by the two negatively related indicators of the number of agricultural machinery service organizations and the number of agricultural technology extension agencies. The results for social services and the use of agricultural machinery technology to promote education suggest that the increase in agricultural mechanization in itself is not through popularization in terms of head counts but has been a gradual process of information dissemination regarding the precision and characteristics of SAM. The results of the structural model are shown in Table 5. The influencing factors are presented in Table 6, including the total effect of the change after considering the indirect effects.

Table 5. Causality and path coefficients of latent variables in SEM.

Path	Path Factor	*t*-Value
AP → Environment	−0.3868	8.5509
AMDL → Environment	0.7745	25.1899
AMIAT → AP	0.4656	9.0505
AMIAT → Environment	0.4313	6.5283
Policy → AP	−0.3460	5.6425
Policy → AMIAT	−0.2000	2.3704
Policy → Environment	0.1782	6.5167
Policy → EP	0.9818	468.6016
EP → AP	0.8638	10.3488
EP → AMDL	0.9905	521.7642
EP → AMIAT	1.1834	14.5986

Note: Economic and population factors (EP), agricultural production (AP), the agricultural mechanization development level (AMDL), and the agricultural machinery industry and agricultural technology (AMIAT).

Table 6. Total effect of SAM in Model B.

Path	Total Path Coefficient	*t*-Value
EP → AP	1.4147	19.1034
EP → Environment	0.7304	21.1399
AMIAT → Environment	0.2512	5.1996
Policy → AP	0.9496	162.2230
Policy → AMIAT	0.9612	146.2765
Policy → AMDL	0.9725	343.0872
Policy → Environment	0.9787	378.2150

Note: Economic and population factors (EP), agricultural production (AP), the agricultural mechanization development level (AMDL), and the agricultural machinery industry and agricultural technology (AMIAT).

The path coefficients in Tables 5 and 6 describe whether there is a causal relationship between a pair of latent variables. If the path coefficient is small, the causal relationship reflected by that path may not exist. Of course, the paths in Table 5 only show the direct effects between the latent variable, that is, the direct impact of the cause variable on the result variable. Table 6 reflects the combination of direct and indirect effects of some of the latent variables.

4.3. Results of Hypothesis Testing

The results of the PLS-SEM analysis showed that agricultural mechanization is directly related to the six dimensions of economy, population, agricultural production, the agricultural machinery industry and agricultural technology, the environment, and policies. It can be seen from Figure 3 that the level of Agricultural Mechanization Development Level (AMDL) is obviously promoted by the economic, population, and policy factors. The economic and population factors are the most critical factors affecting SAM. The effect of the agricultural machinery industry and agricultural technology (AMIAT) on SAM is not statistically obvious.

Secondly, the effects of policy on agricultural machinery (mainly referring to the agricultural machinery purchase subsidy policy) are not significant, but the path regression coefficient of the overall impact of the effect is 0.9725, and that is a strong positive correlation. In terms of agricultural production, the positive effect of economic factors on agricultural production input and output is obvious. The direct effect of policy factors on agricultural production is negatively correlated, but the total effect is still a relatively large positive correlation. The effects of the agricultural machinery industry and agricultural science and technology on agricultural production were also quantified in this article, showing a certain degree of persuasive power. The impact of agricultural mechanization on agricultural production is not significant. In terms of economic and population factors, apart from the strong influence of policy factors, no other path of influence is significant. The factors of the agricultural machinery industry and agricultural science and technology are similar to agricultural production factors: economic factors and policy factors are significantly affected by these factors. Finally, among the directly related factors, the two factors with the greatest effect on the environment are the AMDL and AMIAT. The path regression coefficient of the impact of agricultural mechanization on the environment is 0.7745, and it shows a strong positive correlation. Although this is not very high, a certain significant correlation has been shown, reflecting the increasing degree of the environmental impact of SAM. The coefficient of the direct impact of the agricultural machinery industry and agricultural science and technology on the environment is not particularly high, but the overall effect is reduced because the impact of agricultural production itself has a relatively strong negative correlation, indicating that when agricultural output is higher, the environmental impact will improve. There is a certain harmony between the two factors. Statistically speaking, this improvement (more than 40%) should be given sufficient attention.

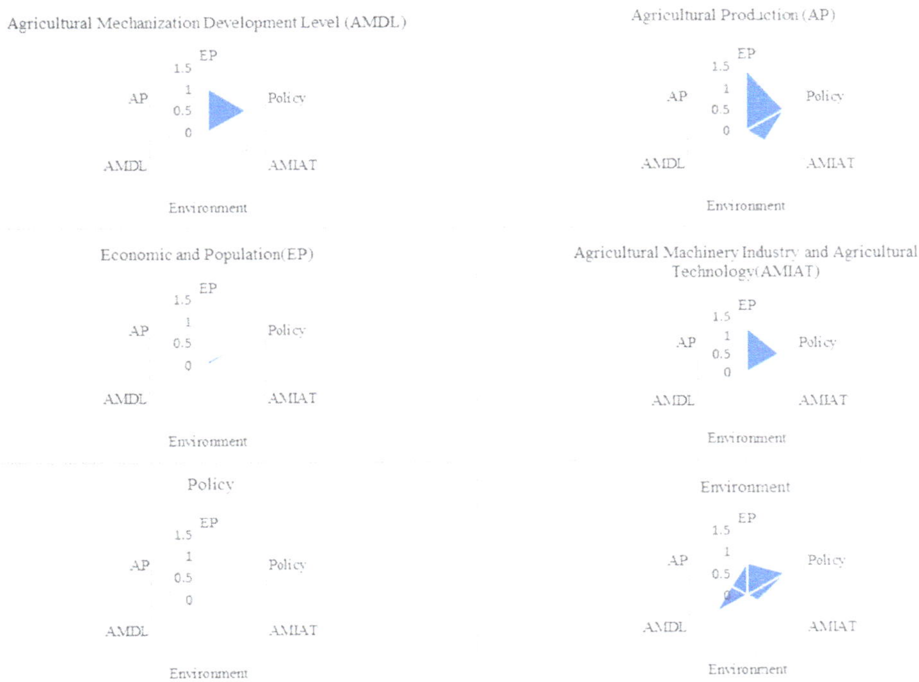

Figure 3. Radar chart of the total effect of the latent variables.

4.4. Discussion

(1) Sustainable Agricultural Mechanization (SAM) is directly affected by economic effects to a high degree. The development of SAM needs the necessary economic input and asset investment for its support. The cost of agricultural machinery and the efficiency of output have a direct impact on farmers' use, making these important factors. The income level of farmers directly affects the farmers' willingness and ability to purchase agricultural machinery, so agricultural mechanization and economics form feedback loops of mutual restriction and mutual promotion. This conclusion has also been confirmed in sub-Saharan Africa, South Asia, and Latin America [12,15,17].

(2) Our results show that SAM requires policy investment support, and the indirect incentives of policy-guided market regulation can still bring great vitality to SAM, but the direct effect of the agricultural machinery policy subsidy on SAM has not been very significant. However, through policy guidance, market regulation of indirect incentives can still bring great vitality to SAM. Policy factors are still the leading factors promoting SAM, and our results corroborate several findings of prior studies [12,38,44]. However, as the degree of marketization deepens and government functions are gradually weakened, whether farmers' willingness to purchase agricultural machinery can be maintained for future development are unclear.

(3) However, our findings are also in contrast with other research results [45]. Our study confirms that the impact of agricultural mechanization on agricultural production is not significant, indicating that, on the one hand, many factors affecting agricultural output, such as climate, natural disasters, markets, and other factors, can strongly affect production results, and that therefore more complex agricultural mechanization is only one of the factors affecting production. On the other hand, one can also see that the current development of mechanization in Hubei is not particularly balanced.

Because of the many lakes in large areas, mechanization is only found in a few areas, and only a few popular crops are grown. There is still a significant gap between Hubei and the provinces and regions with a high degree of SAM, such as Heilongjiang, Henan, and Jiangsu Provinces.

(4) Environmental factors are influenced by several other factors comprehensively. The past period of high economic growth has been accompanied by high pollution and high consumption issues, which can be seen very clearly here. Adjusting the relative balance of agricultural development and the ecological environment, which must be a non-negotiable part of SAM, requires attention. At the same time, our results also show that through the improvement of agricultural machinery technology and agricultural science and technology input, accompanied by the effective improvement of agricultural output mode and processes, the environment can also play a significant role in achieving the sustainable green development of agriculture [11].

So far, we have clarified the relationships, influence paths, and intensity of the interactions among several factors affecting SAM, analyzed the factors influencing SAM within the overall and structural relationships, and clearly showed the mechanisms and quantity of the internal influencing relationships. Among these key elements of SAM, the total effect of the two aspects of policy and economic factors is consistent with the actual situation [7], the relationship between the level of agricultural mechanization and agricultural production is worth exploring, and the environment is the result of comprehensive action. In addition, we also suggest that we should focus on improving the system of laws and regulations regarding agricultural mechanization and that we should standardize the production, sales, use, and service of agricultural machinery. New studies should be pursued to actively explore policies and measures to promote the development of agricultural mechanization, strengthen administrative laws regarding agricultural machinery, and perfect the operation mechanism of regulatory supervision of agricultural machinery. Multiple measures should be used to strengthen public legislation and education and to enhance the legislative awareness and ideas of the public to meet the objective requirements of building a modern agricultural system and sustainable development.

5. Conclusions

Although some studies have investigated the factors affecting agricultural mechanization in China, relatively few have involved systematic and structured econometric research. Based on historical data and the status quo of the development of agricultural mechanization in Hubei, this study used a partial least squares–structural equation model (PLS-SEM) framework to conduct a reasonable modeling analysis based on 28 measurement indicators and determined the relationships and paths of the factors affecting Sustainable Agricultural Mechanization (SAM) in Hubei. The measurement results provided solid support for most of our hypotheses and effectively verified and supplemented the corresponding qualitative research: SAM is directly affected by economic effects to a high degree, and environmental factors are comprehensively influenced by several other factors. In addition, some influencing relationships are presented in the form of quantitative results for the first time, such as the total effect of policy on the agricultural mechanization level and the path coefficient of the impact of agricultural mechanization on the environment.

Our finding relies on data we collected. Different data-collection methods, data facticity, and limitations of data may result in greater deviation from our results and the interpretation of our results to formulate our conclusions. Therefore, future research that could enrich our understanding of China's SAM could potentially proceed with longer-term empirical research. Meanwhile, the internet and big data technology can be used to monitor SAM in real time to reflect instantaneous developments and changes in SAM in response to different factors. These represent improvements that future studies can be undertaken to develop a more in-depth understanding of other economic, policy, and environmental factors impacting the adoption of Sustainable Agricultural Mechanization by producers in China and beyond.

Author Contributions: Conceptualization, Z.L. and M.Z.; methodology, Z.L., M.Z. and H.H.; software, Z.L. and J.F.; validation Z.L.; formal analysis, Z.L.; investigation, Z.L. and H.H.; resources, Z.L.; data curation, Z.L. and J.F.; writing—original draft preparation, Z.L.; writing—review and editing, Z.L. and Y.Y.; visualization, Z.L.; supervision, Z.L. and Y.Y.; project administration, Y.Y.; funding acquisition, Z.L. All authors have read and agreed to the published version of the manuscript.

Funding: This research was supported by the National Natural Science Foundation of China (71503095), a guiding project of humanities and social sciences of the Hubei Provincial Department of Education (15G028).

Institutional Review Board Statement: Not applicable.

Informed Consent Statement: Not applicable.

Data Availability Statement: The data used in this study are available from the authors on reasonable request.

Acknowledgments: We appreciate the advice of the reviewers very much.

Conflicts of Interest: The authors declare no conflict of interest.

Appendix A

Table A1. Descriptive statistics of the variables.

Variable Codes	Variable	N	Minimum	Maximum	Mean	Std. Deviation
s1	Gross Domestic Product of the region	17	5633.24	45,828	23,168.34	13,167.26
s2	Agricultural investment in fixed assets	17	46.24	650.49	293.2665	201.8406
s3	Number of employees in agriculture, forestry, animal husbandry, and fishery	17	863	1105	939.2941	88.02457
s4	The proportion of rural labor force with junior high school education or above	17	0.67	0.76	0.722941	0.024438
s5	The sown area of food crops	17	3817	4852	4346.647	353.1728
s6	The sales output value of agricultural machinery industry	17	887	4735	2745.824	1202.251
s7	The contribution rate of agricultural science and technology progress	17	39.64	61	52.5288	6.2936
s8	Informationization level	17	1.09	34.32	16.8453	12.3385
s9	Total investment in agricultural mechanization	17	9.1	54.9	35.1059	14.5315
s10	Farm machinery purchase cost	17	7.9	50.9	30.6	12.6586
s11	Fixed base price index of mechanized farm tools	17	99.1	107.7	102.1765	2.3720
s12	Number of service organizations of agricultural mechanization	17	3.4	219	45.3471	67.6336
s13	Number of trainees in agricultural mechanization	17	98,832	560,816	381,649.5	181,420.3
s14	Machine sowing area	17	233.66	3362.6	1565.45	1085.071
s15	Machine farming area	17	2015.93	6127.6	4602.291	1526.089
s16	Machine collecting area	17	1263.46	4633.2	3227.768	1161.949
s17	Total power of agricultural machinery	17	1768.6	4626.1	3541.795	924.7491

Table A1. *Cont.*

Variable Codes	Variable	N	Minimum	Maximum	Mean	Std. Deviation
s18	Comprehensive operation rate of main crops cultivation and harvesting	17	42	71.3	58.2059	9.4999
s19	Number of farm machinery households	17	94.6	237.3	211.5353	37.8286
s20	The original value of agricultural machinery	17	98.57	550	319.9506	144.1551
s21	Total profit of agricultural machinery	17	28.22	103.44	70.3544	23.7995
s22	Per capita net income of farmers	17	2890.01	16,390.86	8791.248	4666.718
s23	Per capita food production	17	349	500	426.7059	47.7417
s24	Agricultural output value	17	921.59	3492.54	2198.772	853.4645
s25	Agricultural diesel consumption	17	41.09	67.33	58.1059	9.3321
s26	Agricultural carbon emissions	17	1833	4598	3226.294	836.1836
s27	Government Agricultural Machinery Policy Subsidies—National	17	0.7	237.54	140.2788	86.7749
s28	Number of agricultural mechanization technology promotion agencies	17	670	1016	832.1765	103.1482

Table A2. Factor load estimation results of the measurement model.

Path	Original Sample Value (O)	T Statistic (|O/STERR|)	Significance	Path	Original Sample Value (O)	T Statistic (|O/STERR|)	Significance
s1 ← EP	0.9810	345.5423	***	s2 ← EP	0.9258	106.6113	***
s10 ← EP	0.9215	880.8130	***	s20 ← EP	0.9929	678.2270	***
s11 ← EP	−0.7261	17.9869	**	s21 ← EP	0.9740	202.7099	***
s12 ← AMDL	−0.7505	26.1794	**	s22 ← EP	0.9599	217.1062	***
s13 ← AMDL	0.9712	255.6121	***	s23 ← AP	0.9943	1125.4639	***
s14 ← AMDL	0.9537	196.9908	***	s24 ← AP	0.9733	329.7788	***
s15 ← AMDL	0.9855	407.7759	***	s28 ← AMDL	−0.6921	14.8183	**
s16 ← AMDL	0.9960	1594.6752	***	s3 ← EP	−0.9494	132.5960	***
s17 ← AMDL	0.9891	1055.2473	***	s4 ← EP	0.7831	35.1248	**
s18 ← AMDL	0.9924	1028.9101	***	s5 ← AP	0.9734	262.2697	***
s19 ← AMDL	0.7595	24.3129	**	s6 ← AMIAT	0.9834	372.2486	***
s25 ← E	0.9885	511.9002	***	s7 ← AMIAT	0.9625	181.3262	***
s26 ← E	0.9885	512.6484	***	s8 ← AMIAT	0.9821	359.1926	***
s27 ← P	1.0000			s9 ← EP	0.9357	101.2481	***

Note: (1) Economic and population factors (EP), agricultural production (AP), the agricultural mechanization development level (AMDL), the agricultural machinery industry and agricultural technology (AMIAT), environment (E), and policy (P). (2) *** indicates a significance level of 1%, ** indicates a significance level of 5%. (3) STERR indicates standard error.

Table A3. Estimation of path coefficients for structural model.

Path	Original Sample Value (O)	T Statistic (∣O/STERR∣)	Significance
AP → Environment	−0.3624	5.3331	*
AMDL → AP	−0.1030	0.9804	–
AMDL→ Environment	0.8246	12.7605	**
AMIAT → AP	0.4739	8.8222	**
AMIAT → AMDL	0.0514	0.7795	–
AMIAT → Environment	0.4446	7.7764	**
Policy → AP	−0.3525	6.0133	**
Policy → AMDL	−0.0507	0.6018	–
Policy → AMIAT	−0.2062	2.2246	*
Policy → Environment	0.2363	5.0714	*
Policy → EP	0.9819	450.3113	***
EP → AP	0.9639	7.5819	**
EP → AMDL	0.9900	7.9282	**
EP → AMIAT	1.1889	13.3998	**
EP → Environment	−0.1449	1.1388	–

Note: (1) Economic and population factors (EP), agricultural production (AP), the agricultural mechanization development level (AMDL), and the agricultural machinery industry and agricultural technology (AMIAT). (2) *** indicates a significance level of 1%, ** indicates a significance level of 5%, and * indicates a significance level of 10%, – indicates failed *t*-test. (3) STERR indicates standard error.

Table A4. Factor load of observed variables in measurement Model B.

Latent Variable	Observation Variables	Factor Load
EP	s1	0.9811
	s2	0.9261
	s9	0.9353
	s10	0.9212
	s11	−0.7265
	s20	0.9929
	s21	0.9738
	s22	0.9601
	s3	−0.9490
	s4	0.7834
AP	s5	0.9734
	s23	0.9943
	s24	0.9733
AMIAT	s7	0.9625
	s8	0.9821
	s6	0.9834
AMDL	s12	−0.7543
	s13	0.9705
	s14	0.9516
	s15	0.9864
	s16	0.9957
	s17	0.9895
	s18	0.9923
	s19	0.7627
	s28	−0.6874
Environment	s25	0.9884
	s26	0.9885
Policy	s27	1

Note: Economic and population factors (EP), agricultural production (AP), the agricultural mechanization development level (AMDL), and the agricultural machinery industry and agricultural technology (AMIAT).

References

1. Sims, B.; Kienzle, J.; Hilmi, M. *Agricultural Mechanization a Key Input for Sub-Saharan African Smallholders*; FAO: Rome, Italy, 2016.
2. FAO. Sustainable Agricultural Mechanization. Available online: http://www.fao.org/sustainableagricultural-mechanization (accessed on 13 February 2022).
3. Sims, B.; Kienzle, J. Sustainable Agricultural Mechanization for Smallholders: What Is It and How Can We Implement It? *Agriculture* **2017**, *7*, 50. [CrossRef]
4. Kobrich, C.; Rehman, T. Sustainable agriculture and the MCDM paradigm: The development of compromise programming models with special reference to small-scale farmers in Chile's VIth region. In Proceedings of the First International Conference on Multiple Objective Decision Support Systems (MODSS) for Land, Water and Environmental Management: Concepts, Approaches and Applications, Honolulu, HI, USA, September 1996; pp. 557–569.
5. FAO. *Agri-Hire in Sub-Saharan Africa Business Models for Investing in Sustainable Mechanization*; FAO: Rome, Italy, 2021.
6. Ministry of Agriculture and Rural Affairs, P.R.C. National Agricultural Mechanization Development Plan in the 14th Five-Year Plan. Available online: http://www.moa.gov.cn/govpublic/NYJXHGLS/202201/t20220105_6386316.htm (accessed on 13 February 2022).
7. Yang, M. Discussion on the development of Agricultural Mechanization in China under the new normal. *Agric. Mach. Qual. Superv.* **2015**, *3*, 4–9.
8. Gowdy, J.; Baveye, P. Chapter 27-An Evolutionary Perspective on Industrial and Sustainable Agriculture. In *Agroecosystem Diversity*; Lemaire, G., Carvalho, P.C.D.F., Kronberg, S., Recous, S., Eds.; Elsevier: Amsterdam, The Netherlands, 2019; pp. 425–433.
9. Mrema, G.; Kienzle, J.; Mpagalile, J. Current status and future prospects of agricultural mechanization in sub-saharan Africa (SSA). *Agric. Mech. Asia Afr. Lat. Am.* **2018**, *49*, 13–30.
10. Yu, J.; Wu, J. The sustainability of agricultural development in China: The agriculture–environment nexus. *Sustainability* **2018**, *10*, 1776. [CrossRef]
11. Jiang, M.; Hu, X.; Chunga, J.; Lin, Z.; Fei, R. Does the popularization of agricultural mechanization improve energy-environment performance in China's agricultural sector? *J. Clean. Prod.* **2020**, *276*, 124210. [CrossRef]
12. Kormawa, P.; Mrema, G.; Mhlanga, N.; Fynn, M.; Kienzle, J.; Mpagalile, J. *Sustainable Agricultural Mechanization: A Framework for Africa*; FAO: Addis Ababa, Ethiopia; AUC: Addis Ababa, Ethiopia, 2018.
13. Daum, T.; Adegbola, Y.P.; Kamau, G.; Daudu, C.; Zossou, R.C.; Crinot, G.F.; Houssou, P.; MOSES, L.; Ndirpaya, Y.; Wahab, A. Impacts of agricultural mechanization: Evidence from four African countries. *Hohenh. Work. Pap. Soc. Inst. Chang. Agric. Dev.* **2020**. [CrossRef]
14. Li, W.; Wei, X.; Zhu, R.; Guo, K. Study on Factors Affecting the Agricultural Mechanization Level in China Based on Structural Equation Modeling. *Sustainability* **2019**, *11*, 51. [CrossRef]
15. Van Loon, J.; Woltering, L.; Krupnik, T.J.; Baudron, F.; Boa, M.; Govaerts, B. Scaling agricultural mechanization services in smallholder farming systems: Case studies from sub-Saharan Africa, South Asia, and Latin America. *Agric. Syst.* **2020**, *180*, 102792. [CrossRef]
16. Baudron, F.; Misiko, M.; Getnet, B.; Nazare, R.; Sariah, J.; Kaumbutho, P. A farm-level assessment of labor and mechanization in Eastern and Southern Africa. *Agron. Sustain. Dev.* **2019**, *39*, 17. [CrossRef]
17. Devkota, R.; Pant, L.P.; Gartaula, H.N.; Patel, K.; Gauchan, D.; Hambly-Odame, H.; Thapa, B.; Raizada, M.N. Responsible Agricultural Mechanization Innovation for the Sustainable Development of Nepal's Hillside Farming System. *Sustainability* **2020**, *12*, 374. [CrossRef]
18. Chen, Y.E.; Chen, W. A Study on the Relationship among Agricultural Mechanization, Industrial Upgrading and Agricultural Carbon Emission—The Empirical Research Based on Dynamic Panel Data Model. *J. Agrotech. Econ.* **2018**, *5*, 122–133.
19. Cui, H.; Zhao, T.; Shi, H. STIRPAT-Based Driving Factor Decomposition Analysis of Agricultural Carbon Emissions in Hebei, China. *Pol. J. Environ. Stud.* **2018**, *27*, 1449–1461. [CrossRef]
20. Cui, Y.; Khan, S.U.; Deng, Y.; Zhao, M.; Hou, M. Environmental improvement value of agricultural carbon reduction and its spatiotemporal dynamic evolution: Evidence from China. *Sci. Total Environ.* **2021**, *754*, 142170. [CrossRef]
21. Chi, Y.; Zhou, W.; Wang, Z.; Hu, Y.; Han, X. The Influence Paths of Agricultural Mechanization on Green Agricultural Development. *Sustainability* **2021**, *13*, 12984. [CrossRef]
22. Guo, X.; Huang, S.; Wang, Y. Influence of Agricultural Mechanization Development on Agricultural Green Transformation in Western China, Based on the ML Index and Spatial Panel Model. *Math. Probl. Eng.* **2020**, *2020*, 6351802. [CrossRef]
23. Liu, D.; Zhu, X.; Wang, Y. China's agricultural green total factor productivity based on carbon emission: An analysis of evolution trend and influencing factors. *J. Clean. Prod.* **2021**, *278*, 123692. [CrossRef]
24. Li, W.; Zhang, P. Relationship and integrated development of low-carbon economy, food safety, and agricultural mechanization. *Environ. Sci. Pollut. Res.* **2021**, *28*, 68679–68689. [CrossRef]
25. Bai, R. Thinking on the development of agricultural mechanization during the '13th Five-year' in China. *J. Chin. Agric. Mech.* **2014**, *35*, 1–5.
26. Lu, B.; Zhang, Z.; Zhu, M.; Chen, Z.; Lv, X. Discrimination and analysis of key influencing factors for agricultural mechanization development. *Trans. Chin. Soc. Agric. Eng. (Trans. CSAE)* **2008**, *24*, 114–117.
27. Yan, Z.; Qi, Y.; Liao, J. Comparative analysis of evaluation methods of agricultural mechanization level in China. *J. Agric. Mech. Res.* **2018**, *10*, 1–5.

28. Daum, T.; Birner, R. Agricultural mechanization in Africa: Myths, realities and an emerging research agenda. *Glob. Food Secur.* **2020**, *26*, 100393. [CrossRef]
29. Kirui, O.K. The Agricultural mechanization in Africa: Micro-level analysis of state drivers and effects. *ZEF-Discuss. Pap. Dev. Policy* **2019**, *60*. [CrossRef]
30. Cossar, F. *Boserupian Pressure and Agricultural Mechanization in Modern Ghana*; Intl Food Policy Res Inst: Washington, DC, USA, 2016; Volume 1528. Available online: https://ssrn.com/abstract=2778742 (accessed on 20 January 2022).
31. Berhane, G.; Dereje, M.; Minten, B.; Tamru, S. *The Rapid–But from a Low Base–Uptake of Agricultural Mechanization in Ethiopia: Patterns, Implications and Challenges*; Intl Food Policy Res Inst: Washington, DC, USA, 2017; Volume 105. [CrossRef]
32. Saliou, I.O.; Zannou, A.; Aoudji, A.K.; Honlonkou, A.N. Drivers of Mechanization in Cotton Production in Benin, West Africa. *Agriculture* **2020**, *10*, 549. [CrossRef]
33. Mottaleb, K.A.; Krupnik, T.J.; Erenstein, O. Factors associated with small-scale agricultural machinery adoption in Bangladesh: Census findings. *J. Rural. Stud.* **2016**, *46*, 155–168. [CrossRef] [PubMed]
34. Ghosh, B.K. Determinants of Farm Mechanisation in Modern Agriculture: A Case Study of Burdwan Districts of West Bengal. *Int. J. Agric. Res.* **2010**, *5*, 1107–1115. [CrossRef]
35. Aryal, J.P.; Sofina, M.; Olaf, E. Understanding factors associated with agricultural mechanization: A Bangladesh case. *World Dev. Perspect.* **2019**, *13*, 1–9. [CrossRef]
36. Aryal, J.P.; Thapa, G.; Simtowe, F. Mechanisation of small-scale farms in South Asia: Empirical evidence derived from farm households survey. *Technol. Soc.* **2021**, *65*, 101591. [CrossRef]
37. Belton, B.; Win, M.T.; Zhang, X.; Filipski, M. The rapid rise of agricultural mechanization in Myanmar. *Food Policy* **2021**, *101*, 102095. [CrossRef]
38. Li, Z.; Zhu, M.; Huang, H. Evaluating the sustainable development of agricultural mechanization in Hubei Province based on fuzzy DEMATEL and ISM combined method. *Trans. Chin. Soc. Agric. Eng.* **2022**, *38*, 51–58.
39. Mamat, T.; Ding, W.; Kasim, E.; Hassan, M. The sustainable development of agricultural mechanization based on combination weighting and ahp. *Chin. J. Agric. Resour. Reg. Plan.* **2018**, *39*, 67–73.
40. WOLD, S. Modeling data tables by principal components and PLS: Class pattern and quantitative predictive relations. *Cemomet. Intell. Lab. Syst.* **1989**, *7*, 53–65. [CrossRef]
41. Fornell, C.; Larcker, D.F. Evaluating structural equation models with unobservable variables and measurement error. *J. Mark. Res.* **1981**, *18*, 39–50. [CrossRef]
42. Ringle, C.M.; Christian, M.; Wende, S.; Becker, J.M. *SmartPLS 3*; SmartPLS GmbH: Boenningstedt, Germany, 2015. Available online: http://www.smartpls.com (accessed on 13 February 2022).
43. Hair, J.F. *Multivariate Data Analysis*; Prentice Hall: Englewood Cliffs, NJ, USA, 2011.
44. Tian, X.; Li, W.; Li, R. The Environmental Effects of Agricultural Mechanization: Evidence from Agricultural Machinery Purchase Subsidy Policy. *Chin. Rural. Econ.* **2021**, *9*, 95–109.
45. Zhou, Z.; Kong, X. Agricultural Machines Subsidy Policy and the Grain Output of China: DID Empirical Analysis Based on National County-Level Data. *China Soft Sci.* **2019**, *4*, 20–32.

sustainability MDPI

Article

The Impact of Government Subsidies on Technological Innovation in Agribusiness: The Case for China

Liping Wu [1], Kai Hu [1], Oleksii Lyulyov [2,3], Tetyana Pimonenko [2,3,*] and Ishfaq Hamid [4]

1 College of Economics and Management, Jiangxi Agricultural University, Nanchang 310045, China
2 Department of Management, Faculty of Applied Sciences, WSB University, 41-300 Dabrowa Gornicza, Poland
3 Department of Marketing, Sumy State University, 40007 Sumy, Ukraine
4 MICA, The School of Ideas, Ahmedabad 380058, India
* Correspondence: tetyana_pimonenko@econ.sumdu.edu.ua

Abstract: With the implementation of the rural revitalization strategy and the promotion of agricultural and rural modernization, the subsidies enjoyed by agricultural enterprises in China are increasing. As a result, the effectiveness of government subsidies for the technological innovation of agricultural enterprises has attracted more and more attention. Based on the perspectives of the whole industry chain of agriculture, forestry, animal husbandry, fisheries, and of processing, manufacturing, circulation, and service, this paper takes the listed agricultural companies from 2007 to 2019 as a research sample and empirically tests the effects and mechanisms of government subsidies on the technological innovation of agricultural enterprises. The study applies the fixed effect and intermediary effect models. The findings show that government subsidies potentially encourage agricultural enterprises to grow more successfully. Moreover, R&D expenditure is essential for enterprise technological innovation and leads to an intermediate impact. At the same time, government subsidies for the technological innovation of agricultural enterprises have a certain heterogeneity between different industries, state-owned enterprises and non-state-owned enterprises, and large enterprises and small and medium-sized enterprises. Therefore, this study argues that the government should continue to raise subsidies. In addition, the subsidies should be "different from enterprise to enterprise", and government subsidy funds should be better supervised to foster agricultural technological innovation properly.

Keywords: industry chain; government grants; technological innovation in agricultural enterprises; R&D investment

Citation: Wu, L.; Hu, K.; Lyulyov, O.; Pimonenko, T.; Hamid, I. The Impact of Government Subsidies on Technological Innovation in Agribusiness: The Case for China. *Sustainability* **2022**, *14*, 14003. https://doi.org/10.3390/su142114003

Academic Editors: Aaron K. Hoshide and Francesco Caputo

Received: 15 September 2022
Accepted: 24 October 2022
Published: 27 October 2022

1. Introduction

The Chinese government accepted the agricultural and rural modernization plan during the 14th Five-Year Plan period (2021–2025) [1]. In consideration of this, innovation was outlined as the core force for agricultural and rural modernization. The innovations in agricultural development are directed at improving the production of agricultural goods. The innovations in rural development allow the improvement of the production of agricultural goods and the education, health, and social infrastructure of rural areas.

Therefore, agricultural enterprises face several types of risks, such as environmental risks and operations risks [2–4]. In addition, agricultural enterprises face the issue of a lack of financing for the implementation of innovations [5–11]. Consequently, this limits the development of agricultural enterprises. Government subsidies in the form of financial aid have been implemented for a long time in China to modernize agricultural and rural development. In this case, government subsidies for agriculture and rural development may be defined as investments [12–18]. Past studies [14,19–21] outline that government subsidies could guide and motivate enterprises to increase R&D investment to implement technological innovation activities. At the same time, the inefficiency of government subsidies could be caused by the adverse selection of the innovation activities of enterprises for subsidies [14]. Adverse selection results in asymmetric information on available options for government subsidies.

Consequently, it could provoke inequalities and gaps in a company's innovation development and cause a decline in their long-term competitiveness [15]. Past research [16] has proven that information asymmetry between the government and enterprises causes subsidies to have a reverse effect. This could limit the achievement of indicated goals in the plan for agricultural and rural modernization during the 14th Five-Year Plan period (2021–2025) [1]. Thus, it is justifiable to analyze how government subsidies affect the technological innovation of agricultural enterprises and their mechanisms of action. It should be noted that in the ongoing economic open system theory [22], the development of all sectors, including agriculture and rural development, should be analyzed in connection with each other. Thus, agriculture is increasingly closely linked to the secondary and tertiary industries and fails to scientifically reflect the value of the whole industrial chain, such as production, processing, circulation, the service of agriculture, forestry, animal husbandry, and fisheries.

This paper focuses on analyzing the impact of government subsidies on agricultural enterprises' technological innovation from the whole industry chain perspective. Such samples allow the modelling of agricultural enterprises' whole and individual behavior. In addition, they allocate and measure the statistical effects that could not be determined based on the data of the individual enterprises. Regarding the standard of the National Bureau of Statistics' "Statistical Classification of Agriculture and Related Industries (2020)" (Order No. 32 of the National Bureau of Statistics) [23], agricultural enterprises are defined as all economic activities formed in the production, processing, manufacturing, service and other links of agriculture, forestry, animal husbandry, and fisheries, as well as relevant enterprises in the secondary and tertiary industries.

Our research aims to fill the following scientific gaps: (1) to develop a methodology to check the link between government subsidies and the technological innovation of agricultural enterprises; (2) to analyze agriculture from the whole industrial chain, and extend the scope of agricultural enterprises to agriculture, forestry, animal husbandry, fisheries production, processing, manufacturing, circulation, service, and other industries; and (3) to develop a methodology to check whether research and development could extend the innovation among agricultural enterprises. The remainder of this paper is divided into the following sections: Section 2 presents empirical evidence from the literature; Section 3 discusses the methodology and data; Section 4 analyzes the findings; and Section 5 considers conclusions and policy implications.

2. Literature Review

2.1. The Relationship between Government Subsidies and Technological Innovation in Agricultural Enterprises

Past research shows that there has been no consensus on the effect of government subsidies on enterprise technology innovation. A few main views constitute these findings. Government subsidies could incentivize enterprises to innovate technologically [3,22–41]. The late economist Kenneth Arrow [24] suggested that the technological innovation of enterprises has a spillover effect. Moreover, the free-riding behavior of other enterprises has seriously hit the enthusiasm of enterprises for independent innovation. This has provoked an insufficient supply of technological innovation. Subsequent research [25–27] has confirmed that government subsidies positively impact companies' performances. One of these studies [25] analyzed 158 listed energy companies in China. In this case, government subsidies for technological innovations negatively impacted company performance in the short term. At the same time, a positive effect was shown in the long term. Other researchers [26] demonstrated that Chinese government subsidies stimulate innovations in environmental management. However, these types of subsidies did not encourage the rapid growth of technological innovations. It should be noted that carbon-free technological innovation could enhance the performance of companies [27]. However, the effect could be different depending on the time and efficiency of management.

Government subsidies can directly provide financial support to agricultural enterprises. As part of the profits of enterprises, government subsidies directly increase enterprises' funds, alleviate the shortage of funds available to agricultural enterprises, improve their enthusiasm to innovate, and solve the spillovers of innovative results. They also reduce the risk caused by the uncertainty of innovation and encourage enterprises to increase R&D investment in technological innovation [28,33]. Secondly, government subsidies send a positive signal of government recognition and reduce the information asymmetry between enterprises and external investors. An enterprise that enjoys this subsidy shows that the government recognizes their development. This proves that the enterprise has strong R&D innovation ability, good innovation projects, and is more willing and capable of technological innovation [3,28–33]. At the same time, scholars [32,33] confirm that government subsidies should be implemented at all levels, from companies to individuals. In this case, government subsidies could positively impact agriculture.

Government subsidies can improve the ability of enterprises to access resources. They can also improve the ability of enterprises to obtain resources by supplementing their innovation resources and enhancing their recruitment of talented workers. Agribusinesses receiving government subsidies send positive signals of good relations with the government, indicating they have sufficient government resources. The government provides an invisible guarantee for agricultural enterprises to make up for the natural weakness of agriculture and attracts banks, venture capitalists, etc., to increase investment. Furthermore, it also increases the attractiveness of enterprises to prospective employees, which improves the overall level of Research and Development (R&D) personnel, and enhances the technological innovation capabilities of enterprises [34].

Past research [35] on strategic emerging industry enterprises found that the impact of financial incentive policies on innovation conforms to an inverted U-shape. In this case, the scholars confirmed that government subsidies stimulate innovation to a certain point, after which efficiency declines. Thus, the government should control monitoring systems for government subsidies. Other research [36] found no significant positive impact of government subsidies on private R&D for small- and medium-sized firms. In summary, agricultural government subsidies increase funds for production and investment, release positive signals to attract more external financing and outstanding human capital, improve the ability of companies to obtain resources, and promote technological innovation for agricultural producers. Thus, we propose our first research hypothesis:

Hypothesis 1 (H1). *Government subsidies can promote technological innovation in agribusiness.*

2.2. Mechanisms of Government Subsidies for Technological Innovation in Agricultural Enterprises

Enterprises with effective Research and Development (R&D) generate technical knowledge, have certain externalities, are easily learned or reproduced, and suffer from market failure. At the same time, the investment, risk, and uncertainty of these activities provokes issues for enterprises, especially agricultural enterprises, in obtaining funds from the capital markets. Nevertheless, based on the importance of R&D and the solutions to market failures, the government should promote enterprises to carry out such innovation [42].

Based on data from Chinese companies, past research [42–44] finds that government subsidies have an incentive effect on companies' R&D activities. Government subsidies can support agricultural enterprises in increasing investment in three ways: by reducing the cost of R&D, reducing the uncertainty of these types of projects, and dispersing subsequent risks. Thus, government subsidies reduce R&D costs. According to the theory of externalities, the externalities of R&D activities lead to the spillover of knowledge, which to a certain extent discourages the enthusiasm of enterprises involved in such research. Government subsidies, as part of corporate profits, reduce the marginal cost of enterprise R&D and then stimulate agricultural enterprises to increase investment. Furthermore, government subsidies reduce uncertainty about projects [45]. This increases the market demand for a project's results and improves the expected return of the enterprise [45]. At the same time, it can also attract more qualified personnel to participate in projects, reducing their uncertainty. Finally, government subsidies can diversify R&D risks. The government provides subsidies and shares information about the project, which can attract external investors and incentivize them to join, reducing the risk of failure for enterprises to a certain extent [46,47].

According to the new economic growth theory, human capital and investment are important factors in promoting economic growth and technological innovation [42]. Enterprises, through R&D activities, improve the stock of human capital and promote enterprise innovation [41]. R&D activities are the most direct source of technological innovation. Enterprises increase investment in activities, generate new knowledge and information, and directly promote technological innovation. Furthermore, this increase in investment enables enterprises to use existing external knowledge better, enhance their knowledge stock, and indirectly promote their innovation capabilities [46]. Thus, past studies [48,49] emphasize that providing an effective R&D policy allows the development of additional advantages. This could be due to the implementation of transborder strategies on knowledge sharing, geographical changes in research developments and innovations, and the international fragmentation of research activities. It has been demonstrated that competitiveness depends on innovative activities [50]. At the same time, lack of labor and financial resources are the biggest limitations to investing in R&D.

Therefore, an increase in investment in R&D can promote technological innovation. Thus, it can be concluded that government subsidies encourage agricultural enterprises to increase investments by reducing the cost of R&D and project uncertainty, as well as helping to disperse production risk. Therefore, we propose our second hypothesis:

Hypothesis 2 (H2). *Government subsidies encourage both investment and technological innovation.*

3. Materials and Methods

3.1. Sample Selection and Data Sources

Taking the A-share (representing publicly listed Chinese companies that trade on Chinese stock exchanges, such as the Shenzhen and Shanghai Stock Exchanges) of listed agricultural companies from 2007 to 2019 as a research sample, this paper no longer limits agriculture to traditional agriculture, forestry, animal husbandry, and fisheries. Instead, it extends it to the perspective of the whole industry chain to the production, processing, manufacturing, circulation, and service of agriculture, forestry, livestock, and fisheries. Drawing from the practice of [44] and referring to the standards of the Statistical Classification of Agriculture and Related Industries (2020) (Order No. 32 of the National Bureau of Statistics) issued by the National Bureau of Statistics and the Guidelines for the Classification of Listed Companies (Revised in 2012) issued by the China Securities Regulatory Commission, agriculture-related industries include agriculture, forestry (A02), animal husbandry (A01 and A03), fisheries (A04), and services related to these natural resource-based industries (A05). Processing and manufacturing in these industries includes food processing (C13) and the manufacture of food (C14), fertilizers and pesticides (C26), and agricultural machinery (C35). Listed agricultural companies are involved in agriculture, forestry, animal husbandry, and fisheries, as well as enterprises in the secondary and tertiary input sectors whose products are essential for firms within these natural resource-based industries. We narrowed down our sample to 177 listed agricultural companies from 194 after removing 17 companies with serious financial risks. Our non-balanced panel data consisted of 2301 enterprises from these 177 companies. The enterprise patent data used in this article come from the China Research Data Service Platform (CNRDS database) [51]. Some of the missing data were provided by searching on the patent website of the State Intellectual Property Office [52]. The screening of listed agricultural companies was mainly based on analyzing enterprises' main business scopes, such as Hexun Network [53] and

Flush Database [54]. The data for the other variables were collected from the CSMAR database [55].

3.2. Variable Settings

Past studies [45,46] demonstrate that patent applications are one of the incentives for developing and implementing technological innovation at companies. In addition, considering the analytical report of World Intellectual Property Indicators 2021 [56], patents guarantee the authorship protection of innovation. Furthermore, patents allow the obtainment of additional revenue for agricultural companies. Considering this, our research used the patent applications of enterprises as the measure of technological innovation ($Patent_{t+1}$). Considering the time lag of technological innovation, the technology innovation level of t + 1 was measured by adding 1 logarithm to the number of patent applications in the t-period based on the methods outlined in [45,46]. The t-period starts with a value of 0 zero.

Government grants were the explanatory variable we evaluated. There are large differences in the amount of government subsidies distributed based on the size of a natural resource-based enterprise. In order to narrow the absolute difference between the data, the logarithm of the government subsidies received by the company in the current year was taken to measure the explanatory variables. Based on other scholars' work on enterprise technological innovation, our research used six control variables that may affect the technological innovation of agricultural enterprises, such as enterprise size, age, asset–liability ratio, growth potential, proportion of fixed assets, concentration of equity, and salary incentives (Table 1). In order to analyze the impact mechanism of government subsidies on technological innovation, we defined investment as an intermediary variable using the logarithm of the company's investment in the current year.

Table 1. Description of the variables and the calculation formula.

Variable	Symbol	Variable Name	Computational Formula
Explained variable	$Patent_{t+1}$	Number of patent applications	ln(1 + t Number of patent applications)
Explanatory variable	SUB	Governmental subsidy	ln(Government subsidy amount)
Control variables	Size	Enterprise scale	The natural logarithm of the company's total market value
	Age	Enterprise age	Sample year minus the year of company establishment
	Debt	Asset–liability ratio	End Liabilities/End Total Assets
	Growth	Growth ability	Increase the rate of business revenue
	Fixasset	The proportion of fixed assets	Net fixed assets/ending total assets
	Share	Equity concentration	The shareholding of the largest shareholder
	Salary	Compensation incentive	ln(Total annual salary of ending directors, supervisors, and senior executives)
Mediating variables	R&D	Research input	The company's R&D investment was logarithmic

3.3. Model Settings

In order to analyze the impact of government subsidies on the technological innovation of agricultural enterprises, we used a basic econometric model specified as:

$$\text{Patent}_{\text{it}+1} = \alpha_0 + \alpha_1\text{SUB}_{\text{it}} + \beta\text{CV}_{\text{it}} + \sum\text{Year} + \sum\text{Ind} + \varepsilon_{\text{it}} \quad (1)$$

where $\text{Patent}_{\text{it}+1}$—technological innovation in the Company$'\text{s}_{\text{t}+1}$ period; α_0—denotes the constant term; SUB_{it}—the government subsidy of the company's t period; CV_{it}—the control variable matrix; ε_{it}—the residual term; i and t—the enterprises and years; and Year and Ind—the fixed effect of the year and industry, respectively.

The two-way fixed-effect model [57] is applied to decrease the impact of the macroeconomic environment and the nature of the industry. However, R&D investment is introduced as the intermediary variable to identify the mechanisms of government subsidies for the technological innovation of agricultural enterprises. Therefore, the following Ordinary Least Square (OLS) econometric models are set up based on model (1) r using methods from [58,59] in order to analyze the intermediary effect of R&D investment.

$$\text{RD}_{\text{it}} = \alpha_0 + \alpha_1\text{SUB}_{\text{it}} + \beta\text{CV}_{\text{it}} + \sum\text{Year} + \sum\text{Ind} + \varepsilon_{\text{it}} \quad (2)$$

$$\text{Patent}_{\text{it}+1} = \alpha_0 + \alpha_1\text{SUB}_{\text{it}} + \alpha_2\text{R\&D}_{\text{it}} + \beta\text{CV}_{\text{it}} + \sum\text{Year} + \sum\text{Ind} + \varepsilon_{\text{it}} \quad (3)$$

where RD_{it} equals the R&D for company i during time period t; SUB_{it} is the government subsidy of the company's t period; and CV_{it} is the control variable matrix with ε_{it} as residual error of the model. Year and Ind are the fixed effect of the year and industry, respectively.

4. Results

4.1. Descriptive Statistics and the Correlation Analysis

The descriptive statistical results of the variables signify that the average number of patent applications is 1.6146, the median is 0.6931, and the maximum and minimum values are 7.3671 and 0 with a standard deviation of 1.3992 (Table 2). Thus, the vast majority of listed agricultural companies have technological innovations but vary greatly. In addition, the average value of government subsidies is 16.3633, the median is 16.4341, the maximum and minimum values are 20.7799 and 8.9227, respectively, and the standard deviation is 1.5357. This suggests that the government subsidies enjoyed by listed agricultural companies are more balanced, but specific differences exist.

Table 2. Descriptive statistical results of the variables.

Variable	Obs	Mean	Standard	Minimum	Median	Maximum
Patent	2301	1.6146	1.3992	0	0.6931	7.3671
SUB	1622	16.3633	1.5357	8.9227	16.4341	20.7799
Size	1680	22.3731	0.9348	19.1148	22.2638	26.3942
Age	2266	14.3350	5.9605	1	14	35
Debt	1697	0.4297	0.2184	0.0084	0.4129	2.0498
Growth	1601	1.2077	37.4913	−0.9913	0.1052	1497.1560
Fixasset	1697	0.3104	0.1619	0.0040	0.2882	0.8491
Share	1697	35.1022	14.2938	4.0800	33.7400	95.9500
Salary	1694	15.0283	0.8552	11.6082	15.0366	17.9634
R&D	1213	16.9112	1.7823	9.6347	17.0643	21.4612

The correlation analysis of the variables is shown in Table 3. Thus, the correlation coefficient between the current government subsidy (SUB) and the next phase of patent applications is 0.423 at the 1% level of significance. The correlation coefficient between the SUB and the intermediary variable for R&D input is significant with a value of 0.384. The correlation coefficient (r) denoting a positive association between R&D and the next phase of patent applications is 0.574, which is also significant at the 1% confidence level. Among the control variable, enterprise size and age are significant and positively correlate with the number of next patent applications.

Table 3. Variable correlation analysis.

Variable	$Patent_{t+1}$	SUB	Size	Age	Debt	Growth	Fixasset	Share	Salary	R&D
$Patent_{t+1}$	1.000									
SUB	0.423 ***	1.000								
Size	0.450 ***	0.441 ***	1.000							
Age	0.327 ***	0.136 ***	0.161 ***	1.000						
Debt	0.115 ***	0.211 ***	−0.053 **	0.012	1.000					
Growth	−0.034	−0.011	−0.031	−0.005	0.000	1.000				
Fixasset	0.028	0.108 ***	−0.004	−0.039	0.121 ***	−0.036	1.000			
Share	−0.112 ***	0.026	0.094 ***	−0.127 ***	−0.110 ***	0.037	0.072 ***	1.000		
Salary	0.534 ***	0.420 ***	0.580 ***	0.252 ***	−0.067 ***	−0.025	0.003	−0.071 ***	1.00	
R&D	0.574 ***	0.384 ***	0.484 ***	0.089 ***	0.039	−0.002	0.007	0.013	0.506 ***	1.000

Note: **, and *** are significant at the 10%, 5%, and 1% levels.

However, equity concentration is significant and negatively correlated (−0.112) with the number of next patent applications. Executive compensation correlates significantly with the number of next patent applications at the 1% significance level with a positive r equal to 0.534. There is a significant correlation between the main variables and further multiple regressions. The absolute value of the correlation coefficient between the main variables is less than 0.5, indicating no limited multicollinearity. Multicollinearity or high degrees of association ($r > 0.7$) between independent variables is problematic since the OLS regression model assumes "independent" impacts of independent variables specified in the model on the dependent variable. Multicollinearity distorts the parameter estimates in the OLS model rendering inferences gleaned from the model results potentially inaccurate.

4.2. Regression Analysis Results

The regression results from empirical tests on the impact of government subsidies on technological innovation in agribusiness using model (1) are shown in Table 4. After the number of patent applications in the current period plus one to take the logarithm and lag one period as the explanatory variable, the enterprise-level variables and the annual and industry fixed effects are gradually controlled. Additionally, the regression coefficient of government subsidies is significantly positive at the 1% confidence level. The findings from column (4) of Table 4 suggest that under the two-way fixed effect of control years and industries, the regression coefficient of SUB is 0.221. The change in government subsidies in the current period is 1%, and the average change in the number of patent applications of enterprises in the next year is 0.221%. This implies that government subsidies promote agricultural innovation, which validates our first hypothesis. Among the other control variables, the regression coefficients of enterprise size, asset–liability ratio, and executive compensation are significantly positive. This indicates that growth in scale results in an increasing level of debt. Furthermore, increases in executive compensation are conducive to increasing patent applications and technological innovation. The regression coefficients of enterprise age and equity concentration are significantly negative. This suggests that the longer the company is established, the higher the equity concentration, the fewer the number of patent applications, and the lower the level of technological innovation.

Table 4. Return results of the impact of government subsidies on technological innovation in agricultural enterprises.

Variable	(1)	(2)	(3)	(4)
	$Patent_{t+1}$	$Patent_{t+1}$	$Patent_{t+1}$	$Patent_{t+1}$
SUB	0.464 *** (17.03)	0.209 *** (7.92)	0.201 *** (7.60)	0.221 *** (8.33)
Size		0.299 *** (6.32)	0.352 *** (6.97)	0.402 *** (8.20)
Age		0.011 * (1.70)	−0.006 (−0.72)	−0.016 * (−1.88)
Debt		0.346 * (1.96)	0.459 ** (2.51)	0.306 * (1.70)
Growth		−0.001 *** (−5.70)	−0.001 *** (−3.60)	0.000 (1.47)
Share		−0.009 *** (−3.41)	−0.010 *** (−3.87)	−0.008 *** (−3.13)
Fixasset		−0.215 (−1.00)	−0.096 (−0.45)	−0.332 (−1.34)
Salary		0.664 *** (12.15)	0.576 *** (10.14)	0.431 *** (7.59)
_cons	−5.768 *** (−13.10)	−18.187 *** (−22.77)	−17.769 *** (−21.27)	−17.733 *** (−23.15)
Year	No	No	Yes	Yes
Industry	No	No	No	Yes
N	1549	1460	1460	1460
R^2	0.179	0.358	0.373	0.452

Note: *, **, and *** are significant at the 10%, 5%, and 1% levels.

4.3. Analysis of the Intermediary Affect Test Results

Empirical testing has verified that government subsidies can promote technological innovation in agribusiness. According to the previous analysis, government subsidies may affect the technological innovation of enterprises by influencing their R&D investment. According to [58], the empirical test is carried out through models (1) and (3), and whether the R&D investment plays an intermediary role according to the regression coefficient and significance level of government subsidies and R&D investment.

Column (1) of Table 5 shows the regression results of model (1). The regression coefficient of government grants is 0.221, which is significant at the 1% confidence level. This implies that the basic variable government grant significantly positively affects the number of patent applications for the interpreted variable. Column (2) shows the regression results of model (2), and the regression coefficient of government subsidy is 0.201, which is also significant at the 1% level. Thus, government subsidies appear to have a significant impact on investment in R&D.

Column (3) in Table 5 summarizes the regression results for model (3). The regression coefficient of government subsidy after adding the intermediary variable R&D investment is still significant, but the coefficient drops from 0.221 to 0.212. This indicates that the positive effect of government subsidies on the number of patent applications is partially absorbed by the R&D investment of the intermediary variable. Thus, R&D investment plays a part in the intermediary effect. The proportion of the intermediary effect to the total effect is 27.56%. Moreover, the government subsidy acts on the level of technological innovation of the enterprise by influencing such investment of the enterprise. Therefore, our second hypothesis is also validated.

Table 5. Test of the intermediary effect of government subsidies affecting the technological innovation in agricultural enterprises.

Variable	(1)	(2)	(3)
	Patent$_{t+1}$	**R&D**	**Patent$_{t+1}$**
SUB	0.221 *** (8.33)	0.201 *** (4.90)	0.212 *** (7.05)
R&D			0.304 *** (11.98)
_cons	Yes	Yes	Yes
Year	Yes	Yes	Yes
Industry	Yes	Yes	Yes
N	1460	1112	1099
R^2	0.452	0.428	0.548

Note: *** is significant at the 10%, 5%, and 1% levels.

4.4. Analysis of Heterogeneity

In order to investigate the heterogeneity of the samples, this paper conducts empirical tests according to the industry, the nature of the enterprise, and the size of the enterprise. Our research analyzes the production, processing, manufacturing, circulation, and service of agriculture, forestry, animal husbandry, and fisheries from the perspective of the whole industrial chain. The nature of the enterprise is according to whether the actual controller of the enterprise is a government department at all levels. If so, it is a state-owned enterprise; otherwise, it is a non-state-owned enterprise. The size of enterprise is divided into large, small, and medium-sized enterprises. The core criteria are the operating income of the enterprise in the current year. If it exceeds RMB200 million, it is a large enterprise; otherwise, it is a small or medium-sized enterprise.

The group regression results (Table 6) show that from the perspective of the industry, the regression coefficient between the government subsidies for the processing of agriculture, forestry, animal husbandry, and fishery products and the manufacturing industry, the number of manufacturing materials in the manufacturing industry, and the number of patent applications in the next period is significantly positive. At the same time, the regression coefficient between the government subsidies for traditional agriculture, forestry, animal husbandry, and fisheries and the number of patent applications in the next period is not significant. Government subsidies for these natural resource-based industries promote technological innovation by these businesses. At the same time, government subsidies for traditional agriculture, forestry, animal husbandry, and fisheries do not significantly affect enterprises' technological innovation. The reason for this may be that agriculture, forestry, animal husbandry, and fisheries are more susceptible to fluctuations in natural factors and market factors. Therefore, despite government subsidies, these subsidies have not substantially improved enterprises' R&D conditions, and their R&D power is insufficient.

4.5. Robustness Test

In order to test the robustness of the results, we used the number of patent grants instead of the number of patent applications as the agent variable of technological innovation. The regression results (Table 7) show that the regression coefficient of the SUB is significantly positive at the 1% level, which is consistent with the results in Table 4. This confirms that the regression results of Table 4 are stable. The conclusions of this study have passed the empirical test, have strong explanatory power, and can be used to guide and encourage technological innovation in agricultural enterprises.

Table 6. Group regression results by industry, enterprise nature, and size.

Variable	(1)	(2)	(3)	(4)	(5)	(6)	(7)
		Grouped by Industry		Grouped by Enterprise Nature		Grouped by Size	
	A	Agr	Ag	St	NSt	L	S/M
	$Patent_{t+1}$	$Patent_{t+1}$	$Patent_{t+1}$	$Patent_{t+1}$	$Patent_{t+1}$	$Patent_{t+1}$	$Patent_{t+1}$
SUB	0.006 (0.14)	0.182 *** (4.84)	0.364 *** (7.38)	0.168 *** (4.78)	0.293 *** (7.41)	0.318 *** (7.47)	0.128 *** (3.77)
Age	0.058 *** (4.28)	−0.025 ** (−2.32)	−0.041 *** (−2.71)	−0.059 *** (−4.57)	−0.001 (−0.11)	−0.022 * (−1.89)	−0.007 (−0.62)
Size	0.393 *** (4.67)	0.419 *** (6.32)	0.327 *** (3.59)	0.412 *** (6.99)	0.248 *** (3.71)	0.443 *** (6.36)	0.388 *** (4.04)
Debt	0.715 ** (2.18)	0.378 (1.43)	0.212 (0.68)	1.080 *** (4.59)	−0.444 * (−1.86)	0.832 *** (2.87)	−0.056 (−0.24)
Growth	−0.000 * (−1.87)	−0.016 ** (−2.38)	−0.088 (−0.75)	0.033 (0.41)	0.001 *** (2.97)	−0.135 * (−1.74)	0.000 (0.65)
Share	−0.015 *** (−3.80)	−0.016 *** (−4.41)	0.021 *** (4.36)	−0.011 *** (−3.49)	−0.009 *** (−2.67)	−0.004 (−1.12)	−0.016 *** (−4.33)
Fixasset	0.774 (1.59)	0.913 *** (2.60)	−2.437 *** (−7.50)	−1.435 *** (−4.67)	0.977 *** (2.64)	−0.761 ** (−2.38)	0.026 (0.07)
Salary	0.262 ** (2.33)	0.510 *** (6.73)	0.702 *** (6.27)	0.296 *** (4.04)	0.520 *** (6.32)	0.510 *** (6.62)	0.238 *** (2.88)
_cons	−12.068 *** (−9.44)	−17.696 *** (−16.82)	−21.382 *** (−11.04)	−14.898 *** (−13.83)	−16.777 *** (−15.90)	−22.729 *** (−20.15)	−12.519 *** (−6.60)
Year	Yes	Yes	Yes	Yes	Yes	Yes	Yes
Industry	Yes	Yes	Yes	Yes	Yes	Yes	Yes
N	346	674	440	726	734	773	687
R^2	0.308	0.431	0.573	0.611	0.434	0.507	0.298

Note: *, **, and *** are significant at the 10%, 5%, and 1% levels; A: agriculture, forestry, animal husbandry and fisheries; Agr: agriculture, forestry, animal husbandry, and fishery products processing and manufacturing industry; Ag: agriculture, forestry, animal husbandry, and fishery means of production manufacturing industry; St: state-owned enterprises; NSt: non-state-owned enterprises; L: large-lot producer; S/M: medium and small-sized enterprises.

Table 7. Summary of Ordinary Least Squares (OLS) regression parameter estimates for technological innovation and research and development.

Variable	OLS	Variable	OLS
SUB	0.175 *** (6.67)	Fixasset	−0.118 (−0.50)
Age	−0.014 * (−1.66)	Salary	0.382 *** (7.18)
Size	0.369 *** (7.80)	_cons	−16.085 *** (−20.67)
Debt	0.290 * (1.78)	Year Industry	Yes Yes
Growth	0.000 * (1.76)	N	1460
Share	−0.005 ** (−2.02)	R^2	0.432

Note: *, **, and *** are significant at the 10%, 5%, and 1% levels.

5. Discussion

Our model results are consistent with the results of [42,43]. At the same time, the findings underline the necessity of government subsidies for technological innovation in agribusiness in China. Firstly, the study found that government subsidies effectively promote technological innovation in agribusiness. Government subsidies affect the technological innovation of enterprises by influencing their R&D investment; that is, the positive effects of government subsidies on the number of patent applications are partially absorbed by the R&D investment of the intermediary variable. Moreover, R&D investment is an intermediary effect that accounts for 27.56% of the total effect. Thirdly, the effects of government subsidies on the technological innovation of agricultural enterprises have a certain heterogeneity. From an industry perspective, government subsidies for processing agriculture, forestry, animal husbandry, and fishery products and manufacturing promote technological innovation in enterprises. However, government subsidies for traditional agriculture, forestry, animal husbandry, and fisheries do not significantly affect these enterprises' technological innovations. In terms of the nature of the enterprises, government subsidies promote the technological innovation of state-owned and non-state-owned enterprises. Their impact on technological innovation for non-state-owned enterprises is greater than it is for state-owned enterprises. In terms of the size of enterprises, government subsidies promote technological innovation for all sizes of companies. The impact of technological innovation is greater for large enterprises than it is for small and medium-sized enterprises.

The results of this study confirm the assumptions that innovations and digital technologies are the core instruments with which to support the sustainable development of agriculture. These findings are consistent with past research [60–62]. At the same time, innovations and digital technologies require sufficient financial resources from the government subsidies that are available to agricultural companies. However, the government should consider all the effects from innovation projects when making decisions on how to allocate government subsidies to innovative agricultural projects. These subsidized projects can positively and/or negatively impact the environment and society. Past research confirms that innovations in water management can provoke the relocation of local people [63–65]. Other researchers have demonstrated that R&D investments in agriculture positively impact farmers and local communities [66–69]. This suggests that the government should balance agricultural productivity and economic profits with minimizing negative environmental impacts (e.g., soil degradation, water and soil pollution, deforestation, etc.) and promoting societal benefits (e.g., healthy diets, community vibrancy, etc.). The following three policy suggestions are put forward based on the above research conclusions: Firstly, the government should continue to increase subsidies. The rural revitalization strategy needs scientific and technological innovation as a support. The core key to agricultural and rural modernization also depends on scientific and technological innovations, which play a pivotal role in agricultural and rural development. As the main body of technological innovation, agricultural enterprises play an important strategic role in agricultural modernization. Studies have shown that government subsidies effectively promote the technological innovation activities of agricultural enterprises. Moreover, our findings confirm that government subsidies are effective options for stimulating innovation in agricultural enterprises. Therefore, the Chinese government should continue to increase agricultural subsidies, such as direct subsidies, tax incentives, and research and development subsidies. The Chinese government should also account for possible negative externalities of subsidized agriculture, including environmental pollution and the forced relocation of entire communities.

Secondly, government subsidies should "vary from enterprise to enterprise". The impact of government subsidies on the technological innovation of agricultural enterprises varies according to the type of industry, the nature of the enterprise, and its size. Government subsidies have a significant role in promoting technological innovation in the processing and manufacturing of agriculture, forestry, animal husbandry, and fishery products. Their impact on technological innovation for non-state-owned enterprises is greater than it is for state-owned enterprises. The impact of technological innovation is greater for large enterprises than it is for small and medium-sized enterprises. Therefore, government departments should be divided into categories. The government's limited subsidy resources should be invested in enterprises with strong technological innovation capabilities. Thus, agricultural processing and manufacturing companies need to be supported with high-quality resources to invest in agricultural enterprises with a strong willingness to adopt innovative technologies.

Thirdly, government subsidy funds need to be better supervised. Government subsidies affect the technological innovation of agricultural enterprises through R&D investment. Therefore, the government should strengthen the supervision of the use of subsidy funds and improve the performance of the use of funds. It is possible to establish and improve a monitoring system covering the whole process and the whole chain of fund allocation, implementation, and supervision. It is necessary to analyze the efficiency of government subsidies. At the same time, the focus is on supervising agricultural enterprises with low R&D investment levels and on encouraging enterprises to increase their investment in innovative, sustainable technologies and processes.

The efficiency of government policy for supporting the innovation implementations in agricultural companies should become an instrument for improving the export structure of agriculture and achieving sustainable development goals. Thus, the agricultural sector is a crucial element of food security. This involves the rational use of limited resources and the implementation of green technologies and energy efficiency innovations while mitigating adverse environmental and community impacts.

6. Conclusions

From the whole industry chain perspective, this paper extended the agricultural scope to the production, processing, manufacturing, circulation, and service of agriculture, forestry, animal husbandry, and fisheries. It empirically tested the effect and influence mechanism of government subsidies on agricultural enterprises' technological innovation by taking the companies listed from 2007 to 2019 as a research sample. We developed Ordinary Least Squares statistical regression models to test these hypotheses.

Despite the valuable findings and practical recommendations, our research has a few limitations. Our analysis focused on China only. At the same time, the globalization and openness of the economy facilitates potential improvements or declines in the competitiveness and sustainability of companies involved in agriculture and agro-forestry. The competitiveness of agricultural businesses also depends on other internal and external factors and should be studied in future investigations. Internal factors include the social responsibility of companies, the education level of managers, technological innovations, etc. External factors include government corruption and quality, sustainable development pathways in the region, geographic characteristics, etc. Innovative agricultural projects that are subsidized by the government can have a wide range of positive and/or negative economic, ecological, and social impacts which warrant further investigation.

Author Contributions: Conceptualization, L.W., K.H., O.L., T.P. and I.H.; methodology, L.W., K.H., O.L., T.P. and I.H.; formal analysis, L.W., K.H., O.L., T.P. and I.H.; investigation, L.W., K.H., O.L., T.P. and I.H.; writing—original draft preparation, L.W., K.H., O.L., T.P. and I.H.; writing—review and editing, L.W., K.H., O.L., T.P. and I.H.; visualization, L.W., K.H., O.L., T.P. and I.H. All authors have read and agreed to the published version of the manuscript.

Funding: This research was funded by a grant from the Ministry of Education and Science of Ukraine, 0121U100468, "Green investing: cointegration model of transmission ESG effects in the chain "green brand of Ukraine—social responsibility of business".

Institutional Review Board Statement: Not applicable.

Informed Consent Statement: Not applicable.

Data Availability Statement: Not applicable.

Conflicts of Interest: The authors declare no conflict of interest.

References

1. Outline of the 14th Five-Year Plan (2021–2025) for National Economic and Social Development and Vision 2035 of the People's Republic of China. Available online: https://www.fujian.gov.cn/english/news/202108/t20210809_5665713.htm (accessed on 20 October 2022).
2. Smith, M. A Real Options Approach To Evaluating Agricultural Investments under Uncertainty: When To Get In and Out of Sugarcane Production. *SocioEconomic Chall.* **2018**, *2*, 21–34. [CrossRef]
3. Didenko, I.; Sidelnyk, N. Insurance Innovations as a Part of the Financial Inclusion. *Bus. Ethics Leadersh.* **2021**, *5*, 127–135. [CrossRef]
4. Sokolov, M.; Mykhailov, A.; Khandurin, D. Distribution of investment resources: Where is agriculture in the Ukraine's economy? *Financ. Mark. Inst. Risks* **2018**, *2*, 38–42. [CrossRef]
5. Sung, B. Do government subsidies promote firm-level innovation? Evidence from the Korean renewable energy technology industry. *Energy Policy* **2019**, *132*, 1333–1344. [CrossRef]
6. Bogachov, S.; Kwilinski, A.; Miethlich, B.; Bartosova, V.; Gurnak, A. Artificial intelligence components and fuzzy regulators in entrepreneurship development. *Entrep. Sustain. Issues* **2020**, *8*, 487–499. [CrossRef]
7. Kuzior, A.; Kwilinski, A.; Hroznyi, I. The factorial-reflexive approach to diagnosing the executors' and contractors' attitude to achieving the objectives by energy supplying companies. *Energies* **2021**, *14*, 2572. [CrossRef]
8. Kwilinski, A.; Dzwigol, H.; Dementyev, V. Model of entrepreneurship financial activity of the transnational company based on intellectual technology. *Int. J. Entrep.* **2020**, *24* (Suppl. S1), 1–5.
9. Butko, M.; Popelo, O.; Pishenin, I. Innovations in Human Resources Management in Eurointegration Conditions: Case for Ukrainian Agro-Industrial Complex. *Mark. Manag. Innov.* **2019**, *2*, 74–82. [CrossRef]
10. Kravchenko, S. Simulation of the national innovation systems development: A transnational and coevolution approach. *Virtual Econ.* **2019**, *2*, 41–54. [CrossRef]
11. Kozmenko, S.; Vasil'yeva, T.; Leonov, S. The structuring of components of the net domestic product according to the innovation criterion. *Innov. Mark.* **2010**, *6*, 30–41. [CrossRef]
12. Czyżewski, B.; Polcyn, J.; Brelik, A. Political orientations, economic policies, and environmental quality: Multi-valued treatment effects analysis with spatial spillovers in country districts of poland. *Environ. Sci. Policy* **2022**, *128*, 1–13. [CrossRef]
13. Guo, D.; Guo, Y.; Jiang, K. Government-subsidized R&D and firm innovation: Evidence from China. *Res. Policy* **2016**, *45*, 1129–1144.
14. Mamuneas, T.P.; Nadiri, M.I. Public R&D policies and cost behavior of the US manufacturing industries. *J. Public Econ.* **1996**, *63*, 57–81.
15. Kiss, L.B. Examination of Agricultural Income Inequality in the European Union. *Bus. Ethics Leadersh.* **2020**, *4*, 36–45. [CrossRef]
16. Lee, E.Y.; Cin, B.C. The effect of risk-sharing government subsidy on corporate R&D investment: Empirical evidence from Korea. *Technol. Forecast. Soc. Change* **2010**, *77*, 881–890.
17. Zhang, H.; Li, L.; Zhou, D.; Zhou, P. Political connections, government subsidies and firm financial performance: Evidence from renewable energy manufacturing in China. *Renew. Energy* **2014**, *63*, 330–336. [CrossRef]
18. China Securities Regulatory Commission. Guidelines for Industry Classification of Listed Companies. 2012. Available online: http://www.csrc.gov.cn/pub/zjhpublic/G00306201/201211/t20121116_216990.htm (accessed on 10 August 2022).
19. Dzwigol, H.; Trushkina, N.; Kwilinski, A. The Organisational and Economic Mechanism of Implementing the Concept of Green Logistics. *Virtual Econ.* **2021**, *4*, 41–75. [CrossRef]
20. Polcyn, J. Determining value added intellectual capital (VAIC) using the TOPSIS-CRITIC method in small and medium-sized farms in selected european countries. *Sustainability* **2022**, *14*, 3672. [CrossRef]
21. Polcyn, J. Eco-efficiency and human capital efficiency: Example of small-and medium-sized family farms in selected european countries. *Sustainability* **2021**, *13*, 6846. [CrossRef]
22. Belle, H. *Perspectives on the World: An Interdisciplinary Reflection*; Worldviews Group, Ed.; VUB Press: Brussels, Belgium; Paul Company Publishers Consortium: Brussels, Belgium, 1995; ISBN 978-90-5487-113-2.
23. National Bureau of Statistics. Statistical Classification of Agriculture and Related Industries 2020. Available online: http://www.stats.gov.cn/english/PressRelease/202201/t20220113_1826284.html (accessed on 10 August 2022).
24. Arrow, K.J. The economic implications of learning by doing. *Rev. Econ. Stud.* **1962**, *29*, 155–173. [CrossRef]
25. Luo, G.; Liu, Y.; Zhang, L.; Xu, X.; Guo, Y. Do governmental subsidies improve the financial performance of China's new energy power generation enterprises? *Energy* **2021**, *227*, 120432. [CrossRef]
26. Ren, S.; Sun, H.; Zhang, T. Do environmental subsidies spur environmental innovation? Empirical evidence from Chinese listed firms. *Technol. Forecast. Soc. Change* **2021**, *173*, 121123. [CrossRef]
27. Li, F.; Xu, X.; Li, Z.; Du, P.; Ye, J. Can low-carbon technological innovation truly improve enterprise performance? The case of Chinese manufacturing companies. *J. Clean. Prod.* **2021**, *293*, 125949. [CrossRef]
28. Chygryn, O.; Krasniak, V. Theoretical and applied aspects of the development of environmental investment in Ukraine. *Mark. Manag. Innov.* **2015**, *3*, 226–234.

29. Kaya, H.D. How Does The Use Of Technology In Entrepreneurial Process Affect Firms' Growth? *SocioEconomic Chall.* **2021**, *5*, 5–12. [CrossRef]
30. Samoilikova, A. Financial Policy of Innovation Development Providing: The Impact Formalization. *Financ. Mark. Inst. Risks* **2020**, *4*, 5–15. [CrossRef]
31. Zhou, Y.; He, Z.; Zhao, S. How do government subsidies affect the strategic choices of enterprises and individuals in agricultural waste recycling? *Sustain. Prod. Consum.* **2021**, *28*, 1687–1698. [CrossRef]
32. Lu, Q.H.; Xu, T. Research on subsidy mechanism in the government-led agricultural supply chain finance. In *E3S Web of Conferences*; EDP Sciences: Les Ulis, France, 2021; Volume 275, p. 01016. [CrossRef]
33. Bin, Z.H.U.; Lulu, L.I. An Empirical Analysis of the Impact of Public Subsidies on Private Enterprise's R&D Investment. *Soc. Chin. J. Sociol./Shehui* **2014**, *34*, 165–186.
34. Lin, B.; Luan, R. Are government subsidies effective in improving innovation efficiency? Based on the research of China's wind power industry. *Sci. Total Environ.* **2020**, *710*, 136339. [CrossRef]
35. Yu, F.; Guo, Y.; Le-Nguyen, K.; Barnes, S.J.; Zhang, W. The impact of government subsidies and enterprises' R&D investment: A panel data study from renewable energy in China. *Energy Policy* **2016**, *89*, 106–113.
36. Jiang, C.; Zhang, Y.; Bu, M.; Liu, W. The effectiveness of government subsidies on manufacturing innovation: Evidence from the new energy vehicle industry in China. *Sustainability* **2018**, *10*, 1692. [CrossRef]
37. Aghion, P.; Cai, J.; Dewatripont, M.; Du, L.; Harrison, A.; Legros, P. Industrial policy and competition. *Am. Econ. J. Macroecon.* **2015**, *7*, 1–32. [CrossRef]
38. Lazzarini, S.G. Strategizing by the government: Can industrial policy create firm-level competitive advantage? *Strateg. Manag. J.* **2015**, *36*, 97–112. [CrossRef]
39. Wang, P.; Dong, C.; Chen, N.; Qi, M.; Yang, S.; Nnenna, A.B.; Li, W. Environmental Regulation, Government Subsidies, and Green Technology Innovation—A Provincial Panel Data Analysis from China. *Int. J. Environ. Res. Public Health* **2021**, *18*, 11991. [CrossRef]
40. Wu, H.; Hu, S. The impact of synergy effect between government subsidies and slack resources on green technology innovation. *J. Clean. Prod.* **2020**, *274*, 122682. [CrossRef]
41. Guo, Y. Signal transmission mechanism of government innovation subsidy and enterprise innovation. *China Ind. Econ.* **2018**, *9*, 98–116.
42. Wu, A. The signal effect of government R&D subsidies in China: Does ownership matter? *Technol. Forecast. Soc. Change* **2017**, *117*, 339–345.
43. Wang, K.H.; Liu, J.L. The dynamic effects of government-supported R&D subsidies: An empirical study on the Taiwan science park. *Asian J. Technol. Innov.* **2009**, *17*, 1–12.
44. Guo, F.; Zou, B.; Zhang, X.; Bo, Q.; Li, K. Financial slack and firm performance of SMMEs in China: Moderating effects of government subsidies and market-supporting institutions. *Int. J. Prod. Econ.* **2020**, *223*, 107530. [CrossRef]
45. Wang, N.; Hagedoorn, J. The lag structure of the relationship between patenting and internal R&D revisited. *Res. Policy* **2014**, *43*, 1275–1285.
46. Yu, M.; Fan, R.; Zhong, H. Chinese industrial policy and corporate technological innovation. *China Ind. Econ.* **2016**, *12*, 5–22.
47. Long, X.; Wang, J. China's patent explosion and its quality implications. *World Econ.* **2015**, *6*, 115–120.
48. Papanastassiou, M.; Pearce, R.; Zanfei, A. Changing perspectives on the internationalisation of R&D and innovation by multinational enterprises: A review of the literature. *J. Int. Bus. Stud.* **2020**, *51*, 623–664.
49. Zhao, S.; Sun, Y.; Xu, X. Research on open innovation performance: A review. *Inf. Technol. Manag.* **2016**, *17*, 279–287. [CrossRef]
50. Paiva, T.; Ribeiro, M.; Coutinho, P. R&D collaboration, competitiveness development, and open innovation in R&D. *J. Open Innov. Technol. Mark. Complex.* **2020**, *6*, 116.
51. The China Research Data Service Platform. Available online: https://www.cnrds.com (accessed on 4 October 2022).
52. The State Intellectual Property Office. Available online: https://english.cnipa.gov.cn (accessed on 4 October 2022).
53. Shenzhen Hexun Huagu Information Technology. Available online: https://www.jiguang.cn/en/about (accessed on 4 October 2022).
54. Economic Database. Royal Flush iFind Database. Available online: http://www.51ifind.com (accessed on 4 October 2022).
55. China Stock Market & Accounting Research Database. Available online: https://cn.gtadata.com (accessed on 4 October 2022).
56. WIPO. World Intellectual Property Indicators 2021. World Intellectual Proper. 2021. Available online: https://www.wipo.int/edocs/pubdocs/en/wipo_pub_941_2021.pdf (accessed on 4 October 2022).
57. Somaini, P.; Wolak, F.A. An algorithm to estimate the two-way fixed effects model. *J. Econom. Methods* **2016**, *5*, 143–152. [CrossRef]
58. Wen, Z.; Ye, B. Analyses of mediating effects: The development of methods and models. *Adv. Psychol. Sci.* **2014**, *22*, 731. [CrossRef]
59. Bilan, Y.; Lyeonov, S.; Stoyanets, N.; Vysochyna, A. The impact of environmental determinants of sustainable agriculture on country food security. *Int. J. Environ. Technol. Manag.* **2018**, *21*, 289–305. [CrossRef]
60. Stoian, M.; Ion, R.A.; Turcea, V.C.; Nica, I.C.; Zemeleaga, C.G. The Influence of Governmental Agricultural R&D Expenditure on Farmers' Income—Disparities between EU Member States. *Sustainability* **2022**, *14*, 10596. [CrossRef]
61. Streimikis, J.; Baležentis, T. Agricultural sustainability assessment framework integrating sustainable development goals and interlinked priorities of environmental, climate and agriculture policies. *Sustain. Dev.* **2020**, *28*, 1702–1712. [CrossRef]

62. Raszkowski, A.; Bartniczak, B. On the Road to Sustainability: Implementation of the 2030 Agenda Sustainable Development Goals (SDG) in Poland. *Sustainability* **2019**, *11*, 366. [CrossRef]
63. Li, K.; Zhu, C.; Wu, L.; Huang, L. Problems caused by the Three Gorges Dam construction in the Yangtze River basin: A review. *Environ. Rev.* **2013**, *21*, 127–135. [CrossRef]
64. Guo, H.; Hu, Q.; Zhang, Q.; Feng, S. Effects of the three gorges dam on Yangtze river flow and river interaction with Poyang Lake, China: 2003–2008. *J. Hydrol.* **2012**, *416*, 19–27. [CrossRef]
65. Wu, J.; Huang, J.; Han, X.; Gao, X.; He, F.; Jiang, M.; Jiang, Z.; Primack, R.B.; Shen, Z. The three gorges dam: An ecological perspective. *Front. Ecol. Environ.* **2004**, *2*, 241–248. [CrossRef]
66. Hurni, H.; Giger, M.; Liniger, H.; Mekdaschi Studer, R.; Messerli, P.; Portner, B.; Schwilch, G.; Wolfgramm, B.; Breu, T. Soils, agriculture and food security: The interplay between ecosystem functioning and human well-being. *Curr. Opin. Environ. Sustain.* **2015**, *15*, 25–34. [CrossRef]
67. Fahmi, Z.; Samah, B.A.; Abdullah, H. Paddy industry and paddy farmers well-being: A success recipe for agriculture industry in Malaysia. *Asian Soc. Sci.* **2013**, *9*, 177. [CrossRef]
68. Stoneham, G.; Eigenraam, M.; Ridley, A.; Barr, N. The application of sustainability concepts to Australian agriculture: An overview. *Aust. J. Exp. Agric.* **2003**, *43*, 195–203. [CrossRef]
69. Alston, J.M. Reflections on agricultural R&D, productivity, and the data constraint: Unfinished business, unsettled issues. *Am. J. Agric. Econ.* **2018**, *100*, 392–413.

sustainability

MDPI

Article

Sustainability Trade-Offs in Media Coverage of Poverty Alleviation: A Content-Based Spatiotemporal Analysis in China's Provinces

Yuting Sun [1,*] and Shu-Nung Yao [2,*]

1 School of Public Administration, Hohai University, Nanjing 210098, China
2 Department of Electrical Engineering, National Taipei University, New Taipei 23741, Taiwan
* Correspondence: 20220019@hhu.edu.cn (Y.S.); snyao@gm.ntpu.edu.tw (S.-N.Y.)

Abstract: Poverty alleviation has always been fundamental for China to achieve the goal of creating a moderately prosperous society. This study conducted a content-based spatiotemporal analysis of media coverage, regression analysis of panel data, and text mining to examine how China's Targeted Poverty Alleviation (TPA) Strategy is characterised by online mainstream media platform. A total of 10,857 articles related to TPA in 31 specific provinces of mainland China were collected and analysed by Natural Language Processing (NLP) analysis. The results of this study indicated that spatiotemporal distribution of TPA coverage was consistent with the typical logic of the Chinese government in policy implementation based on spatial and social marginalisation. Media attention on TPA is influenced by economic, environmental, and community sustainability indicators, reflecting the sustainability trade-offs in TPA-related media coverage. The keywords embedded in media coverage indicated that agricultural product promotion in extremely impoverished areas and the experiences of economically developed agricultural areas were essential for poverty eradication. Keywords emphasise top-down administrative-led poverty governance for extremely impoverished areas and local autonomy for relatively impoverished areas. This study provides perspectives for antipoverty governance and media empowerment in the postpoverty era in China.

Keywords: content-based analysis; media coverage; spatiotemporal distribution; sustainability trade-offs; Targeted Poverty Alleviation Strategy

Citation: Sun, Y.; Yao, S.-N. Sustainability Trade-Offs in Media Coverage of Poverty Alleviation: A Content-Based Spatiotemporal Analysis in China's Provinces. *Sustainability* **2022**, *14*, 10058. https://doi.org/10.3390/su141610058

Academic Editor: Aaron K. Hoshide

Received: 29 June 2022
Accepted: 12 August 2022
Published: 14 August 2022

1. Introduction

Poverty may involve a lack of opportunities, empowerment, or security, malnutrition, or poor health [1]. Definitions of poverty are diverse. Traditional approach views poverty as either low utility or shortfall in primary goods, resources, or income; Capability approach proposed views poverty as the deprivation of basic capabilities from realising their full potential [2–4]. Absolute poverty describes a condition characterised by severe deprivation of basic human needs, which depends not only on income but also on access to services [5]. Relative poverty as a standard is measured in terms of the society in which an individual lives and which therefore differs between countries and over time. Marginalisation is a root cause of poverty in many cases and is synonymous with poverty [6–8]. Poverty is unevenly distributed from the spatial dimension of marginalisation. In rural areas, especially those with a poor ecological environment, a remote geographical location, and inadequate public services and facilities, diverse forms of poverty are evident [9,10]. The social dimension of marginalisation emphasises the importance of the marginalisation process for marginalised groups [11–13]. As a marginalised group, low-income individuals in disadvantaged villages face difficulties participating in various aspects of life.

Both developing and developed countries today are experiencing poverty in different forms and to varying degrees. Much of the research on poverty in developed countries tends to predominantly focus on urban poverty and relative poverty [14], while studies on poverty in developing countries are mostly restricted to rural poverty and extreme poverty [15]. This is because the majority of the population of developing countries live in the rural areas, and they have less access to the socio-economic and infrastructural facilities than their urban counterparts [16]. Poverty governance is essentially a multi-faceted process of using political power, exercising political authority, mobilizing political resources, running political institutions, and gaining political legitimacy [17]. In developed countries, poverty reduction is regarded as a poverty governance model. Developing countries, due to potentially weak political and administrative areas of governance hardly achieved sustainable rapid growth for a long time in reducing poverty [18–20]. China has a large poor population and has contributed to most of the world poverty reduction [21]. It is essential to examine poverty governance in China due to its diversity.

Poverty reduction related to human prosperity is one goal of sustainable development [22]. Sustainable development is a balance between economic, social, and ecological goals [23]. Previous evidence shows that, over the past two centuries, economic development has resulted in a sharp decrease in absolute poverty worldwide [24]. Poverty alleviation faces trade-offs among economic, environmental, and community/social sustainability. Especially in China, due to the long-standing urban-rural dual system, rural multidimensional poverty contributed 80% of the national multidimensional poverty. Poverty mainly occurs in rural areas with a very large peasant population of rural hukou [25]. To eliminate rural poverty and realise rural revitalisation, poverty reduction must be adapted to the social and environmental conditions in different regions of China, and the social process of poverty alleviation must be adjusted to and embedded in local areas. Consequently, it is essential to address poverty alleviation trade-offs among sustainability dimensions and ensure the balanced and full development between urban and rural areas.

Poverty alleviation is a vital goal for China to create a moderately prosperous society. Since the 1980s, the Chinese government has been committed to poverty reduction [26]. In 2013, the Targeted Poverty Alleviation (TPA) Strategy was set forth by Chinese President Xi Jinping as a departure from previous poverty alleviation strategies; he sought to move from the transfusion type of poverty alleviation to haematopoietic poverty alleviation [26]. Since then, poverty alleviation measures have been applied with the aim of establishing of a long-term system for reducing poverty. After eight years of continual effort, China finally eradicated absolute poverty in 2020 [27].

Mass media is considered as a major force against poverty [28]. Some research has examined the media role influencing policymakers to set the agenda for poverty alleviation programs and the effect of the influence of media coverage of poverty alleviation programmes on the people [29]. With the development of Internet technologies and horizontal and vertical strategic advancements in the field of media integration, diversified-level mainstream news sources gathered in online media platforms. Understanding TPA-related spatial distributions and temporal trends and identifying the logic of online media related to TPA are paramount for sustainable long-term poverty eradication. Therefore, a discussion of online mainstream media coverage in this crucial stage of poverty alleviation is essential, and this discussion is expected to inspire future media coverage strategies in the postpoverty era.

Previous research employed text mining approach to examine policy documents of the poverty reduction strategy. Smith-Carrier and Lawlor [30] used corpus linguistics and critical discourse analysis to study a poverty reduction strategy implemented in Ontario, Canada. Some researchers examined specific TPA programs of China such as projects on power generation [31–33] from policy documents and e-commerce poverty alleviation from social media platform [34]. These studies mainly focused on policy texts and few studies have adopted text mining to examine how China's TPA Strategy is characterised by online media. Furthermore, when considering the amount of media coverage and meaning of media content as assessments of media attention, few studies have focused on sustainability trade-offs of media attention on TPA and how China's TPA Strategy is being characterised by the media.

Based on above research gaps, in this study, content analysis was initially applied to investigate the spatial and temporal distribution of TPA-related coverage. Then, according to economic, environmental, and social sustainability, a panel data econometrics regression model with the above three dimensions of sustainability and sum of TPA-related coverage on mainland China's provinces from 2017 to 2020 was established to explore the sustainability trade-offs in TPA-related media attention. Finally, a text mining approach was employed to explore keywords in the media coverage, and the mechanism through which TPA is represented by the media. The research questions investigated in this study are described as follows: (1) What dynamic patterns of temporal and spatial distributions are represented in TPA-related coverage? (2) What indicators of sustainability dimensions affect TPA-related coverage? (3) How are dynamic patterns of temporal and spatial distributions of keywords embedded in media storytelling related to TPA? The results of this study can contribute to a deeper understanding of antipoverty governance in rural China through media empowerment.

2. Literature Review

2.1. Marginality and Poverty

Marginality is primarily defined and described using two major conceptual frameworks, namely spatial and societal dimensions [35]. The spatial dimension of marginality is primarily based on physical location and distance from centres of development; such locations lie at the edge of or are poorly integrated into a system [36,37]. This concept is used to gain insights into the influence of physical locations and distance on the livelihoods of individuals and groups. The societal dimension focuses on understanding the underlying causes of exclusion, inequality, social injustice, and the spatial segregation of people by demography, religion, culture, social structure, economics, and politics. Such factors are related to access to resources by individuals and groups [36–42].

The concept of marginality, which is multidimensional and multidisciplinary, generally integrates geographical or physical locations with sociocultural, political, and economic spheres where disadvantaged people struggle to gain access (societal and spatial) to resources and fully participate in social life [35,37,43]. Marginality is a social construct, and social and political forces are regarded as the core determinants of marginality [44]. Marginalisation is a social structure and the result of conscious actions by social actors. When marginalisation becomes a part of rules, it constitutes the condition and medium of individual actions in social structures [45].

Marginalisation is a root cause of poverty in many cases and is synonymous with poverty [6–8]. The main reasons for such difficulties lie in the internal cohesion and local mechanism of marginalised areas. Villagers may reject outsiders and lack trust in foreign poverty alleviation personnel and technology. Moreover, villages lack effective integration mechanisms for different stakeholders. The aforementioned phenomena affect the development and effects of poverty alleviation governance in villages.

Demarginalisation is a process with low and high marginality on two ends of a continuum [46]. Although the development of technical tools helps people in spatially marginalised areas to improve their lives gradually, social difficulties persist [47]. Demarginalisation efforts must be adapted to the environmental and social conditions of different regions, and the social process embedded in local areas must be accordingly adjusted. Poverty alleviation is a poverty governance model in developed countries. Some countries (e.g., Turkey, German, Greece, etc.) established extensive participation and negotiation mechanisms with stakeholders [21]. Japan paid attention to youth education and participation in sustainable development knowledge, as well as promoting increased awareness of, commitment to, and ownership of the agenda [22,48]. Good governance associated with the buzzwords of "participation" and "empowerment" provides a sense of purpose and hope for equality of opportunity in the fast-moving world of development policy [49]. As a consequence, a sustainable poverty alleviation strategy focuses on improving the lives of the poor and vulnerable through strengthening the social dimension of demarginalisation [50].

2.2. News Media's Role in Poverty Alleviation and Discussing Poverty-Related News

News media are regarded as crucial components in the fight against poverty. They call to action with strong involvement from civil society to policies and programmes, thus setting national agendas and ultimately motivating the implementation of policy-making processes [51]. Therefore, policy debates through media can help democratise policies and increase awareness of poverty alleviation policies, ultimately strengthening the advocacy and demand for improved poverty alleviation programmes. Reports on TPA reports are vital for the promotion of TPA activities. Thus, how media portray the fight against poverty has a considerable effect on the public's and policymakers' expectations of poverty alleviation processes and eventually determines whether and to what extent stakeholders and social actors participate in poverty alleviation.

In an English-language context, previous studies [52–55] centred on news framing of poverty coverage in mass media. This type of coverage mostly focuses on the "consumption of the poor suffering" from the perspective of "the other" [52] rather than on the plight of low-income individuals [53]. Poverty coverage by mass media is mostly attributed at the societal level rather than the personal level; thus, the audience may form the belief that governments should take responsibility to alleviate poverty [54,55]. Although some researchers have employed content analysis to examine policy texts focusing on certain TPA projects in China, such as policy texts on photovoltaic power [32,33], few studies have focused on the media coverage of TPA. Given the importance of rural and national representations for fighting poverty in mainland China, the main aim of this study was to determine if and how online mainstream media give voice to or represent the values of rural Chinese people in discussions of poverty alleviation strategies in mainland China.

2.3. TPA policy and Regional Profiles in a Chinese Context

China's national poverty reduction programmes have experienced a three-stage evolution: from region-focused targeting (*quyu miaozhun*) during 1986–2000 to village-focused targeting (*zhengcun tuijin*) during 2001–2010 and then to household-focused targeting (*jingzhun daohu*) during 2011–2020 [25]. The TPA Strategy was on the agenda of China's 13th Five-Year Plan for Economic and Social Development [26]. Essentially, TPA involves accurately identifying low-income individuals, accurately allocating poverty alleviation funds, accurately formulating poverty alleviation measures, accurately implementing poverty alleviation strategies in appropriate areas, and accurately assessing poverty alleviation results. The aim of TPA is to alter poverty alleviation methods from being material centric to being region and household centric [56]. Furthermore, TPA policy focused on economic, environmental, and social sustainability. In 2015, ten projects of the TPA were issued including vocational education and training, helping cadres' residency in impoverished villages, microfinance, ex situ poverty alleviation relocation (ESPAR), e-commerce, tourism, photovoltaic power generation, papyrifera planting, entrepreneurship training of rich leaders, and leading enterprises driving poverty alleviation [26]. In addition, to strengthen the community-level poverty alleviation capacity, China's governments at various levels have dispatched human resources to poverty-stricken villages. These targeted poverty alleviation measures are calling for the establishment of a long-term system to better carry out efforts.

As to TPA implemented regions, in 2012, the State Council Leading Group Office of Poverty Alleviation and Development in China identified impoverished counties for poverty alleviation and development. In total, 832 counties across 22 provinces were identified as national-level impoverished counties according to the criteria of farmers' per capita net income and the size of the poor population. The aforementioned poverty-stricken areas are located in ecologically vulnerable zones with poor living conditions, frequent natural disasters, defective economic foundations, poor infrastructure, and insufficient public services [57]. Before extreme poverty was eradicated by the end of 2020, people experiencing poverty were primarily concentrated in designated poverty-stricken areas, especially in the most impoverished areas of minority provinces, autonomous regions, and remote regions in western China, including several prefectures in Tibet, Xinjiang, Gansu, Sichuan, and Yunnan [57]. After eight years of continual effort, on 23 November 2020, China announced that it had eliminated absolute poverty nationwide by uplifting all of its citizens beyond the absolute poverty line of 2300 RMB per year (2010 constant prices) set in 2012, or less than a dollar per day poverty line [25,27]. In total, 832 impoverished counties, 128,000 impoverished villages, and 98.99 million impoverished people have been lifted out of poverty [27].

3. Research Methodology and Data

3.1. Content Analysis and Natural Language Processing

Conventional content analysis can provide an integrative perspective of a text and its related context for researchers to understand social phenomena in a subjective yet scientific manner [58,59]. Content analysis involves the thematic categorisation of words to reveal the content and context of the language used [60]. The frequency of appearance of thematically categorised words can provide an objective means of gauging the salience of certain concepts in a corpus [61]. In this study, content analysis was conducted to examine media trends as well as identify and analyse observable semantic data pertaining to TPA-related articles.

Natural language processing (NLP) involves an automatic analysis of human language and aims to address complexity and multiple connotations. In text mining, NLP is employed to understand data as though a human coder is reading the relevant text [62]. NLP enables the identification of relevant information within a text from a large corpus, which can assist researchers in making large data sets manageable and enhancing the trustworthiness of analysis results [63]. The text mining techniques used in the current study were correspondence analysis and Word2Vector analysis.

Content analysis combined with NLP may help extract meaning from data and enhance the inferences that researchers can make from a given text [63]. Currently, this method has been widely used in web data mining, search engines, geopolitical events, sentiment analysis, and social media content [64–66]. This present study combined content analysis and NLP in qualitative data analysis to provide a deeper understanding of texts.

3.2. Data Collection and Related Analysis

3.2.1. Media Attention: Poverty Alleviation Coverage

The present study's research sample was obtained from NTV (www.ntv.cn, accessed on 1 January 2021), which is a mainstream media convergence platform including state-controlled and local news sources concerned with three issues, namely agriculture, rural areas, and farmers, distinguished from social media, bloggers, or independent journalists. The term 'fupin' (poverty alleviation) was used to search for articles about poverty alleviation. The concept of TPA was proposed in 2013, and absolute poverty was eradicated in China in 2020. All included articles were published between July 2017 and December 2020. The article content and related news information were collected using the R crawling package (rvest and httr) to extract textual information [67]. Data collection involved the retrieval of 10,857 articles and related information, including article headings, URLs, publication dates, news sources, and news content. Advanced parsing techniques were used to remove redundant segments (e.g., pictures and short videos) of data that might have biased the results [68]. Furthermore, ethical issues for data mining were considered in relation to individual privacy [69]; hence, neither personal data nor behavioural data were revealed in this study.

3.2.2. Indicators of Sustainability Dimensions Affecting Media Attention

Previous studies indicated that specific indicators quantitatively assess the economic, environmental, and social sustainability [70–73]. Indicators used in economic sustainability assessments includes profitability [70], farm revenues and household income [71], crop yield [72] as well as several inputs and outputs such as farm productivity and technical efficiency [72,73]. Environmental indicators include pesticide use, greenhouse gas emissions, biodiversity, water pollution, soil quality, and land conservation [73,74]. The social dimension is associated with the broader society indicators, such as vitality of rural areas and contribution to local residents [73]. In this study, indicators were selected and constructed from related studies. In addition, researchers also took data accessibility and openness into consideration. Thus, the indicators chosen in this study were those similar or relevant with sustainability dimensions which can be found in the National Bureau of Statistics of China.

Specifically speaking, first, "agricultural gross domestic product (GDP)" represents agricultural productivity; "effective irrigated area" reflects technical efficiency; and "per capita disposable income of rural residents" represents rural household income. These three variables were employed to assess economic sustainability. Second, "pollution control investment" indicates pollution treatment and "affected area of crop" indicates disaster risk resistance capacity. These two variables were used to evaluate environmental dimension. Third, "rural doctors and medical workers" reveals the input of rural community healthcare resources; and "numbers of receiving social relief in rural areas" reveals equity and poverty alleviation efficiency of rural communities. These two variables were selected to measure social dimensions. The list of the variables for the years 2017–2020 are shown in Table 1. Then a panel data approach was selected for the analysis, and annual data for 31 provinces in mainland China were taken from the National Statistical Yearbook 2021 Indicators. For 31 provinces and 4 years, there were 124 observations, which was a suitable number to proceed with quantitative data analysis.

Table 1. Sustainability indicators used as variables in regression analysis.

Dimensions	Variables	Description	Unit
Media attention	Media coverage	Numbers of articles related to poverty alleviation	Piece
Rural economic sustainability	Agriculture GDP	Agricultural gross domestic product (GDP) as an indicator of agricultural productivity	100 million Yuan (RMB)
	Effective Irrigated Area	Technical efficiency of farmland	Thousand Hectares
	Rural Residents' Income	Per capita disposable income (PCDI) of rural residents	Yuan (RMB)
Environmental sustainability	Pollution Control Investment	Capital input of pollution treatment	10,000 Yuan (RMB)
	Affected Area of Crop	Disaster Risk resistance capacity of agriculture	Thousand Hectares
Social/Community sustainability	Rural Doctors and Medical Workers	Input of rural community healthcare resources	10,000 people
	Numbers of Receiving Social Relief in Rural Areas	Social relief stand for equity and poverty alleviation efficiency of rural community	10,000 people

3.3. *Category Building of Media Coverage and Intercoder Reliability*

Numerous key steps in content analysis are required to enable valid and reliable inferences to be derived from data [63]. Each coverage should be coded with a region based on which place of TPA experience it reported. Through observation before coding, researchers found place names such as XX province, XX city, XX town and XX township, etc., mostly occurred in news title and news content. As news title and news source are short, region coding of each coverage was easier to identify. Thus, a dictionary of geographical names was firstly constructed. Furthermore, news sources were mainly named by the form of "place + newspaper/media/broadcasting ... ", and local media always reported local TPA experience. Thus, a local media dictionary was then constructed to detect regions of news sources. Specifically speaking, in the dictionary of geographical names, 31 provincial-level administrative regions in mainland China comprised the largest geographical unit in mainland China. This unit included the geographical names of 22 provinces, 5 autonomous regions, and 4 municipalities directly under the control of the central government. The names of prefecture-level cities were provided in the second unit; district and county names were provided in the third unit; and town names were provided in the fourth and smallest unit. Village names were not included in the dictionary. In the local media dictionary, names of local media were included in the fifth and as a unit to identify the region each coverage belongs to through detecting news sources. If coverage can't be identified from news title and news source, content of coverage can be as judgment. For news content that focuses nationwide TPA experience but not related to a specific place, researchers manually classified it as nationwide/others. Finally, researchers established a table presenting the classification of regional categories. The categorisation scheme applied in this study involved manual coding and a data-driven approach (Tables 2 and 3).

Table 2. Categorisation scheme and 2019 agricultural land use in Chinese provinces.

Regions Categories		Agricultural Land in 2019	
Regions	**31 Provincial-Level Administrative Regions**	**Area (100,000 km²)**	**Percent (%)**
Seven regions		**696.68**	**72.49%**
Southwest	Yunnan, Guizhou, Sichuan, Tibet, Chongqing	196.51	20.45%
Northwest	Shaanxi, Gansu, Ningxia, Qinghai, Xinjiang	168.02	17.48%
North	Hebei, Shanxi, Inner Mongolia, Beijing, Tianjin	120.93	12.58%
Northeast	Heilongjiang, Jilin, Liaoning	69.29	7.21%
Eastern	Shandong, Jiangsu, Shanghai, Zhejiang, Fujian, Anhui, Jiangxi	58.84	6.12%
Central	Henan, Hubei, Hunan	44.59	4.64%
South	Hainan, Guangxi, Guangdong	38.50	4.01%
Nationwide/Other regions			
Extremely impoverished areas		**633.51**	**65.91%**
Southwest	Yunnan, Guizhou, Sichuan, Tibet, Chongqing	196.51	20.45%
Northwest	Shaanxi, Gansu, Ningxia, Qinghai, Xinjiang	168.02	17.48%
North	Hebei, Shanxi, Inner Mongolia	119.20	12.40%
Northeast	Heilongjiang, Jilin	57.08	5.94%
Eastern	Anhui, Jiangxi	23.86	2.48%
Central	Henan, Hubei, Hunan	44.59	4.64%
South	Hainan, Guangxi	24.25	2.52%
Relatively impoverished areas		**63.17**	**6.57%**
North	Beijing, Tianjin	1.73	0.18%
Northeast	Liaoning	12.21	1.27%
Eastern	Shandong, Jiangsu, Shanghai, Zhejiang, Fujian	34.97	3.64%
South	Guangdong	14.26	1.48%

Note: Agricultural land includes cultivated land, orchards, forestland, pasture, and other agricultural land [75,76].

Table 3. Coding scheme: take Gansu and Jiangsu as two coding examples.

Province	Prefecture-Level City	County-Level Administrative Unit	Street/Sub-district/Town/Xiang	Community/Village	News Sources
Gansu (Extremely impoverished)	Lanzhou, Tianshui, Baiyin, Jinchang, Jiayuguan, Wuwei, Qingyang, Pingliang, Zhangye, Jiuquan, Dingxi, Longnan, Linxia hui, Linxia Hui Autonomous Prefecture, ***Gannan Tibetan Autonomous Prefecture***	***Hezuo county level city***	Dangzhou street etc.	Zhihema community etc.	Gansu People, Gansu China, Gansu Daily, Gansu Economic Daily, Gannan Headline, Silkroad Pearl net, New Tianshui, Western Business Daily etc.
			Nawu town etc.	Duoheer village etc.	
			Kajiaman Xiang etc.	Haikeer village etc.	
		Lintan county, Zhuoni county, Diebu county, Zhouqu county, ***Xiahe county***, Maqu county, Luqu county	Lapuleng town etc.	Tawa Community etc.; Jiujia village etc.	
			Damai xiang etc.	Shantang village etc.	
Jiangsu (Relatively impoverished)	Nanjing, Suzhou, Wuxi, ***Changzhou***, Zhenjiang, Nantong, Yangzhou, Taizhou, Xuzhou, Lianyungang, Huaian, Yancheng, Suqian	***Jintan district***, Liyang county level city, etc.	Xicheng street	Hebin community etc.; Chengnan village etc.	Jiangsu China, Jiangsu Broadcasting Corporation, Nanjing Broadcasting System, Nanjing Daily, Yangtse Evening Post, Jiangnan Times, Wuxi Daily, etc.
			Jincheng town etc.	Wuxing community etc.; Nanyao village etc.	
			Hualuogeng high-tech sub-district etc.	Yaotang village etc.	

Note: As several County-level administrative units in one Prefecture-level city in each Chinese province, we take place names highlighted in bold and italics as sub-leveled examples.

News articles were coded by two doctoral students in mainland China who had sufficient understanding of China's TPA context and were familiar with content analysis approaches. Two coders were trained using a random sample of 20% of all coded articles. When different codes were applied, the two coders held discussions and selected the most suitable code. First, a reliability pretest involving Holsti's formula [77] indicated an average agreement of 0.87 among the classification of regional categories. Subsequently, Krippendorff's alpha [78] yielded an average reliability value of 0.93 among regional categories; this value is considered acceptable.

3.4. Relevant Analyses and Research Process

First, spatial distribution of media coverage as a proportion of total coverage was calculated by province units from 2017 to 2020. R packages such as Remap, baidumap, and ggplot2 facilitated data extraction and output visualisation. Second, the temporal distribution of media coverage related to different provinces was determined for each month from 2017 to 2020. Term frequency (TF) refers to how often a term appears in the corpus, and inverse document frequency (IDF) decreases the weight of commonly used words and increases the weight of words that are less commonly used in a corpus. The term frequency-inverse document frequency (TF-IDF) statistic is used to measure how important a word is to a document in a corpus [79,80].

An analysis based on TFs can determine whether a data set can capture differences in language over various years. Such an analysis involves determining which words are more or less likely to be used during a certain period by using the log odds ratio; then, the results are assessed as a descriptive base. The number of times each word is used over 2 years is counted, and the log odds ratio for each word is then calculated [80,81]. The formula for the log odds ratio is as follows:

$$\text{Log odds ratio} = \ln\left(\frac{\left[\frac{n+1}{\text{total}+1}\right]\text{pre year}}{\left[\frac{n+1}{\text{total}+1}\right]\text{later year}}\right) \quad (1)$$

where n is the number of times that a given word is used in each period, and total indicates the total number of words in each period.

In terms of panel data, the analysis first formulates the generalized functional model in the sense of [82] as media coverage being a function of Agriculture GDP, effective irrigated area, rural residents' income, pollution control investment, affected area of crop, rural doctors and medical workers, and numbers of receiving social relief in rural areas. The model is stated in the following equation:

$$\begin{aligned} \text{Media coverage}_{(it)} &= \text{Intercept}_{(it)} + \text{GDP}_{(it)} + \text{Effective area}_{(it)} + \text{Income}_{(it)} + \text{pollution invement}_{(it)} \\ &+ \text{affected area}_{(it)} + \text{healthcare resources}_{(it)} + \text{social relief}_{(it)} \end{aligned} \tag{2}$$

Two indices are included in this equation, t for the time series and i for the cross-sections. The dependent variable of this model is media coverage in logarithmic terms. Except for the variable of rural doctors and medical workers, other independent variables are included in logarithmic (log) terms to decrease multicollinearity. In order to avoid pseudo regression, the ADF–Fisher test of the panel unit root test is conducted, and results indicates that all variables were stationary, as the *p*-values were less than 0.05. Then, multicollinearity test and heteroscedasticity test are separately conducted to examine the OLS model. Regarding the multicollinearity test, all variance inflation factor (VIF) values corresponding to the independent variables are below 10, and we could conclude that the model does not have collinearity problems. Regarding the heteroscedasticity test, *p*-value is 0.31 > 0.05 in White test. Thus, the errors have equal variance across the range of the dependent variable and the OLS regression analysis is efficient.

After these preliminary considerations, the approach proposes a panel fixed effects regression (FE), a panel random effects regression (RE), and a panel pool (POOL) regression to be compared for selection. A rule of thumb for the Hausman test indicates that a *p*-value (probability) smaller than 0.05 would indicate the selection of the fixed effects regression model, whereas the opposite would indicate the selection of the random effects regression model. F-test indicates that a $p < 0.05$ would achieve the selection of the fixed effects regression model, whereas the opposite would select the pool effects regression model; BP-test indicates that $p < 0.05$ would select the fixed effects regression model, whereas the opposite would choose the pool effects regression model. In this case, the pool effects regression model should be chosen according to the results of Hausman test, F-test, and BP-test (Table 4).

Table 4. Related tests for panel data regression analysis.

Test	Purpose	Value	Result
Hausman test	FE vs. RE	$\chi^2(7) = 17.073, p = 0.017$	FE
F-test	FE vs. POOL	$F(30,86) = 1.302, p = 0.173$	POOL
Breusch-Pagan test (BP-test)	RE vs. POOL	$\chi^2(1) = 0.640, p = 0.212$	POOL

3.5. Data Processing

Several text mining techniques were used to ensure the accuracy of the analysis results. First, the Jieba system was used [83] for the segmentation of Chinese words and sentences in all articles. Second, a simplified Chinese stop words corpus was imported. This corpus excluded high-frequency numbers and letters used in texts in such a manner that would prevent the meaningful interpretation of results. Third, compound words and specific expressions were detected and combined. Fourth, alternative spellings were accounted for, and names with two and more words were combined into one word so that they would not be counted separately. After the aforementioned processing steps, additional stop words such as the time expressions year and hour, directions such as under and above, and other similar terms were manually excluded according to a common interpretation of the Mandarin language [69]. After data cleaning, 6,162,276 words were used for further text analysis. Statistics analysis in this study was conducted in R [83].

4. Results and Discussion

4.1. Spatiotemporal Distribution of Media Coverage on Poverty Alleviation

Figure 1 shows the spatial distribution of media coverage on TPA as a proportion of total coverage during 2017–2020. In this figure, a darker hue represents a higher media focus on poverty alleviation. Media attention on poverty alleviation is represented by obvious spatial clusters distributed in 31 provinces and cities in China.

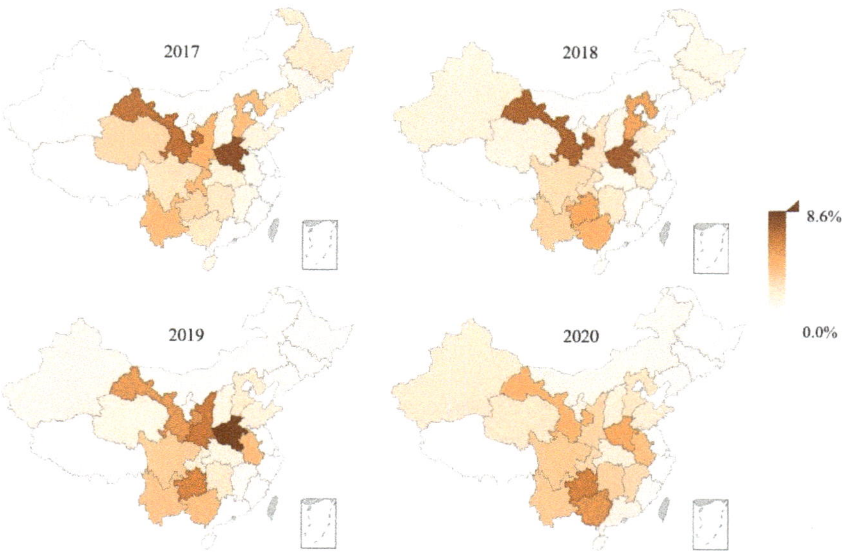

Figure 1. Spatial distribution of media coverage on TPA as a proportion of total coverage in a given year during 2017–2020.

Media attention was mainly distributed on the central, northwest, and southwest regions of China and neighbouring provinces. Substantial differences existed between eastern and western regions, and attention was higher in inland areas than in coastal areas. Studies have divided the total land area of mainland China into an eastern part and a western part by using the Hu Huanyong Line, with 96% of the population of China living in the eastern part, which covers 36% of the total land area of China [84]. Rural low-income individuals have been unevenly distributed in China over the past decade, with 16.4% of them living in northwestern regions and 83.6% of them living in southeastern regions [85,86]. This result indicated that provinces closer to or crossing the Hu Huanyong Line, such as Shaanxi, Henan, and Gansu, had high media coverage; provinces far away from the Hu Huanyong Line, such as the eastern provinces of Xinjiang and Tibet, had the lowest media coverage. This finding is consistent with mainland China's poverty distribution.

In terms of dynamic spatial patterns, media coverage of certain regions expanded yearly. Such coverage concentrated on certain provinces in 2017 and then expanded to cover most regions in China by 2020. In line with the yearly progress on poverty alleviation in China, media focus on regional poverty alleviation has undergone a marked shift from impoverished regions to the non-impoverished and those emerging out of poverty and from regional targeted coverage to widespread national coverage.

With regard to the media coverage of specific regions, Henan, which is a province located in central China, received the most coverage each year. The northwestern province Gansu and southwestern provinces such as Guangxi, Sichuan, Guizhou, and Yunnan also received substantial media attention from 2017 to 2020. Media attention on northeast

China declined from 2017 to 2020. Similarly, Shanxi and Inner Mongolia, as central impoverished provinces with 36 and 31 impoverished counties, respectively, received little attention over a long period. In eastern China, Jiangxi and Anhui, which are relatively impoverished compared with other provinces this region, attracted media attention, which increased yearly.

The results indicate that poverty is dynamic, complex, and multidimensional, and it can have different characteristics in different geographical regions. A lack of natural endowments, poor geographic conditions, and a fragile ecological environment place areas at a major competitive disadvantage [87]. These factors are the main drivers of China's persistent poverty and are reflected in media coverage on poverty alleviation.

4.2. Monthly Distribution of News Coverage of Poverty Alleviation

Figure 2 presents the monthly distribution of provinces' news coverage. In terms of temporal distribution, the second halves of the years 2017, 2018, and 2020 are represented by darker hues than are their first halves, which indicates that media attention was higher in the second halves of these years. Since 17 October 2014 has been designated as National Poverty Alleviation Day; thus, media coverage on poverty alleviation increased every October. TPA activities in the first half of the year were examined at the end of the year. Furthermore, the highest number of poverty-alleviation-related coverage was nationwide, which indicated that the trends of media coverage for TPA closely follow political affairs and important sessions of the Communist Party of China (CPC), such as the 19th National Congress in October 2017, the Third Plenary Session of the 19th Central Committee in January 2018, the Political Bureau Meeting of the CPC Central Committee in July 2018, and the two sessions held in May 2020. After the top-down policy deployment of TPA in late 2017 and 2018, each province had to transform national policy documents into concrete actions for implementation in 2019. When progress in alleviating COVID-19 outbreaks was achieved in March 2020, media attention shifted to the struggle to end poverty. The aforementioned findings indicate that the temporal trends of media coverage were in line with China's TPA mechanism.

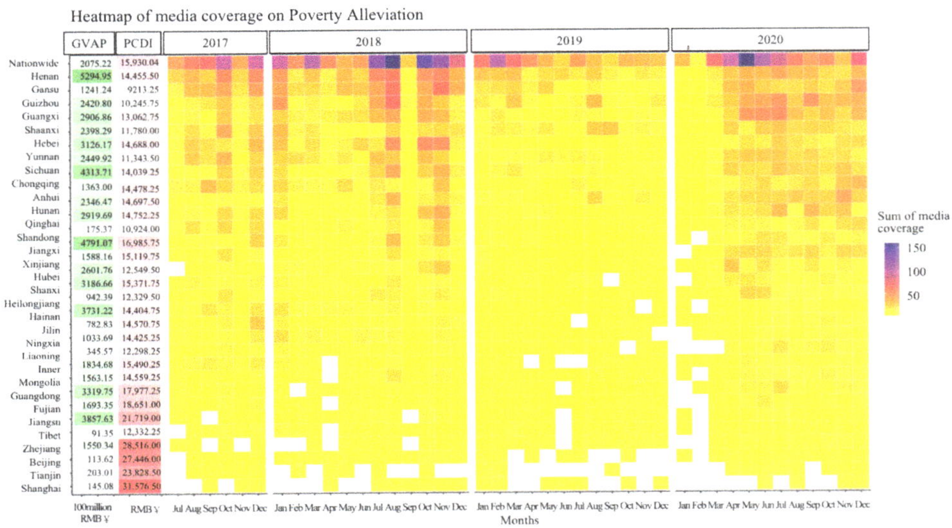

Figure 2. 2017–2020 Heatmap of the monthly distribution of TPA coverage in different provinces. Notes: AGDP is the short of "Agricultural gross domestic product (GDP)", the value of AGDP is the average value from 2017 to 2020; PCDI is the short of "Per Capita Disposable Income of Rural residents", the value of PCDI in Figure 2 is the average value from 2017 to 2020 [87].

Of all the provinces, Henan received the most media attention on TPA because Henan is located in the area of three mountains and one beach and has the largest rural population in China, which reached 45.11 million people in 2020 according to the National Bureau of Statistics of China [87]. Furthermore, agriculture is a crucial contributor to poverty reduction [88,89]. As a large agricultural province and a major source of high-quality agricultural products in China, the value that Henan adds to this primary industry is second only to that added by Shandong and Sichuan [87]. According to the average value of "Agricultural gross domestic product (AGDP)" from 2017 to 2020 in Figure 2, provinces with high agricultural GDP generally received high media attention, while the "Per Capita Disposable Income of rural residents (PCDI)" in these high agricultural GDP provinces is low. As agricultural-dominated provinces with low levels of economic development are the focus of TPA policies, this result also reveals that media attention on poverty issues is related to those agricultural provinces with low economic development.

Media attention focused first on areas of extreme poverty in the central and western regions of China. Attention was then given to Gansu, Guizhou, Guangxi, Shaanxi, Hebei, Yunnan, and Sichuan. Apart from Hebei and Shaanxi, the other five provinces are currently among the seven provinces with the most poverty in China and include national-level impoverished counties. The aforementioned results indicate that impoverished regions and those with and large agrarian populations are factors attracting media attention to TPA. Xinjiang and Ningxia are provinces that have yet to eradicate poverty but lack media attention. Media attention on Ningxia was biased in the first three months of 2020.

From 2017 to 2020, provinces and cities such as Shanghai, Beijing, Tianjin, Jiangsu, Zhejiang, Tibet, Guangdong, and Liaoning lacked media coverage for several months. Tibet was the first area to eradicate poverty among areas with national-level poverty in December 2019. The remaining seven cities and provinces are located in developed areas along the eastern coast [57]. Therefore, media coverage was in line with areas' poverty level and the time at which poverty was eradicated. Particular focus was given to areas that had not eradicated poverty. The more difficult was poverty alleviation, and the longer it took, the more media coverage an area received. Provinces with relatively low poverty levels in eastern China exhibited the lowest media coverage among all the regions.

4.3. Indicators Affecting News Coverage of Poverty Alleviation

Through looking into the results of the pool effects regression model as shown in Table 5, all seven variables significant positively or negatively influenced media coverage which indicates that media attention is related to three dimensions of sustainability. As to the absolute values of the co-efficient among independent variables, TPA-related coverage mainly focused on rural economics sustainability, followed by rural community sustainability, and the least on environmental indicators.

Specifically regarding economic sustainability, "agricultural gross domestic product (GDP)" positively influenced media coverage where a 1% increase in agricultural GDP results in a 0.339% increase in media coverage. The relation between per capita disposable income of rural residents and media coverage was negative, indicating that an increase of 1% in the disposable income of rural residents decreased media coverage by 0.890%. Furthermore, the "effective irrigated area" also negatively influenced media coverage where a 1% increase in "effective irrigated area" decreased media coverage by 0.417%. These results reflected that high media attention on TPA was related to those agricultural provinces with low income and low agricultural irrigation technology.

In terms of environmental sustainability, "pollution treatment" negatively influences media coverage where a 1% increase in pollution results in a 0.167% drop in media coverage. High capital input for pollution treatment is associated with those highly developed areas with serious industrial pollution or dominated by industrialization. "Affected area of crop" positively influences media coverage where a 1% increase in affected crop area boosts media coverage by 0.117%. Combined with our result on economic sustainability, TPA-related coverage focused more on agriculture than industry, especially those with weak resistance to weather agricultural risk.

Table 5. Pool effect regression for sustainability indicators influencing media coverage (2017–2020).

Variables	Coef	Std. Err	t	p
Intercept	5.864	0.979	5.992	0.000 ***
Rural Economic Sustainability				
Agriculture GDP	0.339	0.103	3.297	0.001 **
Rural resident Per Capita Disposable Income	−0.890	0.254	−3.507	0.001 **
Effective Irrigated Area	−0.417	0.114	−3.669	0.000 ***
Environmental Sustainability				
Pollution Control Investment	−0.167	0.04	−4.172	0.000 ***
Affected Area of Crop	0.117	0.046	2.528	0.013 *
Social/Community-level Sustainability				
Rural doctors and medical workers	0.293	0.112	2.616	0.010 *
Receiving Social Relief in Rural Areas	0.037	0.01	3.631	0.000 ***
Adjusted R^2	0.74			
F-statistic	F (7,116) = 51.134			
p	$p = 0.000$			

Notes: * $p < 0.05$, ** $p < 0.01$, *** $p < 0.001$.

Regarding social and community sustainability, "rural doctors and medical workers" positively influenced media coverage and ten thousand medical workers increase rural community media coverage by 0.037%. Social relief played a positive role in media coverage. For every 1% change in the number of those receiving social relief in rural areas, the media coverage positively changed by 0.293%. These results indicated that the role of media in public welfare with reporting is more associated with the input of rural community resources and solving social inequity.

4.4. Analysis of Annual Differences in Keywords Used in Media Reports

Yearly comparisons of six groups based on keywords' log odds ratios were conducted, and the top 25 keywords are presented in Figure 3. The larger the absolute value of the log odds ratio, the better the keyword can represent the prominent feature of this category. Figure 3 indicated that among the common keywords of the two years, the greater the absolute value of log ratio, the more important it can represent in that year. The closer the log ratio is to 0, the more it represents that the prominence of the word in the two years is basically the same.

Keywords in media coverage between 2017 and 2019 were different, and keywords in 2020 were also different from those in the previous three years. The keywords in 2017 were mainly related to state organs and major national conferences, such as the Ministry of Agriculture, the CPC, the Agricultural Fair, the Central Committee of the CPC, and the 18th and 19th National Congresses of the CPC. The number of keywords related to state organs used in 2017 was 1–3 times higher than that used in 2018, 5–10 times higher than that used in 2019, and over 10 times higher than that used in 2020. The number of keywords related to major national conferences used in 2017 was 1–3 times higher than that used in 2018 and 2.5–4 times higher than that used in 2020. The number of keywords related to major national conferences used in 2017 was not markedly different to that used in 2019. Policy-oriented keywords appeared frequently in 2017, and the coverage of such keywords decreased year on year. Furthermore, compared with the keywords used in 2018, 2019, and 2020, the keywords used in 2017 focused more on the names of agricultural products, planting, and breeding and animal husbandry. These keywords were combined with keywords such as "exhibition" and "agricultural fair", which indicated that media attention focused on TPA policies and the promotion of TPA-related agricultural products.

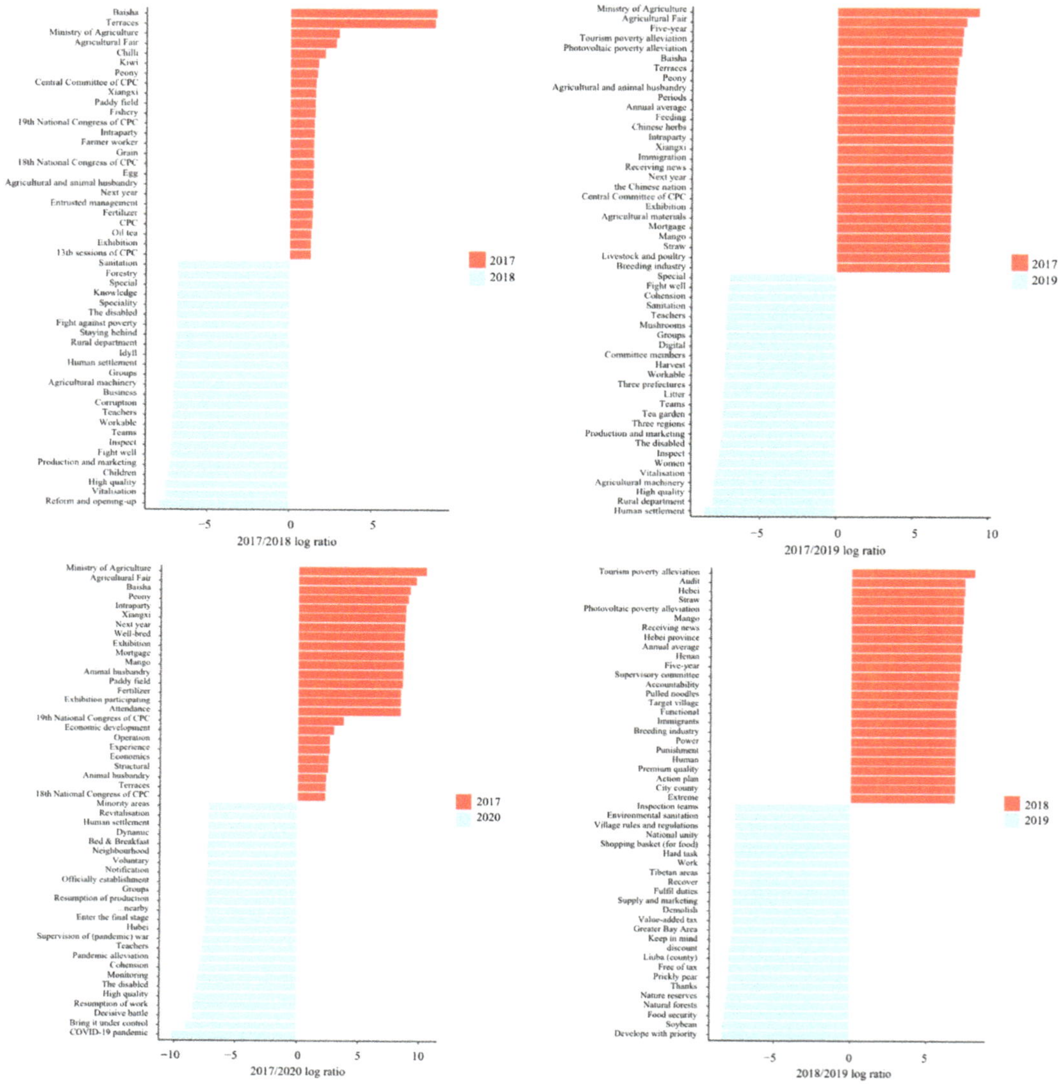

Figure 3. *Cont.*

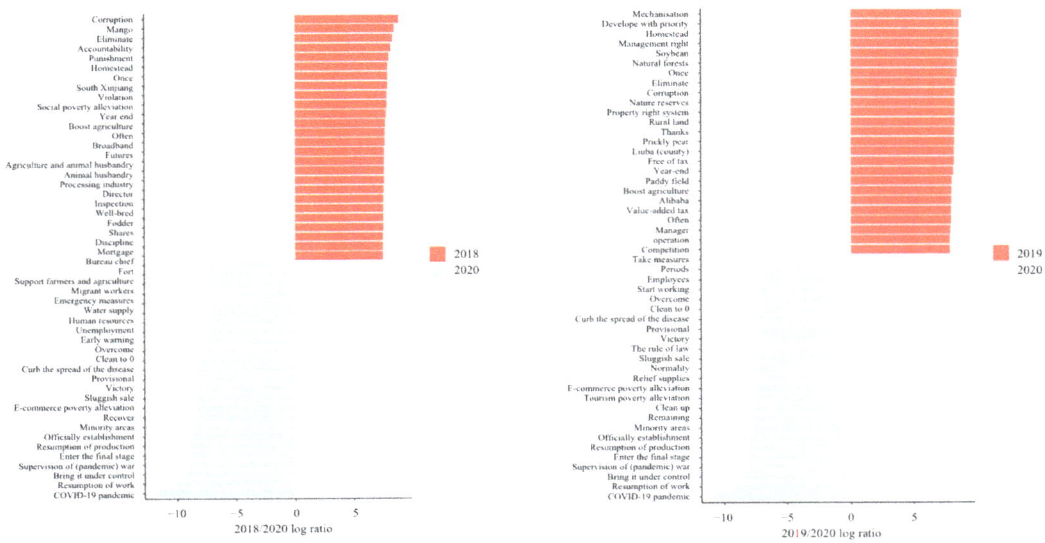

Figure 3. 2017–2020 Pairwise comparisons of keyword log ratios in different years.

In 2018, keywords related to vulnerable groups, such as "children", "staying behind", and "disabled individuals" were similar to those in 2017. Keyword similarities were also noted in relation to volunteering (e.g., "teachers", "teams", and "rural departments") and eradicating poverty (e.g., "fight well", "inspect", "engagement", and "workable"). Negative words such as "corruption" appeared in media articles, which indicated that some problems could not be ignored in terms of poverty alleviation. Keywords in 2019 were related to impoverished areas (e.g., "three districts" and "three states"), vulnerable groups (e.g., "the disabled" and "women"), and those assisting with poverty alleviation ("rural department", "team", "committee members", and "teachers"). Moreover, some keywords were related to new digital technologies and tools that can help low-income individuals in agricultural settings (e.g., "harvest of tea garden" and "mushrooms") and improve welfare in human settlements. In 2020, which was the final year in the fight for poverty alleviation, the COVID-19 pandemic became an obstacle to achieving poverty alleviation goals. Words related to the pandemic were the most prominent that year, which reflected the urgency of pandemic alleviation (e.g., "bring it under control"). Keywords such as "revitalisation" and "human settlement" were less prominent; keywords related to the top-down decisions and policies related to economic activities, such as "resumption of work", "resumption of production", and "resumption of normal life", were more prominent. Moreover, phrases such as "decisive battle" and "supervision of war" reflected the determination to persevere in the "war" against the COVID-19 pandemic in China.

Keywords related to poverty alleviation actions, such as "audit", "supervisory committee", "accountability", and "punishment" were more prominent in 2020 than in 2019. Similarly, keywords such as "corruption", "accountability", "punishment", "violation", "inspection", "discipline", and "director" were more salient in 2018 than in 2020. The aforementioned results indicated that political elites often abused their power for personal gain and violated the law, and such activities affected poverty alleviation work. Media attention in 2018 mainly focused on reflections on TPA mechanisms the exposing of related problems.

Keywords related to natural ecological environment, such as "natural forest", "protected land", "environmental sanitation", and "Tibetan area", appeared more in 2019 than in 2018, which indicated the increased focus on the fragility of the natural environment.

When combined with media coverage of contiguous areas with substantial poverty, such as the "Three Regions" and "Three Prefectures", which were used in the previous comparison involving the year 2017, media coverage in 2019 was concentrated on the spatial dimensions of ecological fragility and poverty alleviation in marginal areas such as "Tibetan areas" and "minority areas".

Compared with the keywords in 2020, those in 2019 were more related to agricultural modernisation and professional management (e.g., "mechanisation", "management right", "property right system", "operation", and "Alibaba e-commerce"). The combination of the aforementioned keywords with the word "digital" highlighted that compared with the media coverage in 2017, that in 2019 was more focused on alleviating poverty among marginalised groups in impoverished areas by using technology and modern poverty alleviation approaches. A trend of large-scale and specialised poverty alleviation measures was identified. In addition to theoretical, organisational, and institutional innovation [72], technological innovation is another means of poverty alleviation with Chinese characteristics.

In contrast to the media coverage in 2018 and 2019, that in 2020 focused on challenges to TPA work related to the COVID-19 pandemic, as indicated by the usage of keywords such as "unsalable", "migrant workers", and "unemployment". The keywords used in 2020 were related to TPA actions, such as "curb the spread of the disease", "resumption of work", "early warning", "water supply", "emergency", and "support farmers and agriculture", and achievement, such as "overcome". Furthermore, discourse on encouragement-based emotions revealed the destiny of communities sparing no effort and uniting social forces to overcome difficulties. Media empowerment is a means of poverty alleviation; it can involve maintaining social stability during the COVID-19 pandemic, stimulating the endogenous self-empowerment of low-income individuals, and promoting changes in cognition and action.

4.5. Regional Differences in Media Coverage Keywords

Term frequency-inverse document frequency (TF-IDF) measures how relevant a word to a certain category of documents, and it is widely used to extract the most representative words as features for text classification. Figures 4 and 5 display the TF-IDF results by region. The greater TF-IDF, the more representative the keyword is to distinguish the regional categories. Figure 4 shows that in each regional category, keywords with greater TF-IDF values are more representative of the region-based TPA coverage. Most of the top 20 keywords for regional poverty alleviation coverage in Figure 4 were nouns; they were related to the main crops or breeding organisms that are suitable for the local climate, soil, and natural conditions. Moreover, keywords related to the local landscapes, buildings, and customs of southwestern minority areas were salient. Agricultural contacts, business opportunities, and local industries for TPA was promoted in eastern China according to locational advantages. The Maritime Silk Road aided local poverty alleviation efforts. Poverty alleviation through business development was highlighted by the media, especially in relation to local primary industries such as arable and pastoral farming. Media coverage also focused on poverty alleviation in the central and western regions of mainland China as well as developing countries worldwide.

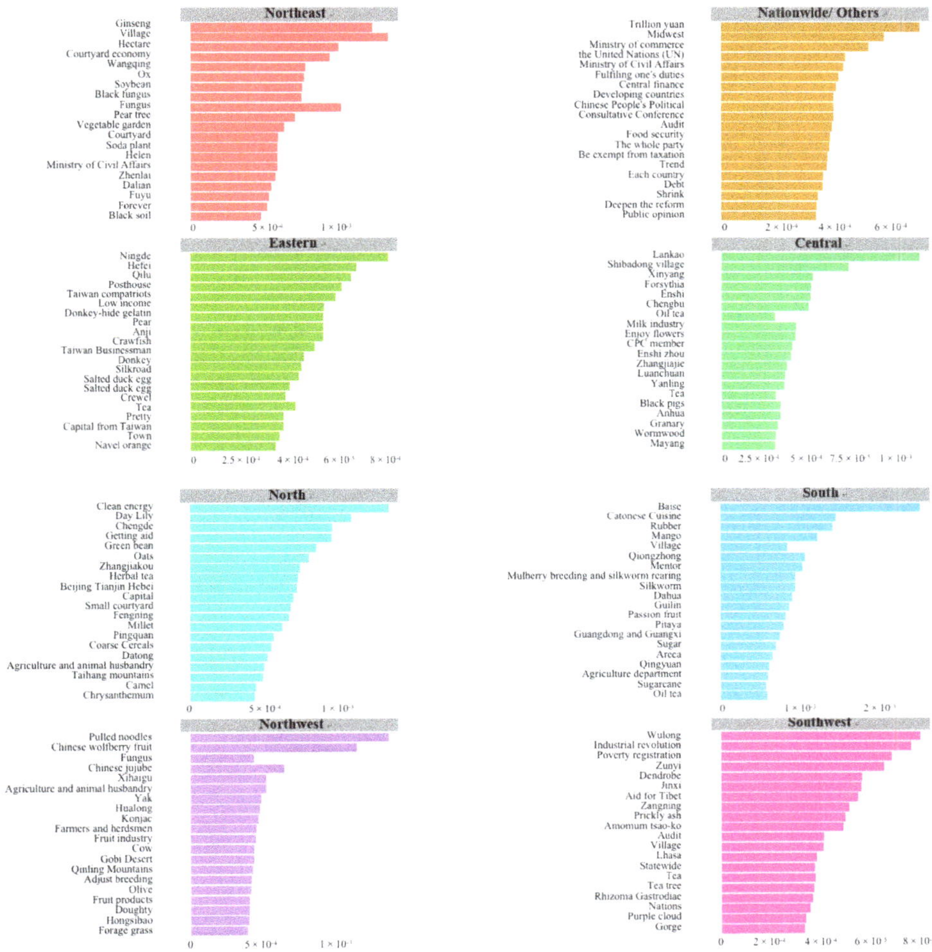

Figure 4. Term frequency-inverse document frequency (TF-IDF) results by region during 2017–2020.

Figure 5 shows that in each regional category, keywords of attributes with greater TF-IDF values can better represent the emotional characteristics of the region-based TPA coverage. As indicated in Figure 5, the top 20 keywords related to characteristics and attributes were extracted; they included adjectives, adverbs, verb, adverbs, idioms, phrases, and other modifiers. First, the number of negative words was small, and they were related to descriptions of spatial marginalisation based on geographic disadvantages. Specifically, the words "drought", "barren", "fragile", and "soil erosion" mainly appeared in media coverage of north-eastern areas, north China, and north-western and south-western regions. Words and phrases such as "live at the mercy of the elements", "make use of local resources", "formidable project", and hardship were prominent in media coverage of north-west and south-west regions, which reflected the harsh environmental conditions in western areas and the hardship of people's lives there. By contrast, numerous words and phrases with rich and diverse meanings, such as "smooth", "prosperous", "rich", "clear", "roomy", "beautiful", "make a difference", "blooming everywhere", "vigorous", "dense" and "progress", indicated that some residents in certain regions had a better life than did those in other regions. These positive adjectives are in contrast to words associated with geographical disadvantages, which indicated that impoverished areas had been improved through TPA projects. Some positive and inspiring words and phrases, such as "heart to heart", "concerted efforts", "pragmatic", and "constant dropping wears away the stone", were also common in the media coverage of various regions. These words and phrases from positive local stories shared by local media are related to initiative and enthusiasm and emphasise the collectivist spirit of solidarity, reflecting a form of discourse empowerment.

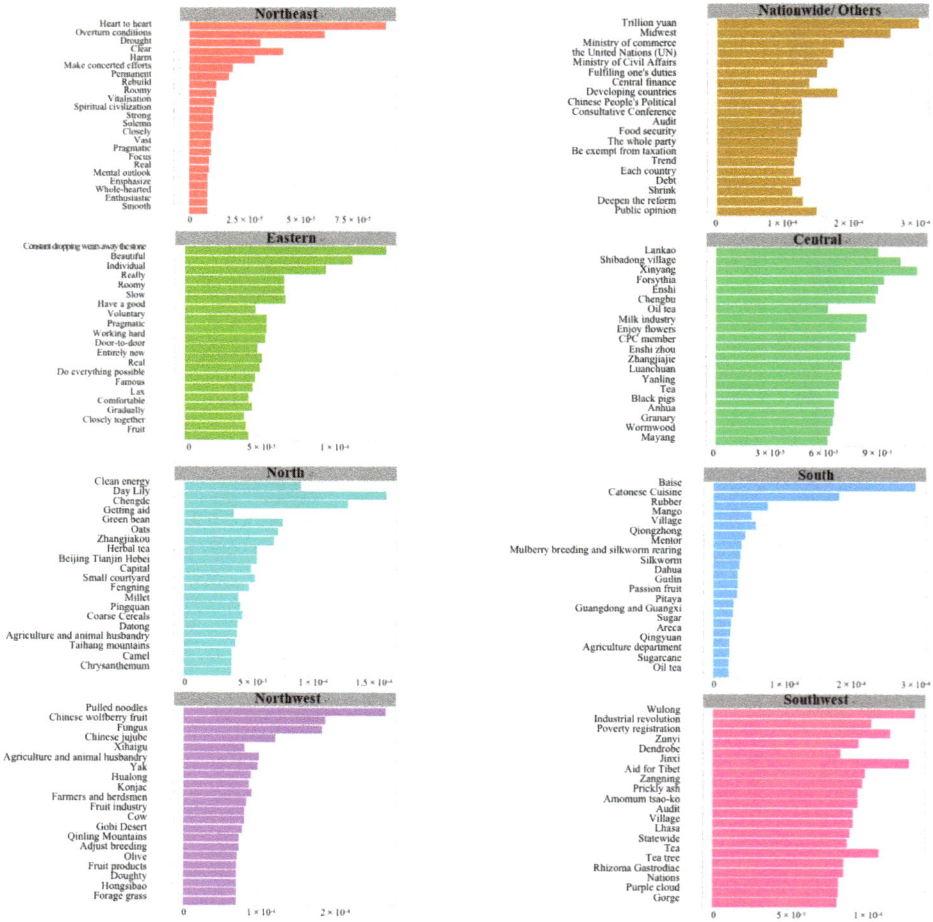

Figure 5. Term frequency-inverse document frequency (TF-IDF) results with characteristics and attributes by region during 2017–2020.

Some areas lack natural endowments and are characterised by poor living conditions, giving residents a substantial competitive disadvantage [57,85]. However, the marginalisation of spatial dimensions can also inspire people to pursue subjective initiatives of poverty alleviation. The core keywords in poverty-alleviation-related media coverage were "increasing people's confidence" and "enriching their knowledge", which reflected the collective consensus and wish for a better life after poverty alleviation.

"Weak" and "lax" were words used to describe the inaction of local party organisations. These words were common in poverty alleviation coverage related to north, south, and east China. Words and phrases related to work style, such as "mendacious", "incorruptibility", and "evil forces", were concentrated in central and southern China, which reflected the shortcomings of bureaucratic systems in achieving local poverty alleviation in these regions. Furthermore, "door to door", "focus", "emphasis", and "refined" were highlighted in the media coverage of the aforementioned regions. The aforementioned results indicated that problems existed in the local grassroot bureaus and targeted nature of TPA work.

In general, in north and west China, poverty alleviation activities were mainly focused on the poverty caused by spatial marginalisation. In central, east, and south China, poverty alleviation work was mainly conducted to overcome the poverty caused by human factors, such as the ineffectiveness of local bureaucrats. Nationwide/other media coverage emphasised that state power can solve problems caused by inappropriate local governance.

4.6. Differences in Keywords Used in Various Regions

Figure 6 shows that among the common keywords of the two regions, the greater the absolute value of log odds ratio, the more important the keyword can represent in that region. The closer the log ratio is to 0, the more it represents that the importance of the keyword in both two regions is basically the same. As depicted in Figure 6, compared with the relatively impoverished areas of the nine eastern provinces, higher media coverage, a higher frequency of certain keywords, and more diverse word types were observed for extremely impoverished areas in the 22 central and western provinces. Keywords related to ESPAR, such as "ethnic groups", "highways", and "autonomous regions", were the most prominent keywords in the 22 central and western provinces. The aforementioned results indicate that media coverage keywords in extremely impoverished areas were based on geographic marginalisation. "Farmhouse touring", "tourism poverty alleviation", "model zone", and words related to crops and livestock (e.g., "oil tea", "pepper", "walnut", "beef cattle", "prickly ash", "pig", and "kiwi fruit") as well as e-commerce poverty alleviation and infrastructure construction (e.g., "[road] hardening", "sanitation", and "green water") indicate that a modern poverty alleviation model was embedded in the extremely impoverished areas. Furthermore, the beneficial effects of poverty alleviation were highlighted through quantifiers and phrases such as "10,000 tonnes", "output value", "net income", and "more than 10,000 yuan" as well as words related to low-income groups' interests such as "10,000 households", "average household", and "net income".

Figure 6. Log ratios of keywords by region during 2017–2020.

Terms related to arable farming were more widely embedded in the media coverage related to poverty alleviation in areas with extreme poverty than in the nationwide coverage or media coverage in other regions. Words related to agricultural crops and related products were more prominent in areas with absolute poverty (e.g., "edible fungi", "oil tea", "pepper", "walnut", "fruit", "prickly ash", "grape", "kiwi fruit", and "mushroom") than in other areas. Words and phrases related to leisure and amusement, such as "picking", "farmhouse touring", "bed & breakfast", "orchard", and "tea garden", were connected with people's livelihood (i.e., "enriching the people"), which indicates that for areas with extreme poverty, poverty alleviation emphasised the revitalisation and activation of rural life. The promotion of administrative forces (words such as "county party committee") was the main driving force of poverty alleviation in extremely impoverished areas. Finally, similar to the keywords related to the media coverage of relatively impoverished areas, words related to profit, achievements, and the interests of low-income households highlighted the effectiveness of poverty-alleviation-related governance.

Lower media coverage and fewer types of keywords were noted for the relatively impoverished areas than for the extremely impoverished areas; however, the meanings of the keywords used in the media coverage of the relatively impoverished areas are somewhat flourish. First, in terms of geographical locations and the effects of climatic and environmental factors (e.g., water, soil, and precipitation), the extremely impoverished provinces in the central and western regions were suited to arable farming and animal husbandry. The provinces located in eastern coastal areas were more suited to the development of the aquaculture industry. Second, some words and phrases such as "support", "counterpart support", "cooperation", "pairing", "hand in hand", "co-construct", and "blood transfusion" (to give aid and support) revealed poverty alleviation to be a common social mission involving solidarity that shaped the collective consensus of communities. In contrast with the action of getting rid of poverty in extremely impoverished areas, the poverty alleviation mechanisms in relatively impoverished areas involved prosperity of industry specialisation and urbanisation (e.g., see words and phrases such as "Academy of Agricultural Sciences", "special commissioners", "Neighbour", and "below the basic living allowance").

In contrast to the nationwide coverage or media coverage in other regions, the media coverage in relatively impoverished areas included keywords related to leisure-oriented arable agricultural activities such as "orchards" and "tea gardens". Media attention was concentrated on the south-eastern areas of China, especially Guangdong and Fujian, which indicated that the media paid more attention to the TPA practices of these areas with developed agricultural industries than it did to those of other areas. Media coverage of relatively impoverished areas was related to local practices and assigned developed areas the social responsibility to pursue the values of "sharing" and "great harmony".

Different from media coverage of extremely impoverished areas, international poverty alleviation strategies and domestic industrial poverty alleviation modes, such as manufacture and finance were highlighted in nationwide/other coverage. Examples of words and phrases used in nationwide/other media coverage are "powerful country", "giant", "developing countries", "whole party", "all nations" and "running a country", and examples of words and phrases related to international commerce are "tax reduction", "fee reduction", "exemption", "value-added tax", "import", "open to the outside world", and "equality". Moreover, words related to bureaucracies such as "Ministry of Civil Affairs", "Ministry and Party Group" and "Ministry of Education" as well as words related to officials" statuses and positions such as "Liu Yongfu" and "Director-general" indicated that poverty alleviation measures at the international and national level were dominated by bureaucracies with a top-down approach.

Compared with media coverage of relatively impoverished areas, state-dominated coverage and coverage of extremely impoverished areas were highlighted more in nationwide and other media coverage. By contrast, local autonomy was more important than administration-led activities in relatively impoverished areas. Nationwide/other media coverage paid attention to international assistance, especially among developing countries, and the coverage of extreme poverty in China was also emphasised. The aforementioned results are consistent with the concept of an "environmentally bundled economic interest", which indicates that China's local governments fulfil the central government's new political mission and satisfy the demand for local economic development [90].

5. Conclusions, Contributions, and Limitations

This study employed regression analysis of panel data and content analysis with a text mining approach to investigate the spatiotemporal distribution and influence factors of Targeted Poverty Alleviation (TPA)-related coverage, and keywords embedded in media coverage. The results of this study revealed that, first, the media coverage was higher in central inland areas than in southeast coastal areas. Each year, more attention was given to relatively impoverished areas than to extremely impoverished areas in nationwide/other media coverage. Second, regarding temporal characteristics, media coverage on poverty alleviation was high in 2017 and 2018, low and scattered in 2019, and high in 2020. The monthly distribution of media coverage was characterised by a midyear peak and a peak at the end of the year. These patterns conform to the schedule of important political events, such as committee meetings. Third, the temporal distribution of media keywords demonstrated that policy propaganda was highlighted in 2017, diversified group characteristics and problem exposure of local TPA practice were emphasised in 2018, digital and modernised practices were prominent in 2019, and positive actions for addressing the poverty caused by the COVID-19 pandemic were highlighted in 2020.

For the purpose of poverty alleviation, there are several trade-offs between economic, environmental, and community/social sustainability. Factors influencing TPA-related coverage are significantly related to key economic, environmental, and community sustainability indicators we evaluated. Specifically, high media attention on TPA is related to those agricultural provinces with low income, low agricultural technology, and weak risk resistance capacity in agriculture rather than those highly developed areas dominated by industrialisation. Furthermore, media as a social power of poverty alleviation tends to be associated more with the input of community resources and addressing social inequity.

Keywords describing extremely impoverished areas emphasised spatial poverty, the welfare of low-income households, the revitalisation of rural areas, and the effectiveness of poverty governance. Keywords related to relatively impoverished areas focused on local autonomy in developing diversified industries, collective responsibility, and the concept of sharing and harmony in rural China. China's local governments developed an approach based on "environmentally bundled economic interests" that simultaneously fulfils the central government's new political mission and local economic development. Finally, the positive discourse in media coverage indicated that TPA stakeholders pursued subjective initiatives. This result is consistent with the finding of Tsai and Liao [91] that poverty alleviation has been promoted with substantial enthusiasm. Consequently, media coverage was consistent with China's poverty alleviation mechanism. The implementation of TPA policies from top to bottom can be promoted effectively, and campaign-style enforcement can be achieved.

Previous policy following a "pollute first, clean up later" linked to the idea of progressive economic development, which led poverty alleviation only focused on agricultural intensification and progressive economics development but with limited understanding of the impacts on ecological dynamics [92]. Environmental degradation leads to lower agricultural outputs and rural incomes [93]. For example, rice yields declined due to increasing fertilizer application rates [94]. In addition, losses of cultivated land transferred to other land uses due to urbanization in some regions [92]. However, compared with 69.10% of agricultural land use composition in 2008 [75], statistics in 2019 is 72.49%; especially cultivated land had increased from 12.80% to 13.30% [76]. Consequently, combined with the results of media attention and media discourse in this study, environmentally bundled economic interests and social participation of current Targeted Poverty Alleviation creates trade-offs and synergetic relationships, which made implementation of TPA effective.

This study contributes to the poverty governance literature through various means. First, in terms of theory, this study not only extends the importance of the marginality theory of poverty governance but also provides a theoretical marginality-based perspective for understanding poverty governance in media settings. Second, this study is the only one to use text mining to examine the representations of Chinese TPA in online media for providing an integrative perspective of TPA media representation. Third, the results of this study are vital for understanding poverty governance in different regions through dynamic patterns of media coverage and keywords related to TPA, especially in regions where poverty alleviation projects and social capital are seldom combined. The results of this study can provide insight for local governments to assess local poverty scenarios and for the central government in terms of allocating resources to poverty alleviation.

Regarding practical implications, this study can inspire new strategies for alleviating poverty in various regions. First, for extremely impoverished areas, this study revealed that the top-down administrative-led approach was used to overcome spatial poverty; after poverty is eradicated, market-oriented operations and full local autonomy should be learned from eastern areas to establish a long-term mechanism for poverty alleviation. Second, the growth of China's agricultural industry is particularly crucial for poverty alleviation. This industry is four times as effective in reducing poverty as secondary and tertiary industries are [85,95]. In this study, the spatial distribution of media attention was focused on large agricultural provinces with large agrarian populations; these provinces are major bases for high-quality agricultural products in China. The keywords embedded in media coverage indicated that throughout all the study years, agricultural product promotion in extremely impoverished areas and the experiences of economically developed agricultural areas were essential for poverty eradication. As technical efficiency is one assessment of agricultural economic sustainability, accordingly, poverty alleviation projects of the agricultural industry should be combined with modernised technological innovations to achieve poverty alleviation with Chinese characteristics.

This study provides strategies related to antipoverty governance for the media in the postpoverty era. First, for extremely impoverished areas, the media should pay more attention to the process of local TPA by comparing regional differences in media ecology in areas with similar natural endowments or geographical proximity to establish a long-term mechanism for the media's representation of TPA approaches. Second, news media is considered a major force against poverty, keywords in media coverage indicated discourse empowerment with the dual function of increasing people's confidence and enriching their knowledge. This coverage focused on local stories on getting rid of poverty and becoming rich. Thus, media discourse should focus on providing detailed microlevel perspectives on local TPA cases instead of brief descriptions. Positive local stories should be shared by local media as a form of discourse empowerment to inspire low-income households to pursue self-empowerment. Third, media coverage should improve public awareness of TPA sustainability and promote an optimal mechanism established for the agricultural sector and the rural community of impoverished areas.

Inevitably, several limitations hindered this study. First, the term frequency-inverse document frequency (TF-IDF) and log odds ratio were used in this study. These parameters focus on keywords embedded in articles but not on topics. In future research, latent Dirichlet allocation, which is a topic modelling algorithm for extracting topics present in sets of patent documents, will be adopted [96]. Subsequently, the trends of different topics represented in media coverage will be explored. Second, although NTV (www.ntv.cn, accessed on 1 January 2021) as a convergence platform integrates mainstream news sources and news from state-controlled media and local media, the expressions of individuals such as bloggers and opinion leaders are lacking. Prospective studies could enable deeper interpretations than the current format affords by including diverse media voices and conducting different analyses. Finally, the TPA approach was proposed in November 2013; however, news articles in this study covered only the three years before China eradicated absolute poverty in November 2020. Thus, NTV does not provide sufficient information on the TPA in China over time.

Author Contributions: Conceptualisation, Y.S.; methodology, Y.S.; software, Y.S.; validation, Y.S.; formal analysis, Y.S.; investigation, Y.S.; resources, Y.S. and S.-N.Y.; data curation, Y.S.; writing—original draft preparation, Y.S. and S.-N.Y.; writing—review and editing, Y.S. and S.-N.Y.; visualisation, Y.S.; project administration, Y.S.; funding acquisition, S.-N.Y. and Y.S. All authors have read and agreed to the published version of the manuscript.

Funding: This research was funded by the Ministry of Science and Technology, Taiwan (grant number 109-2221-E-305-011-MY2), Fundamental Research Funds for the Central Universities, Hohai University, China (grant number B220201066), and University System of Taipei Joint Research Program (grant number USTP-NTPU-TMU-111-02).

Institutional Review Board Statement: Not applicable.

Informed Consent Statement: Not applicable.

Data Availability Statement: The data presented in this study are available on request from the corresponding authors.

Acknowledgments: Special thanks to Aaron K. Hoshide for his contribution in highly promoting the review process of the earlier version of this article. The authors would also like to thank the anonymous reviewers for their constructive comments.

Conflicts of Interest: The authors declare no conflict of interest.

References

1. World Bank. *World Development Report 2001*; Oxford University Press: Oxford, UK, 2001.
2. Sen, A. *Inequality Reexamined*; Oxford University Press: Oxford, UK, 1992.
3. Sen, A. *Development as Freedom*; Oxford University Press: Oxford, UK, 1999.
4. Sen, A. *Commodities and Capabilities*; Oxford University Press: Oxford, UK, 1999.
5. United Nations. *The Copenhagen Declaration and Programme of Action*; World Summit for Social Development: New York, NY, USA, 1995.
6. Bird, K.; Shepherd, A. Livelihoods and chronic poverty in semi-arid Zimbabwe. *World Dev.* **2003**, *31*, 591–610. [CrossRef]
7. Goodhand, J. Enduring disorder and persistent poverty: A review of the linkages between war and chronic poverty. *World Dev.* **2003**, *31*, 629–646. [CrossRef]
8. Green, M.; Hulme, D. From correlates and characteristics to causes: Thinking about poverty from a chronic poverty perspective. *World Dev.* **2005**, *33*, 867–879. [CrossRef]
9. Deller, S. Rural poverty, tourism and spatial heterogeneity. *Ann. Tour. Res.* **2010**, *37*, 180–205. [CrossRef]
10. Dou, H.; Ma, L.; Liu, S.; Fang, F. Identification of rural regional poverty type based on spatial multi-criteria decision-making: Taking Gansu Province, an underdeveloped area in China, as an example. *Environ. Dev. Sustain.* **2021**, *24*, 3439–3460. [CrossRef]
11. Dain, S. *Conceptual and Operational Framework on Exclusion in Social Protection for Health*; Pan American Health Organization: Washington, DC, USA, 2003.
12. Larsen, J.E. *Who Cares About and for Marginal People*; Danish Social Science Research Council: Copenhagen, Denmark, 2002.
13. Perlman, J. Metamorphosis of Marginality: Rio's Favelas 1969–2002. Available online: https://www.kas.de/c/document_library/get_file?uuid=dd6076a1-2db1-53b6-253e-dba86fe53fa2&groupId=252038 (accessed on 7 May 2002).
14. Desmond, M.; Western, B. Poverty in America: New directions and debates. *Annu. Rev. Sociol.* **2018**, *44*, 305–318. [CrossRef]
15. Falola, T.; Odey, M.O. (Eds.) *Poverty Reduction Strategies in Africa*; Routledge: London, UK, 2018.
16. Chaudhry, I.S.; Malik, S.; Ashraf, M. Rural poverty in Pakistan: Some related concepts, issues and empirical analysis. *Pak. Econ. Soc. Rev.* **2006**, *44*, 259–276.
17. Wang, Z.; Guo, S. Politics of Poverty Governance: An Introduction. *J. Chin. Political Sci.* **2022**, *27*, 205–219. [CrossRef] [PubMed]
18. Grindle, M.S. Good enough governance: Poverty reduction and reform in developing countries. *Governance* **2004**, *17*, 525–548. [CrossRef]
19. Davis, T.J. Good governance as a foundation for sustainable human development in sub-Saharan Africa. *Third World Q.* **2017**, *38*, 636–654. [CrossRef]
20. Singh, P.K.; Chudasama, H. Evaluating poverty alleviation strategies in a developing country. *PLoS ONE* **2020**, *15*, e0227176. [CrossRef] [PubMed]
21. Fang, Y.; Zhang, F. The future path to China's poverty reduction: Dynamic decomposition analysis with the evolution of China's poverty reduction policies. *Soc. Indic. Res.* **2021**, *158*, 507–538. [CrossRef] [PubMed]

22. Cheng, Y.; Liu, H.; Wang, S.; Cui, X.; Li, Q. Global action on SDGs: Policy review and outlook in a post-pandemic Era. *Sustainability* **2021**, *13*, 6461.
23. Redclift, M. The multiple dimensions of sustainable development. *Geography* **1991**, *76*, 36–42.
24. Malerba, D. The trade-off between poverty reduction and carbon emissions, and the role of economic growth and inequality: An empirical cross-country analysis using a novel indicator. *Soc. Indic. Res.* **2020**, *150*, 587–615. [CrossRef]
25. Li, X.; Gao, Q.; Tang, J. Who are identified as poor in rural China's targeted poverty alleviation strategy? Applying the multidimensional capability approach. *J. Chin. Political Sci.* **2022**, *27*, 221–246. [CrossRef]
26. Zhou, Y.; Guo, Y.; Liu, Y.; Wu, W.; Li, Y. Targeted poverty alleviation and land policy innovation: Some practice and policy implications from China. *Land Use Policy* **2018**, *74*, 53–65. [CrossRef]
27. The State Council Information Office of the People's Republic of China. *Poverty Alleviation: China's Experience and Contribution*; The State Council Information Office of the People's Republic of China: Beijing, China, 2021.
28. La Ferrara, E. Mass media and social change: Can we use television to fight poverty? *J. Eur. Econ. Assoc.* **2016**, *14*, 791–827. [CrossRef]
29. Anorue, L.I.; Obayi, P.M.; Chima, O.A. The influence of media coverage of government programmes on Poverty Alleviation in Nigeria. *Int. J. Arts Humanit.* **2011**, *3*, 27–38.
30. Smith-Carrier, T.; Lawlor, A. Realising our (neoliberal) potential? A critical discourse analysis of the poverty reduction strategy in Ontario, Canada. *Crit. Soc. Policy* **2017**, *37*, 105–127. [CrossRef]
31. Cheng, X.; Chen, J.; Jiang, S.; Dai, Y.; Zeng, J.; Shuai, C.; Liu, G. Pursuing sustainable development goals: A review of renewable energy and poverty alleviation nexus. *Environ. Dev.* **2021**, *40*, 100679. [CrossRef]
32. Geall, S.; Shen, W.; Gongbuzeren. Solar energy for poverty alleviation in China: State ambitions, bureaucratic interests, and local realities. *Energy Res. Soc. Sci.* **2018**, *41*, 238–248. [CrossRef]
33. Zhang, H.; Xu, Z.; Sun, C.; Elahi, E. Targeted poverty alleviation using photovoltaic power: Review of Chinese policies. *Energy Policy* **2018**, *120*, 550–558. [CrossRef]
34. Yan, J.; Zhang, Y.; Chen, L.; Huang, L.; Huang, Y. The Impact of Video Information and Publisher's Characteristics in Tik Tok Platform on the Spreading Effect of Poverty Alleviation by E-Commerce. In *International Conference on Management Science and Engineering Management*; Springer: Cham, Switzerland, 2020; pp. 763–777.
35. Gurung, G.S.; Kollmair, M. *Marginality: Concepts and Their Limitation*, 2nd ed.; NCCR North-South: Bern, Switzerland, 2007.
36. Leimgruber, W. *Between Global and Local: Marginality and Marginal Regions in the Context of Globalization and Deregulation*; Ashgate: London, UK, 2004.
37. Sommers, L.M.; Mehretu, A.; Pigozzi, B.W. Towards typologies of socioeconomic marginality: North/South comparisons. In *Marginality in Space-Past, Present and Future: Theoretical and Methodological Aspects of Cultural, Social and Economic Parameters of Marginal and Critical Region*; Jussila, H., Majoral, R., Mutambirwa, C.C., Eds.; Ashgate: Farnham, UK, 2019; pp. 7–24.
38. Brodwin, P. *Marginality and Cultural Intimacy in a Transnational Haitian Community*; No. 91; University of Wisconsin-Milwaukee: Milwaukee, WI, USA, 2001.
39. Darden, J.T. Blacks and other racial minorities: The significance of color in inequality. *Urban Geogr.* **1989**, *10*, 562–577. [CrossRef]
40. Gans, H.J. From "underclass" to "undercaste": Some observations about the future of the post-industrial economy and its major victims. In *Urban Poverty and the Underclass: A Reader*; Mingione, E., Ed.; Blackwell: Oxford, UK, 1996; pp. 141–152.
41. Hoskins, I. Combining work and care for the elderly: An overview of the issues. *Int. Labour Rev.* **1993**, *132*, 347–369.
42. Massey, D. *Space, Place and Gende*; Polity Press: Cambridge, UK, 1994.
43. Andersen, J.; Larsen, J.E. Gender, poverty and empowerment. *Crit. Soc. Policy* **1998**, *18*, 241–258. [CrossRef]
44. Cullen, B.T.; Pretes, M. The meaning of marginality: Interpretations and perceptions in social science. *Soc. Sci. J.* **2000**, *37*, 215–229. [CrossRef]
45. Bourdieu, P. *The Weight of the World: Social Suffering in Contemporary Society*; Polity Press: Cambridge, UK, 1999.
46. Leimgruber, W.; Chang, C.Y.D. (Eds.) Rural space between marginality and centrality. Approaches to marginality issues in rural areas. In *Rural Areas between Regional Needs and Global Challenges*; Springer: Cham, Switzerland, 2019; pp. 3–14.
47. Mehta, M. *Cultural Diversity in the Mountains: Issues of Integration and Marginality in Sustainable Development*; Consultation on the Mountain Agenda: Lima, Peru, 1995.
48. Government Japan. *Japan's Voluntary National Review Report on the Implementation of the Sustainable Development Goals*; Government Japan: Tokyo, Japan, 2017.
49. Cornwall, A.; Brock, K. What do buzzwords do for development policy? *A critical look at 'participation,''empowerment'and 'poverty reduction.' Third World Q.* **2005**, *26*, 1043–1060.
50. World Bank. *The World Bank: Annual Report 2015*; The World Bank: Washington, DC, USA, 2015.
51. Buccini, G.; Harding, K.L.; Hromi-Fiedler, A.; Pérez-Escamilla, R. How does "Becoming Breastfeeding Friendly" work? A programme impact pathways analysis. *Matern. Child Nutr.* **2019**, *15*, e12766. [CrossRef] [PubMed]
52. Wasserman, E. Ethics of poverty coverage. *J. Mass Media Ethics* **2013**, *28*, 138–140. [CrossRef]
53. Roth, M. Lack of Media Coverage on Poverty. Available online: https://www.borgenmagazine.com/lack-media-coverage-poverty/ (accessed on 6 August 2014).
54. Kim, S.H.; Carvalho, J.P.; Davis, A.C. Talking about poverty: News framing of who is responsible for causing and fixing the problem. *J. Mass Commun. Q.* **2010**, *87*, 563–581. [CrossRef]
55. Kim, S.H.; Shanahan, J.; Choi, D.H. TV news framing supports societal poverty solutions. *Newsp. Res. J.* **2012**, *33*, 101–112. [CrossRef]

56. Li, X.Y.; Tang, Y. Practice and optimization suggestion of precise poverty alleviation in Jiangxi province under the rural revitalization strategy. In Proceedings of the 6th International Conference on Management Science and Management Innovation, Changsha, China, 17–18 May 2019; pp. 94–99.
57. Qi, X.; Ye, S.; Xu, Y.; Chen, J. Uneven dynamics and regional disparity of multidimensional poverty in China. *Soc. Indic. Res.* **2021**, *159*, 169–189. [CrossRef]
58. Hsieh, H.F.; Shannon, S.E. Three approaches to qualitative content analysis. *Qual. Health Res.* **2005**, *15*, 1277–1288. [CrossRef]
59. Zhang, Y.; Wildemuth, B.M. Qualitative analysis of content. In *Applications of Social Research Methods to Questions in Information and Library Science*; Wildemuth, B.M., Ed.; Libraries Unlimited: Westport, CT, USA, 2009; pp. 308–319.
60. Weber, R.P. *Basic Content Analysis*, 2nd ed.; Sage: London, UK, 1990.
61. Haider, A.S. Syrian-Lebanese relations: A corpus-based critical discourse analysis of Bashar Al-Assad's speeches and interviews. *Dirasat Hum. Soc. Sci.* **2019**, *46*, 551–570.
62. Yu, C.H.; Jannasch-Pennell, A.; DiGangi, S. Compatibility between text mining and qualitative research in the perspectives of grounded theory, content analysis, and reliability. *Qual. Rep.* **2011**, *16*, 730–744. [CrossRef]
63. Renz, S.M.; Carrington, J.M.; Badger, T.A. Two strategies for qualitative content analysis: An intramethod approach to triangulation. *Qual. Health Res.* **2018**, *28*, 824–831. [CrossRef]
64. Hu, H.; Ge, Y.; Hou, D. Using Web Crawler Technology for Geo-Events Analysis: A Case Study of the Huangyan Island Incident. *Sustainability* **2014**, *6*, 1896–1912. [CrossRef]
65. Tao, Y.; Zhang, F.; Shi, C.; Chen, Y. Social media data-based sentiment analysis of tourists' air quality perception. *Sustainability* **2019**, *11*, 5070. [CrossRef]
66. Han, Y.; Lee, H. Lifestyle Experiences: Exploring Key Attributes of Lifestyle Hotels Using Instagram User-Created Contents in South Korea. *Sustainability* **2021**, *13*, 2591. [CrossRef]
67. Bauer, C.; Scharl, A. Quantitative evaluation of Web site content and structure. *Internet Res.* **2000**, *10*, 31–44. [CrossRef]
68. Scharl, A.; Dickinger, A.; Weichselbraun, A. Analyzing news media coverage to acquire and structure tourism knowledge. *Inf. Technol. Tour.* **2008**, *10*, 3–17. [CrossRef]
69. Sun, Y.; Liang, C.; Chang, C.C. Online social construction of Taiwan's rural image: Comparison between Taiwanese self-representation and Chinese perception. *Tour. Manag.* **2020**, *76*, 103968. [CrossRef]
70. Testa, R.; Foderà, M.; Di Trapani, A.M.; Tudisca, S.; Sgroi, F. Choice between alternative investments in agriculture: The role of organic farming to avoid the abandonment of rural areas. *Ecol. Eng.* **2015**, *83*, 227–232. [CrossRef]
71. Chopin, P.; Dore', T.; Guinde', L.; Blazy, J.M. MOSAICA: A multi-scale bioeconomic model for the design and ex ante assessment of cropping system mosaics. *Agric. Syst.* **2015**, *140*, 26–39. [CrossRef]
72. Sydorovych, O.; Raczkowski, C.W.; Wossink, A.; Mueller, J.P.; Creamer, N.G.; Hu, S.; Bell, M.; Tu, C. A technique for assessing environmental impact risks of agricultural systems. *Renew. Agric. Food Syst.* **2009**, *24*, 234–243. [CrossRef]
73. Sidhoum, A.A.; Dakpo, K.H.; Latruffe, L. Trade-offs between economic, environmental and social sustainability on farms using a latent class frontier efficiency model: Evidence for Spanish crop farms. *PLoS ONE* **2022**, *17*, e0261190. [CrossRef]
74. Dempsey, N.; Bramley, G.; Power, S.; Brown, C. The social dimension of sustainable development: Defining urban social sustainability. *Sustain. Dev.* **2011**, *19*, 289–300. [CrossRef]
75. He, C.; Huang, Z.; Wang, W. Land use changes and economic growth in China. *Land Lines* **2012**, *24*, 14–19.
76. Ministry of Natural Resources of the People's Republic of China. The Third National Land Resource Survey 2019. Available online: https://gtdc.mnr.gov.cn/shareportal#/special/statisticalReport (accessed on 27 August 2021).
77. Holsti, O.R. *Content Analysis for the Social Sciences and Humanities*; Addison Wesley: Boston, MA, USA, 1969.
78. Krippendorff, K. *Content Analysis: An Introduction to Its Methodology*, 2nd ed.; Sage: London, UK, 2004.
79. Nieman-Gonder, J.; Islam, S.; Zhu, X.; Lobo, S.; Anderson, S.; Scalisi, L.; Moncada, D. Customer reviews: Motivated by error-free service or successful recovery. *J. Organ. Psychol.* **2018**, *18*, 41–72.
80. Silge, J.; Robinson, D. *Text Mining with R: A Tidy Approach*; O'Reilly Media: Sebastopol, CA, USA, 2017.
81. Sciandra, A. COVID-19 outbreak through Tweeters' words: Monitoring Italian social media communication about COVID-19 with text mining and word embeddings. In Proceedings of the 2020 IEEE Symposium on Computers and Communications (ISCC), Rennes, France, 7–10 July 2020; pp. 1–6.
82. Erkut, B. Renewable Energy and Carbon Emissions: New Empirical Evidence from the Union for the Mediterranean. *Sustainability* **2022**, *14*, 6921. [CrossRef]
83. R Development Core Team. *R: A Language and Environment for Statistical Computing*; R Foundation for Statistical Computing: Vienna, Austria, 2022; Available online: http://www.R-project.org/ (accessed on 10 March 2022).
84. Hu, H.Y. The distribution of population in China. *Acta Geogr. Sin.* **1935**, *2*, 32–74.
85. Liu, Y.; Liu, J.; Zhou, Y. Spatio-temporal patterns of rural poverty in China and targeted poverty alleviation strategies. *J. Rural. Stud.* **2017**, *52*, 66–75. [CrossRef]
86. Zhou, L.; Xiong, L.Y. Natural topographic controls on the spatial distribution of poverty-stricken counties in China. *Appl. Geogr.* **2018**, *90*, 282–292. [CrossRef]
87. National Bureau of Statistics of China. *China Statistical Yearbook 2021*; China Statistics Press: Beijing, China, 2021.
88. Christiaensen, L.; Demery, L.; Kuhl, J. The (evolving) role of agriculture in poverty reduction: An empirical perspective. *J. Dev. Econ.* **2011**, *96*, 239–254. [CrossRef]

89. Guo, Y.; Zhou, Y.; Liu, Y. Targeted poverty alleviation and its practices in rural China: A case study of Fuping county, Hebei Province. *J. Rural. Stud.* **2022**, *93*, 430–440. [CrossRef]
90. Wang, J.H.; Tseng, S.W.; Zheng, H. The paradox of small hydropower: Local government and environmental governance in China. *J. Dev. Stud.* **2015**, *51*, 1475–1487. [CrossRef]
91. Tsai, W.H.; Liao, X. Mobilizing cadre incentives in policy implementation: Poverty alleviation in a Chinese county. *China Inf.* **2019**, *34*, 45–67. [CrossRef]
92. Zhang, K.; Dearing, J.A.; Dawson, T.P.; Dong, X.; Yang, X.; Zhang, W. Poverty alleviation strategies in eastern China lead to critical ecological dynamics. *Sci. Total Environ.* **2015**, *506*, 164–181. [CrossRef]
93. Liu, L. Environmental poverty, a decomposed environmental Kuznets curve, and alternatives: Sustainability lessons from China. *Ecol. Econ.* **2012**, *73*, 86–92. [CrossRef]
94. Guo, J.H.; Liu, X.J.; Zhang, Y.; Shen, J.L.; Han, W.X.; Zhang, W.F.; Christie, P.; Goulding, K.W.T.; Vitousek, P.M.; Zhang, F.S. Significant acidification in major Chinese croplands. *Science* **2010**, *327*, 1008–1010. [CrossRef] [PubMed]
95. Ravallion, M.; Chen, S. China's (uneven) progress against poverty. *J. Dev. Econ.* **2007**, *82*, 1–42. [CrossRef]
96. De Clercq, D.; Wen, Z.; Song, Q. Innovation hotspots in food waste treatment, biogas, and anaerobic digestion technology: A natural language processing approach. *Sci. Total Environ.* **2019**, *673*, 402–413. [CrossRef] [PubMed]

sustainability

MDPI

Article

Structural Evolution and Sustainability of Agricultural Trade between China and Countries along the "Belt and Road"

Lunzheng Zhou and Guangji Tong *

College of Economics and Management, Northeast Forestry University, Harbin 150040, China; 1611886506@nefu.edu.cn

* Correspondence: tonggj63@nefu.edu.cn; Tel.: +86-135-0361-0923

Abstract: Enhancing trade in agricultural products between China and countries along the "Belt and Road" (B&R) will help strengthen China's food security and promote global, sustainable economic development. Based on the agricultural trade data between China and B&R countries from 2001 to 2019, we used the TII index, the HHI index, and the social network analysis method to calculate the trade structure of agricultural products between China and B&R countries, in terms of plane structure and spatial network structure, and analyzed the influencing factors of their spatial network structure. The results show that China's agricultural trade with B&R countries is highly concentrated in terms of regions and types, the import trade is decentralized, while the export trade is concentrated, and the regions with high trade intensity are mainly concentrated in the countries in close proximity. China's agricultural trade network with B&R countries has become increasingly close, and China has a significant presence in trade networks. The trade network shows four major segments, and the internal and external trade of each segment has become increasingly close. Water resources, geographical location, transportation, trade agreements, and trade structure are the main influencing factors in the trade network between China and B&R countries. Our findings provide useful insights for informed decision-making in the development of international agricultural sustainable cooperation strategies.

Keywords: "Belt and Road"; agriculture products; trade structure; social networks; QAP

Citation: Zhou, L.; Tong, G. Structural Evolution and Sustainability of Agricultural Trade between China and Countries along the "Belt and Road". *Sustainability* **2022**, *14*, 9512. https://doi.org/10.3390/su14159512

Academic Editor: Aaron K. Hoshide

Received: 27 June 2022
Accepted: 30 July 2022
Published: 3 August 2022

1. Introduction

Food security is the foundation of national development. Under the rigid constraints of natural resources, the establishment of the carbon peaking and carbon neutrality goals has brought new challenges to China's domestic agricultural production. The global spread of COVID-19 [1], the Russia–Ukraine conflict, and climate change have exacerbated agricultural supply risks. How to ensure the sustainable trade of Chinese and global agricultural products in the surging world economic tide has become a major issue that both domestic and foreign practitioners and academics must reconsider.

Currently, the impact of world agricultural trade on global food security has risen from 9% to 17%, and the food exports of many developed agricultural countries have made up for the food shortages of most countries [2]. Integrated planning and the full utilization of both domestic and international markets and resources to enable the sustainable production and consumption of global agricultural products is the only choice for ensuring food security in China and globally [3]. In recent years, the No. 1 document of the Central Government has repeatedly mentioned the strengthening of international cooperation in agriculture in B&R countries, expanding diversified import channels, and expanding the exports of superior agricultural products. In 2022, the No. 1 central document emphasized the optimization of the agricultural product trade layout, and the diversification strategy of agricultural product import. At present, the focus of China's agricultural imports has shifted to B&R countries [4], but to achieve coordinated and sustainable development, fundamentally speaking, this still depends on whether the structure of agricultural trade between China and the B&R countries is reasonable. In particular, the structuring of the agricultural trade relationship to be based on comparative advantage, and the empirical investigation of related issues, are of great significance for ensuring the effective supply of agricultural products in China and the world.

The current research on agricultural trade between China and B&R countries has yielded many results, but it still needs to be further expanded and improved. We examined the literature and found that in terms of cooperation objects, this mostly involves the analysis of China and some of the countries or regions along the B&R; this lacks a holistic analysis of the region, which mainly includes China and ASEAN [5], China and Central Asia [6,7], China and South Asia [8], China and the "21st Century Maritime Silk Road" [9,10], China and Russia [11,12], China and Central and Eastern Europe [13,14], China and Southeast Asia [15], China and the countries along the Silk Road Economic Belt [16–18], China and Africa [19], and China and Ukraine [20].

In terms of trade categories, they are mostly focused on a certain type of agricultural trade, and they lack an overall exploration of agricultural trade. This mainly includes dairy products [21], corn [22], aquatic products [23,24], apples [25], grains [26–29], and fruits [30]. In terms of analyzing data, many studies only explore the trade network structure under cross-sectional data at certain time points, which cannot reflect the dynamic evolution of trade in a specific time context. Wei (2019) analyzed the structure, association characteristics, and strategy choices of agricultural trade networks between China and B&R countries, using cross-sectional data in 2017 [31]. Zhan (2019) analyzed the competitiveness and complementarity of agricultural trade networks of B&R countries in 2007 and 2015 [32]. Su (2019) analyzed the structure and cooperation trends of agricultural trade networks between China and B&R countries in 2012 and 2016 [33].

Although some scholars have explored the structure of agricultural trade between China and B&R countries from a holistic perspective over the past two years, most of them have adopted an index approach to explore the structure of flat trade and to analyze its influencing factors. For example, Yang et al. (2021) analyzed the evolution of agricultural trade characteristics between China and B&R countries [34]. Although all of the agricultural products of the B&R countries were studied, the countries and agricultural products within the B&R countries were not discussed separately, and all the countries and agricultural products along B&R countries were treated as a single whole, without discussing the internal structure reflecting the trade. Liu et al. (2021) explored the countries of China's agricultural trade with B&R countries in 2018, using descriptive analysis, but they took a static trade situation of one year as the sample of the study, they did not analyze the evolution of trade dynamically from the time series; they only used a descriptive analysis of plane structure, and they did not use index and spatial analysis methods [35]. Sun (2021) used the index method to investigate the intra-industry trade of agricultural products and its influencing factors between China and B&R countries, which included all B&R countries and all types of agricultural products, but they only explored the flat structure. While the definition of agricultural products in Sun's paper was based on the World Trade Organization (WTO) Agricultural Agreement + Fishery Products, this paper uses the United Nations International Trade Standards Classification, with its different focuses [36].

This paper explores the structural evolution of agricultural trade between China and B&R countries, from both a planar and a spatial perspective. The study differs from the existing literature in four ways. First, instead of limiting the scope to certain countries or certain agricultural products, all countries and agricultural products along the B&R countries are used as samples to classify and compare the agricultural trade relations between China and B&R countries, which is conducive to grasping the current situation of trade cooperation in terms of levels and varieties within the overall sample, and exploring the directions for sustainable cooperation. Second, the study is not limited to a single point in time; the data chain covers four time points, namely, China's accession to the WTO, the global financial crisis, the introduction of the "Belt and Road" initiative, and the latest data; it explores the dynamic evolution of trade relations between China and B&R countries.

It is useful for China and B&R countries to forecast the future direction of sustainable cooperation based on past bilateral trade dynamics in the current situation of increased world uncertainty. Third, the literature has used descriptive analysis, index analysis, and social network analysis separately, but the comprehensive planar analysis can highlight the accuracy of trade data and the continuity of the time series, while the spatial analysis can better reflect the relationship between countries, between countries and small groups, and between countries and the whole. Combining the advantages of the two analysis methods, we choose to adopt the dual perspective of planar and spatial research. Fourth, drawing on the research techniques of Li (2017) [37], the competition index, CS, was improved and the competitive advantage index, CN, was constructed, which provides a methodological improvement for measuring the spatial analysis of multinational cooperation network development and is conducive to providing analytical tools for scholars studying trade cooperation.

2. Materials and Methods

2.1. Research Methods

To quantify the planar and spatial network structure of agricultural trade between China and B&R countries, we used a combination of descriptive statistics and the index method to quantify the planar structure [38–40]. The trade index was selected as the trade intensity index and the export concentration index. We used the social network analysis method [29,31,36,37,41,42] to quantify the spatial network structure.

2.1.1. Planar Structure Quantification Method

1. Trade Intensity Index (TII). TII index was proposed by Brown (1949) [43] and was later improved and refined by Kojima (1964) [44]. TII index measures the ratio of a country's exports to a trading partner country, to that country's total exports to that trading partner country's total imports, as a share of total world imports; this is often used in inter-country trade interdependence analysis, with the following formula.

$$TII_{ab} = \frac{X_{ab}/X_a}{M_b/M_w} \quad (1)$$

In the formula, a, b, w represent country a, b, and the world market, respectively, TII_{ab} represents the trade combination degree of country a and b, X_{ab} represents the export volume of country a to country b, X_a represents the total export volume of country a, M_b represents the total import volume of country b, and M_w represents the total import volume of country. When $TII_{ab} > 1$, this indicates that a and b have a close trade relationship. When $TII_{ab} < 1$, it indicates that the trade relationship between a and b is loose.

2. Export Concentration Index (HHI). HHI index, also known as the Hirschman index, is used to measure the degree of concentration of a country or region in terms of the types of products exported [45]. The formula is as follows.

$$HHI = \sqrt{\sum_{k=1}^{n} (X_k/X)^2} \quad (2)$$

In the formula, X_k is the export volume of k products of the country, and X is its total export volume. The value range of HHI is $\left[\frac{1}{\sqrt{n}}, 1\right]$. The smaller the value is, the more fragmented the country's export product structure is; likewise, the larger the value, the more concentrated the export product structure.

2.1.2. Quantification Method of Spatial Structure

1. Origin of social network approach. The application of social network analysis in economics is mainly inspired by the sociologist Granovetter (1985) [46], who argues that the key to many high-transaction-cost behaviors in the real economy that are still

traded through the market is that both buyers and sellers are embedded in a long-term network of business relationships, i.e., both buyers and sellers are unwilling to lose the trust relationship they have built up in mutual transactions, and the whole system. The whole system is constantly adaptive through mutual coordination and information exchange. This means that the real economic system has the essential characteristics of a social network. After that, social network analysis has gradually received the attention of economists and has been widely used in many fields such as industrial economics, finance, and international trade. With the development of economic globalization, the close economic ties between countries make global trade relations an organic whole, and the growing international trade is becoming the key to shape the global economic and political landscape. The adoption of social network analysis method to study the characteristic laws of international trade system has become an emerging research direction.

2. Standard construction of social network methods. The social network analysis method regards trading countries as points, and the resulting trade relations as connecting lines, and analyzes the structural characteristics of trade networks according to the connections between nodes in the network. The two–two relationship conditions are different and can be constructed into different trade networks; according to the import–export and competitive advantage of a two–two relationship, two different trade relationship networks can be constructed, so as to reflect the trade prospects between trading countries more comprehensively, where the two–two import-export relationship is reflected by the bilateral trade volume. For the competitive advantage relationship, based on the method of Li Jing and Chen Ni et al. (2017) [29], the trade competition index is improved into the trade competition difference index based on the comparative advantage theory, which is the competitive advantage index. The original trade competition index CS formula is as follows.

$$CS = 1 - \frac{1}{2}\sum |a_i^n - a_j^n| \tag{3}$$

where i and j represent countries, n represents industries, a_i^n represents the comparative advantage of industry n in country i, and a_j^n represents the comparative advantage of industry n in country j. The closer the comparative advantage of two countries, the smaller $\frac{1}{2}\sum |a_i^n - a_j^n|$ becomes, and the larger the CS. The improvement in this paper is to consider industry n as a single industry, and the comparative advantage is specified as the NRCA index, as in Equation (4).

$$CN = 1 - \frac{1}{2}|NRCA_i^n - NRCA_j^n| \tag{4}$$

$$RCA_{ij} = \frac{X_{i_j}/X_{tj}}{X_{i_w}/X_{tw}} \tag{5}$$

$$NRCA_{ij} = \frac{RCA_{ij} - 1}{RCA_{ij} + 1} \tag{6}$$

where: X_{i_j} represents the export value of i products of Country j, X_{tj} represents the export volume of products of Country j, X_{i_w} represents the export value of i products in the world, and X_{tw} stands for the world export of goods.

According to Formulas (5) and (6), when $RCA_{ij} = \pm\infty$, $NRCA_{ij} = 1$; when $RCA_{ij} = 0$, $NRCA_{ij} = -1$, and $NRCA_{ij} \in [-1, 1]$, it can be deduced that $-2 \leq NRCA_i^n - NRCA_j^n \leq 2$. We obtain $0 \leq 1 - \frac{1}{2}|NRCA_i^n - NRCA_j^n| \leq 1$, and $CN \in [0, 1]$. Therefore, if NRCA is taken as an independent variable, the domain of CN index is [−1,1] and the range is [0,1]. According to the theory of comparative advantage put forward by Ricardo, international trade is based on the relative difference in production technology and the resulting relative cost of production. Every country should concentrate on producing and exporting products with "comparative advantage" and importing products with "comparative disadvantage". The NRCA index represents the trade competitiveness index of a country. One country exports products with relative advantages, while the other exports products with relative disadvantages. The absolute value of the NRCA index is the convergence point of the interests of the trade between the two countries, namely, the competitive advantage. According to Formula (4), if the set $|\mathrm{NRCA}_i^n - \mathrm{NRCA}_j^n| \geq \mathrm{A}$, there are comparative advantages for trade between the two countries, so that $1 - \frac{1}{2}|\mathrm{NRCA}_i^n - \mathrm{NRCA}_j^n| \leq 1 - \frac{1}{2}\mathrm{A}$, launch $\mathrm{CN} \leq 1 - \frac{1}{2}\mathrm{A}$, $|\mathrm{NRCA}_i^n - \mathrm{NRCA}_j^n| \geq \mathrm{A}$ set up, which sets up $\mathrm{CN} \leq 1 - \frac{1}{2}\mathrm{A}$, and the two countries have the competitive advantage.

3. Analysis of the density of trade networks (Dn). Dn Trade network density reflects the sparseness of trade relationships between countries.

$$\mathrm{Dn} = \mathrm{L}/[\mathrm{N} \times (\mathrm{N} - 1)] \tag{7}$$

In the formula, N is the number of countries in the trade network, where the number of countries in that trade network that meet the criteria is L. The value range of Dn is $[0, 1]$.

The larger the Dn, the greater the number of important trade relationships in the network and the higher the trade density.

4. Analysis of the centrality of trade networks. De is the relative degree centrality, which measures a country's position and role in the overall network. NC is the relative degree centrality index, which measures the centrality of the entire network.

$$\mathrm{De} = \mathrm{n}/(\mathrm{N} - 1) \tag{8}$$

$$\mathrm{NC} = \sum_{i=1}^{M} (\mathrm{Max}(\mathrm{De}) - \mathrm{De_i}) \ / \ (\mathrm{M} - 2) \tag{9}$$

In the formula, n denotes the number of countries in network trade with which a country has significant trade relations, N denotes the maximum possible number of directly connected countries, and M denotes the number of countries in the trade network. The value range of De is $[0, 1]$. The larger the De, the more central a country is in the network, the more "influence" it has in the network, and the more it can influence other countries. The value range of NC is $[0, 1]$. The larger the NC, the greater the degree to which the network is built around a point or points in the network, and the more concentrated the trade.

5. Block model analysis. Block model analysis is a network location analysis model proposed by White et al. (1976) [47]. According to the block model theory, using the CONCOR method in Ucinet 6, a trade network can be divided into several plates to reveal the trade relations between the inside and the outside of the plates, revealing the roles and functions of each economic segment and its member countries in international trade. In this paper, referring to the classification method of Li Jing et al. (2017) [37], the economic plates are classified into four major categories. One is the internal type, if the plate has many internal relationships and few or no external relationships; two is the outward type, if the plate has few or no internal relationships and many external relationships; three is the eclectic type, if there are many internal relationships and also many external relationships; and four is the isolated type, if there are few or no both external and internal relationships.

2.1.3. Analysis Methods of Spatial Network Influence Factors

After analyzing the characteristics of the spatial network of agricultural trade, it is necessary to analyze what factors affect the spatial network of agricultural trade between China and B&R countries. In order to avoid the problems of multicollinearity and spurious regression in social network analysis, the study combines QAP correlation analysis and QAP regression analysis. These analyses were based on the research methods of Liu (2007) [48], Li (2014) [49], and Ma (2016) [42]. The framework for analyzing the occurrence of agricultural trade in a country in this paper consists of three parts: agricultural production—transportation of agricultural products—intercountry trade. In the agricultural production stage, the main factors affecting agricultural production are arable land, water, and seeds. The main influencing factors in the transportation stage are distance and means of transportation, and the main influencing factors in the trade stage are economic distance between two countries, cultural differences, trade structure, and trade agreements. Combining the above analysis and referring to the existing research results, 10 indicators are selected to characterize the corresponding influencing factors [31,50–55].

1. Agricultural resource endowment: We use the absolute value matrix of the difference in per capita water resources (PCWR), the absolute value matrix of the difference in per capita arable land area (PCLA), and the absolute value matrix of the difference in the share of investment in scientific research in each country (SCI), to represent the effects of water, arable land, and seeds, respectively.

2. Agricultural product transportation: We consider whether the two countries are bordering each other to indicate the trade distance. If the two countries are bordering, it will be recorded as 1, otherwise it will be 0, for constructing the distance matrix (DIS). Agricultural products belong to the large volume of low-value goods; the two countries trade in order to save costs, and generally use railroad or waterway transportation. We adopt the absolute value of the difference between the railroad length of each country matrix (TRA), indicating the convenience of transportation in each country.

3. Trading between two countries: The economic distance between countries will affect agricultural trade. We combine two ways of representing economic distance in the existing literature, namely, the absolute value of the difference between the total economic output value of each country (DGDP), and the economic distance matrix (DE) of two countries. The formula for calculating the economic distance between two countries in the DE matrix is: $DE_{ij} = \frac{(PGDP_i - PGDP_j)^2}{GDP_i * GDP_j}$, where DE_{ij} denotes the economic distance between country i and country j, and PGDP and GDP are the GDP per capita and GDP, respectively. The trade agreement facilitates international trade between the two countries and is recorded as 1 if both countries are members of the trade agreement; otherwise, it is 0. The trade agreement matrix (TA) is constructed. As the cultural factor, language is the tool of communication between two countries. If an official language is used in both countries, it is recorded as 1; otherwise, it is 0. The cultural matrix (CUL) is constructed. For trade structure, this paper uses the share of an agricultural export in a country's total trade to represent the trade structure and to construct the trade structure matrix (IS). F is used to represent the trade network of US $100 million between China and B&R countries in 2019, and then the model is constructed as follows:

$$F = f(PCWR, PCLA, SCI, DIS, TRA, DGDP, DE, TA, CUL, IS) \tag{10}$$

In the formula, the GDP, population, scientific research expenditure ratio, water resources, arable land resources, and railroad length of each country are obtained from the World Bank database, trade agreements and official languages are obtained from the official websites of each country, and geographical distances are obtained from Google Maps.

2.2. Description of Study Subjects and Data

2.2.1. Definition of the Research Area

Since the "Belt and Road" is an open international economic cooperation region, the academic community has not precisely defined the distribution range. This paper refers to the definition methods of scholars [22,26], and in view of the availability of trade data, the B&R countries are divided into six regions and 60 countries. The specific regions are: ① Mongolia and Russia; ② Central Asia, including Kazakhstan, Kyrgyzstan, Tajikistan, Uzbekistan, and Turkmenistan; ③ Southeast Asia, including Vietnam, Laos, Cambodia, Thailand, Malaysia, Singapore, Indonesia, Brunei, the Philippines, Myanmar, and Timor-Leste; ④ South Asia, including India, Pakistan, Bangladesh, Afghanistan, and Nepal; ⑤ Western Asia and the Middle East, including Turkey, Iran, Syria, Iraq, Saudi Arabia, Qatar, Bahrain, Kuwait, Lebanon, Oman, Yemen, Jordan, Israel, Palestine, Armenia, Georgia, Azerbaijan, and Egypt; ⑥ Central and Eastern Europe, including Poland, the Czech Republic, Slovakia, Hungary, Slovenia, Croatia, Romania, Bulgaria, Serbia, Montenegro, Bosnia and Herzegovina, Albania, Estonia, Lithuania, Latvia, and Ukraine.

2.2.2. Agricultural Product Scoping and Data Sources

According to the United Nations Standard International Trade Classification (SITC Rev.3), the definition of agricultural products includes four categories and 22 chapters of agricultural products. The four categories of agricultural products are 0, 1, 2, and 4. Category 0 is food and live things, including 10 chapters; category 1 is beverages and tobacco, including two chapters of agricultural products; category 2 is non-edible raw materials (except fuel), including seven chapters of agricultural products, except 27 and 28; and category 4 is animal and vegetable oils, and fats and waxes, including three chapters of agricultural products. In order to study the change of trade structure after China's accession to the WTO, this paper selects the data related to China's agricultural trade with B&R countries from 2001 to 2019 for analysis, and the data are obtained from the UN COMTRADE database.

3. Results

3.1. Planar Structure Analysis

3.1.1. Trade Type Structure

According to Table 1, the import and export of agricultural products categories between China and B&R countries are highly concentrated. Agricultural products are classified using SITC into 22 chapters, of which the first 10 chapters account for more than 80% of the total proportion of trade, so that more than 80% of the total trade is concentrated in 45% of the agricultural product categories. Among them, from the time series, the share of China's agricultural products' import categories showed a decentralized trend, decreasing from 94.94% in 2009 to 90.37% in 2019, while China's agricultural products export categories showed a relatively concentrated trend, increasing from 82.54% in 2001 to 90.02% in 2019. In terms of specific types of agricultural products, China's agricultural imports to B&R countries are more evenly concentrated into five categories: 02 (dairy products and poultry eggs), 23 (crude rubber), 05 (vegetables and fruits), 42 (fixed vegetable fats and oils), and 03 (fish, crustaceans, mollusks and aquatic invertebrates, and their products), while China's exports to B&R countries are highly concentrated into two categories of agricultural products: 05 (vegetables and fruits) and 03 (fish, crustaceans, mollusks, and aquatic invertebrates and their products), with five products accounting for approximately 45% of the total in recent years.

Table 1. Trade structure of specific types of agricultural products.

China's Import of Agricultural Products to B&R Countries (%)							China's Export of Agricultural Products to B&R Countries (%)						
SITC	**2001**	**2009**	**2013**	**2017**	**2018**	**2019**	**SITC**	**2001**	**2009**	**2013**	**2017**	**2018**	**2019**
02	28.28	17.93	18.61	18.38	17.76	15.37	05	25.40	46.03	42.68	47.66	44.35	45.82
23	12.61	17.11	18.52	23.55	18.17	15.07	03	5.79	12.21	17.88	13.84	13.14	13.57
05	7.27	10.74	12.07	10.78	12.49	14.28	26	9.19	6.59	7.13	6.73	7.28	6.09
42	9.55	24.72	16.40	12.82	11.74	12.77	07	4.51	4.28	3.54	4.50	4.83	5.18
03	12.03	8.39	5.71	5.88	9.12	12.47	29	5.00	5.73	4.99	5.16	5.63	5.12
25	14.74	6.04	5.32	7.16	8.22	7.18	04	14.26	2.51	2.87	1.89	2.57	3.40
04	2.05	1.37	3.53	6.55	6.33	5.75	06	3.42	3.70	3.38	3.39	3.56	3.35
26	3.11	5.54	9.91	3.03	2.78	2.61	22	4.24	1.89	2.00	2.57	3.12	2.75
09	1.24	2.32	2.16	2.21	2.46	2.59	08	2.65	3.47	2.64	2.81	3.20	2.66
07	0.71	0.78	1.21	1.42	1.75	2.28	12	8.08	3.28	2.84	2.43	2.07	2.08
Total	91.59	94.94	93.44	91.78	90.82	90.37	Total	82.54	89.69	89.95	90.98	89.75	90.02

3.1.2. Trade Region Structure

According to Table 2, China's agricultural trade with B&R countries is highly concentrated, with imports tending to be decentralized, and exports tending to be concentrated.

In terms of the top 10 import and export trade shares overall, China's agricultural imports from B&R countries fell from 93.78% in 2001 to 89.68% in 2019, indicating that the effect of China's diversified import strategy has emerged, while China's agricultural exports to B&R countries rose from 66.51% in 2001 to 78.76% in 2019, and have been concentrated overall. Specifically, from the top 10 countries in import and export trade, 8 of them rank in the top 10 countries for both import and export, namely, Thailand, Indonesia, Russia, Vietnam, Malaysia, India, the Philippines, and Myanmar, indicating that China has close trade ties in agricultural products with B&R countries, for both importing and exporting. From the perspective of individual import and export trade countries, the ranking of China's agricultural trade with B&R countries has almost always tended to stabilize, with Thailand, Indonesia, and Russia holding the top three in the import ranking, and the top three in the export ranking being Vietnam, Thailand, and Malaysia.

Table 2. Regional distribution structure of agricultural trade products.

Distribution of China's Agricultural Products Imports from B&R Countries (%)							Distribution of China's Agricultural Products Exports to B&R Countries (%)						
Country	**2001**	**2009**	**2013**	**2017**	**2018**	**2019**	**Country**	**2001**	**2009**	**2013**	**2017**	**2018**	**2019**
Thailand	18.41	18.04	22.38	24.29	22.89	21.01	Vietnam	4.61	10.04	12.81	19.92	21.58	20.67
Indonesia	23.26	18.89	17.34	20.04	18.60	17.72	Thailand	4.86	8.98	13.85	13.10	13.29	13.60
Russia	28.46	22.95	13.72	16.93	18.97	16.79	Malaysia	16.19	12.59	13.95	10.13	9.69	10.87
Vietnam	3.07	4.93	8.71	11.29	11.49	10.64	Indonesia	12.05	10.83	9.47	10.29	9.70	9.46
Malaysia	14.20	19.74	13.50	9.75	8.11	7.56	Philippine	6.01	7.12	7.40	8.43	8.08	7.51
India	2.21	4.90	8.59	2.70	3.43	5.65	Russia	10.52	11.44	10.72	8.11	7.96	6.90
Ukraine	0.05	0.07	1.86	2.60	3.06	4.93	Turkey	1.04	2.47	1.89	2.06	2.36	2.58
Philippine	1.91	1.43	1.73	1.92	2.25	2.11	Myanmar	2.24	0.78	1.22	2.05	2.24	2.50
Laos	0.13	0.39	1.46	1.17	1.43	1.68	Bengal	0.82	1.91	1.43	1.49	1.76	2.41
Myanmar	2.08	1.97	2.37	1.01	1.07	1.59	India	8.15	5.16	3.64	3.17	2.40	2.25
Total	93.78	93.30	91.68	91.70	91.30	89.68	Total	66.51	71.33	76.38	78.76	79.06	78.76

3.1.3. Trade Intensity Structure

In order to analyze the structure of agricultural trade intensity between China and B&R countries as a whole, this paper divides B&R countries into six regions. As shown in Table 3, trade intensity is greater in regions where China is close to B&R countries, such as Mongolia and Russia, Southeast Asia, and Central Asia, which border China and are ranked in the top three in terms of trade intensity. From the general trend, China's trade intensity with B&R countries tends to disperse, among which China's trade intensity with Mongolia and Russia tends to weaken, and its trade intensity with Central Asia, Southeast Asia, and South Asia increases, while the trade intensity of B&R countries with China tends to strengthen slightly, but remains stable overall. From the comparison of trade intensity between China to B&R countries and B&R countries to China, the overall trade intensity of China to B&R countries is higher than the trade intensity of B&R countries to China, indicating that China is more dependent on B&R countries, especially neighboring countries, while B&R countries are less dependent on China's trade.

3.1.4. Trade Concentration Structure

According to Figure 1, the concentration of China's import trade to B&R countries has a tendency to decrease, while the concentration of the export trade has a tendency to increase. In terms of agricultural trade types, the concentration of agricultural import types from China to B&R countries tends to decline, from 0.39 in 2001 to 0.33 in 2019, while the concentration of export types from China to B&R countries tends to rise, from 0.35 in 2001 to 0.49 in 2019. In terms of agricultural trade regions, the regional concentration of China's imports to B&R countries tends to decline in recent years, from 0.73 in 2017 to

0.67 in 2019, while the regional concentration of China's exports to B&R countries tends to increase, from 0.55 in 2008 to 0.70 in 2019. From the value of concentration, the regional concentration curve is always above the category concentration, and the trade concentration between China and the countries (regions) along the B&R is higher than the category trade concentration.

Table 3. Structure of agricultural product trade intensity.

China's Agricultural Trade Intensity with B&R Countries							B&R Countries Agricultural Trade Intensity with China						
Region	**2001**	**2009**	**2013**	**2017**	**2018**	**2019**	**Region**	**2001**	**2009**	**2013**	**2017**	**2018**	**2019**
Mongolia	73.95	141.87	104.38	125.07	97.01	95.84	Mongolia	9.10	5.03	2.56	3.86	3.55	3.40
Central Asia	1.18	2.02	1.66	2.65	2.33	2.85	Central Asia	1.80	1.48	1.72	1.51	1.43	1.48
Southeast Asia	1.70	3.27	3.59	4.84	4.38	4.85	Southeast Asia	2.55	2.38	2.05	2.68	2.43	2.49
South Asia	0.93	1.30	1.00	1.09	1.06	1.25	South Asia	0.44	0.89	0.84	0.52	0.59	0.97
Western Asia and Middle East	0.33	0.65	0.48	0.70	0.65	0.70	Western Asia and Middle East	0.18	0.10	0.10	0.20	0.20	0.33
Central and Eastern Europe	0.03	0.04	0.03	0.03	0.03	0.03	Central and Eastern Europe	0.17	0.05	0.16	0.30	0.30	0.44

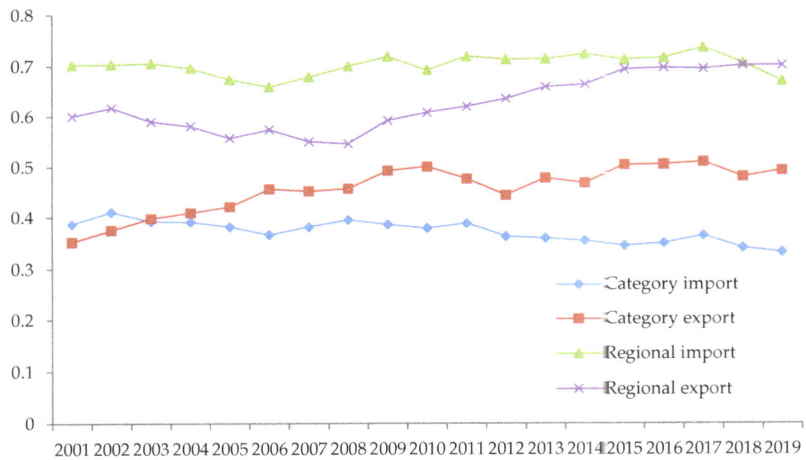

Figure 1. Product categories and regional concentration of agricultural trade.

3.2. Spatial Network Structure Analysis

In order to reflect the closeness of the trade relations, in this paper, referring to the method of Li Jing et al. (2017) [37], the import–export relationship is divided into US $10 million and US $100 million categories, and the existence of significant trade relations is judged if the trade volume between the two countries meets the classification criteria. If the CN index is less than 0.7 and 0.8, a significant competitive advantage relationship is indicated. In order to reflect the evolution of the agricultural trade network between China and B&R countries from 2001 to 2019 based on the cross-sectional analysis of the trade network analysis as a single year, considering the financial crisis in 2008 and the "Belt and Road" initiative proposed by China for the first time in 2013, and other important nodes due to the delayed impact of the financial crisis, which generally only appeared in 2009, we selected four different years, 2001, 2009, 2013, and 2019, as representatives, and constructed 16 trade networks according to the time dimension and the degree of trade relations (Figures 2–17).

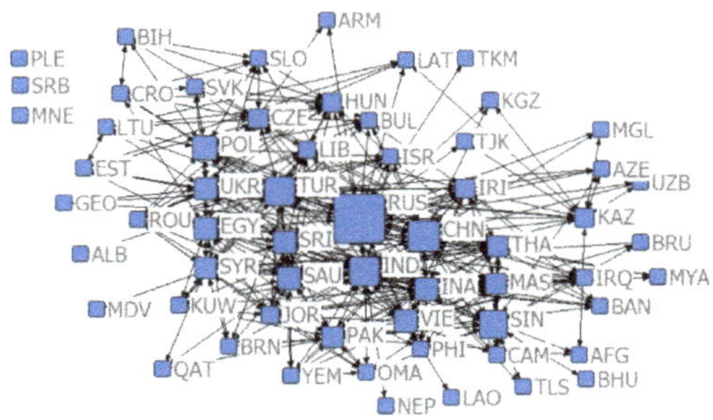

Figure 2. Relative degree centrality network of the trade volume of US $10 million in 2001.

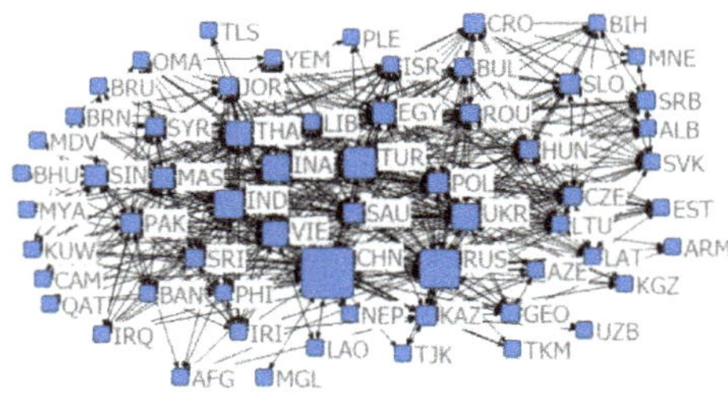

Figure 3. Relative degree centrality network of the trade volume of US $10 million in 2009.

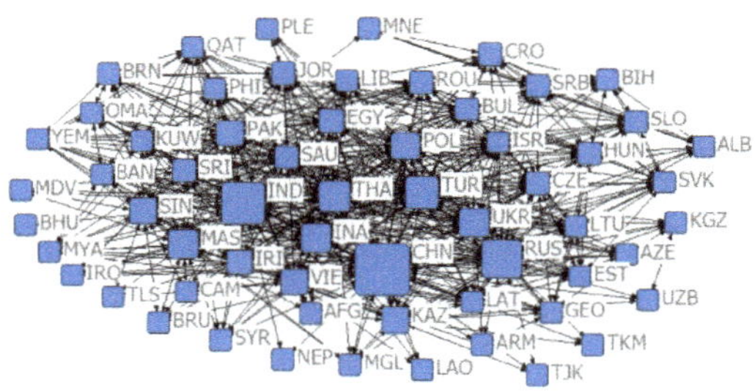

Figure 4. Relative degree centrality network of the trade volume of US $10 million in 2013.

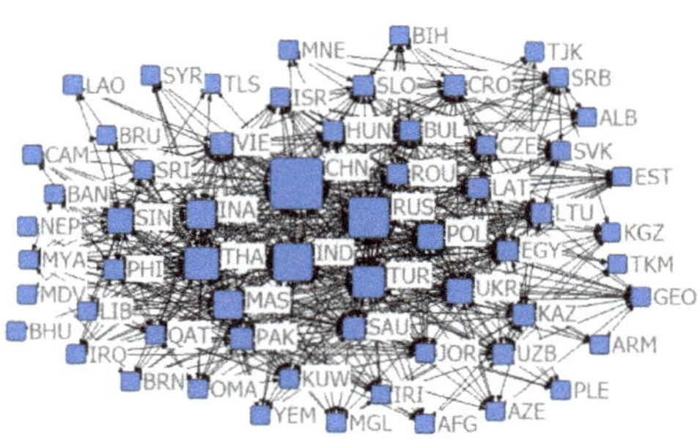

Figure 5. Relative degree centrality network of the trade volume of US $10 million in 2019.

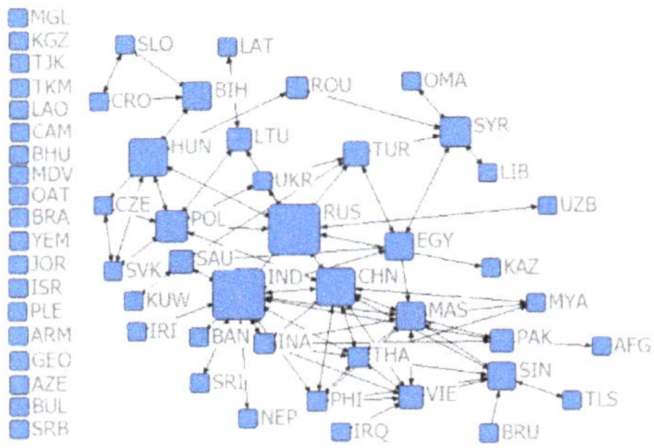

Figure 6. Relative degree centrality network of the trade volume of US $100 million in 2001.

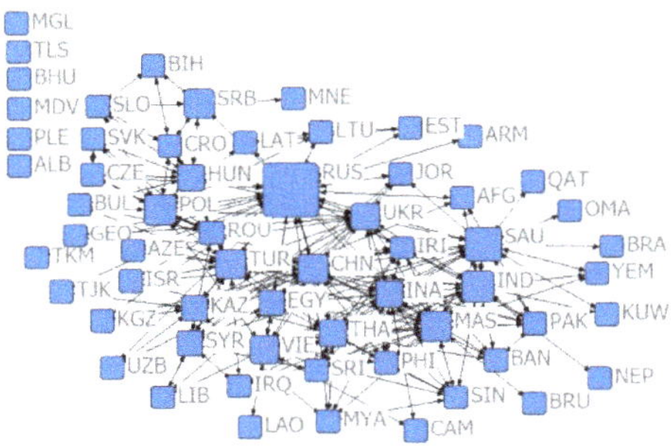

Figure 7. Relative degree centrality network of the trade volume of US $100 million in 2009.

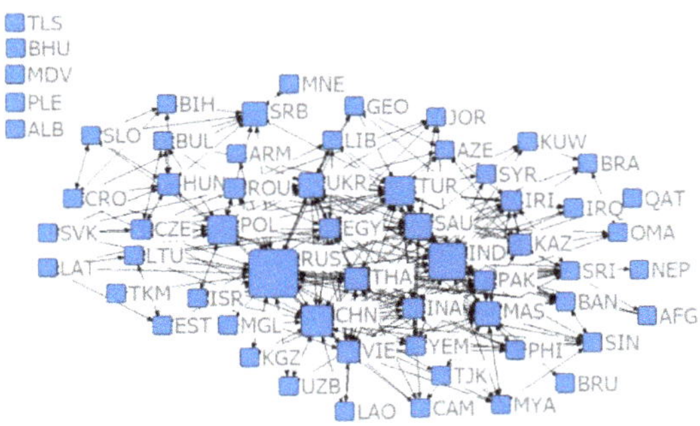

Figure 8. Relative degree centrality network of the trade volume of US $100 million in 2013.

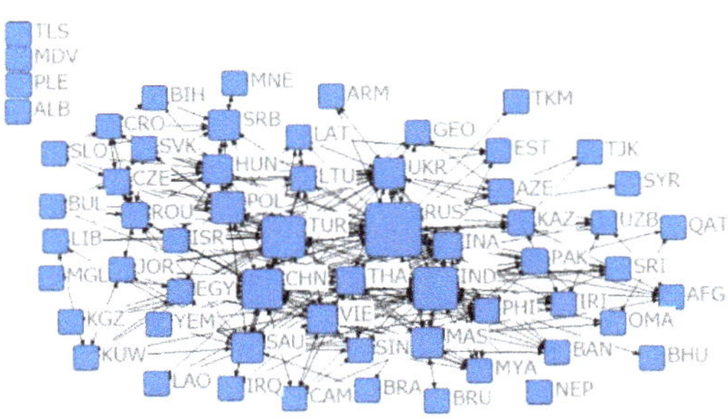

Figure 9. Relative degree centrality network of the trade volume of US $100 million in 2019.

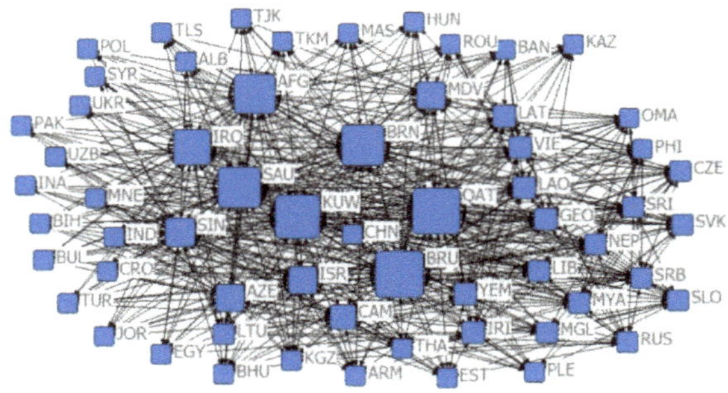

Figure 10. The 2001 competition index CN = 0.7 relative degree centrality network.

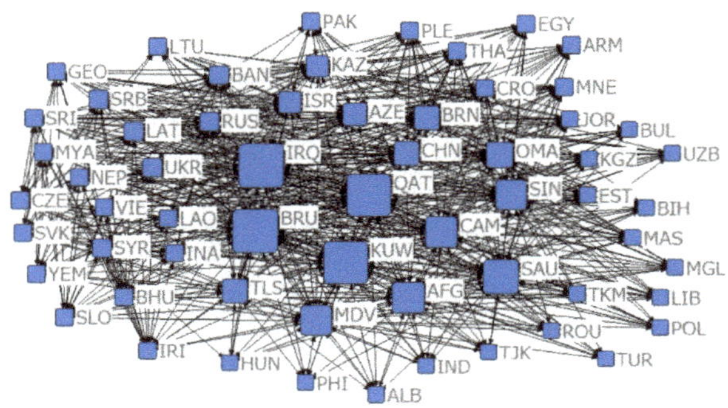

Figure 11. The 2009 competition index CN = 0.7 relative degree centrality network.

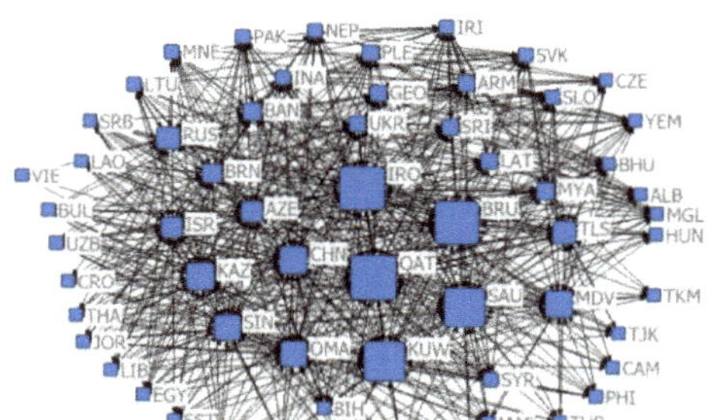

Figure 12. The 2013 competition index CN = 0.7 relative degree centrality network.

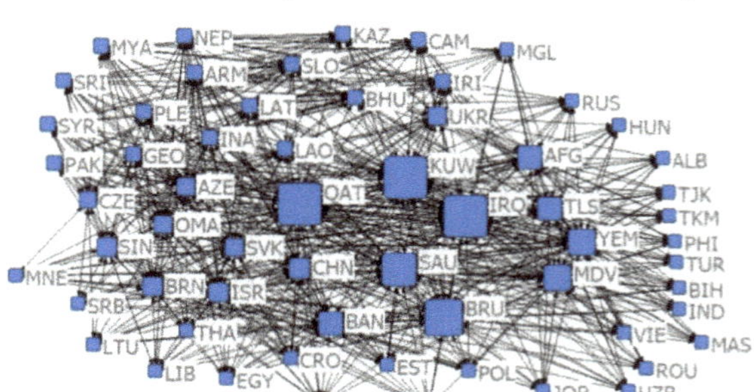

Figure 13. The 2019 competition index CN = 0.7 relative degree centrality network.

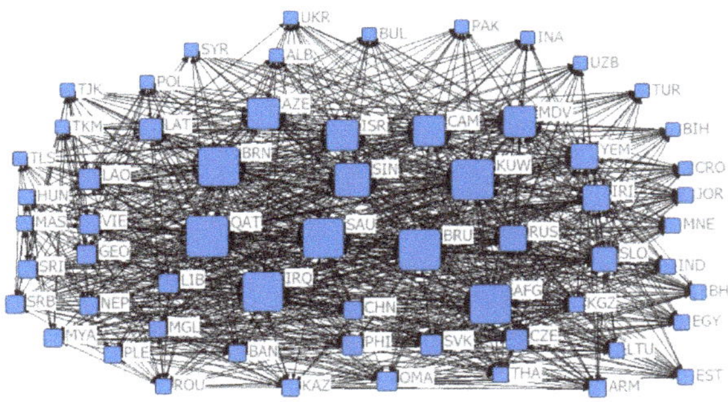

Figure 14. The 2001 competition index CN = 0.8 relative degree centrality network.

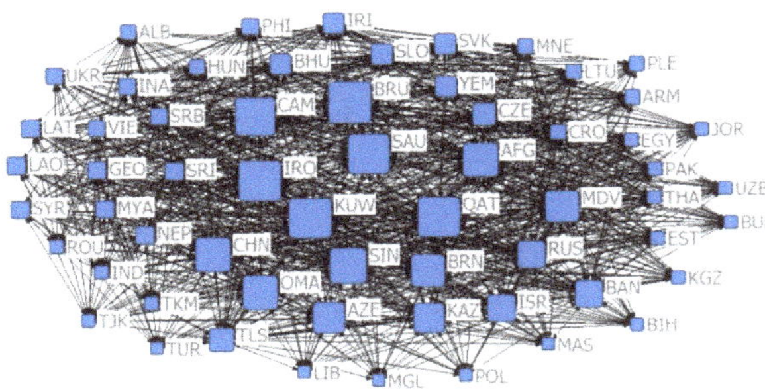

Figure 15. The 2009 competition index CN = 0.8 relative degree centrality network.

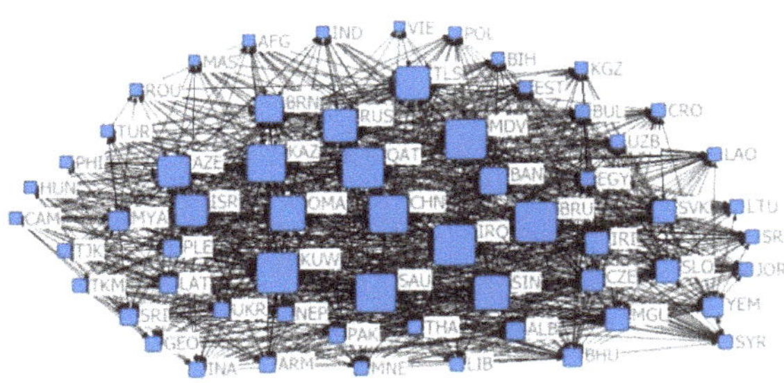

Figure 16. The 2013 competition index CN = 0.8 relative degree centrality network.

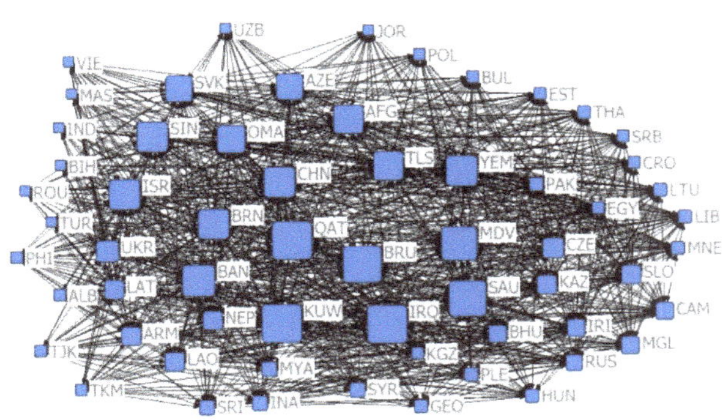

Figure 17. The 2019 competition index CN = 0.8 relative degree centrality network.

3.2.1. Analysis of the Density of Trade Networks

Combining Figures 2–17 and Table 4 in terms of the network density, Dn for the US $10 million criterion ranged from 0.1785 to 0.3251 over the four years of the study sample, tending to increase overall and decreasing slightly in recent years. Dn in the US $100 million criterion increases from 0.0404 to 0.1505 year by year, and Dn in 2019 is 3.52 times that of 2001, indicating a rapid increase in trade relations reaching the US $100 million criterion. In general, China's agricultural trade relations with B&R countries achieve faster growth, but the US $10 million criterion relationship declines slightly, the US $100 million criterion relationship gradually increases, and trade tends to be concentrated.

Table 4. Network density of 16 trade relations of China with B&R countries.

Year	Trade Volume (US $ Million)	CN Index	Network Density (Dn)	Significant
2001	10		0.1785	653
	100		0.0404	148
		0.7	0.3221	1179
		0.8	0.4975	1821
2009	10		0.2809	1028
	100		0.1121	410
		0.7	0.3582	1311
		0.8	0.5339	1954
2013	10		0.3251	1190
	100		0.1421	520
		0.7	0.3339	1222
		0.8	0.5063	1853
2019	10		0.3178	1163
	100		0.1505	551
		0.7	0.3509	1284
		0.8	0.5426	1986

With the CN = 0.7 criterion, the four-year Dn ranged from 0.3221 to 0.3509, with growth rates of 11.21%, −6.78%, and 5.09% for the four time points, indicating that the development space of agricultural trade between China and B&R countries has experienced the process of "rise–fall–rise". In 2013, the "Belt and Road" initiative injected new impetus to the agricultural trade of B&R countries. With the CN = 0.8 criterion, Dn ranged from 0.4975 to 0.5426, with growth rates of 7.32%, −5.175%, and 7.17% for the four time points, indicating that, when lowering the competitive advantage criterion, the development space of the agricultural trade of the B&R countries is also in the trend of "rise–fall–rise". Comparing the CN = 0.7 and CN = 0.8 criteria, it is found that the "Belt and Road" initiative provides development opportunities for countries with different levels of competitive advantage. Compared with the CN = 0.7 criterion, Dn in 2019 is the largest under the CN = 0.8 criterion, indicating that the countries with weaker comparative advantages in agricultural trade have achieved the best development in history after the introduction of the "Belt and Road" initiative.

3.2.2. Analysis of the Centrality of Trade Networks

Combining Figures 2–17 and Table 5, in terms of the De values, China, Russia, India, Turkey, Thailand, and Malaysia occupy the main centrality positions in the standard condition of trade volume, and they are the most influential countries in the agricultural trade network between China and B&R countries. Over the four years of the study sample, under the criterion of US $10 million, China's centrality rankings are 2, 1, 1, and 1, respectively, indicating that under the criterion of small and medium trade, China has the highest centrality and the greatest influence among B&R countries. Under the US $100 million criterion, China's De rankings over the four years are 2, 2, 2, and 2, respectively, indicating that among B&R countries, under the large trade volume criterion, China's influence is firmly in second place, after Russia. Under the criterion of CN = 0.7, China's De rankings over the four years are 31, 11, 10, and 15, respectively, indicating that, under the condition of greater competitive advantage, China's agricultural trade development space is generally improving. Under the criterion of CN = 0.8, China's De ranking over the four years are 29, 10, 11, and 11, respectively, which indicates that China's trade development space is steadily increasing and staying stable under the smaller trade competitive advantage. The criterion of CN = 0.7 and CN = 0.8 indicates that China's trade development space with B&R countries is huge.

Table 5. Network centricity of trade relations of China with B&R countries.

Year	Trade Volume (US $ Million)	CN Index	Top Five Countries in Terms of Relative Degree of Centrality (De %)	China Ranking	NC (%) Index
2001	10		Russia (34), China (31), India (27), Turkey (27), Saudi Arabia (24)	2	40
	100		India (13), China (10), Malaysia (10), Russia (9), Thailand (8)	2	18.16
		0.7	Brunei (51), Qatar (51), Kuwait (50), Saudi Arabia (47), Bahrain (46)	31 (15)	54.55
		0.8	Afghanistan (57), Qatar (55), Kuwait (53), Saudi Arabia (52), Brunei (51)	29 (29)	46.58
2009	10		China (49), Russia (46), Turkey (44), India (38), Ukraine (37)	1	55.11
	100		Russia (30), China (22), India (21), Turkey (20), Malaysia (19)	2	40.11
		0.7	Brunei (49), Iraq (48), Qatar (48), Kuwait (48), Saudi Arabia (40)	11 (31)	47.20
		0.8	Iraq (54), Brunei (54), Qatar (54), Kuwait (54), Afghanistan (50)	10 (44)	37.63
2013	10		China (54), India (46), Russia (46), Turkey (44), Thailand (41)	1	59.1
	100		Russia (35), China (29), India (29), Turkey (26), Ukraine (24)	2	45.45
		0.7	Iraq (49), Qatar (49), Brunei (48), Kuwait (44), Saudi Arabia (45),	10 (33)	49.75
		0.8	Iraq (55), Kuwait (55), Qatar (55), Brunei (54), Saudi Arabia (50)	11 (43)	42.12
2019	10		China (54), Russia (49), India (47), Turkey (45), Thailand (42)	1	59.66
	100		Russia (37), China (33), Turkey (30), India (29), Ukraine (24)	2	48.16
		0.7	Iraq (51), Qatar (52), Kuwait (52), Brunei (47), Saudi Arabia (42)	15 (29)	53.28
		0.8	Iraq (57), Kuwait (57), Qatar (57), Brunei (55), Maldives (51)	11 (44)	41.78

According to Table 5, from the relative degree centrality index, under the US $10 million criterion, the NC index between China and B&R countries grew from 40% to 59.66% in the four years of the simulation, with an overall growth rate of 49.15% and annual growth rates of 37.78%, 7.24%, and 0.95%, indicating that the centrality of agricultural trade between China and B&R countries tends to be concentrated, and that the countries have closer trade relations, but the centrality tends to slow down. Under the US $100 million criterion, the NC index between China and B&R countries grew from 18.16% to 48.16% in the four years of the simulation and annual growth rates of 120.87%, 13.31%, and 5.96%.

Compared with the US $10 million criterion, the US $100 million criterion centered on a smaller base but a faster growth rate, proving that China has relatively fewer countries, with a trade volume exceeding US $100 million along the B&R countries, but this closeness is growing faster. Under the criterion of CN = 0.7, the NC index between China and B&R countries ranged from 47.20% to 58.28% in the four years of the simulation and annual growth rates of −13.47%, 5.40%, and 17.15%, indicating that after the financial crisis, the central tendency of China and B&R countries with large competitive advantages has increased, and the trade exchanges tended to be closer. Under the criterion of CN = 0.8, the NC index between China and B&R countries ranged from 37.63% to 41.78% in the four years of the simulation and annual growth rates of 19.21%, 11.93%, and 0.81%, indicating that the trade center potential of China and B&R countries with smaller competitive advantages in the "Belt and Road" initiative tended to disperse, and the countries with room for trade development were more widely distributed.

3.2.3. Analysis of the Trade Block Model

According to the analysis in Table 6, the first plate is an internal type plate. The second plate is an internal type plate. Third plate is an isolated plate and the fourth plate is an internal plate. According to the above analysis, it can be concluded that in the early stage of China's accession to WTO, the agricultural trade between China and B&R countries is in an internal block or isolated state; a "small circle" trade. Similarly, the first, second, third, and fourth plates in 2009 are, respectively, the internal plate, isolated plate, isolated plate, and isolated plate. In 2013, the first, second, third, and fourth plates are the simultaneous plate, the internal plate, the simultaneous plate, and the internal plate, respectively. In 2019, the first, second, third, and fourth plates are the simultaneous plate, internal plate, internal plate, and internal plate, respectively.

Table 6. Trade volume (US$ 100 million) in 2001, 2009, 2013, and 2019 between China and B&R country plates (%).

Year	Plate	Plate 1	Plate 2	Plate 3	Plate 4	Year	Plate	Plate 1	Plate 2	Plate 3	Plate 4
2001	1	21.60	2.10	0	1.00	2009	1	30.40	8.70	0	3.70
	2	2.10	18.70	0	8.60		2	8.70	7.10	0	0.70
	3	0	0	0	0		3	0	0	0	0
	4	1.00	8.60	0	30.00		4	3.70	0.70	0	2.10
2013	1	54.10	12.70	14.40	1.90	2019	1	59.10	20.60	12.60	7.00
	2	13.60	4.40	1.20	0		2	20.60	4.50	7.90	1.40
	3	14.70	1.20	14.00	4.90		3	12.90	7.90	7.80	7.40
	4	2.30	0	4.90	19.70		4	7.00	1.40	7.40	27.30

Due to space limitations, only the country distributions for 2001 and 2019 are listed. According to the CONCOR method analysis results, in 2001, the countries in the first plate are China, nine countries in Southeast Asia, five countries in South Asia, and four countries in Western Asia and the Middle East. The countries in the second plate are Russia, two countries in Central Asia, three countries in Western Asia and the Middle East, and seven countries in central and Eastern Europe. The countries in the third plate are Mongolia, three countries in Central Asia, two countries in Southeast Asia, three countries in South Asia, seven countries in Western Asia and the Middle East, and six countries in central and Eastern Europe. The countries in the fourth plate are four countries in Western Asia and the Middle East, and three countries in Central and Eastern Europe. In 2019, the countries in the first plate are China, Russia, 10 countries in Southeast Asia, 3 countries in South Asia, and 2 countries in Western Asia and the Middle East. The countries in the second plate are four countries in South Asia, seven countries in Western Asia and the Middle East, and one country in Central and Eastern Europe. The countries in the third plate are Mongolia, four countries in Central Asia, five countries in Western Asia and the Middle East, and six countries in Central and Eastern Europe. The countries in the fourth plate are one in Southeast Asia, one in South Asia, four in Western Asia and the Middle East, and nine in Central and Eastern Europe. China always belonged to the first plate, and most of the countries in the first plate belonged to China's neighboring countries. The trade density of the first plate continued to increase from 2001 to 2019, rising from 0.216 to 0.591, indicating that the trade of countries in the first plate became increasingly close.

According to the analysis in Table 7, the first, second, third, and fourth plates in 2001 were, respectively, the export-oriented plate, dual-oriented plate, dual-oriented plate, and the dual-oriented plate. In 2009, they were the export-oriented plate, take into account plate, take into account plate, and the take into account plate. In 2013, they were the export-oriented plate, multi-faceted plate, multi-faceted plate, and the internal plate. In 2019, they were the simultaneous plate, simultaneous plate, simultaneous plate, and the simultaneous plate. Due to space limitations, only the country distributions in 2001 and 2019 are listed. Countries belonging to the first plate in 2001, included China, Russia, one in Southeast Asia, four in West Asia and the Middle East, and three in Central and Eastern Europe. The countries belonging to the second plate numbered three in Southeast Asia and eight in West Asia and the Middle East. The countries belonging to the third plate were Mongolia, and two in Central Asia, five in Southeast Asia, five in South Asia, three West Asia and the Middle East, and four in Central and Eastern Europe. Finally, the countries in the fourth plate numbered three in Central Asia, two in Southeast Asia, three in South Asia, three in West Asia and the Middle East, and nine in Central and Eastern Europe. The countries belonging to the first segment in 2019 were China, one in Southeast Asia, one in South Asia, eight in West Asia and the Middle East, and two in Central and Eastern Europe. The countries in the second segment were Mongolia and Russia, three in Central Asia, four in Southeast Asia, two in South Asia, one in West Asia and the Middle East, and five in Central and Eastern Europe. The countries in the third segment cinluded two in Central Asia, one in Southeast Asia, one in South Asia, two in West Asia and Middle East, and seven in Central and Eastern Europe. The countries in the fourth segment included four in Southeast Asia, four in South Asia, seven in West Asia and the Middle East, and four in Central and Eastern Europe. From 2001 to 2019, China was in the first plate, which changed from an export-oriented plate in 2001 to a balanced plate in 2019, indicating that the internal and external trade space of the plate gradually increased, and that the agricultural trade between China and B&R countries has broad prospects. From the perspective of the overall trade sector, the second, third, and fourth trade sectors had a large space for trade development. Although the fourth sector was transformed into an internal sector in 2013, and the foreign trade was not close enough, after the "Belt and Road" Initiative was proposed, the four sectors were all balanced sectors in 2019. This shows that the trade potential of agricultural products between China and B&R countries is huge.

Table 7. Trade volume (CN = 0.7) in 2001, 2009, 2013, and 2019 between China and B&R country plates (%).

Year	Plate	Plate 1	Plate 2	Plate 3	Plate 4	Year	Plate	Plate 1	Plate 2	Plate 3	Plate 4
2001	1	0	38.90	75.40	0.50	2009	1	0.80	31.50	66.70	99.70
	2	38.90	4.40	99.50	79.10		2	31.50	0	0	65.80
	3	75.40	99.50	1.80	11.00		3	67.90	0	0	10.20
	4	0.50	79.10	11.20	0		4	99.00	65.80	10.20	2.20
2013	1	0	40.50	94.60	72.00	2019	1	14.10	29.50	86.30	99.00
	2	40.50	0	55.20	0		2	30.00	0	1.20	56.30
	3	96.00	55.90	0	5.90		3	86.30	1.20	0	13.40
	4	72.00	0	5.90	0		4	99.00	56.30	12.90	3.30

3.3. Analysis of Spatial Network Influencing Factors

3.3.1. QAP Correlation Analysis

Using the QAP correlation analysis method and Ucinet 6 software, 5000 random permutations are selected to obtain the results of correlation analysis between China's agricultural trade network F and each influencing factor with B&R countries in 2019, where P represents the probability that the random correlation coefficient is greater or less than the actual value. This is shown in Table 8. The agricultural trade network F is not significantly correlated with the per capita arable land resource matrix (PCLA), scientific research expenditure share (SCI), and economic distance (DE), and is significantly correlated with the per capita water resource matrix (PCWR), geographic location matrix (DIS), transportation (TRA), total economic output (DGDP), trade agreement matrix (TA), culture matrix (CUL), and trade structure matrix (IS). We tentatively determine that the geographical distance, transportation, GDP difference, trade agreements, cultural differences, and trade structure differences between the two countries are the main factors affecting the agricultural trade network between China and B&R countries.

Table 8. QAP correlation analysis results of F and influencing factors of China with B&R countries.

Variable	Actual Correlation Coefficient	Significance Level	Mean of Correlation Coefficient	Standard Deviation	Minimum	Maximum	$P \geq 0$	$P \leq 0$
PCWR	−0.045	0.046	0.001	0.072	−0.123	0.268	0.955	0.046
PCLA	0.043	0.231	<0.001	0.068	−0.147	0.299	0.231	0.770
SCI	0.055	0.184	−0.001	0.065	−0.138	0.268	0.184	0.816
DIS	0.313	<0.001	<0.001	0.026	−0.083	0.101	<0.001	1.000
TRA	0.362	<0.001	<0.001	0.071	−0.153	0.286	<0.001	1.000
DGDP	0.253	0.002	0.001	0.074	−0.103	0.275	0.002	0.998
DE	−0.015	0.400	0.001	0.039	−0.077	0.210	0.601	0.400
TA	0.225	<0.001	<0.001	0.051	−0.122	0.204	<0.001	1.000
CUL	0.103	0.012	−0.001	0.040	−0.090	0.163	0.012	0.989
IS	−0.107	0.020	<0.001	0.072	−0.139	0.294	0.980	0.020

3.3.2. QAP Regression Analysis

Based on the correlation analysis results in Table 9, we included the matrices of PCWR, DIS, TRA, DGDP, TA, CUL, and IS, which were significantly correlated with F, in the QAP regression analysis, and set the number of random permutations to 5000. The results are shown in Table 9. As seen from the table, (1) PCWR is significantly correlated with the trade network between China and B&R countries, with a regression coefficient of −0.000002, and a negative regression coefficient, indicating that the smaller the difference in water resources, the closer the trade in agricultural products. (2) The correlation of DIS to China's trade network with B&R countries is significant, with a regression coefficient of 0.360666, indicating that countries with closer geographical locations are more likely to trade agricultural products, which also verifies the fact that China has close agricultural trade with neighboring countries such as Russia, Thailand, and Vietnam. (3) The correlation of TRA with the trade network between China and B&R countries is significant, with a regression coefficient of 0.053962, indicating that the more convenient the domestic and international transportation, the greater the possibility of agricultural trade. (4) DGDP is not significantly correlated with China's trade network with B&R countries, indicating that although GDP is correlated with agricultural trade between countries, countries with a large difference in GDP between the two countries do not necessarily have close agricultural trade. (5) TA is significantly correlated with the trade network between China and B&R countries, indicating that a trade agreement between two or more countries can facilitate bilateral or multilateral trade exchanges. (6) The correlation of CUL with the trade network between China and B&R countries is significant, indicating that the two countries have the same or similar culture, which is conducive to the economic and trade exchanges between the two countries and the expansion of bilateral agricultural trade. (7) IS is significantly correlated with the trade network between China and B&R countries, with a negative coefficient of −0.000983, indicating that the smaller the gap in the agricultural trade structure, the closer the bilateral trade between the countries.

Table 9. QAP regression analysis results of China with B&R countries.

Variable	Normalized Regression Coefficient	Probability of Significance	$P \geq 0$	$P \leq 0$
PCWR	−0.000002	0.015 **	0.986	0.015
DIS	0.360666	<0.001 ***	<0.001	1.000
TRA	0.053962	0.001 ***	0.001	1.000
DGDP	0.0000001	0.440	0.440	0.560
TA	0.179529	0.001 ***	0.001	1.000
CUL	0.098974	0.036 **	0.036	0.965
IS	−0.000983	0.007 ***	0.994	0.007
Sample size	3660	R^2 = 0.240	Adjusted R^2 = 0.238	

Note: ***, ** indicate that the coefficients passed the significance tests of 1%, 5%.

4. Discussion

4.1. Comparison to Prior Studies

The main objective of our paper is to explore the factors influencing the evolution and trade space of agricultural trade between China and B&R countries, from a planar and spatial perspective. Our results confirm that China's agricultural trade with B&R countries has become increasingly close and highly concentrated, with high intensity areas being mainly concentrated in closer or neighboring countries, and that the trade development space shows a "rise-fall-rise " trend, with China remaining at the "power" core of the trade network, and with the centralization of trade in the B&R countries tending to be concentrated. The agricultural trade network between China and B&R countries can be divided into four major segments, with increasingly close internal and external trade in each segment. The trade network linkages are mainly influenced by water resources, geographic location, railroad convenience, trade agreements, and trade structure. The findings of this paper are consistent with previous studies on agricultural trade between China and B&R countries. For example, Yu (2016) found that China's total bilateral trade with eight South Asian countries has quadrupled, and that China's total agricultural imports from South Asia are greater than its total agricultural exports, with a widening deficit [8]. Zhan (2018) pointed out that the network density of agricultural export relations, competitive relations, and complementary relations among B&R countries is increasing day by day [32]. Su (2019) argued that the density of spatially linked networks of agricultural trade in China and B&R countries is high, and that China is at the center of this spatially linked network [33]. As expected, the empirical results of this paper show that China's agricultural trade with B&R countries is getting closer and closer, and that China is gradually becoming the center of agricultural trade. Our findings are almost consistent with those of the aforementioned scholars.

However, our study is somewhat different from previous studies, from the following perspectives. In analyzing the comprehensiveness of agricultural products and time chains, we have used all the total agricultural products and the time of WTO accession as samples, which can provide China with an overall perspective and ability to grasp the dynamics of agricultural cooperation in B&R countries; this is conducive to making comprehensive and sustainable decisions. For example, Li (2018) takes aquatic products as a sample and concludes that China is located in the middle and high end of the regional value chain of B&R countries, and has the ability to dominate the regional value chain [23], which is similar to the conclusion reached in this paper. However, the conclusion of this paper has a greater generalization of agricultural products and is more conducive to national agricultural sector decision-making.

Second, we use a combination of planar- and spatial-shaped analysis, taking into account the quantity, specific agricultural product dynamics, and spatial national agricultural trade dynamics, each focusing on the other and complementing each other. For example, both Chen (2019) and Su (2019) analyzed the relationship model of food and agricultural trade networks between China and B&R countries from a spatial perspective [26,33], which can reflect that the density of food trade networks among B&R countries is increasing. However, they did not compare with the flat volume structure reference, and could not produce accurate figures.

Third, we improved the competitive index CS and constructed the competitive advantage index CN in the research technique of Li et al. (2017) [37], which provides a method for scholars to measure the spatial relationship. For example, Wei (2018) refers to Li Jing's (2017) method of screening nodes, and uses the total agricultural import and export trade of more than US $100 million, and unilateral agricultural imports of more than US $10 million as the criteria for trade flows, without further constructing the network model with the criteria of comparative advantage to derive new trade structure information [31].

Finally, regarding measurements of the influencing factors of spatial association relationships, for example, previous studies found that Wei's (2018) "proximity effect", FTAs, differences in consumer population base, and differences in total economic size all enhance the association relationships of agricultural trade between countries [31]. However, we found that trade linkages are influenced by water resources, geographic location, railroad accessibility, trade agreements, and trade structure. There were also both overlapping elements and new elements that can provide a reference for China to select sustainable cooperation partners.

4.2. Sustainability Implications

First, the types of agricultural products traded and the import and export areas between China and B&R countries are highly concentrated. China has close trade with countries that are geographically close, so China has to maintain friendly relations with neighboring countries, such as Russia, India, Vietnam, Thailand and other large agricultural countries, and actively sign trade agreements. It can not only save China's trade costs and improve its own agricultural products supply security capacity, but also promote the differentiated division of labor in agriculture between China and B&R countries, improve labor productivity, increase the export of agricultural products with comparative advantages, and achieve sustainable mutual benefits.

Second, large agricultural countries such as China, Russia, and India occupy a dominant position, and the trade network can be divided into four major segments, with increasingly close trade within and outside each segment. Therefore, China should stabilize production and trade with large agricultural countries such as Russia and India to reduce the overall trade risk and enhance sustainable agricultural production and trade. China should also actively use its dominant position as a large agricultural trading country to guide other strong agricultural trade countries to play a greater role in emphasizing their own advantages, supporting countries with weaker trade by providing agricultural production factors, and promoting sustainable cooperation among countries.

Last, the relationship between China and the agricultural trade network of B&R countries is mainly influenced by water resources, geographical location, railroad convenience, trade agreements and trade structure. Therefore, China can, based on the conditions of natural resources, public facilities, and trade agreements of the B&R countries, reduce unnecessary waste of agricultural resources and the environment in the context of the global goal of achieving carbon peaking and carbon neutrality. China should also strengthen infrastructure development and natural resource advantages of neighboring countries to complement each other, reduce trade costs and increase agricultural productivity, and improve environmentally friendly and sustainable production between countries. Additionally, China should develop an agricultural production policy within China that suits its own resource environment and trade structure, so that both domestic and international agricultural exports and imports can take advantage of their comparative advantages and form a sustainable domestic production and international trade relationship.

4.3. Limitations and Future Research Directions

There are also some shortcomings to our research. First, when discussing the classification of specific agricultural products, this paper only explores in the plane structure, and it hardly shows the spatial structure. Therefore, although the current situation of the cooperation of specific classified agricultural products with B&R countries can be understood from the perspective of a single Chinese country, the analysis of the agricultural products classified in the spatial scope of B&R countries is lacking. The results of our study need to be supplemented and improved again by subsequent studies. Second, when discussing the factors influencing the spatial structure of trade, we only list 10 indicators, due to the length and the availability of the data, and there are other important influencing factors to be explored to further improve the indicator system for promoting bilateral trade. Third, we assumed only two significant indicators, namely the US $10 million and US $100 million markers, since other data were not available for the countries and years we analyzed. Fourth, we constructed indicators of comparative competitive advantage in trade without constructing indicators of trade complementarity. The conclusions drawn can only reflect a situation of competitive advantage in trade.

There are three future research directions. First, our research here mainly explores the agricultural trade structure and its influencing factors between China and B&R countries from a plane and spatial perspective. However, with the establishment of AFTA, CEFTA, and RCEP, the policies between countries are very different, and the global dual carbon initiative goals are included in the influence of the factors. These factors enrich the system of indicators affecting sustainable trade structure and help to quantify the effect of regional cooperation and the reference direction of future sustainable cooperation in the face of COVID-19, the Russia–Ukraine conflict, and climate change. Second, the spatial structure of this paper only explores the overall agricultural trade structure from different markers, so we can take global bulk agricultural products, such as soybean, wheat, corn, and rice, as the research objects, and explore the agricultural trade structure between China and B&R countries, which is conducive to exploring the targets of trade-led sustainable cooperation from the perspective of specific agricultural products. Third, future research can construct trade complementarity indexes. International trade not only has the theory of comparative advantage, but complementarity is also one of the important theories for promoting the development of international trade, and so it is beneficial to expand the criteria of trade cooperation and explore the trade network space from multiple perspectives to provide a rich reference for international agricultural sustainable cooperation.

5. Conclusions

From the perspective of plane and space, this paper analyzes the plane structure and the spatial network structure, and the influencing factors of agricultural trade between China and the B&R countries from 2001 to 2019. This paper takes all agricultural products and all countries along the B&R as research samples, classifies countries and products, and uses plane and spatial perspectives to conclude that China's agricultural product trade with B&R countries tends to be decentralized in terms of import types and regions, and concentrated in terms of export types and regions, the regions with high trade intensity are mainly concentrated in close proximity or in neighboring countries, and that trade relations are getting closer. China has always had a greater influence in the trade network, and the trade centrality of the B&R countries tends to be concentrated. China's agricultural trade network with the B&R countries can be divided into four major segments, with increasingly close internal and external trade in each segment. The results of this paper are beneficial for

China and B&R countries to provide a basis for making decisions on trade cooperation from the perspective of agriculture as a whole, to promote global agricultural cooperation, and to facilitate global agricultural cooperation, the flow of global agricultural factors, world food production, and the establishment of a reference method to study global trade structure.

Author Contributions: Conceptualization: L.Z. and G.T.; methodology: L.Z. and G.T.; software: L.Z.; validation: L.Z. and G.T.; formal analysis: L.Z.; investigation: L.Z.; data curation: L.Z. and G.T.; writing—original draft preparation: L.Z.; writing—review and editing: L.Z. and G.T.; visualization: L.Z. and G.T.; funding acquisition: L.Z. and G.T. All authors have read and agreed to the published version of the manuscript.

Funding: The research was funded by Heilongjiang Provincial Philosophy and Social Science Office, grant number 18JLD310.

Institutional Review Board Statement: Not applicable.

Informed Consent Statement: Not applicable.

Data Availability Statement: The data presented in this study are available on request from the corresponding author.

Acknowledgments: The authors would like to thank the anonymous reviewers for their constructive comments and valuable suggestions on this article.

Conflicts of Interest: The authors declare no conflict of interest.

References

1. Glogovețan, A.I.; Dabija, D.C.; Fiore, M.; Pocol, C.B. Consumer Perception and Understanding of European Union Quality Schemes: A Systematic Literature Review. *Sustainability* **2022**, *14*, 1667. [CrossRef]
2. Guo, J.; Mao, K.; Yuan, Z.; Qin, Z.; Xu, T.; Bateni, S.M.; Zhao, Y.; Ye, C. Global Food Security Assessment during 1961–2019. *Sustainability* **2021**, *13*, 14005. [CrossRef]
3. Majerova, J.; Sroka, W.; Krizanova, A.; Gajanova, L.; Lazaroiu, G.; Nadanyiova, M. Sustainable Brand Management of Ali-mentary Goods. *Sustainability* **2020**, *12*, 556. [CrossRef]
4. Du, Z.X.; Gao, M.; Han, L. Study on the impact of supply side import end change on China's food security. *Chin. Rural Econ.* **2021**, *1*, 15–30.
5. Zhao, Y.L.; Lin, G. Analysis of bilateral agricultural trade flow and trade potential between China and 10 ASEAN countries—A study based on Trade Gravity Model. *J. Int. Trade* **2008**, *12*, 69–77.
6. Li, T.; Li, Y.X. Analysis on complementarity of agricultural trade between China and five Central Asian countries. *J. Int. Trade* **2011**, *1*, 53–62.
7. Zhu, X.X.; Li, Y.X. Analysis on the competitiveness and complementarity of agricultural trade between China and five Central Asian countries. *Int. Econ. Trade Res.* **2011**, *27*, 17–22.
8. Yu, M.Z.; Liang, Y.F.; Gao, Y. One belt, one road strategy and the competition and complementarity of agricultural products trade between China and South Asia. *Issues Agric. Econ.* **2016**, *37*, 83–94.
9. Geng, Z.Z.; Xiao, H.F. Analysis on the trade characteristics of agricultural products between China and countries along the "21st century Maritime Silk Road". *Issues Agric. Econ.* **2016**, *37*, 81–88.
10. Yang, F.M.; Tian, Y.Y. Research on agricultural trade between China and countries along the "21st century Maritime Silk Road"—From the perspective of competitiveness, complementarity and trade potential. *Mod. Econ. Res.* **2018**, *8*, 54–65.
11. Tong, G.J.; Shi, L. Empirical analysis of Sino Russian agricultural trade based on intra industry. *Issues Agric. Econ.* **2017**, *38*, 89–100.
12. Bai, X.B.; Xu, Z.; Zhou, Y.H. Analysis of the characteristics of China-Russia agricultural trade and prospects of cooperation. *Russ. Stud.* **2021**, *4*, 176–196.
13. Liu, C.P.; Xiao, H.F. A study on the causes of agricultural trade growth between China and 16 countries in central and Eastern Europe—An Empirical Analysis Based on CMS model. *J. Agrotech. Econ.* **2018**, *9*, 135–144.
14. Yao, C.M.; Xu, X.K.; Wang, Y. Dynamic decomposition of the causes of agricultural trade growth between China and Eastern European countries. *Stat. Deci.* **2021**, *37*, 155–159.
15. An, X.N.; Xin, L. Analysis of the current situation and potential of agricultural trade between China and Southeast Asia based on the "One Belt, One Road" initiative. Chin. *J. Agri. Res. Reg.* **2019**, *40*, 118–127.
16. Wang, R.; Wen, H.D. Research on China's export potential of agricultural products to countries along the "Silk Road Economic Belt"—An Empirical Analysis Based on Stochastic Frontier gravity model. *J. Agrotech. Econ.* **2016**, *10*, 116–126.
17. Wang, L.; Liu, S.G.; Duan, P.L.; Yin, P. Structural characteristics of agricultural products trade network in countries along the Silk Road Economic Belt. *Econ. Geogra.* **2019**, *39*, 198–206.

18. Yang, J.Q. Spring Festival Study on the export trade potential of China's agricultural products from countries along the "Silk Road Economic Belt"—An analytical framework based on TPI and extended stochastic frontier gravity model. *J. Int. Trade* **2020**, *6*, 127–142.
19. Ya, Z.; Pei, K. Factors Influencing Agricultural Products Trade between China and Africa. *Sustainability* **2022**, *14*, 5589. [CrossRef]
20. Zhao, X.; Sun, Z.L. Analysis of the prospects of agricultural trade between China and Ukraine in the context of "Belt and Road". *World Agric.* **2021**, *3*, 90–99.
21. Manitra, R.A.G. Competitiveness and trade potential of India's dairy industry. *Food Policy* **2006**, *31*, 216–227.
22. Zhao, M.Z. Study on the dependence of corn on the international market—Based on the comparative analysis of four food crops. *J. Int. Trade* **2015**, *9*, 109–121.
23. Li, C.; Wang, L.Y. China's leading Belt and Road, and other countries' regional value chain construction: Take aquatic products export trade as an example. *Macroeconomics* **2018**, *9*, 72–84.
24. Zhang, Y.; Zhao, L.; Chen, Y.S. Belt and Road strategy of aquatic products export trade in China: A case study of Shandong province. *J. Xiamen Univ.* **2018**, *4*, 135–144.
25. Zhang, F.H.; Zhang, J.G.; Song, X.L.; Pang, G.J. Comparative analysis of Belt and Road, and the other countries' export market in China and the United States. *J. Agrotech. Econ.* **2019**, *1*, 135–144.
26. Chen, Y.W.; Li, E.L. "Belt and Road" national grain trade network spatial pattern and its evolution mechanism. *Progress Geogra.* **2019**, *38*, 1643–1654. [CrossRef]
27. Sun, Z.L.; Jia, X.L.; Li, X.D. Belt and Road, China's grain trade trend and the estimation of its virtual cultivated land flow. *J. Huazhong Agric. Univ.* **2019**, *1*, 24–32.
28. Han, D.; Li, G.S.; Zhong, Y. A comparison of grain competitiveness and factors influencing grain trade between China and countries along the "Belt and Road". *J. Jiangxi Univ. Finance Econ.* **2020**, *4*, 76–92.
29. Han, D.; Li, G.S. Evolution of Food Trade Patterns and Influence Mechanisms between China and Countries Along the "Belt and Road"—Based on Social Networking Perspective. *Issues Agric. Econ.* **2020**, *8*, 24–40.
30. Peng, S.G.; Zhou, Y.H. China's export growth Belt and Road country – from the perspective of three yuan margin. *Issues Agric. Econ.* **2021**, *4*, 132–144.
31. Wei, S.H. China's Agricultural Trade with "Belt and Road" Countries: Network Structure, Linkage Characteristics and Strategic Options. *Issues Agric. Econ.* **2018**, *11*, 101–113.
32. Zhan, M.H. Belt and Road, the competition and complementarity of agricultural products trade in the countries along the belt—Based on social network analysis method. *Issues Agric. Econ.* **2018**, *2*, 103–114.
33. Su, X.; Zhang, H. China's Belt and Road, and other countries' agricultural trade network structure and cooperation situation. *Reform.* **2019**, *7*, 96–110.
34. Yang, Y.W.; Dong, W.L.; Yang, J. The evolution of agricultural trade characteristics between China and countries along the Belt and Road. *Soc. Sci.* **2021**, *1*, 67–73.
35. Liu, Y.J.; Liu, Y.T. Evolution and optimization strategies of agricultural trade pattern between China and countries along the "Belt and Road". *Zhejiang Acad. J.* **2021**, *4*, 113–122.
36. Sun, Z.L.; Li, X.D.; Li, S.J. Research on intra-industry trade of agricultural products between China and countries along the Belt and Road and its influencing factors. *J. Huazhong Agri. Univ.* **2021**, *1*, 57–68+176.
37. Li, J.; Chen, N.; Wan, G.H.; Chen, S. Belt and Road, the state of trade, the complementary relationship and dynamic change of goods trade – Based on network analysis method. *J. Manag. World* **2017**, *4*, 10–19.
38. Lin, D.Y.; Zhu, J. Study on the comparative advantage and import market structure of main grain varieties in China. *World Econ. Stud.* **2015**, *2*, 115–126.
39. Zheng, L.; Song, Z.Y.; Liu, W.D.; Liu, Y. Analysis of trade pattern and trade structure in Western China. *Geographical Res.* **2015**, *34*, 1933–1942.
40. Sun, D.S.; Su, J.X.; Li, N.H.; Zhang, L. Study on the impact of Sino US trade friction on Sino US agricultural trade structure. *Issues Agric. Econ.* **2021**, *1*, 95–106.
41. Li, J.; Chen, S.; Wan, G.H.; Fu, C.M. Spatial correlation of China's regional economic growth and its explanation—Based on network analysis method. *Econ. Res. J.* **2014**, *49*, 4–16.
42. Ma, S.Z.; Ren, W.W.; Wu, G.J. Characteristics of a country's agricultural trade network and its impact on the division of labor in the global value chain—From the perspective of social network analysis. *J. Manag. World* **2016**, *3*, 60–72.
43. Baster, A.S.J.; Brown, A.J. Applied Economics: Aspects of the World Economy in War and Peace. *Economica* **1949**, *16*, 78. [CrossRef]
44. Kojima, K. The Pattern of International Trade Among Advanced Countries. *Hitotsubashi J. Econ.* **1964**, *5*, 16–36.
45. Chen, Y.S.; Chang, K.C. The relationship between a firm's patent quality and its market value—The case of US pharmaceutical industry. *Technol. Forecast. Soc. Chang.* **2010**, *77*, 20–33. [CrossRef]
46. Granovetter, M. Economic Action and Social Structure: The Problem of Embeddedness. *Am. J. Sociol.* **1985**, *91*, 481–510. [CrossRef]
47. Harrison, C.W.; Scott, A.B.; Ronald, L.B. Social Structure from Multiple Networks. I. Block-models of Roles and Positions. *Am. J. Sociol.* **1976**, *81*, 730–780.
48. Liu, J. QAP: A method to measure the relationship between "relationships". *Chin. J. Sociol.* **2007**, *4*, 164–174.
49. David, K.; Wilbur, I.; Smith, K.O.; John, A.C. A Gravity Model Analysis of the Benefits of Economic Integration in the Pacific Rim. *J. Econ. Integra.* **1999**, *14*, 347–367.

50. Cemal, A.; Jun, F. Regional Blocs and Agricultural Trade Flow: The Case of ASEAN. Japan Int. Res. *Cent. Agric. Sci.* **2008**, *42*, 115–121.
51. Liu, H.M.; Li, G.J.; Wang, K.Q. Study on Influencing Factors of international trade of Agricultural Virtual Water in China—Analysis Based on gravity model. *J. Manag. World* **2010**, *9*, 76–87.
52. Zhong, Z.H.; Qin, C.L. Trade Network Structure of "Belt and Road" and Its Influencing Factors—A Study Based on Network Analysis Method—Based on network analysis method. *Int. Econ. Trade Res.* **2017**, *33*, 16–28.
53. Yao, X.; Pu, Y.; Wu, G.; Wang, B.; Wang, L. China's industrial integration and status along the "Belt and Road": Industry comparison, regional differences and correlation factors. *Econ. Res. J.* **2019**, *54*, 172–186.
54. Feng, L.Y.; Xu, H.L.; Wu, G.; Zhao, Y.; Xu, J.L. Exploring the structure and influence factors of trade competitive advantage network along the Belt and Road. *Phys. Stat. Mech. Appl.* **2020**, 559. [CrossRef]
55. Vătămănescu, E.M.; Alexandru, V.A.; Mitan, A.; Dabija, D.C. From the Deliberate Managerial Strategy towards International Business Performance: A Psychic Distance vs. Global Mindset Approach. *Syst. Res. Behav. Sci.* **2020**, *3*, 374–387. [CrossRef]

sustainability

MDPI

Article

Agricultural Support Policies and China's Cyclical Evolutionary Path of Agricultural Economic Growth

Xiangdong Guo [1,*], Pei Lung [2], Jianli Sui [3], Ruiping Zhang [1] and Chao Wang [1]

1 School of Economics and Management, Beijing Jiaotong University, Beijing 100044, China; rpzhang@bjtu.edu.cn (R.Z.); chaowang@bjtu.edu.cn (C.W.)
2 The Daniels College of Business, University of Denver, Denver, CO 80208, USA; Pei.lung@du.edu
3 Center for Quantitative Economics, Jilin University, Changchun 130012, China; jlsui@163.com
* Correspondence: xiangdong.guo@bjtu.edu.cn; Tel.: +86-10-5168-8044

Abstract: Due to the weak nature of agricultural production, governments usually adopt supportive policies to protect food security. To discern the growth of agriculture from 2001 to 2018 under China's agricultural support policies, we use the nonlinear MS(M)-AR(p) model to distinguish China's agricultural economic cycle into three growth regimes—rapid, medium, and low—and analyze the probability of shifts and maintenance among the different regimes. We further calculated the average duration of each regime. Moreover, we calculated the growth regime transfers for specific times. In this study, we find that China's agricultural economy has maintained a relatively consistent growth trend with the support of China's proactive agricultural policies. However, China's agricultural economy tends to maintain a low-growth status in the long-term. Finally, we make policy recommendations for agricultural development based on our findings that continue existing agricultural policies and strengthen support for agriculture, forestry, and animal husbandry.

Keywords: agricultural development; agricultural economic cycle; agricultural policies

Citation: Guo, X.; Lung, P.; Sui, J.; Zhang, R.; Wang, C. Agricultural Support Policies and China's Cyclical Evolutionary Path of Agricultural Economic Growth. *Sustainability* **2021**, *13*, 6134. https://doi.org/10.3390/su13116134

Academic Editor: Aaron K. Hoshide

Received: 14 April 2021
Accepted: 27 May 2021
Published: 29 May 2021

1. Introduction

Agriculture not only affects the macroeconomics of a country but is also associated with the food security and employment issues of a country, particularly for developing countries. With the majority of countries in the world still in the developing stage and a very high proportion of the population still underdeveloped, development remains a central topic in the world economy. As early as 1946, economists Burns and Mitchell emphasized that economic growth can be effectively guided only by a thorough analysis of the mechanisms of change inherent in economic growth [1].

The agricultural surplus theory considers a highly developed agricultural economy as the fundamental condition for macroeconomic development. For this reason, the Chinese government introduced beneficial agricultural policies every year since 2004 to support agricultural development. China's economy has been increasing with a double digit high growth rate in the past few decades. The growth rate has only slowed down slightly in recent years, but it is still one of the fastest growing economies in the world. Within this historical context, what are the dynamics that drive China's agricultural economic growth? In this paper, we investigate the true underlying dynamics of China's agricultural economy during this period of time, and summarize its experience to incentivize further growth of China's agricultural economy.

The available literature focuses mainly on the factors influencing the growth of agriculture. Even though agriculture may grow rapidly in the short term, long-term growth is constrained by factors such as overconsumption of natural resources and environmental pollution [2]. The role of water use in driving agricultural growth in different regions of China was based on a panel vector autoregressive model [3]. Infrastructure development has a catalytic effect on China's agricultural GDP [4]. Agricultural production and development can also be influenced by other factors. For example, in Nigeria rainfall, currency exchange rate, and food exports are the most important factors driving agricultural output. Food importation, diversion of funds for agriculture, and low penetration of agricultural technology were identified as the major constraints to agricultural development in Nigeria [5].

Other studies reach the same results, in which increasing agricultural research and development expenditures will support agricultural economic growth [6–9]. However, increasing fiscal spending on agriculture, while promoting agricultural growth, can also impact the quality of agroecosystems [10]. Soil and water conservation has a significant impact on the per capita income of rural households in China [11].

To achieve sustainable long-term growth in agriculture, we need to strengthen reforms and innovation in the rural economic system [12]. In the long run, a land system that is compatible with the country's macroeconomic condition has a positive contribution to China's agricultural economic growth [13,14]. Studies in Commonwealth of Independent States (CIS) countries have shown that policy factors such as land reform contribute, to some extent, to the growth of agricultural production [15]. Studies in the EU countries show that the average farm income is already close to the average non-farm income, thanks to the Common Agricultural Policy (CAP) support [16]. Agricultural policies are important for poverty reduction and agricultural development [17–22]. On the contrary, distorted agricultural policies can hinder its agricultural development [23]. Effective monetary and fiscal policies can boost agricultural growth over time [24–26].

The studies above focus on the different factors that are related to agricultural growth, but few scholars have analyzed the historical trajectory of agricultural growth in depth. As a reflection of China's economic growth, agricultural growth has shown up-and-down cycles over the past 20 years. This is despite China's high rate of macroeconomic growth and strong support from the government's pro-agricultural policies.

Recent studies on agricultural economic cycles in Spain, Cuba, and the United Kingdom identified the causative factors driving these cycles. Studies in the Spanish regional economy show that the agricultural production cycle is constrained by the natural environment and ecological conditions, meanwhile, at the same time, benefitted from rapid economic and social development and globalization [27]. The history of agricultural development in Cuba indicates that, according to the theory of the adaptive renewal cycle, the process of agricultural development is nonlinear and is divided into four stages: Growth, maturation, collapse, and transformation [28]. Studies on the United Kingdom agriculture shows that agriculture sustainably reinforces cultural management and ecosystems and will affect cultural service assets in a broad sense [29].

Since 1952, China's agricultural economic cycle has fluctuated several times, with three classical economic cycles and three growth cycles. The agricultural economies have achieved growth and development in the midst of cyclical fluctuations, and responded to the economic policies and institutional reforms in different economic periods [30]. Changes in China's agricultural policies are the main cause of agricultural fluctuations. Institutional factors are important causes of the cyclical fluctuations of the agricultural economy [31]. The pricing mechanism and land system can promote the change of the agricultural economy to a high-growth state [32].

The magnitude of fluctuations in China's agricultural economic cycle has declined significantly and China's agricultural development has gradually stabilized from the 1980s. Overall, China's agricultural economic cycle has a high frequency of fluctuations with small magnitude [33]. China's agricultural economy is characterized by significant inertia during low-growth rates. When the agricultural growth rate is relatively high, its risk of shocks is higher, as well. When the growth rate is relatively low, its uncertainty is relatively low [34]. Agricultural economic cycles have certain spatial correlations and will amplify agricultural economic fluctuations through cyclical spatial spillover, forming cyclical synergistic effects [35]. Technological and institutional innovations in China should emphasize more on sustainable development which considers the relationship between agriculture and the environment, rather than setting inconsistent and sometimes incompatible policy goals from different perspectives [36].

Agricultural policies implemented by the Chinese government have raised farmers' income levels and contributed to long-term food security goals. However, such policies have also led to a price gap between domestic and international markets for agricultural products leading to a sharp increase in agricultural imports and the accumulation of large stocks [37]. Previous experience in developing agriculture through institutional reform, technological change, market reform, and investment in agriculture remains the key to future success in ensuring food security and sustainable agriculture growth in China [38]. The true family farm of moderate "small and precise" scale, which has emerged quite widely in China over the past 30 years, can chart a more sustainable way forward for Chinese agriculture [39].

Agricultural economic growth should not come at the cost of damage to natural resources and pollution of the ecological environment, but should focus on the coordination and balance between the short- and long-term [2]. Environmentally friendly technological innovation is a long-term driving force for both the development and sustainable growth of agricultural economies. On the whole, every 1% increase in environmentally friendly agricultural technology innovation causes a 0.375% increase in agricultural economic growth, while every 1% increase in the extent of environmentally friendly technology diffusion causes a 0.542% increase in agricultural economic growth [40].

Attention should be paid to the phenomenon of decreasing ecological land use in the agro-pastoral zone, and the land use structure should be adjusted to provide good ecological conditions for the sustainable development of the agricultural economy in the agro-pastoral zone [41]. Soil and water conservation can contribute to agricultural economic growth and rural poverty reduction in China. Soil quality and capital inputs are now more important than farmland size and agricultural labor in poverty reduction and economic growth. Governments and farmers need to prioritize investments in soil and water conservation to boost the agricultural economy and reduce rural poverty [11]. As the use of linear measures does not allow for effective measurement of the characteristics of economic cycles, such as those found in agriculture, scholars have proposed various nonlinear econometric models to characterize the variability of economic cycles in detail, such as the Smoothed Migration Autoregressive (STAR) model, the Markov transfer (MS) model, and Threshold Autoregressive (TAR) models [42–44]. In recent years, researchers have measured the economic cycles of South Africa, Brazil, Turkey, etc. using Markov regime transfer models [45–47].

On this basis, in the existing studies on the fluctuation dynamics of the Chinese economic cycle, scholars have quantitatively measured the economic cycle based on different forms of Markov transfer (MS) models to analyze the dynamic characteristics of the economic cycle when it varies across regimes [48–50]. Sui Jianli and Song Diandian first used a Markovian regime shift model to study the Chinese agricultural economic cycle, but he chose a two-regimes Markovian shift model that could only distinguish the Chinese agricultural economic cycle into a high-growth and a low-growth regime [51].

This paper follows the previous research path but differs in that the time domain of this paper is placed in the specific context of the Chinese government's annual agricultural policies to support agricultural development from the early 21st century. Our research focuses on China's agricultural economic development under the support of agricultural policies. In this paper, we construct a Markov transfer model with three regions to analyze the agricultural development in China since 2001 and use it to initially assess the effects of agricultural policies during this period.

In addition, a nonlinear MS (M)-Autoregressive Process (AR) (p) model with "mean form" and "intercept form" was created, and the growth rate of the overall agricultural product is included in our study. The nonlinear model with the "mean form" focuses on capturing the trajectory of the mean levels as they shift across time, while the nonlinear model with the "intercept form" is able to track the dynamic path of time series means since they smoothly transition over time as they shift by regime. The model is able to more accurately and sharply capture the cyclical fluctuations in the mean of each variable's time series data. Drawing on the ideas of Krolzig, this paper provides a quantitative demonstration of the growth dynamics in the context of policy support for China's agricultural economy [52]. The specific contribution of this paper to the scientific literature on agricultural development is to empirically highlight the importance of agricultural policy for agricultural development.

The Markov transfer model has the advantage of being able to accurately distinguish the dynamic changes between different variables, which is different from other models. Therefore, the Markov transfer model is used in this paper to study the growth of China's agricultural economy. In order to accurately understand the cyclical evolutionary path of the agricultural economy, this paper takes not only the total agricultural output as the object of study, but also the total output of agriculture, forestry, fishery, and animal husbandry sub-sectors of Chinese agriculture.

2. Methods and Materials

2.1. Nonlinear MS (M)-AR (p) Model Construction and Model Estimation

In this paper, we model the growth rates of China's agriculture, forestry, livestock, fisheries, and overall agriculture, respectively, and its regime shifts. The growth rate is measured as y_t and the regime shift in the growth rate is measured as s_t. We consider a linear p-order Autoregressive Process (AR) as the starting point for the nonlinear model:

$$y_t = v + A_1 y_{t-1} + \ldots + A_p y_{t-p} + u_t \tag{1}$$

where v is chosen as the intercept term. In addition, this paper requires the necessary assumptions regarding the smoothness of the y_t time series. In the equation $1 - A_1 L - A_2 L^2 - \cdots - A_p L^p = 0$ of the lag operator L, we assumed that the characteristic roots of the equation are located outside the unit circle. We also assumed that the error term u_t of Equation (1) follows the standard normal distribution, i.e., $u_t \sim NID(0, \Sigma)$. Based on these assumptions, the model form presented in Equation (1) is the "intercept form" of the classical AR (p) model. The "mean value form" of the linear p-order AR (p) model is presented as follows:

$$y_t - \mu = A_1(y_{t-1} - \mu) + \ldots + A_p(y_{t-p} - \mu) + u_t \tag{2}$$

In this paper, μ is defined as the mean of the time series y_t for each variable. We can clearly see that the linear AR (p) models in "intercept form" and "mean form" constructed in the previous section have limitations in detailing the possible y_t nonlinear features in the time series and cannot successfully capture the "structural mutations" embedded in the time series. In light of this, we explored in depth the "structural mutation" phenomenon in China.

This paper follows the approach of Hamilton and Krolzig [43,52] to add random s_t variables to the time series y_t to deeply explore the potential nonlinear "Markovian shifts" in China's agricultural economic growth process, where M different regimes s_t can be characterized, $s_t \in \{1, \ldots, M\}$. By introducing s_t into the time series data generation process, we are able to more accurately examine the dynamic changes in the nonlinear AR (p) model. At the same time, this paper further assumes s_t that it is possible for the Markov process to be followed to traverse all M regimes, based on which, the specific transfer matrix can be expressed in the following form:

$$P = \begin{bmatrix} P_{11} & P_{12} & \cdots & P_{1M} \\ P_{21} & P_{22} & \cdots & P_{2M} \\ \vdots & \vdots & \ddots & \vdots \\ P_{M1} & P_{M2} & \cdots & P_{MM} \end{bmatrix} \tag{3}$$

In Equation (3), $p_{ij} = \Pr(s_{t+1} = j | s_t = i)$, $\Sigma_{j=1}^{M} p_{ij} = 1, \quad \forall i, j \in \{1, \ldots, M\}$.

In the following sections of this paper, we construct nonlinear MS (*M*)-AR (*p*) models with reference to the "mean-form" linear AR (*p*) models. The MSM (*M*)-AR (*p*) model containing the variable parameter function $\mu(s_t)$ can be constructed by introducing the regime state variable s_t into the mean μ shown in Equation (2).

$$y_t - \mu(s_t) = A_1[y_{t-1} - \mu(s_{t-1})] + \ldots + A_p[y_{t-p} - \mu(s_{t-p})] + u_t, \quad u_t \sim NID(0, \Sigma) \quad (4)$$

It is known that

$$\mu(s_t) = \begin{cases} \mu_1, & s_t = 1 \\ \vdots & \vdots \\ \mu_M, & s_t = M \end{cases} \quad (5)$$

The variable-parameter functions $A_1(s_t), \ldots, A_p(s_t), \Sigma(s_t)$, and, $v(s_t)$ have very similar expressions to $\mu(s_t)$ as defined above and, therefore, will not be repeated in the following section.

The average duration $D[s_t(i)]$ with regime variables s_t using the following formula:

$$D[s_t(i)] = E[s_t = i] = \frac{1}{1 - p_{ii}}, i = 1, 2, 3 \quad (6)$$

The unique approach of this paper is to create a nonlinear MS (M)-AR (p) model that includes both "mean form" and "intercept form." The nonlinear MS (M)-AR (p) model is further explored using the Expectation Maximization (EM) algorithm and the Maximum Likelihood (ML) technique [43,52].

When we use these nonlinear MS (M)-AR (p) models for economic analysis, we need to first verify the stationarity of the variable data. In this paper, we will use ADF (Augmented Dickey–Fuller) test, PP (Phillips–Perron) test, and KPSS (Kwiatkowski–Phillips–Schmidt–Shin) test to check the stability of the variable data. In addition, for the multiple nonlinear models created in this paper, it is necessary to calculate the AIC (Akaike Information Criterion), HQ (Hannan Quinn), and SC (Schwarz Criterion) values under different model settings according to the AIC information criterion, HQ information criterion, and SC information criterion to analyze the reliability and validity of the model.

2.2. Data Selection for China's Agricultural Economic Growth

Based on the quarterly data of China's gross product of agriculture, forestry, livestock, fishery, and overall gross agricultural product from Q1 2001 to Q1 2018, this paper further calculated the quarterly data of the growth rate of each variable to examine the cyclical dynamic change process of China's agricultural economic growth in detail. The growth rate is calculated by the year-on-year method and takes into account price inflation. The data in this paper were obtained from the China Economy Internet (CEI) data (http://db.cei.cn, accessed on 24 May 2021) and the China Statistical Yearbook.

To further explore the dynamic paths of the growth rates of China's gross product of agriculture, forestry, livestock, fishery, and overall agricultural product over time, this paper uses the H-P filtering technique [52,53] to capture the *trend component* and the *volatility component* of the time series of each variable to provide a clearer picture of the dynamic evolution of the aggregate value of each variable within the time domain under study. Specifically, the "trend component" can clearly depict the trend state and the change process of each variable time series over a long period of time. The "volatility component" can more carefully depict the fluctuation magnitude and uncertainty of each variable time series data in different economic periods. The "volatility component" can provide a more detailed picture of the volatility and uncertainty of the time series of each variable in different economic periods.

3. Results

3.1. Growth Rate Dynamic Trajectory Analysis

In this paper, we first depict the time fluctuation paths of the growth rates of the gross product value in China's agriculture, forestry, livestock, fishery, and agriculture industries. We can see that, on the one hand, the growth rates of the gross product value of China's agriculture, livestock, fishery, and agriculture time series have generally similar time dynamic trajectories, while the growth rates of China's forestry industry alone show relatively different trend changes (Figures 1–5). On the whole, the fluctuations and oscillations of the time series of the growth rates of the gross product value of China's agriculture, forestry, fishery, and agriculture are relatively small, while the fluctuations of the growth rate of the gross product value of China's livestock are more drastic—showing steep fluctuations with steep increases and decreases.

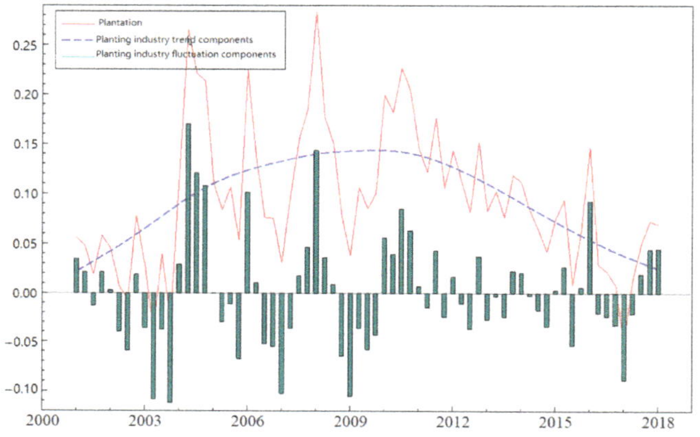

Figure 1. Time series of the growth rate of the gross agriculture product.

The "trend component" of the growth rate of China's gross agriculture product shows that China's agriculture started to achieve significant growth in the early 21st century and showed an increasing trend year by year, reaching its highest "peak" roughly around 2010 and declining from 2010–2018 (Figure 1). In addition, looking at the "fluctuating components" of the growth rate of China's gross agriculture product depicted in Figure 1, the growth rate of China's gross agriculture product has been characterized by weak fluctuations since 2010, with more significant fluctuations clustering in the sample interval before 2010. However, in recent years, the volatility of the growth rate of China's gross agriculture product has increased.

The "trend component" of the growth rate of China's gross forestry product as depicted shows that China's gross forestry product has generally grown steadily over the time horizon selected for this paper, rising year by year at the beginning of the 2000s and peaking in 2003 (Figure 2). Subsequently, the growth rate of the gross forestry product moves down from the "peak" at a very slow pace. In the rest of the sample period, the fluctuation of the growth rate of China's gross forestry product is small, and, especially in recent years, the fluctuation of the growth rate of China's gross forestry product is extremely weak.

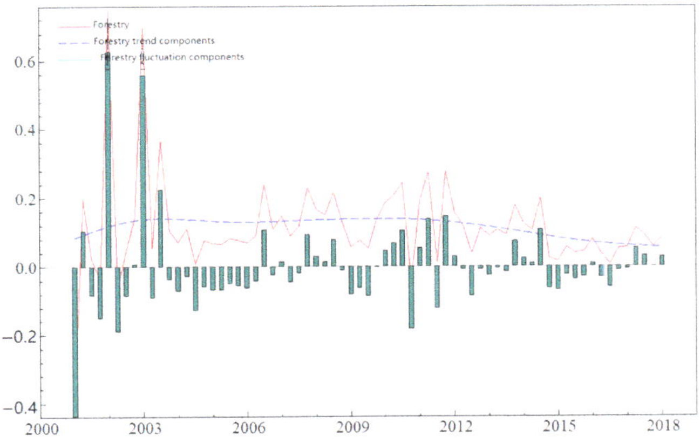

Figure 2. Time series of growth rate of the gross forestry product.

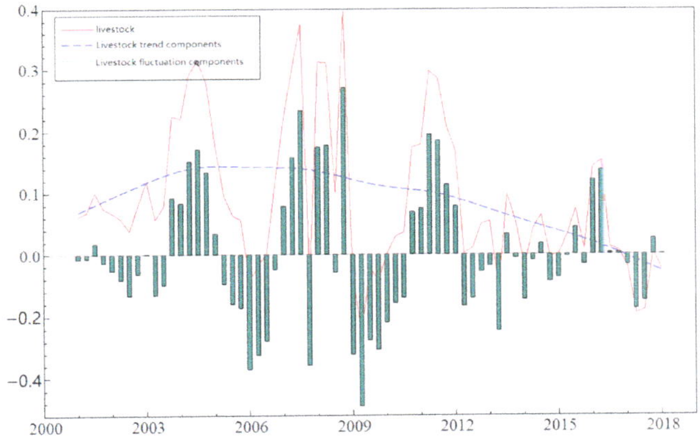

Figure 3. Time series of growth rate of the gross livestock product.

The "trend component" shown in Figure 3 shows that the growth rate of China's gross livestock product changes slowly and appears to hover between "peaks" and "troughs" several times. The growth rate of the gross livestock product in general gradually shifts downward, particularly in recent years. Industry livestock growth rates are lower than the initial levels during the early 21st century.

At the same time, the "fluctuation component" in Figure 3 reflects that the time series of the growth rate of China's gross livestock product contains significant fluctuation clustering characteristics, and shows higher fluctuation than the growth rate of gross product of agriculture, forestry, and fishery industries. In other words, there is relatively more volatility and uncertainty in the time series of the growth rate of China's gross product value of livestock. However, it is also clear that the volatility of the growth rate of China's gross livestock product has significantly decreased in the recent years. In recent years, the Chinese government has implemented policies to curb the development of animal husbandry in order to protect the environment, resulting in negative growth in the animal husbandry industry.

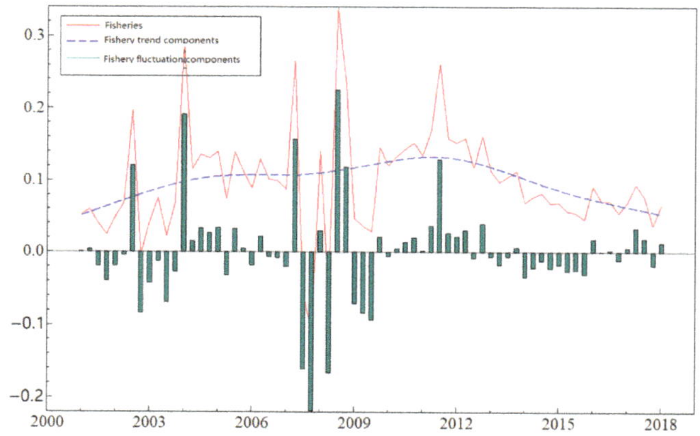

Figure 4. Time series of growth rate of the gross fishery product.

The dynamic trajectory of the growth rate of China's gross fishery product over time is shown in Figure 4. It can be seen from the "trend component" of the figure that, during the initial period in 2001, the growth rate of gross fishery product was at a low level. Subsequently, the growth rate of the gross fishery product shows a "cyclical-like" pattern of gradually climbing from a lower growth level to a higher "peak" level, and then slowly falling back to a lower "trough" level. In recent years, the growth rate of China's gross fishery product has been moving downward. In addition, the "volatility component" of the time series of the gross fishery product growth rate shows that during the global financial crisis from 2007 to 2010, China's gross fishery product growth rate exhibited a highly volatile clustering characteristic. However, in recent years, the volatility of China's gross fishery product growth rate has significantly weakened.

The growth rate of China's gross agricultural product for all four agricultural sectors aggregated together is depicted in Figure 5. Since 2001, this aggregated growth rate has shown a gradual increase in development momentum and, after reaching the highest "peak" in succession, has been declining. In addition, looking at the "fluctuating component" of the growth rate of China's agricultural product depicted in Figure 5, we can see that, since 2001, China's gross agricultural product has shown more obvious fluctuations and a certain clustering of fluctuations. During the subsequent period of 2012–2018, the volatility of the growth rate of China's gross agricultural product decreased.

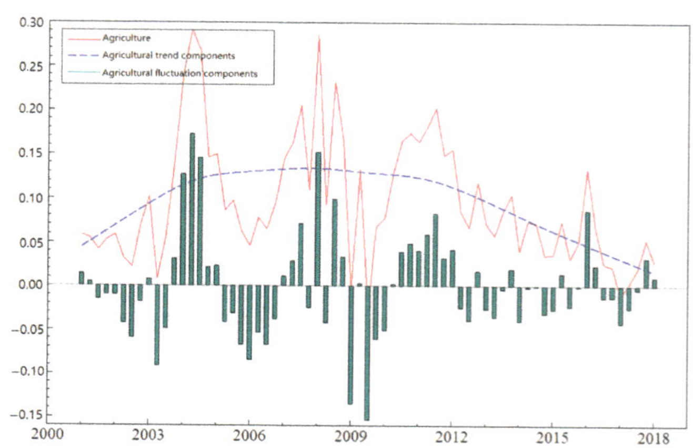

Figure 5. Time series of growth rate of the gross agricultural product.

3.2. The Parameter Estimation of the MSM (M)-AR (p) Model

The growth rates of China's gross product of agriculture, forestry, livestock, fishery, and agriculture, depicted in Figures 1–5, can be used to make a preliminary visual judgment of the long-term dynamic changes in China's agricultural economy. The next part of this paper is based on a nonlinear MS (M)-AR (p) model with time series data on the growth rates of China's gross product of agriculture, forestry, livestock, fishery, and agriculture. The results show that the growth rates of China's gross product of agriculture, forestry, livestock, fishery, and overall agriculture were stationary at the 5% significance level, while the growth rates of all variables were first-order single integers at the 1% significance level.

The study shows that the AIC, HQ, and SC values of the time series of the growth rate of China's gross agriculture product were the smallest when the model was set to the MSM (3)-AR (3) form. The AIC, HQ, and SC values of the time series of the growth rate of China's gross forestry product were the smallest when the model was set to the MSM (3)-AR (1) form. The AIC, HQ, and SC values of the time series of the growth rate of China's livestock, fishery, and China's agricultural output are all minimized when the model is set in the form of MSM (3)-AR (4). Thus, it is reasonable and reliable to use the MSM (3)-AR (p) model constructed in this paper to investigate the dynamic evolution of China's agricultural economy in terms of the growth region system and its changing dynamics.

The parameter estimation results of the MSM (M)-AR (p) model calculated by different variables are presented in Tables 1 and 2, respectively. The results of the t-statistical test indicate that all values were significant at the 1% or 5% level, except for the growth rate of the total fishery output in the zone system 1 ($s_t = 1$). This indicates that the model we chose is appropriate. The mean μ estimates of the growth rates of the gross product of China's agriculture, forestry, livestock, fishery, and agricultural industries are all consistent with the parameter constraints $\mu_1 < \mu_2 < \mu_3$ (Tables 1 and 2). Therefore, the growth rates of each variable are considered as the low-growth regime, medium-growth regime, and rapid-growth regime for regime 1, regime 2, and regime 3, respectively.

Table 1. MSM (3)-autoregressive process (AR) (p) model parameter estimation results.

Regime	Growth Rate of Gross Agriculture Product		Growth Rate of Gross Forestry Product		Growth Rate of Gross Livestock Product		Growth Rate of Gross Fishery Product	
	Estimated Value	Standard Deviation	Estimated Value	Standard Deviation	Estimated Value	Standard Deviation	Estimated Value	Standard Deviation
μ_1	0.0793 **	0.0362	0.0520 ***	0.0158	0.0277 **	0.0125	0.0217	0.0201
μ_2	0.1334 ***	0.0510	0.1254 ***	0.0130	0.0809 **	0.0343	0.0802 ***	0.0144
μ_3	0.2057 ***	0.0392	0.7102 ***	0.0501	0.2812 ***	0.0261	0.1685 ***	0.0205
A_1	0.3598 **	0.1643	−0.1339 **	0.0668	0.3666 ***	0.0732	−0.3219 ***	0.1190
A_2	0.0973	0.1808	–	–	−0.1270	0.0829	−0.5408 ***	0.1085
A_3	0.1980 *	0.1185	–	–	0.2104 ***	0.0728	−0.3489 ***	0.1116
A_4	–	–	–	–	−0.4658 ***	0.0684	−0.6732 ***	0.1074

Note: "***", "**", and "*" indicate significance at the 1%, 5%, and 10% levels, respectively.

Table 2. Estimation results for each parameter of the MSM (3)-AR (p) model.

Regime	Growth Rate of Gross Agricultural Product	
	Estimated Value	Standard Deviation
μ_1	0.0617 **	0.0268
μ_2	0.0764 ***	0.0274
μ_3	0.1600 ***	0.0251

Table 2. *Cont.*

Regime	Growth Rate of Gross Agricultural Product	
	Estimated Value	Standard Deviation
A_1	0.2081 **	0.0906
A_2	0.7292 ***	0.0975
A_3	0.2134 **	0.0990
A_4	−0.5313 ***	0.0868

Note: "***" and "**" indicate significance at the 1% and 5% levels, respectively.

3.3. The Dynamic Shift Probabilities of China's Agricultural Growth Regime

Based on the time series data of the growth rates of total output value of Chinese agriculture, forestry, livestock, fishery and overall agricultural in China, we calculated the dynamic shift probabilities of China's agricultural economic growth regimes using Equation (3). The results are presented in Tables 3–7, respectively. We can see that the probability of staying at a low-growth rate for the agriculture sector is 0.9133 ($p_{11} = 0.9133$), and the probability of shifting to the medium or fast growth range is very low, corresponding to A and B, which respectively, indicates that the agriculture sector is likely to maintain a low-growth state (Table 3). The probability of staying in the medium-growth range for agriculture is only 0.0548, while the probability of shifting to the low-growth range is as high as 0.9452 ($p_{21} = 0.9452$), and there is almost no possibility of shifting to the high-growth range. This reinforces the high probability of maintaining a low-growth rate for the agriculture sector.

The probability of maintaining a high-growth range for the agriculture is relatively high at 0.5588 ($p_{33} = 0.5588$), but the probability of falling back to a medium-growth state is also relatively high (0.4421). Since the probability of agriculture maintaining the medium-growth state is very low and the probability of shifting to the low-growth state is very high 0.9452 ($p_{21} = 0.9452$), the probability of falling back from the fast-growth state to the medium-growth state through the medium-growth state to the low-growth state is also very high. Again, the result shows that the probability of maintaining the low-growth rate in the agriculture sector is high.

Overall, the probability of staying at the low-growth state for the agriculture sector is very high, while the probability of shifting to the medium or fast rate range is very low. Moreover, the probability of falling from a medium to a low-speed state and of falling from a fast state to a low-speed state through a medium-speed state is very high. Therefore, agriculture in China clearly tends to maintain a low-growth rate.

Table 3. Transfer probability matrix of the regime for the growth rate of the gross agriculture product.

	Low-Growth Regime	Medium-Growth Regime	Rapid-Growth Regime
Low-growth regime	0.9133	<0.0001	0.0867
Medium-growth regime	0.9452	0.0548	<0.0001
Rapid-growth regime	<0.0001	0.4421	0.5588

The probability of forestry maintaining the low and medium speed states are both very high at 0.8937 ($p_{11} = 0.8937$) and 0.9774 ($p_{22} = 0.9774$), respectively, and the probability of shifting to other states is very small (Table 4). The probability of maintaining forestry in the fast-growth state is close to zero, while the probability of shifting to the low and medium speed states is very high, 0.6109 ($p_{31} = 0.6109$) and 0.3891 ($p_{32} = 0.3891$), respectively. This indicates that forestry will likely maintain a low to medium-growth state and has the highest probability of staying in the medium-growth state.

Table 4. Transfer probability matrix of the regime of the growth rate of the gross forestry product.

	Low-Growth Regime	Medium-Growth Regime	Rapid-Growth Regime
Low-growth regime	0.8937	0.0007	0.1056
Medium-growth regime	0.0226	0.9774	<0.0001
Rapid-growth regime	0.6109	0.3891	<0.0001

The probability of maintaining a low-growth state in the livestock sector is very high 0.9063 ($p_{11} = 0.9063$), while the probability of shifting to either the medium or high growth range is low 0.0937 ($p_{12} = 0.0937$) and close to zero, respectively (Table 5). Therefore, it is relatively easy for the livestock industry to maintain a low-growth range. At the same time, the probability that the livestock sector maintains a fast-growth state is relatively high 0.6609 ($p_{33} = 0.6609$), but the probability of maintaining a medium-growth range is very low.

Table 5. Transfer probability matrix of the regime of the growth rate of the gross livestock production.

	Low-Growth Regime	Medium-Growth Regime	Rapid-Growth Regime
Low-growth regime	0.9063	0.0937	<0.0001
Medium-growth regime	0.3023	0.0066	0.6911
Rapid-growth regime	0.1461	0.1930	0.6609

The probability of shifting China's total fishery output from the low-growth range to the medium-growth range and the fast-growth range is $p_{12} = 0.5515$ and $p_{13} = 0.4464$, respectively. The probability of maintaining a particular growth rate is highest sustaining fast-growth ($p_{33} = 0.8026$) followed by keeping at medium ($p_{22} = 0.7345$) and low ($p_{11} = 0.0020$) growth (Table 6). It can be seen that the probability of maintaining the total fishery output value in the low-growth range is very low, and it is easy to climb from the low-growth state to the medium- and high-growth state. There is a high probability of maintaining the medium-growth state and high-growth state, and it is easy to maintain the medium- and high-growth state, so China's fishery industry has a very good developmental trend.

The probability of transferring China's total fishery output value from the medium-growth regime back to the low-growth regime is $p_{21} = 0.2646$, and the probability of transferring from the fast-growth regime to the medium-growth regime is $p_{32} = 0.1974$. The probability of shifting from the "fast-growth regime" to the "medium-growth regime" is close to zero. Therefore, the probability of shifting from a higher growth state to a lower growth state is low, which indicates a more stable development trend of fisheries.

Table 6. Transfer probability matrix of regimes for the growth rate of the gross fishery product.

	Low-Growth Regime	Medium-Growth Regime	Rapid-Growth Regime
Low-growth regime	0.0020	0.5515	0.4464
Medium-growth regime	0.2646	0.7345	0.0009
Rapid-growth regime	<0.0001	0.1974	0.8026

From the results of the transfer probability matrix of China's gross agricultural product, the maintaining probabilities of the low-, medium-, and rapid-growth regime of the gross agricultural product are $p_{11} = 0.8425$, $p_{22} = 0.5909$, and $p_{33} = 0.8020$, respectively (Table 7). This indicates that China's gross agricultural product does not easily change its growth status when it is in different regimes, and it has certain inertia characteristics. When China's gross agricultural product is in the low-growth regime, it does not easily change to a higher growth rate, but when the level of agricultural development increases significantly, the gross agricultural product easily stays in the medium-growth regime and the rapid-growth regime and does not decline significantly.

It is clear that the probability of gross agricultural product climbing from the low-growth regime to the medium-growth regime is low ($p_{12} = 0.1575$), and essentially does little to possibly jump from the low-growth regime to the rapid-growth regime, while the probability of transferring the gross agricultural product from the medium-growth regime to the rapid-growth regime is relatively high ($p_{23} = 0.3999$). Thus, when the gross agricultural product is in the low-growth regime, it is difficult to achieve a significant increase due to the limitation of resources and technical facilities. When the gross agricultural product is in the medium-growth regime, the existing capital and technology advantages can be fully utilized to achieve a smooth transition from the medium-growth regime to the rapid-growth regime. The rise from the medium-growth regime to the rapid-growth regime is smooth.

The probability of change of China's gross agricultural product falling back from the medium-growth regime is $p_{21} = 0.0092$, while the probability of change from the fast-growth regime to the low-growth regime is relatively low ($p_{31} = 0.1980$). At the same time, the probability of change of gross agricultural product falling back from the rapid-growth regime to the medium-growth regime is close to zero. Thus, it is clear that the probability of a decline in gross agricultural product from the medium-growth regime is not high, and the gross agricultural product has a relatively stable development trend. In addition, when the gross agricultural product is in the rapid-growth regime, it does not decrease to the medium-growth regime. Therefore, the national strategy of modernizing agricultural development has achieved significant results during the sample period.

The overall improvement of China's agricultural production conditions and mechanization level in recent years has laid a good foundation for the development of agricultural facilities. Coupled with the Chinese government's rational planning and allocation of resources and active financial support policies, China's agricultural production modernization level has maintained steady and rapid growth. China's existing agricultural policies have achieved positive results.

Table 7. Transfer probability matrix of growth rate regime of China's gross agricultural product.

	Low-Growth Regime	Medium-Growth Regime	Rapid-Growth Regime
Low-growth regime	0.8425	0.1575	<0.0001
Medium-growth regime	0.0092	0.5909	0.3999
Rapid-growth regime	0.1980	<0.0001	0.8020

3.4. Estimated Average Duration of Each Regime of China's Agricultural Economy

We calculated the average durations of the growth rates of China's agriculture, forestry, livestock, fisheries, and overall agriculture for each growth rate (Table 8). Combining the maintenance probabilities given in Tables 3–7, and considering the average duration presented in Table 8, we further examined the dynamic characteristics of the variables in different regimes. The probability that China's gross agriculture product is in the low-growth regime is $p_{11} = 0.9133$ maintained with an average duration of 11.54 quarters. This indicates that the lower level of development of the agriculture is more inclined to maintain a low rate of growth and will not make a significant leap to a state of rapid growth.

At the same time, the probability of maintaining a fast-growth state for the gross agriculture product is $p_{33} = 0.5588$, and 2.27 quarters of average duration. The gross agriculture product can be stabilized in the rapid-growth regime with a moderate probability of maintenance in the short-term. When the gross agriculture product is in the medium-growth regime, the maintenance probability is $p_{22} = 0.0548$ with an average duration of 1.06 quarters. The probability and average duration of sustaining agriculture in the medium-growth regime are low. Overall, China's agriculture is most likely to remain in the low-growth regime and is the least likely to be in the medium-growth regime among all regimes.

The probability that China's gross forestry product is in the low-growth regime is $p_{11} = 0.8937$ with an average duration of 9.41 quarters and has the inertia characteristic of maintaining a low-growth rate. The probability of maintaining the medium-growth rate of forestry product is higher $p_{22} = 0.9774$ with an average duration of 44.19 quarters, and the gross forestry product tends to maintain the medium-growth regime. The probability of maintaining the forestry product in the rapid-growth regime is close to zero with an average duration of 1.00 quarter. On balance, the gross forestry product tends to maintain a medium-growth rate over a longer time horizon, with a possible shift to a low-growth regime but a difficult shift to a rapid-growth regime.

The probabilities of maintaining medium- and fast-growth in the gross livestock product are $p_{22} = 0.0066$ and $p_{33} = 0.6609$ with 1.01 and 2.95 quarters of average duration, respectively. The probability that the gross product of livestock is in the low-growth regime is the highest among all regimes ($p_{11} = 0.9063$) and has an average duration of 10.67 quarters. Overall, the gross product of the livestock sector is more likely to remain in the low-growth regime, the least likely to remain in the medium-growth regime, and moderately likely to be in the rapid-growth regime.

The probability of maintaining low growth in China's gross fishery product was the lowest among all regimes ($p_{11} = 0.0020$) and only showed an average duration of 1.00 quarter. The probabilities of maintaining medium- and rapid-growth rates were $p_{22} = 0.7345$ and $p_{33} = 0.8026$, with 3.77 and 5.07 quarters average duration, respectively. This indicates that the gross fishery product is more likely to achieve medium- and high-growth rates and has the highest probability of being in a fast-growth state and the lowest probability of being in a low-growth state.

Table 8. Estimated average duration of each regime (quarters).

Regime	Gross Agriculture Product Growth Rate	Gross Forestry Product Growth Rate	Gross Livestock Product Growth Rate	Gross Fishery Product Growth Rate	Gross Overall Agricultural Product Growth Rate
Low-growth regime	11.54	9.41	10.67	1.00	6.35
Medium-growth regime	1.06	44.19	1.01	3.77	2.44
Rapid-growth regime	2.27	1.00	2.95	5.07	5.05

The average duration of the gross agricultural product in the low-growth regime was relatively long, with an average duration of 6.35 quarters, while the average duration in the rapid-growth regime was also relatively long (Table 8). However, the average duration in the medium-growth regime was relatively short, at 2.44 quarters. At the same time, the probability that China's gross agricultural product will remain in the low-growth regime and rapid-growth regime was high ($p_{11} = 0.8425$, $p_{33} = 0.8020$), and the probability of medium-growth was lower ($p_{22} = 0.5909$). This indicates that China's gross agricultural product is more likely to be in the low-growth regime and rapid-growth regime, and less likely to be in the medium-growth regime.

3.5. Filter Probability and Smoothing Probability

In the next section, the dynamic paths of the time series trajectories of the growth rates of each variable are clarified by depicting the smoothed probabilities of the gross product value of China's agriculture, forestry, livestock, fishery, and overall agriculture industries in each growth regime. In the smoothed probabilities of the low-growth regime (probability of $s_t = 1{:}\Pr(s_t = 1|\xi_{t|t}) > 0.5$), medium-growth regime (probability of: $s_t = 2{:}\Pr(s_t = 2|\xi_{t|t}) > 0.5$), and rapid-growth regime (probability of $s_t = 3{:}$ $\Pr(s_t = 3|\xi_{t|t}) > 0.5$), $\xi_{t|t}$ refers to all information sets based on past *t* periods.

The growth rates of China's gross product of agriculture, forestry, livestock, fishery, and overall agricultural industries changed frequently and dynamically among various regimes during the development of China's agricultural economic growth, and the dynamic paths of the growth rates of each variable in each regime will be further detailed in this paper. Within the sample interval studied, the gross agriculture product was in the rapid-growth regime during (a) Q2 to Q4 2004, (b) Q1 2006, (c) Q4 2007 to Q1 2008, and (d) Q1 to Q4 2010 (Figure 6). Looking back at history, it is easy to see that these particular timeframes correspond to the launch of relevant policies in China (Appendix A, Table A1). For example in 2004, the first "Document No. 1" of the century was issued, which introduced the "three subsidies" policy, namely "direct subsidies for farmers, subsidies for good seeds, and subsidies for the purchase of agricultural machinery, and minimum purchase price for rice."

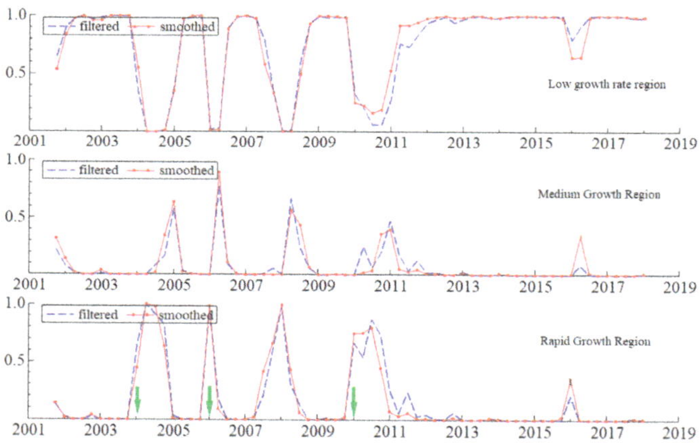

Figure 6. Filter probability and smoothing probability of gross agriculture product located in different growth regimes. Note: The green arrows mark the times when important policies were introduced.

In 2006, China completely abolished all agricultural taxes except the tobacco tax nationwide. In 2009, the Chinese government implemented the property rights policy of giving farmers fuller and more secure rights to contracted land management, and indicated that the existing land contract relationship would remain stable and unchanged for a long time. This injected confidence and vitality into the agricultural economy. In 2010, the government clearly proposed to give full play to the effective allocation of resources to promote agricultural development, and to improve the efficiency of agricultural development by improving the allocation of resource factors.

Under a series of effective policy measures, China's agriculture has achieved remarkable results and has been in the rapid-growth regime for a long period of time. However, China's gross agriculture product still fell into the medium-growth regime during Q1 2005, Q2 2006, and Q2 2008. It was in the low-growth rate during the periods from (a) Q4 2001 to Q1 2004, (b) Q2 to Q4 2005, (c) Q3 2006 to Q3 2007, (d) Q3 2008 to Q4 2009, and (e) Q1 2011 to Q1 2018.

Overall, we found that China's gross agriculture product generally fluctuated between the rapid-growth regime and the medium-growth regime but was able to maintain a relatively stable growth rate when it was in the low-growth regime. However, when in the low-growth regime, the growth rate remained relatively stable and had a more consistent smoothing probability. In the time domain selected in this paper, the gross agriculture product in the low-growth regime basically coincided with the occurrence of natural disasters, such as floods and snowstorms.

This judgment is consistent with the phenomena revealed in Tables 3 and 8. That is, it is difficult for China's gross agriculture product to remain in the rapid-growth regime or medium-growth regime in general, but it tends to stay in the low-growth regime. Unlike prior literature, the model in this paper captures the rapid-growth regime of Chinese agriculture around 2006 and 2008, suggesting that the model in this paper is more accurate in modeling real fluctuations in economic development.

We can see from Figure 7 that in the early part of this century, China's agricultural development, including forestry, was greatly affected by the *soft landing* of China's macroeconomics. In the case of forestry, the gross forestry product showed more dramatic fluctuations at the beginning of the 20th century, with more frequent cyclical fluctuations. Coupled with the sudden impact of natural disasters on forestry development, the process of change in gross forestry product showed a direct fall from the rapid-growth regime to the low-growth regime and was less often in the medium-growth regime.

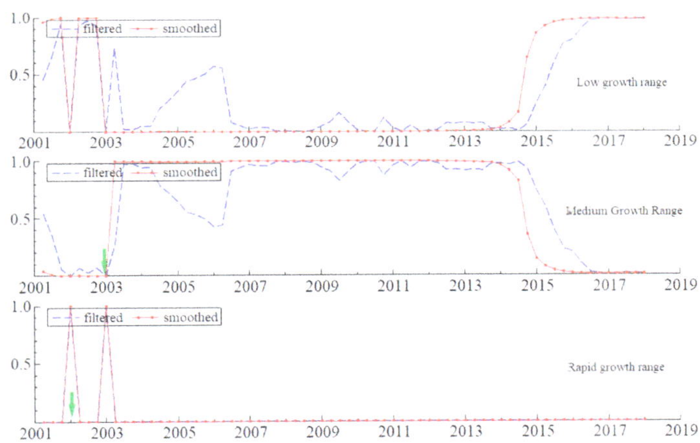

Figure 7. Filter probability and smoothing probability of gross forestry product located in different growth regimes. Note: The green arrows mark the times when important policies were introduced.

During the periods from Q2 2001 to Q4 2001 and Q2 2002 to Q4 2002, China's gross forestry product was in the low-growth regime, and, during Q1 2002 and Q1 2003, China's gross forestry product was in the rapid-growth regime. In 2002, the Chinese government promulgated the *Regulations on Returning Farmland to Forestry* to promote the development of forestry, which solved the inherent problems in forestry development while expanding the area of forestland and laid the policy foundation for sustainable development of forestry (Appendix A, Table A1).

In 2003, the Chinese government further promulgated the *Decision on Accelerating the Development of Forestry*, which provided concrete measures for forestry development by adjusting the structure of forestry industry and strengthening the construction of forestry bases (refer to the attachment for details). This series of policies and measures strongly promoted the development of forestry and the process of forestry modernization. The gross forestry product entered the medium-growth regime from Q2 2003 to Q3 2014.

However, China's gross forestry product was in the low-growth regime from Q4 2014 to Q1 2018. From the smoothed probability time dynamics trajectory shown in Figure 7, whether in the low-growth regime, medium-growth regime or rapid-growth regime, the smoothed probability of China's gross forestry product in different regimes were all close to the 1.00 level. This shows that the risk prevention and control of China's gross forestry product was more effective.

At the beginning of this century, the gross livestock product, which has a certain scale but is still immature in general, was basically in the low-growth regime due to the influence of the soft landing of China's economy (Figure 8). Several factors caused China's grow livestock product to drop from a rapid-growth to a low-growth regime. In 2004, the No. 1 Document encouraged the continuous improvement of feed, technology, equipment, and other inputs. In addition, there was increasing industry maturity in the processing of both dairy and meat products.

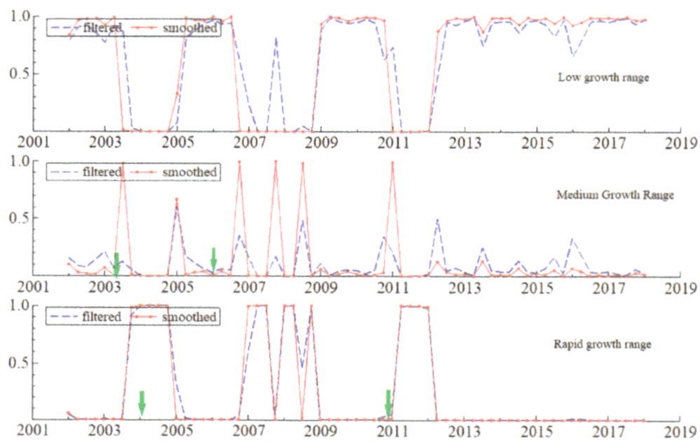

Figure 8. Filter probability and smoothing probability when the gross livestock product is in different growth regimes. Note: The green arrows mark the times when important policies were introduced.

In 2006, under the influence of the policy of abolishing agricultural tax and benefiting farmers, China's gross livestock product stepped into the rapid-growth regime again, fluctuating back and forth until falling back to a low-growth regime in 2009. In 2011, stimulated by the policy of accelerating the construction of water resources and infrastructure, it again entered the rapid-growth regime. From 2001 to 2012, there was an interaction between the rapid-growth regime and the low-growth regime, with a long-term low-growth regime after 2012. Unlike previous studies, our model sensitively captures the rapid-growth regime of China's livestock industry from (a) 2003–2005, (b) 2007–2008, (c) 2008–2009, and (d) 2010–2012, indicating that our model can more accurately simulate agricultural economic development.

China's fisheries industry was also affected by the *soft landing* of China's economy and was developing at a low rate in the early 2000s. However, fisheries entered a rapid-growth phase starting in 2004 under the influence of the first Document No. 1 (Figure 9). In 2008, the Chinese government introduced policies to strengthen the safety of fishery production, improve regulatory efforts, strengthen fishery production measures, and optimize departmental cooperation to ensure safe fishery production, while establishing a long-term mechanism to maintain sustainable fishery development (Appendix A, Table A1).

This series of comprehensive requirements as well as specific initiatives have greatly improved the pattern of fishery production, and under the new management mechanism, China's gross fishery production climbed to the medium-growth regime during the periods from Q3 2008 to Q1 2009 and from Q2 2010 to Q4 2012, while China's gross fishery production was in the rapid-growth regime during the periods from Q3 2008 to Q1 2009 and from Q2 2010 to Q4 2012. In general, China's fishery industry has been developing at a medium to high speed for a long time and has achieved a very good development trend.

The possibility of China's fishery product moving into a low-growth regime still requires special attention.

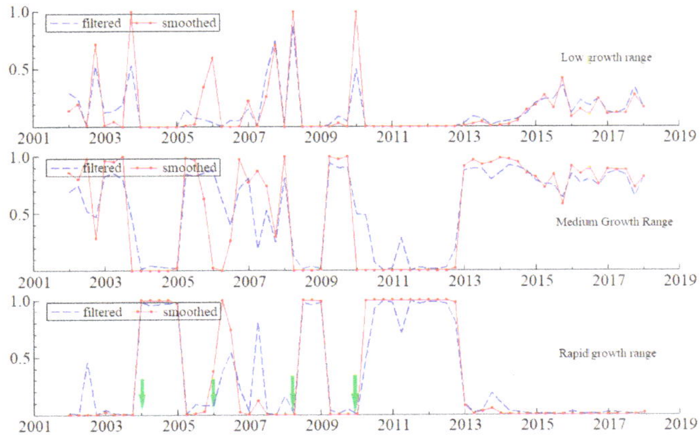

Figure 9. Filter probability and smoothing probability when the gross fishery product is in different growth regimes. Note: The green arrows mark the times when important policies were introduced.

Finally, we can see the basic overview of China's gross agricultural product in the low-growth regime, medium-growth regime, and rapid-growth regime. Specifically, China's agricultural economy was in the rapid-growth regime for 24 quarters in stages throughout the sample period (Figure 10). Looking back at history, during the period when China's gross agricultural product was in rapid growth, the Chinese government had major policies benefiting agriculture (Appendix A, Table A1).

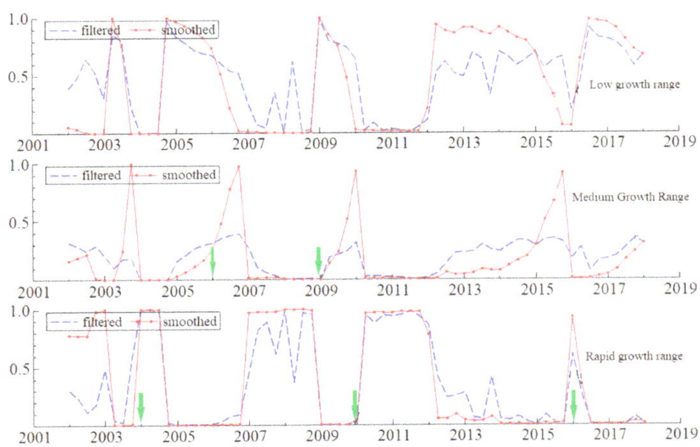

Figure 10. Filter probability and smoothing probability when the overall agricultural product is in different growth regimes. Note: The green arrows mark the times when important policies were introduced.

In 2004, the Chinese government issued the first No. 1 Document, which introduced direct subsidies for farmers, subsidies for good seeds and agricultural machinery, and subsidies for the purchase of agricultural machinery, and a minimum purchase price for rice, which greatly motivated farmers. In 2006, the agricultural tax regulations were abolished. The agricultural tax regulations, which had been implemented for nearly 50 years in New

China, became historical records, and the system of taxing farmers by the area of land, which had lasted for 2600 years, was retired from the historical precedence.

The burden of farmers nationwide was reduced by CNY 133.5 billion per year, and the per capita burden was reduced by about CNY 140. After 2009, the Chinese government implemented the property rights policy of giving farmers fuller and more secure rights to contracted land management and indicated that the existing land contracting relationship would remain stable and unchanged for a long time. In 2010, the Chinese government made it clear that it was committed to giving full play to the role of efficient allocation of resources in promoting agricultural development, and to achieving increased efficiency in agricultural development by improving the allocation of resource factors. This injected confidence and vitality into the agricultural economy, and China's agricultural economy once again entered a high-speed growth regime. As the above-mentioned policy dividend receded, China's total agricultural output value shifts to a low-growth regime after a brief transition to a medium-growth regime.

Overall, China's gross agricultural product generally fluctuated frequently between the rapid-growth regime and the low-growth regime. The medium-growth regime occurred over shorter time periods. After 2012, except for 2016 when it was in the medium-high-growth range for a short time, China's gross agricultural product was in the low-growth regime for a long time and was able to maintain relative stability (Tables 7 and 8).

The Chinese government has introduced beneficial agricultural policies every year since 2004. However, the growth of China's agricultural economy fluctuated between fast and low growth. After 2012, despite active fiscal policy, and, except for one quarter of fast growth in 2016, it has been in a low growth trend for a long time. This fully illustrates the vulnerability of agricultural economic growth and indicates that the government should pay attention to agricultural development in the long-term and increase policy support.

Comparing Figures 6–10 with Figures 1–5, we can see that the changes in the total output value of the agriculture, forestry, livestock, fishery, and overall agriculture industries show some similarities. Specifically, the time range of the "fast-growth regime" for each industry shown in Figures 5–8 corresponds to the period when the "fluctuation component" is relatively strong as shown in Figures 1–5, and the time range of the low-growth regime for each industry shown in Figures 6–10 corresponds to the period when the "fluctuation component" is relatively calm as shown in Figures 1–5. This implies that the risk of shocks is higher when the growth rate of the total output of each industry was relatively high in the agriculture, forestry, livestock, fishery, and overall agriculture industries and less uncertain when the growth rate of the total output of each industry was relatively low.

4. Discussion

4.1. Contrast to Prior Studies

Since China is still a developing country, the level of development of China's rural areas still lags behind that of the developed world, and agricultural production is still the main source of income for many Chinese farmers. Development remains a major issue for the Chinese government now and in the future. Therefore, in order to better understand the implications of our results, we need to compare and contrast our results to previous studies.

First, based on the above-mentioned empirical findings, we found that China's agricultural economy maintained a relatively good development with the support of the benefit agriculture policy during the sample period, i.e., the probability of maintaining China's agricultural economy in the fast-growth regime was $p_{33} = 0.8020$ and had an average duration of 5.05 quarters, which is only slightly lower than the average duration of 6.65 quarters in the low-growth regime. This indicates that China's agricultural economy developed relatively well during periods following the agricultural support policies, which is consistent with prior studies [13–15].

Second, we found that the agricultural economy tended to maintain a low-growth rate, with the highest maintenance probability of $p_{11} = 0.8425$ and the longest average duration of 6.35 quarters, which is consistent with the results of another study [34], demonstrating that China's agricultural economy is not easy to move into the expansion phase of its economic cycle. Our results are also consistent with research showing that maintaining rapid long-term growth in the agricultural economy is not easy due to factors such as excessive consumption of natural resources and environmental pollution [2].

A possible reason for this phenomenon is that the law of diminishing marginal returns is particularly evident in traditional agriculture due to natural conditions. Since the amount of resources invested in agriculture, such as land, is fixed, increasing the labor force will not increase the agricultural output significantly, resulting in slow agricultural growth. This pattern is particularly evident in developing countries where technological progress in agriculture is slow. Another possible reason is that since the demand for agricultural products typically lacks elasticity, an increase in the supply of agricultural products will cause the prices of agricultural products to fall, and hence an increase in production will not lead to an increase in income, which will, in turn, lead to a slow growth in agricultural output.

Third, based on the results of our empirical tests, we distinguished three growth statuses of China's agricultural economy—low, medium, and rapid—and we show their specific transfer paths in Figures 6–10. Specifically, China's agricultural economy was in the rapid-growth regime for 24 quarters in stages throughout the sample period. These rapid- growth states are clearly synchronized with the timing of important agricultural policies. Our analysis suggests that the possible reasons for the shifts to the rapid-growth regime are the introduction of major pro-agricultural policies by the Chinese government.

For example, in 2004, the Chinese government issued the first "Document No. 1" in the time period under examination, which introduced "direct subsidies for farmers, subsidies for good seeds and agricultural machinery purchases, and a minimum purchase price for rice," that greatly motivated farmers to produce more agricultural products. China's total agricultural output value also entered a high-growth zone. In 2006, when the agricultural tax regulations were abolished, China changed from an agricultural tax-raising country to an agricultural subsidy country, and the total agricultural output value entered a high-growth zone at the same time. In 2009, the Chinese government implemented the property rights policy of "granting farmers more complete and more secure rights to contracted land management and the existing land contract relationship should remain stable and unchanged for a long time."

From 2009 to 2010, a series of major initiatives for the benefit of farmers were proposed, including improving the policy system for the benefit of farmers, focusing on promoting the allocation of resources to rural areas, promoting the transformation of agricultural development, and improving the level of modern agricultural equipment. It also includes accelerating the improvement of rural people's livelihood, narrowing the gap between the development of urban and rural public utilities, coordinating the reform of urban and rural areas, and enhancing the vitality of agricultural and rural development. This was intended to enhance the vitality of agricultural and rural development, to promote urbanization actively and steadily, with the development of small and medium-sized cities and small towns as the focus, and to deepen the reform of the household registration system, etc. This has injected confidence, capital, technology, and vitality into China's agricultural economy, which has once again stepped into a high-speed growth regime. The above results support the findings of Qiao et al. [12], Jin and Deininger [13], Deininger et al. [14], and Xi et al. [32] that demonstrated agricultural growth is influenced by policies and institutions.

In addition, our research results show that China's agriculture, forestry, and animal husbandry tended to grow at a low rate. It was in the low-growth regime for a long time after 2011. The recent low-growth rate of the Chinese agricultural economy has been previously documented [34]. Policymakers should pay attention to this issue and continue to rely on agricultural support policies to avoid food shortages in China as well as adverse macroeconomic outcomes.

In this paper, the causal explanation of the correlation between agricultural economic cycles and government support policies is based on the common sense judgment of temporal synchronization and is not the result of empirical studies constructed using specialized models, such as Data Envelopment Analysis (DEA), Common Agricultural Policy Regionalized Impact (CAPRI), Propensity Score Matching (PSM), and other modeling methods. Hence, the generalizability of our results may be limited. Using specialized policy efficiency assessment models to evaluate the efficiency of agricultural support policies in China may be a fruitful, future research direction.

4.2. Policy Recommendations

Since China is still a developing country. The level of development of China's rural areas still lags behind that of the developed world, and agricultural production is still the main source of income for many Chinese farmers. Development remains a major issue for the Chinese government now and for the foreseeable future. Therefore, based on the results of our study, we propose the following policy recommendations.

First, there is a large body of theoretical and empirical research that shows a strong correlation between agricultural policy and agricultural development. China's current agricultural policies have achieved relatively good results. Therefore the current agricultural policy in the form of the annual Central Government Document No. 1 should be continued.

Secondly, agriculture, forestry, animal husbandry, and fisheries each have their own industry characteristics. Past research results show that the impacts of policies are rather limited with the exception of fisheries. Therefore, the formulation of agricultural policies in line with the development characteristics and stages of each industry may achieve better policy results.

Finally, China is a vast country with a wide range of regional development levels, and the geographical characteristics of agricultural development vary. Therefore policymakers should consider this situation and delegate policy-making authority to provincial governments or lower institutions. This is especially true for autonomous regions, autonomous prefectures, and even autonomous counties that are less developed. These institutions may be able to develop policies in their own regions that are more in line with the level of local agricultural development. Financial and policy supports from the central government could be channeled to allow for more local control if policy makers can be convinced that this improves productivity. Thus, it may be possible to promote the development of Chinese agriculture better than the current grand unified agricultural policy.

Because this study is not an assessment of the effectiveness of current agricultural policies in China, the above policy recommendations may be somewhat biased. Our next research aims to specifically assess the policy efficiency of China's current agricultural policies. Therefore it may be possible to make more nuanced policy recommendations that are more in line with China's agricultural development after completing this additional research.

5. Conclusions

Based on the above findings and discussion, it can be concluded that China's agricultural economy has achieved relatively good development in the context of the country's rapid macroeconomic development and agricultural policies, especially the fishery industry, which has been able to maintain medium to high-growth rates. However, in the long-term, China's agricultural economy tends to maintain a low-growth rate. Since 2011, China's agriculture, forestry, and livestock industries have mostly maintained low-growth rates. In order to maintain agricultural development in China, the Chinese government should continue its current agricultural support policies, especially increasing support for the agriculture, forestry, and livestock industries. Future research should focus on using specialized policy assessment models to evaluate the efficiency of agricultural support policies in China.

Author Contributions: Conceptualization, X.G., J.S., R.Z., and C.W.; methodology, X.G., J.S., and P.L.; formal analysis, X.G., J.S., and C.W.; writing—original draft preparation, X.G. and J.S.; writing—reviewing and editing, X.G., P.L., and R.Z.; supervision, P.L. and C.W. All authors have read and agreed to the published version of the manuscript.

Funding: This research received no external funding.

Institutional Review Board Statement: Not applicable for studies not involving humans or animals.

Informed Consent Statement: Not applicable for studies not involving humans.

Data Availability Statement: The data presented in this study are available in article.

Acknowledgments: We gratefully acknowledge Yu Zhu for his comments and suggestions for improvement, and Yudong Chen for checking the paper. Similarly, we also thank Qiusheng Zhang, Li Li, Changsheng Gao, Zisheng Fang, Xinsheng Deng, Qing Chang, Dehong Liu, and Zizheng Wu for their advice and help with the paper. Finally, we are particularly grateful to Aaron K. Hoshide for his editorial comments and to the two anonymous reviewers for their review comments.

Conflicts of Interest: The authors declare no conflict of interest.

Appendix A

Table A1. Summary of the Chinese government's main agricultural support policies from 2002–2018.

Year	Summary of Major Agricultural Policies
2002	Regulations on Restoring Farmland to Forest Returning farmland to forest is to protect and improve the ecological environment in the west, the sloping arable land that is prone to soil erosion and the arable land that is prone to land sanding, stop cultivation in a planned and step-by-step manner. In line with the principle that it is appropriate to plant trees, irrigation, grass, and the combination of trees, irrigation, and grass, planting forests and grasses according to local conditions to restore forest and grass vegetation. The state implements the system of funds and food subsidies for returning farmland to forest, the state provides appropriate subsidized food, seedling planting fees, and cash (living expenses) subsidies to those returning farmland to forest for a certain period of time without compensation according to the approved area of returning farmland to forest. The Yellow River Basin and the northern region, each mu of fallow land is subsidized with 100 kg of raw food and CNY 20 of cash per year, and at least 8 years for ecological forest, 5 years for economic forest, and 2 years for grass. Each mu of fallow land and forest able wasteland subsidies has a seedling reforestation fee of CNY 50.
2003	The decision of the State Council of the Communist Party of China (CPC) Central Committee on Accelerating the Development of Forestry Stabilize the land contract relationship and allow the reasonable transfer of land use rights under the principle of law, voluntarily and with compensation. Development of private fast-growing forests and its related industries, the implementation of agroforestry compound management, turning potential resource advantages into realistic economic advantages, to achieve the effect of increasing farmers' income, agricultural efficiency and environmental improvement. Encourage pulp and paper enterprises to cross-region, cross-sector, cross-ownership to establish a paper forest base. In the forestry policy, the first should be clear property rights, with the responsibility to the household. In accordance with the principle of "who makes who has, who operates who gains," further extend the use of forest land, allowing inheritance, transfer, mortgage, lease, and access to the secondary market flow. Second, to relax logging restrictions and revitalize forest assets.
2004	Opinions of the State Council of the Central Committee of the Communist Party of China on Several Policies to Promote Increased Incomes of Farmer On 31 December 2003, the "Opinions of the State Council of the Central Committee of the Communist Party of China on Several Policies to Promote Increased Incomes for Farmers" was issued and published on 9 February 2004. A prominent problem of agricultural and rural development at that time was the difficulty of increasing farmers' income. The income gap between urban and rural residents widened from 1.8:1 in the 1980s to 3.1:1, and the difficulty of increasing farmers' income not only restricted the development of the rural economy but also affected the growth of the national economy as a whole. The "Opinions" proposed to "adhere to the 'more to give, less to take, let live' policy, adjust the agricultural structure, expand employment of farmers, accelerate scientific and technological progress, deepen rural reform, increase agricultural investment, strengthen support for agricultural protection, and strive to achieve a relatively rapid growth in farmers' income, as soon as possible to reverse the income gap between urban and rural residents The trend of widening income gap between urban and rural residents." The document contains 22 articles, putting forward a series of high gold content, pointing to clear and practical policy measures.

Table A1. *Cont.*

Year	Summary of Major Agricultural Policies
2005	Opinions of the State Council of the Central Committee of the Communist Party of China on Several Policies to Further Strengthen Rural Work and Improve Comprehensive Agricultural Production Capacity In February 2005, the "Opinions of the State Council of the Central Committee of the Communist Party of China on a number of policies to further strengthen rural work to improve the comprehensive production capacity of agriculture" was issued. The document pointed out the current and future period, to strengthen the construction of agricultural infrastructure, accelerate the progress of agricultural science and technology, improve the comprehensive production capacity of agriculture, as a major and urgent strategic task, and effectively grasp it. Additionally, stressed that we should "strictly protect the arable land as the basis, to strengthen the construction of agricultural water conservancy as the focus, to promote scientific and technological progress as a support, to improve the service system as a guarantee, and strive to make efforts in the next few years, so that the material and technical conditions of agriculture significantly improved, land product rate and labor productivity significantly increased, the overall efficiency and competitiveness of agriculture significantly enhanced."
2006	Several Opinions of the State Council of the CPC Central Committee on Promoting the Construction of a New Socialist Countryside On 21 February 2006, the "Opinions of the State Council of the Central Committee of the Communist Party of China on Promoting the Construction of a New Socialist Countryside" was released. The document points out that building a new socialist countryside is a major historical task in China's modernization process. Only by developing the rural economy, building a good home for farmers, and enabling them to lead a prosperous life can we ensure that all people share the fruits of economic and social development and continuously expand domestic demand and promote sustainable development of the national economy. The full text of the document contains 32 articles, eight parts. The document emphasizes the need to adhere to the development of the rural economy as the center, to further liberate and develop the rural productive forces. Adhere to the "more to less to take the live" approach, focusing on the "more to" efforts to mobilize a wide range of forces to participate.
	Decision of the Standing Committee of the National People's Congress on the Abolition of the Regulations of the People's Republic of China on Agricultural Tax On 29 December 2005, the 19th meeting of the Standing Committee of the 10th National People's Congress decided to repeal the Regulations of the People's Republic of China on Agricultural Tax from 1 January 2006. As a result, the state no longer levies a separate tax on agriculture.
2007	Several Opinions of the State Council of CPC Central Committee on Actively Developing Modern Agriculture and Solidly Promoting the Construction of a New Socialist Countryside On 29 January 2007, the "CPC Central Committee and State Council on the active development of modern agriculture to solidly promote the construction of a new socialist countryside of a number of opinions" was published. The document clearly pointed out that the construction of a new socialist countryside should put the construction of modern agriculture in the first place. The construction of the new socialist countryside has received an enthusiastic response from the grassroots. However, there are some deviations in practice. Emphasis on the new rural construction should put the development of modern agriculture in the first place, which is conducive to the serious implementation of the spirit of the Fifth Plenary Session of the Sixteenth Central Committee around the world, the new socialist countryside construction to move forward solidly and healthily. The document proposes to equip agriculture with modern material conditions, transform agriculture with modern science and technology, upgrade agriculture with modern industrial systems, promote agriculture with modern forms of operation, lead agriculture with modern development concepts, develop agriculture with training new farmers, improve the level of agricultural water conservancy, mechanization and information technology, improve land product rate, resource utilization and agricultural labor productivity, improve agricultural quality, efficiency and competitiveness.

Table A1. *Cont.*

Year	Summary of Major Agricultural Policies
2008	Several Opinions of the CPC Central Committee and State Council on Effectively Strengthening Agricultural Infrastructure Construction to Further Promote Agricultural Development and Increase Farmers' Income On 30 January 2008, the "Opinions of the Central Committee of the Communist Party of China and the State Council on Effectively Strengthening Agricultural Infrastructure to Further Promote Agricultural Development and Increase Farmers' Income" was announced. That year's Central Document No. 1 deepened the requirements of last year's Central Document No. 1 on developing modern agriculture as the primary task of new rural construction, grasping the key link between maintaining economic stability and promoting agricultural development, and also taking into account the work of all aspects of rural areas. There are more than 40 policy requirements and measures in the whole document, among which the agriculture and farmers can be directly benefited as "three obvious," "three adjustments," "four increases," "four improvements," "four increase," and "two substantial." Embodies the benefits to farmers to gradually increase, with the growth of national financial resources to the "three rural" support to further increase the requirements.
	General Office of the State Council on strengthening the work of fishery production safety notice Comprehensively implement the production safety responsibility system for fisheries, further strengthen safety management and supervision, increase investment, improve infrastructure, improve technical equipment, improve laws and regulations, and constantly improve the safety quality of practitioners and disaster prevention and avoidance capabilities, and strive to build a long-term mechanism for fisheries production safety, effectively curb fishing safety accidents, effectively protect the lives and property of the people, and promote the safe development of the fisheries economy. Expansion of a number of new safety from the wind, supporting the perfect fishing port, so that the national coastline within an average of 200 km above the first-class fishing port is able to provide services for 45% of the marine fishing vessels. Key fishing ports were equipped with safety monitoring equipment, the construction of marine fisheries ship management dynamic monitoring system, fishing vessel safety equipment testing, and inspection base and fisheries crew training base, so that serious fishing vessel accidents are significantly control. By 2015, the formation of a more complete fishery production safety support and security system, fisheries safety supervision and disaster prevention and mitigation capabilities are significantly enhanced, the quality of practitioners have improved to a certain extent, the fisheries safety production situation has improved significantly. Specific measures include: Strengthening the construction of fishing port safety infrastructure, vigorously improve the quality of fishing vessel safety, and actively promote the construction of fisheries safety communication network, and strive to improve fisheries safety technology and equipment. Increase financial investment, improve the production of fishery safety laws and regulations and systems, promote scientific and technological advances in fisheries safety, improve fisheries safety risk protection mechanism, etc.
2009	Several Opinions of the CPC Central Committee and State Council on Promoting Stable Development of Agriculture and Sustained Incomes of Farmers in 2009 The 2009 No. 1 Document "the CPC Central Committee and the State Council on 2009 to promote the stable development of agriculture and farmers continue to increase income of a number of opinions" presents four new highlights. First, the support for farmers to grow food has been increased again. Including increased investment in agricultural infrastructure and scientific and technological services, and increased direct subsidies to agriculture. Second, to increase efforts to solve the problem of employment of migrant workers. The document proposes urban and rural infrastructure construction and new public welfare jobs, as much as possible to use more migrant workers. To take work for food and other ways to guide farmers to participate in agricultural and rural infrastructure construction. Third, the rural livelihood construction focus on the rural power grid construction, rural road construction, rural drinking water safety project construction, rural biogas construction, rural housing renovation, and other five areas. Fourth, the transfer of agricultural land emphasizes further regulation. For adhering to the basic rural management system, the 2009 No. 1 Document first emphasized the implementation and protection of farmers' land rights and interests, focusing on two aspects of work: The ownership of collectively owned land is further defined clearly and its rights and interests are protected. The contracted land plots are identified, registered, and certified.

Table A1. *Cont.*

Year	Summary of Major Agricultural Policies
2010	Several Opinions of the State Council of the CPC Central Committee on Increasing the Efforts to Coordinate Urban and Rural Development to Further Strengthen the Foundation of Agricultural and Rural Development In early 2010, the "Opinions of the State Council of the Central Committee of the Communist Party of China on Increasing the Efforts to Integrate Urban and Rural Development to Further Strengthen the Foundations of Agricultural and Rural Development" was released, further improving and strengthening good policies for the "three rural areas" on the basis of maintaining policy continuity and stability, and putting forward a series of new major principles and measures, including improve the policy system for a strong agricultural policy system, promote the allocation of resources to rural areas, improve the level of modern agricultural equipment, promote the transformation of agricultural development, accelerate the improvement of rural people's livelihood, narrow the gap between urban and rural public utilities development, coordinate urban and rural reform, enhance the vitality of agricultural and rural development. Strengthen the construction of rural grass-roots organizations, and consolidate the Party's ruling base in rural areas. The document places special emphasis on institutional innovation to promote the development of urbanization. Propose to actively and steadily promote urbanization, improve the level of urban planning and development quality, to strengthen the development of small and medium-sized cities and small towns as the focus. Deepen the reform of the household registration system, accelerate the implementation of policies to relax the conditions for settling in small and medium-sized cities and small towns, especially in counties and central towns, and promote the eligible agricultural transfer population to settle in cities and towns and enjoy the same rights and interests as local urban residents.
2011	The CPC Central Committee and State Council Decision on Accelerating the Reform and Development of Water Resources On 29 January 2011, the "CPC Central Committee and the State Council on the decision to accelerate the reform and development of water conservancy" was released, the theme of accelerating the reform and development of water conservancy. This is the first time in 62 years since the founding of new China, the central document on the work of water resources for comprehensive deployment. The document proposes to put the work of water resources on a more prominent position, focusing on accelerating the construction of agricultural water conservancy, to promote the leapfrog development of water resources. Proposes to strive for 5 to 10 years of efforts to fundamentally reverse the situation of the obvious lag in water conservancy construction.
2012	The CPC Central Committee and State Council on Accelerating Agricultural Science and Technology Innovation to Continuously Enhance the Ability to Ensure the Supply of Agricultural Products The "Opinions of the State Council of the CPC Central Committee on Accelerating Agricultural Science and Technology Innovation to Continuously Enhance the Supply Assurance Capability of Agricultural Products," released in February 2012, highlights the deployment of agricultural science and technology innovation and makes the promotion of agricultural science and technology innovation the focus of the work of the "three rural areas." Two of the most popular policies: One is the public, basic, social discourse on agricultural science and technology. The other is the township agricultural personnel salaries and wages to be linked to the average income of local institutions. The grassroots agricultural extension system reform and construction demonstration county project basically covers all agricultural counties, agricultural technology extension institutions conditions, and construction projects cover all townships.
2013	Several Opinions of the CPC Central Committee and State Council on Accelerating the Development of Modern Agriculture and Further Enhancing the Vitality of Rural Development On 31 January 2013, the "Opinions of the State Council of the Central Committee of the Communist Party of China on Accelerating the Development of Modern Agriculture and Further Enhancing the Vitality of Rural Development" was released. The document makes comprehensive arrangements for "accelerating the development of modern agriculture and further enhancing the vitality of rural development," requiring that we must respond to the changes in the stage, follow the laws of development, enhance the sense of worry, and make national efforts to persistently strengthen agriculture, benefit the countryside, and enrich farmers. In accordance with the work objectives of ensuring supply and income, reform and innovation to add vitality, increase the efforts of rural reform, policy support, science and technology-driven efforts.

Table A1. *Cont.*

Year	Summary of Major Agricultural Policies
2014	Several Opinions of the CPC Central Committee and State Council on Comprehensively Deepening Rural Reform and Accelerating Agricultural Modernization In January 2014, the "Opinions of the State Council of the Central Committee of the Communist Party of China on Comprehensively Deepening Rural Reform and Accelerating Agricultural Modernization" was released. It is pointed out that to comprehensively deepen rural reform, we should adhere to the direction of socialist market economy reform, handle the relationship between the government and the market, and stimulate rural economic and social vitality. Encourage exploration and innovation, and promote land transfer in an orderly manner under the premise of protecting the interests of farmers. Tailor the reform to local conditions, proceed in a gradual and orderly manner, and do not engage in a "one-size-fits-all" approach or pursue a one-step solution. Allow the adoption of different, transitional systems and policy arrangements. To integrate urban and rural linkages, give farmers more property rights, promote equal exchange of urban and rural factors and balanced allocation of public resources, so that farmers can participate equally in the modernization process and share the fruits of modernization.
2015	Several Opinions of the CPC Central Committee and State Council on Increasing Reform and Innovation to Accelerate the Construction of Agricultural Modernization The 2015 Central Document No. 1, "Opinions of the State Council of the Central Committee of the Communist Party of China on Increasing Reform and Innovation to Accelerate the Construction of Agricultural Modernization," points out that we should actively adapt to the new normal of economic development, follow the general requirements of stabilizing food and increasing income, improving quality and efficiency, and driving innovation, continue to comprehensively deepen rural reform, comprehensively promote the construction of the rule of law in rural areas, and promote the simultaneous development of new industrialization, information technology, urbanization and agricultural modernization. Efforts to improve food production capacity to explore new potential, in the optimization of agricultural structure to open up new ways, in the transformation of agricultural development to seek new breakthroughs, in the promotion of farmers to obtain new results on income, in the construction of a new countryside to take new steps to provide strong support for sustainable and healthy economic and social development.
2016	Several Opinions of the State Council of the Central Committee of the Communist Party of China on Implementing the New Concept of Development and Accelerating Agricultural Modernization to Achieve the Goal of Comprehensive Well-off In January 2016, the "Opinions of the State Council of the Central Committee of the Communist Party of China on Implementing the New Concept of Development and Accelerating Agricultural Modernization to Achieve the Goal of Overall Well-off" was released. All regimes and departments are required to firmly establish and thoroughly implement the development concept of innovation, coordination, green, openness and sharing, vigorously promote agricultural modernization, and ensure that hundreds of millions of farmers will join the people of the country to move into an overall well-off society. The document proposes to use the new concept of development to crack the "three rural" new problems, thicken the advantages of agricultural and rural development, increase innovation-driven efforts to promote the supply-side structural reform of agriculture, accelerate the transformation of agricultural development, to maintain stable development of agriculture, and farmers continue to increase income.

Table A1. *Cont.*

Year	Summary of Major Agricultural Policies
2017	Several Opinions of the State Council of the Central Committee of the Communist Party of China on Deepening the Structural Reform on the Supply Side of Agriculture and Accelerating the Cultivation of New Dynamic Energy for Agricultural and Rural Development In February 2017, the "CPC Central Committee and State Council on deepening the supply-side structural reform of agriculture to accelerate the cultivation of new dynamics of agricultural and rural development of a number of opinions" was released. The document clearly points out that the main line of work for the current and future period of the "three rural areas" should be to deeply promote the structural reform of the agricultural supply side. After years of efforts, China's agricultural and rural development has entered a new historical stage. The main contradiction in agriculture from the total lack of structural contradictions, highlighted by the stage of oversupply and undersupply coexist, the main aspect of the contradiction in the supply side. Urgent requirements to further promote the supply-side structural reform of agriculture, accelerate the cultivation of new dynamics of agricultural and rural development. The document points out that to promote the supply-side structural reform of agriculture, based on ensuring national food security, closely around the changes in market demand, to increase farmers' income, to ensure effective supply as the main goal, to improve the quality of agricultural supply as the main direction, to institutional reform and institutional innovation as the fundamental way. Additionally, stressed that the supply-side structural reform of agriculture is a long-term process, we must face the difficulties and challenges, try to reduce the cost of reform, and actively prevent the risk of reform. At the same time, the further implementation of the rural land three rights of separation, to encourage the appropriate scale of land transfer.
2018	Opinions of the State Council of the Central Committee of CPC on Implementing the Strategy of Rural Revitalization On 2 January 2018, the "Opinions of the State Council of the Central Committee of the Communist Party of China on the Implementation of the Rural Revitalization Strategy" was released. Around the implementation of the rural revitalization strategy, the document plans a series of major initiatives and is the top-level design for the implementation of the rural revitalization strategy. The document has two important features: First, it is comprehensive in management. It makes comprehensive arrangements for the overall promotion of rural economy, politics, culture, society, ecological civilization, and the construction of the Party. The second is to manage the long-term. In accordance with the strategic arrangement proposed by the 19th Party Congress of building a moderately prosperous society and achieving the second century goal in two stages, the document deploys the three stages of implementing the rural revitalization strategy according to the principle of "far and coarse, near and fine."

References

1. Burns, A.F.; Wesley, C.M. *Measuring Business Cycles*; National Bureau of Economic Research: Cambridge, MA, USA, 1946; pp. 384–417.
2. Fan, S. Public Expenditures, Growth, and Poverty: Lessons from Developing Countries. *Asean Econ. Bull.* **2009**, *26*, 235. [CrossRef]
3. Pan, D.; Ying, R. Research on the relationship between water resources and agricultural economic growth in China—Based on panel VAR model. *China Popul. Resour. Environ.* **2012**, *22*, 161–166.
4. Wu, Q.; Zhou, X.; Feng, Z. The impact of infrastructure on agricultural economic growth—A study based on inter-provincial panel data of China from 1995–2010. *China Econ. Issues* **2015**, *3*, 29–37.
5. Ofana, O.G.; Effiong, C.E.; Eko, E.O. Constraints to agricultural development in Nigeria. *Int. J. Dev. Econ. Sustain.* **2016**, *4*, 19–33.
6. Sylwester, K. R&D and economic growth. *Knowl. Soc.* **2001**, *13*, 71–84. [CrossRef]
7. Andersen, M.A. Public investment in U.S. agricultural R&D and the economic benefits. *Food Policy* **2015**, *51*, 38–43. [CrossRef]
8. Wang, J. Research on the Interactive Relationship between China's Agricultural Research Investment and Agricultural Eco-nomic Growth. *Agric. Technol. Econ.* **2009**, *1*, 103–109.
9. Yang, C.; Zhang, J.; Zhao, K. Demonstration on the relationship between agricultural science and technology resources and agricultural economic development. *China Popul. Resour. Environ.* **2011**, *21*, 113–118.
10. Kirchner, M.; Schmidt, J.; Kindermann, G.; Kulmer, V.; Mitter, H.; Prettenthaler, F.; Rüdisser, J.; Schauppenlehner, T.; Schönhart, M.; Strauss, F.; et al. Ecosystem services and economic development in Austrian agricultural landscapes—The impact of policy and climate change scenarios on trade-offs and synergies. *Ecol. Econ.* **2015**, *109*, 161–174. [CrossRef]

11. Abdul-Rahim, A.S.; Sun, C.; Noraida, A.W. The Impact of Soil and Water Conservation on Agricultural Economic Growth and Rural Poverty Reduction in China. *Sustainability* **2018**, *10*, 4444. [CrossRef]
12. Qiao, Z.; Jiao, F.; Li, N. Changes in China's Rural Economic System and Agricultural Growth: An Empirical Analysis of China's Agricultural Growth from 1978 to 2004. *Econ. Res.* **2006**, *7*, 73–82.
13. Jin, S.; Klaus, D. Land rental markets in the process of rural structural transformation: Productivity and equity impacts from China. *J. Comp. Econ.* **2009**, *37*, 629–646. [CrossRef]
14. Deininger, K.; Jin, S.; Xia, F.; Huang, J. Moving off the farm: Land institutions to facilitate structural transformation and agricultural productiv-ity growth in China. *World Dev.* **2014**, *59*, 505–520. [CrossRef]
15. Lerman, Z. Land reform, farm structure, and agricultural performance in CIS countries. *China Econ. Rev.* **2009**, *20*, 316–326. [CrossRef]
16. Guth, M.; Smędzik-Ambroży, K.; Czyżewski, B.; Stępień, S. The Economic Sustainability of Farms under Common Agricultural Policy in the European Union Countries. *Agriculture* **2020**, *10*, 34. [CrossRef]
17. Susilastuti, D. Poverty Reduction Models: Indonesian Agricultural Economic Approach. *Eur. Res. Stud. J.* **2017**, *20*, 164–176. [CrossRef]
18. Boonyanam, N. Agricultural economic zones in Thailand. *Land Use Policy* **2020**, *99*, 102774. [CrossRef]
19. Chaudhary, D. Agricultural Policies and Rural Development in Nepal: An Overview. *Res. Nepal J. Dev. Stud.* **2018**, *1*, 34–46. [CrossRef]
20. Chanie, A.M.; Pei, K.Y.; Lei, Z.; Zhong, C.B. Rural Development Policy: What does Ethiopia Need to Ascertain from China Rural Development Policy to Eradicate Rural Poverty? *Am. J. Rural. Dev.* **2018**, *6*, 79–93. [CrossRef]
21. Corral, S.; Díaz, A.S.; Monagas, M.D.C.; García, E.C. Agricultural Policies and Their Impact on Poverty Reduction in Developing Countries: Lessons Learned from Three Water Basins in Cape Verde. *Sustainability* **2017**, *9*, 1841. [CrossRef]
22. Pernechele, V.; Jean, B.; Léopold, G. Agricultural policy incentives in sub-Saharan Africa in the last decade (2005–2016). *Monit. Anal. Food Agric. Policies (MAFAP) Synth. Study* **2018**, 2143-2019-4790. [CrossRef]
23. Grabowski, R. Agricultural distortions and structural change. *J. Asian Econ.* **2013**, *24*, 17–25. [CrossRef]
24. Akbar, M.; Jamil, F. Monetary and fiscal policies' effect on agricultural growth: GMM estimation and simulation analysis. *Econ. Model.* **2012**, *29*, 1909–1920. [CrossRef]
25. Fang, H. The impact of monetary supply shocks on the agricultural economy under the conditions of an open economy—Based on the causal graph method. *China Rural Econ.* **2011**, *8*, 41–53.
26. Chen, D. Monetary policy, inflationary pressure and agricultural product prices. *China Soft Sci.* **2014**, *7*, 185–192.
27. Sánchez-Picón, A.; Aznar-Sánchez, J.A.; García-Latorre, J. Economic cycles and environmental crisis in arid southeastern Spain. A historical perspective. *J. Arid Environ.* **2011**, *75*, 1360–1367. [CrossRef]
28. Palma, I.P.; Toral, J.N.; Vázquez, M.R.P.; Fuentes, N.F.; Hernández, F.G. Historical changes in the process of agricultural development in Cuba. *J. Clean. Prod.* **2015**, *96*, 77–84. [CrossRef]
29. Chatterton, J.; Graves, A.; Audsley, E.; Morris, J.; Williams, A. Using systems-based life cycle assessment to investigate the environmental and economic impacts and benefits of the livestock sector in the UK. *J. Clean. Prod.* **2015**, *86*, 1–8. [CrossRef]
30. Ge, Y.; Zhao, R.; Liu, D. Causal factors of cyclical fluctuations in China's agricultural economy and their transmission mechanisms. *Agric. Technol. Econ.* **1998**, *1*, 46–49.
31. Guo, W. A study of agricultural volatility in China. *Rural Econ. Soc.* **1991**, *2*, 1–9.
32. Xi, L.; Peng, K. Analysis of China's rural economic system changes and agricultural cyclical growth. *China Popul. Resour. Environ.* **2010**, *20*, 123–129.
33. Sun, X. Analysis of the characteristics of my country's agricultural cycle fluctuations. *Learn. Explor.* **2011**, *2*, 168–169.
34. Sui, J.; Cai, Q. Identification of the path evolution of China's agricultural business cycle: Practice and experience since the reform and opening up. *China Rural Econ.* **2016**, *9*, 30–43.
35. Cheng, S.; López, M.A. Synergy analysis of agricultural economic cycle fluctuation based on ant colony algorithm. *Open Phys.* **2018**, *16*, 978–988. [CrossRef]
36. Yu, J.; Wu, J. The Sustainability of Agricultural Development in China: The Agriculture–Environment Nexus. *Sustainability* **2018**, *10*, 1776. [CrossRef]
37. Hejazi, M.; Marchant, M.A. China's evolving agricultural support policies. *Choices* **2017**, *32*, 1–7.
38. Huang, J.; Guolei, Y. Understanding recent challenges and new food policy in China. *Glob. Food Secur.* **2017**, *12*, 119–126. [CrossRef]
39. Huang, P.C.C. Is the family farm the approach for agricultural development in China? *Open Times* **2014**, *2*, 176–194.
40. Yao, Y.; Chen, W.; Li, X. Research on the relationship between environmentally friendly agricultural technology innovation and agricultural economic growth. *China Popul. Resour. Environ.* **2014**, *24*, 122–130.
41. Wang, L. Impact of ecological land use change on agricultural economy in the farming pastoral ecotone of northern China. *China Rural Econ.* **2009**, *4*, 80–85.
42. Skalin, J.; Teräsvirta, T. Another look at Swedish business cycles, 1861–1988. *J. Appl. Econ.* **1999**, *14*, 359–378. [CrossRef]
43. Hamilton, J.D. Analysis of time series subject to changes in regime. *J. Econ.* **1990**, *45*, 39–70. [CrossRef]
44. Leamer, E.; Potter, S. *A Nonlinear Model of the Business Cycle*; Federal Reserve Bank of New York: New York, NY, USA, 2003.
45. Moolman, E. A Markov switching regime model of the South African business cycle. *Econ. Model.* **2004**, *21*, 631–646. [CrossRef]

46. de Medeiros, O.R.; Sobral, Y.D. A Markov Switching Regime Model of the Brazilian Business Cycle. *SSRN Electron. J.* **2007**. [CrossRef]
47. Dicle, O. Sectoral Business Cycle Asymmetries and Regime Shifts: Evidence from Turkey. *Appl. Econ. J.* **2019**, *26*, 59–80.
48. Liu, J.; Sui, J.; Yan, C. My country's economic cycle wave dynamics and economic policy orientation under the financial crisis. *China Ind. Econ.* **2009**, *8*, 37–46.
49. Chen, L.; Liu, H. Research on the Asymmetry and Persistence of my country's Business Cycle Fluctuations. *Econ. Res.* **2007**, *4*, 43–52.
50. Liu, J.; Sui, J. A Test of the Relationship between the Uncertainty of China's Currency Growth and Economic Growth (1980–2008). *Chin. Soc. Sci.* **2010**, *4*, 74–86, 221–222.
51. Sui, J.; Song, D. Analysis of the changes in the dynamic path of China's agricultural economic growth. *Learn. Explor.* **2014**, *6*, 106–109.
52. Krolzig, H.-M. *Markov-Switching Vector Auto Regressions: Modelling, Statistical Inference, and Application to Business Cycle Analysis*; Springer Science & Business Media: Berlin, Germany, 2013; Volume 454.
53. Hodrick, R.J.; Prescott, E.C. Postwar U.S. Business Cycles: An Empirical Investigation. *J. Money Crédit Bank.* **1997**, *29*, 1–16. [CrossRef]

sustainability

MDPI

Article

Land Use Changes in the Teles Pires River Basin's Amazon and Cerrado Biomes, Brazil, 1986–2020

Aline Kraeski [1], Frederico Terra de Almeida [1,2,*], Adilson Pacheco de Souza [1,2], Tania Maria de Carvalho [2], Daniel Carneiro de Abreu [2,3], Aaron Kinyu Hoshide [3,4] and Cornélio Alberto Zolin [5]

1 Environmental Sciences, Federal University of Mato Grosso, Sinop 78557-267, MT, Brazil
2 Institute of Agrarian and Environmental Sciences, Federal University of Mato Grosso, Sinop 78557-287, MT, Brazil
3 AgriSciences, Federal University of Mato Grosso, Sinop 78555-267, MT, Brazil
4 College of Natural Sciences, Forestry and Agriculture, The University of Maine, Orono, ME 04469, USA
5 EMBRAPA Agrossilvipastoril, Brazilian Agricultural Research Corporation, Sinop 78550-970, MT, Brazil
* Correspondence: fredterr@gmail.com; Tel.: +55-(66)-99995-1315

Abstract: The Teles Pires River basin in Brazil's center-west has recently expanded agricultural economic development at the expense of both the Amazon rainforest and Cerrado savannah. We evaluated these changes occurring in this basin over the last 34 years. Maps were generated to determine changes in land use classifications between 1986, 1991, 1996, 2000, 2005, 2011, 2015, and 2020. The supervised classification of Landsat 5 and 8 images used the maximum likelihood algorithm. Satellite spatial data on land use downloaded from the United States Geological Survey were validated according to 1477 locations, where our research team categorized land use in the field during 2020. The growth in agricultural crops (+643%) and pasture (+250%) from 1986 to 2020 were detrimental to natural areas, such as the rainforest and savannah. The percentage increase in the agricultural areas between the evaluated years peaked around 1996 and stabilized in 2020 at 40% of the Teles Pires River basin's land area. Land use change patterns were related to political/economic events in Brazil, forest/pasture conversions until 2011, and the change from pasture to crops from 2011 to 2020. There was greater intensity in the changes in the upper Teles Pires River basin toward the south, which expanded northward over time. Sustainable agricultural intensification is needed in such stabilized, frontier areas.

Keywords: agricultural frontier; Brazil; land conversion; land use; southern Amazon; supervised classification; Teles Pires River; territorial dynamics

Citation: Kraeski, A.; de Almeida, F.T.; de Souza, A.P.; de Carvalho, T.M.; de Abreu, D.C.; Hoshide, A.K.; Zolin, C.A. Land Use Changes in the Teles Pires River Basin's Amazon and Cerrado Biomes, Brazil, 1986–2020. *Sustainability* **2023**, *15*, 4611. https://doi.org/10.3390/su15054611

Academic Editors: Pablo Peri and Luca Salvati

Received: 28 December 2022
Revised: 21 February 2023
Accepted: 24 February 2023
Published: 4 March 2023

1. Introduction

Humans historically have dominated their geographical surroundings, interacting, and modifying it according to their interests. Unfortunately, this often causes degradation of the natural resources due to the different forms of environmental impacts [1]. An important example involves changes in land cover, which, when associated with the absence of conservation practices, generate impacts such as a reduction in water supply [2,3] resulting from changes in the hydrological cycle [4,5]. In Brazil, this has involved deforestation of the Amazon rainforest and habitat conversion of the Cerrado savannahs for agricultural use [6,7]. Natural habitat areas are typically converted to pasture for extensive livestock (e.g., *Bos taurus* Nelore beef breed) grazing and/or the export of commodity crops, such as soybeans (*Glycine max* L.), maize (*Zea mays* L.), and cotton (*Gossypium* sp.). These recent land use conversions and the resulting agricultural economic development have been driven by international demand for food and livestock feed [6] Continued habitat conversions and/or changes in the climate could make the sole reliance on rain-fed agriculture more challenging if precipitation continues to decline in the Amazon [8].

In order to meet the growing needs of humans (e.g., food) and mitigate the environmental impacts of agricultural production by more efficiently using natural resources, it is essential to adopt planning policies that integrate environmental, social, and economic aspects [9]. Monitoring changes in land classifications (i.e., classes), as a function of different land uses, has become an effective tool to support land use planning. In particular, the use of technologies such as remote sensing, which can currently count on a variety of sensors operating at different scales, allow for the acquisition of information on land use classifications over large areas at low cost [10,11].

Data resulting from remote sensing can be easily processed using geoprocessing programs that make it possible to carry out different types of operations to generate information. In studies that aim to characterize land use and occupation, the application of the image classification method [12] stands out. This method consists of labeling the image pixels according to their spectral characteristics, using, for this purpose, mathematical techniques that perform the pattern recognition resulting in thematic maps [13].

Due to the global importance and recognition for the ecological services it offers, the Amazon has been monitored for decades using remote sensing and image classification products. This monitoring has sought to evaluate changes to this rich biome in order to help implement public policies for Amazon conservation. In a study carried out in the Colombian Amazon, Landsat images and supervised classification were used to map the changes in land cover and identify locations affected by deforestation over a period of sixteen years, between 2000 and 2016 [14]. Other researchers have dedicated themselves to mapping and monitoring forest changes in the Brazilian Amazon in the state of Pará from 2000 to 2019, using multitemporal remote sensing data and machine learning classification [15]. In the upper Teles Pires River basin's transition zone between the Amazon and Cerrado biomes, previous research mapped the spatial and temporal dynamics of land use from 1986 to 2014, using Landsat images and supervised classification. The results showed an intense reduction in native vegetation as a result of agricultural expansion [16].

Several other studies around the world have already employed remote sensing to identify the forms of appropriation of spaces and changes in landscapes. Changes in land use in poor areas of China were mapped between 2013 and 2018 [17]. Another study monitored land cover changes in a district of India between 1990 and 2010 [18]. Other studies focus on mapping specific targets, such as that conducted by researchers [19], who monitored the urban spatial–temporal dynamics in Nagpur, India, between 1991 and 2010. In Brazil, one of the nationwide actions for mapping land cover and land use is the MapBiomas project, aimed at the conservation of the different Brazilian biomes, which has generated a historical series of annual maps from an initiative involving a collaborative network of specialists [20].

Our present study focused on mapping the changes that occurred in the Teles Pires River basin (Figure 1). The largest part of the Teles Pires River basin is located in the state of Mato Grosso, Brazil. The Mato Grosso state is characterized by great socioeconomic and ecological diversity, where the Pantanal, Amazon and Cerrado biomes share space [21]. Both the Amazon and Cerrado biomes are found within the Teles Pires River basin. The region has been in full economic development mode, driven by industrial agricultural exports. In the last decade, Mato Grosso has installed large hydroelectric projects, which has resulted in profound changes to the landscape, pointing to the need to monitor such changes in view of the possible impacts on the environment. Despite this region's importance in the national and international context, the region lacks continuous monitoring of the changes in land use resulting from agricultural expansion by large farming enterprises. The goals of our research were to expand the knowledge about the land use dynamics in the region in order to better manage the water resources and plan economic activities. The objective of our study was to evaluate the changes in land use in the Teles Pires River basin over a 34-year period from 1986 to 2020.

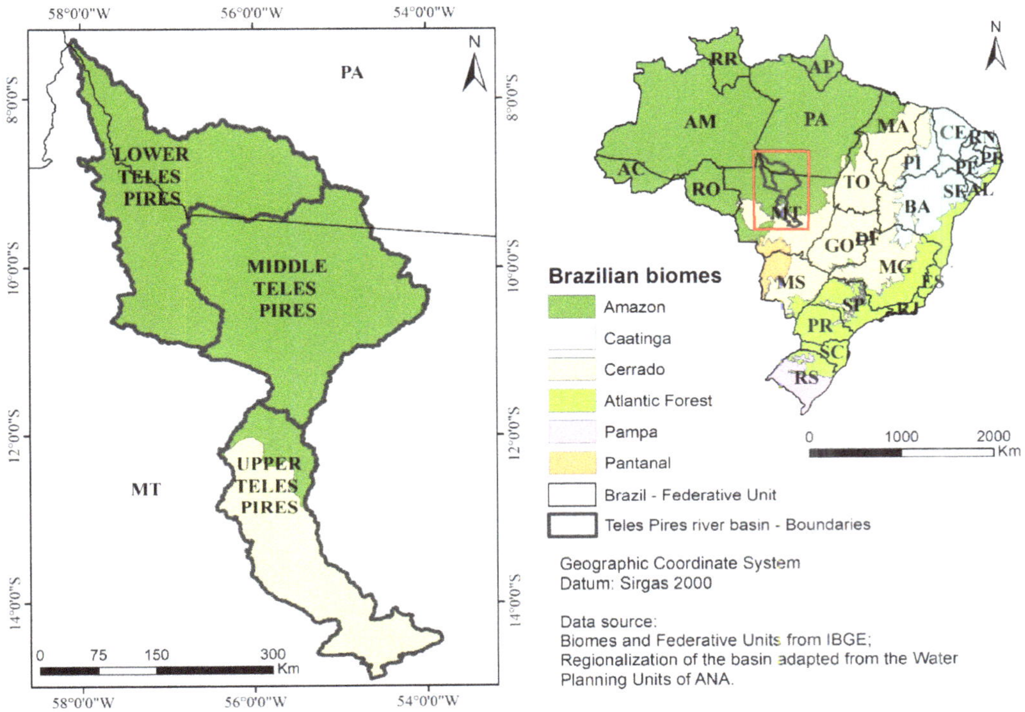

Figure 1. Location and regions of Teles Pires River basin in Brazil.

2. Using Remotely Sensed Data for Conservation

Despite the current abundance of research that generates information on land occupation and use, globally, there is still a lack of adequate information that allows for assessing the intensity of the changes that occur in terms of land use. This makes it impossible to estimate and evaluate the effects, impacts, or potential expansion of agricultural production in ecologically diverse biomes [22]. Gradually, this deficiency has been overcome thanks to the use and improvement of technologies such as remote sensing, which facilitates the acquisition of information in large areas and allows for the collection of historical data [23].

The longest record of orbital images of the Earth's surface already covers five decades and corresponds to the Landsat mission, which is a partnership between the National Aeronautics and Space Administration (NASA) and the United States Geological Survey (USGS), having launched the first satellite in 1972, the Landsat 1 [24]. Over the years, the Landsat mission has outdone itself with the launch of satellites carrying increasingly innovative sensors. Currently, the Landsat series includes the Landsat 8 and 9 satellites, the last one having launched in September 2021 [25]. The Landsat data set currently provides a global basis for monitoring the changes in environments due to the expansion of human occupation, data that are freely available to the public and open source [24].

Through the use of software and geoprocessing tools, the data obtained through remote sensing, such as satellite images, can be treated and processed, allowing for the extraction of information of interest and the generation of products, such as thematic maps. One of the most used processes in the generation of thematic maps from remote sensing images is so-called image classification, which can also be distinguished as being either supervised or unsupervised. The unsupervised classification is characterized by not requiring prior knowledge of the study area, as the algorithm examines the unknown pixels of the image and divides them into different classes. Meanwhile, the supervised classification demands knowledge of the study area, as it is up to the analyst to select the training pixels representative of each class, so that the algorithm performs the classification [26]. One of the most popular supervised classification methods used is the maximum likelihood algorithm [18].

The maximum likelihood algorithm is based on the probability that each pixel in the image belongs to each of the classes identified by the analyst during training. The maximum likelihood algorithm then assigns pixels to the class with the highest likelihood of being in a particular classification [27]. For this, it evaluates the variance and covariance of the spectral response of the training class when classifying the unknown pixels, generating accurate results, as it is based on statistical parameters [28].

With rapid technological advancement occurring in this area, several new methodologies have been developed and applied in satellite image processing [29], such as machine learning techniques [30], the random forest algorithm [31], neural networks, and others [32]. Even so, classic methods such as maximum likelihood continue to be used due to their easy access and availability in various software, and when well executed they result in land use and occupation data of satisfactory accuracy [33].

In order to measure the quality of the information generated in the classification of images, it is essential to validate the results, identifying the accuracy of the mapping. This can be easily obtained by comparing a set of classified pixels with terrestrial truth data [12]. The most common way of representing the accuracy of the classification of remote sensing data has been the use of an error matrix, which is the basis for a series of statistical analyzes, such as general accuracy and the kappa index [34]. Together, the error matrix and the kappa index have come to represent the standard way of evaluating the accuracy of image classification [12]. Following such verification to improve data accuracy, such remotely sensed data become more reliable for use. Thus, maps resulting from image classification can be used to follow-up and monitor changes in land use and occupation, especially in areas threatened by environmental degradation, such as tropical rainforests and savannahs.

In tropical regions, monitoring human intervention in environments is essential, given the intense pace of the conversion of the natural areas into arable areas. About half of the world's remaining tropical rainforests are in Latin America. The tropical rainforests in both Central America and South America are also experiencing the world's highest rates of deforestation, largely driven by large-scale commercial agricultural production [35].

Based on mapping global deforestation footprints between 2001 and 2015 [36], tropical forests are under increasing threat. This is especially the case in tropical countries, such as Brazil, with high historical deforestation footprints that have allowed for the production and export of agricultural commodities, such as cattle, soybeans, coffee, cocoa, and wood, to other countries. Thus, it is possible to associate spatial patterns of deforestation with global supply chains [37]. Massive investments have been made to support the production of export commodities in tropical countries, resulting in high rates of deforestation [38]. This activity requires significant conversion of the land for use. Governments sometimes see these investments as beneficial, by improving the use of land seen as idle, disregarding that many of these lands are occupied by traditional peoples and ignoring the ecological importance of natural systems [38].

3. Materials and Methods

3.1. Study Area

The Teles Pires River basin is located between latitudes 7°16′47″ and 14°55′17″ south and longitudes 53°49′46″ and 58°7′58″ west, occupying a territorial area of 141,524 square kilometers in the Brazilian states of Mato Grosso and Pará, within the limits of the Legal Amazon (Figure 1). Integrated into the Amazon hydrographic region, its main river is the Teles Pires, which together with the Juruena River is responsible for the formation of the Tapajós River, one of the main tributaries of the right bank of the Amazon River. The wide latitudinal extension of the basin causes it to have a diversity of environments, allowing for easy regional classification into the upper, middle, and lower Teles Pires (Figure 1). The lower and middle portions of the Teles Pires River basin are characterized by the presence of the Amazon biome, while the upper Teles Pires is marked by the Cerrado biome. The predominant climate in the area, according to Köppen's classification, is Aw tropical climate with dry winter and rainy summer, present in the entire upper and part of the middle Teles Pires, while the rest of the basin has an Am humid tropical climate with a short dry season and more precipitation [39].

3.2. Spatial Data Sources

The area of study was delimited using the ArcGis 10.1 software, the ArcHydro extension, and the Digital Elevation Model (DEM) from the Brazilian Agricultural Research Corporation, also known as Embrapa [40]. We used Shuttle Radar Topography Mission (SRTM) data with a spatial resolution of 90 m and the drainage network from the Brazilian Institute of Geography and Statistics [41]. The ArcGis spatial data used a continuous database of the Brazilian territory using a scale of 1:250,000. These were the only spatial data sources with complete coverage for the region.

The DEM was reconditioned by imposing the drainage pattern by following an automated ArcHydro Tools algorithm in the ArcGis software (AGREE), a method that deepens the DEM in order to coincide with the vector hydrography and assists in the hydrological analyses in areas with low topographic differences [42]. Soon after, the correction of the depressions was performed and the flow direction was obtained by the eight direction pour point model method, which assumes that the water flows to one of the eight neighboring cells according to the greatest slope, used for generating the accumulated flow, and calculation of the upstream cells that drain to each cell of the raster. Spatial data in ArcGis are either shape or gridded (i.e., raster). The raster drainage network was generated using the threshold of 250 cells of accumulated flow to define the drainage. The outlet of the basin was identified, located at the coordinates 7°20′58″ south latitude and 58°7′57″ west longitude, and the area was delimited from this point.

3.3. Mapping Land Use

The data used were orbital images from the Thematic Mapper and Operational Land Imager sensors, Landsat 5 and 8, respectively. These were obtained free of charge as Level 1 products from the Earth Explorer platform of the United States Geological Survey or USGS [43]. The images are high quality for temporal analysis and have a spatial resolution of 30 m for both sensors, while the radiometric resolution is 8 bits for the Thematic Mapper sensor and 16 bits for the Operational Land Imager sensor. The ArcGis software version 10.1 and the ENVI software were used in the processing steps.

In order to characterize the changes in land use gradually, we used an average interval of five years between the classifications. The factors influencing our choice of years were the availability and quality of the data and the absence of apparent clouds. This resulted in the use of data for 1986, 1991, 1996, 2000, 2005, 2011, 2015, and 2020 (Table 1). We used data from the dry season in the region between the months of May and September, when there is low cloudiness, which favors the classification of images.

All the images were subjected to conversion from digital number (DN) to reflectance at the top of the atmosphere using the radiometric calibration module in ENVI and atmospheric correction with the FLAASH (Fast Line-of-sight Atmospheric Analysis of Spectral Hypercubes) module, based on the MODTRAN (Moderate Resolution Atmospheric Transmission) radiation transfer model, with the application of the tropical atmospheric model, resulting in surface reflectance values. Thereafter, all the processing steps were carried out in ArcGis using the spectral bands of red, near infrared, and medium infrared (measured in micrometers or $\mu m = 1 \times 10^{-9}$ km), corresponding to bands 3 (0.63–0.69 μm), 4 (0.76–0.90 μm), and 5 (1.55–1.75 μm) of the Thematic Mapper (TM) sensor and 4 (0.64–0.67 μm), 5 (0.85–0.88 μm), and 6 (1.57–1.65 μm) of the Operational Land Imager (OLI) sensor. Mosaics were created and the red, green, blue (RGB) color composition was generated using the combination of bands 5, 4, and 3 for the TM, and 6, 5, and 4 for the OLI, which facilitates the interpretation of images, favoring the identification of vegetated areas [44].

Table 1. Orbit/point and day/month of images used for mapping land use and occupation in the Teles Pires River watershed.

Orbit/Point	Year							
	1986	1991	1996	2000	2005	2011	2015	2020
225/70	29 June	29 July	11 Aug.	21 July	17 June	20 July	16 Aug.	10 June
226/66	8 Sept.	18 June	1 July	28 July	10 July	12 Aug.	7 Aug.	3 July
226/67	8 Sept.	20 July	1 July	10 June	10 July	12 Aug.	7 Aug.	3 July
226/68	7 Aug.	20 July	1 July	10 June	26 July	12 Aug.	7 Aug.	3 July
226/69	7 Aug.	20 July	30 May	10 June	26 July	13 Sept.	7 Aug.	3 July
226/70	7 Aug.	20 July	1 July	10 June	26 July	11 July	7 Aug.	3 July
227/66	27 June	27 July	6 June	17 June	17 July	18 July	14 Aug.	10 July
227/67	27 June	27 July	6 June	17 June	17 July	3 Aug.	14 Aug.	10 July
227/68	27 June	27 July	6 June	1 June	17 July	3 Aug.	14 Aug.	10 July
227/69	27 June	27 July	6 June	1 June	17 July	3 Aug.	14 Aug.	10 July
228/65	5 Aug.	18 July	15 July	10 July	8 July	25 July	5 Aug.	17 July
228/66	5 Aug.	18 July	15 July	26 July	8 July	25 July	5 Aug.	17 July
228/67	5 Aug.	18 July	31 July	26 July	8 July	25 July	5 Aug.	15 June
229/65	28 Aug.	26 Aug.	23 Aug.	17 July	15 July	1 Aug.	12 Aug.	22 June
Sensor	TM	TM	TM	TM	TM	TM	OLI	OLI

TM = Thematic Mapper sensor; OLI = Operational Land Imager sensor.

Supervised classification was then used for identifying the classes present in the area and selecting representative samples, which consisted of the training phase of the classifier. At this stage, techniques for the visual interpretation of the images, based on elements such as color, texture, shape, and pattern, were employed, creating interpretation keys for the features [45]. In order to obtain better results in view of the variability of the spectral responses identified for some classes, resulting from the different stages in the development of the vegetation or crops, these classes were divided into subclasses in the training and classification steps, and the sample sets by classes/subclasses had a value greater than 1000 pixels.

The algorithm adopted for the classification was the maximum likelihood, which among the conventional classifiers is efficient in the classification of medium-resolution images [18,28]. From the information determined during training, the algorithm calculates the probability of the pixels of the image belonging to each class and assigns them to the one with the highest probability [27]. After classification, the subclasses were grouped, resulting in the classes summarized in Table 2.

The classifications were subjected to post-processing with the application of a filter that eliminates isolated pixels on the map, replacing their values based on contiguous neighbors, using the majority filter tool. The files were finally converted into vector format and the areas occupied by each class/year were calculated. In order to check for regional differences, the data were segmented based on the water planning units established by the National Water Agency (ANA), which divides the basin into the upper, middle, and lower Teles Pires [46]. To detect the area conversions that occurred, a method for comparing different classifications was used, based on the area of polygons in 1986 and 2020. Using the intersection tool in the ArcGis software, the two maps were crossed (1986 and 2020), making it possible to identify the areas that remained in the same land use and occupation class during the period, and the areas that were converted to other classes. Such data were compiled using cross tabulation resulting in an area conversion matrix, similar to that produced by [17]. The area conversion matrix was used to identify the land use classes with the most significant modifications. Thus, changes in land use were compared between all the mapped time periods.

Table 2. Description of the land use classes mapped for the Teles Pires River basin in Brazil.

Land Use Classes	Description
Water	Surfaces covered by water, encompassing the water bodies of the basin.
Forest	Area of tree forest vegetation with high density of trees.
Cerrado	Area of vegetation with predominance of shrub stratum, showing variations with areas of low-density forest formation.
Pasture	Areas covered by natural or planted perennial forage intended for cattle grazing.
Crops	Areas intended for the cultivation of food crops, fibers, and agribusiness commodities.
Burned area	Surfaces that have undergone recent burning processes, with evidence of the affected areas.
Other area	Formed by the junction of land occupation classes with reduced spatial cover in the basin, including urban areas, mining areas, sandbanks, and rock formations.

3.4. Mapping Validation

Information from ground truth was acquired to estimate the accuracy of the classifications. During field expeditions in 2020, 1477 points distributed throughout the area were collected (Figure 2), specifically 42 for water, 350 for forest, 77 representing Cerrado, 621 for pasture, 323 in crops, and 64 others. The 2020 images used in the classification were dated between 10 June and 17 July 2020, and field visits were carried out between 26 July and 28 September 2020.

In order to assess the mapping accuracy of all classified years and obtain better distribution of samples and more significant sample sets, samples were also selected through visual identification using high-resolution historical images from the Google Earth platform and Landsat images with dates close to those of the classifications. For the years 1986, 1991, and 1996, due to the absence of high-resolution images, sampling was performed exclusively with Landsat images. To ensure total independence in the validation, samples coinciding with the areas used in the classify training were not used.

Each point collected was associated with the class present at the site and its geographic coordinates, and these data were compared with those of the corresponding classifications, enabling the construction of error matrices. These were then used to calculate the overall, producer, and user accuracy indices [34], and the kappa index [47].

Using the same sample set of points created in the previous step, an agreement analysis was used between the classification results and the data set produced by the MapBiomas project [48]. Confusion matrices were based on the methodology from Shimabukuro et al. 2020 [31]. This type of matrix shows the classification accuracy of remotely sensed spatial data (e.g., MapBiomas) compared to ground truth classification from our ground surveys (Figure 2), as well as from other geographic and spatial sources. Differences in legend between the mapping in this study and MapBiomas were adjusted according to Table 3. As there is no class corresponding to the burned area in MapBiomas, samples from this category were not used in the analysis. The area values of the MapBiomas mapping per class were also quantified for comparison with the areas obtained in the mapping of this study.

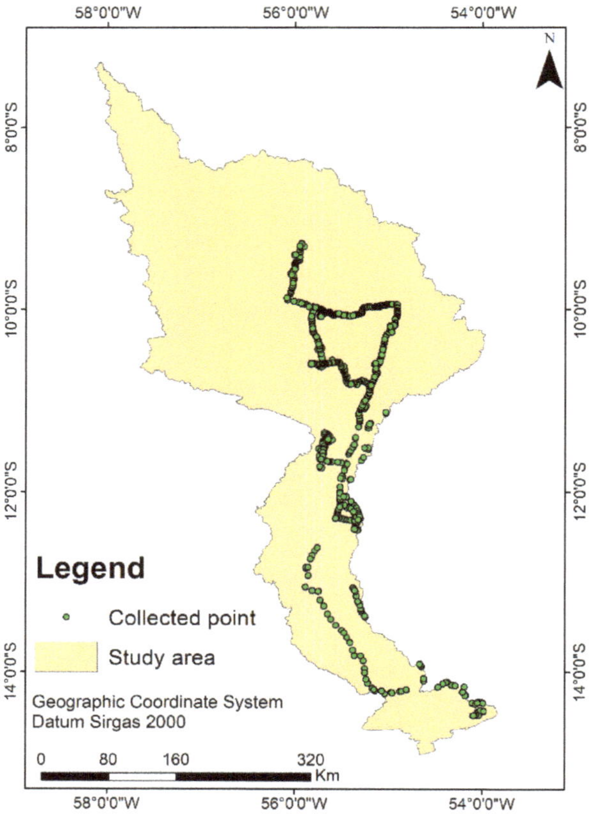

Figure 2. Distribution of points collected in the field in Brazil's Teles Pires River basin in 2020.

Table 3. Legend adjustments of classes mapped by the MapBiomas project with those mapped in this study.

Classification	MapBiomas
Water	River, lake, and ocean.
Forest	Forest training and silviculture.
Cerrado	Savanna formation, grassland formation, and other non-forest natural formations.
Pasture	Pasture.
Crops	Crops, temporary crops, sugarcane, soybeans, other temporary crops, and cotton.
Burned area	-
Other area	Urbanized area, other non-vegetated areas, and mining.

4. Results

The Teles Pires River basin has an area of 141,524 square kilometers (km^2), of which 34,453 km^2 corresponds to the region of the upper Teles Pires, 55,890 km^2 to the middle Teles Pires, and 51,181 km^2 to the lower Teles Pires. These regions represent 24.34%, 39.49%, and 36.16% of the basin, respectively. Figures 3 and A1 show the maps obtained in the classification of land use for the basin between 1986 and 2020. The results of the accuracy indices generated in the validation step are shown in Tables 4, 5, A1 and A2.

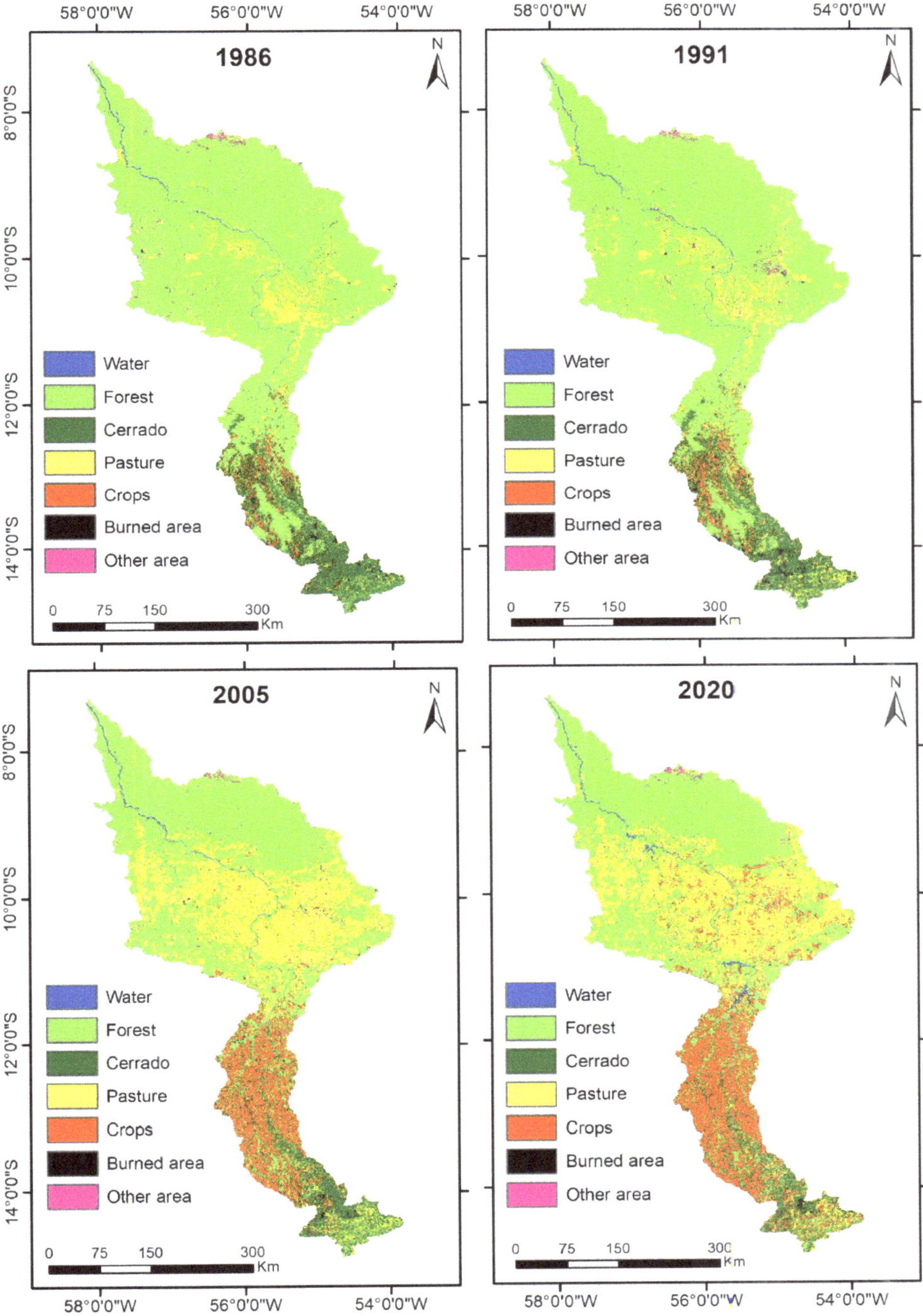

Figure 3. Maps of land use in the Teles Pires River basin during 1986, 1991, 2005, and 2020.

Table 4. Results of the accuracy indices of land use classifications for all mapped years for Brazil's Teles Pires River basin based on Google Earth and Landsat data.

	1986	1991	1996	2000	2005	2011	2015	2020
Kappa Index	0.96	0.96	0.97	0.97	0.97	0.97	0.97	0.97
Overall Accuracy (%)	98.43	97.97	98.73	98.03	98.24	98.49	98.04	98.46
Producer Accuracy (%)								
Water	100	100	100	100	100	100	100	100
Forest	100	100	100	100	100	99.99	100	99.98
Cerrado	93.27	94.46	93.44	91.06	89.93	92.56	91.97	95.96
Pasture	98.58	98.65	99.21	97.55	98.29	98.71	97.90	97.67
Crops	99.46	93.09	98.45	99.16	97.79	98.53	97.06	97.30
Burned area	100	100	100	99.93	100	100	99.57	100
Other area	75.66	79.45	86.67	71.70	79.14	74.60	74.72	83.50
User Accuracy (%)								
Water	100	100	100	100	100	100	100	100
Forest	98.75	98.88	98.77	98.18	98.54	98.81	98.77	99.18
Cerrado	100	100	100	100	100	100	100	99.95
Pasture	96.23	92.73	97.84	98.65	97.64	98.78	95.45	97.72
Crops	89.94	92.62	92.39	93.23	95.14	93.22	94.76	92.51
Burned area	100	100	100	100	100	100	100	100
Other area	100	100	100	100	100	100	100	100

Table 5. Confusion matrix between classification and field data for the year 2020.

2020	Earthly Truth								
	Water	Forest	Cerrado	Pasture	Crops	Burned Area	Other Area	Total	UA (%)
Water	**42**	0	0	0	0	0	0	42	100.00
Forest	0	**346**	7	1	1	0	0	355	97.46
Cerrado	0	2	**59**	0	0	0	0	61	96.72
Pasture	0	0	7	**545**	33	0	0	585	93.16
Crops	0	2	4	75	**289**	0	24	394	73.35
Burned area	0	0	0	0	0	**0**	0	0	-
Other area	0	0	0	0	0	0	**40**	40	100.00
Total	42	350	77	621	323	0	64	1477	
PA (%)	100.00	98.86	76.62	87.76	89.47	-	62.50		
Overall accuracy (%):	89.44			Kappa Index:		0.85			

PA = producer accuracy; UA = user accuracy.

The overall accuracy of the classifications based on Google Earth and Landsat data varied between 97.97 and 98.73%, while the kappa index values were between 0.96 and 0.97 (Table 4), classified as excellent according to the evaluation proposed by [49]. The general accuracy obtained for 2020, considering only the data collected in the field, was 89.44%, and the kappa index was 0.85 (Table 5), still considered excellent for being in the range of 0.80 and 1 [49]. The agreement between the mapping generated in this study and that of the MapBiomas project also had a satisfactory result, with overall accuracy between 91.60 and 96.56%, and a kappa index between 0.83 and 0.94 (Tables A1 and A2).

In general, producer accuracy was superior for the classifications for natural areas, such as water and forest as well as for burned areas, indicating that they are more likely to be correctly classified. The user accuracy was higher for the classes for water, the Cerrado biome, burned areas, and other areas, indicating higher probability of the classified areas actually representing these categories in the field. Table 6 presents the areas for land use and occupation obtained from these classifications, in addition to those for crops and pasture.

Changes over time for these classifications are shown in Figure 4 (1986, 1991, 2005, and 2020) and Figure A2 (1996, 2000, 2011, and 2015), respectively.

Table 6. Areas corresponding to the classes of land use and occupation in Brazil's Teles Pires River basin mapped between 1986 and 2020.

Land Use Classes		1986	1991	1996	2000	2005	2011	2015	2020	Percent Change 1986–2020
Water	km^2	686.6	762.6	747.1	718.8	724.7	737.2	924.8	1159.1	
	% of total	0.49	0.54	0.53	0.51	0.51	0.52	0.65	0.82	68.82
	% change	-	11.07	−2.03	−3.79	0.82	1.73	25.45	25.34	
Forest	km^2	114,444.0	111,470.4	100,091.7	96,777.9	83,218.4	80,710.7	79,330.5	78,979.0	
	% of total	80.87	78.77	70.73	68.39	58.81	57.03	56.06	55.81	−30.99
	% change	-	−2.60	−10.21	−3.31	−14.01	−3.01	−1.71	−0.44	
Cerrado	km^2	13,069.5	11,923.2	11,416.0	8381.7	7927.5	5805.7	5196.4	5356.1	
	% of total	9.24	8.43	8.07	5.92	5.60	4.10	3.67	3.78	−59.02
	% change	-	−8.77	−4.25	−26.58	−5.42	−26.77	−10.49	3.07	
Pasture	km^2	8839.7	11,658.0	20,248.2	23,533.1	32,385.7	34,423.2	34,643.2	30,988.6	
	% of total	6.25	8.24	14.31	16.63	22.89	24.33	24.48	21.90	250.56
	% change	-	31.88	73.69	16.22	37.62	6.29	0.64	−10.55	
Crops	km^2	3244.0	4098.7	8150.7	11,194.6	16,177.6	19,201.6	20,787.8	24,105.9	
	% of total	2.29	2.90	5.76	7.91	11.43	13.57	14.69	17.03	643.09
	% change	-	26.35	98.86	37.35	44.51	18.69	8.26	15.96	
Burned area	km^2	708.4	902.0	509.9	406.2	412.3	145.6	125.9	163.8	
	% of total	0.50	0.64	0.36	0.29	0.29	0.10	0.09	0.12	−76.87
	% change	-	27.33	−43.47	−20.34	1.50	−64.69	−13.53	30.10	
Other area	km^2	522.8	699.4	350.0	502.5	667.2	489.4	505.1	757.9	
	%	0.37	0.49	0.25	0.36	0.47	0.35	0.36	0.54	44.98
	% change	-	33.78	−49.96	43.57	32.78	−26.65	3.21	50.05	

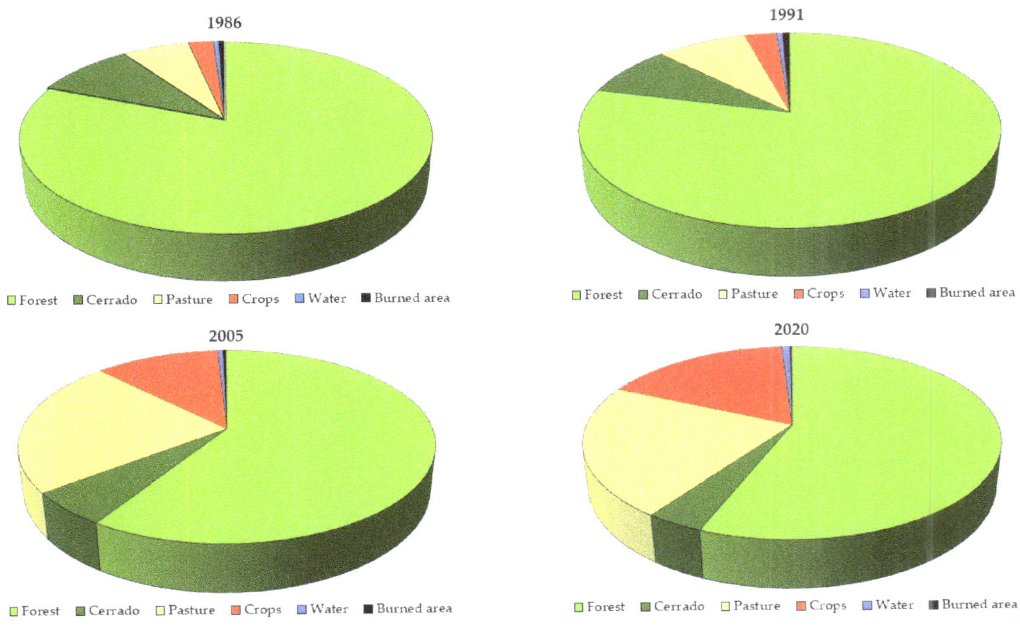

Figure 4. Change in land use classification percentages for the Teles Pires River basin for 1986, 1991, 2005, and 2020.

Our land use classifications when compared to the ground truth data we collected had a relatively lower accuracy (289/394 = 73.35%) for crops compared to other land use categories, which all had >93% accuracy (Table 5). The times that we misclassified crops as something else was highest for the pasture (75), other area (24), Cerrado (4), and forest (2), land use classes (Table 5). Our land use classifications using Landsat data in 2020 for crops was (3280/3463 = 94.72%) (Table A2). Misclassification as something other than crops were for other areas (143), and pasture (38). Misclassification of crops with some other land use ranged from 89.39% in 1986 (Table A1) to 95.14% in 2005 (Table A2).

Our results indicated growth of areas occupied by agricultural areas (i.e., pasture and crops), which led to the reduction of native vegetation (i.e., forest and Cerrado savannah) with recent stabilization to ~60% native vegetation and ~40% crops for the Teles Pires River basin (Figure 5a). Figure 5b presents the changes between the time periods evaluated. The percent growth in agricultural areas increased drastically by 80% from 1991 to 1996, and then by 40% from 2000 to 2005, while more recently declining by −0.6% from 2015 to 2020. This recent minor decline in total agricultural area is driven by the −10.55% drop in pasture, which takes up more land area than crops which actually increased from 2015 to 2020 (Table 6). The percentage reduction in native vegetation has been less drastic, with greater declines from 1991 to 1996 (−9.6%) and from 2000 to 2005 (−13.3%), with stabilization to around zero with −0.2% growth from 2015 to 2020 (Figure 5b). This has predominantly been driven by reductions in forest, since the Cerrado increased slightly by 3.05% from 2015 to 2020 (Table 6). Between 2015 and 2020, both declines in the increase in agricultural areas and the decrease in native vegetation to near 0% (Figure 5b) does not mean that there was no expansion of agriculture or deforestation. Rather this indicates a relative stabilization of the recent rates of agricultural expansion and deforestation.

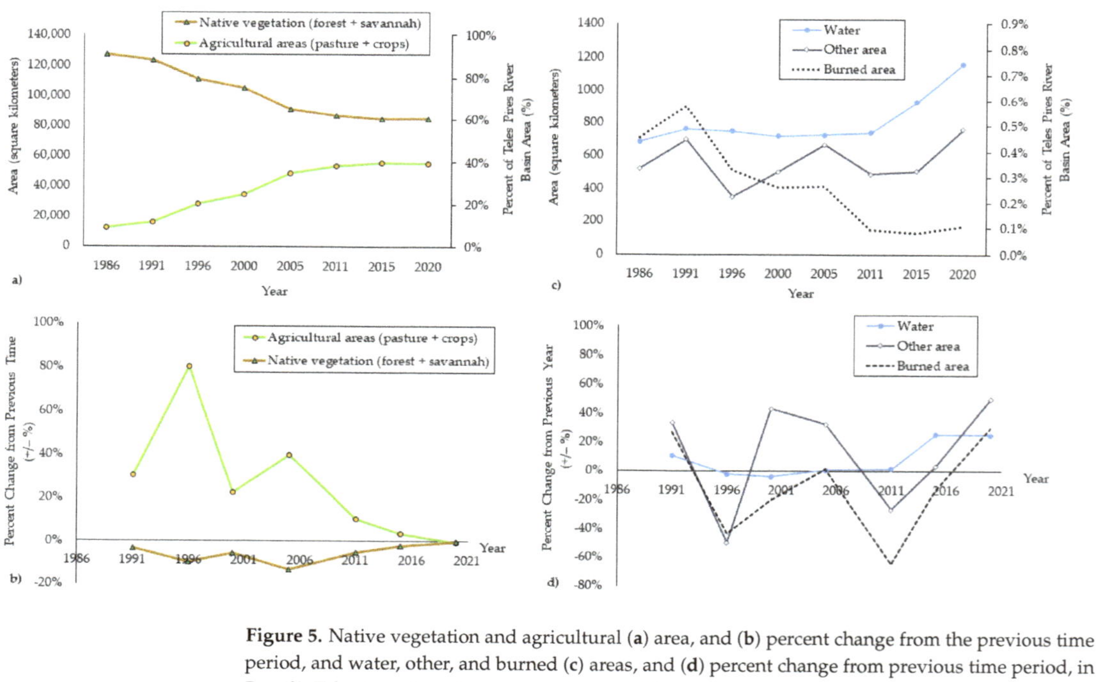

Figure 5. Native vegetation and agricultural (**a**) area, and (**b**) percent change from the previous time period, and water, other, and burned (**c**) areas, and (**d**) percent change from previous time period, in Brazil's Teles Pires River basin from 1986 to 2020.

The regionalization of land use and occupation show differences in the occupation patterns of the different parts of the Teles Pires River basin (Figure 6). Forest was the predominant class in the basin in both years, but showed great variation of the area in the period, from 114,400 to 78,900 km^2. More than 80% of the entire mapped forest was located in the regions of the lower and middle Teles Pires, which in 1986 had forest areas corresponding to 49,200 and 49,300 km^2, respectively. The lower Teles Pires had the smallest decline in forest areas, still showing about 43,300 km^2 in 2020. In the upper Teles Pires, forests share the space with areas of the Cerrado, which also corresponds to the indigenous environment of this sub-region.

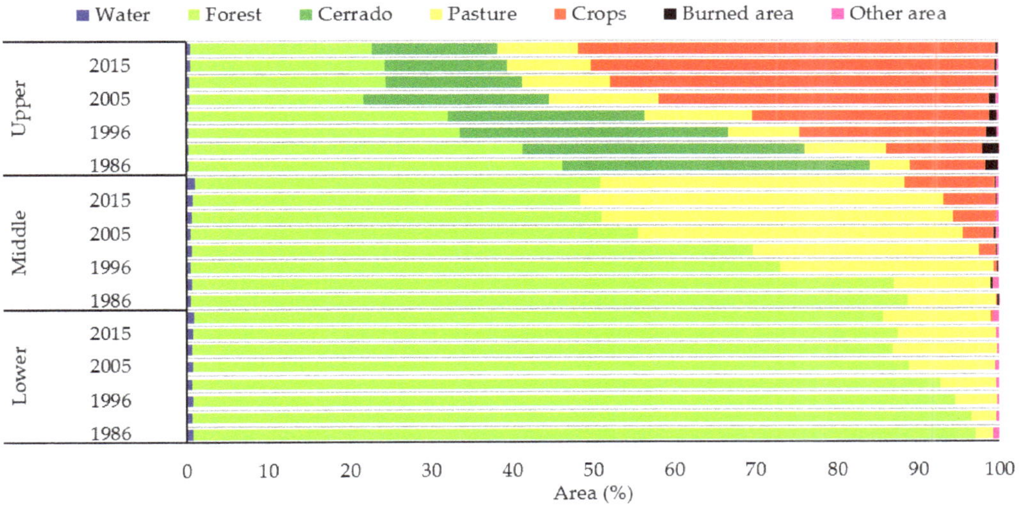

Figure 6. Evolution of land use in Brazil's upper, middle, and lower Teles Pires regions in percent (%) of areas occupied per class.

In the upper Teles Pires, forest areas showed variation between 15,700 and 7700 square kilometers (km^2) of occupied area during the period, while the Cerrado areas varied between 13,000 and 5300 km^2, the latter represented the class with the greatest percentage reduction in our study. Considering the forest and Cerrado classes together, in the upper Teles Pires, there was an area variation from 28,800 to 13,000 km^2 between 1986 and 2020, corresponding to 84% and 38% of the region, respectively. This was the part of the Teles Pires River basin with the lowest percentage of natural vegetation in 2020.

Crops was the land use class with the greatest growth in the period. In 1986, it occupied about 2% of the basin and increased to about 16% in 2020, from 3200 to 24,100 km^2 of the area. Most of the area mapped as agriculture is concentrated in the upper Teles Pires, a region that in 2020 was more than 70% agriculture. Crops already occupy approximately 50% of the upper Teles Pires. In the middle Teles Pires, crops have increased, occupying about 10% of the region in 2020, whereas in the lower Teles Pires crops are less than 0.5% of the land area (Figure 6). Pasture represents the predominant form of agricultural land use in the basin, covering more than 8000 km^2 in 1986, corresponding to 6% of the total area of the basin. Pastures occupied 31,000 km^2 in 2020, 22% of the land area, decreasing from the peak values of 32,385.7 and 34,643.2 km^2 that it occupied between 2005 and 2015 (Table 6). It is in the middle Teles Pires that most of the pastures mapped are concentrated, with more than 20,000 km^2 of pasture areas in 2020, equivalent to 37% of the land area in this region.

Water also increased between 2011 and 2020, from 737 to 1159 square kilometers, a 57% growth in surface area over this 9-year period (Figure 5c). The percentage increase from one time period to the next was more variable for other area and burned area (Figure 5d). Formed by the junction between areas of minority territorial occupation, the classification of other area also showed an increase, primarily from the growth of urban areas in 23 municipalities within the basin. Mining areas were mostly located on the banks of large tributaries of the Teles Pires River, especially in the middle portion of the basin. Burned areas could not be analyzed. Mapping burned areas requires consideration of the temporal distribution of wildfires, using images with dates that correspond to the end of the wildfire season taking place during the dry season, with 80 to 90% of fires occurring between the months of June and October [50].

The area conversions that occurred between land use classes were detected using a method for comparing different maps, based on the classifications of 1986 and 2020, and then the area conversion matrix presented in Table 7 was generated. This made it possible to identify the losses experienced by each of the classes, as well as the allocation classes of these areas. In the matrix, the values in the diagonal cells indicate the area that remained in the same class of land use and occupation between 1986 and 2020, while the other values indicate the changes that occurred in the area. The rows in Table 7 identify the losses for each land use class, while the columns correspond to the gains and their origin.

Table 7. Matrix of conversion of areas in square kilometers (km^2) for land use classifications in the Teles Pires River basin between 1986 and 2020.

		Columns = Gains for Each Class between 1986 and 2020								
	Class	**Water**	**Forest**	**Cerrado**	**Pasture**	**Crops**	**Burned Area**	**Other Area**	**Total**	**Losses (1986–2020)**
Rows = Losses for each class between 1986 and 2020	Water	**523.50**	145.76	4.04	4.29	2.00	0.82	6.11	686.52	163.03
	Forest	537.86	**76,550.19**	925.63	23,232.59	12,694.87	61.91	421.31	114,424.40	37,874.17
	Cerrado	26.89	1012.66	**4011.00**	2081.09	5822.54	87.63	20.17	13,061.98	9050.97
	Pasture	40.20	1132.88	195.12	**5055.31**	2296.68	8.63	108.04	8836.86	3781.55
	Crops	4.57	21.85	131.03	245.18	**2812.59**	2.99	24.41	3242.62	430.04
	Burned area	10.49	33.88	85.85	167.17	404.29	**1.55**	4.93	708.16	706.61
	Other area	15.56	74.18	1.10	199.58	59.03	0.30	**172.71**	522.46	349.76
	Total	1159.07	78,971.40	5353.77	30,985.21	24,092.00	163.83	757.68		
	Gains (1986–2020)	635.57	2421.21	1342.77	25,929.90	21,279.41	162.28	584.98		

Bold numbers on the diagonal indicate the area in each class that was maintained in 1986 and 2020.

The conversion matrix makes it evident that the greatest losses of Cerrado and forest areas resulted from their conversion to crops and pasture. The forest had the highest loss of 37,900 km^2, where 23,200 km^2 was converted to pasture, and 12,700 km^2 transitioned to crops. Forest losses were also recorded due to the conversion to water. The Cerrado had the highest proportion of losses compared to the pre-existing area, with 69% being converted to other uses, representing 9000 km^2 of the lost area, of which 5800 km^2 was converted to crops and 2100 km^2 was opened to pasture. Losses of pasture area were also recorded during this period (1986 to 2020), and these were mainly due to the conversion to crops. As the main area conversions identified in the basin were from forest to crops (F/CP), forest to pasture (F/P), Cerrado to crops (C/CP), Cerrado to pasture (C/P), and pasture to crops (P/CP), these conversions were compared for all the yearly intervals mapped (Figure 7).

Forest areas showed greater conversion to pasture for all the evaluated time periods, with higher values recorded between 2000 and 2005, when more than 10,000 km^2 was converted. This period was also responsible for the greatest conversion of forest into crops. The replacement of the Cerrado areas with pasture and crops varied between the periods. From 1986 to 1991 and from 2000 to 2015, the conversion of Cerrado to pasture was higher, whereas from 1991 to 2000 the conversion of Cerrado to crops was higher, with the peak conversion recorded between 1991 and 1996. The dynamics of pasture conversion to crops was characterized by progressive growth, representing the most significant type of conversion in more recent periods.

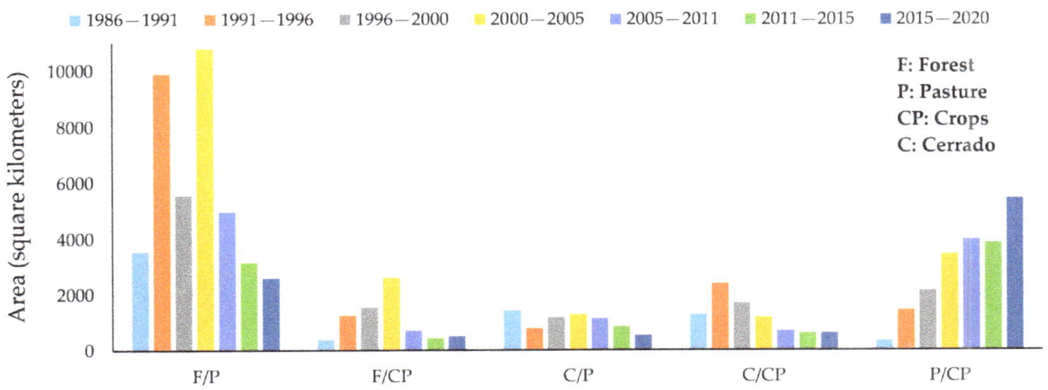

Figure 7. Conversion of areas between forest (F), pasture (P), crops (CP), and Cerrado (C) land use classes between the years mapped in the Teles Pires River basin.

Figure 8 shows the values of the conversion of pasture to crops over the years, mapped for all three sub-regions of the Teles Pires River basin. In the upper Teles Pires, this form of conversion from pasture to crops increased, peaking between 2000 and 2005, when about 53% of the pastures in the region were converted to crops. This type of land use conversion subsequently decreased. In the middle Teles Pires, the conversion of pastures to crops has increased, with a conversion of 3900 km^2 recorded between 2015 and 2020, representing 16% of the pre-existing pastures in 2015. In the lower Teles Pires, this type of conversion is still a recent phenomenon.

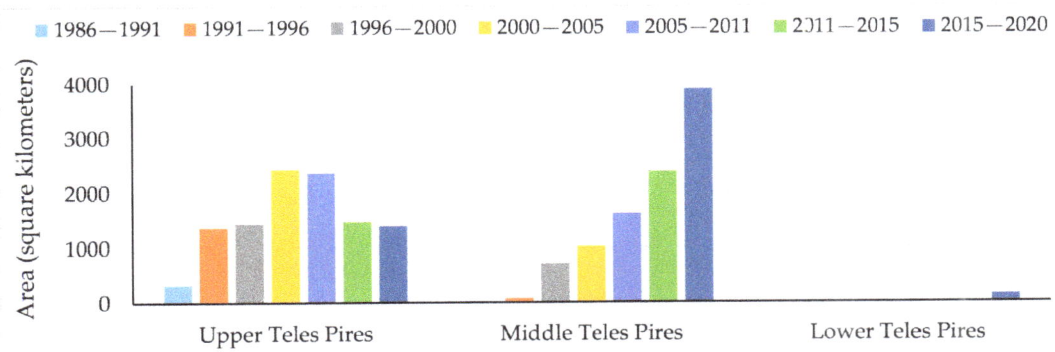

Figure 8. Pasture areas converted to crops in the different regions of the Teles Pires River basin.

5. Discussion

5.1. Potential Misclassifications and Comparisons to Previous Studies

The occupation of states belonging to the Legal Amazon was historically marked by policies to encourage land occupation [51]. This occurred through a colonization model that led to disorderly occupation of the region, contributing to high rates of deforestation [52]. In our study of the Teles Pires River basin, which is located in the states of Mato Grosso and Pará, our results indicate a significant reduction in natural areas between 1986 and 2020, a period marked by intense occupation of the region. One of the greatest challenges encountered during the study is the presence of Cerrado and transition areas, which results in difficulties in the mapping of these areas due to the different plant physiognomies and the natural dynamics shown by this classification, leading to varied spectral responses, which hinders the correct mapping of these features [53]. This is demonstrated by the omission of significant areas of the Cerrado, erroneously classified as forest.

Another class that also had areas omitted during the classification of images was the other land use class, which was confused with crops and pasture. The other class encompassed categories of territorial occupation that were spatially limited and had mostly heterogeneous spectral response, such as urban and mining areas. The maximum likelihood classifier assumes that the training data are representative and normally distributed, but highly heterogeneous response classes do not have this type of data distribution, leading to low efficacy of the classification. This is one of the main limitations of the classification method we adopted [54].

When comparing the data obtained in the present study with the areas mapped by the MapBiomas project's collection 4, for the Teles Pires River basin area, similar classes were identified and some of these showed very similar results, corroborating our results. Considering the forest cover classes of MapBiomas, values between 116,100 and 79,200 km^2 were found for the basin area for 1986 and 2015, respectively. Values very close to those found in our study were also documented by MapBiomas for the years 1991, 1996, 2011, and 2015. The areas of crops, represented in the MapBiomas by three categories ((1) Pasture, (2) Annual crops, and (3) Perennial crops), ranged from 3 to 14% between 1986 and 2015, and the percentages of pasture varied between 6 and 24% between 1986 and 2015, respectively. The areas occupied by water in the Teles Pires River basin, according to MapBiomas data, were between 0.5% and 0.6% between 1986 and 2015. These values are consistent with our results.

In addition to pointing out the accelerated decline of the natural areas in the basin, our research shows that most of the area losses have been occurring in the upper and middle regions of the Teles Pires, which are most impacted by human occupation. Meanwhile, the lower Teles Pires still conserves much of its natural areas and stands out for covering protected areas corresponding to portions of the Juruena National Park, and Munduruku and Kayabi indigenous lands. This highlights the role of protected areas in curbing deforestation in the Legal Amazon [55,56]. The greatest losses of natural area were recorded in the upper Teles Pires, giving the region the status of strongly altered area, due to the high percentage of the area that is already suffering from deforestation [57].

The Teles Pires River basin has more significant losses of natural areas to the south, upper Teles Pires, which evolve in the north direction, toward the middle and lower Teles Pires. Other researchers [58] have already reported the occurrence of this phenomenon in the region where the changes caused by deforestation expand from south to north in the state of Mato Grosso, destroying areas of primary vegetation for agricultural production in the central areas of the state and for livestock farming to the north.

Crops currently represents the predominant land use in the upper Teles Pires. This is consistent with [16], who found that 47.8% of the land area in the region was for agricultural use in 2014. The middle part of the basin is characterized by wide anthropic occupation of the areas by pasture. Livestock farming was the primary driver of deforestation in the municipality of Paranaíta, with an area corresponding to 6445.05 km^2, located in the Teles Pires River basin between the middle and lower portions, with 32.73% of the municipality occupied by pasture in 2016 [52]. This result is similar to the pattern found in the middle Teles Pires region in the present study.

The significant growth of the areas occupied by water observed in the basin can be explained by the construction of four large dams for the installation of hydroelectric power plants (HPPs) in the course of the Teles Pires River in recent years. This has created lakes upstream of the Teles Pires, São Manoel, Colíder, and Sinop HPPs, built between 2011 and 2020. The installation of these HPPs results in an increase in water surfaces and generally also contributes to the reduction of native vegetation [59]. Native vegetation was suppressed in the Teles Pires River basin during the installation of the Teles Pires

HPP between the municipalities of Paranaíta, Mato Grosso, and Jacareacanga, Pará, with negative impacts such as deforestation and fragmentation of the areas [52].

The construction of dams in the Amazon generates impacts that can spread on a local, regional, and global scale. Dams modify the natural flow of the watercourse, changing upstream environments from lotic to lentic, which affects water quality and the transport of sediments and nutrients [60]. These changes can reduce local biodiversity, being especially harmful to migratory and endemic aquatic species, and favoring generalist species [61]. The impacts of dams go far beyond directly influencing rivers. Sometimes dams displaced people to make way for the development and this can increase deforestation from the opening of new roads [62]. Dams with large hydroelectric plants located in tropical regions can also generate significant emissions of greenhouse gases, such as carbon dioxide and methane [63].

As for the conversion of areas in the basin, there is a clear, intense replacement of natural, Cerrado and forest areas with agricultural and pasture areas, which have been pointed out as the main drivers of deforestation in the region for a long time, both by direct conversion and indirect conversion through the displacement of the forms of land use [64–66]. The Cerrado, identified in this study as the class with the highest proportion of area losses, has had its destruction documented for years [67,68]. The Cerrado is the most threatened biome in Brazil, with deforestation rates higher than those of the Amazon [69,70], due to more scarce protection policies [70]. In addition to the high losses of natural areas, another evident phenomenon of the conversion in the Teles Pires River basin is the conversion of pastures, resulting from their transformation into agricultural areas. This occurrence in the state of Mato Grosso has also been reported by other researchers [69,71].

Our results suggest that more ground truthing is needed to confirm whether the misclassification of crops is as prevalent in other areas in Brazil outside of the Teles Pires River basin. Improved ground truthing using real-world data collected in the field is clearly needed since our land use classifications using Landsat data from 2020 had a much greater classification accuracy for crops at 94.72% when compared to MapBiomas (Table A2), versus only 73.35% accuracy when validating against ground truth data collected in the field (Table 5). While MapBiomas is annually produced, other land use data sets, such as TerraClass which is released biennially, may be able to distinguish more accurately between forest and successional stages of forest regrowth, as well as between agriculture and pasture [20]. Thus, TerraClass could be used for future validations.

5.2. *Agricultural Development Policies and Future Sustainable Intensification*

In the Brazilian Legal Amazon, deforestation rates have followed the political and economic scenario, and the results found here reflect this pattern. For instance, high conversion of forest areas were recorded between 1991 and 1996, which may have been associated with high deforestation rates recorded in the Amazon after the implementation of the Real Plan in 1994 [72]. Peak deforestation occurred between 2000 and 2005, which may be associated with the increase in prices of agricultural commodities, especially soybean [73]. In this context, policies to reduce deforestation, such as changes made to the Forest Code [74], also stand out, and more recently deforestation rates of the Legal Amazon have slowed due to policies to combat deforestation [65].

In the Teles Pires River basin, high values of forest conversion to crops were recorded between 1991 and 2005, with a peak recorded in the last five years of this period. Between 2001 and 2006, soybean plantations expanded in the Amazon and record deforestation rates were found, with the occurrence of direct conversion of forests to agricultural production [68]. The subsequent reduction in this conversion is associated with the Soy Moratorium, an agreement signed aimed at reducing deforestation caused by the expansion of soybeans in the Brazilian Amazon. Launched in 2006, the Soy Moratorium involved civil society organizations and companies linked to the soybean industry committing to not buying soybean grown on deforested land after July 2006 [75,76]. This triggered the expansion of soybean production to pasture areas, which originated from previously deforested areas [70]. Our results show that in the Teles Pires River basin, the conversion of pastures to crops increased, especially in the middle part of the basin since the mid-1990s, with higher conversion rates recorded after 2005, corresponding to the period following the adoption of the Soy Moratorium.

It is more common for agricultural production to expand through the conversion of pastures, with the establishment of these pastures to newly deforested areas [76]. Thus, agricultural expansion of Brazil's beef industry, historically used for holding claim to land, has indirectly caused deforestation [68]. This has occurred in the Teles Pires River basin, where pastures continue to expand through the occupation of natural areas. When analyzing changes in land use and occupation between 1986 and 2014 in the upper Teles Pires, [16] reported this pattern of replacement of natural areas first with pastures and then with crops, which corroborates the occurrence of the high values of pasture conversion to crops found in our study. Such pasture to crops conversions may be even greater, since the interval between mapped years may have been insufficient to portray such dynamics. It is worth pointing out that in this region, commodity cropping is already the dominant land use, covering about half of the region's total area in 2020, while pasture areas have declined. Based on the high percentages of pasture conversion to crops found in the upper Teles Pires region, this type of conversion tends to expand in the north direction of the basin, to the middle and lower Teles Pires regions, where there is a large number of pasture areas available for conversion.

The prevalence of conversions of areas of natural vegetation to pasture and agriculture can cause changes in soil cover resulting in potential negative environmental impacts, such as soil degradation and changes to the physical properties of the soil [69,77,78], as well as changes in water availability and quality [56,79,80]. Therefore, the need to evaluate the occurrence of such impacts in the Teles Pires River basin is evident. Improved organization of production and optimization of the use of the natural resources can improve economic development of the Mato Grosso state in Brazil, while adopting better management practices to help conserve river basins. These practices include sustainable intensification strategies for Brazil's beef and commodity crop industries to increase production on an agricultural land base that has stabilized (Figure 5a). Mato Grosso's beef production can potentially increase on the same pasture area by supplementing pasture with grain, reseeding degraded pastures [81], integrating cattle with crops [82], and reducing the time to slaughter [83]. Increasing commodity crop productivity can be accomplished by better hybrid development, especially for maize versus soybeans [84], as well as irrigation during the dry season, to allow for three cropping seasons per year compared to the current two seasons [85].

5.3. Policy Implications

Since deforestation is the main threat to biodiversity, mapping and quantifying changes in land use and occupation is necessary for understanding the dynamics of the landscape to adequately manage development through improved decision-making [18]. Thus, remotely sensed satellite data and maps developed over time can help public policy makers identify critical locations and potential environmental fragility and where efforts to contain deforestation should be prioritized [86]. Only through efficient land governance will it be possible to reduce deforestation in areas such as the Amazon [87]. Such governance must encompass rules and processes that inform decision makers about land use and control, how decisions are implemented, and how antagonistic interests in land use can be resolved. It is worth highlighting the importance of understanding the link between deforestation and global food supply chains in order to create better regulatory policies to protect tropical forests in biodiversity hotspots [36]. The conservation of tropical forests requires comprehensive and long-term solutions, understanding potential socio-ecological trade-offs, and ensuring a balance between land use, environmental goals, and sustainable development [38].

6. Conclusions

Agriculture (e.g., commodity crops) has been the classification with greatest growth in the Teles Pires River basin and, despite the current policies to curb deforestation, it continues to expand by incorporating anthropic areas already consolidated, through the conversion of pasture areas, which has led to their displacement to new areas, maintaining the continuity of deforestation in the basin, even if at lower rates. High values of direct conversion of natural areas to crops were recorded until the year 2005, from which this type of conversion decreased and the growth of agricultural areas through the conversion of pastures began to prevail. In recent years, the middle Teles Pires has stood out for having high occurrences of this type of conversion.

The changes that occurred in the Teles Pires River basin may be related to the expansion of areas for crops and pasture, while the form and intensity of area conversions between the analyzed years accompanied regional trends, denoting the strong influence of Brazil's economic and political dynamics. The different sub-regions of the basin have experienced different stages of land conversion, with the upper Teles Pires toward the south of the basin, showing a higher degree of anthropic alteration. This conversion has expanded northwards over time. The northernmost part of the basin, the lower Teles Pires, continues to have the highest percentage of natural areas. Current land use is characterized by commodity agriculture in the upper Teles Pires, pasture in the middle Teles Pires, and native forest in the lower Teles Pires. The construction of infrastructure for implementing hydroelectric projects along the course of the Teles Pires River in the last decade has also contributed to the reduction of natural areas in this river basin.

Author Contributions: Data collection, writing, methodology, formal analysis, figures—A.K., F.T.d.A. and T.M.d.C.; data collection, review, editing, supervision and financial support—A.P.d.S.; methodology, review, editing—C.A.Z.; review, editing and financial support, figures—A.K.H. and D.C.d.A. All authors have read and agreed to the published version of the manuscript.

Funding: This study was financed by the Coordenação de Aperfeiçoamento de Pessoal de Nível Superior—Brasil (CAPES) and the Agência Nacional de Águas e Saneamento Básico (ANA), Finance Code—001 and Process 88887.144957/2017-00. The authors wish to thank the Conselho Nacional de Desenvolvimento Científico e Tecnológico (CNPq) for their support with scientific initiation grants and a productivity grant (Process 308784/2019-7).

Institutional Review Board Statement: Not applicable.

Informed Consent Statement: Not applicable.

Data Availability Statement: Study data can be obtained by request to the corresponding author or the second author, via email. It is not available on the website as the research project is still under development.

Acknowledgments: The authors also thank all the students and professors in the Tecnologia em Recursos Hídricos no Centro-Oeste" research group (Available online: dgp.cnpq.br/dgp/espelhogrupo/2399343537529589 accessed on 3 March 2023). We thank three anonymous reviewers whose comments and edits substantially improved the quality of this work.

Conflicts of Interest: The authors declare no conflict of interest. Supporting entities had no role in the design of the study; in the collection, analyses, or interpretation of data; in the writing of the manuscript, or in the decision to publish the results.

Appendix A

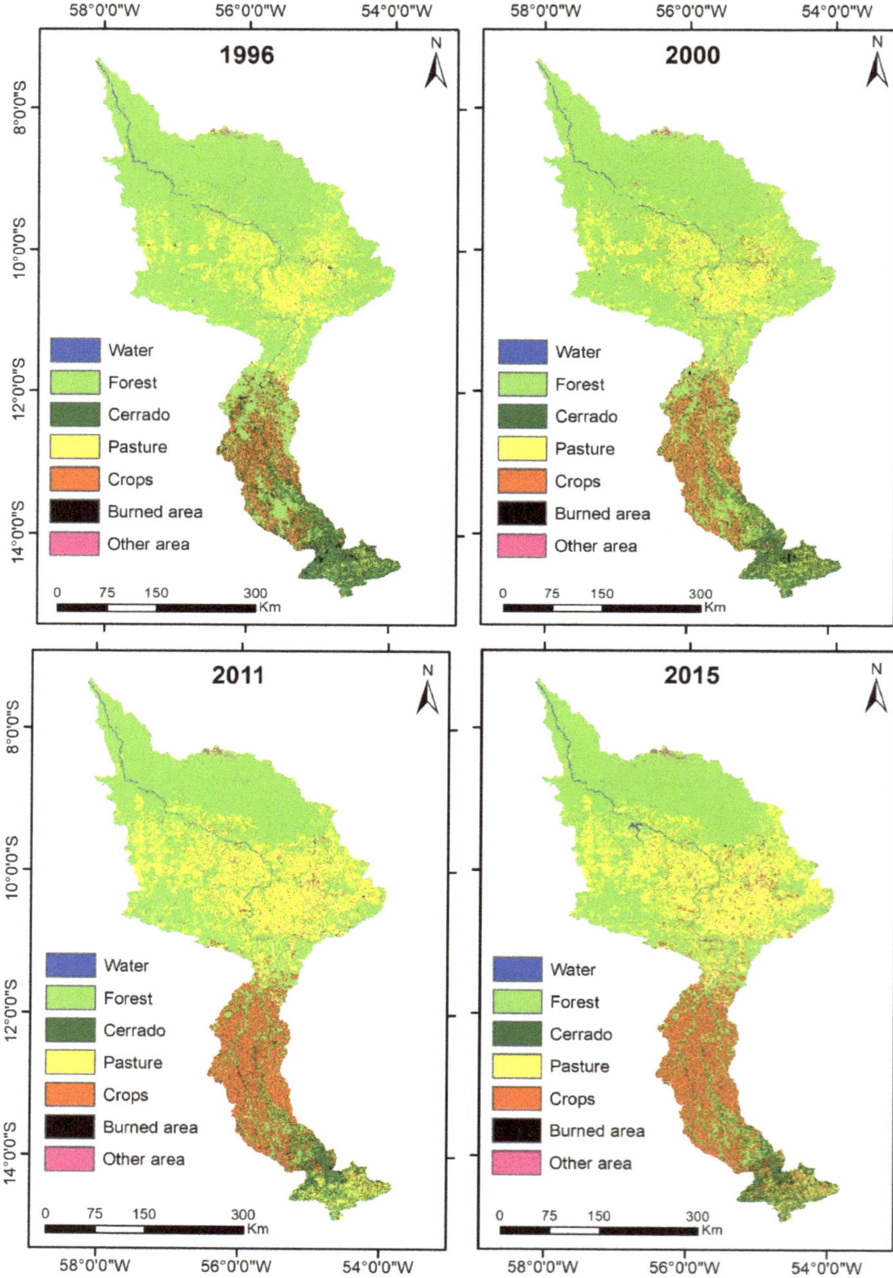

Figure A1. Maps of land use in the Teles Pires River basin during 1996, 2000, 2011, and 2015.

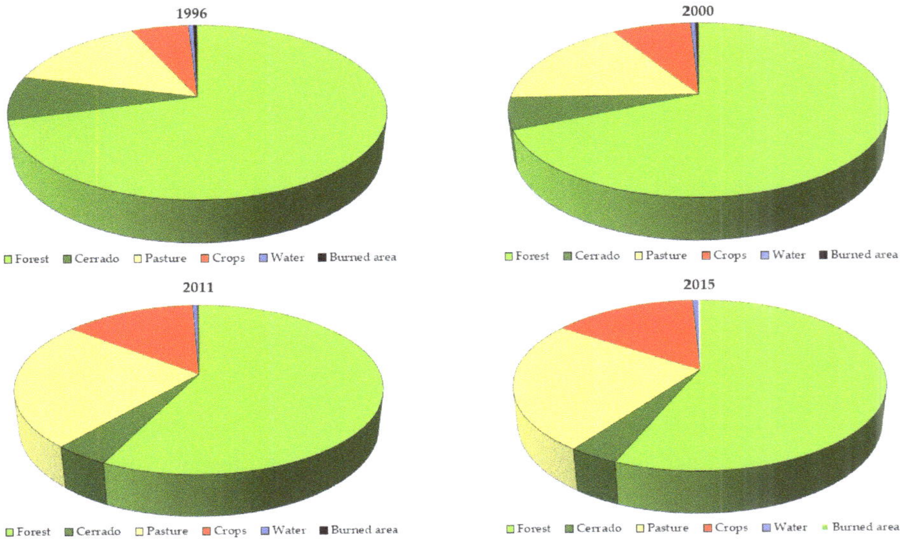

Figure A2. Change in land-use classification percentages for Teles Pires River basin for 1996, 2000, 2011, and 2015.

Table A1. Confusion matrix between the mapping in this article and the MapBiomas project for the years 1986, 1991, 1996, and 2000.

	MapBiomas							
1986	Water	Forest	Cerrado	Pasture	Crops	Other area	Total	UA (%)
Water	**1419**	0	0	0	0	0	1419	100.00
Forest	0	**20,023**	12	0	0	0	20,035	99.94
Cerrado	0	1281	**2199**	0	0	0	3480	63.19
Pasture	0	0	3	**1455**	8	44	1510	96.36
Crops	0	0	0	26	**1475**	149	1650	89.39
Other area	21	0	22	159	28	**373**	603	61.86
Total	1440	21,304	2236	1640	1511	566	28,697	
PA (%)	98.54	93.99	98.35	88.72	97.62	65.90		
Overall accuracy (%):	93.89			Kappa Index:		0.87		
1991	Water	Forest	Cerrado	Pasture	Crops	Other area	Total	UA (%)
Water	**1870**	1	0	0	0	0	1871	99.95
Forest	0	**18,678**	5	0	0	0	18,683	99.97
Cerrado	0	1120	**2461**	0	0	0	3581	68.72
Pasture	0	0	0	**3087**	216	12	3315	93.12
Crops	0	0	0	45	**2939**	158	3142	93.54
Other area	9	0	42	433	12	**335**	831	40.31
Total	1879	19,799	2508	3565	3167	505	31,423	
PA (%)	99.52	94.34	98.13	86.59	92.80	66.34		
Overall accuracy (%):	93.47			Kappa Index:		0.89		
1996	Water	Forest	Cerrado	Pasture	Crops	Other area	Total	UA (%)
Water	**1306**	0	0	0	0	0	1306	100.00
Forest	0	**21,211**	1	26	0	0	21,238	99.87
Cerrado	0	2229	**1491**	0	0	0	3720	40.08
Pasture	0	1	0	**4030**	17	71	4119	97.84
Crops	0	0	0	32	**831**	31	894	92.95
Other area	0	0	34	165	80	**443**	722	61.36
Total	1306	23,441	1526	4253	928	545	31,999	
PA (%)	100.00	90.49	97.71	94.76	89.55	81.28		
Overall accuracy (%):	91.60			Kappa Index:		0.83		

Table A1. *Cont.*

	MapBiomas							
2000	Water	Forest	Cerrado	Pasture	Crops	Other area	Total	UA (%)
Water	**2049**	1	0	0	0	0	2050	99.95
Forest	0	**19,774**	9	0	0	0	19,783	99.95
Cerrado	0	1320	**2355**	0	0	0	3675	64.08
Pasture	0	0	0	**4791**	64	27	4882	98.14
Crops	0	0	9	113	**4121**	176	4419	93.26
Other area	14	5	0	85	22	**401**	527	76.09
Total	2063	21,100	2373	4989	4207	604	35,336	
PA (%)	99.32	93.72	99.24	96.03	97.96	66.39		
Overall accuracy (%):	94.78			Kappa Index:		0.92		

PA = producer accuracy; UA = user accuracy.

Table A2. Confusion matrix between the mapping in this article and the MapBiomas project for the years 2005, 2011, 2015, and 2020.

	MapBiomas							
2005	Water	Forest	Cerrado	Pasture	Crops	Other area	Total	UA (%)
Water	**737**	0	0	0	0	0	737	100.00
Forest	0	**20,805**	77	0	0	0	20,882	99.63
Cerrado	0	1302	**1422**	1	0	0	2725	52.18
Pasture	0	0	0	**4110**	117	11	4238	96.98
Crops	0	0	0	72	**3931**	129	4132	95.14
Other area	14	0	0	38	34	**445**	531	83.80
Total	751	22,107	1499	4221	4082	585	33,245	
PA (%)	98.14	94.11	94.86	97.37	96.30	76.07		
Overall accuracy (%):	94.60			Kappa Index:		0.90		
2011	Water	Forest	Cerrado	Pasture	Crops	Other area	Total	UA (%)
Water	**803**	0	0	0	0	0	803	100.00
Forest	0	**23,060**	107	0	0	0	23,167	99.54
Cerrado	0	904	**2519**	0	0	0	3423	73.59
Pasture	0	2	214	**4157**	46	6	4425	93.94
Crops	0	0	0	57	**3081**	167	3305	93.22
Other area	17	2	0	43	10	**436**	508	85.83
Total	820	23,968	2840	4257	3137	609	35,631	
PA (%)	97.93	96.21	88.70	97.65	98.21	71.59		
Overall accuracy (%):	95.58			Kappa Index:		0.92		
2015	Water	Forest	Cerrado	Pasture	Crops	Other area	Total	UA (%)
Water	**1160**	0	0	0	0	0	1160	100.00
Forest	0	**19,801**	55	0	0	0	19,856	99.72
Cerrado	0	1116	**1668**	0	0	11	2795	59.68
Pasture	0	0	8	**3720**	138	44	3910	95.14
Crops	0	0	0	77	**4242**	146	4465	95.01
Other area	10	0	15	50	51	**468**	594	78.79
Total	1170	20,917	1746	3847	4431	669	32,780	
PA (%)	99.15	94.66	95.53	96.70	95.73	69.96		
Overall accuracy (%):	94.75			Kappa Index:		0.91		
2020	Water	Forest	Cerrado	Pasture	Crops	Other area	Total	UA (%)
Water	**1541**	0	0	0	0	0	1541	100.00
Forest	0	**19,907**	100	1	0	0	20,008	99.50
Cerrado	0	144	**3917**	205	0	0	4266	91.82
Pasture	0	0	4	**3864**	470	0	4338	89.07
Crops	2	0	0	38	**3280**	143	3463	94.72
Other area	64	10	4	0	0	**737**	815	90.43
Total	1607	20,061	4025	4108	3750	880	34,431	
PA (%)	95.89	99.23	97.32	94.06	87.47	83.75		
Overall accuracy (%):	96.56			Kappa Index:		0.94		

PA = producer accuracy; UA = user accuracy.

References

1. Verma, A.K. Sustainable Development and Environmental Ethics. *Int. J. Environ. Sci.* **2019**, *10*, 1–5. Available online: https://ssrn.com/abstract=3689046 (accessed on 1 July 2022).
2. Araújo Neto, J.R.; Andrade, E.M.; Palácio, H.A.Q.; Sales, M.M.; Maia, A.R.S. Influence of land use/occupation on water quality in the Trussu river valley, Ceará, Brazil. *Rev. Ciênc. Agron.* **2017**, *48*, 59–69. [CrossRef]
3. Shi, P.; Zhang, Y.; Li, Z.; Li, P.; Xu, G. Influence of land use and land cover patterns on seasonal water quality at multi-spatial scales. *Catena* **2017**, *151*, 182–190. [CrossRef]
4. McMillan, H.; Montanari, A.; Cudennec, C.; Savenije, H.; Kreibich, H.; Krueger, T.; Liu, J.; Mejia, A.; Van Loon, A.F.; Aksoy, H.; et al. Panta Rhei 2013–2015: Global perspectives on hydrology, society and change. *Hydrol. Sci. J.* **2016**, *61*, 1174–1191. [CrossRef]
5. Van Loon, A.F.; Rangecroft, S.; Coxon, G.; Naranjo, J.A.B.; Van Ogtrop, F.; Van Lanen, H.A.J. Using paired catchments to quantify the human influence on hydrological droughts. *Hydrol. Earth Syst. Sci.* **2019**, *23*, 1725–1739. [CrossRef]
6. Joly, C.A.; Scarano, F.R.; Bustamante, M.; Gadda, T.M.C.; Metzger, J.P.W.; Seixas, C.S.; Ometto, J.P.H.B.; Pires, A.P.F.; Boesing, A.L.; Sousa, F.D.R.; et al. Brazilian assessment on biodiversity and ecosystem services: Summary for policy makers. *Biota Neotrop.* **2019**, *19*, e20190865. [CrossRef]
7. Silva, C.M. Entre Fênix e Ceres: A grande aceleração e a fronteira agrícola no Cerrado. *Var. Hist.* **2018**, *34*, 409–444. [CrossRef]
8. Lathuillière, M.J.; Coe, M.T.; Johnson, M.S. A review of green- and blue-water resources and their trade-offs for future agricultural production in the Amazon Basin: What could irrigated agriculture mean for Amazonia? *Hydrol. Earth Syst. Sci.* **2016**, *20*, 2179–2194. [CrossRef]
9. Butler, D.; Ward, S.; Sweetapple, C.; Astaraie-Imani, M.; Diao, K.; Farmani, R.; Fu, G. Reliable, resilient and sustainable water management: The Safe & SuRe approach. *Glob. Chall.* **2016**, *1*, 63–77. [CrossRef]
10. Rogan, J.; Chen, D. Remote sensing technology for mapping and monitoring land-cover and land-use change. *Prog. Plann.* **2004**, *61*, 301–325. [CrossRef]
11. Liping, C.; Yujun, S.; Saeed, S. Monitoring and predicting land use and land cover changes using remote sensing and GIS techniques—A case study of a hilly area, Jiangle, China. *PloS ONE* **2018**, *13*, e0200493. [CrossRef]
12. Rwanga, S.S.; Ndambuki, J.M. Accuracy Assessment of Land Use/Land Cover Classification Using Remote Sensing and GIS. *Int. J. Geosci.* **2017**, *8*, 611–622. [CrossRef]
13. Mahmon, N.A.; Ya'acob, N.; Yusof, A.L. Differences of image classification techniques for land use and land cover classification. In Proceedings of the 2015 IEEE 11th International Colloquium on Signal Processing & Its Applications (CSPA), Kuala Lumpur, Malaysia, 6–8 March 2015; pp. 90–94. [CrossRef]
14. Murad, C.A.; Pearse, J. Landsat study of deforestation in the Amazon region of Colombia: Departments of Caquetá and Putumayo Remote Sens. *Appl. Soc. Environ.* **2018**, *11*, 161–171. [CrossRef]
15. Brovelli, M.A.; Sun, Y.; Yordanov, V. Monitoring Forest Change in the Amazon Using Multi-Temporal Remote Sensing Data and Machine Learning Classification on Google Earth Engine. *ISPRS Int. J. Geo-Inf.* **2020**, *9*, 580. [CrossRef]
16. Zaiatz, A.P.S.R.; Zolin, C.A.; Vendrusculo, L.G.; Lopes, T.R.; Paulino, J. Agricultural land use and cover change in the Cerrado/Amazon ecotone: A case study of the upper Teles Pires river basin. *Acta Amazon.* **2018**, *48*, 168–177. [CrossRef]
17. Ge, Y.; Hu, S.; Ren, Z.; Jia, Y.; Wang, J.; Liu, M.; Zhang, D.; Zhao, W.; Luo, Y.; Fu, Y.; et al. Mapping annual land use changes in China's poverty-stricken areas from 2013 to 2018. *Remote Sens. Environ.* **2019**, *232*, e111285. [CrossRef]
18. Rawat, J.S.; Kumar, M. Monitoring land use/cover change using remote sensing and GIS techniques: A case study of Hawalbagh block, district Almora, Uttarakhand, India. *Egypt. J. Remote Sens. Space Sci.* **2015**, *18*, 77–84. [CrossRef]
19. Kar, R.; Obi Reddy, G.P.; Kumar, N.; Singh, S.K. Monitoring spatio-temporal dynamics of urban and peri-urban landscape using remote sensing and GIS—A case study from Central India. *Egypt. J. Remote Sens. Space Sci.* **2018**, *21*, 401–411. [CrossRef]
20. Neves, A.K.; Korting, T.S.; Fonseca, L.M.G.; Escada, M.I.S. Assessment of TerraClass and MapBiomas data on legend and map agreement for the Brazilian Amazon biome. *Acta Amazon.* **2020**, *50*, 170–182. [CrossRef]
21. Souza, A.P.; Mota, L.L.; Zamadei, T.; Martim, C.C.; Almeida, F.T.; Paulino, J. Climate classification and climatic water balance in Mato Grosso state, Brazil. *Nativa.* **2013**, *1*, 34–43. Available online: https://periodicoscientificos.ufmt.br/ojs/index.php/nativa/article/view/1334 (accessed on 1 July 2022). [CrossRef]
22. Kuemmerle, T.; Erb, K.; Meyfroidt, P.; Müller, D.; Verburg, P.H.; Estel, S.; Haberl, H.; Hostert, P.; Jepsen, M.R.; Kastner, T.; et al. Challenges and opportunities in mapping land use intensity globally. *Curr. Opin. Environ. Sustain.* **2013**, *5*, 484–493. [CrossRef] [PubMed]
23. Souza, S.O. Geotechnologies applied to time-space analysis of land use and occupation in coastal plain of Caravelas (BA). *Goiano Bull. Geogr.* **2015**, *35*, 71–89. [CrossRef]
24. Wulder, M.A.; Roy, D.P.; Radeloff, V.C.; Loveland, T.R.; Anderson, M.C.; Johnson, D.M.; Healey, S.; Zhu, Z.; Scambos, T.A.; Pahlevan, N.; et al. Fifty years of Landsat science and impacts. *Remote Sens. Environ.* **2022**, *280*, 113195. [CrossRef]
25. Showstack, R. Landsat 9 Satellite Continues Half-Century of Earth Observations: Eyes in the sky serve as a valuable tool for stewardship. *BioScience* **2022**, *72*, 226–232. [CrossRef]
26. Jog, S.; Dixit, M. Supervised classification of satellite images. In Proceedings of the 2016 Conference on Advances in Signal Processing (CASP), Pune, India, 9–11 June 2016; pp. 93–98. [CrossRef]
27. Ali, M.Z.; Qazi, W.; Aslam, N. A comparative study of ALOS-2 PALSAR and landsat-8 imagery for landcover classification using maximum likelihood classifier. *Egypt. J. Remote Sens. Space Sci.* **2018**, *21*, 329–330. [CrossRef]

28. Shalaby, A.; Tateishi, R. Remote sensing and GIS for mapping and monitoring land cover and land-use changes in the Northwestern coastal zone of Egypt. *Appl. Geogr.* **2007**, *27*, 28–41. [CrossRef]
29. Asokan, A.; Anitha, J.; Ciobanu, M.; Gabor, A.; Naaji, A.; Hemanth, D.J. Image Processing Techniques for Analysis of Satellite Images for Historical Maps Classification—An Overview. *Appl. Sci.* **2020**, *10*, 4207. [CrossRef]
30. Souza, C.M., Jr.; Z.Shimbo, J.; Rosa, M.R.; Parente, L.L.; A.Alencar, A.; Rudorff, B.F.T.; Hasenack, H.; Matsumoto, M.; G.Ferreira, L.; Souza-Filho, P.W.M.; et al. Reconstructing Three Decades of Land Use and Land Cover Changes in Brazilian Biomes with Landsat Archive and Earth Engine. *Remote Sens.* **2020**, *12*, 2735. [CrossRef]
31. Shimabukuro, Y.E.; Arai, E.; Duarte, V.; Dutra, A.C.; Cassol, H.L.G.; Sano, E.E.; Hoffmann, T.B. Discriminating Land Use and Land Cover Classes in Brazil Based on the Annual PROBA-V 100 m Time Series. *IEEE J. Sel. Top. Appl. Earth Obs. Remote Sens.* **2020**, *13*, 3409–3420. [CrossRef]
32. Gavade, A.B.; Rajpurohit, V.S. Systematic analysis of satellite image-based land cover classification techniques: Literature review and challenges. *Int. J. Comput. Appl.* **2019**, *43*, 514–523. [CrossRef]
33. Opedes, H.; Mücher, S.; Baartman, J.E.M.; Nedala, S.; Mugagga, F. Land Cover Change Detection and Subsistence Farming Dynamics in the Fringes of Mount Elgon National Park, Uganda from 1978–2020. *Remote Sens.* **2022**, *14*, 2423. [CrossRef]
34. Congalton, R.G. A review of assessing the accuracy classifications of remotely sensed data. *Remote Sens. Environ.* **1991**, *37*, 35–46. [CrossRef]
35. Pendrill, F.; Persson, U.M. Combining global land cover datasets to quantify agricultural expansion into forests in Latin America: Limitations and challenges. *PLoS ONE* **2017**, *12*, e0181202. [CrossRef]
36. Hoang, N.T.; Kanemoto, K. Mapping the deforestation footprint of nations reveals growing threat to tropical forests. *Nat. Ecol. Evol.* **2021**, *5*, 845–853. [CrossRef]
37. Pendrill, F.; Persson, U.M.; Godar, J.; Kastner, T. Deforestation displaced: Trade in forest-risk commodities and the prospects for a global forest transition Environ. *Res. Lett.* **2019**, *14*, 055003. [CrossRef]
38. Davis, K.F.; Koo, H.I.; Dell'Angelo, J.; D'Ororico, P.; Estes, L.; Kehoe, L.; Kharratzadeh, M.; Kuemmerle, T.; Machava, D.; Rodrigues Pais, A.J.; et al. Tropical forest loss enhanced by large-scale land acquisitions. *Nat. Geosci.* **2020**, *13*, 482–488. [CrossRef]
39. Dubreuil, V.; Pechutti, F.K.; Planchon, O.; Sant'anna Neto, J.L. The types of annual climates in Brazil: An application of the classification of Köppen from 1961 to 2015. *Confins* **2018**, *37*. [CrossRef]
40. Brazilian Agricultural Research Corporation—Embrapa. Brazil in Relief. Available online: https://www.cnpm.embrapa.br/projetos/relevobr/ (accessed on 1 July 2022).
41. Brazilian Institute of Geography and Statistics (Instituto Brasileiro de Geografia e Estatística or IBGE). Statistics and Geosciences Downloads. Available online: https://downloads.ibge.gov.br/index.htm (accessed on 1 July 2022).
42. Ettritch, G.; Hardy, A.; Bojang, L.; Cross, D.; Bunting, P.; Brewer, P. Enhancing digital elevation models for hydraulic modelling using flood frequency detection. *Remote Sens. Environ.* **2018**, *217*, 506–522. [CrossRef]
43. United States Geological Survey (USGS). EarthExplorer. Available online: http://earthexplorer.usgs.gov/ (accessed on 1 July 2022).
44. Santos, L.A.C.; Batista, A.C.; Neves, C.O.M.; Carvalho, E.V.; Santos, M.M.; Giongo, M. Multi-temporal analysis of land use and cover in nine municipalities in the south of Tocantins using Landsat images. *Rev. Agroambiente* **2017**, *11*, 111–118. [CrossRef]
45. Furtado, L.G.; Morales, G.P.; Silva, D.F.; Pontes, A.N. Land use and land cover transformations in the Murucupi river basin, Barcarena, Pará. *Rev. Bras. Geogr. Física* **2020**, *13*, 2340–2354. [CrossRef]
46. National Water Agency (Agência Nacional de Águas e Saneamento Básico or ANA). Dados Abertos da Agência Nacional de Águas e Saneamento Básico. Available online: https://dadosabertos.ana.gov.br (accessed on 1 July 2022).
47. Cohen, J.A. Coefficient of agreement for nominal scales. *Educational and Psycol. Meas.* **1960**, *20*, 37–46. [CrossRef]
48. MapBiomas. Annual Mapping of Land Cover and Land Use in Brazil. Available online: http://mapbiomas.org (accessed on 1 February 2023).
49. Landis, J.R.; Koch, G.G. The measurement of observer agreement for categorical data. *Biometrics* **1977**, *33*, 159–174. [CrossRef] [PubMed]
50. Shimabukuro, Y.E.; Miettinen, J.; Beuchle, R.; Grecchi, R.C.; Simonetti, D.; Achard, F. Estimating burned area in Mato Grosso, Brazil, using an object-based classification method on a systematic sample of medium resolution satellite images. *IEEE J. Select. Top. Appl. Earth Observ. Rem. Sens.* **2015**, *8*, 4502–4508. [CrossRef]
51. Oliveira, N.A. "Winning the West": Amazônia Legal Brasileira and the case of Nova Xavantina/MT. *Front. Rev. Hist.* **2015**, *17*, 248–272. Available online: https://1library.org/document/zgg09p2z-conquistando-oeste-amazonia-legal-brasileira-caso-nova-xavantina.html (accessed on 1 July 2022).
52. Silva, M.; Deluski, E.C.; Santos, S.K.F.; Claudino, W.V.; Silva, E.P. Use of geotechnologies in the dynamics of soil occupation in the Municipality of Paranaita-MT. *Agr. Acad.* **2018**, *5*, 334–346. [CrossRef]
53. Batista, F.R.Q.; Nogueira, S.H.; Ferreira, L.G. Mapping of Cerrado Phytophysiognomies by Remote Sensing: Challenges and Possibilities. In *Brazilian Symposium on Remote Sensing (SBSR), 19, 2019, Santos*; Electronic analytics; INPE: São José dos Campos, Brazil, 2019; pp. 403–406. Available online: http://urlib.net/rep/8JMKD3MGP6W34M/3TUPA2H (accessed on 1 July 2022).
54. Phiri, D.; Morgenroth, J. Developments in Landsat land cover classification methods: A review. *Remote Sens.* **2017**, *9*, 967. [CrossRef]

55. Santos, F.A.A.; Rocha, E.J.P.; Santos, J.S. Dynamics of Landscape and its Environmental Impacts in the Amazon. *Rev. Bras. Geogr. Física* **2019**, *12*, 1794–1815. [CrossRef]
56. Santos, V.; Laurent, F.; Abe, C.; Messner, F. Hydrologic response to land use change in a large basin in Eastern Amazon. *Water* **2018**, *10*, 429. [CrossRef]
57. Zeilhofer, P.; Alcantara, L.H.; Fantin-Cruz, I. Effects of deforestation on spatio-temporal runoff patterns in the upper Teles Pires watershed, Mato Grosso, Brazil. *Rev. Bras. Geogr. Física* **2018**, *11*, 1889–1901. [CrossRef]
58. Yoshikawa, S.; Sanga-Ngoie, K. Deforestation dynamics in Mato Grosso in the southern Brazilian Amazon using GIS and NOAA/AVHRR data. *Int. J. Remote Sens.* **2011**, *32*, 523–544. [CrossRef]
59. Diniz, M.B.; Oliveira Junior, J.N.; Neto, N.T.; Diniz, M.J.T. Causes of deforestation in the Amazon: An application of the Granger causality test on the main sources of deforestation in the municipalities of the Brazilian legal Amazon. *Nova Econ.* **2009**, *19*, 121–151. [CrossRef]
60. Timpe, K.; Kaplan, D. The changing hydrology of a dammed Amazon. *Sci. Adv.* **2017**, *3*, e1700611. [CrossRef]
61. Lees, A.C.; Peres, C.A.; Fearnside, P.M.; Schneider, M.; Zuanon, J.A.S. Hydropower and the future of Amazonian biodiversity. *Biodivers. Conserv.* **2016**, *25*, 451–466. [CrossRef]
62. Winemiller, K.O.; McIntyre, P.B.; Castello, L.; Fluet-Chouinard, E.; Giarrizzo, T.; Nam, S.; Baird, I.G.; Darwall, W.; Lujan, N.K.; Harrison, I.; et al. Balancing hydropower and biodiversity in the Amazon, Congo, and Mekong. *Science* **2016**, *351*, 128–129. [CrossRef] [PubMed]
63. Faria, F.A.M.; Jaramillo, P.; Sawakuchi, H.O.; Richey, J.E.; Barros, N. Estimating greenhouse gas emissions from future Amazonian hydroelectric reservoirs Environ. *Res. Lett.* **2015**, *10*, 124019. [CrossRef]
64. Gollnow, F.; Lakes, T. Policy change, land use, and agriculture: The case of soy production and cattle ranching in Brazil, 2001-2012. *Appl. Geogr.* **2014**, *55*, 203–211. [CrossRef]
65. Gusso, A.; Ducati, J.R.; Bortolotto, V.C. Analysis of soybean cropland expansion in the southern Brazilian Amazon and its relation to economic drivers. *Acta Amazon.* **2017**, *47*, 281–292. [CrossRef]
66. Simões, R.; Picoli, M.C.A.; Camara, G.; Maciel, A.; Santos, L.; Andrade, P.R.; Sánches, A.; Ferreira, K.; Carvalho, A. Land use and cover maps for Mato Grosso State in Brazil from 2001 to 2017. *Sci. Data* **2020**, *7*, 1–10. [CrossRef]
67. Klink, C.A.; Machado, R.B. Conservation of the Brazilian Cerrado. *Conserv. Biol.* **2005**, *19*, 707–713. [CrossRef]
68. Gibbs, H.K.; Rausch, L.; Munger, J.; Schelly, I.; Morton, D.C.; Noojipady, P.; Soares Filho, B.; Barreto, P.; Micol, L.; Walker, N.F. Brazil's Soy Moratorium. *Science* **2015**, *347*, 377–378. [CrossRef]
69. Hunke, P.; Roller, R.; Zeilhofer, P.; Schröder, B.; Mueller, E.N. Soil changes under different land-use in the Cerrado of Mato Grosso, Brazil. *Geoderma Reg.* **2015**, *4*, 31–43. [CrossRef]
70. Picoli, M.C.A.; Rorato, A.; Leitão, P.; Camara, G.; Maciel, A.; Hostert, P.; Sanches, I.D. Impacts of public and private sector policies on soybean and pasture expansion in Mato Grosso-Brazil from 2001 to 2017. *Land* **2020**, *9*, 20. [CrossRef]
71. Cohn, A.S.; Gil, J.; Berger, T.; Pellegrina, H.; Toledo, C. Patterns and processes of pasture to crop conversion in Brazil: Evidence from Mato Grosso State. *Land Use Policy* **2016**, *55*, 108–120. [CrossRef]
72. Fearnside, P.M. Deforestation in Brazilian Amazonia: History, Rates, and Consequences. *Conserv. Biol.* **2005**, *19*, 680–688. [CrossRef]
73. Ferreira, M.D.P.; Coelho, A.B. Recent Deforestation in the States of the Legal Amazon: An analysis of the contribution of agricultural prices and government policies. *Rev. Econ. Sociol. Rural* **2015**, *53*, 93–108. [CrossRef]
74. Pacheco, R.; Rajão, R.; Soares-Filho, B.; Hoff, R.V.D. Regularization of Legal Reserve Debts: Perceptions of rural producers in the state of Pará and Mato Grosso in Brazil. *Ambient. Soc.* **2017**, *20*, 181–200. [CrossRef]
75. Rudorff, B.F.T.; Adami, M.; Aguiar, D.A.; Moreira, M.A.; Mello, M.P.; Fabiani, L.; Amaral, D.F.; Pires, B.M. The soy moratorium in the Amazon biome monitored by remote sensing images. *Remote Sens.* **2011**, *3*, 185–202. [CrossRef]
76. Spera, S.A.; Cohn, A.S.; VanWey, L.K.; Mostarda, J.F.; Rudorff, B.F.; Risso, J.; Adami, M. Recent cropping frequency, expansion, and abandonment in Mato Grosso, Brazil had selective land characteristics. *Environ. Res. Lett.* **2014**, *9*, 064010. [CrossRef]
77. Zilverberg, C.J.; Heimerl, K.; Schumacher, T.E.; Malo, D.D.; Schumacher, J.A.; Johnson, W.C. Landscape dependent changes in soil properties due to long-term cultivation and subsequent conversion to native grass agriculture. *Catena* **2018**, *160*, 282–297. [CrossRef]
78. Dionizio, E.A.; Costa, M.H. Influence of land use and land cover on hydraulic and physical soil properties at the Cerrado Agricultural Frontier. *Agric.* **2019**, *9*, 24. [CrossRef]
79. Scanlon, B.R.; Jolly, I.; Sophocleous, M.; Zhang, L. Global impacts of conversions from natural to agricultural ecosystems on water resources: Quantity versus quality. *Water Resour. Res.* **2007**, *43*, W03437. [CrossRef]
80. Borella, D.R.; de Souza, A.P.; de Almeida, F.T.; de Abreu, D.C.; Hoshide, A.K.; Carvalho, G.A.; Pereira, R.R.; da Silva, A.F. Dynamics of Sediment Transport in the Teles Pires River Basin in the Cerrado-Amazon, Brazil. *Sustainability* **2022**, *14*, 16050. [CrossRef]
81. Pedrosa, L.M.; Hoshide, A.K.; Abreu, D.C.; de Molossi, L.; Couto, E.G. Financial transition and costs of sustainable agricultural intensification practices on a beef cattle and crop farm in Brazil's Amazon. *Renew. Agric. Food Sys.* **2019**, *36*, 26–37. [CrossRef]
82. Molossi, L.; Hoshide, A.K.; Pedrosa, L.M.; Oliveira, A.S.; Abreu, D.C. Improve pasture or feed grain? Greenhouse gas emissions, profitability, and resource use for Nelore beef cattle in Brazil's Cerrado and Amazon biomes. *Animals* **2020**, *10*, 1386. [CrossRef] [PubMed]

83. Skidmore, M.E.; Sims, K.M.; Rausch, L.L.; Gibbs, H.K. Sustainable intensification in the Brazilian cattle industry: The role for reduced slaughter age. *Environ. Res. Lett.* **2022**, *17*, 064026. [CrossRef]
84. Pinheiro, D.T.; Santos, D.M.S.; Martins, A.R.R.; da Silva, W.M.; de Araújo, C.V.; de Abreu, D.C.; Hoshide, A.K.; Molossi, L.; de Oliveira, R.A. Closing the gap: Sustainable intensification implications of increased corn yields and quality for second-crop (*safrinha*) in Mato Grosso, Brazil. *Sustainability* **2021**, *13*, 13325. [CrossRef]
85. Da Silva, W.M.; Bianchini, A.; Amorim, R.S.S.; Couto, E.G.; dos Santos Weber, O.L.; Hoshide, A.K.; Pereira, P.S.X.; Cremon, C.; de Abreu, D.C. Soil efflux of carbon dioxide in Brazilian Cerrado wheat under variable soil preparation and irrigation. *Agriculture* **2022**, *12*, 163. [CrossRef]
86. Kraeski, A.; Almeida, F.T.; Carvalho, T.M.; Souza, A.P. Identification of land use conflicts in Permanent Preservation Area in a Brazilian Amazon sub-basin. *Soc. E Nat.* **2022**, *35*, e65951. [CrossRef]
87. Reydon, B.P.; Fernandes, V.B.; Telles, T.S. Land governance as a precondition for decreasing deforestation in the Brazilian Amazon. *Land Use Policy* **2020**, *94*, 104313. [CrossRef]

sustainability MDPI

Article

Estimating Suspended Sediment Concentration Using Remote Sensing for the Teles Pires River, Brazil

Rhavel Salviano Dias Paulista [1], Frederico Terra de Almeida [2,*], Adilson Pacheco de Souza [2], Aaron Kinyu Hoshide [3,4], Daniel Carneiro de Abreu [2,3], Jaime Wendeley da Silva Araujo [2] and Charles Campoe Martim [5]

1 Environmental Sciences, Federal University of Mato Grosso, Sinop 78557-287, MT, Brazil
2 Institute of Agrarian and Environmental Sciences, Federal University of Mato Grosso, Sinop 78557-287, MT, Brazil
3 AgriSciences, Institute of Agrarian and Environmental Sciences, Federal University of Mato Grosso, Avenida Alexandre Ferronato, 1200, Sinop 78555-267, MT, Brazil
4 College of Natural Sciences, Forestry and Agriculture, University of Maine, Orono, ME 04469, USA
5 Postgraduate Program in Environmental Physics, Federal University of Mato Grosso, Cuiabá 78060-900, MT, Brazil
* Correspondence: fredterr@gmail.com or frederico.almeida@ufmt.br; Tel.: +55-66-99995-1315

Abstract: Improving environmental sustainability involves measuring indices that show responses to different production processes and management types. Suspended sediment concentration (SSC) in water bodies is a parameter of great importance, as it is related to watercourse morphology, land use and occupation in river basins, and sediment transport and accumulation. Although already established, the methods used for acquiring such data in the field are costly. This hinders extrapolations along water bodies and reservoirs. Remote sensing is a feasible alternative to remedy these obstacles, as changes in suspended sediment concentrations are detectable by satellite images. Therefore, satellite image reflectance can be used to estimate SSC spatially and temporally. We used Sentinel-2 A and B imagery to estimate SSC for the Teles Pires River in Brazil's Amazon. Sensor images used were matched to the same days as field sampling. Google Earth Engine (GEE), a tool that allows agility and flexibility, was used for data processing. Access to several data sources and processing robustness show that GEE can accurately estimate water quality parameters via remote sensing. The best SSC estimator was the reflectance of the B4 band corresponding to the red range of the visible spectrum, with the exponential model showing the best fit and accuracy.

Keywords: Amazonia; Google Earth Engine; hydro-sedimentology; reflectance; satellite imagery

Citation: Paulista, R.S.D.; de Almeida, F.T.; de Souza, A.P.; Hoshide, A.K.; de Abreu, D.C.; da Silva Araujo, J.W.; Martim, C.C. Estimating Suspended Sediment Concentration Using Remote Sensing for the Teles Pires River, Brazil. *Sustainability* **2023**, *15*, 7049. https://doi.org/10.3390/su15097049

Academic Editors: Majid Mohammadian and Ozgur Kisi

Received: 30 January 2023
Revised: 18 April 2023
Accepted: 20 April 2023
Published: 23 April 2023

1. Introduction

The search for environmentally sustainable production processes requires establishing parameters to evaluate production systems such as those related to forestry and agriculture. These factors can enable development while mitigating adverse environmental impacts. Among them, suspended sediment concentration in water bodies is important for evaluating such land use and its impacts on water resources. Suspended sediment concentration (SSC) is relevant to assess the quality of water bodies, as it is directly connected to the morphology of channels and silting processes in reservoirs [1,2]. Although erosion and particle sedimentation are natural phenomena, anthropogenic actions can enhance these processes, causing loss of water quality, silting up of water bodies, and reducing the useful life of reservoirs [3,4]. Although traditional methods for acquiring sedimentological data are reliable, alternative techniques are needed to speed up and improve SSC quantification, making it less costly [5]. Another problem of traditional methods is the difficulty in establishing continuous observations, impairing long-term SSC assessments. Furthermore, the site-specific nature of sampling makes these methods even more difficult to extrapolate values to other locations along the water body [4].

Remote sensing is an alternative for SSC determination. At first, its application was limited to oceans or large water bodies, but as higher spatial-resolution orbital sensors emerged, remote sensing technology could be used for smaller bodies of water [6]. As SSC increases, the surface reflectance of water bodies also rises, enabling the detection of its variations along a water body [7]. One advantage of sedimentological monitoring using satellite images is the potential to monitor sediment inflow sites in reservoirs, which are hard to monitor [3]. Therefore, satellite imagery can be used to monitor several points of water bodies simultaneously, including hard-to-reach areas. Satellites can provide temporal resolution at a daily time scale which can vary depending on the type of sensor used. Another benefit of remote sensing is the ease of accessing data from services such as the U.S. Geological Survey (USGS)-Earth Explorer and the Copernicus Science Hub. These platforms provide free satellite images to the scientific community. Another important platform is Google Earth Engine (GEE) which, in addition to making data available, allows data processing which has improved access to high-performance computing [8,9].

There are three main remote sensing approaches for water quality characterization, namely: empirical, semi-empirical, and analytical approaches [10]. In an empirical approach, simple or multiple regressions are performed between reflectance values from satellite images and water quality parameters. In a semi-empirical one, however, the spectral behavior of parameters studied must be measured, in loco or in the laboratory, with specific spectral bands being selected to capture such response. Finally, an analytical approach is a physical evaluation in which specific inherent optical properties are studied throughout the entire water column, and, subsequently, these characteristics are related to the apparent optical properties of water quality parameters.

Several authors have used satellite images to quantify suspended sediment concentration (SSC) [11–16]. Using the empirical approach, previous research found strong relationships between SSC and the Normalized Difference Water Index (NDWI) radiometric index for the Teles Pires River [17]. Another study applied this method to two different orbital platforms, Landsat and Sentinel-2 A and B, and evaluated several spectral bands and radiometric indices, finding good SSC estimators for the Doce River in southeast Brazil [18]. Both authors faced the problem of synchronizing the dates of satellite images with those for field sampling. Therefore, they opted to insert images from days after and before sampling, even though there may have been variations in flow rates. Two studies have successfully applied the semi-empirical method with good results relating to responses obtained by spectrometers in the field with images from the Multi-Spectral Instrument (MSI) and Moderate Resolution Imaging Spectroradiometer (MODIS) sensors [19,20]. Another study applied this analytical process to understand the behavior of the optical properties of water along the Amazon basin [4]. Using the same procedure, other researchers built models to estimate suspended matter in Lake Frisian in the northeastern part of The Netherlands [21].

Regardless of the approach used, satellite imagery reflectance values need to be referring to water quality parameters and not to noise caused by the atmosphere. The author emphasized that there is a need for atmospheric calibrations, which seek to recover surface reflectance by subtracting the atmospheric contributions from the reflectance at the top of the atmosphere captured by the orbital sensors [6]. With the spread and popularization of remote sensing data, atmospheric calibration algorithms were developed to minimize such noise; among them is the atmospheric corrector Dark Spectrum Fitting [9]. Dark Spectrum Fitting (DFS) was originally developed for images with metric resolution, but satisfactory results have been obtained when implementing DFS in images from Landsat and Sentinel-2 satellites, which have decametric resolution [22]. DFS is contained within ACOLITE, a multi-sensor atmospheric calibration processor developed by the Royal Belgian Institute of Natural Sciences (RBINS) for aquatic satellite imagery applications.

Another relevant factor is sun glint, which is the reflection of direct sunlight towards remote sensors' field-of-view, preventing proper capture of reflectance from water bodies [23]. Sun glint is a recurrent problem in studies of water surfaces via remote sensing, and some authors have been studying and proposing solutions [24–26]. One solution was implemented in an ACOLITE processor using a sun glint extraction method for the short wave infrared region (SWIR) based Multi-Spectral Instrument (MSI) sensor [27]. In another study, sensors with the SWIR band facilitate atmospheric calibration because water reflectance is equal to zero at this wavelength and under ideal conditions [28]. Moreover, the use of the SWIR band as a parameter for sun glint extraction improves the accuracy of remotely sensed data [27]. By synchronizing field data with satellite passes, remote-sensing products can be validated to assess the presence of sediments [10].

The goal of our research was to evaluate the Google Earth Engine (GEE) platform, both in terms of manipulation and image/data processing, through a semi-automatic method. In doing so, we have three objectives. The first objective of our research was to evaluate empirical models using spectral bands and radiometric indices obtained through images from Sentinel-2 A and B sensors. Second, we sought to estimate the suspended sediments concentration (SSC) in the Teles Pires River basin in Brazil. The third objective of our study was to optimize monitoring mechanisms that help in the management of watersheds and the implementation of public policies.

2. Materials and Methods

2.1. Study Area

Remotely sensed data were used, and field sampling was conducted in the Teles Pires River. The Teles Pires River basin has an area of 141,524 square kilometers and is located between south latitudes 7°16′47″ to 14°55′17″ and west longitudes 53°49′46″ to 58°7′58″. The Teles Pires River basing watershed covers both of the Brazilian states of Mato Grosso and Pará and is within the Amazon Hydrographic Region. Due to its great length, the Teles Pires river basin is commonly divided into upper, medium, and lower regions. In the lower and medium Teles Pires, the predominant biome is the Amazon forest, while the Cerrado (i.e., savannah) is the predominant biome in the upper basin further south. The basin has two climates according to the Köppen's classification. In the upper region and most of the medium basin, it is classified as Aw, which stands for tropical, with dry winters and rainy summers. Meanwhile, the area near the Teles Pires River's mouth, where it merges with the Amazon River, has an Am climate, which is a humid tropical climate with short dry seasons and higher amounts of rainfall [29]. The Teles Pires River currently houses four hydroelectric plants (HEP), namely the São Manuel, Teles Pires, Colíder Pesqueiro do Gil, and the Sinop plants (Figure 1).

The São Manuel hydroelectric plant is located in the lower basin, at coordinates 9°11′11.06″ South and 57°3′1.13″ West and has an installed capacity of 700 megawatts (MW) and a reservoir area of 66 square kilometers (km^2). The Teles Pires hydroelectric plant is located at coordinates 9°21′04″ South and 56°46′39″ West and has an installed capacity of 1820 MW and a reservoir area of 150 km^2. The Colíder Pesqueiro do Gil plant is located at coordinates 10°59′06.62″ South and 55°45′52.06″ West with an installed capacity of 300 MW and a reservoir area of 114.9 km^2. Finally, the Sinop hydroelectric plant (11°16′1″ South, 55°27′14″ West) has an installed capacity of 401.88 MW and a reservoir area of 342 km^2 [30].

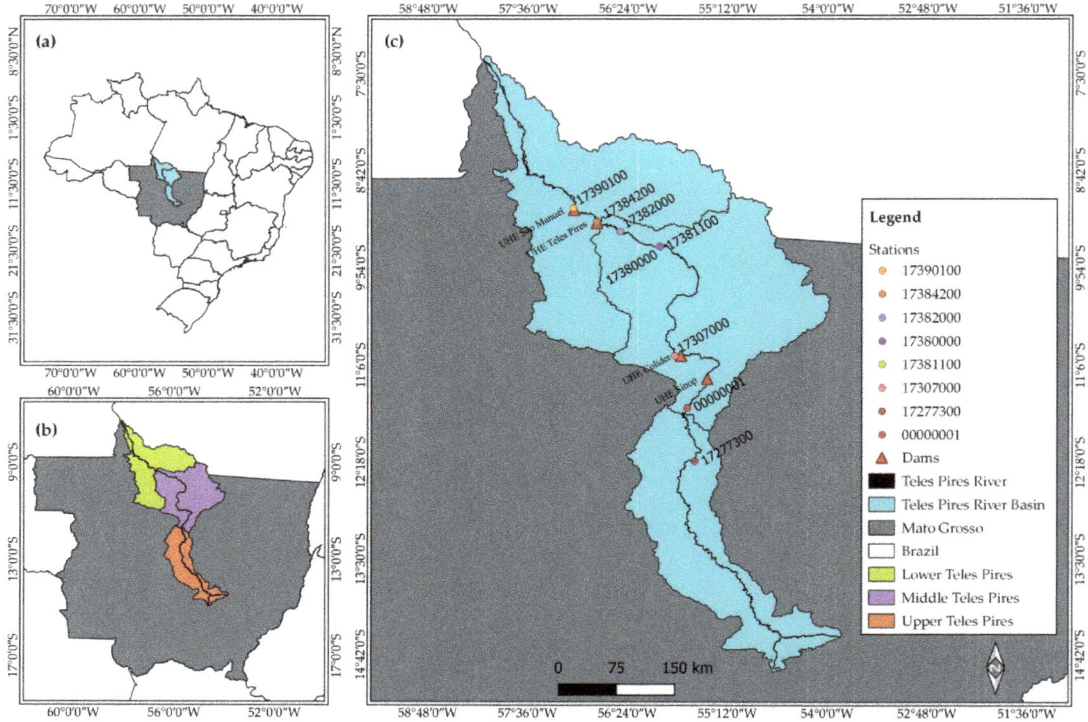

Figure 1. Maps showing the study area and dams for hydroelectric stations: (**a**) river basin highlighted (in blue), inserted in the state of Mato Grosso (in grey) in Brazil, (**b**) Teles Pires river basin upper, middle, and lower divisions from north to south and (**c**) location of dams/stations and in-field sampling locations used for field research with respective codes.

2.2. Suspended Sediment Concentration Data

Suspended sediment concentration (SSC) data used in this study were obtained from two sources. The first data source is the authors themselves, who performed five data collection campaigns on the same day the Sentinel-2 A and B satellites passed within a section near the Sinop power plant reservoir. The second data source is the National Agency for Water and Basic Sanitation (Agência Nacional das Águas e Saneamento Básico—ANA) [31]. A joint resolution known as ANA/ANEEL n° 03/2010 establishes the monitoring of several hydrological parameters by power utilities that operate the hydroelectric complexes, one of which is a hydro-sedimentological survey. In addition to the operators, the Brazilian Geological Survey (Serviço Geológico do Brasil—CPRM) also monitors this parameter within a section of the Teles Pires River. Under these conditions, all stations in operation between 2016 and 2021 that had solid discharge data were selected. Table 1 shows the information on the section of the Teles Pires River studied by the authors and on the seven stations used in this study that were obtained from ANA that met our requirements.

For solid discharge collection, both authors and HEP operators used the equal width increment (EWI) sampling method with vertically integrated samples [5]. According to Oliveira Carvalho 2008 [5], the EWI represents the average suspended sediment concentration (SSC) in the river section studied. In this method, the section is divided into verticals with the same distance between them. In each vertical, subsamples that integrate the SSC from the surface to the river bottom are collected. All the subsamples are gathered to form a single sample capable of representing the average SSC in the investigated section. This is commonly called a composite sample. Here, each composite sample is the sum of the water sample volumes acquired in each vertical (Figure 2).

Table 1. Characteristics and location of the hydro-sedimentological monitoring stations used in sampling of the Teles Pires River, Mato Grosso state, Brazil.

Code	Name	Operator	Location	
			South	West
00000001	Section Curio	Authors	11°37′35″	55°41′23.37″
17277300	HEP Sinop upstream 1	HEP Sinop	12°17′17″	55°36′03″
17390100	HEP São Manuel downstream 1	HEP São Manuel	09°09′55″	57°03′39″
17307000	HEP Colíder Pesqueiro do Gil	HEP Colíder	10°58′59″	55°46′06″
17380000	Downstream the mouth of Peixoto de Azevedo	CRRM	09°38′26″	56°01′10″
17381100	HEP Teles Pires upstream 2	HEP Teles Pires	09°38′23″	56°01′09″
17382000	HEP Teles Pires upstream 1	HEP Teles Pires	09°27′11″	56°29′32″
17384200	HEP Teles Pires downstream	HEP Teles Pires	09°19′52″	56°46′41″

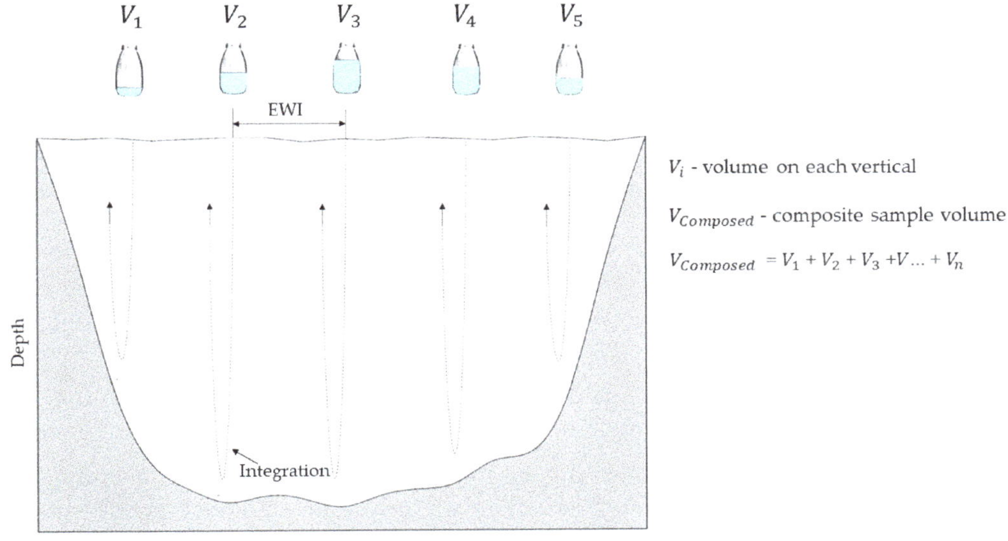

Figure 2. Description and details of the equal width increment (EWI) method.

With regard to operators, ANA/ANEEL resolution n° 03/2010 requires a quarterly collection frequency, comprising four annual measurements in periods of drought and flood, as well as rising and receding waters. Since stations downstream of the mouth of the Peixoto de Azevedo (17380000) and the second upstream station of the Teles Pires River (17381100) are at a similar location (Table 1), this provided a greater volume of data for each location. The raw data used in this study can be found summarized in Appendix A.

2.3. Remote Sensing Data

Remote sensing images were processed in Google Earth Engine (GEE). GEE is a cloud computing platform with a vast orbital data collection allowing for remote sensing data to be processed [8]. GEE is a tool that allows for more open access to high-performance computing and can process large amounts of remote sensing data. We used the Python-based package *geemap* [32] to manipulate and visualize GEE data. One of the collections available in GEE is called COPERNICUS, which comprises images from Multi-Spectral Instrument (MSI) sensor on board the Sentinel-2 A and B satellite.

This satellite is part of the COPERNICUS earth monitoring program, which is coordinated by the European Commission (EC) and the European Space Agency (ESA). The MSI sensor has a 5-day time resolution and 13 spectral band spans. However, we only used bands B3, B4, and B8, which correspond to green, red, and infrared bands, respectively. Such bands have a 10-m spatial resolution. Band B3 is centered at the 560-nanometer (nm) wavelength and 35-nm width. Band B4 is centered at the 665-nm wavelength and 30-nm width, while band B8 is centered at the 842-nm wavelength and 20-nm width [33]. The main reason for choosing the sensor was its time resolution, which has a 5-day revisit interval. This increases the chances of capturing images on the same day as field data sampling. Bands were chosen based on the reflectance bands that capture the range of suspended sediment concentration [34]. Low suspended sediment concentration (SSC) increases green reflectance, while high SSC increases red and infrared reflectance values.

In addition to testing bands alone, two radiometric indexes that work within these spectral bands were also tested. These were the Normalized Difference Vegetation Index and Normalized Difference Water Index. The Normalized Difference Vegetation Index (NDVI) is the normalized ratio between red and near-infrared bands [35]. We compared the physical behavior of healthy and unhealthy plants within the electromagnetic spectrum to determine vegetation status. The other index, the Normalized Difference Water Index (NDWI), is similar to the NDVI and uses the green instead of red band in order to outline aquatic environment features and highlight water bodies in satellite images [36]. NDWI ranges from −1 to 1, with positive values corresponding to aquatic environments.

Satellite images are available in collections, each corresponding to a processing level. We used the second level of processing; that is, products were geometrically corrected and calibrated for the atmosphere to present surface reflectance. However, GEE images were used only for exploratory analysis, identifying cloud-free images with all quality parameters. Using station coordinates and in situ collection dates, an algorithm was developed via the GEE platform, which verified Sentinel-MSI sensor images for the same days of in situ collections. Of the 122 dates, the algorithm verified 31 images corresponding to collection days. From visual inspection, clouds were verified in 16 images, with only 15 images not having cloud obstruction and fitting all quality parameters. The sum of these images combined with those obtained during the in situ campaigns totaled 20 images. Of these 20 images, 14 images were reserved for the creation of the model, with the remaining 6 images used for application of the model. The choice of images for model application took into account the representativeness of the Teles Pires River, selecting at least one image for each studied station. Figure 3 shows the satellite images used for this study, dates, and respective tiles, as well as identifies which images were used to create and validate the models that were developed.

2.4. Atmospheric Calibration

Atmospheric calibration seeks to recover surface reflectance by extracting atmospheric contributions from reflectance at the top of the atmosphere [6]. To do so, we used the Dark Spectrum Fitting (DSF) algorithm proposed by a previous study [9]. The DSF is implemented in the ACOLITE processor and was initially developed for metric-resolution satellite images. However, good results can be obtained by applying DSF to decametric-resolution satellite images, where DSF is based on two assumptions for atmospheric contribution estimation. The first assumption is that the atmosphere is homogeneous and image reflectance is constant, or in part of it since the algorithm allows selection of specific regions for correction. The second assumption is the occurrence of near-zero reflectance values in at least one of the sensor bands, allowing estimation of atmospheric reflectance [22].

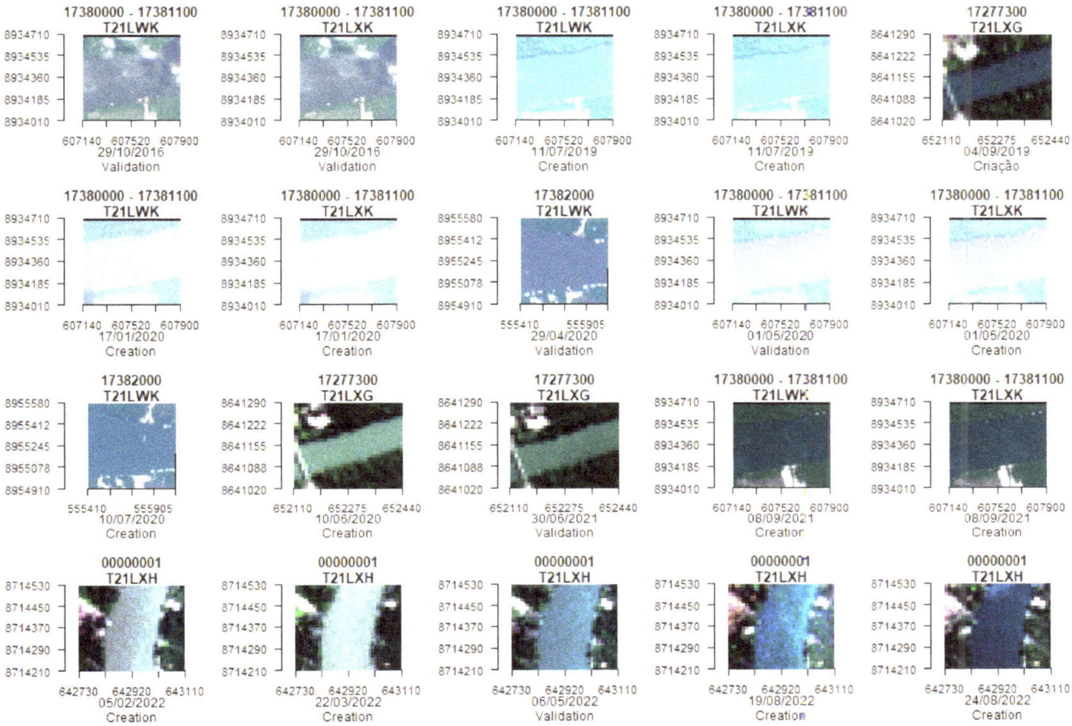

Figure 3. Images from Sentinel-2A and B satellites at different sections along the Teles Pires River, Mato Grosso state, Brazil.

Based on these assumptions, top-of-atmosphere reflectance (ρ_T) is corrected for gas–water interface [37]. After this, dark-object reflectance ($\rho_{\_dark}$) is defined by using the regression intercept containing the lowest reflectance values in each band. Aerosol thickness is calculated by comparing the $\rho_{\_dark}$ values found in the previous step with aerosol models created using a Look Up Table (LUT). This table is generated using the 6SV model [38] and simulates spectral curves that vary with aerosol optical thickness (τ_a) at 550 nm. For each aerosol model contained in the LUT, the band that provides the smallest τ_a is chosen. A combination of a band and model with the smallest τ_a is finally used for calibration, and the parameters required are imported from the LUT.

Another relevant factor in the study of aquatic environments using remote sensing is direct sunlight reflection to the sensor's field of view, known as sun glint. This phenomenon produces anomalies in reflectance capture from water quality parameters by orbital sensors [23]. The ACOLITE processor incorporates a method where a GRS algorithm is used to correct sun glint contamination in SWIR band sensors [27]. The algorithm estimates the bidirectional reflectance distribution factor (BRDF) for air–water interface at the SWIR band and propagates it to visible and near-infrared (NIR) bands [23]. The images selected during verification of intersection with field data were re-downloaded using the Sentinel hub library at processing level 1C. Changes in image acquisition sources occur because the ACOLITE requires auxiliary files for calibration. After atmospheric calibration, images were imported into GEE for data processing.

2.5. Data Processing

After importing images to Google Earth Engine (GEE), pixel values were extracted by a semi-automatic process. For all stations, we manually drew a perpendicular line between river banks. From this point on, the entire process was automated. First, a 30-m buffer zone was established on the drawn line, and then a cut was made. To extract only water-related pixels, we applied the modified normalized difference water index (MNDWI), a radiometric index, which replaces the NIR band with the middle-infrared (MIR) band in the Normalized Difference Water Index (NDWI). This alteration provides less interference from anthropic objects and, due to greater energy absorption of MIR by aquatic environments, increases efficiency in delineating water bodies [39]. Like the NDWI, the index varies from −1 to 1, with 0 being the threshold between aquatic and non-aquatic environments and positive values corresponding to water. In this way, a vector mask based only on pixels greater than zero was created. A new buffer was established to minimize margin effects, reducing the area by 10 m. For pixel value extraction, a 10 × 10 m sample mesh was created, considering spatial resolution of images and avoiding unnecessary interpolations. Since EWI sampling represents average suspended sediment concentration (SSC), pixels contained within the mask were averaged for statistical analysis. Below, Figure 4 describes the entire process, while Figure 5 shows the study design.

All statistics were performed using R 4.0.2 open-source statistical software. First, data normality was verified by the Shapiro–Wilk test at 5% significance ($p = 0.05$). A Spearman correlation matrix was used to assess which variables had the highest correlation with SSC data, and two regression models were later adjusted. The HydroGOF package was used to implement statistical parameters that measure model efficiency [40]. Estimated data were compared with observed data using as parameters mean absolute error (*MAE*; Equation (1)), root mean squared error (*RMSE*; Equation (2)), bias (*BIAS*; Equation (3)), Willmott concordance index (*d*; Equation (4)), the Nash–Sutcliffe efficiency index (*NSE*; Equation (5)), and the mean relative error (*MRE*; Equation (6)).

$$MAE = \frac{1}{N}\sum_{i=1}^{N}|O_i - P_i| \tag{1}$$

$$RMSE = \lceil \frac{1}{N}\sum_{i=1}^{N}(O_i - P_i)^2 \rceil^{0.5} \tag{2}$$

$$BIAS = \frac{1}{N}\sum_{i=1}^{N}(O_i - P_i) \tag{3}$$

$$d = 1 - \lceil \frac{\sum_{i=1}^{N}(P_i - O_i)^2}{\sum_{i=1}^{N}(|P_i - O| + |O_i - O|)^2} \rceil \tag{4}$$

$$NSE = 1 - \frac{\sum_{i=1}^{N}(O_i - P_i)^2}{\sum_{i=1}^{N}(O_i - O)^2} \tag{5}$$

$$MRE = \frac{1}{N}\left(\sum_{i=1}^{N}\frac{(P_i - O_i)}{O_i} \times 100\right) \tag{6}$$

where P_i is the estimated suspended sediment concentration (SSC) estimated (milligrams (mg)/liter (L)), O_i is the observed SSC (mg/L), O is the average of the observed SSC values (mg/L), and N is the number of values in the sample. MAE and RMSE were used to evaluate variation of errors in SSC estimates for spectral bands B3 and B4. Both indices showed perfect model fit when their values were equal to zero [41]. Bias was used to verify whether the model underestimates or overestimates. Positive values indicate underestimation, while negative values mean overestimation. The Willmott concordance index (d) was applied to evaluate model prediction performance with values ranging from 0 to 1. Here, a value of d = 1 indicates perfect agreement. Model performance was established using the Nash–Sutcliffe efficiency (NSE) index, where NSE = 1 means a perfect fit of data by the model, NSE > 0.75 suggests an adequate model, NSE between 0.36 and 0.75 indicates a satisfactory model, and NSES < 0.36 indicates an unsatisfactory model [42]. The mean relative error compares the observed values with the estimated ones and expresses the differences in percentages. This parameter is useful when dealing with values with high or low scales.

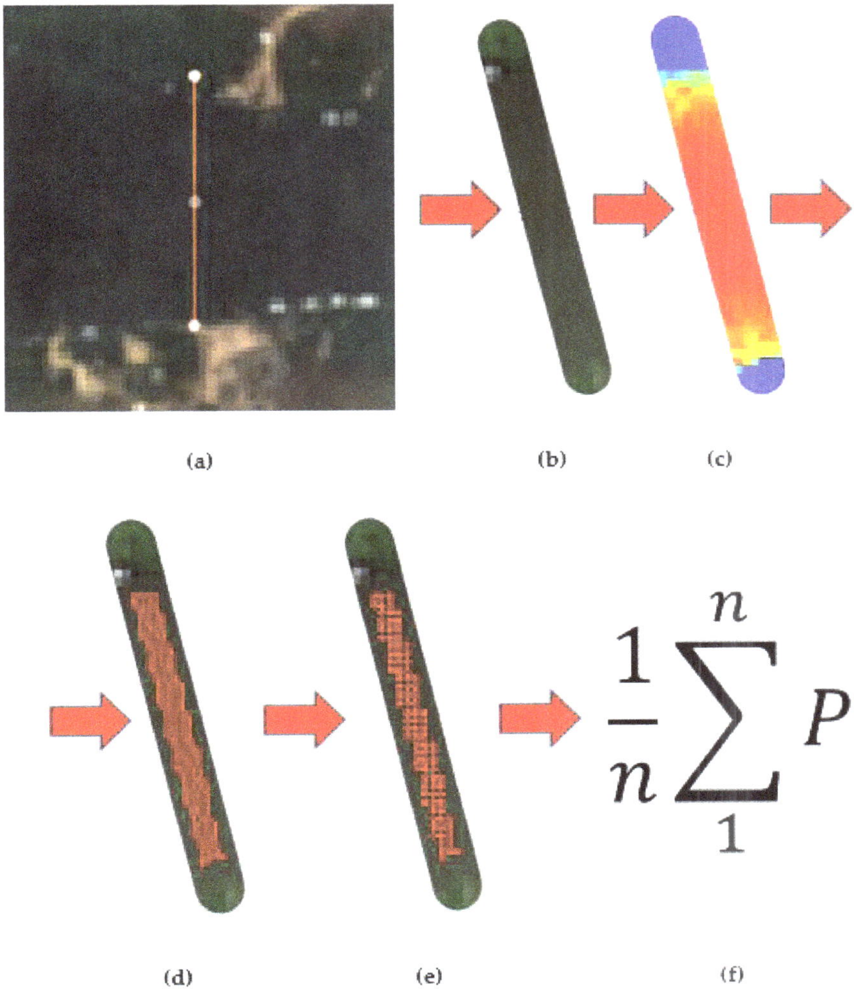

Figure 4. Data processing with (**a**) vector drawn manually from one bank to the other, (**b**) 30-m (m) buffer application from the line and cut, (**c**) MNDWI index application to extract only the aquatic environment, (**d**) 10-m buffer application, (**e**) Sample grid of 10 × 10 m, and (**f**) average of values captured by the sampling grid.

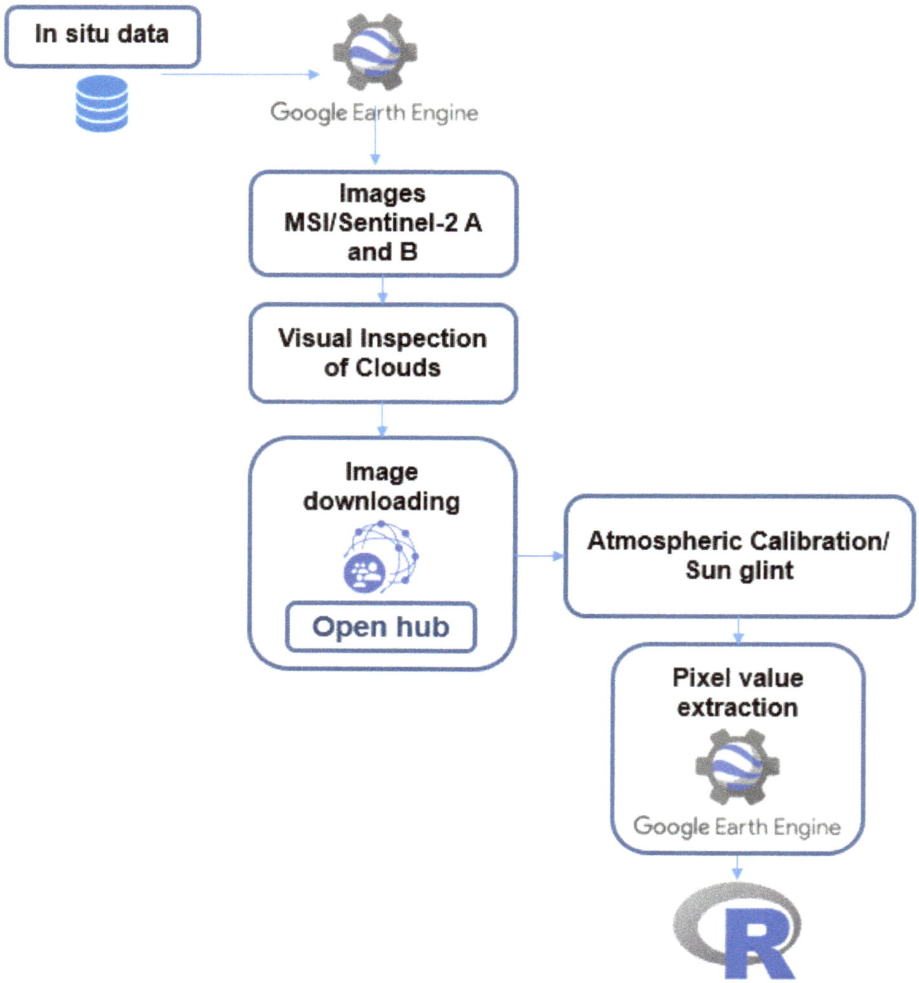

Figure 5. Flowchart showing the methods used, from SSC data, satellite image selection, atmospheric calibration, and data processing, using a semi-automatic method.

3. Results

Table 2 shows the results from all processing and indicates which data (spectral bands and radiometric indices) were used to create and validate suspended sediment concentration (SSC) estimation models. The values of spectral bands B3 and B4 and the Normalized Difference Vegetation Index (NDVI) showed normality with $p = 0.105$, $p = 0.283$, and $p = 0.059$, whereas the in-situ SSC, B8, and Normalized Difference Water Index (NDWI) showed no normality with $p = 0.002$, $p = 0.001$, and $p = 0.024$ (Table 3). Of the analyzed variables, band B4 had the highest correlation coefficient with Rho = 0.726, and B3 had Rho = 0.677 (Table 3). Therefore, there is a strong correlation between the red and green spectrums and SSC. SCC was significantly different for B3, B4, and B8 spectral bands but not for NDWI and NDVI (Table 3). Two regression models were fit to the B3 and B4 bands, which were linear (Figure 6) and exponential (Figure 7).

Table 2. Data from Agência Nacional de Águas' (ANA's) hydro-sedimentological stations specified as codes, sediment collection dates, and images with their function of use in the models, reflectance values from different spectral bands, radiometric normalized difference indexes of the Sentinel satellite imagery for water (NDWI) and vegetation (NDVI), and suspended sediment concentrations (SSC).

Code	Date	Function	Spectral Band B3	Spectral Band B4	Spectral Band B8	NDWI	NDVI	SSC
17381100	29 October 2016	Validation	0.053753	0.038952	0.034259	0.250606	−0.08914	17.55
17381100	29 October 2016	Validation	0.053747	0.038953	0.034259	0.250555	−0.08915	17.55
17380000	11 July 2019	Creation	0.033997	0.018504	0.015644	0.37387	−0.08958	5.10
17380000	11 July 2019	Creation	0.033996	0.018503	0.015605	0.374827	−0.09063	5.10
17277300	4 September 2019	Creation	0.067924	0.053814	0.052215	0.130831	−0.0152	8.91
17381100	17 January 2020	Creation	0.04712	0.04482	0.067682	−0.17882	0.2029	9.82
17381100	17 January 2020	Creation	0.047095	0.044796	0.06767	−0.17899	0.20307	9.82
17382000	29 April 2020	Validation	0.03022	0.02803	0.013714	0.382123	−0.34932	9.39
17381100	1 May 2020	Creation	0.029089	0.025905	0.014683	0.327065	−0.27623	6.74
17381100	1 May 2020	Validation	0.029093	0.02591	0,.014682	0.327161	−0.27635	6.74
17277300	10 June 2020	Creation	0.051608	0.039467	0.017445	0.495145	−0.38752	14.42
17382000	10 July 2020	Creation	0.033754	0.018002	0.017074	0.329243	−0.02757	5.05
17277300	30 June 2021	Validation	0.050261	0.038579	0.02081	0.414598	−0.29942	5.00
17381100	8 September 2021	Creation	0.027345	0.010911	0.021023	0.133084	0.329158	6.28
17381100	8 September 2021	Creation	0.027333	0.010893	0.021005	0.133293	0.329555	6.28
00000001	5 February 2022	Creation	0.055942	0.060056	0.029518	0.309608	−0.34133	17.42
00000001	22 March 2022	Creation	0.05877	0.063464	0.046997	0.11156	−0.14925	24.37
00000001	6 May 2022	Validation	0.048097	0.03775	0.023068	0.352535	−0.24231	11.63
00000001	19 August 2022	Creation	0.037834	0.023956	0.041534	−0.04737	0.269369	6.53
00000001	24 August 2022	Creation	0.036101	0.016822	0.013989	0.443149	−0.09479	5.58

Table 3. Tests for normality, correlation, and significant differences [a] of suspended sediment concentrations (SSC) for the spectral bands and radiometric indices selected.

Element	Normality (Shapiro–Wilk)	Correlation (*Rho*)	Significance (*p*-Value)
Suspended sediment concentration	0.002 **	-	-
Spectral Band			
B3	0.105 *	0.677	0.001 **
B4	0.283 **	0.760	0.001 ***
B8	0.001 **	0.359	0.017 **
Radiometric Index			
Normalized Difference Water Index (NDWI)	0.024 *	−0.097	0.683
Normalized Difference Vegetation Index (NDVI)	0.059 *	−0.276	0.239

[a] Differences significant at a confidence level of 0.1 *, 0.05 **, and 0.01 ***, respectively.

In both models, band B4 had the best coefficients of determination (R^2), which were 0.5849 for the linear and 0.7883 for the exponential; therefore, the exponential model provides the best fit for band B4. In band B3, the same behavior occurred, with the linear fit having an R^2 of 0.4972 and the exponential an R^2 of 0.7003. Table 4 shows the evaluation of the linear and exponential models for bands B3 and B4. For band B4, the exponential model showed the lowest mean absolute error (MAE) and root mean square (RMSE) indices. For band B3, the MAE index indicated smaller errors for the exponential model, while the RMSE index indicated the linear model as more accurate. The BIAS showed that only the linear model with B3 underestimated the suspended sediment concentration (SSC). The Willmott concordance index (d) found the highest concordance for the estimator for the B4 band, with the exponential model presenting a value closer to 1. The same combination also presented the best result for the Nash–Sutcliffe efficiency index (NSE) (0.75), which indicated that the model was satisfactory. For the two bands studied, the exponential model generated the lowest mean relative errors; in particular, band B4 showed the best results for this parameter in both models. Figure 8 expresses the comparison between observed and estimated data for the validation set.

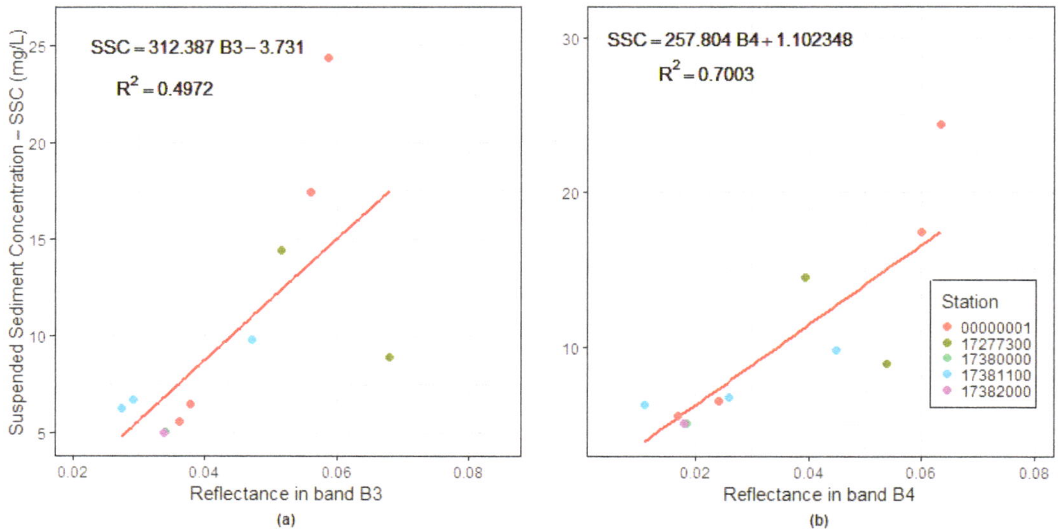

Figure 6. Linear models: with (**a**) linear fit between B3 band (green) and suspended sediment concentration (SSC) and (**b**) linear fit between B4 band (red) and SSC.

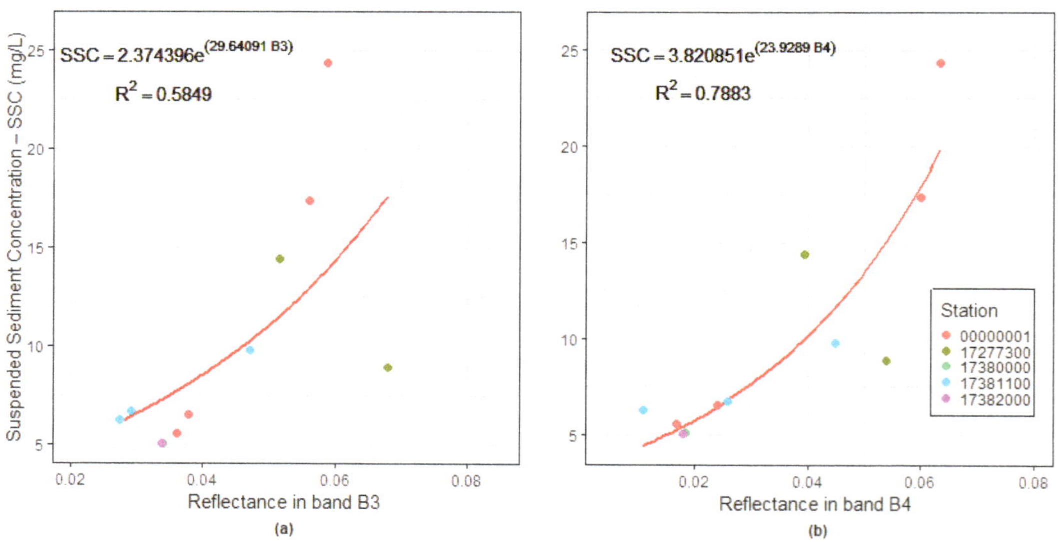

Figure 7. Linear and exponential models for fit between (**a**) band B3 (green) and suspended sediment concentration (SSC) and between (**b**) band B4 (red) and SSC.

Table 4. Statistical evaluation of the models adjusted for spectral bands and suspended sediment concentrations (SSC).

	Spectral Band B3		Spectral Band B4	
Statistical Parameter	**Linear Model**	**Exponential Model**	**Linear Model**	**Exponential Model**
Mean absolute error (MAE) (mg/L)	2.825193	2.7046560	2.251899	1.8859822
Root mean squared error (RMSE) (mg/L)	3.881253	4.1712430	2.996691	2.7348836
BIAS	-3.55×10^{-15}	0.5354025	-8.2462×10^{-16}	−0.3289959
Willmott's concordance (d) index	0.8011445	0.7504934	0.9041224	0.9139091
Nash–Sutcliffe efficiency (NSE) index	0.497190	0.4192480	0.7002607	0.7503466
Mean relative error (%)	30.48	26.15	23.54	18.15%

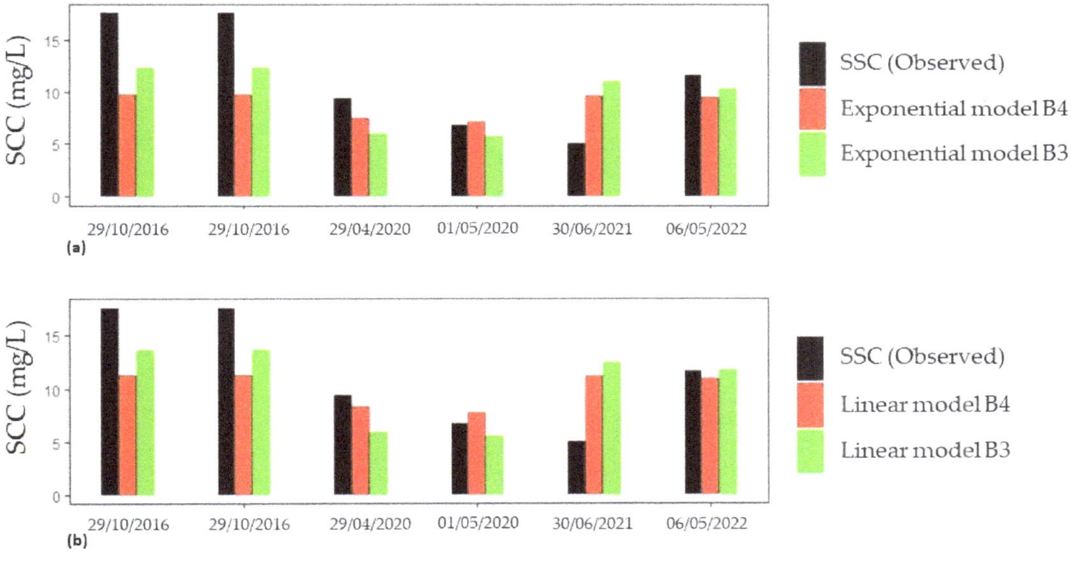

Figure 8. Comparison between exponential and linear models: (**a**) Exponential models B3 and B4 (**b**) Linear models B3 and B4.

4. Discussion

4.1. Comparisons to Previous Research

Data on in situ suspended sediment concentration (SSC), Normalized Difference Water Index (NDWI), and spectral band B8 were not significant for a normal distribution of the data. This may be due to the limitation of observations to the intersection of in situ/satellite collections. Both spectral bands B3 and B4 had significantly normal data distributions. The latter spectral band B4 was the best estimator for both the linear and exponential models. This result is in line with that of Marinho et al. [19], who found strong relationships between SSC and spectral band B4 (red) of the Multi-Spectral Instrument (MSI) sensor used for remote sensing of the Negro River in the northern Amazon forest in Brazil. Moreover, Santos et al. [20] reached the same conclusion for the Purus River in Brazil's eastern Amazon forest using the MODIS sensor. This was also confirmed by Lobo et al. [15] for the Tapajós River in the eastern part of the Brazilian Amazon using the Landsat-MSS/TM/OLI sensors.

These studies corroborated our results, where the spectral band B4 (red) explained suspended sediments better, even at low concentrations. However, Jensen [34] makes a point that more collecting data on chlorophyll content may be needed to improve the accuracy of the models that were used in all of these studies. This author explained that such behavior is due to a link between reflectance shift from red and infrared to green with chlorophyll content in the aquatic environment at low sediment concentrations. Thus, the hypothesis that the Teles Pires River has low chlorophyll contents would explain why band B4 (red) was more efficient in estimating SSC, even at low suspended sediment concentrations. However, to validate this claim, chlorophyll concentrations along the Teles Pires River should be measured in the future.

Radiometric indexes had no correlation with SSC data, especially the Normalized Difference Water Index (NDWI), contradicting the findings of Simões et al. [17]. This difference may have been due to the use of images from days after and before the days of collection in the field. Another hypothesis was launched by McFeeters [36], who had already warned that the broadband aspect of NDWI would increase efficiency in estimating general turbidity. However, such efficiency would be lost when the index is related to isolated variables, such as chlorophyll and suspended sediments. Krug and Noernberg [43] applied NDWI to determine bathymetry and detected that the index also suffers interference from depth variations. Since Simões et al. [17] used the index in two fixed sections, depth was constant, and thus NDWI values were mostly affected by changes in suspended material, which could explain the divergence between the results. The weak correlation of the Normalized Difference Vegetation Index (NDVI) with suspended sediment concentration (SSC) can be explained by very low reflectance values in spectral bands B4 and B8. The genesis of the index indicates that satisfactory results are linked to significant differences between the reflectance values of study objects [35]. SSC in the Teles Pires River provides small reflectance in the red region. Such spectral behavior, combined with high infrared absorption by water, makes the index ineffective in detecting sediments.

The mathematical model that presented the best fit and accuracy was the exponential model, which presented the smallest errors. The combination of the exponential model with the reflectance of the B4 band showed the best results in all the statistical indices that we evaluated. Some stretches studied are at overlapping points of satellite images, which provided two images for the same day, reinforcing the model. The average SSC for the Teles Pires River is approximately 11 mg/L, and the SSC values used to create the model ranged from 5.05 to 24.37 mg/L, demonstrating the feasibility of the model for this water body. When evaluating the application of the model to the validation dataset, the model was able to detect the variation in sediment concentration along the section. Figure 9 demonstrates the potential of this methodology/model in the spatialization of information, as it converts specific SSC information into regional SSC information.

The exponential model had the best fit and accuracy, showing the lowest errors regarding the observed data. Some sections are at overlapping points of satellite images, which provided two images for the same day, reinforcing the test of the model. The average SSC for the Teles Pires River was about 11 milligrams (mg)/liter (L), with values used for model creation varying from 5.05 to 24.37 mg/L. Therefore, the model is feasible for sampling in this water body. No measurements could be evaluated for the last quarter of 2019 or 2020. This situation is related to the rainy season, during which clouds are common in satellite images for the study region. This difficulty can be minimized by using other orbital sensors, increasing the likelihood of capturing images without clouds, as was done in other studies [15,18]. Another solution was proposed by researchers who used unmanned aircraft with onboard spectral sensors during field collections [44,45]. When using the proper method, this approach would reduce impacts from the atmosphere and give some freedom for planning field sampling.

The Google Earth Engine (GEE) tool proved to be flexible in processing images. By applying this tool, several process settings are allowed without requiring an excessive amount of labor. With a few lines of code, we could quickly and efficiently verify the adequacy, occurrence, or lack of occurrence of any image without clouds or noise for the

53 dates of our field sampling. The processing power and availability of remote sensing data make GEE a robust tool for studying water quality parameters.

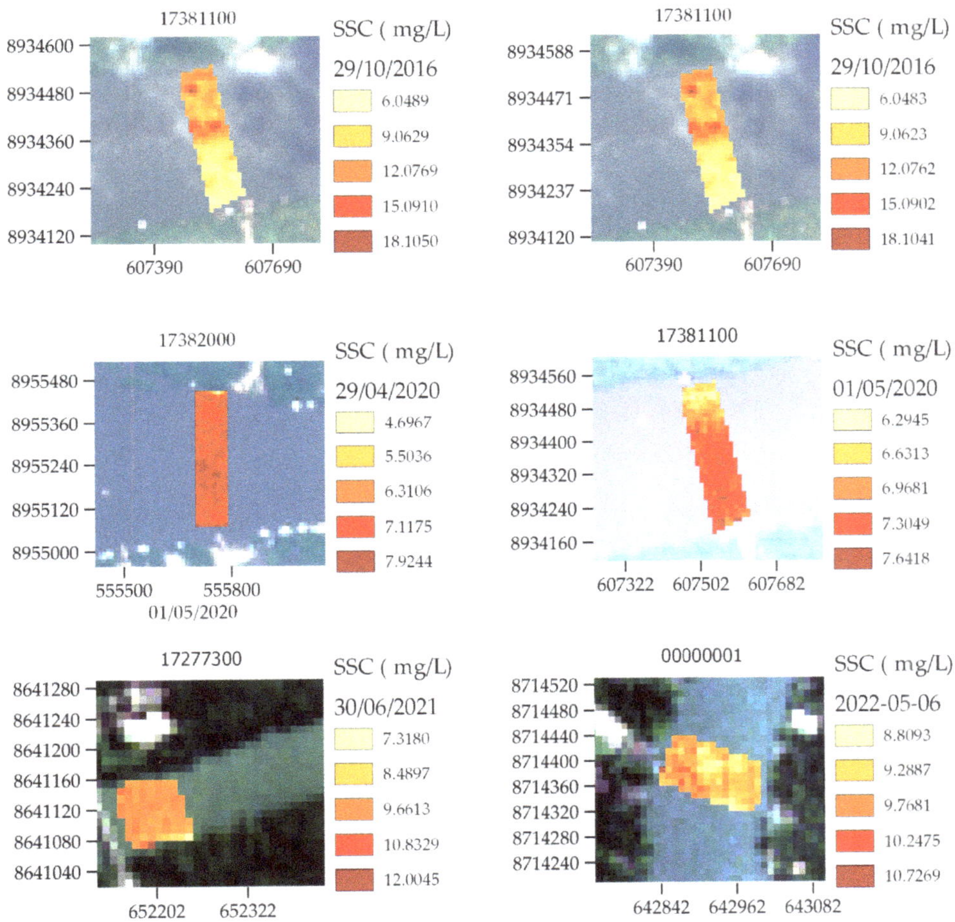

Figure 9. Application of the B4 exponential model on the set of images for validation.

4.2. *Sustainable Agricultural Development Implications and Public Policy Recommendations*

In addition to impacting the useful life of the reservoirs, estimating and/or monitoring suspended sediment concentration (SSC) sheds light on the erosive processes in hydrographic basins. Erosive processes can generate several consequences and vary according to the characteristics of the basin, such as vegetative cover, topography, precipitation regime, and soil properties [46]. In particular, water erosion leads to soil degradation by reducing nutrients and organic matter and favors the transport of fertilizers and pesticides, directly impacting economic activities in these basins [47]. The intensification of erosion can cause damage to the environment and agricultural production affecting food security [48]. Erosion is also intensified in hydrographic basins with a predominance of cultivated or deforested areas [49]. Thus, obtaining SSC data and, consequently, erosion data are essential for the proper management of soil and water in such watersheds.

Recent measurements of suspended sediment concentration (SSC) in the Teles Pires river basin show that SSC is higher in those parts of the basin that have more agricultural land and urban areas [50]. Therefore, it is important to develop public policies that can engage agricultural producers and urban municipalities in the proposition, evaluation, and codification of such policies to encourage the reduction of SSC in the Teles Pires River and other rivers in Brazil. Despite the importance of monitoring solids in water resources, public policies focused on SSC have been limited to monitoring the areas drained from the dams of Hydroelectric Projects. This policy was established by ANA/ANEEL resolution n° 03/2010. However, there is no specific limit to the concentration of solids in watercourses, possibly due to the enormous cost of the necessary measures, a more detailed understanding of erosion processes, and due to poor soil management and use.

Our research raises the possibility of establishing models for estimating the concentration of suspended sediments (SSC) with greater spatiality and temporality. This estimation is possible throughout the water body studied and even over time. Therefore, it is possible to determine the causes and effects of erosion along water resources. With more in-depth studies in the future, different drivers of erosion can be more clearly connected to how they contribute to soil loss and sediment suspension in water bodies. For example, previous research verified how different tributaries contribute to erosion in urban reservoirs located in Campo Grande, midwest Brazil [51]. Understanding where to limit erosion can not only benefit urban areas but also reduce environmental impacts in agricultural areas in Brazil and beyond.

5. Conclusions

Remote sensing is an efficient tool to estimate suspended sediment concentration (SSC) since, even with low sediment concentrations from samples taken from the Teles Pires River, the method can detect SSC variations in the sections studied. The isolated bands provided stronger correlations when compared with radiometric indices, and the B4 band (red) was the best SSC estimator. When the exponential model was used, this band obtained the best results with a coefficient of determination equal to $R^2 = 0.7883$ and the Nash-Sutcliffe index with ENS = 0.7503, indicating the feasibility of the model for the Teles Pires River. Google Earth Engine (GEE) has clearly made the exploratory analysis more agile and efficient for image acquisition and method application. For future work, the use of sensor harmonization products such as The Harmonized Landsat Sentinel-2 (HLS) may increase the probability of occurrence of orbital images on the same day of in situ collections, in addition to minimizing efforts in pre-processing. Another approach that could involve this type of study would be the execution of punctual samples of SSC, allowing for the segmentation of variation of the SSC along the section. Moreover, the application of sediment source fingerprint (SSF), associated with land use analysis, can also be used to develop public policies to encourage more sustainable agricultural development.

Author Contributions: Data collection, writing, methodology, formal analysis, figures and tables—R.S.D.P. and F.T.d.A.; review, editing, supervision—F.T.d.A.; review, editing, supervision, and financial support—A.P.d.S., A.K.H., and D.C.d.A.; data collection—J.W.d.S.A. and C.C.M. All authors have read and agreed to the published version of the manuscript.

Funding: This study was financed by the Coordenação de Aperfeiçoamento de Pessoal de Nível Superior—Brasil (CAPES) and the Agência Nacional de Águas e Saneamento Básico (ANA), Finance Code—001 and Process 88887.144957/2017-00. The authors also thank the Coordination for the Improvement of Higher Education Personnel—Brazil (CAPES) for supporting the graduate scholarship and the National Council of Scientific and Technological Development (CNPq) for supporting scientific initiation scholarships, as well as the AgriScience project.

Institutional Review Board Statement: Not applicable.

Informed Consent Statement: Not applicable.

Data Availability Statement: Study data can be obtained by request to the corresponding author or the first author via e-mail. It is not available on the website as the research project is still under development.

Acknowledgments: The authors thank the Coordenação de Aperfeiçoamento de Pessoal de Nível Superior—Brasil (CAPES) and the Agência Nacional de Águas e Saneamento Básico (ANA), Finance Code—001 and Process 88887.144957/2017-00. The authors also thank the Coordination for the Improvement of Higher Education Personnel—Brazil (CAPES) for supporting the graduate scholarship and the National Council of Scientific and Technological Development (CNPq) for the support with scientific initiation scholarships, as well as the AgriScience Project. The authors also thank all the students and professors of the Tecnologia em Recursos Hídricos no Centro-Oeste research group (dgp.cnpq.br/dgp/espelhogrupo/2399343537529589, accessed 29 January 2023).

Conflicts of Interest: The authors declare no conflict of interest. Supporting entities had no role in the design of the study; in the collection, analyses, or interpretation of data; in the writing of the manuscript, or in the decision to publish the results.

Appendix A

Table A1. Data used in the study, such as the code of the field measurement station (Code) and its description (Name), the concomitant date of the field collection and the satellite image (Date), and the concentration of suspended sediments (SSC).

Code	Name	Date	SSC
17307000	HEP Colíder Pesqueiro do Gil	16 April 2019	15.40
17307000	HEP Colíder Pesqueiro do Gil	17 July 2019	2.40
17307000	HEP Colíder Pesqueiro do Gil	22 June 2021	3.30
17307000	HEP Colíder Pesqueiro do Gil	24 September 2021	2.00
17390100	HEP São Manuel downstream 1	5 February 2016	24.24
17390100	HEP São Manuel downstream 1	25 June 2016	12.61
17390100	HEP São Manuel downstream 1	2 September 2016	10.78
17390100	HEP São Manuel downstream 1	15 December 2016	25.64
17390100	HEP São Manuel downstream 1	15 February 2017	25.43
17390100	HEP São Manuel downstream 1	22 May 2017	10.96
17390100	HEP São Manuel downstream 1	7 August 2017	8.95
17390100	HEP São Manuel downstream 1	6 November 2017	9.95
17390100	HEP São Manuel downstream 1	31 January 2018	14.11
17390100	HEP São Manuel downstream 1	28 June 2018	12.55
17390100	HEP São Manuel downstream 1	3 September 2018	9.41
17390100	HEP São Manuel downstream 1	5 November 2018	10.35
17390100	HEP São Manuel downstream 1	20 February 2019	26.98
17390100	HEP São Manuel downstream 1	23 May 2019	11.43
17390100	HEP São Manuel downstream 1	21 August 2019	9.18
17390100	HEP São Manuel downstream 1	20 November 2019	14.56
17390100	HEP São Manuel downstream 1	4 February 2020	13.78
17390100	HEP São Manuel downstream 1	8 October 2020	11.21
17390100	HEP São Manuel downstream 1	26 November 2020	11.40
17390100	HEP São Manuel downstream 1	9 March 2021	10.11
17390100	HEP São Manuel downstream 1	23 May 2021	8.75
17390100	HEP São Manuel downstream 1	2 July 2021	9.25
17390100	HEP São Manuel downstream 1	2 December 2021	9.51
17390100	HEP São Manuel downstream 1	10 December 2021	10.60

Table A1. *Cont.*

Code	Name	Date	SSC
17277300	HEP Sinop upstream 1	5 March 2019	36.10
17277300	HEP Sinop upstream 1	13 June 2019	10.66
17277300	HEP Sinop upstream 1	4 September 2019	8.91
17277300	HEP Sinop upstream 1	11 December 2019	42.17
17277300	HEP Sinop upstream 1	18 March 2020	34.59
17277300	HEP Sinop upstream 1	10 June 2020	14.42
17277300	HEP Sinop upstream 1	9 September 2020	6.77
17277300	HEP Sinop upstream 1	15 October 2021	2.00
17277300	HEP Sinop upstream 1	30 June 2021	5.00
17277300	HEP Sinop upstream 1	4 December 2021	12.00
17277300	HEP Sinop upstream 1	2 April 2021	25.56
17277300	HEP Sinop upstream 1	12 December 2020	17.36
17381100	HEP Teles Pires upstream 2	3 February 2016	29.22
17381100	HEP Teles Pires upstream 2	10 May 2016	16.97
17381100	HEP Teles Pires upstream 2	19 July 2016	13.31
17381100	HEP Teles Pires upstream 2	29 October 2016	17.55
17381100	HEP Teles Pires upstream 2	25 January 2017	19.93
17381100	HEP Teles Pires upstream 2	17 April 2017	13.77
17381100	HEP Teles Pires upstream 2	27 July 2017	12.30
17381100	HEP Teles Pires upstream 2	30 October 2017	12.01
17381100	HEP Teles Pires upstream 2	24 January 2018	17.81
17381100	HEP Teles Pires upstream 2	15 April 2018	12.15
17381100	HEP Teles Pires upstream 2	5 July 2018	9.60
17381100	HEP Teles Pires upstream 2	31 October 2018	14.31
17381100	HEP Teles Pires upstream 2	18 January 2019	9.43
17381100	HEP Teles Pires upstream 2	24 April 2019	4.91
17381100	HEP Teles Pires upstream 2	20 July 2019	6.29
17381100	HEP Teles Pires upstream 2	31 October 2019	6.52
17381100	HEP Teles Pires upstream 2	17 January 2020	9.82
17381100	HEP Teles Pires upstream 2	1 May 2020	6.74
17381100	HEP Teles Pires upstream 2	11 July 2020	56.40
17381100	HEP Teles Pires upstream 2	16 September 2020	5.64
17381100	HEP Teles Pires upstream 2	30 January 2021	12.66
17381100	HEP Teles Pires upstream 2	13 May 2021	9.64
17381100	HEP Teles Pires upstream 2	8 September 2021	6.28
17381100	HEP Teles Pires upstream 2	24 November 2021	10.96
17381100	HEP Teles Pires upstream 2	22 October 2020	4.28
17382000	HEP Teles Pires upstream 1	3 February 2016	25.13
17382000	HEP Teles Pires upstream 1	13 May 2016	17.26
17382000	HEP Teles Pires upstream 1	15 July 2016	12.79

Table A1. *Cont.*

Code	Name	Date	SSC
17382000	HEP Teles Pires upstream 1	21 October 2016	14.04
17382000	HEP Teles Pires upstream 1	28 January 2017	18.49
17382000	HEP Teles Pires upstream 1	15 April 2017	16.44
17382000	HEP Teles Pires upstream 1	24 July 2017	10.39
17382000	HEP Teles Pires upstream 1	28 October 2017	11.43
17382000	HEP Teles Pires upstream 1	26 January 2018	17.08
17382000	HEP Teles Pires upstream 1	16 April 2018	15.50
17382000	HEP Teles Pires upstream 1	7 July 2018	10.04
17382000	HEP Teles Pires upstream 1	2 November 2018	18.02
17382000	HEP Teles Pires upstream 1	17 January 2019	11.43
17382000	HEP Teles Pires upstream 1	23 April 2019	7.18
17382000	HEP Teles Pires upstream 1	23 July 2019	5.97
17382000	HEP Teles Pires upstream 1	1 November 2019	3.75
17382000	HEP Teles Pires upstream 1	21 January 2020	5.87
17382000	HEP Teles Pires upstream 1	29 April 2020	9.39
17382000	HEP Teles Pires upstream 1	10 July 2020	5.05
17382000	HEP Teles Pires upstream 1	17 September 2020	3.12
17382000	HEP Teles Pires upstream 1	22 October 2020	3.48
17382000	HEP Teles Pires upstream 1	30 January 2021	12.92
17384200	HEP Teles Pires downstream	11 February 2016	10.35
17384200	HEP Teles Pires downstream	26 June 2016	10.85
17384200	HEP Teles Pires downstream	20 July 2016	11.62
17384200	HEP Teles Pires downstream	2 November 2016	11.42
17384200	HEP Teles Pires downstream	27 January 2017	13.50
17384200	HEP Teles Pires downstream	18 April 2017	10.41
17384200	HEP Teles Pires downstream	28 July 2017	8.59
17384200	HEP Teles Pires downstream	3 November 2017	9.26
17384200	HEP Teles Pires downstream	29 January 2018	10.02
17384200	HEP Teles Pires downstream	17 April 2018	8.93
17384200	HEP Teles Pires downstream	3 July 2018	11.30
17384200	HEP Teles Pires downstream	1 November 2018	12.66
17384200	HEP Teles Pires downstream	16 January 2019	2.63
17384200	HEP Teles Pires downstream	22 April 2019	4.17
17384200	HEP Teles Pires downstream	19 July 2019	11.88
17384200	HEP Teles Pires downstream	1 November 2019	2.64
17384200	HEP Teles Pires downstream	20 January 2020	3.27
17384200	HEP Teles Pires downstream	28 April 2020	4.19
17384200	HEP Teles Pires downstream	9 July 2020	5.88
17384200	HEP Teles Pires downstream	18 September 2020	2.91

Table A1. *Cont.*

Code	Name	Date	SSC
17384200	HEP Teles Pires downstream	21 October 2020	0.45
17384200	HEP Teles Pires downstream	1 February 2021	6.20
17380000	Downstream the mouth of Peixoto de Azevedo	24 August 2016	5.80
17380000	Downstream the mouth of Peixoto de Azevedo	30 November 2016	19.90
17380000	Downstream the mouth of Peixoto de Azevedo	2 May 2017	8.00
17380000	Downstream the mouth of Peixoto de Azevedo	24 July 2017	3.10
17380000	Downstream the mouth of Peixoto de Azevedo	31 October 2017	6.10
17380000	Downstream the mouth of Peixoto de Azevedo	20 July 2018	4.60
17380000	Downstream the mouth of Peixoto de Azevedo	25 October 2018	9.80
17380000	Downstream the mouth of Peixoto de Azevedo	6 May 2019	13.30
17380000	Downstream the mouth of Peixoto de Azevedo	11 July 2019	5.10
17380000	Downstream the mouth of Peixoto de Azevedo	30 October 2019	13.80
17380000	Downstream the mouth of Peixoto de Azevedo	16 November 2021	26.80
17380000	Downstream the mouth of Peixoto de Azevedo	27 April 2022	16.30
17380000	Downstream the mouth of Peixoto de Azevedo	26 July 2022	8.90
00000001	Section Curio	19 August 2022	6.53
00000001	Section Curio	5 February 2022	17.42
00000001	Section Curio	24 August 2022	10.74
00000001	Section Curio	22 March 2022	24.37
00000001	Section Curio	6 May 2022	11.63

References

1. Aldrees, A.; Bakheit Taha, A.T.; Mustafa Mohamed, A. Prediction of Sustainable Management of Sediment in Rivers and Reservoirs. *Chemosphere* **2022**, *309*, 136369. [CrossRef] [PubMed]
2. Grant, G.E. The Geomorphic Response of Gravel-Bed Rivers to Dams: Perspectives and Prospects. In *Gravel-Bed Rivers: Processes, Tools, Environments*; John Wiley and Sons: Hoboken, NJ, USA, 2012; pp. 165–181. ISBN 9780470688908.
3. Chelotti, G.B.; Martinez, J.M.; Roig, H.L.; Olivietti, D. Space-Temporal Analysis of Suspended Sediment in Low Concentration Reservoir by Remote Sensing. *Rev. Bras. Recur. Hídr.* **2019**, *24*. [CrossRef]
4. Martinez, J.M.; Espinoza-Villar, R.; Armijos, E.; Moreira, L.S. The optical properties of river and floodplain waters in the Amazon River Basin: Implications for satellite-based measurements of suspended particulate matter. *J. Geophys. Res. Earth Surf.* **2015**, *120*, 1274–1287. [CrossRef]
5. de Oliveira Carvalho, N. *Practical Hydrosedimentology*, 2nd ed.; Interciência: Rio de Janeiro, Brazil, 2008; pp. 1–599.
6. Pahlevan, N.; Mangin, A.; Balasubramanian, S.V.; Smith, B.; Alikas, K.; Arai, K.; Barbosa, C.; Bélanger, S.; Binding, C.; Bresciani, M.; et al. ACIX-Aqua: A global assessment of atmospheric correction methods for Landsat-8 and Sentinel-2 over lakes, rivers, and coastal waters. *Remote Sens. Environ.* **2021**, *258*, 112366. [CrossRef]
7. Ritchie, J.C.; Zimba, P.V.; Everitt, J.H. Remote Sensing Techniques to Assess Water Quality. *Photogramm. Eng. Remote Sens.* **2003**, *69*, 695–704. Available online: https://www.asprs.org/wp-content/uploads/pers/2003journal/june/2003_jun_695-704.pdf (accessed on 10 October 2022). [CrossRef]
8. Gorelick, N.; Hancher, M.; Dixon, M.; Ilyushchenko, S.; Thau, D.; Moore, R. Google Earth Engine: Planetary-Scale Geospatial Analysis for Everyone. *Remote Sens. Environ.* **2017**, *202*, 18–27. [CrossRef]
9. Vanhellemont, Q.; Ruddick, K. ACOLITE for Sentinel-2: Aquatic Applications of MSI Imagery. In Proceedings of the Living Planet Symposium Proceedings, Prague, Czech Republic, 9–13 May 2016; Ouwehand, L., Ed.; Volume 740, p. 55, ISBN 978-92-9221-305-3.
10. Giardino, C.; Bresciani, M.; Stroppiana, D.; Oggioni, A.; Morabito, G. Optical Remote Sensing of Lakes: An Overview on Lake Maggiore. *J. Limnol.* **2014**, *73*, 201–214. [CrossRef]
11. De Oliveira Fagundes, H.; de Paiva, R.C.D.; Fan, F.M.; Buarque, D.C.; Fassoni-Andrade, A.C. Sediment Modeling of a Large-Scale Basin Supported by Remote Sensing and in-Situ Observations. *Catena* **2020**, *190*, 104535. [CrossRef]
12. Doxaran, D.; Froidefond, J.-M.; Lavender, S.; Castaing, P. Spectral Signature of Highly Turbid Waters Application with SPOT Data to Quantify Suspended Particulate Matter Concentrations. *Remote Sens. Environ.* **2002**, *81*, 149–161. [CrossRef]

13. Espinoza Villar, R.; Martinez, J.M.; le Texier, M.; Guyot, J.L.; Fraizy, P.; Meneses, P.R.; de Oliveira, E. A study of sediment transport in the Madeira River, Brazil, using MODIS remote-sensing images. *J. S. Am. Earth Sci.* **2013**, *44*, 45–54. [CrossRef]
14. Lin, H.; Yu, Q.; Wang, Y.; Gao, S. Assessment of the potential for quantifying multi-period suspended sediment concentration variations using satellites with different temporal resolution. *Sci. Tot. Environ.* **2022**, *853*, 158463. [CrossRef] [PubMed]
15. Lobo, F.L.; Costa, M.P.F.; Novo, E.M.L.M. Time-series analysis of Landsat-MSS/TM/OLI images over Amazonian waters impacted by gold mining activities. *Remote Sens. Environ.* **2015**, *157*, 170–184. [CrossRef]
16. Montanher, O.C.; Novo, E.M.L.M.; Barbosa, C.C.F.; Rennó, C.D.; Silva, T.S.F. Empirical Models for Estimating the Suspended Sediment Concentration in Amazonian White Water Rivers Using Landsat 5/TM. *Int. J. Appl. Earth Obs. Geoinf.* **2014**, *29*, 67–77. [CrossRef]
17. Simões, J.M.; Terra De Almeida, F.; Uliana, E.M.; Gonçalves Vendrusculo, L. Uso de Sensoriamento Remoto No Monitoramento de Sólidos Em Suspensão Do Reservatório de Uma Usina Hidrelétrica Na Transição Cerrado-Amazônia. *Gaia Scientia* **2021**, *15*, 69–88. Available online: https://www.alice.cnptia.embrapa.br/alice/bitstream/doc/1138716/1/AP-Uso-sensorimento-remoto-2021.pdf (accessed on 10 October 2022).
18. Aires, U.R.V.; da Silva, D.D.; Fernandes Filho, E.I.; Rodrigues, L.N.; Uliana, E.M.; Amorim, R.S.S.; de Ribeiro, C.B.M.; Campos, J.A. Modeling of Surface Sediment Concentration in the Doce River Basin Using Satellite Remote Sensing. *J. Environ. Manag.* **2022**, *323*, 116207. [CrossRef]
19. Marinho, R.R.; Harmel, T.; Martinez, J.M.; Junior, N.P.F. Spatiotemporal Dynamics of Suspended Sediments in the Negro River, Amazon Basin, from in Situ and Sentinel-2 Remote Sensing Data. *ISPRS Int. J. Geoinf.* **2021**, *10*, 86. [CrossRef]
20. Dos Santos, A.L.M.R.; Martinez, J.M.; Filizola, N.P.; Armijos, E.; Alves, L.G.S. Purus River Suspended Sediment Variability and Contributions to the Amazon River from Satellite Data (2000–2015). *CR Geosci.* **2018**, *350*, 13–19. [CrossRef]
21. Dekker, A.G.; Vos, R.J.; Peters, S.W.M. Comparison of Remote Sensing Data, Model Results and In Situ Data for Total Suspended Matter TSM in the Southern Frisian Lakes. *Sci. Total Environ.* **2001**, *268*, 197–214. [CrossRef]
22. Vanhellemont, Q. Adaptation of the Dark Spectrum Fitting Atmospheric Correction for Aquatic Applications of the Landsat and Sentinel-2 Archives. *Remote Sens. Environ.* **2019**, *225*, 175–192. [CrossRef]
23. Tavares, M.H.; Lins, R.C.; Harmel, T.; Fragoso, C.R.; Martínez, J.M.; Motta-Marques, D. Atmospheric and Sunglint Correction for Retrieving Chlorophyll-a in a Productive Tropical Estuarine-Lagoon System Using Sentinel-2 MSI Imagery. *ISPRS J. Photogramm. Remote Sens.* **2021**, *174*, 215–236. [CrossRef]
24. Ottaviani, M.; Spurr, R.; Stamnes, K.; Li, W.; Su, W.; Wiscombe, W. Improving the Description of Sunglint for Accurate Prediction of Remotely Sensed Radiances. *J. Quant. Spectrosc. Radiat. Transf.* **2008**, *109*, 2364–2375. [CrossRef]
25. Kutser, T.; Vahtmäe, E.; Paavel, B.; Kauer, T. Removing Glint Effects from Field Radiometry Data Measured in Optically Complex Coastal and Inland Waters. *Remote Sens. Environ.* **2013**, *133*, 85–89. [CrossRef]
26. Harmel, T.; Chami, M. Estimation of the Sunglint Radiance Field from Optical Satellite Imagery over Open Ocean: Multidirectional Approach and Polarization Aspects. *J. Geophys. Res. Oceans* **2013**, *118*, 76–90. [CrossRef]
27. Harmel, T.; Chami, M.; Tormos, T.; Reynaud, N.; Danis, P.A. Sunglint Correction of the Multi-Spectral Instrument (MSI)-SENTINEL-2 Imagery over Inland and Sea Waters from SWIR Bands. *Remote Sens. Environ.* **2018**, *204*, 308–321. [CrossRef]
28. Gao, B.C.; Montes, M.J.; Li, R.R.; Dierssen, H.M.; Davis, C.O. An Atmospheric Correction Algorithm for Remote Sensing of Bright Coastal Waters Using MODIS Land and Ocean Channels in the Solar Spectral Region. *IEEE Trans. Geosci. Remote Sens.* **2007**, *45*, 1835–1843. [CrossRef]
29. Dubreuil, V.; Fante, K.P.; Planchon, O.; Neto, J.L.S. Os Tipos de Climas Anuais No Brasil: Uma Aplicação Da Classificação de Köppen de 1961 a 2015. Confins 2018. ISSN: 1958–9212. Available online: https://doaj.org/article/7b1a63c089d3444f961c85dec0ddd4ff (accessed on 24 January 2023).
30. Gallardo, A.L.C.F.; da Silva, J.C.; Gaudereto, G.L.; Sozinho, D.W.F. A Avaliação de Impactos Cumulativos No Planejamento Ambiental de Hidrelétricas Na Bacia Do Rio Teles Pires (Região Amazônica). *Desenvolvimento Meio Ambiente* **2017**, *43*, 53818. [CrossRef]
31. Agência Nacional de Águas (ANA). HIDROWEB v3.2.7. Available online: https://www.snirh.gov.br/hidroweb/apresentacao (accessed on 1 January 2021).
32. Wu, Q. Geemap: A Python Package for Interactive Mapping with Google Earth Engine. *J. Open Source Softw.* **2020**, *5*, 2305. [CrossRef]
33. Drusch, M.; del Bello, U.; Carlier, S.; Colin, O.; Fernandez, V.; Gascon, F.; Hoersch, B.; Isola, C.; Laberinti, P.; Martimort, P.; et al. Sentinel-2: ESA's Optical High-Resolution Mission for GMES Operational Services. *Remote Sens. Environ.* **2012**, *120*, 25–36. [CrossRef]
34. Jensen, J.R. *Sensoriamento Remoto Ambiente: Uma Perspectiva Em Recursos Terrestres*; Pearson Prentice Hall: Upper Saddler River, NJ, USA, 2009; pp. 1–661. ISBN 978-85-60507-06-1. Available online: https://www.scribd.com/document/456069018/Sensoriamento-Remoto-do-Ambiente-Uma-Perspectiva-em-Recursos-Terrestres-John-R-Jensen-pdf (accessed on 22 January 2021).
35. Rouse, R.W.H.; Haas, J.A.W.; Deering, D.W. Monitoring vegetation systems in the Great Plains with ERTS. 3rd ERTS-1 Symposium, Vol. 1, Sect. A, Goddard Space Flight Center, NASA, 1 January 1974. Available online: https://ntrs.nasa.gov/citations/19740022614 (accessed on 16 February 2021).
36. McFeeters, S.K. The Use of the Normalized Difference Water Index (NDWI) in the Delineation of Open Water Features. *Int. J. Remote Sens.* **1996**, *17*, 1425–1432. [CrossRef]

37. Gordon, H.R.; Brown, J.W.; Evans, R.H. Exact Rayleigh Scattering Calculations for Use with the Nimbus-7 Coastal Zone Color Scanner. *Appl. Opt.* **1988**, *27*, 862–871. [CrossRef]
38. Kotchenova, S.Y.; Vermote, E.F.; Matarrese, R.; Klemm, F.J. Validation of a Vector Version of the 6S Radiative Transfer Code for Atmospheric Correction of Satellite Data. Part I: Path Radiance. *Appl. Opt.* **2006**, *45*, 6762–6774. [CrossRef] [PubMed]
39. Xu, H. Modification of Normalised Difference Water Index (NDWI) to Enhance Open Water Features in Remotely Sensed Imagery. *Int. J. Remote Sens.* **2006**, *27*, 3025–3033. [CrossRef]
40. Zambrano-Bigiarini, M. HydroGOF: Goodness-of-Fit Functions for Comparison of Simulated and Observed Hydrological Time Series. 2020. Available online: https://github.com/hzambran/hydroGOF (accessed on 10 October 2022).
41. Moriasi, D.N.; Arnold, J.G.; van Liew, M.W.; Bingner, R.L.; Harmel, R.D.; Veith, T.L. Model Evaluation Guidelines for Systematic Quantification of Accuracy in Watershed Simulations. *Trans ASABE* 1983, 50, 885–900. Available online: https://pubag.nal.usda.gov/catalog/9298 (accessed on 6 September 2022).
42. Van Liew, M.W.; Veith, T.L.; Bosch, D.D.; Arnold, J.G. Suitability of SWAT for the Conservation Effects Assessment Project: Comparison on USDA Agricultural Research Service Watersheds. *J. Hydrol. Eng.* **2007**, *12*, 173–189. [CrossRef]
43. Krug, L.A.; Noernberg, M.A. Extração de Batimetria por Sensoriamento Remoto de Áreas Rasas dos Sistemas Estuarinos do Estado do Paraná. In Proceedings of the Anais XII Simpósio Brasileiro de Sensoriamento Remoto, INPE, Goiânia, Brasil, 16–21 April 2005; p. 3077. Available online: https://www.scribd.com/document/485410630/3077-pdf# (accessed on 10 October 2022).
44. Olmanson, L.G.; Brezonik, P.L.; Bauer, M.E. Airborne Hyperspectral Remote Sensing to Assess Spatial Distribution of Water Quality Characteristics in Large Rivers: The Mississippi River and Its Tributaries in Minnesota. *Remote Sens. Environ.* **2013**, *130*, 254–265. [CrossRef]
45. Kwon, S.; Shin, J.; Seo, I.W.; Noh, H.; Jung, S.H.; You, H. Measurement of Suspended Sediment Concentration in Open Channel Flows Based on Hyperspectral Imagery from UAVs. *Adv. Water Resour.* **2022**, *159*, 104076. [CrossRef]
46. Dos Santos, F.M.; de Souza, N.P.; de Oliveira, R.P.; di Lollo, J.A. Using the SWAT Model to Identify Erosion Prone Areas and to Estimate Soil Loss and Sediment Transport in Mogi Guaçu River Basin in Sao Paulo State, Brazil. *Catena* **2023**, *222*, 106872. [CrossRef]
47. Riquetti, N.B.; Mello, C.R.; Leandro, D.; Guzman, J.A.; Beskow, S. Assessment of the Soil-Erosion-Sediment for Sustainable Development of South America. *J. Environ. Manag.* **2022**, *321*, 115933. [CrossRef]
48. Jin, F.; Yang, W.; Fu, J.; Li, Z. Effects of Vegetation and Climate on the Changes of Soil Erosion in the Loess Plateau of China. *Sci. Tot. Environ.* **2021**, *773*, 145514. [CrossRef]
49. Da Cunha, E.R.; Santos, C.A.G.; da Silva, R.M.; Panachuki, E.; de Oliveira, P.T.S.; de Oliveira, N.S.; dos Falcão, K.S. Assessment of Current and Future Land Use/Cover Changes in Soil Erosion in the Rio Da Prata Basin (Brazil). *Sci. Total Environ.* **2022**, *818*, 151811. [CrossRef]
50. Borella, D.R.; de Souza, A.P.; de Almeida, F.T.; de Abreu, D.C.; Hoshide, A.K.; Carvalho, G.A.; Pereira, R.R.; da Silva, A.F. Dynamics of Sediment Transport in the Teles Pires River Basin in the Cerrado-Amazon, Brazil. *Sustainability* **2022**, *14*, 16050. [CrossRef]
51. Ferreira, M.E.A.; Zanoni, D.A.; Carvalho, G.A.; Anache, J.A.A.; Oliveira, P.T.S.; Sobrinho, T.A. Fingerprinting Sediment Origin of the Silting Process of Urban Reservoirs. *Sustainability* **2023**, *15*, 1745. [CrossRef]

sustainability

MDPI

Article

Estimating Sugarcane Yield in a Subtropical Climate Using Climatic Variables and Soil Water Storage

Jessica Lima Viana [1,*], Jorge Luiz Moretti de Souza [2], Aaron Kinyu Hoshide [1,3], Ricardo Augusto de Oliveira [4], Daniel Carneiro de Abreu [1,5] and Wininton Mendes da Silva [6]

1 AgriSciences, Universidade Federal de Mato Grosso, Caixa Postal 729, Sinop 78550-970, MT, Brazil
2 Departamento de Solos e Engenharia Agrícola (DSEA), Campus Universitário de Curitiba, Universidade Federal do Paraná, Rua dos Funcionários, 1540, Curitiba 80035-050, PR, Brazil
3 College of Natural Sciences, Forestry and Agriculture, The University of Maine, Orono, ME 04469, USA
4 Departamento de Fitotecnia e Fitossanidade (DFF), Campus Universitário de Curitiba, Universidade Federal do Paraná, Rua dos Funcionários, 1540, Curitiba 80035-050, PR, Brazil
5 Instituto de Ciências Agrárias e Ambientais (ICAA), Campus Universitário de Sinop, Universidade Federal do Mato Grosso, Avenida Alexandre Ferronato, 1200, Sinop 78550-728, MT, Brazil
6 Empresa Mato-Grossense de Pesquisa, Assistência e Extensão Rural (EMPAER-MT), Centro Político Administrativo, Cuiabá 78049-903, MT, Brazil
* Correspondence: jessica.limavp@gmail.com; Tel.: +55-66-9966-6763

Abstract: Brazil is the largest producer of sugarcane (*Saccharum* spp.) in the world, and this crop's response to climate and soil water storage is essential for optimal management and genetic/yield improvements. The objective of our study was to build a multivariate model to estimate sugarcane yield in the subtropical conditions of the northwestern Paraná region using climatic and soil water storage variables. Observed yield data was used from experiments conducted at the Experimental Station of the Sugarcane Genetic Improvement Program of the Universidade Federal do Paraná. The sugarcane varieties RB72454, RB867515, RB966928, and RB036066 were analyzed in the 1998–2006, 2008, 2018 and 2019 harvest years. Stepwise multiple linear regression analysis with repeated cross-validation was developed to estimate sugarcane yield given climate and soil water storage variables for crop growth phases. The accumulated degree days in Phases I and II and soil water storage in Phase II of development significantly impacted sugarcane yield. The multiple linear regression model, with accumulated degree days and soil water storage in Phases I and II of development, successfully predicted sugarcane yield for analyzed varieties. Sugarcane production models like the one we developed can improve crop management for greater sustainability and climate change adaption in Brazil and other areas.

Keywords: agrometeorological modeling; multiple linear regression; statistical model; sugarcane; yield prediction

Citation: Viana, J.L.; de Souza, J.L.M.; Hoshide, A.K.; de Oliveira, R.A.; de Abreu, D.C.; da Silva, W.M. Estimating Sugarcane Yield in a Subtropical Climate Using Climatic Variables and Soil Water Storage. *Sustainability* **2023**, *15*, 4360. https://doi.org/10.3390/su15054360

Academic Editor: Cláudio Moisés Santos e Silva

Received: 3 February 2023
Revised: 22 February 2023
Accepted: 26 February 2023
Published: 28 February 2023

1. Introduction

Brazil is the largest producer of sugarcane (*Saccharum* spp.) in the world. The estimated production for the 2021–2022 harvest is 592 million metric tons [1]. However, the volume of raw material harvested this year was 9.5% less than the preceding 2020–2021 harvest due to drought during the production cycle and low temperatures in June and July, including frosts in some production areas [1]. The estimated production of the crop in the South Region of Brazil is 31.9 million metric tons, which was 6.7% lower compared to the previous year's harvest [1]. Climatic conditions during the sugarcane production cycle interferes with production. The identification of climate and soil factors that have a significant effect on the yield of sugarcane is fundamental to predicting production, successfully managing, selecting varieties, and suggesting alternatives for genetic improvement of the crop [2,3].

Sugarcane is a perennial crop, and its development is influenced by edaphoclimatic conditions [4,5]. Sugarcane production in Brazil expanded from 2000 to 2013; however,

88% of the increase was due to increase in production area and only 12% occurred due to increased productivity [6]. More than 90% of all sugarcane is grown without irrigation in Brazil, so the crop depends on rainfall to meet the sugarcane's water demand during growth, development, and production [6].

The availability of water in the soil during the vegetative growth phase is essential for sugarcane to express its productive potential, being the period in which the greatest water demand of the crop occurs [7,8]. Sugarcane stalk yields were 140% higher when grown under full irrigation compared to rain-fed situations, due to adequate water availability throughout the growth period, which makes nutrients available for the roots [9]. Extreme air temperatures, water deficit or surplus, and nitrogen deficiency interfere with the efficiency of solar radiation use and canopy expansion of sugarcane and, consequently, cane yield [10]. Therefore, the availability of water in the soil and climatic factors, such as solar radiation, influence the physiological processes of sugarcane.

In this context, the development and use of predictive models provide interesting alternative to help understand the responses of various crops to different environmental conditions in which they are grown, as well as in estimating their productivity [2]. The development of multiple linear regression models allows the identification of the variables that most interfere with crop yield and their estimation. Agrometeorological models, using as input monthly data of air temperature, precipitation, water deficiency and surplus, potential and actual evapotranspiration, water storage in the soil, and incident, global solar irradiation from the previous year were used by researchers [11] to estimate the sugarcane yields and quality. These yields were measured in metric tons per hectare, while sugarcane quality was measured as total recoverable sugar.

Multivariate regression models are an alternative to identify the variables that most influence sugarcane yield [11–13]. Multiple linear regression models estimating sugarcane yields with multiple variables have shown adequate performance. For example, precipitation, degree days, and negative degree days of the five months prior to harvest were used in multiple linear regression analysis to predict sugarcane yield [12]. Another multiple linear regression analysis verified the influence of climatic variables and available water in the soil on sugarcane productivity during El Niño and La Niña events in the state of Paraná during the summer, fall, winter, and spring [13].

Several researchers have sought to improve the genetics and productivity of sugarcane at the Experimental Station of the Sugarcane Genetic Improvement Program (PMG-CA) at the Federal University of Paraná (UFPR) as well as at the Interuniversity Network for the Development of Sugarcane Energy (RISESA) in Brazil [14,15]. Obtaining information on sugarcane yield in relation to soil and climate variables in long-term experiments can support sustainable agricultural management practices to ensure the growth, development, and production of sugarcane. In view of the context presented, the goal of the present study was to develop growth-stage (phenological phases I, II, and III) specific crop production models in order to better estimate sugarcane yield under subtropical soil and climate conditions. In order to achieve this goal, our research objectives were to (1) develop such growth-stage specific regression models for sugarcane specifically using climatic and soil water storage variables and to (2) validate our regression models with long-term sugarcane production data collected at PMGCA in Brazil.

2. Materials and Methods

2.1. Study Location and Agrometerological Data

Our research was developed in the Laboratory of Modeling of Agricultural Systems (LAMOSA) at the Setor de Ciências Agrárias (SCA) at the Universidade Federal do Paraná (UFPR). Analyses were carried out with crop data from experiments conducted at the Experimental Station of the Sugarcane Genetic Improvement Program (PMGCA) at UFPR and at the Interuniversity Network for the Development of Sugarcane Energy (a.k.a., RIDESA). The PMGCA is located at 22°58′ south latitude, 52°28′ west longitude and is at 470 m of average altitude in the municipality of Paranavaí, in the northwest region of the state of Paraná, Brazil. The climate of the region according to the Köppen classification is Cfa (subtropical climate), with average annual air temperature between 22.1 and 23 °C with average annual precipitation between 1400 and 1600 mm [16]. The agrometeorological weather station data used was obtained from the Paraná Rural Development Institute located in Paranavaí, Paraná state, Brazil for the period 1997 to 2019. Weather station variables used for modeling were incident solar radiation (Rs) measured in megajoules (MJ) per square meter (m^2) per day, rainfall (R) in millimeters, maximum temperature (T_{max}), minimum temperature (T_{min}), and average temperature (T_{avg}) all measured in °C, relative humidity (RH) as a percent (%), as well as wind speed (u) in meters per second [17].

2.2. Sugarcane Data

The observed sugarcane yields measure in metric tons per hectare were obtained from experiments conducted in the area of the Paranavaí Experiment Station, RIDESA, for the first ratoon cane crops (Table 1). Sugarcane yield data was available for the years 1998 to 2006, 2008, 2018, and 2019. The experimental station's experiments involved genetic improvement of sugarcane. From 1999 to 2005, experiments were conducted with other varieties. In this experimental area, sugarcane improvement research has historically been conducted and sugarcane varieties are grown in-field for evaluation. Sugarcane is typically planted in July with annual harvesting over a few years. Sugarcane harvesting in these experiments is done manually and without burning prior to harvest. Harvesting without burning leaves plant residues on the soil as straw and can improve yield and sugar content under conditions of water deficiency [18].

Table 1. Sugarcane yield measured in metric tons (t) of thatch per hectare (ha) for the varieties RB72454, RB867515, RB966928, and RB036066 from the Interuniversity Network for the Development of Sugarcane Energy in Paranavaí, Paraná state, Brazil [19].

	Cultivar's Observed Sugarcane Yield (t/ha)				
Year	**RB72454**	**RB867515**	**RB966928**	**RB036066**	**Average**
1998	130.604	-	-	-	130.604
1999	130.163	-	-	-	130.163
2000	98.988	-	-	-	98.988
2001	140.830	-	-	-	140.830
2002	127.940	-	-	-	127.940
2003	115.794	-	-	-	115.794
2004	141.600	99.304	113.417	-	118.107
2005	65.997	98.910	107.333	-	90.747
2006	164.190	134.173	-	-	149.182
2008	67.917	-	-	-	67.917
2018	-	80.715	152.570	101.533	111.606
2019	-	91.468	105.792	85.608	94.289

2.3. Adjustment of the Multiple Linear Regression Model

Multiple linear regression analyses using the stepwise method were carried out specifying sugarcane yield as the dependent variable in the models. Several independent variables were tested for use in our multiple linear regression models. These variables included soil water storage (SWS) to a depth of 0.6 m down in the soil profile. SWS is measured in centimeters. Weather station data used in regression models were reference evapotranspiration (ET) and daily rainfall (R) measured in millimeters per day, average relative humidity (RH) as a percentage (%), and incident solar radiation (R_s) and radiation balance at the surface (R_n), both measured in megajoules (MJ) per square meter (m^2) per day. Temperature (°C) variables included daily maximum (T_{max}), minimum (T_{min}), and average (T_{avg}) air temperatures as well as accumulated degree days (ADD). Total values were used for R and ADD, while median values were used for SWS, ET, RH, Rn, Rs, T_{max},

T_{min}, and T_{avg}). Median values were used since some variables (SWS, R, ADD, Rs, Rn) did not show normal distribution.

Sugarcane yields were modeled for three different phenological phases of ratoon sugarcane [20]. Phase I runs from July to October for a total of 93 days when sprouting and intense tillering occur. Phase II goes from October through March over 160 days where sugarcane has growth in stature. Phase III is the last phase running from March to July over 112 days where there is decrease in growth and accumulation of sucrose.

Soil water storage was simulated with the HYDRUS-1D software where HYDRUS numerically solves Richards' equation for water flow in saturated and unsaturated media [21]. The van Genuchten–Mualem model [22] was incorporated into HYDRUS to determine the relationship between hydraulic conductivity, volumetric moisture, and soil matric potential. The meteorological variables that were inputted into HYDRUS were rainfall, incident solar radiation, maximum and minimum air temperature, relative humidity, and wind speed. Reference evapotranspiration was calculated with the Penman–Montheith equation [23], which was incorporated into the program.

The accumulated degree days was calculated as the sum of degree days (*DD*) during the crop cycle according to past research [24]. The basal temperatures were considered equal to 19.0 °C, 23.5 °C, and 18.5 °C in sugarcane development stages I, II, and III, respectively The *DD* for these three stages are specified as:

$$DD = (T_{ni} - Tb_i) + \frac{(T_{xi} - T_{ni})}{2} \; for \; T_{ni} > Tb_i \quad (1)$$

$$DD = \frac{(T_{xi} - Tb_i)^2}{2 \cdot (T_{xi} - T_{ni})} \; for \; T_{ni} < Tb_i \quad (2)$$

$$DD = 0 \; for \; T_{xi} < Tb_i \quad (3)$$

where Tb_i is the lower base temperature of the crop, T_{xi} is maximum temperature of the *i*th day, and T_{ni} is minimum temperature of the *i*th day, all measured in °C.

2.4. Statistical Analyses

The associations between observed (Y) and estimated ($\hat{Y}$) values of sugarcane yield were evaluated with the error and coefficients of determination, calculated in the R software package, version 4.1.0 [25]. These associations included the mean absolute error (*MAE*) and the root mean square error (*RMSE*) measured in metric tons per hectare, and the coefficient of determination (R^2) which is dimensionless:

$$MAE = \frac{\sum_{i=1}^{n} \mid \hat{Y}_i - Y_i \mid}{n} \quad (4)$$

$$RMSE = \sqrt{\frac{\sum_{i=1}^{n} (\hat{Y}_i - Y_i)^2}{n}} \quad (5)$$

$$R^2 = \frac{(\sum_{i=1}^{n} (\hat{Y}_i - \overline{Y}) (Y_i - \overline{Y}))^2}{(\sum_{i=1}^{n} (\hat{Y}_i - \overline{Y})^2) (\sum_{i=1}^{n} (Y_i - \overline{Y})^2)} \quad (6)$$

where $\hat{Y}_i$ is the *i*th value of the estimated variable, Y_i is the *i*th value of the observed variable, n equals the number of observed data points, and $\overline{Y}$ is the mean of the values of the observed variable. Note that *MAE* and *RMSE* values equal or close to zero indicate perfect fit to the data [26].

Multiple linear regression analyses were developed with the stepwise method and repeated cross-validation for 3 repetitions with k-fold (K independent groups) equal to 5. The stepwise method is used to select the best subset of variables that results in the model with the lowest cross-validation prediction error [27]. In the k-fold cross-validation method the data is randomly divided into k groups of equal size, uses "k − 1" groups for model fitting and the first group for validation, The process is repeated k times, with a different group being used for validation. The validation error is calculated with the average of the testing errors [27]. The samples were divided into 70% for parameterization and 30% for validation, using the caret and leaps packages of R based on the methodology from previous studies [28,29].

Other statistical tests were used such as analysis of variance, Shapiro–Wilk test for normality, test for multicollinearity using the variance inflation factor, and the Breusch–Pagan test for homoscedasticity. The statistical performance of multiple linear regression models and independent variable coefficients were evaluated using a confidence level of $\alpha < 0.05$. Regression models estimated sugarcane yields where were then compared to observed yield data from field experiments at the Experimental Station of the Sugarcane Genetic Improvement Program at the Universidade Federal do Paraná in Paraná state, Brazil. This contrast regressed observed yields against regression estimated yields where perfect fit of data ($R^2 = 1$) would have coordinates all fall on a positively sloped 45-degree angle line from the origin. Regression models were run in R [25].

3. Results

3.1. Agrometerological Data

The trend of climatic variables and soil water storage in the phenological phases of sugarcane and years used in the multiple linear regression analysis is summarized in Figures A1–A10. The variables for soil water storage (SWS), evapotranspiration (ET), solar radiation (R_s), surface radiation balance (R_n), average daily rainfall (R), maximum temperature (T_{max}), minimum temperature (T_{min}), and average temperature (T_{avg}) showed higher magnitudes in Phase II during sugarcane development. The highest sugarcane yields are typically associated with high average temperatures and high and uniform precipitation during the full vegetative growth phase [8]. In our study, the highest average air temperatures and rainfall coincided with Phase II of sugarcane stalk growth. The only exception was in 2004 where for rainfall during Phase II was lower than rainfall in Phase III.

The sugarcane crop cycle should coincide with ideal weather conditions [7] such as water demand and availability. Prior research found daily reference evapotranspiration of 5.24 ± 1.27 mm per day during the vegetative growth phase of sugarcane, a period in which the highest water demand of sugarcane variety RB867515 occurred in the microregion of Teresina, Piauí state, Brazil [7]. In the climatic conditions of Paranavaí, Paraná state, Brazil where the experimental data that we used were measured, the highest reference evapotranspiration (ET) was also observed in Phase II of sugarcane development. Rainfall irregularities in a tropical climate can provide distinct responses in the development and productivity of sugarcane varieties in crop cycles such as those for the varieties RB93509 and RB931003 in the Coastal Tablelands region of Alagoas in northeastern Brazil, where rainfall was also irregular [30].

3.2. Multiple Linear Regressions

The root mean square error (RMSE) was used to select the optimal model during the multiple linear regression analyses with repeated cross-validation. The lowest RMSE corresponded to the model with two predictors (Figure 1). The accumulated degree days (ADD, °C) and soil water storage (SWS, centimeters) in Phases I and II of sugarcane development constituted the best model fit with two predictors (Figure 2). The regression model coefficients of ADD for sugarcane growth Phase I and SWS for growth Phase II showed statistical significance ($p < 0.05$) (Table 2).

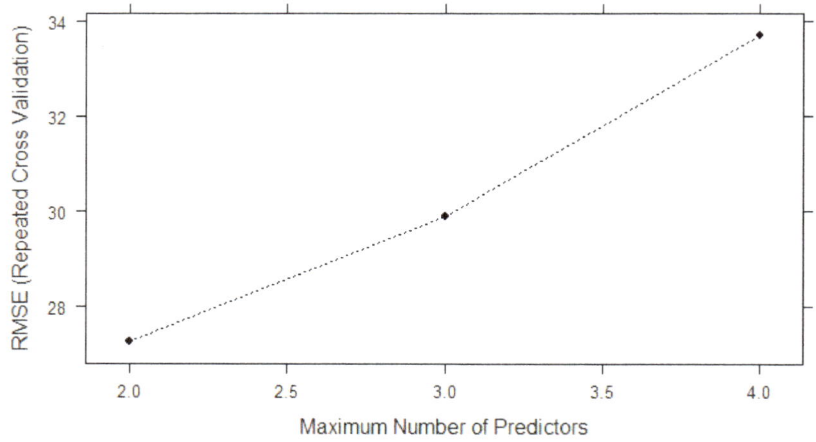

Figure 1. RMSE of repeated cross-validation to select the maximum number of predictors for multiple linear regression models of sugarcane yield estimation (TCH, tons of stalk per hectare), of ratoon cane, varieties RB72454, RB867515, RB966928, and RB036066.

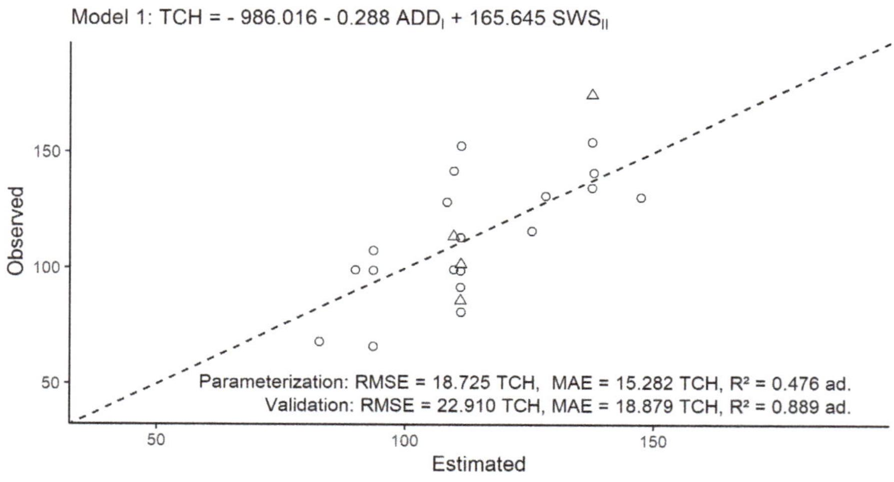

Figure 2. Parameterization and validation of the multiple linear regression model for estimating yield (TCH in metric tons of stalk per hectare) of ratoon cane, varieties RB72454, RB867515, RB966928, and RB036066 (O: parameterization and Δ: validation).

The coefficient of determination (R^2) explained 48% and 89% of the variation in sugarcane yield during model parameterization and validation, respectively. Varietal difference also interfered in the analyses, similar to prior research [31]. In our study, the sugarcane cultivar RB72454 had good agricultural productivity in any type of soil with medium maturity and high sucrose content. This cultivar was grown at the research site from 1998 to 2008. The second, third, and fourth sugarcane cultivars have been grown more recently in 2018 and 2019. The second variety evaluated was RB867515, which had high sucrose content and high agricultural productivity and is recommended for areas with low fertility and sandy and water-constrained soils. The third variety was RB966928, with high agricultural production, early to medium maturity, medium sucrose content, and a high degree of resistance to major diseases. Finally, the RB036066 sugarcane cultivar had high agricultural production, medium maturity, and wide adaptability and production stability. In the multiple linear regression analysis with repeated cross-validation for variety RB72454, the lowest RMSE also corresponded to the model with two predictors (Figure 3). The multiple linear regression analysis was performed only for variety RB72454, due to the greater number of 11 observations of sugarcane yield to allow for the parameterization and validation processes to be performed (Table 1).

Table 2. Multiple linear regression coefficients fitted to sugarcane yield (metric tons per hectare) for varieties RB72454, RB867515, RB966928, and RB036066, in the municipality of Paranavaí, northwestern region of Paraná State.

Coefficient [1]	Estimated	Standardized Estimated	Error [2]	t [3]	Pr(>∣t∣) [4]	$R^2_{ajust.}$ [5]
α	−986.016	∣0.000∣	412.180	−2.392	0.029 *	——
ADD_I	−0.288	∣−0.536∣	0.098	−2.928	0.010 *	0.410
SWS_{II}	165.645	∣0.524∣	57.961	2.858	0.011 *	——

[1] α is the regression coefficient, which represents the intercept; ADD_I is accumulated degree days in Phase I (°C); SWS_{II} is soil water storage up to 60 cm in Phase II. [2] Error is standard error (variable unit). [3] The t is the student's *t*-test values area dimensionless. [4] Pr(>∣t∣) is significance probability; * significant at 5% probability. [5] $R^2_{ajust.}$ is the adjusted coefficient of determination which is dimensionless. The *p*-value = 0.00572 is for the F-statistic.

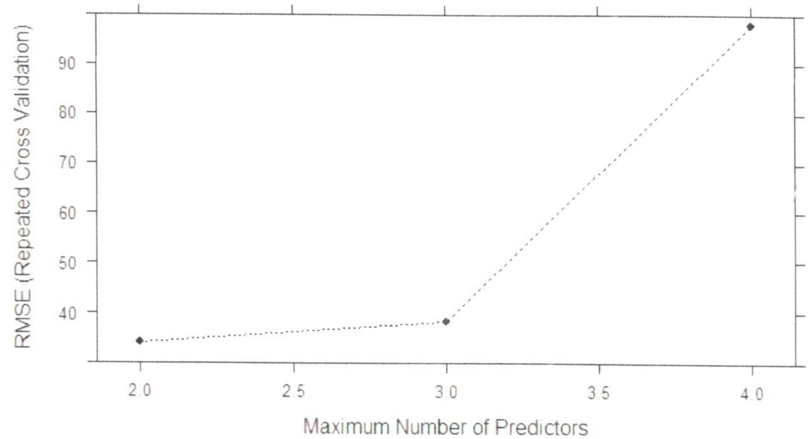

Figure 3. RMSE of repeated cross-validation to select the maximum number of predictors of multiple linear regression models for estimating yield (metric tons of stalk per hectare) of ratoon cane, variety RB72454.

The variables accumulated degree days (ADD_{II}) and soil water storage (SWS_{II}) in phenological phase II of ratoon cane were selected in the analysis and constituted the best model with two predictors for variety RB72454 (Figure 4). The regression coefficients of variables ADD_{II} and SWS_{II} showed statistical significance ($p < 0.05$) in the multiple linear regression model (Table 3).

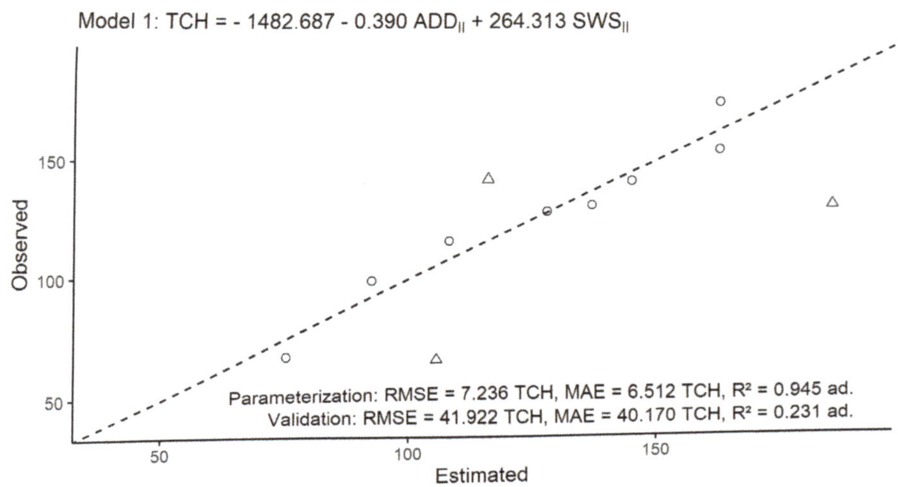

Figure 4. Parameterization and validation of the multiple linear regression model for estimating yield (TCH, tons of stalk per hectare) of ratoon cane, for the variety RB72454 (O: parameterization and Δ: validation).

Table 3. Multiple linear regression coefficients fitted to sugarcane yield (metric tons of per hectare) for variety RB72454 in the municipality of Paranavaí, northwestern region of Paraná State.

Coefficient [1]	Estimated	Standardized Estimated	Error [2]	t [3]	Pr(>|t|) [4]	$R^2_{ajust.}$ [5]
α	−1482.687	|0.000|	345.400	−4.292	0.008 *	——
ADD_{II}	−0.390	|−0.713|	0.058	−6.746	0.001 *	0.922
SWS_{II}	264.313	|0.593|	47.120	5.610	0.002 *	——

[1] The α is the regression coefficient, which represents the intercept; ADD_{II} is the accumulated degree days (°C), in Phase II; SWS_{II} is water storage in the soil up to 60 cm in Phase II. [2] Error is the standard error (unit of the variable). [3] The t is the student's *t*-test values (dimensionless). [4] Pr(>|t|) is the significance probability; * significant at 5% probability. [5] $R^2_{ajust.}$ Is the adjusted coefficient of determination, which is dimensionless. The *p*-value = 0.0007204 is for the F-statistic.

The coefficient of determination (R^2) explained 95% and 23% of the yield variation of variety RB72454 in the parameterization and validation of the model, respectively. The model fitted for variety RB72454 during validation showed the highest root mean square error (RMSE) and mean absolute error (MAE) and least explained variation. However, the effect of the selected variables should be considered due to statistical significance [12]. The climatic variations during the analyzed years is a factor that interfered in the explained variation of the model for variety RB72454 (Pr(>F) < 0.01; Figures A1–A10), as well as other factors that were not considered such as diseases and pests.

The accumulated degree days (ADD_{II}) and soil water storage (SWS_{II}) during Phase II of sugarcane development had an influence on the yield of sugarcane variety RB72454. The accumulated degree days in Phase I (ADD_{I}) and the soil water storage in Phase II (SWS_{II}) had significant effect on the yield of all four sugarcane varieties RB72454, RB867515, RB966928, and RB036066 under the climatic conditions in Paranavaí in Paraná state, Brazil. The identification of climate and soil variables that influence sugarcane yield is indispensable for complex models using a large number of parameters. Crop growth models [2] are fundamental for genetic improvement and identifying drivers for incremental improvements in crop yield. The results obtained in the experiments developed at the Experimental Station of the Sugarcane Genetic Improvement Program at the Universidade Federal do Paraná and at the Interuniversity Network for the Development of Sugarcane Energy can be references for future agricultural planning of sugarcane.

Past research has found that the variables selected in sugarcane modeling suggest that precipitation in the first, second, fourth, and fifth months and degree days in the fourth month prior to harvest had a significant effect on sugarcane yield during the 1999–2000, 2000–2001, and 2001–2002 harvests in the municipality of Pontal, São Paulo [12]. Precipitation was the variable with the most impact on sugarcane yield prediction models, confirming the importance of soil moisture for sugarcane production. Water stress during sugarcane crop development restricts physiological processes, such as its cell division and elongation, consequently reducing aboveground dry biomass [32]. Under severe water stress conditions such as between $-1500 < \psi < -1100$ kilopascals (kPa) measured with the weighing method measuring volumetric moisture versus matric potential, the daily transpiration rate of sugarcane was reduced by approximately 73% compared to plants grown under full water availability. With severe water stress and high values of global solar radiation, leaf temperature reached up to 6.6 °C above air temperature [33].

Research results reaffirm the importance of air temperature and soil water availability for sugarcane yield. For example, the average minimum temperature in spring, sum of precipitation in winter, sum of excess soil water in fall and summer, and the sum of soil water deficiency during summer explained 98% of the variation in sugarcane average annual yields (adjusted $R^2 = 0.982$) in Paranavaí, Paraná state, Brazil [13]. The accumulated degree days is a parameter used in simulation models of both sugarcane growth and production [34] and variation of aerial dry matter accumulation [35]. The model established in our study using accumulated degree days and soil water storage confirmed the importance of both variables in determining sugarcane yields due to their influence on the growth and development of the crop.

The climatic variables and water storage in the soil during the development phases for sugarcane can be used to predict the sugarcane yields. It is important to analyze and build models for each sugarcane variety since sugarcane cultivars respond differently to climatic and soil conditions such as water deficit [36], full irrigation [9], and intercepted photosynthetic radiation [37]. The number of observations is also an important factor in proposing models that explain sugarcane yield. As verified in our study, the reduced number of observations for sugarcane varieties RB867515, RB966928, and RB036066 limited the parameterization and validation of each variety.

The magnitude of the explanatory variables used to constitute models also Influences the ability to predict sugarcane yield. Previous research was unable to establish a predictive model for sugarcane yields using rainfall, degree days, and negative degree days of the five months prior to harvest [12]. The inclusion of other months in the multiple regression analysis would be an alternative, especially for months of vegetative growth. However, with these variables, two models were established to predict sugarcane maturity. During the maturation phase of sugarcane, the most relevant variables were found to be soil moisture and air temperature, with synergistic effects of the combination of both of these factors [38]. Temperatures lower than 20°C decreased growth and increased sucrose accumulation [39,40]. Future research should focus on modeling morphological variables for sugarcane yield models such as dry mass of the plant area and roots, plant height, diameter, and leaf area in relation to climate and soil variables during the phenological phases of the crop. This will help to better understand and identify relevant variables that most interfere in or enhance the development of and, consequently, the yield of sugarcane.

4. Discussion

4.1. Application to Previous Sugarcane Modeling

Our development of a sugarcane growth-stage-specific yield model can be integrated into other models that are more process based and not growth stage specific. These process-based models include DSSAT-Canegro [41,42] and APSIM-Sugar [43,44], specifically for the APSIM-Sugar model's specifications for sugarcane plant transpiration efficiency and supply of water to plant roots [45]. Our results were consistent with other studies that have used these models to estimate sugarcane yields based off of historical yield data in the major production region for sugarcane in southeastern Brazil. For example, our sugarcane total yield range of 80 to 140 metric tons (t)/hectare (ha) (Figure 2) was similar to Marin et al. (2012) [41], who estimated yield ranges of 96 to 129 t/ha for mid-century forecasting. In drier regions of Brazil, sugarcane yields and simulated yields are lower. For example, Dias et al. (2019) [10] modeled sugarcane yields ranging from 40 to 70 t/ha for the northeastern Brazilian state of Piauí in the arid Caatinga biome.

4.2. Sustainable Agricultural Implications

The sustainability of sugarcane in Brazil has both short-term and long-term challenges. In the short run, this involves having the crop play a role in conservation of natural areas such as on-farm required reserves of natural vegetation as well rehabilitating degraded livestock pastures. In the long run, land conservation and preservation are also linked to climate change and reduction in national greenhouse gas (GHG) emissions. Better simulation models for sugarcane can help balance crop productivity with agro-environmental goals. These improved models also allow for better forecasting of both future climate change impacts on sugarcane productivity as well as evaluating current and future crop systems with lower GHG emissions.

Recent sugarcane expansion in Brazil has occurred, for the most part, on land that has already been deforested [46], for example, for the Cerrado savannah and Atlantic Forest biomes [47]. Therefore, improving productivity and sustainability in sugarcane has more indirect effects on land-use change since improving productivity potentially frees up land to grow other crops and agricultural plantings (e.g., reseeded pasture) that are more likely to be introduced following deforestation and native vegetation conversion to agriculture. Although sugarcane is typically established on land that has already been cleared, it is not clear if this would indirectly drive Amazon deforestation for extensive pastured beef cattle due to conversion of pasture to sugarcane further to the southeast of Brazil [48].

Among the main results of this study and our research [49], soil water storage (SWS) can be modified using soil management practices. Sandy soils present differences in water retention due to the particle size composition of the sand fraction, which is an inherent soil characteristic [50,51]; organic matter content [52,53]; as well as land use and management [54]. Sugarcane is more susceptible to water stress during drought due to its shallow root system. Sugarcane water stress is especially severe in sandy soils due to low soil water retention, high soil density, high hydraulic conductivity, and greater root length density and root volume in the upper soil (0 to 0.2 m) layer [51].

According to Cherubin et al., 2022, the crop residue left behind as a soil surface mulch after green harvest of sugarcane can improve moisture retention [55]. Sandy soils require more mulching for adequate soil health for higher yielding sugarcane crops compared to clayey soils [56]. Such crop residues increased root mass and productivity of sugarcane even during drought in a Latosol with a clayey texture in a subtropical climate when 10 t/ha was applied [57]. However, Gallo et al., 2023 [58] demonstrated with field-study validated geospatial modeling in southeastern São Paulo state that unless other soil conservation methods (e.g., reduced tillage, contours, and terraces) were used then straw could not be removed.

In order to improve the physical and hydric conditions of the soil, using sustainable management systems is recommended. Sustainable management practices include continuous harvest without burning and leaving crop residues such as straw on the soil surface following harvest [57]. Another sustainable practice is crop rotation where such crop diversification of agro-ecosystem species can boost soil organic matter and biological activity [59]. This can increase water retention in the soil and reduce compaction in sandy soils. Soil compaction in sandy soils occurs due to heavy and intense traffic during agricultural machinery operations, especially during harvest [60]. Sugarcane yields have declined over the past decade due to this soil compaction from heavy equipment used for green harvesting [55]. However, Souza et al. (2020) found no compaction in sandy soils for sugarcane in Paraíba state, Brazil [61].

Climate change is projected to have adverse impacts on Brazil's sugarcane industry. Duden et al., 2021 [62] used the PCR-GLOBWB spatial terrestrial hydrology model to predict water deficit especially in center-west Brazil with projected increased sugarcane production for ethanol over the next decade. In São Paulo state, the transition to green harvest from burning prior to harvest has resulted in reduction in both total and per hectare greenhouse gas (GHG) emissions attributed to burning, fuel and input use, as well as from sugarcane crop and processing residues [63]. The APSIM-Sugar model forecasts under both RCP8.5 (business as usual) and RCP4.5 (greenhouse gas emission reductions) suggest that rainfed sugarcane yields would decline compared to the past ~40 years due to water stress. Even irrigated yields showed slight decline, but these projected changes were highly variable toward the end-of-21st century [44].

In the face of climate change and intensive land use in conventional systems, the adoption of sustainable management systems is critical for sugarcane cultivation [56]. Sugarcane shows no economies of scale due to lower yields from more industrial harvesting for large-sized farms, so the most profitable farm size at this point is for medium-sized farms [64]. Ruan et al., 2018 [65] use the APSIM-Sugar model to predict sugarcane yields under future climate change with results that suggest future sugarcane yields could be higher in southern China. However, this is challenging since China is third in 2020 global sugarcane production behind Brazil and India at only 14.28% of Brazil's sugarcane production [66]. Therefore, the best opportunity to balance global sugar needs with environmental sustainability may very well rest with Brazil.

5. Conclusions

In the present study, we investigated the main climatic variables and soil water storage characteristics influential in the development phases for sugarcane. Phase I is the sprouting and intense tillering period. Phase II is where the sugarcane crops grows and increases in height. Finally, Phase III is where there is a reduction in growth accompanied by sucrose accumulation. These three phases of sugarcane growth can be separately modeled in order to predict the yield of sugarcane in a subtropical climate grown in sandy soil. The accumulated degree days in Phases I and II and the soil water storage in Phase II of development exerted significant effects on sugarcane yield. Our multiple linear regression model with accumulated degree days (ADD_I) and water storage in the soil (SWS_{II}) in Phases I and II of development allowed for improved predictive capacity of sugarcane yield for the varieties that we analyzed. The model uses data that can be obtained during the sugarcane growing cycle and is an alternative resource to support production decisions related to soil and sugarcane plant management. Our model can also help support the adoption of sustainable management of sugarcane as well as yield forecasting. Our sugarcane production models for these crop development phases can be integrated in the future with whole-growth-stage models such as DSSAT-Canegro and APSIM-Sugar, with calibration of the coefficients, in order to better model future sustainable agricultural pathways for sugarcane both in Brazil and more globally.

Author Contributions: Conceptualization, J.L.V. and J.L.M.d.S.; methodology, J.L.V.; software, J.L.V.; validation, J.L.V.; formal analysis, J.L.V.; investigation, J.L.V.; data curation, J.L.V.; writing—original draft preparation, J.L.V.; J.L.M.d.S. and A.K.H.; writing—review and editing, J.L.V.; J.L.M.d.S., R.A.d.O., A.K.H., D.C.d.A. and W.M.d.S.; supervision, J.L.M.d.S.; project administration, R.A.d.O.; resources, A.K.H. All authors have read and agreed to the published version of the manuscript.

Funding: This research received no external funding.

Institutional Review Board Statement: Not applicable.

Informed Consent Statement: Not applicable.

Data Availability Statement: Study data can be obtained by request to the corresponding author or the first author via e-mail. It is not available on the website, as the research project is still under development.

Acknowledgments: To Capes for the financial support with a doctorate, to RIDESA for the availability of the experimental area and contribution to the research, to the Post-Graduate Program in Soil Science (UFPR), and the Modeling of Agricultural Systems Laboratory (LAMOSA)/SCA/UFPR.

Conflicts of Interest: The authors declare no conflict of interest. Supporting entities had no role in the design of the study; in the collection, analyses, or interpretation of data; in the writing of the manuscript, or in the decision to publish the results.

Appendix A

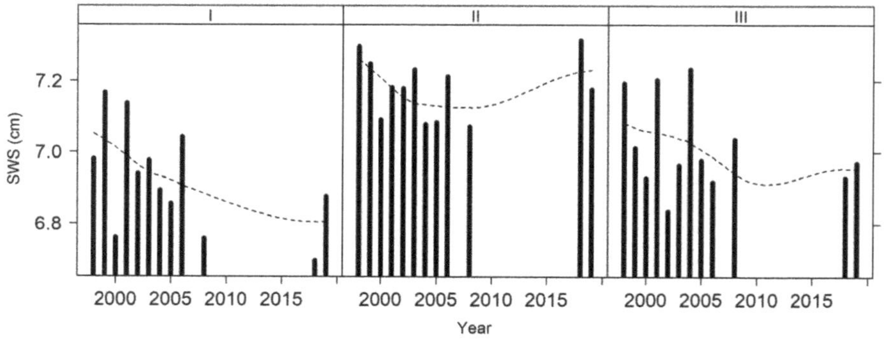

Figure A1. Climate variables for soil water storage (SWS in centimeters (cm)) in the sugarcane development phases I, II, and III (median) and cycle in the municipality of Paranavaí in the northwestern region of Paraná state, Brazil (dashed line corresponds to fitted trend).

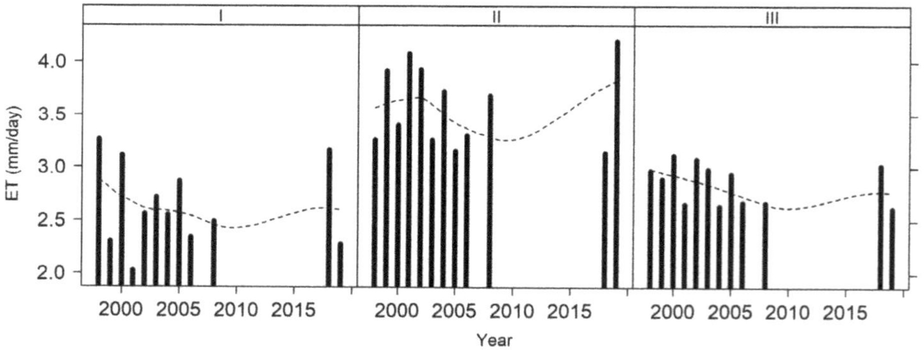

Figure A2. Climate variables for reference evapotranspiration (ET in millimeters (mm)/day) in the sugarcane development phases I, II, and III (median) and cycle in the municipality of Paranavaí in the northwestern region of Paraná state, Brazil (dashed line corresponds to fitted trend).

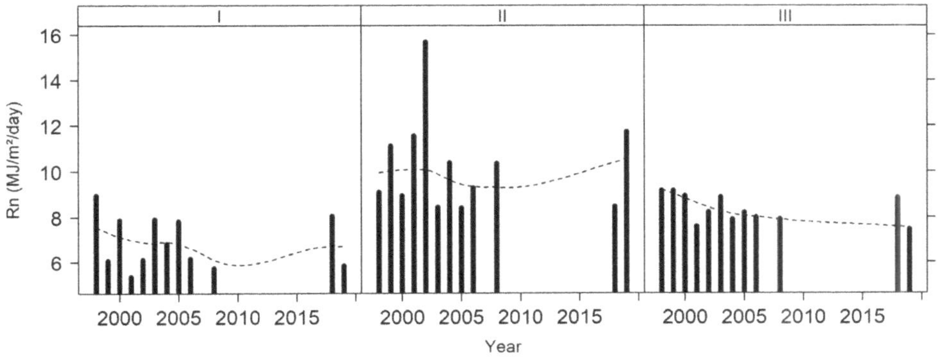

Figure A3. Climate variables for surface radiation balance (Rn in megajoules (MJ)/square meter (m^2)/day) in the sugarcane development phases I, II, and III (median) and cycle in the municipality of Paranavaí in the northwestern region of Paraná state, Brazil (dashed line corresponds to fitted trend).

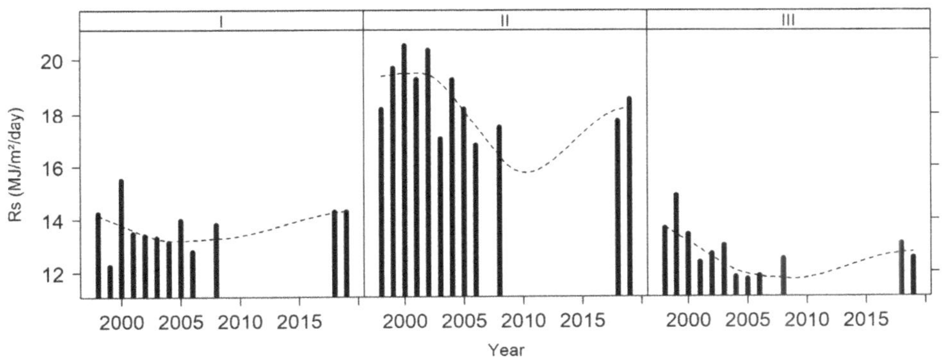

Figure A4. Climate variables for solar radiation (Rs in MJ/m^2/day) in the sugarcane development phases I, II, and III (median) and cycle in the municipality of Paranavaí in the northwestern region of Paraná state, Brazil (dashed line corresponds to fitted trend).

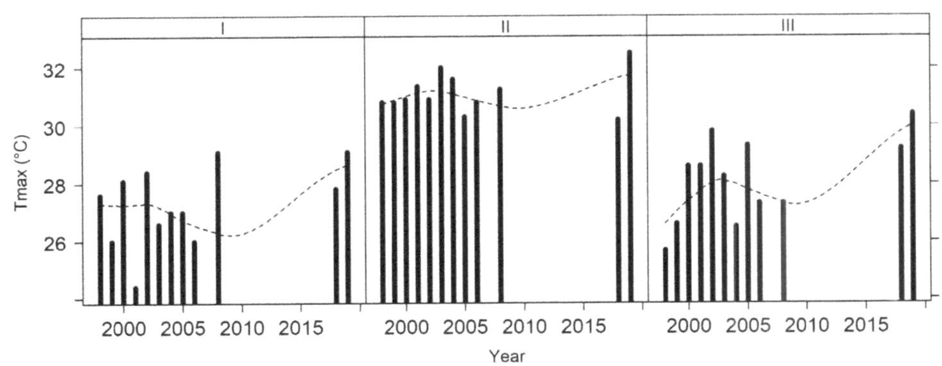

Figure A5. Climate variables for maximum temperature (Tmax in °C) in the sugarcane development phases I, II, and III (median) and cycle in the municipality of Paranavaí in the northwestern region of Paraná state, Brazil (dashed line corresponds to fitted trend).

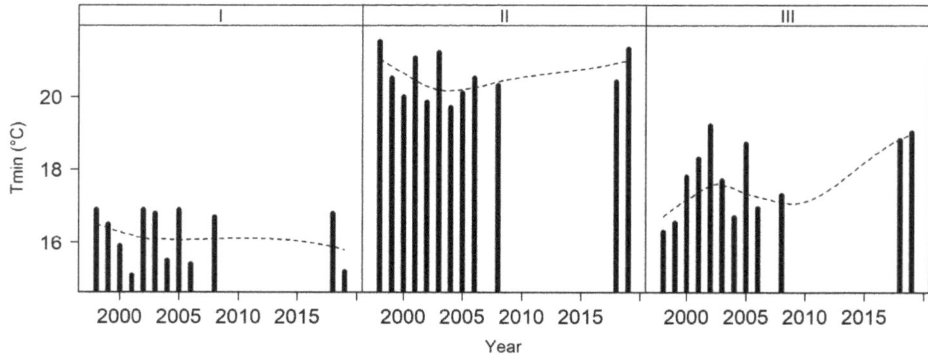

Figure A6. Climate variables for minimum temperature (Tmin in °C) in the sugarcane development phases I, II, and III (median) and cycle in the municipality of Paranavaí in the northwestern region of Paraná state, Brazil (dashed line corresponds to fitted trend).

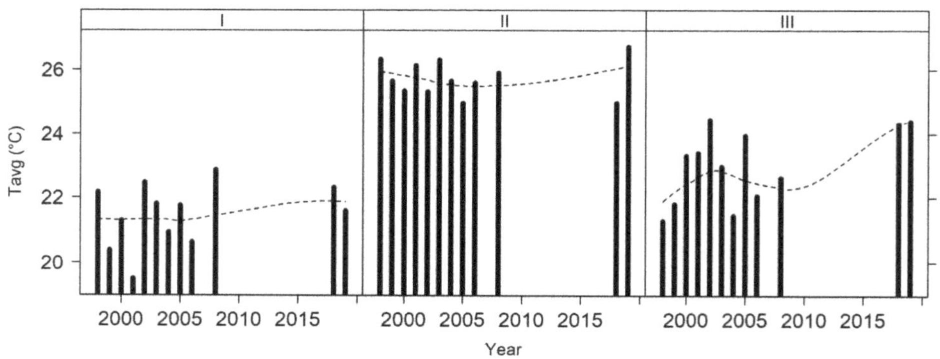

Figure A7. Climate variables for average temperature (Tavg in °C) in the sugarcane development phases I, II, and III (median) and cycle in the municipality of Paranavaí in the northwestern region of Paraná state, Brazil (dashed line corresponds to fitted trend).

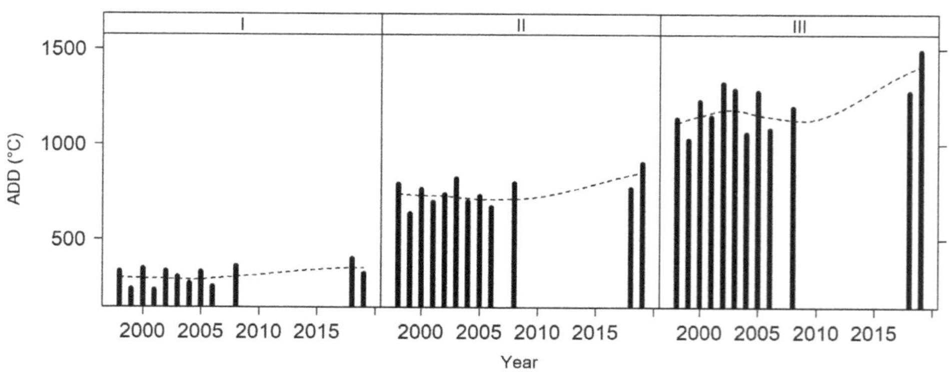

Figure A8. Climate variables for accumulated degree days (ADD in °C) in the sugarcane development phases I, II, and III (accumulated) and cycle in the municipality of Paranavaí in the northwestern region of Paraná state, Brazil (dashed line corresponds to fitted trend).

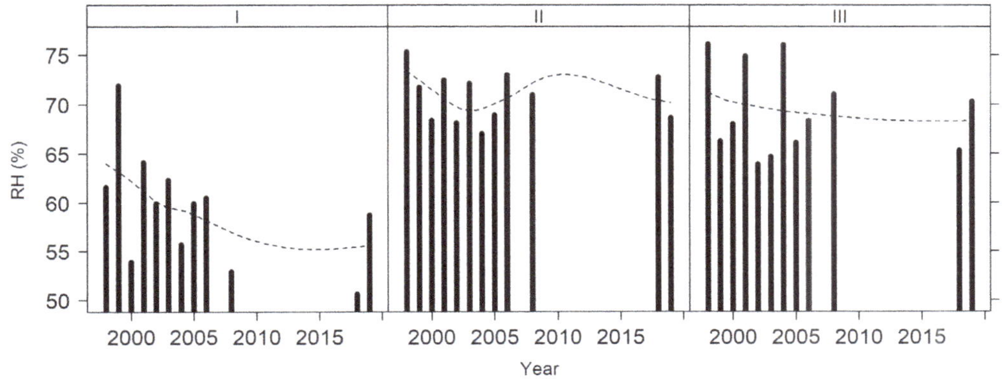

Figure A9. Climate variables for relative humidity (RH as %) in the sugarcane development phases I, II, and III (median) and cycle in the municipality of Paranavaí in the northwestern region of Paraná state, Brazil (dashed line corresponds to fitted trend).

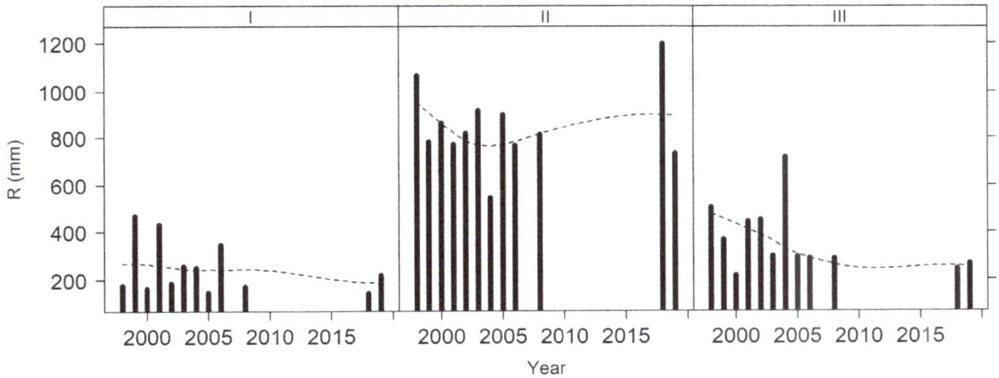

Figure A10. Climate variables for average daily rainfall (R in millimeters (mm)) in the sugarcane development phases I, II, and III (accumulated) and cycle in the municipality of Paranavaí in the northwestern region of Paraná state, Brazil (dashed line corresponds to fitted trend).

References

1. CONAB (Companhia Nacional de Abastecimento). Acompanhamento da Safra Brasileira de Cana-de-Açúcar, Safra 2021/22—Segundo Levantamento, Brasília. Available online: https://www.conab.gov.br/info-agro/safras/cana (accessed on 27 April 2021).
2. Scarpari, F.V.; van Lier, Q.J.; Corrêa, S.T.R.; Barros, A.H.C.; Marin, F.R.; Nassif, D.S.P. Modelos de crescimento da cana-de-açúcar e sua parametrização—Revisão. *Rev. Agric.* **2012**, *87*, 66–80. [CrossRef]
3. Daros, E.; Oliveria, R.A.; Zambon, J.L.C.; Bespalhok Filho, J.C.; Brasileiro, B.P.; Ido, O.T.; Ruaro, L.; Weber, H. RB036066—A sugarcane cultivar with high adaptability and yield stability to Brazilian South-Central region. *Crop Breed. Appl. Biotechnol.* **2018**, *18*, 325–329. [CrossRef]
4. Manhães, C.M.C.; Garcia, R.F.; Francelino, F.M.A.; Francelino, H.O.; Coelho, F.C. Fatores que afetam a brotação e o perfilhamento da cana-de açúcar. *Vértices* **2015**, *17*, 163–181. [CrossRef]
5. Marin, F.R.; Carvalho, G.L. Spatio-temporal variability of sugarcane yield efficiency in the state of São Paulo, Brazil. *Pesqui. Agropecu. Bras.* **2012**, *47*, 149–156. [CrossRef]
6. Marin, F.R.; Martha, G.B., Jr.; Cassman, K.G.; Grassini, P. Prospects for Increasing Sugarcane and Bioethanol Production on Existing Crop Area in Brazil. *BioScience* **2016**, *66*, 307–316. [CrossRef] [PubMed]
7. Andrade, A.S., Jr.; Noleto, D.H.; Bastos, E.A.; Moura, M.S.B.; Anjos, J.C.R. Demanda hídrica da cana-de-açúcar, por balanço de energia, na microrregião de Teresina, Piauí. *Agrometeoro, Passo Fundo* **2017**, *25*, 229–238. [CrossRef]

8. Araújo, R.M. Influência das Condições Climáticas no Crescimento e Desenvolvimento da Cana-de-Açúcar na Região sul do Brasil. Ph.D. Dissertation, Universidade Federal do Paraná, Setor de Ciências Agrárias, Programa de Pós-Graduação em Agronomia—Produção Vegetal, Curitiba, PR, Brazil, 14 March 2019.
9. Oliveira, E.C.A.; Freire, F.J.; Oliveira, A.C.; Simões Neto, D.E.; Rocha, A.T.; Carvalho, L.A. Produtividade, eficiência de uso da água e qualidade tecnológica de cana-de-açúcar submetida a diferentes regimes hídricos. *Pesqui. Agropecu. Bras.* **2011**, *46*, 617–625. [CrossRef]
10. Dias, H.B.; Inman-Bamber, G.; Bermejo, R.; Sentelhas, P.C.; Christodoulou, D. New APSIM-Sugar features and parameters required to account for high sugarcane yields in tropical environments. *Field Crops Res.* **2019**, *235*, 38–53. [CrossRef]
11. Marcari, M.A.; Rolim, G.S.; Aparecido, L.E.O. Agrometeorological models for forecasting yield and quality of sugarcane. *Aust. J. Crop Sci.* **2015**, *9*, 1049–1056.
12. Scarpari, M.S. Modelos Para a Previsão da Produtividade da Cana-de-Açúcar (*Saccharum* spp.) Através de Parâmetros Climáticos. Master's Thesis, Escola Superior de Agricultura "Luiz de Queiroz", Universidade de São Paulo, Piracicaba, SP, Brazil, 13 September 2002.
13. Andretta, R.L. Influência de Variáveis Climáticas, da Água Disponível no solo, e dos Eventos El Niño e La Niña na Produtividade da Cana-de-Açúcar no Estado do Paraná. Master's Thesis, Universidade Federal do Paraná, Setor de Ciências Agrárias, Programa de Pós-Graduação em Agronomia—Produção Vegetal, Curitiba, PR, Brazil, 29 February 2012.
14. Berton, G.S.; Oliveira, R.A.; Daros, E.; Zambon, J.L.C.; Bespalhok Filho, J.C.; Brasileiro, B.P.; Ido, O.T.; Ruaro, L.; Weber, H. RB036091—An early-maturing sugarcane cultivar for the Central South of Brazil. *Crop Breed. Appl. Biotechnol.* **2020**, *20*, e28852029. [CrossRef]
15. Borella, J.; Trautenmüller, J.W.; Brasileiro, B.P.; Oliveira, R.A.; Bespalhok Filho, J.C. Logistic model to selection of energy cane clones. *Cien. Rural.* **2020**, *50*, e20190750. [CrossRef]
16. Nitsche, P.R.; Caramori, P.H.; Ricce, W.S.; Pinto, L.F.D. *Atlas Climático do Estado do Paraná*; Instituto Agronómico do Paraná: Londrina, PR, Brazil, 2019; pp. 1–216, ISBN 978-85-88184-58-3.
17. Instituto de Desenvolvimento Rural do Paraná (IDR-Paraná)—IAPAR-EMATER. Available online: https://www.idrparana.pr.gov.br/ (accessed on 31 January 2023).
18. Aquino, G.S.; Medina, C.C.; Silvestre, D.A.; Gomes, E.C.; Cunha, A.C.B.; Kussaba, D.A.O.; Almeida, L.F.; Shahab, M.; Santiago, A.D. Straw removal of sugarcane from soil and its impacts on yield and industrial quality ratoons. *Sci. Agric.* **2018**, *75*, 526–529. [CrossRef]
19. Rede Interuniversitária Para o Desenvolvimento do Setor Sucroenergético (RIDESA). Available online: https://www.ridesa.com.br/ (accessed on 31 January 2023).
20. Gurski, B.C.; Souza, J.L.M.; Gerstemberg, E.; Oliveira, R.A. Water requirements and restrictions to sugarcane in cane plants and ratoon cane cycles in Southern Brazil. *Acta Agron.* **2020**, *69*, 135–143. [CrossRef]
21. Simunek, J.; Sejna, M.; Saiato, H.; Sakai, M.; van Genuchten, M.T. *The HYDRUS-1D Software Package for Simulating the One-Dimensional Movement of Water, Heat, and Multiple Solutes in Variably-Saturated Media: Version 4.08*; University of California Riverside: Riverside, CA, USA, 2009; pp. 1–330.
22. Van Genuchten, M.T. A closed-form equation for predicting the hydraulic conductivity of unsaturated soils. *Soil Sci. Soc. Am. J.* **1980**, *44*, 892–898. [CrossRef]
23. Allen, R.G.; Pereira, L.S.; Raes, D.; Smith, M. *Crop Evapotranspiration: Guidelines for Computing Crop Water Requirements*; Irrigation and Drainage Paper, 56; Food and Agriculture Organization of the United Nations: Rome, Italy, 1998; pp. 1–297, ISBN 92-5-104219-5.
24. Villa Nova, N.A.; Pedro, M.J., Jr.; Pereira, A.R.; Ometto, J.C. Estimativa de graus-dia acumulados acima de qualquer temperatura base em função das temperaturas máxima e mínima. *Ciênc. Terra* **1972**, *30*, 1–8.
25. R Core Team. R: A language and environment for statistical computing. R Foundation for Statistical Computing, Vienna, Austria. 2020. Available online: https://www.R-project.org/ (accessed on 3 January 2021).
26. Mangiafico, S.S. Summary and Analysis of Extension Program Evaluation in R, Version 1.18.1. 2016. Available online: https://rcompanion.org/handbook/ (accessed on 10 January 2021).
27. James, G.; Witten, D.; Hastie, T.; Tibshirani, R. *An Introduction to Statistical Learning: With Applications in R*, 2nd ed.; Springer: Cham, Switzerland, 2014; Available online: https://www.statlearning.com/ (accessed on 2 January 2022).
28. Lumley, T. Leaps: Regression Subset Selection. R Package Version 3.1, Based on Fortran Code by Miller, A. 2020. Available online: https://CRAN.R-project.org/package=leaps (accessed on 5 February 2021).
29. Kuhn, M. Classification and Regression Training [R Package Caret Version 6.0-90]. 2021. Available online: https://CRAN.R-project.org/package=caret (accessed on 5 February 2021).
30. Abreu, M.L.; Silva, M.A.; Teodoro, I.; Holanda, L.A.; Sampaio Neto, G.D. Crescimento e produtividade de cana-de-açúcar em função da disponibilidade hídrica dos Tabuleiros Costeiros de Alagoas. Bragantia. *Premissas Campinas* **2013**, *72*, 262–270. [CrossRef]
31. Daros, E.; Oliveira, R.A.; Barbosa, G.V.S. *45 Anos de Variedades RB de Cana-de-Açúcar: 25 Anos de Ridesa*, 1st ed.; Curitiba: Graciosa, Portugal, 2015; pp. 1–156.
32. Inman-Bamber, N.G. Sugarcane water stress criteria for irrigation and drying off. *Field Crops Res.* **2004**, *89*, 107–122. [CrossRef]

33. Trentin, R.; Zolnier, S.; Ribeiro, A.; Steidle Neto, A.J. Transpiração e temperatura foliar da cana-de-açúcar sob diferentes valores de potencial matricial. *Eng. Agrícola Jaboticabal* **2011**, *31*, 1085–1095. [CrossRef]
34. Marin, F.R.; Jones, J.W. Process-based simple model for simulating sugarcane growth and production. *Sci. Agric.* **2014**, *71*, 1–16. [CrossRef]
35. Batista, E.L.S.; Zolnier, S.; Ribeiro, A.; Lyra, G.B.; Silva, T.G.F.; Boehringer, D. Modelagem do crescimento de cultivares de cana-de-açúcar no período de formação da cultura. *Rev. Bras. Eng. Agrícola Ambient.* **2013**, *17*, 1080–1087. [CrossRef]
36. Holanda, L.A.; Santos, C.M.; Sampaio Neto, G.D.; Sousa, A.P.; Silva, M.A. Variáveis morfológicas da cana-de-açúcar em função do regime hídrico durante o desenvolvimento inicial. *Irriga. Botucatu* **2014**, *19*, 573–584. [CrossRef]
37. Ferreira, R.A., Jr.; Souza, J.L.; Lyra, G.B.; Teodoro, I.; Santos, M.A.; Porfirio, A.C.S. Crescimento e fotossíntese de cana-de-açúcar em função de variáveis biométricas e meteorológicas. *Rev. Bras. Eng. Agrícola Ambient.* **2012**, *16*, 1229–1236. [CrossRef]
38. Cardozo, N.P.; Sentelhas, P.C. Climatic effects on sugarcane ripening under the influence of cultivars and crop age. *Sci. Agric.* **2013**, *70*, 449–456. [CrossRef]
39. Scarpari, M.S.; Beauclair, E.G.F. Sugarcane maturity estimation through edaphic-climatic parameters. *Sci. Agric.* **2004**, *61*, 486–491. [CrossRef]
40. Scarpari, M.S.; Beauclair, E.G.F. Physiological model to estimate the maturity of sugarcane. *Sci. Agric.* **2009**, *66*, 622–628. [CrossRef]
41. Marin, F.R.; Jones, J.W.; Singels, A.; Royce, F.; Assad, E.D.; Pellegrino, G.Q.; Justino, F. Climate change impacts on sugarcane attainable yield in southern Brazil. *Clim. Chang.* **2013**, *117*, 227–239. [CrossRef]
42. Jones, M.R.; Singels, A.; Chinorumba, S.; Patton, A.; Poser, C.; Singh, M.; Martiné, J.F.; Christina, M.; Shine, J.; Annandale, J.; et al. Exploring process-level genotypic and environmental effects on sugarcane yield using an international experimental dataset. *Field Crops Res.* **2019**, *244*, 107622. [CrossRef]
43. Inman-Bamber, N.G.; McGlinchey, M.G. Crop coefficients and water-use estimates for sugarcane based on long-term Bowen ration energy balance measurements. *Field Crops Res.* **2003**, *83*, 125–138. [CrossRef]
44. Dias, H.B.; Sentelhas, P.C.; Inman-Bamber, G.; Everingham, Y. Sugarcane yield future scenarios in Brazil as projected by the APSIM-Sugar model. *Ind. Crops Prod.* **2021**, *171*, 113918. [CrossRef]
45. Inman-Bamber, N.G.; Jackson, P.A.; Stokes, C.J.; Verrall, S.; Lakshmanan, P.; Basnayake, J. Sugarcane for water-limited environments: Enhanced capability of the APSIM sugarcane model for assessing traits for transpiration efficiency and root water supply. *Field Crops Res.* **2016**, *196*, 112–123. [CrossRef]
46. Leal, M.R.L.V.; Duft, D.G.; Hernandes, T.A.D.; Bordonal, R.O. Brazilian Sugarcane Expansion and Deforestation. In Proceedings of the 25th European Biomass Conference and Exhibition, Stockholm, Sweden, 12–15 June 2017; pp. 1476–1483. [CrossRef]
47. Hernandes, T.A.D.; de Oliveira Bordonal, R.; Duft, D.G.; Leal, M.R.L.V. Implications of regional agricultural land use dynamics and deforestation associated with sugarcane expansion for soil carbon stocks in Brazil. *Reg. Environ. Chang.* **2022**, *22*, 49. [CrossRef]
48. Follador, M.; Soares-Filho, B.S.; Philippidis, G.; Davis, J.L.; de Oliveira, A.R.; Rajão, R. Brazil's sugarcane embitters the EU-Mercosur trade talks. *Sci. Rep.* **2021**, *11*, 13768. [CrossRef] [PubMed]
49. Viana, J.L.; de Souza, J.L.M.; Auler, A.C.; de Oliveira, R.A.; Araújo, R.M.; Hoshide, A.K.; de Abreu, D.C.; da Silva, W.M. Water Dynamics and Hydraulic Functions in Sandy Soils: Limitations to Sugarcane Cultivation in Southern Brazil. *Sustainability* **2023**, *submitted*.
50. Parahyba, R.B.V.; Araújo, M.S.B.; Almeida, B.G.; Rolim Neto, F.C.; Sampaio, E.V.S.B.; Caldas, A.M. Water retention capacity in Arenosols and Ferralsols in a semiarid area in the state of Bahia, Brazil. *An. Acad. Bras. Ciênc.* **2019**, *91*, e20181031. [CrossRef] [PubMed]
51. Viana, J.L. Atributos Físico-Hídricos do Solo, Modelagem do Desenvolvimento Radicu-Lar e Rendimento da Cana-de-Açúcar nas Condições Edafoclimáticas da Região Noroeste do Paraná. Ph.D. Thesis, Universidade Federal do Paraná, Setor de Ciências Agrárias, Programa de Pós-Graduação Ciência do Solo, Curitiba, PR, Brazil, 22 February 2022.
52. Rawls, W.J.; Pachepsky, Y.A.; Ritchie, J.C.; Sobecki, T.M.; Bloodworth, H. Effect of soil organic carbon on soil water retention. *Geoderma* **2003**, *116*, 61–76. [CrossRef]
53. Auler, A.C.; Romaniw, J.; Sá, J.C.M.; Pires, L.F.; Hartman, D.C.; Inagaki, T.M.; Rosa, J.A. Improvement on soil structure and water retention after application of industrial organic waste as a crop fertilizer. *J. Soils Sediments* **2020**, *20*, 2771–2783. [CrossRef]
54. Fidalski, J.; Tormena, C.A.; Alves, S.J.; Auler, P.A.M. Influência das frações de areia na retenção e disponibilidade de água em solos das Formações Caiuá e Paranavaí. *Rev. Bras. Ciência Solo* **2013**, *37*, 613–621. [CrossRef]
55. De Oliveira Bordonal, R.; Carvalho, J.L.N.; Lal, R.; de Figueiredo, E.B.; de Oliveira, B.G.; La Scala, N., Jr. Sustainability of sugarcane production in Brazil. A review. *Agron. Sustain. Dev.* **2018**, *38*, 13. [CrossRef]
56. Cherubin, M.R.; Bordonal, R.O.; Castioni, G.A.; Guimarães, E.M.; Lisboa, I.P.; Moraes, L.A.A.; Menandro, L.M.S.; Tenelli, S.; Cerri, C.E.P.; Karlen, D.L.; et al. Soil health response to sugarcane straw removal in Brazil. *Ind. Crops Prod.* **2021**, *163*, 113315. [CrossRef]
57. Aquino, G.S.; Medina, C.C.; Porteira, A.L., Jr.; Santos, L.O.; Cunha, A.C.B.; Kussaba, D.A.O.; Santos, J.H., Jr.; Almeida, L.F.; Santiago, A.D. Sistema Radicular e Pro-Dutividade de Soqueiras de Cana de Açúcar sob Diferentes Quantidades de palhada. *Pesqui. Agropecu. Bras.* **2015**, *50*, 1150–1159. Available online: https://seer.sct.embrapa.br/index.php/pab/article/view/21795 (accessed on 25 January 2023). [CrossRef]

58. Gallo, B.C.; Magalhães, P.S.G.; Demattê, J.A.M.; Cervi, W.R.; Carvalho, J.L.N.; Barbosa, L.C.; Belinasso, H.; de Mello, D.C.; Veloso, G.V.; Alves, M.R.; et al. Soil Erosion Satellite-Based Estimation in Cropland for Soil Conservation. *Remote Sens.* **2023**, *15*, 20. [CrossRef]
59. Ambrosano, E.J.; Cantarella, H.; Ambrosano, G.M.B.; Schammas, E.A.; Dias, F.L.F.; Rossi, F.; Trivelin, P.C.O.; Muraoka, T.; Azcón, R. Produtividade da cana-de-açúcar após o cultivo de leguminosas. *Bragantia* **2011**, *70*, 810–818. [CrossRef]
60. Souza, G.S.; Souza, Z.M.; Silva, R.B.; Barbosa, R.S.; Araújo, F.S. Effects of traffic control on the soil physical quality and the cultivation of sugarcane. *Rev. Bras. Cienc. Solo* **2014**, *38*, 135–146. [CrossRef]
61. Da Cunha Souza, M.; de Oliveira, F.P.; Silva, J.R.; Martins, A.F.; da Silva, P.L.F. Spatial Variability of Resistance to Penetration in Soil Under Sugarcane Crops with Different Harvest Methods. *Rev. Caatinga* **2020**, *33*, 480–489. [CrossRef]
62. Duden, A.S.; Verweij, P.A.; Kraak, Y.V.; van Beek, L.P.H.; Wanders, N.; Karssenberg, D.J.; Sutanudjaja, E.H.; van der Hilst, F. Hydrological impact of ethanol-driven sugarcane expansion in Brazil. *J. Environ. Manag.* **2021**, *282*, 111942. [CrossRef] [PubMed]
63. Perillo, L.I.; de Oliveira Bordonal, R.; de Figueiredo, E.B.; Moitinho, M.R.; Aguiar, D.A.; Rudorff, B.F.T.; Panosso, A.R.; La Scala, N., Jr. Avoiding burning practice and its consequences on the greenhouse gas emission in sugarcane areas southern Brazil. *Environ. Sci. Pollut. Res.* **2022**, *29*, 719–730. [CrossRef]
64. Patino, M.T.O.; de Amorim, F.R.; de Andrade, A.G.; Alam, M.J.; Solfa, F.D.G. Costs of Agronomic Practices: Profitability at Different Scales of Sugarcane Production in Brazil. *Int. J. Bus. Adm.* **2022**, *13*, 32–43. [CrossRef]
65. Ruan, H.; Feng, P.; Wang, B.; Xing, H.; O'Leary, G.J.; Huang, Z.; Guo, H.; Liu, D.L. Future climate change projects positive impacts on sugarcane productivity in southern China. *Eur. J. Agron.* **2018**, *96*, 108–119. [CrossRef]
66. World Population Review. Sugar Producing Counties 2023. Available online: https://worldpopulationreview.com/country-rankings/sugar-producing-countries (accessed on 2 February 2023).

sustainability

MDPI

Article

Using the GeoWEPP Model to Predict Water Erosion in Micro-Watersheds in the Brazilian Cerrado

Wellington de Azambuja Magalhães [1,*], Ricardo Santos Silva Amorim [2], Maria O'Healy Hunter [1], Edwaldo Dias Bocuti [1], Luis Augusto Di Loreto Di Raimo [1], Wininton Mendes da Silva [3], Aaron Kinyu Hoshide [4,5] and Daniel Carneiro de Abreu [5,6]

[1] Curso de Doutorado Programa de Pós-Graduação em Agricultura Tropical, Universidade Federal de Mato Grosso, UFMT, Cuiabá 78060-900, MT, Brazil

[2] Departamento de Engenharia Agrícola, Universidade Federal de Viçosa, DEA/CCA/UFV, Viçosa 36570-000, MG, Brazil

[3] Empresa Mato-Grossense de Pesquisa, Assistência e Extensão Rural (EMPAER-MT), Centro Político Administrativo, Cuiabá 78049-903, MT, Brazil

[4] College of Natural Sciences, Forestry and Agriculture, The University of Maine, Orono, ME 04469, USA

[5] AgriSciences, Universidade Federal de Mato Grosso, Caixa Postal 729, Sinop 78550-970, MT, Brazil

[6] Instituto de Ciências Agrárias e Ambientais (ICAA), Universidade Federal do Mato Grosso, Campus Universitário de Sinop, Avenida Alexandre Ferronato, 1200, Sinop 78550-728, MT, Brazil

* Correspondence: wellingtonagro@gmail.com; Tel.: +55-65-99961-7964

Abstract: The GeoWEPP model has estimated water and soil losses caused by erosion at the watershed level in different parts of the world. However, this model was developed and its parameters have been adjusted for temperate climates, which are different from tropical climates such as those found in Brazil. Our study evaluated the performance of the GeoWEPP model in estimating soil erosion in three micro-watersheds in the Cerrado (i.e., savannah) of southeastern Mato Grosso state, Brazil. Major land uses modeled were soybean and corn cultivation, traditional pasture, and native vegetation. Input parameters for the GeoWEPP model involved climate, soil, land use and management, and topography. GeoWEPP was calibrated with input parameters for soil erodibility specified as interrill and rill soil erosion, soil critical shear stress, and saturated hydraulic conductivity obtained experimentally and estimated by internal routine equations of the GeoWEPP model. Soil losses observed in micro-watersheds with agriculture, pasture, and native vegetation were 0.11, 0.06, and 0.10 metric tons per hectare per year, respectively. GeoWEPP best modeled soil erosion for native vegetation and pasture, while over-estimating that for crops. Surface runoff was best modeled for crops versus native vegetation and pasture. The GeoWEPP model performed better when using soil erodibility input parameters.

Keywords: environmental impact; soil conservation; water erosion prediction; watershed; WEPP parameters

Citation: Magalhães, W.d.A.; Amorim, R.S.S.; Hunter, M.O.; Bocuti, E.D.; Di Loreto Di Raimo, L.A.; da Silva, W.M.; Hoshide, A.K.; de Abreu, D.C. Using the GeoWEPP Model to Predict Water Erosion in Micro-Watersheds in the Brazilian Cerrado. *Sustainability* **2023**, *15*, 4711. https://doi.org/10.3390/su15064711

Academic Editor: Adriano Sofo

Received: 2 February 2023
Revised: 24 February 2023
Accepted: 1 March 2023
Published: 7 March 2023

1. Introduction

Land use and occupation have undergone several transformations in recent decades, especially in emerging countries such as Brazil, as well as more globally. Studies on soil erosion have been essential for the management and conservation of the environment in general, especially in agricultural environments with higher erosion potential. Prior research has estimated soil loss in the order of 820 million metric tons per year in Brazil, considering not only annual crops but areas cultivated with pastures and perennial crops [1]. Another study estimated soil losses in Brazil to be around 616.5 million metric tons per year with annual crops alone [2]. Water erosion and nutrient runoff into water bodies are typical problems of agricultural activities [3]. Accelerated soil erosion has a global effect, and impacts on the emission of greenhouse gases such as carbon dioxide (CO_2), methane

(CH_4), and nitrous oxide (N_2O) were responsible for 10 to 15% of global anthropogenic emissions in 2019 [4].

Erosion is the main cause of soil degradation in Brazil and the world, and studies aimed at minimizing the adverse impacts of erosion by guiding producers and technicians are increasingly important. Obtaining regional data, as well as methodological approaches, facilitates the adoption of sustainable agricultural practices, especially in biomes such as the Cerrado (i.e., savannah), where agricultural systems have been intensified in recent years in Brazil. Due to the great diversity and complexity of interactions of factors that govern the erosion process, which include combinations of climate, topographic relief, vegetative cover, and soil hydraulic properties, modeling is an essential tool to obtain both a quantitative and consistent approximation of soil erosion process and sediment transport rates under a wide variety of conditions [5]. In addition, such modeling can be used to define the most appropriate land use and management for each location [6–8].

One type of soil erosion model is GeoWEPP, which has been applied in several locations around the world to estimate water and soil loss [9]. This model was developed and calibrated for temperate climate conditions, which are very different from tropical climates. Therefore, developing research that aims to calibrate GeoWEPP to Brazilian edaphoclimatic conditions before being extensively used for erosion prediction is of paramount importance. In Brazil, some studies have already been developed using the WEPP model to estimate soil and water losses under certain soil management conditions. However, most of these studies have used this model to estimate soil losses without comparing model estimates with experimental data [10–13].

The Cerrado and southern Amazon regions in Brazil are characterized by intensive land use for agriculture and livestock. The micro-watersheds used in the present study predominantly had three types of land use. There were native vegetation, pasture, and annual cultivation of commodity crops (Figure 1). Major commodity crops in the region include soybean (*Glycine max* L.) during the wet season, followed by corn (*Zea mays* L.) in the wet/dry seasons (Santa Fe system). The growing of corn or corn under-seeded with pasture grasses such as *Brachiaria* sp. immediately following soybeans has become more prevalent in Brazil. In addition to these cropping sequences involving corn, soybeans can also be followed immediately by cotton (*Gossypium* sp.) as a sustainable intensification strategy to increase crop production on the same land area [14]. Although there has been a shift from conventional to no-till systems for soybeans, corn, and cotton, which can reduce erosion but increase reliance on agro-chemicals such as glyphosate (i.e., Roundup®) [15], conventional tillage is still used on ~40% of Brazil's total crop area [16]. Therefore, it is still important to validate soil erosion models for Brazil to complement field research in conventional tillage.

The GeoWEPP model is based on the physical principles of the erosion process, showing applicability in the simulation of the erosion process for different edaphoclimatic conditions [17]. It is essential that the model be calibrated for the Brazilian edaphoclimatic conditions, since the development of a model is quite costly in terms of time and necessary resources, due to both data collection and the application of guidelines involved in the process [18,19]. The study and application of tools such as GeoWEPP in developing countries such as Brazil can facilitate the mapping of areas sensitive to erosion. This can identify areas better suited for re-seeding perennial crops (e.g., pasture), Amazon re-forestation, and/or re-habilitation of Cerrado biome savannahs and other biomes in Brazil.

The development of this work is a sequence of studies and results that were being carried out in the micro-basins, involving hydrological monitoring techniques and mathematical modeling of sediment production and surface runoff. This study aimed to evaluate the performance of the GeoWEPP model in estimating sediment production when applied to three micro-watersheds of the Cerrado in Mato Grosso state, Brazil. These micro-watersheds are characterized by agricultural production of soybeans (*Glycine max* L.), maize (*Zea mays* L.), extensive pasture (*Brachiaria* sp.), and native vegetation. The specific objectives of our research were to (1) evaluate the dynamics of sediment production related to different land uses over the period of study between 2013 and 2015 and to (2) determine the accuracy of the GeoWEPP model in predicting soil losses when applied to these three watersheds characterized by different soil types and land use.

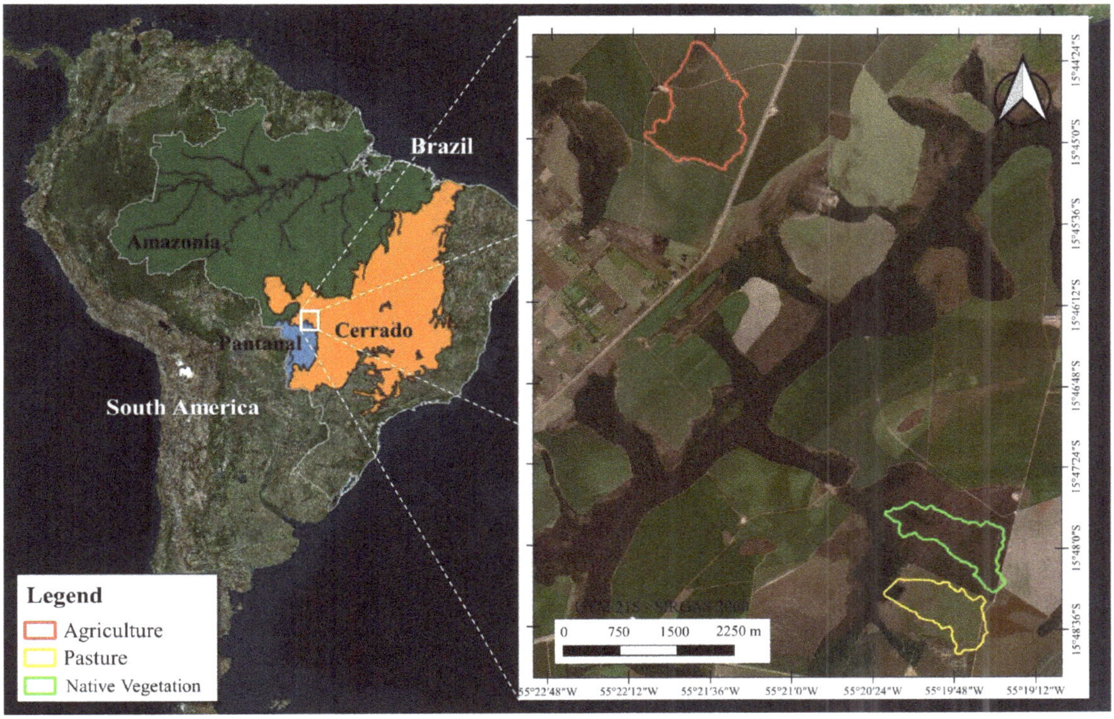

Figure 1. Location of study areas in Campo Verde, Mato Grosso state, Brazil.

2. Materials and Methods

2.1. Study Area

Our study was carried out in three micro-watersheds located in the municipality of Campo Verde, Mato Grosso state, Brazil (Figure 1). The research site is located in the Cerrado (i.e., savannah) biome in Brazil. These micro-watersheds are located within the Rio das Mortes watershed, one of the main tributaries of the Araguaia River. The regional climate is classified as tropical Aw, according to the Köppen classification, alternately wet and dry, with temperatures ranging between 18 and 24 °C. The mean annual precipitation is 1400 mm, varying between 800 and 1600 mm, with 70 to 80% of annual rainfall falling during the rainy season (October to May). The rainy season occurs from October to April and the dry season from May to September. The research area is located in the extreme northwest of the Paraná basin, belonging to the Cachoeirinha Formation. The lithology of the area has unconsolidated gravels, practically those that are monomictic, composed of >90% quartz pebbles that are well rounded with high sphericity. This formation is also characterized by sandy and clayey lenses, some strongly cemented by iron oxides. The reddish to yellowish colors from limonitization often form true laterized crusts [20]. These micro-watersheds were selected because they represented the most common forms of land use in Brazil's Cerrado biome, namely agricultural crops and pasture, as well as native vegetation kept as reserves (Figure 1).

The three selected micro-watersheds had specific characteristics for each of the three predominant land use types. First, commodity agriculture involved annual soybean and corn cultivation under direct sowing in the Santa Fe system for at least 10 years. Second, pasture was extensive with traditional livestock such as Nelore beef cattle (*Bos indicus*) with a stocking density of one animal unit (AU = 450 kg) per hectare. Finally, native vegetation was regenerated in native Cerrado forest, also more than 10 years old. This type of native vegetation is considered to be a reference for an environment that has not been altered for agricultural use.

2.2. Soil Characterization and Soil Sampling

The existing soil classes in the micro-watersheds are Latossolos (Oxisols) in agricultural areas and Neossolos (Entisols) for both pasture and native Cerrado, as defined by the Brazilian Soil Classification System [21]. Our study was carried out in two stages. The first stage involved sampling of soils in the three micro-watersheds. To carry out the experimental analyses of soil and water loss, three representative sampling points were selected within the watersheds. One point under pasture use, one under agricultural use, and one point with native vegetation. The criteria used for determining sampling locations were local relief, soil color and texture, history of use, and occupation. In each of the watersheds, soil samples were collected undisturbed using an auger and a kopeck ring (from the Soltest company). Soil was sampled down to three layers in the upper soil profile at 0 to 0.20, 0.2 to 0.4, and 0.4 to 0.6 m depth in a sampling grid of 200 × 120 m. The samples were sent for analysis to the Soil Physics Laboratory at the Faculty of Agronomy and Animal Science (FAAZ) at the Universidade Federal de Mato Grosso (UFMT) campus in Cuiabá, Mato Grosso state, Brazil. A more detailed description of soil characteristics in these watersheds can be found in Bocuti et al., 2016 [22] and Nóbrega et al., 2020 [23].

These data, associated with the physiographic characterization of the watersheds, were used to run the GeoWEPP model. The data collected also describe the morphometric characteristics and hydrological dynamics of these watersheds [18]. The second stage of our study involved physical and physical–hydric characterization of the study areas with the determination of the effective hydraulic conductivity (K_e), interrill erodibility (K_i), and rill erodibility (K_r) to calibrate the GeoWEPP model. The methodology proposed by the Agricultural Research Service of the United States Department of Agriculture (USDA) Agricultural Research Service (ARS) was used to determine K_e, K_i, and K_r [24] for modeling by the GeoWEPP Software [9] compatible with the ArcGIS 10.4 version.

2.3. Soil Erosion Field Measurements

Between 2014 and 2016, our research team went to the experimental field site to determine the calibration parameters for rill erodibility (Kr) and soil critical shear stress (τ_c) (Figure 2), in addition to the determination of effective hydraulic conductivity (Ke) and interrill erodibility (Ki) parameters (Figure 3). The furrow erodibility (Kr) and the critical shear stress (τ_c) of the soil were determined from tests with the application of different flows in preformed furrows in freshly prepared soil with 9 m of length and 0.46 m of depth, with width and slope equal to the site conditions (Figure 3). During the tests, the surface velocity of the flow and the discharge rate of the furrows were measured.

Figure 2. Preparation of areas for determination of erodibility in the furrow (Kr): (**a**,**b**) in the watershed under pasture, (**c**) in the micro-basin under native vegetation, and (**d**) in the micro-basin under agriculture.

Figure 3. Installation of limiters in the Cerrado micro-basin (**a**) and determination of the average precipitation intensity (**b**) applied at the end of the Ke tests.

The effective hydraulic conductivity (Ke) of the soil was determined in the five study areas with three replications (two in the watersheds with agriculture, two in the pasture, and one in the native vegetation). We used a rainfall simulator following methods outlined in Bocuti et al., 2019 [25] (Figure 2). Interrill erodibility (Ki) in each area was determined in three experimental plots demarcated with galvanized sheets, following the methodology proposed by Elliot 1989 [23]. In the experimental plot, artificial rain was applied, and the Ki was calculated from the equation proposed by Foster 1982 [26] to estimate the sediment release rate in the interrill areas.

Surface runoff samples were also collected to determine the amount of suspended soil particles. Immediately after the start of surface runoff and when the flow rate became approximately constant, the geometry of the furrow and the height of the flow were determined. These data made it possible to determine the wetted area and perimeter, and consequently the hydraulic radius of the cross-section of the flow. For each type of soil in the studied areas, a graph of the average soil shear rate versus the average shear stress of the flow was generated, and a linear equation was adjusted to the set of points. The critical shear stress of the soil was considered to be the one in which soil loss was zero. Soil erodibility was obtained by using the slope of the trend line.

2.4. GeoWEPP Setup and Application

GeoWEPP integrates the WEPP model and the Topography ParameteriZation (TOPAZ) software into the ArcGIS 10.4® software [27] to predict sediment production and surface runoff at a watershed scale. The necessary input files for climate, slope, soil, and land use and management were generated in WEPP, and the topographic data were parameterized by TOPAZ based on a digital elevation model. A survey of altimetry was completed in the study region to delineate the catchment area of each basin using a precision GPS (TOPCON RTK Hiper Lite). With this survey, a digital elevation model (DEM) was generated, with 5×5 m pixels. Finally, the watershed was generated by GIS functions in ArcInfo, which is the command-line sub-package of ArcGIS. A database on WEPP extensions had to be structured for both modules. In general terms, the database was segmented by climate, soils, land use and management, and topography based on methods outlined by Maalim et al., 2013 [28]. A flowchart outlining all the geo-spatial processing steps is shown in Figure 4.

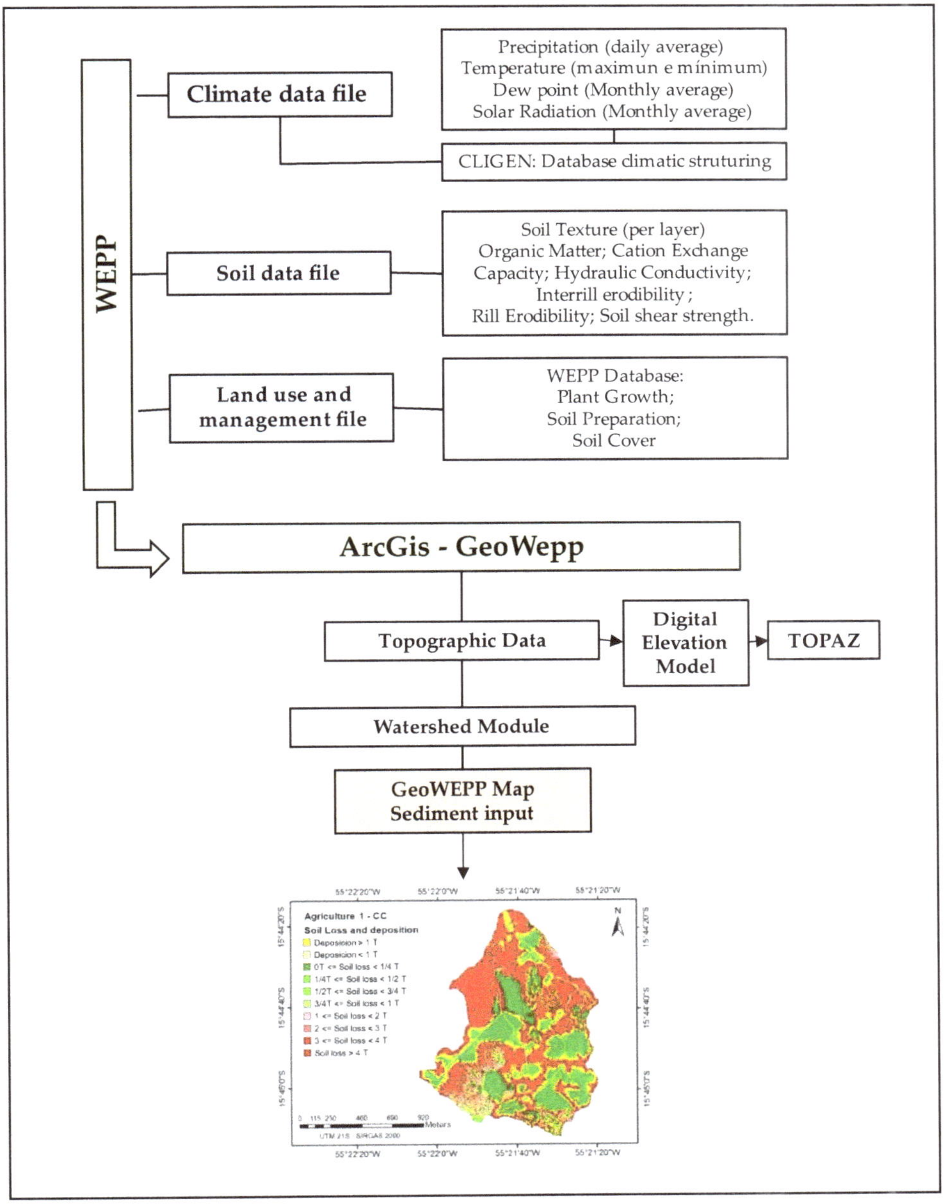

Figure 4. Methodology flowchart for modeling and preparing soil loss maps for different environments studied.

2.5. *WEPP Input Data Preparation*

2.5.1. Climate File

The daily input climate data file for the GeoWEPP model was generated using the CLIGEN software, version 4.3 [29]. The statistical parameters of precipitation, temperatures, and solar radiation were obtained from climatological stations we installed near the study areas. These climatological stations were connected to data loggers that stored the information between 2012 and 2015 at 10 min intervals.

2.5.2. Topography and Land Cover

The digital elevation model (DEM) of the study watersheds was prepared using the "Topo to raster" tool in ArcGIS v.10.1, with a spatial resolution of 5 × 5 m. GeoWEPP integrates the WEPP model and the Topography ParameteriZation (TOPAZ) software into the ArcGIS 10.4® software to predict sediment production and surface runoff at a watershed scale. The necessary input files for climate, slope, soil, and land use and management were generated in WEPP. The topographic data were parameterized by TOPAZ based on the DEM. Finally, the watershed was generated by GIS functions in ArcInfo. The digital elevation model (DEM) of the study watersheds was prepared using the "Topo to raster" tool in ArcGIS v.10.1, with a spatial resolution of 5 × 5 m.

GeoWEPP integrates the WEPP model and the Topography ParameteriZation (TOPAZ) software into the ArcGIS 10.4® software to predict sediment production and surface runoff at a watershed scale. The necessary input files for climate, slope, soil, and land use and management were generated in WEPP. The topographic data were parameterized by TOPAZ based on the DEM. Finally, the watershed was generated by GIS functions in ArcInfo.

2.5.3. Soil File

The soil file that composes the physical attributes such as texture, soil saturated hydraulic conductivity, interrill erodibility, rill erodibility, and critical shear stress were generated by the internal routine equations of the WEPP model. Subsequently, the data determined in the field were inserted into the model, and new files were generated to be modeled in the GIS GeoWEPP interface. The experimental works were carried out in each of the pedological units of the micro-watersheds contemplated in the present study. Plots to determine hydraulic conductivity (K_e), interrill erodibility (K_i), rill erodibility (K_r), and critical shear stress (τ_c) were installed for each pedological unit, with three replications performed for each parameter, following the WEPP methodology proposed by the United States Department of Agriculture (USDA) Agricultural Research Service (ARS) [23] and by [30]. Conventional soil tillage in the area consisted of removing existing plant residues followed by one plowing and two harrowing operations.

2.5.4. Management File

The land use and management file was generated from the management module integrated into the WEPP interface. The model program files were compiled based on local information available at each micro-watershed (native vegetation, pasture, and agriculture). Other input parameters were used from the WEPP database [24]. Thus, the output files were generated for each cropping system based on time frames that can be set by the user [9], which in our case was over the three years of field data collected.

2.5.5. Precipitation, Surface Runoff, Runoff Coefficient, and Soil Losses in Micro-Watersheds

Total precipitation and its temporal distribution were recorded in a pulse rain gauge installed in the experimental area and connected to a datalogger that stored the number of pulses in the collector every ten minutes. This enabled the evaluation of soil and water losses due to the different durations and intensities of precipitation. Triangular metal spillways were installed in each micro-watershed control section to allow all water in the course to pass through an area with known dimensions. The height (h) of the water depth that passes over the spillway sill needs to be known in order to calculate the flow rate.

A Hydrolab DS5X multiparameter probe, capable of estimating the hydraulic head of the course as a function of the hydrostatic pressure and determining the amount of sediment in suspension from the field calibration curve, was used to constantly monitor the hydraulic head above the spillway sill. The probes were installed two meters upstream of the spillway so that the measured hydraulic head would not be influenced by the proximity of the spillway, where the water depth is lower due to the energy line. Height measurements were performed every ten minutes with data stored in the instrument itself.

An automatic water sample collector (Hach-Lange IL 2000-D with 24 bottles) was installed in each flow rate monitoring section whenever there was a rain event. The samples were used to determine the concentrations of sediments by the direct method and were used to prepare the calibration curve of the turbidity sensor. Field collections were carried out every 15 and 30 days during the rainy and dry seasons, respectively. The Kindsvater–Shen equation and its respective calibration adjustment functions were used to quantify the flow of each micro-watershed.

2.6. Statistical Analysis and Modeling

GeoWEPP performance and accuracy were evaluated by comparing the annual amounts of soil loss observed in the field with those estimated by the model and specific rainfall events. Data analysis and modeling were carried out using GeoWEPP both with and without calibration. When calibration was used, the maps of soil loss and surface runoff of the micro-watersheds were generated by inserting the Ki, Kr, Ke, and τc data determined within the micro-watersheds. During calibration, all other components such as climate, soil, land use and management, and topography were also inserted into the model. If calibration was not used, maps of soil loss and surface runoff of the micro-watersheds were generated without the insertion of Ki, Kr, Ke, and τc data. Under this condition, the model itself estimated the parameter values to generate the maps. Thus, only the digital elevation map and the climate, soil, land use and management, and topographic data were inserted into the model.

The evaluation of the efficiency of models was performed using the following three statistics. The first statistic was the root mean square error (RMSE):

$$RMSE = \sqrt{\frac{\sum_{i=1}^{n}\left(Y_{obsj} - Y_{estj}\right)^2}{n}} \quad (1)$$

where Y_{obs} is the observed value, Y_{est} is the estimated value, and n is the total number of pairs of observed and estimated values for the ith treatment and jth repetition. The second statistic was the Willmott concordance index (d):

$$d = 1 - \frac{\sum_{i=1}^{n}\left(Y_{estj} - Y_{obsj}\right)^2}{\sum_{i=1}^{n}\left(\left|Y_{estj} - Y\right| + \left|Y_{obsj} - Y\right|\right)^2} \quad (2)$$

where d is the concordance index, and Y is the mean of observed values. The third statistic used was the Nash–Sutcliffe (NS) coefficient:

$$NS = 1 - \frac{\sum_{i=1}^{n}\left(Y_{obsj} - Y_{estj}\right)^2}{\sum_{i=1}^{n}\left(Y_{obsj} - Y\right)^2} \quad (3)$$

with similar parameters as RMSE and the Willmott concordance index.

3. Results and Discussion

3.1. Soil Characterization

Table 1 shows the soil physical attributes of the studied micro-watersheds. All three micro-watersheds with pasture and native vegetation have the same type of soil in common, Quartzarenic Neosol. This soil type is characterized by sand in the soil profile. The micro-watershed with agriculture had a higher clay content with an oxidic characteristic, which is common in tropical soils.

Table 1. Soil attributes at a depth of 0 to 20 cm within agriculture, pasture, and native vegetation micro-watersheds in the municipality of Campo Verde, Mato Grosso state, Brazil.

Soil Property	Unit	Agri-Culture 1	Agri-Culture 2	Pasture 1	Pasture 2	Native Vegetation
Clay	grams (g)/kilogram (kg)	518	512	31	496	140
Sand	g/kg	300	355	926	946	748
Total organic carbon	g/kg	33.80	29.40	0.56	0.40	14.40
Total porosity	%	56.60	61.71	40.23	43.54	51.51
Aggregate stability index (ASI)	%	93.45	86.93	78.59	26.59	96.06
Mineralogy	n/a	Gibbsite Quartz Kaolinite	Gibbsite Quartz Kaolinite	Gibbsite Quartz	Gibbsite Quartz	Gibbsite Quartz Goethite
Slope	%	5.33	1.60	13.6	3.70	5.55

3.2. WEPP Input Data Parameters

The values for hydraulic conductivity (Ke), interrill soil erodibility (Ki), rill soil erodibility (Kr), and runoff critical shear stress (τc) differed between the watersheds. There were differences between the parameters measured at the sampling points within the watersheds, especially when comparing agriculture and pasture micro-watersheds. Compared to the samples from the agriculture micro-watershed, pasture and native vegetation micro-watersheds had higher Ke, lower Ki, greater Kr, and lower τc (Table 2).

Table 2. Hydraulic conductivity (K_e), interrill soil erodibility (K_i), rill soil erodibility (K_r), and runoff critical shear stress (τ_c) for the different study areas in the micro-watersheds, Campo Verde, Mato Grosso state, Brazil.

Micro-Watershed Land Cover	Hydraulic Conductivity (K_e)	Interrill Soil Erodibility (K_i)	Rill Soil Erodibility (K_r)	Runoff Critical Shear Stress (τ_c)
	Millimeters/Hour	Kilograms/(m × m^3)	Seconds/ Meter	Newtons/ Square Meter
Agriculture 1	30.63 ± 6.76	$2.44 \times 10^{-5} \pm 0.83 \times 10^4$	0.2781	0.001483
Agriculture 2	24.49 ± 3.12	$13.2 \times 10^{-5} \pm 18.2 \times 10^4$	0.1927	0.042942
Pasture 1	81.52 ± 15.80	$1.56 \times 10^{-5} \pm 2.04 \times 10^4$	5.2539	0.000949
Pasture 2	109.94 ± 19.34	$2.47 \times 10^{-5} \pm 4.14 \times 10^4$	11.2723	0.000428
Native vegetation	48.31 ± 13.41	$0.86 \times 10^{-5} \pm 0.21 \times 10^4$	1.4383	0.002991

The K_e and K_i data were reprinted/adapted with permission from Bocuti 2016 [22].

3.3. GeoWEPP Model Results

Table 3 shows the sediment production of the watersheds estimated by the GeoWEPP model with the insertion of parameters. Spatial maps of soil depositions and losses are presented for agricultural areas (Figure 5), pasture (Figure 6), and for native vegetation (Figure 7). When simulated without inserting the Ki, Kr, Ke, and τc parameters, the model was unable to accurately estimate the observed soil loss values in the micro-watersheds. However, with the insertion of parameters, soil loss values for micro-basins with pasture and native vegetation were close to the observed values (Table 3). For estimating surface runoff, the model is sensitive only to the hydraulic conductivity parameter (Ke), while the sediment production and soil loss are sensitive to all parameters (Kr, Ke, τc, and Ki) [31].

Table 3. Sediment production simulated by GeoWEPP for micro-watersheds in Campo Verde, Mato Grosso state, Brazil.

Micro-Watershed Land Cover	Observed				Simulated by GeoWEPP Model	
	Total Water Depth Precip-itated (Wdp)	Surface Runoff (SR)	SR Coefficient	Soil Loss	Soil Loss w/NO Calibration Parameters	Soil Loss w/Calibration Parameters
	millimeters/year		%	metric tons/hectares/year	metric tons/hectares/year	metric tons/hectares/year
Agriculture 1	1640.8	141.3	8.61	0.11	2.71	5.78
Agriculture 2					2.42	6.16
Pasture 1	1616.8	477.8	29.55	0.06	0.28	0.03
Pasture 2					0.44	0.03
Native vegetation	1579.3	361.8	22.91	0.10	0.21	0.10

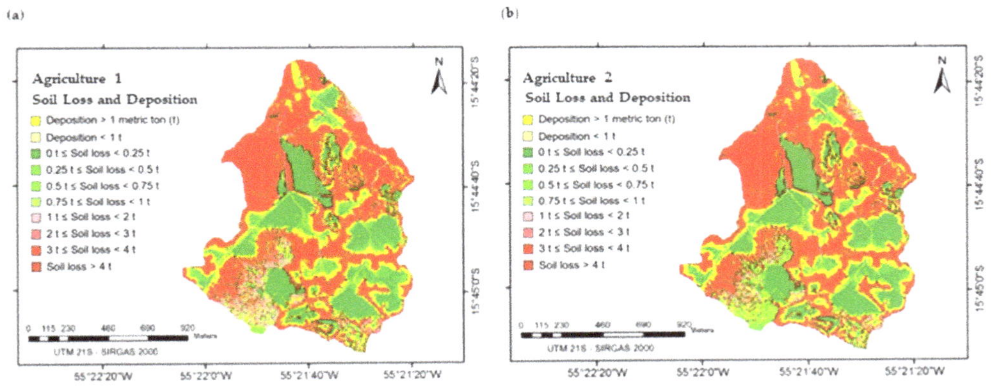

Figure 5. Soil gains/losses generated by GeoWEPP for the agricultural cultivation micro-watershed, considering the calibration parameters in Table 2, for sampling points for (**a**) Agriculture 1 and (**b**) Agriculture 2 in the municipality of Campo Verde, Mato Grosso, Brazil.

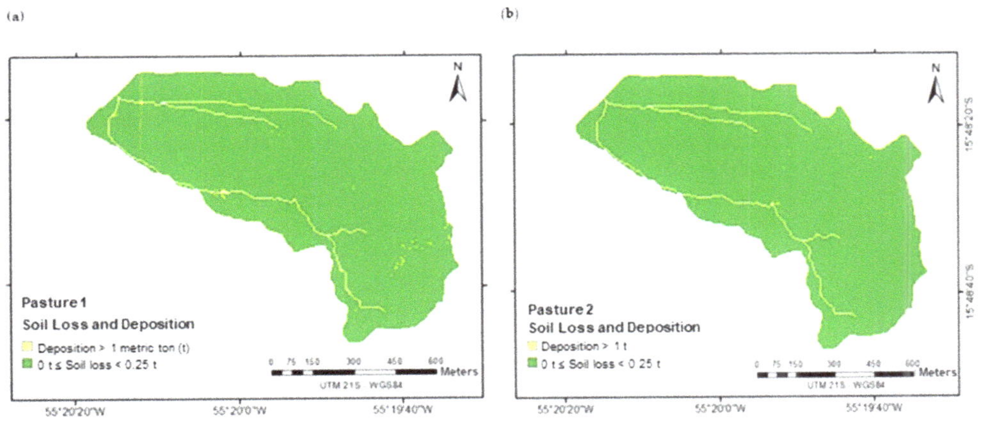

Figure 6. Soil gains/losses generated by GeoWEPP fothe micro-watershed for pasture, considering the calibration parameters in Table 2, for the sampling points for (**a**) Pasture 1 and (**b**) Pasture 2 in the municipality of Campo Verde, Mato Grosso, Brazil.

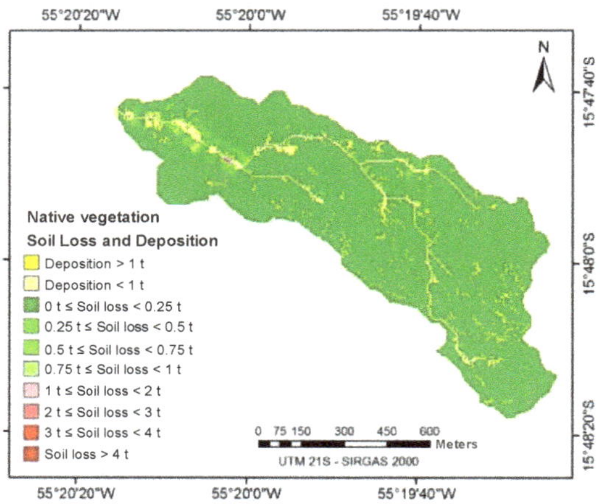

Figure 7. Soil gains/losses generated by GeoWEPP for the micro-watershed for native vegetation in the municipality of Campo Verde, Mato Grosso state, Brazil.

According to previous research (Supplementary Materials), the mean runoff rates were lower for areas with higher total sand content due to their higher infiltration capacity and water percolation in these soils [26]. The areas under pasture and native vegetation have a sandy texture, with approximately 3 and 14% clay content, respectively. The area under agricultural production has a clayey texture and, therefore, a higher microporosity and lower macroporosity, leading to less permeability/movement of water in the soil. The model accurately estimated soil losses in the micro-watershed with native vegetation, with values similar to those observed for pasture. The most significant water and soil losses occurred in the micro-watershed with agricultural crops, indicating higher susceptibility to soil loss even after inserting values for interrill erosion (K_i), rill erosion (K_r), saturated hydraulic conductivity (K_e), and soil critical shear stress (K_e). However, the surface runoff values decreased significantly. It can be seen that the insertion of Ki, Kr, Ke, and τc data was essential for the main changes that occurred in the modeling of soil loss within micro-basins with pasture and native vegetation, since these parameters are accurately modeled.

The accuracy of the GeoWEPP model in predicting the conditions of sandy soil in the micro-basin with pasture may assist research and extension technicians in predicting erosion and recommending conservation practices. Livestock activity is the main driver of land use change of native vegetation to pasture in the world. In Brazil alone, 60 million hectares have been transformed in this way over the last 33 years [32]. Currently, about 58.8% (equivalent to 97.7 million hectares) of pasture show some degree of degradation [33]. Considering that the GeoWEPP model performed well in estimating soil loss for the area under pasture, it can contribute to making Brazilian meat production more sustainable. Brazil has extensive areas of pasture devoted to extensive grazing of beef cattle, mainly in the Amazon biome but also the Cerrado (i.e., savannah) biome.

The GeoWEPP model overestimated soil losses for agricultural crops with parameterization for the agriculture micro-watershed. The model was developed, and its parameters were adjusted, for temperate climate conditions, which are very different from tropical conditions. Among the three micro-watersheds, crop areas were dominated by a clayey Latossolo Vermelho-Amarelo (Oxisol). In a study of the same micro-watershed, a previous study determined that this soil is basically composed of gibbsite, quartz, and goethite [27], characterized by a granular macrostructure with small loose granules, lower soil density, higher proportion of large pores, and higher permeability [34]. This implies these soils are less erodible compared to Oxisols with kaolinitic mineralogy, for example [35]. A prior study estimating the interrill erodibility of Oxisols in Rio Grande do Sul found that the WEPP model overestimated experimental values by 2.4 to 3.6 times [36]. These researchers attributed this overestimation to the equations used in the WEPP model. American soils are generally sandier, and WEPP was developed based on soils subject to a lower weathering rate compared to Brazilian Oxisols.

Although the GeoWEPP model overestimated the values of soil loss in the micro-watershed with agriculture, it can be seen that the practices used in this area have contributed to the maintenance and conservation of soil and water. The observed values of soil loss in this system are 0.11 metric tons/hectare/year, which is considered below the threshold considered suitable for tropical soils. The existence of terracing in the agricultural areas we modeled may also have contributed to the WEPP model not being able to successfully estimate soil loss. This occurs because terraces smaller than 5 m may not have been captured by the digital elevation model (DEM) during mapping.

Our erosions measurements in the field for crops and pasture were consistent with another recent study measuring erosion with simulated rainfall in both the Renato sub-basin and Caiabi sub-basin in the middle and upper regions of Teles Pires River basin in Mato Grosso state, Brazil, respectively. These sub-basins are in the Amazon-Cerrado transition zone, which is north of where we conducted our experiment in the southern Mato Grosso state Cerrado biome. The types of soil in this Teles Pires study's experimental area were all Latisols (versus Oxisols). For crops with crop residue, these researchers measured erosion ranging from 0.035 to 0.102 metric tons (t)/hectare (ha) [37]. This is slightly less than what we measured at 0.11 t/ha (Table 3). The erosion we measured for pasture at 0.06 t/ha (Table 3) was within the range measured for erosion in pasture (0.0101 to 0.087 t/ha) by Alves et al. 2023 [37].

Some studies have observed that anthropogenic factors such as land use and management have more influence on soil loss than precipitation and topography [38]. This is consistent with the results from our study, as evidenced by the hydrological dynamics and surface runoff in the micro-watersheds. A study in the same micro-watersheds we studied that were in pasture and native vegetation found changes in the balance of carbon and nutrients, which were attributed to anthropic factors [39].

The WEPP model indicated lower soil losses in areas with pasture and native vegetation. This result was already expected since the experimental results obtained in the field showed high hydraulic conductivity. The experimental units with pasture had 81.52 and 109.94 mm (mm) of effective soil hydraulic conductivity, while the area with native vegetation had 48.31 mm [25]. Thus, the calibrated model gave these areas the lowest potential to generate soil losses, especially when considering land use and vegetative cover factors. Hydraulic conductivity is one of the parameters in which the GeoWEPP model is more sensitive [40], which can considerably affect model results, as lower water infiltration corresponds to a higher flow rate and greater sediment transport capacity [41].

The characteristics observed in the micro-watersheds we modeled indicate the extent to which soil erosion is triggered since these characteristics describe the processes related to morphometric characteristics and land use and management. These processes are measured by the amount of material removed from the soil per unit area and per unit time. Such surface runoff happens when the water volume is higher than the infiltration limit, which leads to runoff [42]. The processes of transport, deposition, and sedimentation of soil particles occur more intensely in naturally exposed land or under intensive management practices such as those for agriculture, which can lead to landscape degradation.

Table 4 shows the values of the root mean square error (RMSE), the concordance index (d), and the Nash–Sutcliffe coefficient (NS). The NS values were negative for all areas. Wilmott's concordance index (d) expresses the accuracy of the estimates in relation to the observed values. It appears that the model presented a higher level of accuracy for the pasture micro-watershed, presenting the value closest to unity (d = 0.56). Thus, it appears that the WEPP model presented more accurate measurements when applied to watersheds with sandy characteristics, which is corroborated by the work of Amorim et al. (2010), in which the worst performances of the WEPP model occurred in plots conducted in Latossolos (Oxisols) [17].

Table 4. Statistical parameters for the evaluation of soil loss estimates by the GeoWEPP model for the Brazilian Cerrado with insertion of calibration parameters.

Micro-Watershed	Statistical Parameters		
	Root Mean Squared Error (RMSE)	Willmott Concordance Index (d)	Nash-Sutcliffe (NS)
Agriculture	0.12	0.27	−3.38
Pasture	0.48	0.59	−0.56
Native vegetation	0.68	0.44	−0.59

Figure 8 shows the relationship between the soil loss estimates using the parameterized GeoWEPP model compared to the same model without the parameterization. Most of the points for the micro-watersheds with pasture and native vegetation are located below the zero deviation line, indicating the tendency for underestimation of soil losses by the model for the land use and management conditions in GeoWEPP simulations. The main changes with the parameterization occurred for agricultural production in the micro-watershed. In general, the USLE, RUSLE, and WEPP erosion models tend to overestimate small events and underestimate large events [43]. This feature is even more evident in the WEPP model [44], and the model may not be able to estimate or correct soil loss values. A possible explanation for unsatisfactory results for model runs simulated by GeoWEPP may be related to the way GeoWEPP analyzes events. GeoWEPP analyzes runoff and sediment generation daily. The response time for a rainfall event may last only a few hours. Thus, the error for events could be minimized if GeoWEPP analyzed the rainfall events with a minimum time step of one hour. The approximations performed for land cover or even the calibration parameters to represent natural phenomena may have caused errors or may not be representing all the phenomena that impact soil erosion.

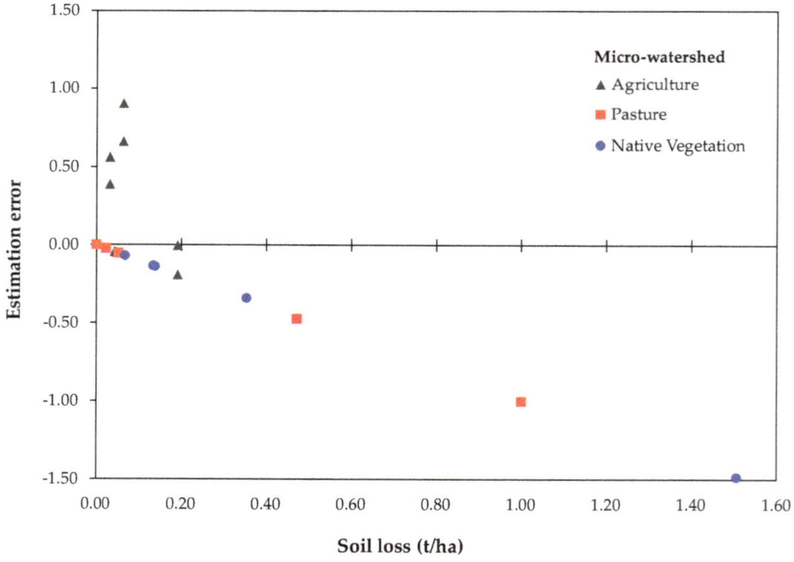

Figure 8. Estimation error of soil loss by the GeoWEPP model versus observed soil losses measured in metric tons (t) per hectare (ha) under land use and management conditions in the micro-watersheds for agriculture, pasture, and native vegetation in Campo Verde, Mato Grosso state, Brazil.

The determination of the minimum length of drainage networks and the minimum critical area by GeoWEPP may also have had a considerable influence on our results. This is because the greater the drainage density, the faster that water and sediments are drained by the basin, which can generate larger peaks. Limited research in Brazil has validated GeoWEPP using calibration parameters from multi-year field data. However, several of these researchers point out that the model has a high level of uncertainty due to the complexity in evaluating soil erosion [45].

It can be Inferred that the inclusion of the calibration parameters for hydraulic conductivity (K_e), interrill erodibility (K_i), rill erodibility (K_r), and critical shear stress (τ_c) alter the flow dynamics and sediment production in the outlet generated by the GeoWEPP model. Therefore, the estimation of these parameters by WEPP can lead to unexpected results. This is because the model has numerous parameters for land use characteristics, but all for American soils under temperate climate conditions.

Prior research found that the use of parameters obtained experimentally improved on average, 306% of the estimates for soil loss, when compared to the use of Ke, Ki, Kr, and τc, generated internally by WEPP. The same occurred for water losses, providing an improvement of 135% [46]. When analyzing the sensitivity of several input parameters of the WEPP model for the Brazilian edaphoclimatic conditions, a previous study verified that the Ke, Ki, Kr, and τc parameters were the ones that presented the greatest improvements within the model [47].

3.4. Future Directions for Sustainable Agricultural Development

The two types of soils represented in our analyses make up over half of the land area of Brazil (Oxisols 38.7 + Entisols 14.5 = 53.2%) [48]. Although the GeoWEPP model validation for crops requires future improvement, the pasture and natural vegetation we validated the model for make up a large percentage of the Cerrado savannah biome at 54.8% for such natural areas and 30.7% for pasture with crops taking up only 13.1% of the land area of the Cerrado biome [49]. Future studies using GeoWEPP can validate the model across a greater variation in types and density of natural vegetation cover since it has been shown in more arid environments such as Brazil's northeast region that denser vegetation cover is associated with lower erosion metrics and greater soil carbon [50]. Natural vegetation buffers in the Cerrado confer ecosystem service benefits such as biodiversity and improved soil and water quality [51].

GeoWEPP can be used to identify priority pasture areas in Brazil for improved soil conservation. Soil erosion is a significant challenge in Brazil, especially in agricultural areas with sandy soils, which can lead to noticeable erosion as rills and gullies even in areas with pasture [52], as well as increased suspended sediments loads in rivers such as the Teles Pires [53]. Brazil's pastures help support ~253 million head of cattle [49]. Using GeoWEPP to model erosion in Brazil's pastures is important due to the predominance of the extensive grazing system in Brazil and the susceptibility of this type of grazing to pasture degradation. Of Brazil's ~850 million hectares or ha (~8.5 million square kilometers or km^2) of total land area, about 21% of this is in pasture or ~179 million ha (~1.79 million km^2). About 35.7% of Brazil's pasture is degraded or 63.7 million ha (637,000 km^2) [54]. Therefore, using GeoWEPP to help rehabilitate degraded pastures is paramount to improving large ruminant livestock productivity and profitability, which can produce more cattle on less land. Such land sparing can allow for the restoration of natural habitats.

4. Conclusions

The GeoWEPP model performs well in estimating soil loss for areas under pasture and native vegetation; however, further study is warranted to increase the model's accuracy in estimating soil loss for areas where there is annual cultivation of commodity crops such as soybeans and maize. The GeoWEPP model performed better when the calibration parameters for both interrill and rill soil erodibility, soil critical shear stress, and saturated hydraulic conductivity were used in the model. The model slightly underestimated soil loss in pasture while accurately modeling this for native vegetation. GeoWEPP overestimated erosion in the micro-watershed we evaluated for agricultural production. In general, the GeoWEPP model can be used in a predictive way if there is an available digital elevation model for the terrain being modeled. This model can facilitate planning by agricultural technicians and by state, university, and independent researchers. As a whole, it is an important predictive tool for soil and environmental management in this region. This is due to GeoWEPP's ability to generate numerous scenarios of land use and management, which can help predict possible impacts from erosion. Future studies should focus on better calibrating the model with commodity cropping systems and expanding the use of this model to other tropical areas in Brazil, as well as around the world.

Supplementary Materials: Related supplementary material can be found in previous research at: (1) https://doi.org/10.5433/1679-0359.2020v41n5Supl1p1909 [19], (2) https://doi.org/10.1371/journal.pone.0179414 [23], (3) https://doi.org/10.19084/RCA18130 [25], (4) http://dx.doi.org/10.1590/1807-1929/agriambi.v24n6p357-363 [41], (5) https://doi.org/10.1016/j.gecco.2019.e00819 [51].

Author Contributions: Conceptualization, W.d.A.M. and R.S.S.A.; methodology, W.d.A.M., W.M.d.S. and E.D.B.; software, W.d.A.M.; validation, W.d.A.M., R.S.S.A. and E.D.B.; formal analysis, W.d.A.M., R.S.S.A., A.K.H. and W.M.d.S.; investigation, E.D.B.; resources, R.S.S.A., E.D.B., M.O.H., A.K.H. and W.d.A.M.; data curation, M.O.H. and D.C.d.A.; writing—original draft preparation, W.d.A.M., W.M.d.S., A.K.H. and M.O'H.H; writing—review and editing, W.d.A.M., A.K.H., D.C.d.A., L.A.D.L.D.R. and W.M.d.S.; visualization, L.A.D.L.D.R., A.K.H.; D.C.d.A. and W.M.d.S.; supervision, M.O.H. and R.S.S.A.; project administration, R.S.S.A. and E.D.B.; funding acquisition, R.S.S.A. All authors have read and agreed to the published version of the manuscript.

Funding: This research was developed in partnership with the CarBioCial project (donation number: 01 LL0902A). The authors are also grateful for financial support from the Mato Grosso State Research Support Foundation (www.fapemat.mt.gov.br, accessed on 5 December 2019), grant number: 335908/2012 and the Brazilian National Council for Scientific and Technological Development (www.cnpq.br, accessed on 11 December 2019), grant number: 481990/2013-5.

Institutional Review Board Statement: Not applicable.

Informed Consent Statement: Not applicable.

Data Availability Statement: Study data can be obtained by request to the corresponding author or the first author via e-mail.

Acknowledgments: The authors would like to thank Fazenda Santa Luzia and Fazenda Rancho do Sol for providing the study areas, and CAPES for granting the doctoral scholarship to the first author. The authors also acknowledge the collaboration of field hosts (Fazendas Santa Luzia and Rancho do Sol) and the field assistance of Túlio G. Santos and Alan R.R. Martin. We thank four anonymous reviewers whose comments and edits improved the quality of our work.

Conflicts of Interest: The authors declare no conflict of interest. Supporting entities had no role in the design of the study; in the collection, analyses, or interpretation of data; in the writing of the manuscript, or in the decision to publish the results.

References

1. Hernani, L.C.; de Freitas, P.L.; Pruski, F.F.; de Maria, I.C.; de Castro Filho, C.; Landers, J.N. A erosão e o seu impacto. In *Uso Agrícola dos solos Brasileiros*; Manzatto, C.V., de Freitas Junior, E., Peres, J.R.R., Eds.; Embrapa Solos: Rio de Janeiro, Brazil, 2002; pp. 47–60. Available online: https://ainfo.cnptia.embrapa.br/digital/bitstream/item/215162/1/A-erosao-e-seu-impacto-2002.pdf (accessed on 9 November 2022).
2. Dechen, S.C.F.; Telles, T.S.; Guimarães, M.D.F.; Maria, I.C.D. Perdas e custos associados à erosão hídrica em função de taxas de cobertura do solo. *Bragantia Camp.* **2015**, *74*, 224–233. Available online: http://www.scielo.br/pdf/brag/v74n2/0006-8705-brag-74-2-224.pdf (accessed on 5 October 2018). [CrossRef]
3. Ellen MacArthur Foundation. Cities and Circular Economy for Food, 2019 Report. Available online: http://www.ellenmacarthurfoundation.org/publications (accessed on 17 September 2022).
4. Lal, R. Soil Erosion and Gaseous Emissions. *Appl. Sci.* **2020**, *10*, 2784. [CrossRef]

5. Bhattarai, R.; Dutta, D. Estimation of Soil Erosion and Sediment Yield Using GIS at Catchment Scale. *Water Resour. Manag.* **2007**, *21*, 1635–1647. [CrossRef]
6. Beven, K. Changing ideas in hydrology—The case of physically based models. *J. Hydrol.* **1989**, *105*, 157–172. [CrossRef]
7. Grayson, R.B.; Moore, I.D.; McMahon, T.A. Physically-based hydrologic modeling: II. Is the concept realist? *Water Resour. Res.* **1992**, *28*, 2659–2666. [CrossRef]
8. Tucci, C.E.M. *Modelos Hidrológicos*, 2nd ed.; Editora da UFRGS; Universidade Federal do Rio Grande do Sul: Porto Alegre, Brazil, 2005; pp. 1–678. ISBN 8570258232.
9. Renschler, C.S.; Flanagan, D.C.; Engel, B.A.; Frankenberger, J.R. GeoWEPP—The Geo-spatial interface for the Water Erosion Prediction Project. In Proceedings of the 2002 ASAE Annual Meeting, Chicago, IL, USA, 18–31 July 2002; American Society of Agricultural and Biological Engineers: St. Joseph, MI, USA, 2002. [CrossRef]
10. Jong Van Lier, Q.; Sparovek, G.; Flanagan, D.C.; Bloem, E.M.; Schnug, E. Runoff mapping using WEPP erosion model and GIS tools. *Comput. Geosci.* **2005**, *31*, 1270–1276. [CrossRef]
11. Ferreira, A.M.; Silva, A.M.; Passos, C.A.; Valentino, C.H.; Gonçalves, F.A.; Menezes, P.H.B.J. Estimativa da erosão hídrica do solo pelo modelo Water Erosion Prediction Project na Sub-Bacia do Córrego do Gigante, sul de Minas Gerais. *Eng. Sanit. Ambient.* **2021**, *26*, 471–483. [CrossRef]
12. Silva, A.M.; Mello, C.R.; Curi, N.; Oliveira, P.M. Simulação da variabilidade espacial da erosão hídrica em uma sub-bacia hidrográfica de Latossolos no sul de Minas Gerais. *Rev. Bras. Ciênc. Solo* **2008**, *32*, 2125–2134. [CrossRef]
13. Cecílio, R.A.; Rodriguez, R.D.G.; Baena, L.G.N.; Oliveira, F.G.; Pruski, F.F. Aplicação dos modelos RUSLE e WEPP para a estimativa da erosão hídrica em microbacia hidrográfica de Viçosa (MG). *Rev. Verde Agroecol. Desenvolv. Sustent.* **2009**, *4*, 39–45. [CrossRef]
14. Spera, S.A.; Cohn, A.S.; VanWey, L.K.; Mustard, J.F.; Rudorff, B.F.; Risso, J.; Adami, M. Recent cropping frequency, expansion, and abandonment in Mato Grosso, Brazil had selective land characteristics. *Environ. Res. Lett.* **2014**, *9*, 064010. [CrossRef]
15. Ofstehage, A.; Nehring, R. No-till agriculture and the deception of sustainability in Brazil. *Int. J. Agric. Sustain.* **2021**, *19*, 335–348. [CrossRef]
16. Fuentes-Llanillo, R.; Telles, T.S.; Junior, D.S.; de Melo, T.R.; Friedrich, T.; Kassam, A. Expansion of no-tillage practice in conservation agriculture in Brazil. *Soil Tillage Res.* **2021**, *208*, 104877. [CrossRef]
17. Pandey, A.; Himanshu, S.K.; Mishra, S.K.; Singh, V.P. Physically based soil erosion and sediment yield models revisited. *Catena* **2016**, *147*, 595–620. [CrossRef]
18. Amorim, R.S.S.; da Silva, D.D.; Pruski, F.F.; de Matos, A.T. Evaluation of the performance of the USLE, RUSLE and WEPP soil erosion models for different edaphoclimatic conditions in Brazil. *Eng. Agric.* **2010**, *30*, 1046–1059. [CrossRef]
19. Magalhães, W.A.; Amorim, R.S.S.; Hunter, M.O.; Bocuti, E.D.; Dos Santos, T.G. Morphometric characteristics and hydrological dynamics in microbasins with different land uses of cerrado soils in Mato Grosso, Londrina (PR). *Semin. Ciênc. Agrár.* **2020**, *41*, 1909–1922. [CrossRef]
20. Weska, R.K. Uma síntese do cretáceo superior mato-grossense. *Rev. Geociênc.* **2006**, *25*, 71–81. Available online: https://www.revistageociencias.com.br/geociencias-arquivos/25_1/7.pdf (accessed on 18 February 2023).
21. Dos Santos, H.G.; Jacomine, P.T.; Dos Anjos, L.H.C.; De Oliveira, V.Á.; Lumbreras, J.F.; Coelho, M.R.; De Almeida, J.A.; de Araujo Filho, J.C.; De Oliveira, J.B.; Cunha, T.J.F. *Brazilian Soil Classification System*, 5th ed.; Embrapa: Brasília, Brazil, 2018; pp. 93–96.
22. Bocuti, E.D. Condutividade Hidráulica Efetiva e Erodibilidade Entressulcos de Solos da Bacia do Rio das Mortes. Master's Thesis, Universidade Federal de Mato Grosso, Cuiabá, Brazil, 2016; pp. 1–76.
23. Nóbrega, R.L.B.; Guzha, A.C.; Torres, G.N.; Kovacs, K.; Lamparter, G.; Amorim, R.S.S.; Couto, E.; Gerold, G. Effects of conversion of native cerrado vegetation to pasture on soil hydro-physical properties, evapotranspiration and streamflow on the Amazonian agricultural frontier. *PLoS ONE* **2020**, *12*, e0179414. [CrossRef]
24. Elliot, W.J. *A Compendium of Soil Erodibility Data from WEPP Cropland Soil Field Erodibility Experiments 1987, 1988*; U.S. Department of Agriculture, Agricultural Research Service: West Lafayette, IN, USA, 1989; pp. 1–291.
25. Bocuti, E.D.; Amorim, R.S.S.; dos Santos, T.G.; Raimo, L.D.L.; Pereira, H.G. Interrill erodibility and its relationship with Cerrado soil attributes. *Rev. Ciênc. Agrár.* **2019**, *42*, 68–78. [CrossRef]
26. Foster, G.R. Modeling the erosion process. In *Hydrologic Modelling of Small Watersheds*; Haan, C.T., Johnson, H.P., Brakensienk, D.L., Eds.; ASAE Monograph, ASAE: St. Joseph, MI, USA, 1982; pp. 296–380.
27. ESRI 2016. ArcGIS Desktop: Release 10.4.1. Redlands, CA: Environmental Systems Research Institute. Available online: https://desktop.arcgis.com/en/quick-start-guides/10.4/arcgis-desktop-quick-start-guide.htm (accessed on 15 December 2016).
28. Maalim, F.K.; Melesse, A.M.; Belmont, P.; Gran, K.B. Modeling the impact of land use changes on runoff and sediment yield in the Le Sueur watershed, Minnesota using GeoWEPP. *Catena* **2013**, *107*, 35–45. [CrossRef]
29. Cligen. Agricultural Research Service, U.S. Department of Agriculture. Available online: https://www.ars.usda.gov/midwest-area/west-lafayette-in/national-soil-erosion-research/docs/wepp/cligen/ (accessed on 15 January 2017).

30. Flanagan, D.C.; Nearing, M.A. *USDA-Water Erosion Prediction Project Hillslope Profile and Watershed Model Documentation*; NSERL Rep. No. 10; USDA-ARS National Soil Erosion Research Laboratory: West Lafayette, IN, USA, 1995. Available online: https://www.ars.usda.gov/midwest-area/west-lafayette-in/national-soil-erosion-research/docs/wepp/wepp-model-documentation/ (accessed on 3 March 2017).
31. Admas, M.; Melesse, A.M.; Abate, B.; Tegegne, G. Soil Erosion, Sediment Yield, and Runoff Modeling of the Megech Watershed Using the GeoWEPP Model. *Hydrology* **2022**, *9*, 208. [CrossRef]
32. Parente, L.; Mesquita, V.; Miziara, F.; Baumann, L.; Ferreira, L. Assessing the pasturelands and livestock dynamics in Brazil, from 1985 to 2017: A novel approach based on high spatial resolution imagery and Google Earth Engine cloud computing. *Remote Sens. Environ.* **2019**, *232*, 111301. [CrossRef]
33. Campos, M.O.; Cerri, C.E.P.; La Scala, N. Atmospheric CO_2, soil carbon stock and control variables in managed and degraded pastures in central Brazil. *Remote Sens. Appl. Soc. Environ.* **2022**, *28*, 100848. [CrossRef]
34. Ferreira, M.M.; Fernandes, B.; Curi, N. Influencia da mineralogía da fração argila nas propriedades físicas de Latossolos da região sudeste do Brasil. *Rev. Bras. Cienc. Solo* **1999**, *23*, 515–524. [CrossRef]
35. Alcantara, G.H. Erodibilidade em Sulcos e Tensão Cisalhante Crítica de Latossolos com Diferentes Teores de Óxidos de Ferro. Master's Thesis, Faculdade Ciências Agrárias e Veterinárias, Universidade Estadual Paulista, São Paulo, Brazil, 2018; 71p.
36. Nunes, M.C.M.; Cassol, E.A. Estimativa da erodibilidade em entressulcos de latossolos do Rio Grande do Sul. *Rev. Bras. Cienc. Solo* **2008**, *32*, 2839–2845. [CrossRef]
37. Alves, M.A.B.; de Souza, A.P.; de Almeida, F.T.; Hoshide, A.K.; Araújo, H.B.; da Silva, A.F.; de Carvalho, D.F. Impact of land use on soil erosion in agricultural frontier areas in the Cerrado-Amazon ecotone, Brazil. *Sustainability*, 2023, *submitted*.
38. González-Arqueros, M.L.; Mendoza, M.E.; Vázquez-Selem, L. Human impact on natural systems modeled through soil erosion in GeoWEPP: A comparison between pre-Hispanic periods and modern times in the Teotihuacan Valley (Central Mexico). *Catena* **2017**, *149 Pt 1*, 505–513. [CrossRef]
39. Nóbrega, R.L.B.; Guzha, A.C.; Lamparter, G.; Amorim, R.S.S.; Couto, E.G.; Hughes, H.J.; Gerold, G. Impacts of land-use and land-cover change on stream hydrochemistry in the Cerrado and Amazon biomes. *Sci. Total Environ.* **2018**, *635*, 259–274. [CrossRef] [PubMed]
40. Nearing, M.A.; Deer-Ascough, L.; Laflen, J.M. Sensitivity analysis of the WEPP hillslope-profile erosion model. *Trans. ASAE* **1990**, *33*, 839–849. [CrossRef]
41. Bocuti, E.D.; Amorim, R.S.S.; Raimo, L.A.D.L.D.; Magalhães, W.D.A.; de Azevedo, E.C. Effective hydraulic conductivity and its relationship with the other attributes of Cerrado soils. *Rev. Bras. Eng. Agric. Ambient.* **2020**, *24*, 357–363. [CrossRef]
42. Pissarra, T.C.T.; Politano, W.; Ferraudo, A.S. Avaliação de características morfométricas na relação solo-superfície da bacia hidrográfica do Córrego Rico, Jaboticabal (SP). *Rev. Bras. Cienc. Solo* **2004**, *28*, 297–305. [CrossRef]
43. Nearing, M.A. Why soil erosion models over-predict small soil losses and under-predict large soil losses. *Catena* **1998**, *32*, 15–22. [CrossRef]
44. Tiwari, A.K.; Risse, L.M.; Nearing, M.A. Evaluation of WEPP and its comparison with USLE and RUSLE. *Trans. ASAE* **2000**, *43*, 1129–1135. [CrossRef]
45. Kinnell, P.I.A. A comparison of the abilities of the USLE-M, RUSLE2 and WEPP to model event erosion from bare fallow areas. *Sci. Total Environ.* **2017**, *596–597*, 32–42. [CrossRef]
46. Gonçalves, F.A. Validação do Modelo WEPP na Predição de Erosão Hídrica para Condição Edafoclimática da Região de VIÇOSA-MG. Ph.D. Thesis, Universidade Federal de Viçosa, Viçosa, Brazil, 2007; pp. 1–116. Available online: https://www.locus.ufv.br/bitstream/123456789/629/1/texto%20completo.pdf (accessed on 31 January 2023).
47. Amorim, R.S.S. Avaliação dos Modelos de Predição da Erosão Hídrica USLE, RUSLE e WEPP para Condições Edafoclimáticas Brasileiras. Ph.D. Thesis, Universidade Federal de Viçosa, Viçosa, Brazil, 2003; pp. 1–120. Available online: https://www.locus.ufv.br/bitstream/123456789/9620/1/texto%20completo.pdf (accessed on 31 January 2023).
48. Guerra, A.J.T.; Fullen, M.A.; Jorge, M.D.C.O.; Alexandre, S.T. Soil Erosion and Conservation in Brazil. *Anu. Inst. Geociênc.* **2014**, *37*, 81–91. [CrossRef]
49. Oliveira, P.T.S.; Godoi, R.D.F.; Colman, C.B.; Motta, J.S.; Sone, J.S.; Almagro, A. Agricultural Land Degradation in Brazil. In *The Handbook of Environmental Chemistry*; Springer: Berlin/Heidelberg, Germany, 2022; pp. 1–31. [CrossRef]
50. Lins, C.M.T.; de Souza, E.R.; Souza, T.E.M.d.S.; Paulino, M.K.S.S.; Monteiro, D.R.; Júnior, V.S.D.S.; Dourado, P.R.M.; Junior, F.E.D.A.R.; da Silva, Y.J.A.; Schaffer, B. Influence of vegetation cover and rainfall intensity on soil attributes in an area undergoing desertification in Brazil. *Catena* **2023**, *221*, 106751. [CrossRef]
51. Nóbrega, R.L.B.; Ziembowicz, T.; Torres, G.N.; Guzha, A.C.; Amorim, R.S.S.; Cardoso, D.; Johnson, M.S.; Santos, T.G.; Couto, E.; Gerold, G. Ecosystem services of a functionally diverse riparian zone in the Amazon-Cerrado agricultural frontier. *Glob. Ecol. Conserv.* **2020**, *21*, e00819. [CrossRef]
52. Bouramtane, T.; Hilal, H.; Rezende-Filho, A.T.; Bouramtane, K.; Barbiero, L.; Abraham, S.; Valles, V.; Kacimi, I.; Sanhaji, H.; Torres-Rondon, L.; et al. Mapping Gully Erosion Variability and Susceptibility Using Remote Sensing, Multivariate Statistical Analysis, and Machine Learning in South Mato Grosso, Brazil. *Geosciences* **2022**, *12*, 235. [CrossRef]

53. Borella, D.R.; de Souza, A.P.; de Almeida, F.T.; de Abreu, D.C.; Hoshide, A.K.; Carvalho, G.A.; Pereira, R.R.; da Silva, A.F. Dynamics of Sediment Transport in the Teles Pires River Basin in the Cerrado-Amazon, Brazil. *Sustainability* **2022**, *14*, 16050. [CrossRef]
54. Polidoro, J.C.; de Freitas, P.L.; Hernani, L.C.; dos Anjos, L.H.C.; Rodrigues, R.D.A.R.; Cesário, F.V.; de Andrade, A.G.; Ribeiro, J.L. Potential impact of plans and policies based on the principles of conservation agriculture on the control of soil erosion in Brazil. *Land Degrad. Dev.* **2021**, *32*, 3457–3468. [CrossRef]

sustainability

MDPI

Article

Effects of Land Use and Cropping on Soil Erosion in Agricultural Frontier Areas in the Cerrado-Amazon Ecotone, Brazil, Using a Rainfall Simulator Experiment

Marco Aurélio Barbosa Alves [1], Adilson Pacheco de Souza [2], Frederico Terra de Almeida [2], Aaron Kinyu Hoshide [3,4], Handrey Borges Araújo [2], Apoliano Francisco da Silva [2] and Daniel Fonseca de Carvalho [5,*]

1 Postgraduate Program in Agronomy-Soil Sciences, Federal Rural University of Rio de Janeiro, Seropédica 23897-045, Rio de Janeiro, Brazil
2 Institute of Agrarian and Environmental Sciences, Federal University of Mato Grosso, Sinop 78557-287, Mato Grosso, Brazil
3 College of Natural Sciences, Forestry and Agriculture, The University of Maine, Orono, ME 04469, USA
4 AgriSciences, Federal University of Mato Grosso, Caixa Postal 729, Sinop 78550-970, Mato Grosso, Brazil
5 Institute of Technology, Federal Rural University of Rio de Janeiro, Seropédica 23897-000, Rio de Janeiro, Brazil
* Correspondence: daniel.fonseca.carvalho@gmail.com; Tel.: +55-(21)-98787-7262

Abstract: Agricultural soils provide ecosystem services, but the removal of natural vegetation reduces water infiltration capacity, increasing surface runoff. Thus, monitoring erosion is critical for sustainable agricultural management. Sediment losses and surface runoff were evaluated using a simulated rainfall of 75 mm/h in areas with crops and pastures in both the Caiabi River and Renato River sub-basins of the Teles Pires River watershed in Mato Grosso State, Brazil. In both the Caiabi and Renato sub-basins, data were collected from 156 observations in the upper, middle, and lower regions where (1) soybeans, (2) maize, and (3) pasture were grown alone, with another crop, or with soil that was scarified. Erosion occurred independent of soil texture and was closely related to the management and use of systems involving fewer crops and more soil scarification, regardless of sub-basin location. In uncovered, scarified soil, the soil losses from erosion were greater compared to covered soil, regardless of sub-basin and sub-basin region. In the Renato River sub-basin, soil losses in cultivated areas not planted with crops but with scarification were 66.01, 90.79, and 60.02 g/square meter in the upper, middle, and lower regions, respectively. Agricultural producers need to increase the planting of crops throughout the year and minimize soil disturbance, which will reduce soil erosion and improve sustainability.

Keywords: Cerrado-Amazon; crops; geoprocessing; GIS; land use; mapping; rainfall simulator; satellite images; soil erosion

Citation: Alves, M.A.B.; de Souza, A.P.; de Almeida, F.T.; Hoshide, A.K.; Araújo, H.B.; da Silva, A.F.; de Carvalho, D.F. Effects of Land Use and Cropping on Soil Erosion in Agricultural Frontier Areas in the Cerrado-Amazon Ecotone, Brazil, Using a Rainfall Simulator Experiment. *Sustainability* **2023**, *15*, 4954. https://doi.org/10.3390/su15064954

Academic Editors: Maurizio Lazzari and Teodor Rusu

Received: 6 January 2023
Revised: 1 March 2023
Accepted: 7 March 2023
Published: 10 March 2023

1. Introduction

Soil is essential for both macro- and microscopic life and provides ecosystem services, ensuring a stock for carbon, nutrient cycling, water retention and infiltration, and food production [1]. However, most soils in Brazil and around the world are compromised due to contamination, pollution, and erosive processes that have contributed to soil degradation [2,3]. In addition to adversely affecting farming production and ecosystem services, degraded soils contribute to global warming and hydrological extremes since they have lower capacities to store carbon and facilitate the infiltration and storage of water [4–6].

Erosion is one of the main causes of soil loss in Brazil and around the world, specifically water erosion [2,7], which is responsible for detaching soil particles, transporting them, and depositing them in areas at lower altitudes. Therefore, soil carbon stocks are exposed and lost through decomposition, mineralization, and transport, while mineral particles silt up rivers, streams, and lakes [8]. While soil compaction, surface crusting, and soil erosion can be reduced by soil organic matter, which can also provide nutrients to plants, soil erosion can also be reduced by vegetative cover [9]. Uncovered and/or badly managed soils accelerate this process since the absence of vegetative cover (e.g., crops) promotes direct exposure to raindrops [10]. Factors such as the duration of rain and the slope of the area are also responsible for increases in soil losses by erosion, as they influence surface runoff [11,12], while soil turning (scarification) promotes the breakdown of aggregates, making the particles more susceptible to the erosive process [13].

In order for Brazil to reduce its soil and water losses, producers need to adopt management techniques such as the no-tillage system, contour lines, and pasture management with rotational grazing. In contrast, areas covered with native forest, crop straw residues, and pasture tend to have less soil loss than areas with exposed and tilled soils [13–15], minimizing the environmental damage from soil erosion. Undisturbed natural vegetation can also minimize soil erosion. For example, in an area with native forest located in the Amazon biome, the measured surface runoff and soil losses were close to zero due to the greater rainfall infiltration capacity of the soil [16]. The removal of natural vegetation from forest areas and its transformation into crops and pastures cause soil and water losses and the subsequent destruction of biodiversity, due to the reduction of carbon stocks [11]. When this removal is accompanied by farming practices involving continuous soil disturbance, such as scarification, these losses are even more marked since soil disruption favors erosive processes [2,17]. Long-term research studies over decades demonstrate that shifting grasslands to progressively more disturbed or scarified (e.g., tilled) cropping systems can decrease soil microbiome diversity, make soil microbiome processes more variable, and increase the prevalence of pathogenic soil organisms [18].

Located in the Cerrado-Amazon ecotone and holding great socioeconomic and environmental importance, the upper and middle Teles Pires regions in Mato Grosso State, Brazil, are agricultural frontiers that have undergone constant native vegetation removal and are susceptible to severe soil, water, and nutrient losses [17]. These areas represent the dynamics of land occupation and land use in the region; soil erosion monitoring is an important tool for decision-making, especially regarding sustainable agricultural production. Due to difficulties in collecting and quantifying the runoff in the experimental plots and the temporal variability of the intensity of natural rainfall, rainfall simulators have been used in different classes, uses, and occupations [12] and in studies about water infiltration into the soil [19], revealing distinct classifications and operational characteristics [20–22]. Additionally, the simulation equipment allows control of the duration and rainfall intensity, and the size and speed of the droplets impacting the soil [20]. Known as sprinkler infiltrometers, these simulators take less time and have a lower cost when conducting field research, compared to experimental natural rainfall plots, and can accurately control water input, thereby reducing the errors associated with natural rainfall variability [23].

The use of rainfall simulators to measure soil erosion in agricultural production systems in the Cerrado-Amazon transition region is unprecedented. Our research can serve as a guide for agricultural producers, ranchers, the public agencies of agrarian policies, and non-governmental agencies seeking to improve the management of soil and water resources in the tropics. The goal of our study is to quantify soil losses when under different agricultural land uses in the Caiabi and Renato River sub-basins of the Teles Pires River watershed in Mato Grosso State, Brazil.

2. Materials and Methods

2.1. Study Area

The Teles Pires River watershed is located in the states of Mato Grosso and Pará, Brazil (Figure 1). Despite being located, in hydrological terms, in the Amazon region, the Teles Pires watershed has variable vegetative cover, with its upper and lower regions in the Cerrado and Amazon biomes, respectively. The upper and middle Teles Pires regions correspond to 26.2 and 57.71% of the basin area; they have a population density of 45.9

and 27.5% of the total population of the basin and are responsible for 66.3 and 18.7% of the gross domestic product (GDP) obtained in the two Teles Pires River areas, respectively. The two regions together represent more than 17% of the GDP of Mato Grosso. Analyses of the soil and water losses were conducted in two drainage sub-basins, the Caiabi River (upper) and Renato River (middle), with drainage areas of approximately 500 and 1450 km^2, respectively (Figure 1).

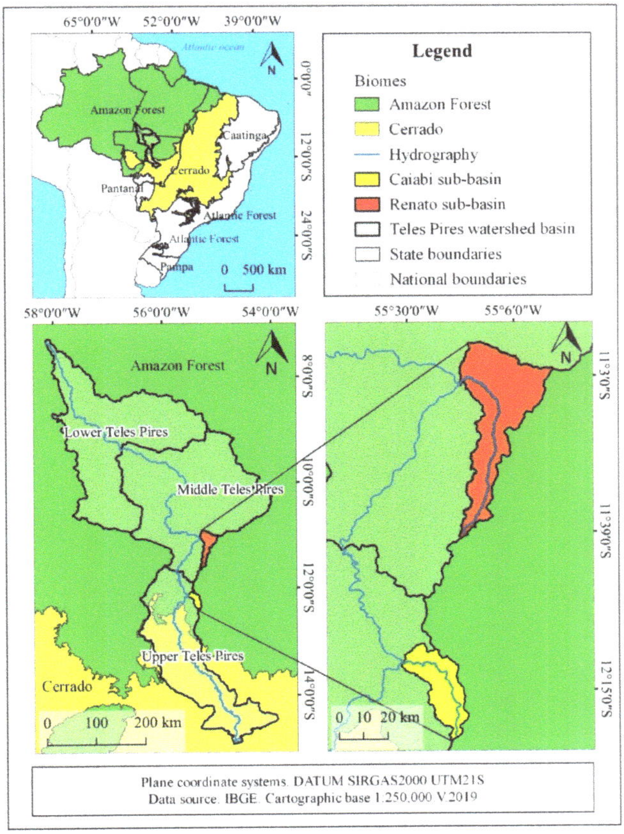

Figure 1. The Teles Pires River watershed and the location of the Caiabi and Renato River basins (data source: Ref. [24]).

Regarding the soils in the Caiabi sub-basin, the most recurrent classes are the inceptisols, oxisols, and entisols [24,25], formed from metasedimentary rocks belonging to the Cuiabá Group and the Raizama and Araras formations (Upper Paraguai Group). Conversely, the Renato sub-basin presents pedological characterization, with ultisols, oxisols, entisols, and Plinthic oxisols [24,25], formed from the granitic and rhyolitic rocks of the Juruena magmatic arc, with several gold occurrences: sandstones from the Dardanelos Formation and the Beneficiente Group, with sandstones, siltstones, and claystones from the Upper Tapajós basin (Capoeiras Formation).

In the Caiabi River sub-basin, located in the Cerrado-Amazon ecotone, monoculture areas of soybean (*Glycine max* L.), immediately followed by maize (*Zea mays* L.) in the same production year succession, are predominant, while in the Renato River sub-basin, there is also a predominance of native Amazon rainforest under forest management. According to the Köppen climate classification model, the climate of the region is the Aw type, considered a tropical wet and dry climate, with a dry period between June and August [26]. The mean annual temperature is 25 °C, with a minimum temperature below 16 °C and a maximum temperature above 34 °C. The mean annual precipitation varies by approximately 1900 mm [27]. In 2019, the percentages (%) of land use and occupation in the Caiabi sub-basin were most for crops (51.96%) followed by water (22%), native forest (31.78%), pastures (3.6%), and burned areas (0.08%), respectively. Likewise, in the Renato River hydrographic sub-basin, these same classes of land use and occupation were highest for native forest (61.32%) followed by water (27%), pastures (9.7%), crops (9.57%), and burned areas (0.18%), respectively (Figure 2).

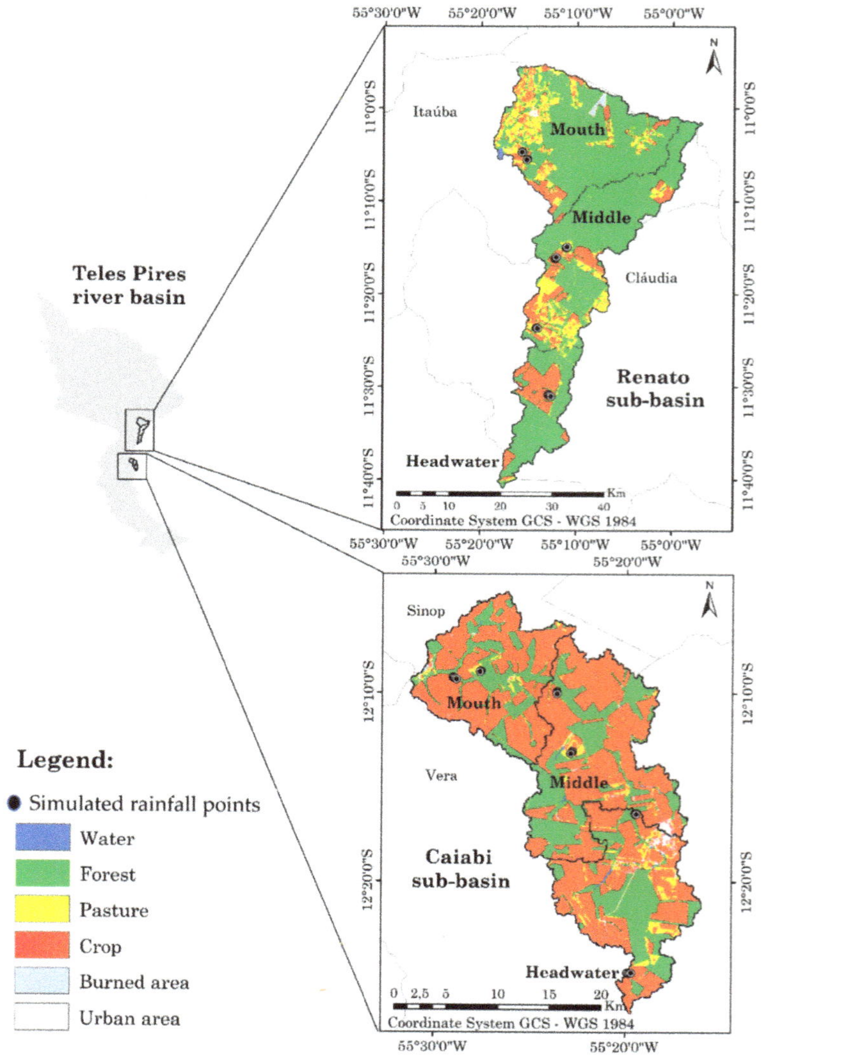

Figure 2. Location of the sampling points of simulated rainfall and land use, along with occupation in the Caiabi and Renato sub-basins, which are tributaries of the Teles Pires river.

Rainfall simulator experimental tests were carried out on 12 farms, two in each region of each sub-basin. In general, in the region each property operates in, only one type of farming activity occurs (either cropping or pasture). In the Caiabi river sub-basin, 84 tests were carried out on six farms, distributed as follows: (1) on the farms occupied by soybeans, only four soil covers were evaluated at each sampling point (soybean + straw, straw only, uncovered soil, and scarified soil), totaling 48 tests; (2) on the farms occupied by pastures, three soil covers were evaluated at each sampling point (pasture, uncovered soil, and scarified soil), totaling 36 tests. In the Renato sub-basin, 72 tests were carried out, since in both of the areas occupied by corn and by pastures, three soil covers were evaluated (with vegetation, uncovered soil, and scarified soil), resulting in 36 tests in areas occupied by corn and 36 tests in areas occupied by pastures. In Figure 2, the four repetitions evaluated in the same farm are not so evident, due to the spatial scale. However, the tests with simulated rainfall were carried out considering a minimum distance of 500 m between repetitions (examples can be seen in Figure 3). The areas selected for carrying out these tests had spent at least 5 years under the same land use.

Figure 3. Locations of the simulated rainfall points in (**a**) Continental Farm (headwater sub-basin region), (**b**) Aremisa III Farm (middle sub-basin region), with land use of maize crops and pasture, in the Renato sub-basin, (**c**) São José Farm, and (**d**) Taguá Farm, both in the middle region of the Caiabi sub-basin. The scales in (**a–d**) are 1:54,200, 1:361,000, 1:55,000, 1:62,300, and 1:90,500 cm, respectively.

The evaluations of soil and water losses occurred for different agricultural crops, depending on the specific cultivation calendar for soybeans and maize in the state of Mato Grosso. Figure 4 summarizes the rainfall in the two hydrographic sub-basins being evaluated, along with the arrangement of crops in the region, regulated by the federal ordinances of the Secretaria de Política Agrícola (SPA) of the Ministério de Agricultura, Pecuária e Abastecimento (MAPA) of Brazil. For soybean cultivation, the Portaria SPA/MAPA 249/2022 legislation [28] recommends sowing soybeans between 1 October and 31 December in the municipalities located in the Caiabi and Renato hydrographic basins. This legislation also establishes a "sanitary void" for soybean cultivation, which is a recommended break in the planting of soybeans after 1 January and before 15 September each year. Therefore, in the region studied, maize is planted as the second crop (*safrinha*), and its planting depends on the sowing time and the growth cycle of the soybean cultivar that is planted. This is regulated by the Agricultural Zoning of Climatic Risk (ZARC), established by Ordinance SPA/MAPA 332/2022 [29], which recommends planting maize between 1 February and 10 March each year.

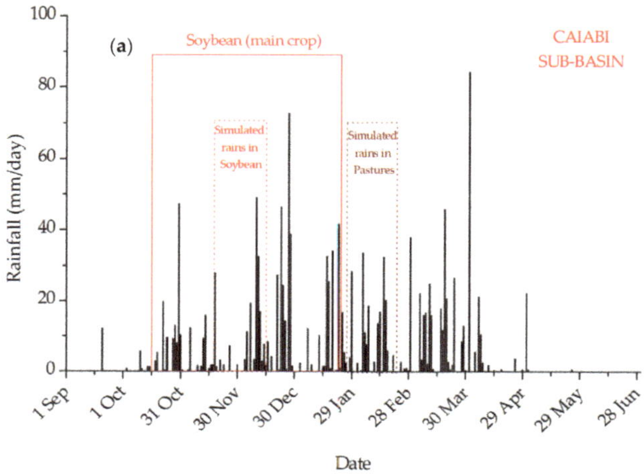

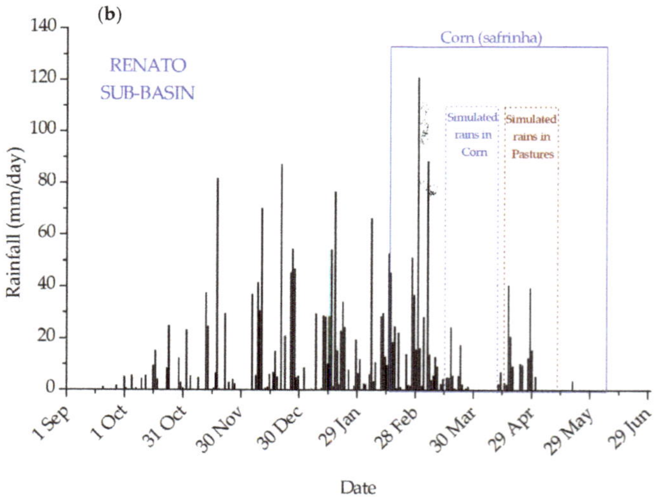

Figure 4. Daily rainfall in (**a**) the Caiabi sub-basin and (**b**) the Renato sub-basin, between 15 September 2020 and 30 June 2021.

2.2. Simulated Rainfall Treatments

Field trials with the InfiAsper rainfall simulator [20] were carried out in pastures and in areas under crop cultivation in the upper, middle, and lower regions of the two watersheds (Figure 5). The simulator operates with two Veejet 80.150 nozzles parallel to each other, positioned 2.3 m from the ground surface, with an average service pressure of 35.6 kPa. The diameter of drops applied by the simulator, considering the different pressure settings and rotation of the obturator disk, was exhaustively measured by Macedo et al. (2021) [22] using the flour method. Alves Sobrinho et al. 2008 [20] also made this assessment, confirming a mean drop diameter of 2.0 mm. Calibration tests were carried out in the laboratory, using a grid with 25 collectors over an area of 50 square centimeters, and uniformity coefficients above 80% were obtained.

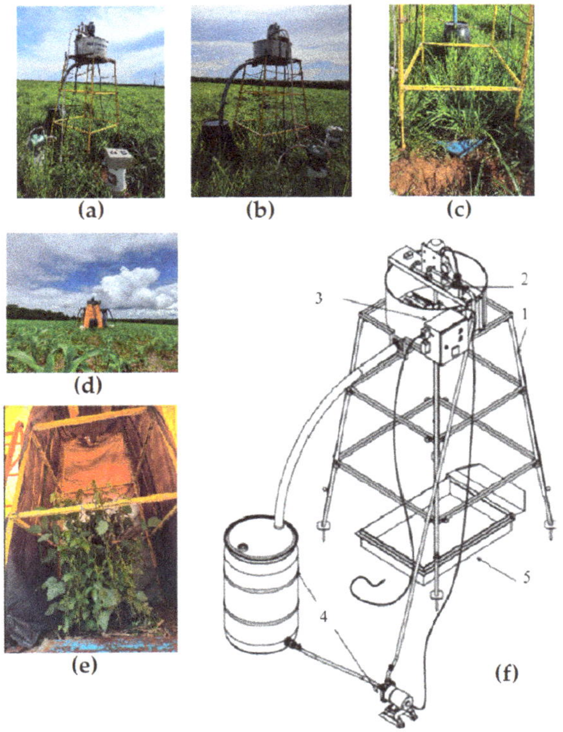

Figure 5. Installation of the *InfiAsper* rainfall simulator (**a**–**d**) and its operation (**e**), with plastic inside the support structure of the *InfAsper* to reduce the effects of wind; scheme of the components of the *InfiAsper* rainfall simulator (**f**) showing: (1) metallic structure, (2) water application unit, (3) control panel, (4) reservoir and water pump, and (5) runoff collector [22]

Due to the regional agricultural calendar, the evaluations in cultivation areas in the Caiabi River basin occurred in soybean plantations at the V7 vegetative stage, with the plants cultivated in maize straw (no-tillage system), whereas in cultivation areas in the Renato River basin, the crops were maize at the V4 vegetative stage after soybean succession. In the two basins, the pasture areas were occupied by *Brachiaria* spp., with an average height of 50 cm. The simulated rainfall was carried out in different areas in the two sub-basins, namely, for soybean (the Caiabi sub-basin) and for maize (the Renato sub-basin). Due to the cultural practices adopted in the region, the amount of remaining soybean straw in the maize crop was very small, when compared with the remaining maize straw, which can influence the timing of soybean planting. For this reason, no simulated rainfall was carried out in those plots with only straw in the Renato sub-basin.

In the cultivation areas of the Caiabi River basin, simulated rainfall was performed considering the following treatments (Figure 6): covered with soybean (plant + straw), only straw, and without crops and scarified at 0.1 m soil depth (4). In the Renato sub-basin, the treatments evaluated were covered with maize, without crops, and scarified. In the pasture areas for both basins, rainfall was assessed with the conditions of soils covered with pasture, uncovered soils, and soils scarified at 10 cm in depth. Simulated rainfall was replicated 4 times per basin region and treatment, totaling 156 tests. The useful area of the simulator plot was 0.7 square meters (m^2) and the average slope of the surface in the field was 3 degrees (Figure 6h). To standardize soil moisture, the plots were dampened before the beginning of the simulated rainfall, according to the methodology described in [12].

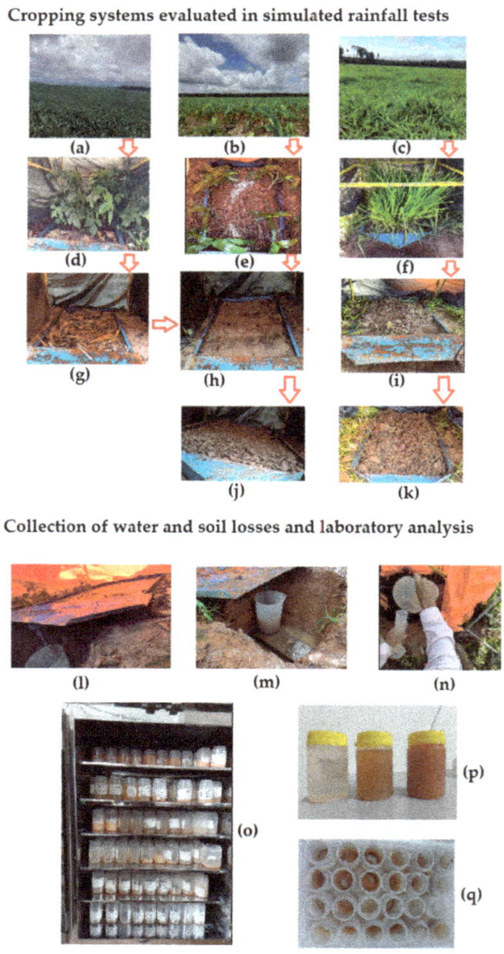

Figure 6. Soil cover treatments with (**a**) soybean, (**b**,**e**) pasture, and (**c**,**f**) maize, soybean + straw (**d**), and straw only (**g**), without crops (**h**,**i**), and scarified (**j**,**k**). The red arrows represent the flowchart of simulated events, as follows: (i) soybean + straw, straw only, without crops and with scarified soil; (ii) maize, without crops and with scarified soil; (iii) pasture, without crops and with scarified soil. In simulated rainfall tests in the different cropping systems, surface runoff was collected every minute (**l**,**m**), measuring the volume of water loss (**n**). Subsequently, the samples were dried in an oven (**o**,**q**); the differences between soil losses in pasture conditions, uncovered soil, and scarified soil for the same sampling point and time after the start of runoff (6 min) can be observed in (**p**).

The rainfall intensity (RI) of the simulated rainfall was defined, based on the intensity–duration–frequency (IDF) relationship created for the study region, according to the authors of [30]. The RI value was approximately 75 mm per hour, considering a return period of 10 years and an average duration of 42 min. After the beginning of the runoff, the material collection was performed at intervals of 1 min using plastic containers, and the runoff volume was measured with a graduated cylinder and then transferred to 0.5-liter (L) jars. The volume of runoff water was sent to the laboratory for the quantification of the water and soils lost through erosion [12]. To identify the amount of sediment in the water, the decantation and subsequent evaporation process of the water was conducted. To achieve this, the jars with the collected material were subjected to a temperature of 55 °C, in a forced circulation ion oven, until reaching a constant dry mass (up to 96 h). After drying, the samples were weighed on an analytical balance for the quantification of sediment.

2.3. Analyses of the Soil Characteristics and Vegetative Cover Dry Matter

For the physical-hydric characterization of the soil, near each point where the simulated rainfall was applied, mini-trenches of 0.4 × 0.4 m were dug for the collection of disturbed and undisturbed soil samples in the 0- to 0.1- and 0.1- to 0.2-m layers in the three regions of each sub-basin. The attributes analyzed in the physical-hydric characterization were granulometry (sand, silt, and clay), bulk density, particle density, total porosity, macro- and microporosity, and hydraulic conductivity. Granulometry was determined by the pipette method, using a sodium hydroxide (NaOH) solution with mechanical agitation for 16 h, based on the principle of Stokes' law. Bulk density was obtained by the graduated cylinder method, using undisturbed samples. In the laboratory, the samples were dried in an oven at 105 °C and then weighed 48 h later [31]. Particle density was determined by the volumetric flask method. Total porosity was obtained via the relationship between the bulk density and particle density in Equation 1 [32]. Macroporosity was obtained by the tensior table, with a tension of 10 kilopascals (kPa), and microporosity was obtained by taking tl difference between total porosity and macroporosity [31]:

$$TPo = 1 - Bd/Pd$$

where TPo equals total porosity, Bd is bulk density, and Pd equals particle density. soil class of all studied areas is latosol [25]. Soil textural distribution particles in the watersheds are shown in Table 1.

Table 1. Soil textural distribution in the hydrographic sub-basins of the Renato and Caiabi R

Cropland Use	Caiabi River Sub-Basin Region *							
	Upper			Middle			Lower	
	Sand	Clay	Silt	Sand	Clay	Silt	Sand	Clay
	(%)							
	0 to 0.1 m in depth							
Cultivated	42.49Ab	27.90Aa	29.61Aa	76.56Aa	17.80Bb	5.64Ab	78.50Aa	15.60Ab
Pasture	49.24Ab	36.10Aa	14.66Ba	49.21Bb	34.60Aa	16.19Aa	84.37Aa	11.00Ab
	0.1 to 0.2 m in depth							
Cultivated	52.40Ab	35.80Ba	11.80Aa	75.75Aa	20.20Bb	4.05Aa	79.29Aa	16.90Ab
Pasture	36.12Bb	49.00Aa	14.88Aa	45.40Bb	43.70Aa	10.90Aa	83.32Aa	11.80Ab
	Renato River Sub-Basin Region *							
	0 to 0.1 m in depth							
Cultivated	75.18Ab	16.20Aab	8.62Aa	82.87Aa	12.90Ab	4.23Aa	73.90Bb	19.40A
Pasture	80.43Aa	15.90Aa	3.67Aa	83.16Aa	12.90Aa	3.94Aa	81.94Aa	14.70F
	0.1 to 0.2 m in depth							
Cultivated	75.88Aab	19.00Aab	5.12Aa	80.52Aa	16.20Ab	3.28Aa	71.98Bb	23.2(
Pasture	75.30Aa	18.30Aa	6.40Aa	81.58Aa	13.40Aa	5.02Aa	79.93Aa	17.7

* Means that are followed by equal uppercase letters in the same column and equal lowerc row do not differ significantly from each other, as established by the Kruskal-Wallis test at a

Vegetation cover was removed after rainfall on the covered ground and before rainfall on the bare ground from each plot. To measure the dry matter content of the vegetation cover, plant materials were collected, identified, and taken to the Hydraulics and Hydrology Laboratory at the Federal University of Mato Grosso. Subsequently, they were dried in an oven with forced air circulation at a temperature of 65 °C for 72 h, until reaching a constant dry mass of around less than 5% moisture. The dry mass was quantified using a thousandth of an analytical scale, while for the straw dry mass, the separation of the remaining soil particles was carried out.

2.4. Experimental and Statistical Design

In cultivated areas in the Caiabi River, the experimental design was conducted in randomized blocks (RBD), in a 3 × 4 factorial scheme, with 3 regions in the basin (upper, middle, and lower) and 4 soil cover/management treatments (soybean + straw, straw, without crops, without crops and scarified). In the Renato River basin, a similar experimental design was used (RDB), but in a 3 × 3 factorial scheme, with 3 regions in the basin and 3 soil cover/management treatments (maize, without crops, without crops and scarified). In pasture areas, regardless of the basin, a 3 × 3 factorial scheme was used, with 3 regions in the basin and 3 soil cover/management treatments (*Brachiaria* spp., without crops, and without crops and scarified). In all the conditions mentioned above, simulated rainfall was replicated 4 times. The repetitions were spaced 50 m apart, ensuring the same level in the toposequence and land use. The slope of the field area in our experiments ranged from 3 to 4 degrees. All variables were subjected to the Kruskal-Wallis test at a 5% probability (test for nonparametric data) using the Statistica program, version 10.0.

3. Results

3.1. Vegetative Cover Dry Matter

The sampled dry-matter contents of soybean and maize (crops), pasture, and straw are contrasted in Table 2. The dry mass of vegetative cover in plots that were subjected to simulated rainfall ranged from 7.26 to 11.91 metric tons/hectare in the Caiabi sub-basin and did not show significant differences in the Kruskal–Wallis test between the hydro-graphic sub-basin regions. The straw represents at least 70% of the dry mass of the vegetative cover (soybean + straw). This vegetative cover condition makes it possible to understand the soil and water losses in the absence of these covers and their relationships with the physical and water characteristics of the soil. No comparisons were made between the two sub-basins because the crops (soybean and maize) and the phenological stages were different.

Table 2. Dry matter (metric tons/hectare) measured for soil-cover treatments of vegetation, straw, and pasture in the experimental plots of simulated rainfall in the hydrographic basins of the Caiabi and Renato rivers.

		Dry Matter (Metric Tons/Hectare)		
Sub-basin	**Region**	**With Crops ***	**Straw**	**Pasture**
Caiabi	Upper	11.91 Aa	8.34 Aa	8.24 Aa
	Middle	10.20 Aa	7.96 Aa	8.90 Aa
	Lower	10.99 Aa	8.95 Aa	7.26 Aa
Renato	Upper	5.21 Ba	-	8.07 Aa
	Middle	6.48 Ba	-	8.29 Aa
	Lower	12.07 Aa	-	8.65 Aa

* "With crops" indicates the presence of soybean and straw in the experimental plots of the Caiabi sub-basin (with only maize for the Renato sub-basin). Differences between means were compared using the Kruskal–Wallis test at 5% probability, where (i) capital letters represent the analysis of the sub-basin regions for the same soil cover, while (ii) lowercase letters represent the analysis of land cover for the same region of the sub-basin.

The Renato River sub-basin has phytophysiological, geological, and pedological characteristics that are different from those found in the Caiabi River basin. Those areas occupied by cultivation had maize at the V4 vegetative stage, while the plants in pasture areas (*Brachiaria* spp.) had an average height of 45 cm. The dry weight obtained for maize varied along the sub-basin, due to the planting density; in both regions, the spacing adopted was 0.5 m between rows. However, the plant populations were 55,000 and 80,000 plants per hectare in the upper/middle and lower regions, respectively. In this sub-basin, there were also no significant differences that were verified by the Kruskal-Wallis test for the dry weight of plants in cultivation, with vegetation (e.g., maize) and pasture areas.

3.2. Caiabi River Sub-Basin

3.2.1. Soil Characteristics in the Caiabi River Sub-Basin

The physical attributes of the soil in the Caiabi river basin varied according to the different regions (Table 3). There was a greater presence of clay in the upper region, while the lower region was characterized by a higher sand content. Although the soil classes are the same, with a predominance of oxisol [25], the lower part of the Caiabi River has a lower altitude, favoring the deposition of sand. The same occurred with bulk density and particle density, with higher values in the lower Caiabi River, since minerals present in the sand fraction have higher densities than clay minerals.

Table 3. Physical and hydric characterization of the Caiabi River sub-basin soils in the pasture and cultivation areas.

Cropland Use	**Caiabi River Sub-Basin Region ***								
	Upper			**Middle**			**Lower**		
	Micro	Macro	**TPo**	**Micro**	**Macro**	**TPo**	**Micro**	**Macro**	**TPo**
	(m^3/m^3)								
	0 to 0.1 m depth								
Cultivated	0.28Aa	0.08Aa	0.36Aa	0.27Aa	0.11Aa	0.38Aa	0.35Aa	0.08Aa	0.43Aa
Pasture	0.27Aa	0.10Aa	0.38Aa	0.35Aa	0.02Aa	0.37Aa	0.29Aa	0.11Aa	0.39Aa
	0.1 to 0.2 m depth								
Cultivated	0.25Aa	0.07Aa	0.32Aa	0.27Aa	0.07Aa	0.34Aa	0.34Aa	0.04Aa	0.38Aa
Pasture	0.26Aa	0.13Aa	0.39Aa	0.32Aa	0.06Aa	0.38Aa	0.29Aa	0.12Aa	0.41Aa
	Pd	**Bd**	**K_0**	**Pd**	**Bd**	**K_0**	**Pd**	**Bd**	**K_0**
	(grams/cm^3)		**(cm/hour)**	**(grams/cm^3)**		**(cm/hour)**	**(grams/cm^3)**		**(cm/hour)**
	0 to 0.1 m depth								
Cultivated	2.14Bb	1.02Bb	1.21Aa	2.54Aa	1.50Aa	1.12Aa	2.52Aa	1.50Aa	1.28Ba
Pasture	2.44Aab	1.41Aa	0.33Bb	2.33Ab	1.58Aa	0.67Bb	2.61Aa	1.58Aa	1.70Aa
	0.1 to 0.2 m depth								
Cultivated	2.30Ab	1.18Ab	0.47Ab	2.54Aab	1.57Aa	1.70Aa	2.63Aa	1.57Aa	1.81Aa
Pasture	2.44Ab	1.40Aa	1.19Bb	2.52Aab	1.57Aa	1.78Aa	2.70Aa	1.57Aa	1.78Aa

* The soil characteristics measured included microporosity (Micro), macroporosity (Macro), total porosity (TPo), particle density (Pd), bulk density (Bd), and hydraulic conductivity (K_0). Means followed by equal uppercase letters in the same column and equal lowercase letters in the same row do not differ significantly from each other, as shown by the Kruskal-Wallis test at a 5% confidence level.

3.2.2. Soil Losses and Surface Runoff in the Caiabi River Sub-Basin

The dry weight of vegetative cover in plots that were subjected to simulated rainfall in the Caiabi River basin ranged from 7.26 to 11.91 metric tons (t) per hectare (Table 1). There were no significant differences between the treatments, using the Kruskal-Wallis test to compare sub-basin regions and land use. This condition allows us to understand the soil and water losses in the absence of these crops and their relationship with the physical and hydric characteristics of the soil (Tables 4 and 5), as well as the surface runoff variable resulting from the simulated rainfall in the Caiabi River sub-basin. There were significant differences in soil scarification for the other soil cover/management conditions in pasture areas. However, there were no significant differences between the positions in the sub-basin for this same use (Table 5).

Table 4. Average values of soil loss (grams/square meter) under different uses, soil cover/management, and regions of the Caiabi River sub-basin.

Cropland Use	Soil Loss (Grams/Square Meter) in Sub-Basin Region *			
	Soil Cover	**Upper**	**Middle**	**Lower**
Cultivation	With other crops	16.2Aa	5.64Aa	12.1Aa
	Straw residue	14.97Aa	8.92Aa	13.6Aa
	Without other crops	31.5Aa	23.6Ba	18.6Aa
	Soil scarified	30.6Aa	27.5Ba	20.50Aa
Pasture	With cultivated	7.09Aa	9.83Aa	2.30Aa
	Without crops	42.8Bb	22.3Aa	16.8Ba
	Soil scarified	35.9Ba	32.7Ba	20.5Ba

* Means followed by equal uppercase letters in the same column and equal lowercase letters in the same row do not differ significantly from each other, according to the Kruskal-Wallis test at a 5% confidence level.

Table 5. Surface runoff (millimeters/hour) under different uses, soil cover/management, and regions of the Caiabi River sub-basin.

Cropland Use	Surface Runoff (millimeters/hour) in Sub-Basin Region *			
	Soil Cover	**Upper**	**Middle**	**Lower**
Cultivated	With other crops	45.30Aa	39.50Aa	26.20Aa
	Straw residue	48.90Aa	44.75Aa	23.22Aa
	Without other crops	49.80Aa	44.70Aa	37.70Aa
	Scarified soil	35.40Aa	33.54Aa	21.70Aa
Pasture	With cultivated	68.60Ba	60.10Ba	61.70Ba
	Without crops	69.60Ba	58.90Ba	66.10Ba
	Scarified soil	30.00Aa	38.50Aa	47.60Aa

* Means followed by equal uppercase letters in the same column and equal lowercase letters in the same row do not differ significantly from each other, according to the Kruskal-Wallis test at a 5% confidence level.

3.3. *Renato River Sub-Basin*

3.3.1. Soil Characteristics in Renato River Sub-Basin

Unlike the Caiabi river basin, the soil attributes of the Renato River basin did not vary by region (Table 6). A similar presence of clay and sand was observed in the upper, middle, and lower parts of the basin. The same occurred with bulk density and particle density, porosity, and hydraulic conductivity, which showed a similar distribution in the different regions. The hydrographic basin of the Renato River is located in the Amazon Forest biome, unlike the Caiabi River basin, which is located predominantly in the Cerrado biome. These two biomes have different geological formations, although the soil classes are the same [25].

Table 6. Physical and hydric characterization of the Renato River sub-basin soils in pasture and cultivation areas.

Cropland Use	Renato River Sub-Basin Region *								
	Upper			Middle			Lower		
	Micro	Macro	TPo	Micro	Macro	TPo	Micro	Macro	TPo
	(m^3/m^3)								
	0 to 0.1 m depth								
Cultivated	0.43Aa	0.09Aa	0.52Aa	0.29Ab	0.08Aa	0.37Ab	0.28Ab	0.09Aa	0.37Ab
Pasture	0.40Aa	0.02Aa	0.42Aa	0.37Aa	0.06Aa	0.43Aa	0.33Aa	0.04Aa	0.37Aa
	0.1 to 0.2 m depth								
Cultivated	0.40Aa	0.07Aa	0.47Aa	0.27Ab	0.08Aa	0.35Aa	0.24Ab	0.11Aa	0.35Aa
Pasture	0.37Aa	0.06Aa	0.43Aa	0.33Aa	0.10Aa	0.43Aa	0.21Ab	0.14Aa	0.35Aa
	Pd	Bd	K_0	Pd	Bd	K_0	Pd	Bd	K_0
	(grams/cm^3)		(cm/hour)	(grams/cm^3)		(cm/hour)	(grams/cm^3)		(cm/hour)
	0 to 0.1 m depth								
Cultivated	2.71Aa	1.57Aa	0.79	2.73Aa	1.53Aa	1.22	2.65Aa	1.56Ba	0.68
Pasture	2.78Aa	1.53Ab	1.22	2.63Aa	1.59Ab	0.57	2.69Aa	1.75Aa	0.90
	0.1 to 0.2 m depth								
Cultivated	2.71Aa	1.66Aa	0.49Bb	2.69Aa	1.64Aa	0.78Ba	2.74Aa	1.71Aa	0.47Bb
Pasture	2.70Aa	1.51Ab	0.78Ac	2.76Aa	1.53Ab	1.30Ab	2.69Aa	1.70Aa	1.65Aa

* Soil characteristics measured included microporosity (Micro), macroporosity (Macro), total porosity (TPo), particle density (Pd), bulk density (Bd), and hydraulic conductivity (K_0). Means followed by equal uppercase letters in the same column and equal lowercase letters in the same row do not differ significantly from each other, according to the Kruskal-Wallis test at a 5% confidence level.

3.3.2. Soil Losses and Surface Runoff in the Renato River Sub-Basin

Significant increases in soil loss were observed under the scarified soil treatment when compared to plots with vegetation, regardless of the sub-basin region and land use (Table 7). This was similar to what was reported in the Caiabi River sub-basin, with higher soil losses for uncovered and scarified soils. Therefore, soil cover and limiting the soil surface turnover and soil exposure in areas of farming expansion are important for soil conservation, regardless of the biome. This sub-basin is characterized by the abundant presence of the Amazon rainforest (Figure 2), unlike the Caiabi River sub-basin, which is located in the Amazon-Cerrado ecotone [33]. Differences in soil losses in pasture areas with cover (vegetation) in different regions of the Renato River sub-basin are related to soil granulometry, since there is, on average, 80% of sand in the lower region (Table 4). In soils with higher sand contents, infiltration tends to be greater than in clayey soils.

Table 7. Average values of soil loss (grams/square meter) under different uses, soil cover/management, and the regions of the Renato River sub-basin region.

Cropland Use	Soil Loss (Grams/Square Meter) in Sub-Basin Region *			
	Condition	Upper	Middle	Lower
Cultivated	With vegetation	4.55Aa	10.20Aa	3.50Aa
	Without other crops	13.20Aa	42.20Ba	10.10Aa
	Scarified soil	66.01Ba	90.79Ca	60.02Ba
Pasture	With vegetation	8.70Ab	5.15Ab	1.01Aa
	Without crops	20.91ABa	25.20ABa	10.07ABa
	Scarified soil	42.01Ba	49.60Ba	17.50Ba

* Means followed by equal uppercase letters in the same column and equal lowercase letters in the same row do not differ significantly from each other, according to the Kruskal-Wallis test at a 5% confidence level.

Regarding surface runoff, a behavioral inversion was observed under scarified soil treatments when compared to other soil cover types, since there was a reduction in the runoff in the uncovered and scarified soils of pasture areas (Table 8). This inversion is the result of the rupture of the surface layers, which are usually compacted in areas of farming and pasture, thus increasing the roughness that limits runoff and promotes water infiltration into the soil. The differences between the average values of surface runoff, especially in the lower regions between the Caiabi and Renato River basins, result from differences in the saturated hydraulic conductivity in the soils (Tables 3 and 6).

Table 8. Surface runoff (millimeters/hour) according to the different uses, soil cover/management, and regions of the Renato River sub-basin.

Cropland Use	Surface Runoff (Millimeters/hour) in Sub-Basin Region *			
	Condition	Upper	Middle	Lower
Cultivated	With vegetation	51.6Aa	51.7Aa	47.7Aa
	Without other crops	59.2Aa	59.6Aa	57.0Aa
	Scarified soil	61.7Aa	61.8Aa	61.0Aa
Pasture	With vegetation	47.2Aa	64.3Ba	58.5Ba
	Without crops	55.2Aa	67.0Ba	59.1Ba
	Scarified soil	34.4Aa	35.0Aa	33.2Aa

* Means followed by equal uppercase letters in the same column and equal lowercase letters in the same row do not differ significantly from each other, according to the Kruskal-Wallis test at a 5% confidence level.

4. Discussion

4.1. Erosion Drivers and Implications

Agricultural crops (soybean + straw, only straw, maize, and *Brachiaria* grass species) are responsible for minimizing the direct impact of raindrops, acting as rain droplet buffers, preventing the disaggregation of particles, and reducing the sediment load in surface runoff [34]. According to the authors of [7], well-managed pastures can be considered sustainable, as they maintain soil quality in terms of the physical, chemical, and biological aspects and prevent erosive processes. In this sense, the pastures of the three sub-basin regions, with an average height of 50 cm, achieved satisfactory phytomass productivity (Table 2). Similar to soybean under a no-tillage scheme, the data are in accordance with the authors of [35], who observed similar results in the Cerrado latosols.

When studying the different levels of cultivated crops, such as soybean, maize, and pasture, prior researchers [12,16] concluded that soil losses increase with the reduction and removal of vegetative cover. Similar results were observed in our study, as soil losses in the Caiabi River sub-basin indicated significant differences between soil treatments with and without vegetative cover. In general, areas covered with vegetation (including straw) provided lower soil losses, revealing the importance of vegetative cover for reducing soil degradation (Table 5).

The occurrence of differences in soil losses for the types of crops and soil scarification demonstrates the need to maintain vegetation cover or straw, regardless of land use (cultivation or pasture), with minimal soil turnover. Due to the distinct physical and hydric characteristics in the sub-basin regions (Tables 3 and 6), a reduction in soil loss was observed from the upper to the lower region for those soils without crops and that were scarified, regardless of land use. The upper region has soils with a predominantly clayey texture, while the middle and lower regions have soils with a more sandy texture (Table 2). In this case, surface sealing may occur in clayey soils, which makes infiltration difficult and promotes an increase in runoff and consequent soil losses, due to the reduction in roughness [36,37]. Moreover, this sub-basin is located in an ecotone area (Cerrado/Amazon rainforest). In this transition area, there are geological, morphological, and pedological variabilities, as well as in the phytophysiognomy of the region [33], affecting the erosive processes along the sub-basin.

In terms of pastures, even though the pasture evaluation did not include soil covered only with straw, soil losses were still reduced with such live vegetative cover. These results are in accordance with [12,14,15], who, while studying simulated rainfall in uncovered soil treatments in different regions of Brazil, observed an intensification of losses due to erosion compared to covered soils. In general, unprotected soils tend to be lost due to the direct impacts caused by raindrops, which provoke detachment and the consequent transport of particles [38].

In addition to vegetative cover, water infiltration into the soil depends on other intrinsic factors, such as texture, porosity, bulk density, and compaction levels, which may compromise hydraulic conductivity (Table 6). Therefore, even in covered soil treatments that reduce soil losses, surface runoff may be high, as observed in this study. In this sense, the lowest values of runoff flow observed in the scarified plots resulted from the roughness formed in the layer that was turned over, along with soil aggregate rupture, which not only facilitates infiltration in areas of low slopes in the short term but also intensifies greater soil losses. Soil turnover in pasture and cultivated areas (e.g., plowing, sub-soiling) disrupts the aggregates, which facilitates this rupture due to rainfall. This facilitates the erosive transport of soil after tillage breaks up the soil layer compacted by animal trampling and wheel slippage from the tractors, planters, sprayers, harvesters, and trucks used during harvesting. Consequently, some researchers [13,39,40] recommend minimal turnover, combined with leaving the straw from crops in-field or using alternate turnovers (e.g., the planting of crops such as maize after the soybean harvest) as ways to minimize soil and water losses by erosion.

The increased values of water loss due to surface runoff in pasture areas with vegetation and without crops (Tables 5 and 8) may be related to precipitation falling directly on the straw, which covers the soil in no-tillage soybean cultivation. This condition may favor water runoff by its running off directly onto the upper surface of the straw. Nevertheless, the water loss depends on fragment sizes, layer height, and straw density, which can reduce infiltration [41]. To reduce soil and water losses, adopting crops with the purpose of protecting the soils and providing better conditions for the use and sustainability of production systems, including good water infiltration, is recommended [42]. Therefore, soil and water conservation management practices should be adopted, such as the use and incorporation of straw to increase infiltration, level terraces, no-tillage systems, minimum tillage, conserved pastures with rotational grazing, and crop rotation. By adopting these practices, it is possible to reduce the exposure and consequent soil loss, in addition to avoiding nutrient and carbon losses, as well as river aggradation [13,38–40,43,44].

According to our results observed in the scarified plots, the vulnerability of soil particles is evident with regard to transport. This confirms greater soil losses in those areas where conventional planting involves traditional management using soil tillage, typically involving one plowing stage and two harrowing stages, in addition to the minimum use of soil cover. Several authors obtained similar results in different regions of Brazil in studies with natural rainfall [39,40,45] and also in studies with simulated rainfall [13].

Raindrops and the surface runoff of water during rain events can lead to soil erosion, with the amount and type of vegetation covering the soil being a significant factor in reducing soil erosion [46]. Soil losses in the two sub-basins that we studied with vegetative cover (pasture, soybean, soybean with straw, and maize) can happen, especially during the early stages of plant development when there is more soil exposure. After the rainfall interception, infiltration, and saturation of the soil surface layers, the water surplus moves depending on the topographic gradients. Therefore, vegetative cover does not entirely eliminate erosion in those areas used by farming production systems. However, it drastically decreases erosion when compared with badly managed and/or unprotected areas, as was shown in the understory of olive orchards, with both the lower-cost natural regeneration of early successional weeds and intentionally planted cover crops in Minas Gerais State, Brazil [47].

Vegetative cover is naturally responsible for protecting the soil from the direct action of rainfall, and it might not necessarily eliminate all losses, except for some cases in native forests. In this context, several studies show the absence of soil losses in areas of preserved native forest or their drastic reduction in comparison with agricultural land use, such as pasture and cultivated crops [17,42,48,49]. In other words, the soil losses observed in this study may be directly related to the conversion of native forests into farming land. Furthermore, they indicate the need for new studies on simulated and natural rainfall in the Teles Pires River sub-basin region and other rivers in the Cerrado and Amazon biomes, with their transitory ecotones.

In agricultural frontiers such as the Teles Pires River basin and, consequently, in the drainage sub-basins studied, soil and water losses lead to numerous environmental problems. These problems include the pollution and contamination of rivers and streams by the transport of chemical products, the deposition of particles that cause aggradation, the exposure of stocked carbon, removal of the surface layer responsible for farming production, damage to cart roads by the formation of gullies, dam bursts, and the destruction of local biodiversity [2]. Thus, an alternative method to circumvent and/or mitigate erosive processes is to adopt conservationist practices, especially those involving vegetative cover, or the combination of such practices with edaphic or mechanical practices, such as terracing, catchment basins, drainage channels, and the construction of dams on the sides of plantations.

4.2. Policy Recommendations

Environmental conservation policies that are already implemented in Brazil have contributed to reducing soil and water losses, including those that encourage the direct planting system (i.e., no-till farming), the recovery of springs and degraded pasture areas, carbon sequestration, and the adoption of agroforestry systems included in the Low Carbon Agriculture Plan, present in Law 12,187 [50] and regulated by Decree 7390 of 9 December 2010 [51]. The protection of native forests is also regulated by federal law (12,651, of 25 May 2012), known as the "Forest Code," which establishes general rules on the protection of native vegetation, including permanent preservation areas, legal reserves, and areas of restricted use [52].

With regard to water resources, the National Policy on Water Resources (Law 9433 of January 8, 1997) has the following main objectives. The first objective is to ensure the necessary availability of water for current and future generations, with adequate quality standards for the respective uses. The second objective is the rational and integrated use of water resources. The third and final objective is to prevent and defend against critical hydrological events, either of natural origin or arising from the inappropriate use of natural resources, and to encourage and promote the capture, preservation, and use of rainwater [53].

In addition to minimizing soil losses, these initiatives contribute to increasing carbon stocks, conserving the biodiversity of biomes, and preserving rivers and lakes from the silting up caused by constant erosion processes [2]. The soil's ground cover, in addition to protecting against the direct impact of raindrops, also protects agroecosystems from wind erosion and solar rays, which affect the soil microbiology [54,55]. From 2009 to 2020, Brazil made progress in achieving the goals of establishing policies for the conservation of natural resources, with a focus on reducing climate change [56]. However, due to recent increases in deforestation, these environmental challenges are omnipresent.

In the present study, the impacts of agricultural land cover were evaluated on soybeans (in the Caiabi River basin) and maize (in the Renato River basin). The maintenance of bare soils, combined with scarification, promotes greater soil loss regardless of the crop and the region of the watershed. According to Borrelli et al. 2017 [2], the presence of cover and the absence of soil disturbance are the quickest ways to conserve pedological and edaphic resources. In this sense, the correct management of the soybean crop with the direct planting system and contour planting are alternative methods capable of stopping or mitigating erosion [10]. The no-till system has been used in Brazil since the 1960s in the southern region of the country; the results point to an increase in the capacity of water infiltration into the soil, with a reduction in surface runoff, in addition to favoring the microbial community, improving the soil structure, and nutrient cycling [2,12,40,44]. Contours or contour planting avoids the formation of preferential lines for surface water flow, minimizing sediment transport.

For maize cultivation, soil and water conservation practices are also important, although, in the state of Mato Grosso, most areas use this crop immediately following soybean cultivation as a second planting (*safrinha*), when the rains are less frequent. Even so, in the months of March and April, rainfall can still be enough (288 and 121 mm/month, respectively [27]) to require attention when it comes to soil conservation. Technical assistance, combined with rural extension, can encourage rural producers to keep residual straw (e.g., stover) on the ground after the maize harvest, protecting the soil during the fallow period of the dry season (June through September). Maize stover can anchor soil prior to planting the soybean crop again at the start of the wet season in October.

Most of the cultivated areas with pastures in Brazil are still degraded or are in the process of degradation. This is due to misuse, such as exceeding the pasture-carrying capacity, a lack of pH correction and fertilization practices, and a lack of planning efforts to avoid erosion [14,15]. Another important factor that must be taken into account in the conservation of pastures is the criterion for animals entering the pasture, which can favor the loss of surface protection, increasing the vulnerability of the soil to erosion [7].

In addition to the existing sustainable agricultural development policies, new policies are needed, with targets that reach all agricultural and livestock producers. In studies in southern Brazil, previous researchers [57] concluded that one of the biggest limitations in combating water erosion is the lack of information on the subject for rural producers. These new policies can be specific for each crop or can be integrated as a sustainability plan for the production systems of soybeans, maize, and pasture. Incentives can be included for crop succession and rotation, intercropping grasses with legumes, and combating degradation with rotational grazing. In an evaluation of erosion processes under different agricultural management scenarios, integrated soil conservation practices were found to have a greater effect on combating soil erosion [57].

Although all the landscapes evaluated in the present study are considered relatively flat, in the state of Mato Grosso, there are agricultural areas on sloping land [14,55]; therefore, new initiatives should pay special attention to those areas with steep slopes above a gradient of 15%. The greater the slope of an area, the greater the potential for soil and water losses [1,58]. One study measured the effect of slopes of 15%, 25%, 35%, and 45% on soil and water losses and concluded that the greater the local slope, the greater the losses of water and soil resources [14]. Consequently, greater soil losses can accelerate the silting up of watercourses, as well as increase the exposure of stored carbon in soils, which can increase CO_2 emissions, thus compromising the biodiversity of biomes [59].

Soils are providers of ecosystem services; when these are compromised, civilizations can be in imminent danger of existential instability. Soil degradation affects the hydrological cycle, compromising food security in the countryside and in urban centers. Therefore, it is necessary to adopt management practices that ensure sustainability, especially in biomes with high levels of deforestation, such as the Cerrado and the Amazon [60]. In this regard, article 225 of the Federal Constitution of Brazil states that "everyone has the right to an ecologically balanced environment, an asset for common use by the people and essential to a healthy quality of life, imposing on the public authorities and the community the duty to defend and preserve it for present and future generations" [61]. The sustainability of natural ecosystems and agricultural production systems is necessary to optimize the use of natural resources linked to soil and water, preserving them for current and future generations.

5. Conclusions

In this study, we evaluate the erosion process in agricultural production systems in the Cerrado-Amazon transition region, using the InfiAsper rainfall simulator. The erosive process is independent of soil texture and is closely related to management and use systems associated with vegetation cover and soil scarification, regardless of the area's position within the sub-basins. The removal of the vegetation cover formed by soybean, maize, and pasture negatively affects soil–water dynamics, with a significant increase in soil losses in all regions of both sub-basins. Areas subjected to soil management with surface scarification (soil turnover) experience greater soil losses, regardless of the sub-basin region and land use. The results indicate the need for agricultural producers and farmers to make use of management practices that prioritize the maximum vegetation cover for crop cultivation and animal husbandry areas, as well as minimal soil turnover, as a path to the sustainability of production systems by reducing erosive processes. Future rainfall simulator studies can quantify changes in soil erosion by (1) using cover crops (e.g., *Crotalara juncea*) following soybeans–maize, (2) growing cotton (*Gossypium* spp.) as a second crop after soybeans in Brazil, and (3) in sugarcane (*Saccharum* spp.) and other large-scale commodity crops in Brazil and around the world, to support more sustainable agricultural systems and development.

Author Contributions: Data collection, writing, methodology, formal analysis, and figures—M.A.B.A. and D.F.d.C.; data collection, review, editing, supervision, and financial support—A.P.d.S. and F.T.d.A.; review, editing, supervision, and financial support—A.K.H.; supervision and data collection, H.B.A. and A.F.d.S. All authors have read and agreed to the published version of the manuscript.

Funding: This study was financed by the Coordenação de Aperfeiçoamento de Pessoal de Nível Superior—Brasil (CAPES) and the Agência Nacional de Águas e Saneamento Básico (ANA), Finance Code—001 and Process 88887.144957/2017-00. The authors wish to thank the Conselho Nacional de Desenvolvimento Científico e Tecnológico (CNPq) for their support with scientific initiation grants and a productivity grant.

Institutional Review Board Statement: Not applicable.

Informed Consent Statement: Not applicable.

Data Availability Statement: Study data can be obtained upon request to the corresponding author or the second author, via e-mail. The data are not available on the website as the research project is still under development.

Acknowledgments: We thank the Federal Rural University of Rio de Janeiro, specifically the PPGA-CS and GPASSA. The authors also thank all the students and professors of the Tecnologia em Recursos Hídricos no Centro-Oeste" research group (dgp.cnpq.br/dgp/espelhogrupo/2399343537529589, accessed on 15 December 2022). We are thankful for the comments and edits from two anonymous reviewers and one MDPI Special Issue editor, which improved the quality of this work.

Conflicts of Interest: The authors declare no conflict of interest. Supporting entities had no role in the design of the study; in the collection, analysis, or interpretation of data; in the writing of the manuscript, or in the decision to publish the results.

References

1. Van Oost, K.; Govers, G.; Desmet, P. Evaluating the effects of changes in landscape structure on soil erosion by water and tillage. *Landsc. Ecol.* **2000**, *15*, 577–589. [CrossRef]
2. Borrelli, P.; Robinson, D.A.; Fleischer, L.R.; Lugato, E.; Ballabio, C.; Alewell, C.; Bagarello, V. An assessment of the global impact of 21st century land use change on soil erosion. *Nat. Commun.* **2017**, *8*, 2013. [CrossRef]
3. García-Ruiz, J.M.; Beguería, S.; Nadal-Romero, E.; González-Hidalgo, J.C.; Lana-Renault, N.; Sanjuán, Y.A. Meta-analysis of soil erosion rates across the world. *Geomorphology* **2015**, *239*, 160–173. [CrossRef]
4. García-González, I.; Hontoria, C.; Gabriel, J.L.; Alonso-Ayuso, M.; Quemada, M. Cover crops to mitigate soil degradation and enhance soil functionality in irrigated land. *Geoderma* **2018**, *322*, 81–88. [CrossRef]
5. Lenka, N.K.; Jaiswal, S.P.; Thakur, J.K.; Lenka, S.; Mandal, A.; Dwivedi, A.K.; Yashona, D.S. Soil degradation effect on soil productivity, carbon pools and soil enzyme activity. *Curr. Sci.* **2017**, *112*, 2434–2439. [CrossRef]

6. Jin, Q.; Wang, C.; Sardans, J.; Vancov, T.; Fang, Y.; Wu, L.; Wang, W. Effect of soil degradation on the carbon concentration and retention of nitrogen and phosphorus across Chinese rice paddy fields chineses. *Catena* **2022**, *209*, e105810. [CrossRef]
7. Galdino, S.; Sano, E.E.; Andrade, R.G.; Grego, C.R.; Nogueira, S.F.; Bragantini, C.; Flosi, A.H. Large-scale modeling of soil erosion with Rusle for Conservationist planning of degraded cultivated Brazilian pastures. *Land Degrad. Develop.* **2016**, *27*, 773–784. [CrossRef]
8. Zambon, N.; Johannsen, L.L.; Strauss, P.; Dostal, T.; Zumr, D.; Cochrane, T.A.; Klik, A. Splash erosion affected by initial soil moisture and surface conditions under simulated rainfall. *Catena* **2021**, *196*, 104827. [CrossRef]
9. Antonella, D.; Maurizio, L.; Mario, C.; Severino, R. Soil Erosion Modelling on Arable Lands and Soil Types in Basilicata, Southern Italy. In *Computational Science and Its Applications—ICCSA 2017, Proceedings of the 17th International Conference, Trieste, Italy, 3–6 July 2017*; Gervasi, O., Murgante, B., Misra, S., Borruso, G., Torre, C.M., Rocha, A.M.A.C., Taniar, D., Apduhan, B.O., Stankova, E., Cuzzocrea, A., Eds.; Part V; Springer International Publishing: Cham, Switzerland, 2017; Volume 5, pp. 57–72. ISBN 978-3-319-62403-7. [CrossRef]
10. Assis, K.G.O.; Silva, Y.J.A.B.; Lopes, J.W.B.; Medeiros, J.C.; Teixeira, M.P.R.; Rimá, F.B.; Singh, V.P. Soil loss and sediment yield in a perennial catchment in southwest Piauí, Brazil. *Environ. Monit. Assess.* **2021**, *193*, 26. [CrossRef] [PubMed]
11. Bourgoin, C.; Betbeder, J.; Le Roux, R.; Gond, V.; Oszwald, J.; Arvor, D.; Baudry, J.; Boussard, H.; Le Clech, S.; Mazzei, L.; et al. Looking beyond forest cover: An analysis of landscape-scale predictors of forest degradation in the Brazilian Amazon. *Environ. Res. Lett.* **2021**, *16*, e114045. [CrossRef]
12. Almeida, W.S.; Panachuki, E.; Oliveira, P.T.S.; Silva, M.R.; Sobrinho, T.A.; Carvalho, D.F. Effect of soil tillage and vegetal cover on soil water infiltration. *Soil Tillage Res.* **2018**, *175*, 130–138. [CrossRef]
13. Panachuki, E.; Dos Santos, M.A.N.; Pavei, D.S.; Alves Sobrinho, T.; Camacho, M.A.; Montanari, R. Soil and water loss in Ultisol of the Cerrado-Pantanal Ecotone under different management systems. *Afr. J. Agric. Res.* **2015**, *10*, 926–932. [CrossRef]
14. Cantalice, J.R.B.; Nunes, E.O.S.; Cavalcante, D.M.; Barbosa, B.; Junior, G.B.; Guerra, S.M.S.; Neto, F.R. Vegetative-hydraulic parameters generated by agricultural crops for laminar flows under a semiarid environment of Pernambuco, Brazil. *Ecol. Indic.* **2019**, *106*, e105496. [CrossRef]
15. Sone, J.S.; Oliveira, P.T.S.; Euclides, V.P.B.; Montagner, D.B.; Araujo, A.R.; Zamboni, P.A.P.; Sobrinho, T.A. Effects of Nitrogen fertilization and stocking rates on soil erosion and water infiltration in a Brazilian Cerrado farm. *Agric. Ecosyst. Environ.* **2020**, *304*, 107159. [CrossRef]
16. Rieger, F.A.; Zolin, C.A.; Paulino, J.; Souza, A.P.; Matos, E.D.S.; Magalhães, C.A.D.S.; Farias Neto, A.L.D. Water erosion on an Oxisol under integrated crop-forest systems in a transitional area between the Amazon and Cerrado biomes. *Rev. Bras. Ciênc. Solo* **2016**, *40*, e0150111. [CrossRef]
17. Lense, G.H.E.; Parreiras, T.C.; Moreira, R.S.; Avanzi, J.C.; Mincato, R.L. Effect of spatial-temporal variation of land use and land cover on soil erosion. *Rev. Caatinga* **2021**, *34*, 90–98. [CrossRef]
18. Neal, A.L.; Hughes, D.; Clark, I.M.; Jansson, J.K.; Hirsch, P.R. Microbiome Aggregated Traits and Assembly Are More Sensitive to Soil Management than Diversity. *mSystems* **2021**, *6*, e01056-20. [CrossRef]
19. de Lima Moraes, A.G.; de Carvalho, D.F.; Antunes, M.A.H.; Ceddia, M.B.; Flanagan, D.C. Steady infiltration rate spatial modeling from remote sensing data and terrain attributes in southeast Brazil. *Geoderma Reg.* **2020**, *20*, e00242. [CrossRef]
20. Alves Sobrinho, T.; Gómez-Macpherson, H.; Gómez, J.A. A portable integrated rainfall and overland flow simulator. *Soil Use Manag.* **2008**, *24*, 163–170. [CrossRef]
21. Iserloh, T.; Ries, J.B.; Arnáez, J.; Boix-Fayos, C.; Butzen, V.; Cerdà, A.; Wirtz, S. European small portable rainfall simulators: A comparison of rainfall characteristics. *Catena* **2013**, *110*, 100–112. [CrossRef]
22. Macedo, P.M.S.; Pinto, M.F.; Alves Sobrinho, T.; Schultz, N.; Coutinho, T.A.R.; Carvalho, D.F. A modified portable rainfall simulator for soil erosion assessment under different rainfall patterns. *J. Hydrol.* **2021**, *596*, e126052. [CrossRef]
23. Carvalho, D.F.; Macedo, P.M.S.; Pinto, M.F.; Almeida, W.S.; Schultz, N. Soil loss and runoff obtained with customized precipitation patterns simulated by InfiAsper. *Int. Soil Water Conserv. Res.* **2022**, *10*, 407–413. [CrossRef]
24. IBGE—Instituto Brasileiro de Geografia e Estatística. Estado do Mato Grosso: Pedologia (Mapa Exploratório de Solos), p1. 2009. Available online: https://mapas.ibge.gov.br/tematicos/solos (accessed on 1 July 2022).
25. Alves, M.A.B.; Borella, D.R.; Luz, C.C.S.; Castagna, D.; Silva, W.C.; Silva, A.P.; Almeida, F.T.; Souza, A.P. Soil classes in the Caiabi and Renato river basins, tributaries of the Teles Pires river in the south of the Amazon. *Nativa* **2022**, *10*, 431–448. [CrossRef]
26. Alvares, C.A.; Stape, J.L.; Sentelhas, P.C.; Gonçalves, J.L.M.; Sparovek, G. Köppen's climate classification map for Brazil. *Meteorol. Zeitschrift.* **2013**, *22*, 711–728. [CrossRef] [PubMed]
27. Souza, A.P.; Mota, L.L.; Zamadei, T.; Martim, C.C.; Almeida, F.T.; Paulino, J. Climate classification and climatic water balance in Mato Grosso state, Brazil. *Nativa* **2013**, *1*, 34–43. [CrossRef]
28. Portaria SPA/MAPA nº 249, de 22 de 4 de julho de 2022-MAPA. Aprova o Zoneamento Agrícola de Risco Climático Para a Cultura da Soja no Estado de Mato Grosso, Ano-Safra 2022/2023. Available online: https://www.gov.br/agricultura/pt-br/assuntos/riscos-seguro/programa-nacional-de-zoneamento-agricola-de-risco-climatico/portarias (accessed on 27 December 2022).

29. Portaria SPA/MAPA nº 332, de 26 de Setembro de 2022-MAPA. Aprova o Zoneamento Agrícola de Risco Climático—ZARC Para a Cultura de Milho 2ª Safra no Estado de Mato Grosso, Ano-Safra 2022/2023. Available online: https://www.gov.br/agricultura/pt-br/assuntos/riscos-seguro/programa-nacional-de-zoneamento-agricola-de-risco-climatico/portarias/safra-vigente/mato-grosso (accessed on 27 December 2022).
30. Sabino, M.; Souza, A.P.; Uliana, E.M.; Lisboa, L.; Almeida, F.T.; Zolin, C.A. Intensity-duration-frequency of maximum rainfall in Mato Grosso State. *Rev. Ambiente Água* **2020**, *15*, e2373. [CrossRef]
31. Teixeira, P.C.; Donagemma, G.K.; Fontana, A.; Teixeira, W.G. *Manual de Métodos de Análise de Solo*, 3rd ed.; Embrapa Solos: Rio de Janeiro, Brazil, 2017; pp. 1–573. Available online: https://www.infoteca.cnptia.embrapa.br/handle/doc/1085209 (accessed on 5 May 2022).
32. Grossman, R.B.; Reinsch, T.G. Bulk Density and Linear Extensibility: Core Method. In *Methods of Soil Analysis*; Dane, J.H., Topp, G.C., Eds.; Part 4, Physical Methods; SSSA: Madison, WI, USA, 2002; pp. 208–228.
33. Wenzel, D.A.; Uliana, E.M.; Almeida, F.T.; Souza, A.P.; Mendes, M.A.D.S.A.; Silva, S.L.G. Physiographic characteristics of the medium and upper Teles Pires River sub-basins, Mato Grosso, Brazil. *Rev. Ciênc. Agroambient.* **2018**, *15*, 123–131. [CrossRef]
34. Uribe, M.; Peñuela, G.A.; Pino, N.J. Megathyrsus maximus and Brachiaria decumbens improve soil characteristics and select promising rhizobacteria during rhizoremediation of petroleum hydrocarbons. *Rhizosphere* **2022**, *22*, e100517. [CrossRef]
35. Bonetti, J.D.A.; Paulino, H.B.; Souza, E.D.D.; Carneiro, M.A.C.; Silva, G.N.D. Influence of the integrated crop-livestock system on soil and soybean and brachiaria yield. *Pesqui. Agropecu. Trop.* **2015**, *45*, 104–112. [CrossRef]
36. Panachuki, E.; Bertol, I.; Alves Sobrinho, T.; Oliveira, P.T.S.D.; Rodrigues, D.B.B. Soil and water loss and water infiltration in red latosol under diferent management systems. *Rev. Bras. Ciênc. Solo* **2011**, *35*, 1777–1786. [CrossRef]
37. Dalla Rosa, J.; Cooper, M.; Darboux, F.; Medeiros, J.C. Influence of tillage systems and simulated rainfall on the processo f surface crust formation. *Rev. Bras. Ciênc. Solo* **2013**, *37*, 400–410. [CrossRef]
38. Almeida, W.S.; Seitz, S.; Oliveira, L.F.C.; Carvalho, D.F. Duration and intensity of rainfall events with the same erosivity change sediment yield and runoff rates. *Int. Soil Water Conserv. Res.* **2021**, *9*, 69–75. [CrossRef]
39. Bertol, I.; Schick, J.; Bandeira, D.H.; Paz-Ferreiro, J.; Vázquez, E.V. Multifractal and joint multifractal analysis of water and soil losses from erosion plots: A case study under subtropical conditions in Santa Catarina highlands, Brazil. *Geoderma* **2017**, *287*, 116–125. [CrossRef]
40. Antoneli, V.; Lenatorvicz, H.H.; Bednarz, J.A.; Pulido-Fernández, M.; Brevik, E.C.; Cerdà, A.; Rodrigo-Comino, J. Rainfall and land management effects on erosion and soil properties in traditional Brazilian tobacco plantations. *Hydrol. Sci. J.* **2018**, *63*, 1008–1019. [CrossRef]
41. Andrade, M.A.; Mello, C.R.D.; Beskow, S. Hydrological simulation in a watershed with predominance of Oxisol in the Upper Grande river region, MG—Brazil. *Rev. Bras. Eng. Agrícola Ambient.* **2013**, *17*, 69–76. [CrossRef]
42. Falcão, K.D.S.; Panachuki, E.; Monteiro, F.D.N.; Silva, R.; Rodrigues, D.B.; Sone, J.S.; Oliveira, P.T.S. Surface runoff and soil erosion in a natural regeneration area of the Brazilian Cerrado. *Int. Soil Water Conserv. Res.* **2020**, *8*, 124–130. [CrossRef]
43. Merten, G.H.; Araújo, A.G.; Biscaia, R.C.M.; Barbosa, G.M.C.; Conte, O. No-till surface runoff and soil losses in southern Brazil. *Soil Tillage Res.* **2015**, *152*, 85–93. [CrossRef]
44. Deuschle, D.; Minella, J.P.N.A.; Hörbe, T.; Londero, A.L.; Schneider, F.J. Erosion and hydrological response in no-tillage subjected to crop rotation intensification in Southern Brazil. *Geoderma* **2019**, *340*, 157–163. [CrossRef]
45. Silva, R.M.; Santos, C.A.G.; dos Santos, J.Y.G. Evaluation and modeling of runoff and sediment yield for different land covers under simulated rain in a semiarid region of Brazil. *Int. J. Sediment Res.* **2018**, *33*, 117–125. [CrossRef]
46. Beniaich, A.; Silva, M.L.N.; Guimarães, D.V.; Avalos, F.A.P.; Terra, F.S.; Menezes, M.D.; Avanzi, J.C.; Cândido, B.M. UAV-based vegetation monitoring for assessing the impact of soil loss in olive orchards in Brazil. *Geoderma Reg.* **2022**, *30*, e00543. [CrossRef]
47. Beniaich, A.; Guimarães, D.V.; Avanzi, J.C.; Silva, B.M.; Acuña-Guzman, S.F.; dos Santos, W.P.; Silva, M.L.N. Spontaneous vegetation as an alternative to cover crops in olive orchards reduces water erosion and improves soil physical properties under tropical conditions. *Agric. Water Manag.* **2023**, *279*, 108186. [CrossRef]
48. Leite, M.H.S.; Couto, E.G.; Amorim, R.S.S.; Costa, E.L.D.; Maraschin, L. Soil and nutriente losses under diferente tillage systems in a clayey oxisol under natural rainfall. *Rev. Bras. Ciênc. Solo* **2009**, *33*, 689–699. [CrossRef]
49. Neves, S.M.A.D.S.; Motinho, M.C.; Neves, R.J.; Soares, E.R.C. Estimation of soil loss for rainfall erosion in the catchment area of Jauru, Mato Grosso State, Brazil. *Soc. Nat.* **2011**, *23*, 423–433. [CrossRef]
50. Brazil. Lei Nº 12.187, 29 December 2009. Institui a Política Nacional Sobre Mudança do Clima—PNMC e dá Outras Providências. Available online: https://www.suape.pe.gov.br/images/publicacoes/Leis/Lei_n._12.187.2009.pdf (accessed on 12 December 2022).
51. Brazil—Decreto No—7.390, 9 December 2010 Regulamenta os Arts. 6o, 11 e 12 da Lei no 12.187, de 29 de Dezembro de 2009, Que Institui a Política Nacional Sobre Mudança do Clima—PNMC, e dá Outras Providências. Available online: https://pesquisa.in.gov.br/imprensa/jsp/visualiza/index.jsp?jornal=1&pagina=4&data=10/12/2010 (accessed on 13 December 2022).
52. Brazil. Lei Nº 12.651, 25 May 2012 Dispõe Sobre a Proteção da Vegetação Nativa; Altera as Leis nºs 6.938, de 31 de Agosto de 1981, 9.393, de 19 de Dezembro de 1996, e 11.428, de 22 de Dezembro de 2006; Revoga as Leis nºs 4.771, de 15 de Setembro de 1965, e 7.754, de 14 de abril de 1989, e a Medida Provisória nº 2.166-67, de 24 de Agosto de 2001; e dá Outras Providências. 38p. Available online: https://www.planalto.gov.br/ccivil_03/_ato2015-2018/2018/lei/l13709.htm (accessed on 12 December 2022).

53. Brazil. Lei N° 9.433, 8 January 1997. Institui a Política Nacional de Recursos Hídricos, Cria o Sistema Nacional de Gerenciamento de Recursos Hídricos, Regulamenta o Inciso XIX do Art. 21 da Constituição Federal, e Altera o art. 1° da Lei n° 8.001, de 13 de março de 1990, Que Modificou a Lei n° 7.990, de 28 de Dezembro de 1989. Available online: https://legislacao.presidencia.gov.br/atos/?tipo=LEI&numero=9433&ano=1997&ato=a12ATVU90MJpWTbaf (accessed on 20 December 2022).
54. Santos, J.P.C.; Caldas, V.I.S.P.; Silva, A.S. Susceptibility to erosion of soils of the Lagos São João Basin, in the State of Rio de Janeiro—Brazil, from the AHP method and multicriteria analysis. *Rev. Bras. Geogr. Fis.* **2019**, *12*, 1415–1430. [CrossRef]
55. Nachtigall, S.D.; Nunes, M.C.M.; Moura-Bueno, J.M.; Lima, C.L.R.D.; Miguel, P.; Beskow, S.; Silva, T.P. Spatial modeling of soil water erosion associated with agroclimatic seasonality in the Southern region of Rio Grande do Sul, Brazil. *Eng. Sanit. Ambient.* **2020**, *25*, 933–946. [CrossRef]
56. TALANOA. A Política Nacional de Mudança do Clima em 2020: Estado de Metas, Mercados e Governança Assumidos na Lei 12.187/2009. Rio de Janeiro, Brasil. 2020. Available online: https://www.oc.eco.br/wp-content/uploads/2020/12/Politica-Nacional-de-Mudanc%CC%A7a-de-Clima.pdf (accessed on 18 December 2022).
57. Didoné, E.J.; Minella, J.P.G.; Evrard, O. Measuring and modelling soil erosion and sediment yields in a large cultivated catchment under no-till of Southern Brazil. *Soil Tillage Res.* **2017**, *174*, 24–33. [CrossRef]
58. Raimo, L.A.D.L.; Amorim, R.S.; Couto, E.G.; Hunter, M.O.; Torres, G.N.; Bocuti, E.D.; Silva, L.D. Natural erosion potential in the Mato Grosso State, Brazil. *Rev. Bras. Eng. Agrícola Ambient.* **2022**, *26*, 655–661. [CrossRef]
59. Schindewolf, M.; Schönke, D.; Amorim, R.S.; Schmidt, J. Effects of contour banks and no-till measures on run-off and sediment yield in Campo Verde Region, Mato Grosso. In *Interdisciplinary Analysis and Modeling of Carbon-Optimized Land Management Strategies for Southern Amazonia*, 1st ed.; Gerold, G., Jungkunst, H.F., Wantzen, K.M., Schönenberg, R., Amorim, R.S.S., Couto, E.G., Madari, B., Hohnwald, S., Eds.; Universitätsverlag Göttingen: Göttingen, Germany, 2014; pp. 85–100. ISBN 978-3-86395-138-2. [CrossRef]
60. Klink, C.A.; Machado, R.B. A Conservação do Cerrado Brasileiro. *Megadiversidade* **2005**, *1*, 147–155. Available online: https://www.researchgate.net/profile/Ricardo-Machado-4/publication/228342037_A_conservacao_do_Cerrado_brasileiro/links/553a78670cf29b5ee4b64c2f/A-conservacao-do-Cerrado-brasileiro.pdf (accessed on 20 December 2022).
61. Brazil. Constituição da República Federativa do Brasil 1988; Senado Federal: Brasília, DF. 2016; pp. 1–496. Available online: https://www2.senado.leg.br/bdsf/bitstream/handle/id/518231/CF88_Livro_EC91_2016.pdf (accessed on 15 December 2022).

sustainability

MDPI

Article

Towards Sustainable Agricultural Development for Edible Beans in China: Evidence from 848 Households

Jiliang Ma [1,†], Jiajia Qu [2,†], Nawab Khan [3,*] and Huijie Zhang [4,*]

1 Institute of Agricultural Economics and Development, Chinese Academy of Agricultural Sciences, Beijing 100081, China; majiliang@caas.cn
2 Institute of Food and Nutrition Development, Ministry of Agriculture and Rural Affairs, Beijing 100081, China; 13298340197@163.com
3 College of Management, Sichuan Agricultural University, Chengdu 611100, China
4 Agricultural Information Institute, Chinese Academy of Agricultural Sciences, Beijing 100081, China
* Correspondence: nawabkhan@stu.sicau.edu.cn (N.K.); zhanghuijie@caas.cn (H.Z.)
† These authors contributed equally to this work.

Abstract: Minor beans other than soybeans or peanuts are edible beans (EBs) that significantly contribute to the Chinese agricultural sector and play a vital role in the sustainability of agricultural production, diversification of food consumption, and income generation for producers. These beans are an important source of protein in a healthy diet, helping to improve national food security. In addition, adjusting and optimizing the industrial structure promotes the sustainable development of agriculture and diversifies staple food crops and introduction of new revenue streams for EB products. The current study examines the responses of mung bean and broad bean producers to environmental and internal input constraints. This study uses the production function with a multilevel mixed-effects method and is based on 848 households from two major EB-producing provinces of China in 2018 and 2019. The results show that local climatic conditions influence planting behavior. These types of beans are considered as a supplement and backup crop to the staple crop. Commercialization encourages cultivation. Producers show variable price responses to output prices, but very strong responses to product costs. Minor bean production is favored by small households because of its low labor intensity. For households growing these beans for consumption, soil fertility and environmental outcomes are improved. Findings from research on planting behavior have strong policy implications for guiding research and development for drought and pest resistance, market monitoring for price stabilization, promoting EB production through low-cost technologies, and encouraging sustainable agriculture.

Keywords: edible beans; planting behavior; high quality; multilevel model; determinants; China

Citation: Ma, J.; Qu, J.; Khan, N.; Zhang, H. Towards Sustainable Agricultural Development for Edible Beans in China: Evidence from 848 Households. *Sustainability* **2022**, *14*, 9328. https://doi.org/10.3390/su14159328

Academic Editor: Aaron K. Hoshide

Received: 22 June 2022
Accepted: 27 July 2022
Published: 29 July 2022

1. Introduction

It has been noted that the national development goal for the next decade is to achieve harmony and unity between humans and nature through eco-friendly development [1]. This requires a multipronged approach: in addition to increasing the supply of safe and high-quality agricultural products, the focus needs to shift from quantity to quality [2], while increasing agricultural incomes by establishing sustainable and efficient agricultural production structures compatible with existing resources [3]. Although edible beans (EBs) account for a small share of agricultural production in China, EBs play an integral role in the sustainability of agricultural production, diversification of food consumption, and income generation for producers [4]. EBs refer to more than 20 kinds of legume crops such as mung bean (MB), adzuki beans, common beans, broad bean (BB), peas, etc., except soybeans and peanuts. They are an important source of fiber and protein for a healthy diet [5]. At present, the per capita per year consumption of EBs is about 1.7 kg, which is expected to expand rapidly with the increase in income and the improvement of nutrition awareness under the strategy of high-quality agricultural development [5].

EBs are also important crops for ecological protection, utilization of idle land, and reduction in disaster damage, which can continuously promote the development of green, low-carbon, and circular agriculture [6]. Therefore, EBs are widely recommended for intercropping or rotation with cereals, root crops, cotton, and fruit saplings. In addition, they are grown on small marginal plots that are less fertile for other crops. These beans are also heat and drought tolerant and have a short growing period, making them popular in all weather conditions. Consequently, during extreme weather events, farmers are drawn to when they quickly replant them to mitigate income losses. EBs are important tools for optimizing crop structure, developing rural areas, reducing poverty, and revitalizing rural areas. Although EBs are widely grown in most parts of China, their production is concentrated in less-developed provinces with large ethnic minority populations and widespread poverty. The provinces with the most EBs are Yunnan, Inner Mongolia, Heilongjiang, Jilin, Sichuan, Guizhou, Chongqing, Shanxi, and Shaanxi, with a planting area of more than 100,000 hectares. In 2018, together this accounted for 65% of the country's total arable land.

According to the investigation by China Agriculture Research System, smallholder production remains the main format of EB production and an important source of household income. The central government has been encouraging adjustments and optimizing the agricultural structure and local brand certification and management to increase high-quality agricultural supply. Examples of EBs serving as a major sector for local poverty reduction include MB from Baicheng City, Jilin Province, the BB from Dali Prefecture, Yunnan Province, and adzuki bean from Kelan County, Shanxi Province. In the case of Kelan County, Shanxi Province, a local noncereal production base was established with adzuki as the leading crop. The production base covers 6400 hectares and 141 administrative villages (of which 90 villages are defined as poor), which benefits 20,485 residents in 7923 households. It was estimated that the average household income increased by 244.98 US$ (1 US$ = 6.8985 yuan on 2019) for 3412 poor households (8854 residents). Fresh BB from Dali Prefecture, Yunnan Province also proves to be a feasible approach for erasing poverty among households living at high altitudes. Despite the strategic importance and growing demand, the domestic cultivation area of EBs had more than halved between 2001 and 2018, from 3.8 million hectares to 1.8 million hectares. The share of EBs in crop cultivation areas also dropped from 3.5 to 1.5%. As a result, bean imports climbed sharply and exacerbated the declining competitiveness of domestic production. There is a rich body of literature on the planting behaviors of cereal farmers, but very few on EB farmers.

The main objective of this study was to analyze EB plantations and their impact on China's sustainable agricultural development (the ratio of EB planted area to total planted area). Based on a survey of 848 households in 2018 and 2019, in two producing areas (MB in Baicheng City, and BB in Dali Prefecture), this study provides an empirical analysis of producers' planting behavior and its determinants. Results from this analysis will contribute to the understanding of farmers' planting behavior and provide evidence in policy recommendations on sectoral development to enhance competitiveness and profit under a high-quality agricultural development strategy. The unique market structure of EBs, different from cereal crops, can also shed insights into other low-volume agricultural products.

2. Literature Review

The development of cash crop production (such as mung and broad beans) to improve household welfare has been at the heart of food policy debates in many developing countries [7,8]. Several studies have shown that cash crop production can be an effective way to improve the household's economic well-being and agricultural development [9,10]. For example, Christiansen et al. [11] studied coffee farmers in Tanzania and found that in the presence of health and drought shocks, coffee farmers remained economically resilient compared to other major crop farmers, suggesting a positive impact on agricultural development and economic well-being of farmers. Similar studies have been found in other developing countries. For example, Kennedy et al. [12] found that participation in a cash crop program resulted in increased household income in six African and Southeast Asian countries (including Gambia, Guatemala, Kenya, Malawi, the Philippines, and Rwanda); later Finnis [13] showed that in southern India, growing cash crops of farmers achieve higher economic and social benefits, and in Malawi, households that choose to grow cash crops also show significantly higher incomes than those that do not use or adopt [14].

Several recognized channels for promoting crop production can improve the economic well-being of households and agricultural development. First, cash crop production can be an effective way to increase family farming income [15–17]. Specialization in cash crop production (relative to staple crops) typically results in higher economic returns per unit of land, including land, water, technology, and, to some extent, labor input. Second, boosting cash crop production can help diversify household livelihoods, thereby further improving household resilience to economic shocks (such as market price shocks) and other climate-related shocks (such as droughts, extreme heat, and cold temperatures). For example, several studies have found that crop diversification, such as intercropping and crop rotation, can increase the resilience of family farming production [18,19]. Third, the benefits of cash crop production also benefit other non-cash-crop farmers through the impact on employment because most cash crop production is labor-intensive [20,21]. Increased labor demand for high-value cash crops is likely to increase average wages for non-cash-crop farmers. In addition, the introduction of cash crop opportunities has shown that households can reduce cash constraints and be able to purchase improved crop production inputs [22]. As a result, their ability to adapt to yield enhancement techniques and agronomic practices is enhanced [23]. This cash income ultimately provides farmers with the opportunity to invest and improve farm management, thereby stimulating agricultural innovation and increasing yields [23–25].

However, there are reasons to question the positive impact of cash crop production on household economic well-being [14,26]. For example, some recent studies have shown that cash crop production has failed to increase the economic well-being of households in some developing countries, especially the poorest households, due to high barriers to entry [27]. They found that promoting cash crop production did little to improve the living standards of the poorest and that these poorest households were often ignored or barred from participating in the production of these cash crops [28]. In China, previous studies on cash crop production have focused on two aspects. A set of the literature focuses on the conceptual and theoretical basis of farmers' economic crop production choices and their operating mechanisms [29]. Other research on cash crop production has generally focused on how cash crop production affects household labor distribution, their subsequent household relocation decisions, and other noneconomic outcomes such as environmental and ecological consequences [30,31]. Although there is ample evidence that cash crop production can be an effective approach to improving household economic well-being [28,32], these observations are mostly correlated. Evidence on its causality is rather limited. It is unclear to what extent and under what conditions cash crop production can achieve desirable results at the micro-household level [14,33].

In addition, farmers' decisions on the production of cash crops (relative to staple crops) are increasingly influenced by perceived risks from climate change [34,35]. In the context of soybean production in China, farmers' perceptions are enhanced due to repeated changes in climate (such as excessive rainfall and flooding in different regions), which directly affects farmers' expected soybean yields and their economic benefits [36]. In response to this heightened perception of climate-related risks and the observed adverse effects on soy production, farmers may consider other alternatives and more resilient crops to address this potential negative impact [37]. As hypothesized by Asrat and Simane [38] and Ojo and Baiyegunhi [39], adaptation to climate change involves a multistep process in which strong perceptions (or strongly perceived changes in climate conditions) must be established, and subsequently, appropriate on-site responses may be initiated for these changes. Several studies have examined this relationship [34,35,40–43] and concluded that household adaptation to climate change behavior is directly related to its perception [44]. However, there is limited research on the combined effects of edible bean (EB) cultivation and its impact on commercial and agriculture development [45] and economic welfare [39]. It can be investigated from the above studies that EBs and crops planting play an important role in increasing farmers' income and agricultural development in China.

3. Data and Methodology

3.1. Survey Design and Data Collection

Mung bean (MB) in Baicheng City, Jilin Province, and broad bean (BB) in Dali Prefecture, Yunnan Province are selected for this study for their significant role in edible bean (EB) production and representativeness. Both locations are important players in national production. Baicheng City MB is a certified geography signature product with an average annual planting area of 80,000 hectares. Annual MB production in Baicheng City reaches 100,000 tons, accounting for 11% of national total production. Approximately half of Baicheng MB is exported, representing above 30% of total exports. A major noncereal wholesale market is located in Baicheng City, facilitating the commercialization of local MB. Yunnan province is the top producer of EBs in the country, representing 16.5% of the national planting area. In Yunnan province, Dali Prefecture is the largest producer of BB, with more than 20% of provincial production coming from this location.

The household-level of data is obtained from households in two consecutive years, 2018 and 2019. A cluster sampling method was used in the survey [46]. Household-level data was obtained from 5 counties/cities in Baicheng City, Jilin Province, and 4 counties/cities in Dali Prefecture, Yunnan Province, covering 32 townships and 66 villages. Among 938 questionnaires distributed, 848 provide valid responses (90%) (see Table 1 for details). Dali Prefecture represents 45.8% of the total sample, while Baicheng City captures the remainder. The distribution of households reflects the geographic concentration of production, with wide cultivation of BB in Dali Prefecture, but a high concentration of MB in Baicheng City (mainly in Tiaonan and Tongyu counties).

Table 1. Samples Location.

Prefecture	Geographical Location	County/City	No. of Households	Share (%)
Dali	Belongs to Yunnan Province, located in the central and western part of China	Dali	106	12.5
		Erdu	103	12.1
		Midu	94	11.1
		Xiangyun	85	10.0
		Subtotal	388	45.8
Baicheng	Belongs to Jilin Province, located in northeast China	Da'an	30	3.5
		Tiaobei	41	4.8
		Tiaonan	146	17.2
		Tongyu	178	21.0
		Zhenglai	65	7.7
		Subtotal	460	54.2
Total			848	100.0

The county-level data on the weather was obtained at the county level from the Chinabric database. The highest and lowest temperature and an average rainfall of Chinabric database at county/city are collected. Weather information in April is collected for MB as they are usually planted in April in Baicheng City and harvested in late September and early October. Dali producers usually plant BB in late September and harvest in April and May of the following year. Hence, weather information in October was collected for Dali Prefecture.

3.2. Empirical Model

This paper uses regression analysis, which is a statistical technique for estimating the relationship that independent variables have on a dependent variable [47]. Regression analysis has many estimation methods to correct for errors. Specifically, a multilevel mixed-effects model was chosen for this study [48,49] because it recognizes the hierarchical nature, clustering, and the structure of the data in this study. Households in the same county tend to be more exposed to the same environmental and social characteristics than households chosen at random from the population. The county difference refers to the resource and environmental endowments difference. Multilevel mixed-effects models accommodate the existence of data hierarchy by introducing separate residual components for each level of the hierarchy. In this study, residuals are assumed to exist at both household and county levels.

The residual variance is partitioned into a between-group component (the variance of county-level residuals) and a within-group component (the variance of the household-level residuals within the same county). Between-group residuals, or county effects, signify unobserved county characteristics that affect household production behavior [50]. County effects are unobservable but can lead to a correlation between EB planting behaviors for households within the same county. Because individual households at the county level are not independently distributed, results from classic Ordinary Least Squares (OLS) are no longer valid [51].

Multilevel mixed-effects models allow both fixed and random effects, especially in the case of non-independence in a hierarchical data structure; they are also called multilevel models. Multilevel mixed-effect models offer several advantages. First, they correct inferences based on observations that are not independent. Ignoring the intercorrelation among households can result in underestimated standard errors of regression coefficients and thus an overstatement of statistical significance. Standard errors for the coefficients of higher-level predictor variables will be the most affected by ignoring the grouping of the county.

Second, multilevel mixed-effects models address individual county-level effects by separating county unobserved conditions from local agronomical practices in obtaining household planting behavior. Moreover, county-level effects are estimated simultaneously with the effects of county-level coefficients. Instead of multiple county dummy variables in a fixed-effects model, this approach allows the estimation of county-level variables of interest (weather indicators in this study), while separating observed (weather) and unobserved (county random effects) county characteristics.

Third, results from the multilevel mixed-effects model can be generalized to a wider population of counties. Unlike a fixed-effects model, inferences are limited to the counties in the sample and cannot be made beyond these groups. By assuming that the random county effects come from a common distribution, planting behaviors obtained from this study can be extrapolated to other counties.

This analysis uses a linear multilevel mixed model with random intercepts. Regression models with one dependent variable and more than one independent variable are called multilinear regression [47]. The regression model can be expressed as

$$y_{ij} = \beta_{0j} + X_{ij}\beta_1 + e_{0ij} \tag{1}$$

$$\beta_{0j} = \beta_0 + \mu_{0j} \tag{2}$$

where $j = 1, 2, \ldots, n$ refers to county, $I = 1, 2, \ldots, m$ refers to households, y_{ij} is the share of EB planting area in total planting area for, x_{ij} are variables affecting EB producer behavior for i-th household of j-th county, β_{0j} is the sum of fixed intercept β_0 and a random intercept for the i-th county μ_{0j}, β_1 is the fixed slope, e_{0ij} is a zero-mean Gaussian error term.

Equation (2) assumes that μ_{0j} and e_{0ij} are unobserved random variables that are independent of each other. Independence of between-group can be measured by ICC (Intra-Class Correlation Coefficient) as the ratio of between-group variance to the total variance. STATA (16th version) software is employed to estimate the results, and Standard Maximum Likelihood (ML) and Restricted Maximum Likelihood (REML) estimation are applied, with the latter providing consistent estimates.

$$ICC = \frac{\sigma^2_{\mu 0}}{\sigma^2_{\mu 0} + \sigma^2_{e0}},$$

where $\sigma^2_{\mu 0}$ is between-group variance and σ^2_{e0} is within-group variance. ICC takes a value between 0 and 1. ICC is close to 1 when between-group variances are large. If ICC is close to 0, within-group variance is large, or observations are independently distributed. The rule of thumb is that mixed-effects models are appropriate when ICC is greater than 0.1.

4. Results and Discussion

Planting behavior is defined as decisions by producers to maximize profit under certain constraints in resources. The existing literature on agricultural planting behavior covers planting behaviors about crop selection and input. Crop planting behaviors involve the willingness to grow [52,53], and planting area and input decision involves the choice of variety [54], application of chemical fertilizer [55], and pesticide and herbicide [56–59]. Referring to the literature, the dependent variable is defined as the share of edible beans (EBs) planning area in household total planting area instead of planting area, which controls the variations of land size across households. The explanatory variables address weather, market and household, and individual characteristics.

Weather is proven to have affected crop production [60–62], planting system, and crop allocation [60,63], as well as producer decision [64,65], and we noticed that farmer's willingness to plant maize is negatively related to average temperate during the growing season but positively related to average rainfall. Interviews with EB farmers in our research regions reveal that farmers pay close attention to temperature and rainfall before EB planting. Hence, average rainfall and the highest and lowest temperature at the beginning of the planting season are included in the analysis.

Most research indicated that the higher the market price, the stronger the willingness to plant [66,67]. On the contrary, other researchers pointed out that to a large extent, the planting behavior did not depend on the price level, and the farmers' response to price fluctuation was more diverse [68]. In addition to prices, the production costs of different crops also affected the behavior of farmers. The higher the production costs, the lower the cost benefit of crops, which directly reduced the production benefits of farmers and the planting behavior [69]. Hence, output prices of last year and production costs are included in the analysis.

In terms of household and individual characteristics, this study selected household size, the share of wage income in household total income, household head age, and education. Larger households are more likely to choose multiple crops to diversify production risks. Wage income is treated as a proxy for non-agricultural income. Household head demographics identify the producer's access to new technology and information [70–73].

4.1. Descriptive Statistics of the Sample

There is substantial variation in planting area: the average mung bean (MB) producer reported a planting area of 2.67 hectares (ha), while the average broad bean (BB) producer only reported 0.14 ha. This regional difference is due to the availability of arable land: the average Baicheng household cultivates 9.63 ha, far greater than the average land size of 0.39 ha in Dali Prefecture. BB production is an integral part of livelihood as its average share in total planting area reached 59% (Table 2).

Table 2. Descriptive statistics of the sample.

Category	Variable	Unit	Mung Bean				Broad Bean			
			M	S.E	Min.	Max.	M	S.E	Min.	Max.
Dependent Variable	Share of edible bean planting area	%	0.38	0.25	0.01	1.00	0.59	0.32	0.01	1.00
Level 1: County/City Level										
Weather	Lowest temperature	°C	−6.06	1.37	−8.50	−4.50	7.83	2.13	4.50	11.50
	Highest temperature	°C	27.90	0.98	26.70	29.30	24.45	0.59	23.70	25.20
	Average rainfall	mm	0.46	0.50	0	1.70	72.21	50.84	9.60	144.8
Level 2: Household Level										
Market	Output price of last year	US$/kg	0.94	0.06	0.82	1.11	0.57	0.11	0.43	0.87
	Average cost	1000 US$/ha	0.63	0.53	0.05	3.83	1.13	0.73	0.02	3.89
Household	Household size		3.46	1.14	1.00	8.00	5.01	1.44	2.00	10.00
	Share of irrigated planting area	%	0.38	0.38	0	1.00	0.78	0.32	0	1.00
	Wage share in total income	%	0.05	0.14	0	0.91	0.46	0.36	0	1.27
Individual	Age	Years	48.27	9.57	21.00	74.00	52.23	9.35	29.00	87.00
	Education	Years	7.96	2.36	0	12.00	8.60	2.31	0	12.00

Note: Weather information in Tiaobei was not available and replaced with Da'an in a similar latitude. Education level: 0 = illiterate, 6 = primary school, 9 = middle high school, 12 = high school graduate and above. Source: Weather data from Chinabric database.

Interviews show that farmers adjust their planting decision based on weather conditions. For instance, farmers in Baicheng City will choose to plant maize if the average rainfall is favorable; otherwise, farmers will choose to either delay maize planting or switch to MB for its drought resistance. There was little difference in temperature between 2018 and 2019 in Baicheng, China, with a slightly warmer temperature in July 2019. The highest temperature was 35.9–38.7 °C in 2018 and 36.1–39.7 °C in 2019. Similarly, the lowest temperature was close to −5.2–−4.6 °C in 2018 and −8.5–−6.3 °C in 2019. However, average rainfall differed remarkably across China in Baicheng City.

Market factors include last year's output prices and average cost. Output prices are the prices sold from the farm gate. The average output price is 0.94 US$/kg for MB and 0.57 US$/kg for BB. The average production cost is 42.40 US$ for MB, and the BB cost almost doubles that of MB at 75.54 US$. Input costs, such as the purchases of seeds, fertilizer, pesticides, and herbicides, accounted for 50.5% of total MB cost, while land rent accounted for another 41.5%. As a result, labor costs only represented 5.1% of the production cost for MB. Input costs were a higher portion of BB production cost at 71.4%. Labor cost is also higher at 12.1%, but land rent was a much smaller part of the total cost at 4.6%. The difference in cost structure can be attributable to the land market and the adoption of mechanization. The land market is very active in Baicheng City, with an average rent of 434.88 US$ per hectare. This leads to a larger land size that supports the use of machines in every stage of MB production, resulting in a much lower dependence on human labor. Dali is the opposite case of smaller plot size from the inactive land market and low adoption of agricultural machinery.

The average size of EB-growing households was four to five persons, with middle school education. The average household head was 48 years old among MB growers. BB growers were older, with the household head at 52 years old. The share of household land that was irrigated was 37.6% of Baicheng because MB is planted in lands without irrigation. BB plots are more likely to be irrigated as 78.3% of household land was irrigated in Dali. MB producers also tend to be more specialized in agriculture with only 5.3% of household income derived from wages. Almost half (46%) of household income came from wages and salaries, indicating that BB-growing households tend to have more diversified income sources and are less dependent on crops for income earning.

Table 3 reports the values of ICC for MB are 0.132 under ML and 0.142 under REML, suggesting that 13.2% of variance comes from between-group components. ICC values are even higher for BB at 0.304 under ML and 0.327 under REML [74]. The results confirm the appropriateness of mixed-effect models in this analysis to avoid inconsistent estimation.

The superiority of mixed-effect models is further confirmed by chi-square values when compared with OLS. Akaike information criterion (AIC) values indicate that ML is the preferred model [75]. Coefficients in Table 3 are used to estimate elasticities of planting behaviors in Table 4, evaluated at the sample mean.

Table 3. Estimation results from multilevel models.

Category	Variable	Mung Bean		Broad Bean	
		ML	REML	ML	REML
Fixed Effects					
Weather	Lowest temperature	0.033 ** (0.016)	0.031 * (0.018)	0.013 (0.041)	0.005 (0.042)
	Highest temperature	0.02 (0.015)	0.021 (0.016)	−0.043 * (0.023)	−0.044 * (0.027)
	Average rainfall	−0.070 ** (0.033)	−0.064 * (0.037)	−0.001 (0.001)	−0.001 (0.001)
Market	Output price	0.069 ** (0.033)	0.067 * (0.037)	−0.076 (0.061)	−0.083 (0.069)
	Average cost	−0.177 *** (0.042)	−0.178 *** (0.043)	−0.088 ** (0.041)	−0.089 ** (0.041)
Household	Household size	−0.041 *** (0.010)	−0.042 *** (0.010)	−0.001 (0.010)	−0.002 (0.010)
	Share of irrigated planting area	−0.048 * (0.027)	−0.047 * (0.028)	0.149 *** (0.048)	0.144 *** (0.048)
	Wage share in total income	0.054 (0.073)	0.056 (0.074)	0.156 *** (0.039)	0.156 *** (0.039)
Individual	Age	0.003 ** (0.001)	0.003 ** (0.001)	0.001 (0.002)	0.001 (0.002)
	Education	−0.002 (0.004)	−0.002 (0.005)	0.001 (0.006)	0.001 (0.007)
	Constant	−0.755 (0.522)	−0.693 (0.567)	0.695 (1.126)	0.931 (1.180)
Random Effects					
	Level 2 Error	0.046 (0.003)	0.047 (0.003)	0.067 (0.005)	0.069 (0.005)
	Level 1 Error	0.001 (0.001)	0.002 (.001)	0.021 (0.010)	0.029 (0.016)
	ICC	0.132	0.142	0.304	0.327
	Observations	460	460	388	388
	Chi-square	118.804 ***	107.193 ***	40.395 ***	38.189 ***
	AIC	−73.258	12.796	106.676	193.105

Note: ***, ** and * indicate 1%, 5% and 10% statistical significance level. Standard errors in parentheses. Estimated parameters are listed above the stand errors in the parentheses.

Table 4. Elasticities of the determinant factors on edible bean planting in China.

Category	Variable	Mung Bean	Broad Bean
Weather	Lowest temperature	1.473	−1.777 *
	Highest temperature	0.528 **	0.172
	Average rainfall	−0.085 **	−0.122
Market	Output price	1.182 **	−0.506
	Average cost	−0.137 ***	−0.078 **
Household	Household size	−0.374 ***	−0.008
	Share of irrigated planting area	−0.048 *	0.197 ***
	Wage share in total income	0.008	0.121 ***
Individual	Household head age	0.382 **	0.088
	Household head education	−0.042	0.015

Note: ***, ** and * indicate 1%, 5% and 10% statistical significance level. Elasticity is an economics concept that measures the responsiveness of one variable to changes in another variable, Elasticity $= \frac{dy}{dx}\frac{\Delta x}{\Delta y}$, in which y is the independent variable and x is a dependent variable.

4.2. Impacts of Weather on Farmers' Planting Behavior

Weather plays an important role in EB producers' decisions. The highest temperature and average rainfall before planting season could substantially affect planting decisions.

A heat wave before planting season is likely to increase MB planting (at a 5% significant level). When the highest temperature in April increases by 1%, the share of MB cultivation would increase by 0.53%. A field study suggests that MB production serves as a strategy to mitigate loss. When it becomes too warm (higher temperature) for maize production, farmers would choose MB as a substitute to decrease crop cultivation loss.

The average rainfall is negatively associated with MB production. A 1% increase in average rainfall in April could lead to a 0.09% drop in MB planting share. This is due to the substitutability between maize and MB. The average profit for MB is about 72.48 US$, slightly higher than that of maize at 65.23 US$. However, farmers prefer maize production because it requires less labor input due to a higher level of mechanization and comes with subsidies from the government. When soil contains more moisture in April, farmers would choose to plant maize for a potential good harvest. However, when average rainfall is low, farmers exhibit a higher propensity for MB, highlighting EBs' role in loss mitigation. The results of Pataczek et al. [76] show that mung bean (Vigna radiata) is gaining attention as a short-season crop that can tolerate dryland conditions (with less rainfall). MB is such a minor crop that dryland smallholder farmers can use to break the downward spiral and increase the profitability and sustainability of their farms. Integration of mung bean in cropping systems may increase the sustainability of dryland production systems. Diversification of local production systems through the inclusion of mung bean as a catch crop provides additional income to farmers and has the potential to improve soil fertility [77].

The lowest temperature affects the planting decisions of BB planting in Dali. The share of BB planting area would drop by 1.78% if the lowest temperature in October falls by 1% (at a 10% significance level), as short episodes of extreme minimum temperature can essentially impede crop development and thus impact crop yield [78]. This is also associated with the growth pattern of BB. Experienced farmers will cut BB planting in a warm October because it could lead to rapid growth and the flowering season could be susceptible to frost and thus lower crop yield. Average rainfall does not appear to impact the planting behavior of BB, mainly due to reliable and sufficient rainfall patterns. Table 2 indicates that average rainfall was 210.2–473.4 mm in October 2018, and 188.4–541.6 mm in October 2019.

4.3. Impacts of Market on Farmers' Planting Behavior

MB farmers are very responsive to output price but not BB. Given that the output price is the price sold from the farm gate, the own-price elasticities we estimated represent the derived (farm gate) supply rather than the market supply [79]. If the output price goes up by 1%, the share of the MB planting area will increase by 1.18% (at a 5% significance level). This phenomenon reflects different levels of commercialization between these crops. MB farmers are active participants in markets and thus are more sensitive to price signals. Additionally, MB farms tend to be much larger at 2.6 hectares, allowing large-scale application of machinery to lower production costs and adjustment in the planting area. On the other hand, BB production is more stable and less responsive to market prices because most of the output is still used for own and local consumption. The lower level of machinery further limits farmers' ability to expand production quickly.

The average cost can negatively impact planting behavior for both EBs. The lower the cost, the higher the planting area. If cost decreases by 1%, the share of the planting area would increase by 0.14 and 0.08% for MB and BB, respectively. Because the majority of the total cost is spent on inputs like seeds, fertilizer, and chemicals, policies targeting lower input costs could promote the production of EBs. In 2018, Yunnan initiated a program to support local green food by providing free fertilizer to lower production costs.

4.4. Impacts of Household and Individual Characteristics on Farmers' Planting Behavior

Household size is negatively associated with EBs planting share, especially in the case of MB, implying that larger households with more arable labor are less inclined to plant

EBs. In Baicheng City, larger households prefer maize for its higher labor productivity from mechanization. In Dali, the BB is less demanding in labor requirement; sometimes farmers choose to skip applying fertilizer or chemicals to cut labor input without sacrificing yield much.

The elasticities of the share of irrigated areas are significant for both MB and BB but of opposite signs due to local agronomical conditions. The share of MB planting area declines when a household has more irrigated plots, with a 1% increase of irrigated land share leading to a 0.37% decrease in MB plant area (at a 10% significance level). The land is largely non-irrigated in Baicheng City, with only 37.6% of the planting area irrigated. Farmers prefer to grow maize on irrigated plots while cultivating MBs in non-irrigated areas. This negative elasticity underscores the key role of MBs in loss mitigation. The impact of irrigation is positive in Dali, partly due to the high availability of irrigated plots (78.3% of the planting area is irrigated).

Wages and salaries are a good proxy of income diversification in rural households. A higher share of wages and salaries in household income indicates a diversified income base and fewer resources dedicated to agricultural production. The elasticity of MB is insignificant but significant for the BB. This is because there are fewer substitute crops in Dali and the flexibility offered by BB production through low labor demand and smaller plots. Diversification from agriculture introduces new income sources for EBs producers. Elasticity associated with household head age is positive and significant for MB with a value of 0.38 at a 5% significance level. Older household heads express a stronger preference for MB for its low input requirement compared to maize.

5. Conclusions and Implications

Few studies have focused on farmers' behavior in growing noncereal crops such as edible beans (EBs), despite their crucial role in the sustainability of agricultural production, the diversification of food consumption, and farmers' income. EB production also has its unique characteristics different from cereals: short growth period, low soil requirements, low yield, less machinery usage, but a high degree of commercialization. Therefore, our research will focus on improving the competitiveness of EBs and improving the allocation of agricultural resources under the eco-friendly development strategy. This paper estimates the planting behavior of two representative edible legumes: mung bean (MB) and broad bean (BB) by estimating a multilevel mixed-effects model. The empirical analysis uses the household-level data of Baicheng City in Jilin Province and Dali Prefecture in Yunnan Province in 2018 and 2019, combined with the county-level meteorological observation data in the month of planting.

This study determines whether or not it is a major factor in the decision to grow EBs. In the case of MB, producers choose to grow MB to mitigate the potential loss of income from unfavorable soils and temperatures. BB farmers are also temperature-sensitive when making planting decisions to ensure high yields. Although the commercialization and specialization of the two types of EB are different, and producers may be less responsive to output prices, they are generally responsive to production costs. Smaller households are showing a higher interest in growing edible pulses due to reduced demand for land and income generation. Although BB is grown in small pieces and eaten more often, its role in increasing nitrogen fertilizer cannot be ignored. Based on the above results, it is believed that under the background of high-quality agricultural development, the eco-friendly development and high-quality development level of the bean industry should start by improving the agricultural development capabilities of the bean industry in order to address climate change. To reduce carbon emissions to address climate change, China's lesser bean producers can improve production efficiency. They can accomplish this by market monitoring and relying on scientific and technological improvements to increase crop productivity per unit input of greenhouse gas emitting inputs (e.g., tractors, processing machinery, etc.).

First, in the process of high-quality agricultural development, to enhance the adaptability of EBs to climate change. It is necessary to increase investment in research and development, carry out the screening of main varieties and key technologies, and enhance stress resistance so that beans are resistant to drought, pests, and other risks. The ecological and cultural value of edible pulses should be further developed. In Dali Prefecture and other key ecological development zones, it is necessary to further explore the ecological service value of beans, improve the organic combination of the beans industry and local culture, and explore new forms of the beans industry. Second, for EBs with a high degree of marketization (such as MB in Baicheng City, Jilin Province), a market price monitoring mechanism should be established, and market supervision of small-scale agricultural products should be strengthened to prevent excessive consumption and reserve as well as stabilize market prices and farmers' production. At the same time, we should encourage the "Internet + Project" development model of EBs, further standardize and guide the EB sales and circulation market, and provide farmers with services such as prenatal markets. Information, quality control in the process of production and post-purchase, and sales through the internet or Internet of Things technology.

Third, it is important to improve quality and efficiency through science and technology, reduce the cost of bean production, and promote the upgrading of the bean industry. Strengthen the promotion and application of new varieties of high-yield and high-quality EBs suitable for mechanized production, and promote mature, high-yield, efficient, simplified, and integrated production technologies characterized by large-scale, standardized, and mechanized production, reducing the production cost of EBs and increasing farmers. The economic income of farmers has improved their enthusiasm to grow EBs.

Finally, it is recommended to establish a demonstration zone for the integration of the bean product industry, develop and improve the bean product industry chain, and closely integrate the bean product industry with the health industry. After starting with implementation of lesser bean industry integration demonstration areas, China can improve integration of bean production and processing, as well as facilitate marketing to the health industry. Future research can enable breakthroughs in the development of processing edible bean products using more eco-friendly processes.

Author Contributions: J.M., J.Q., N.K. and H.Z. developed and outlined this concept, including the method and approach to be used; J.M., J.Q., N.K. and H.Z. developed and outlined the manuscript; J.M., J.Q., N.K. and H.Z. contributed to the methodology and revision of this manuscript; J.M., J.Q. and N.K. wrote the article. All authors have read and agreed to the published version of the manuscript.

Funding: This research was funded by the China Agriculture Research System of MOF and MARA-Food Legumes (CARS-08) and the National Natural Science Foundation of China (No. 71904190).

Institutional Review Board Statement: Not applicable.

Informed Consent Statement: Not applicable.

Data Availability Statement: Not applicable.

Conflicts of Interest: The authors declare that they have no conflict of interest.

Abbreviations

AD	Agricultural Development
M	Mean
MIN	Minimum
MB	Mung bean
OLS	Ordinary Least Squares
ICC	Intra-Class Correlation Coefficient
AIC	Akaike Information Criterion
CSM	Cluster Sampling Method
EB	Edible bean

S.E	Standard Error
MIX	Maximum
ML	Maximum Likelihood
REML	Restricted Maximum Likelihood
BB	Broad bean
χ^2	Chi-Square
N	Number of Samples

References

1. Fang, C.; Wang, Z.; Liu, H. Beautiful China Initiative: Human-nature harmony theory, evaluation index system and application. *J. Geogr. Sci.* **2020**, *30*, 691–704. [CrossRef]
2. Xu, B.; Chen, W.; Zhang, G.; Wang, J.; Ping, W.; Luo, L.; Chen, J. How to achieve green growth in China's agricultural sector. *J. Clean. Prod.* **2020**, *271*, 122770. [CrossRef]
3. Lange, L.; Connor, K.O.; Arason, S.; Bundgård-Jørgensen, U.; Canalis, A.; Carrez, D.; Gallagher, J.; Gøtke, N.; Huyghe, C.; Jarry, B. Developing a sustainable and circular bio-based economy in EU: By partnering across sectors, upscaling and using new knowledge faster, and for the benefit of climate, environment & biodiversity, and people & business. *Front. Bioeng. Biotechnol.* **2021**, *8*, 619066. [PubMed]
4. Charrua, A.B.; Havik, P.J.; Bandeira, S.; Catarino, L.; Ribeiro-Barros, A.; Cabral, P.; Moldão, M.; Romeiras, M.M. Food security and nutrition in mozambique: Comparative study with bean species commercialised in informal markets. *Sustainability* **2021**, *13*, 8839. [CrossRef]
5. Li, L.; Yang, T.; Liu, R.; Redden, B.; Maalouf, F.; Zong, X. Food legume production in China. *Crop J.* **2017**, *5*, 115–126. [CrossRef]
6. Yu, T.; Mahe, L.; Li, Y.; Wei, X.; Deng, X.; Zhang, D. Benefits of Crop Rotation on Climate Resilience and Its Prospects in China. *Agronomy* **2022**, *12*, 436. [CrossRef]
7. Weber, M.T.; Staatz, J.M.; Holtzman, J.S.; Crawford, E.W.; Bernsten, R.H. Informing food security decisions in Africa: Empirical analysis and policy dialogue. *Am. J. Agric. Econ.* **1988**, *70*, 1044–1052. [CrossRef]
8. Tankari, M.R. Cash crops reduce the welfare of farm households in Senegal. *Food Secur.* **2017**, *9*, 1105–1115. [CrossRef]
9. Huang, C. Research on China's path of development-oriented poverty reduction: Review and prospect. *China Agric. Univ. J. Soc. Sci. Ed.* **2016**, *33*, 5–17.
10. Yu, F.; Chen, L. Interactive Development-oriented Poverty Reduction Model for Bijie Experimental Region under the Guidance of Scientific Outlook on Development. *Asian Agric. Res.* **2013**, *5*, 21.
11. Christiaensen, L.J.; Demery, L.; Khl, J. *The Role of Agriculture in Poverty Reduction: An Empirical Perspective*; World Bank Publications: Washington, DC, USA, 2006; Volume 4013.
12. Kennedy, E.; Bouis, H.; Von Braun, J. Health and nutrition effects of cash crop production in developing countries: A comparative analysis. *Soc. Sci. Med.* **1992**, *35*, 689–697. [CrossRef]
13. Finnis, E. Why grow cash crops? Subsistence farming and crop commercialization in the Kolli Hills, South India. *American Anthropologist* **2006**, *108*, 363–369. [CrossRef]
14. Masanjala, W.H. Cash crop liberalization and poverty alleviation in Africa: Evidence from Malawi. *Agric. Econ.* **2006**, *35*, 231–240. [CrossRef]
15. Nguyen, C.V. Measuring the impact of cash crops on household expenditure and poverty in rural Viet Nam. *Asia-Pac. Dev. J.* **2012**, *16*, 87–112. [CrossRef]
16. Zhang, S.; Ma, J.; Zhang, L.; Sun, Z.; Zhao, Z.; Khan, N. Does Adoption of Honeybee Pollination Promote the Economic Value of Kiwifruit Farmers? Evidence from China. *Int. J. Environ. Res. Public Health* **2022**, *19*, 8305. [CrossRef]
17. Khan, N.; Ray, R.L.; Zhang, S.; Osabuohien, E.; Ihtisham, M. Influence of mobile phone and internet technology on income of rural farmers: Evidence from Khyber Pakhtunkhwa Province, Pakistan. *Technol. Soc.* **2022**, *68*, 101866. [CrossRef]
18. Barbieri, P.; Pellerin, S.; Nesme, T. Comparing crop rotations between organic and conventional farming. *Sci. Rep.* **2017**, *7*, 13761. [CrossRef]
19. Khan, N.; Ray, R.L.; Kassem, H.S.; Hussain, S.; Zhang, S.; Khayyam, M.; Ihtisham, M.; Asongu, S.A. Potential role of technology innovation in transformation of sustainable food systems: A review. *Agriculture* **2021**, *11*, 984. [CrossRef]
20. Poulton, C.; Tyler, G.; Hazell, P.; Dorward, A.; Kydd, J.; Stockbridge, M. All-Africa review of experiences with commercial agriculture. In *Background Paper for the Competitive Commercial Agriculture in Sub-Saharan Africa (CCAA) Study*; World Bank: Washington, DC, USA, 2008.
21. Irz, X.; Lin, L.; Thirtle, C.; Wiggins, S. Agricultural productivity growth and poverty alleviation. *Dev. Policy Rev.* **2001**, *19*, 449–466. [CrossRef]
22. Govereh, J.; Jayne, T.S. Cash cropping and food crop productivity: Synergies or trade-offs? *Agric. Econ.* **2003**, *28*, 39–50. [CrossRef]
23. Meng, L.; Christopher, G.; Wanglin, M.; Jiang, W. Impact of cash crop cultivation on household income and migration decisions: Evidence from low-income regions in China. *J. Integr. Agric.* **2020**, *19*, 2571–2581.
24. Khan, N.; Ray, R.L.; Kassem, H.S.; Ihtisham, M.; Siddiqui, B.N.; Zhang, S. Can Cooperative Supports and Adoption of Improved Technologies Help Increase Agricultural Income? Evidence from a Recent Study. *Land* **2022**, *11*, 361. [CrossRef]

25. Khan, N.; Ray, R.L.; Sargani, G.R.; Ihtisham, M.; Khayyam, M.; Ismail, S. Current progress and future prospects of agriculture technology: Gateway to sustainable agriculture. *Sustainability* **2021**, *13*, 4883. [CrossRef]
26. Orr, A. 'Green Gold'? Burley tobacco, smallholder agriculture, and poverty alleviation in Malawi. *World Dev.* **2000**, *28*, 347–363. [CrossRef]
27. Kuma, T.; Dereje, M.; Hirvonen, K.; Minten, B. Cash crops and food security: Evidence from Ethiopian smallholder coffee producers. *J. Dev. Stud.* **2019**, *55*, 1267–1284. [CrossRef]
28. Ding, J.; Wang, Z.; Liu, Y.; Yu, F. Rural households' livelihood responses to industry-based poverty alleviation as a sustainable route out of poverty. *Reg. Sustain.* **2020**, *1*, 68–81. [CrossRef]
29. Liu, M.-Y.; Feng, X.-L.; Wang, S.-G.; Zhong, Y. Does poverty-alleviation-based industry development improve farmers' livelihood capital? *J. Integr. Agric.* **2021**, *20*, 915–926. [CrossRef]
30. Su, S.; Zhou, X.; Wan, C.; Li, Y.; Kong, W. Land use changes to cash crop plantations: Crop types, multilevel determinants and policy implications. *Land Use Policy* **2016**, *50*, 379–389. [CrossRef]
31. Li, J.; Zhang, Z.; Jin, X.; Chen, J.; Zhang, S.; He, Z.; Li, S.; He, Z.; Zhang, H.; Xiao, H. Exploring the socioeconomic and ecological consequences of cash crop cultivation for policy implications. *Land Use Policy* **2018**, *76*, 46–57. [CrossRef]
32. Lei, Z.; Yingyu, H.; Changling, Z.; Wei, L.; Liang, H.; Anhong, G.; Lu, C. The construction and application of assessing index to crop growing condition. *J. Appli. Meteor.* **2019**, *30*, 543–554.
33. Jones, S.; Gibbon, P. Developing agricultural markets in Sub-Saharan Africa: Organic cocoa in rural Uganda. *J. Dev. Stud.* **2011**, *47*, 1595–1618. [CrossRef]
34. Mtambanengwe, F.; Mapfumo, P.; Chikowo, R.; Chamboko, T. Climate change and variability: Smallholder farming communities in Zimbabwe portray a varied understanding. *Afr. Crop Sci. J.* **2012**, *20*, 227–241.
35. Ajuang, C.O.; Abuom, P.O.; Bosire, E.K.; Dida, G.O.; Anyona, D.N. Determinants of climate change awareness level in upper Nyakach Division, Kisumu County, Kenya. *SpringerPlus* **2016**, *5*, 1015. [CrossRef]
36. Fosu-Mensah, B.Y.; Vlek, P.L.; MacCarthy, D.S. Farmers' perception and adaptation to climate change: A case study of Sekyedumase district in Ghana. *Environ. Dev. Sustain.* **2012**, *14*, 495–505. [CrossRef]
37. Quiggin, J.; Adamson, D.; Chambers, S.; Schrobback, P. Climate change, uncertainty, and adaptation: The case of irrigated agriculture in the Murray–Darling Basin in Australia. *Can. J. Agric. Econ. Rev. Can. D'agroeconomie* **2010**, *58*, 531–554. [CrossRef]
38. Asrat, P.; Simane, B. Farmers' perception of climate change and adaptation strategies in the Dabus watershed, North-West Ethiopia. *Ecol. Processes* **2018**, *7*, 7. [CrossRef]
39. Ojo, T.; Baiyegunhi, L. Climate change perception and its impact on net farm income of smallholder rice farmers in South-West, Nigeria. *J. Clean. Prod.* **2021**, *310*, 127373. [CrossRef]
40. Maddison, D. *The Perception of and Adaptation to Climate Change in Africa*; World Bank Publications: Washington, DC, USA, 2007; Volume 4308.
41. Hassan, R.M.; Nhemachena, C. Determinants of African farmers' strategies for adapting to climate change: Multinomial choice analysis. *Afr. J. Agric. Resour. Econ.* **2008**, *2*, 83–104.
42. Idrisa, Y.; Ogunbameru, B.; Ibrahim, A.; Bawa, D. Analysis of awareness and adaptation to climate change among farmers in the Sahel Savannah agro-ecological zone of Borno State, Nigeria. *Int. J. Environ. Clim. Chang.* **2012**, *2*, 216–226. [CrossRef]
43. Huong, N.T.L.; Yao, S.; Fahad, S. Assessing household livelihood vulnerability to climate change: The case of Northwest Vietnam. *Hum. Ecol. Risk Assess. Int. J.* **2019**, *25*, 1157–1175. [CrossRef]
44. Fahad, S.; Wang, J. Climate change, vulnerability, and its impacts in rural Pakistan: A review. *Environ. Sci. Pollut. Res.* **2020**, *27*, 1334–1338. [CrossRef]
45. Fierros-Gonzalez, I.; Lopez-Feldman, A. Farmers' perception of climate change: A review of the literature for Latin America. *Front. Environ. Sci.* **2021**, *9*, 205. [CrossRef]
46. Grosh, M.; Glewwe, P. *Designing Household Survey Questionnaires for Developing Countries*; World Bank: Washington, DC, USA, 2000.
47. Uyanık, G.K.; Güler, N. A study on multiple linear regression analysis. *Procedia-Soc. Behav. Sci.* **2013**, *106*, 234–240. [CrossRef]
48. Rabe-Hesketh, S.; Skrondal, A. Multilevel modelling of complex survey data. *J. R. Stat. Soc. Ser. A (Stat. Soc.)* **2006**, *169*, 805–827. [CrossRef]
49. Reidsma, P.; Oude Lansink, A.; Ewert, F. Economic impacts of climatic variability and subsidies on European agriculture and observed adaptation strategies. *Mitig. Adapt. Strateg. Glob. Chang.* **2009**, *14*, 35–59. [CrossRef]
50. Raudenbush, S.W.; Bryk, A.S. *Hierarchical Linear Models: Applications and Data Analysis Methods*; Sage: Thousand Oaks, CA, USA, 2002; Volume 1.
51. Skrondal, A.; Rabe-Hesketh, S. *Generalized Latent Variable Modeling: Multilevel, Longitudinal, and Structural Equation Models*; Chapman and Hall/CRC: Boca Raton, FL, USA, 2004.
52. Peng, L.; Zhou, X.; Tan, W.; Liu, J.; Wang, Y. Analysis of dispersed farmers' willingness to grow grain and main influential factors based on the structural equation model. *J. Rural. Stud.* **2020**, *93*, 375–385. [CrossRef]
53. Bergtold, J.S.; Fewell, J.E.; Duffy, P.A. *Farmers' Willingness to Grow Cover Crops: Examining the Economic Factors of Adoption in Alabama*; Agricultural and Applied Economics Association: Denver, CO, USA, 2010.
54. Zheng, S.; Wang, Z.; Song, S. Farmers' behaviors and performance in cooperatives in Jilin Province of China: A case study. *Soc. Sci. J.* **2011**, *48*, 449–457. [CrossRef]

55. Han, H.-Y.; Zhao, L.-G. Farmers' character and behavior of fertilizer application-evidence from a survey of Xinxiang County, Henan Province, China. *Agric. Sci. China* **2009**, *8*, 1238–1245. [CrossRef]
56. Zhao, L.; Wang, C.; Gu, H.; Yue, C. Market incentive, government regulation and the behavior of pesticide application of vegetable farmers in China. *Food Control* **2018**, *85*, 308–317. [CrossRef]
57. Bagheri, A.; Emami, N.; Damalas, C.A. Farmers' behavior towards safe pesticide handling: An analysis with the theory of planned behavior. *Sci. Total Environ.* **2021**, *751*, 141709. [CrossRef]
58. Jallow, M.F.; Awadh, D.G.; Albaho, M.S.; Devi, V.Y.; Thomas, B.M. Pesticide risk behaviors and factors influencing pesticide use among farmers in Kuwait. *Sci. Total Environ.* **2017**, *574*, 490–498. [CrossRef] [PubMed]
59. Damalas, C.A. Farmers' intention to reduce pesticide use: The role of perceived risk of loss in the model of the planned behavior theory. *Environ. Sci. Pollut. Res.* **2021**, *28*, 35278–35285. [CrossRef] [PubMed]
60. Wang, Y.; Tan, Z.; Sun, G. The impact of climate change on the potential suitable distribution of major crops in Zambia and the Countermeasures. In Proceedings of the International Conference on Computer and Computing Technologies in Agriculture; (CCTA), 16–19 September 2014, Beijing, China; pp. 460–472.
61. Yi, F.; Sun, D.; Zhou, Y. Grain subsidy, liquidity constraints and food security—Impact of the grain subsidy program on the grain-sown areas in China. *Food Policy* **2015**, *50*, 114–124. [CrossRef]
62. Ma, J.; Maystadt, J.-F. The impact of weather variations on maize yields and household income: Income diversification as adaptation in rural China. *Glob. Environ. Chang.* **2017**, *42*, 93–106. [CrossRef]
63. Liu, Z.; Yang, P.; Tang, H.; Wu, W.; Zhang, L.; Yu, Q.; Li, Z. Shifts in the extent and location of rice cropping areas match the climate change pattern in China during 1980–2010. *Reg. Environ. Chang.* **2015**, *15*, 919–929. [CrossRef]
64. Cui, X. Climate change and adaptation in agriculture: Evidence from US cropping patterns. *J. Environ. Econ. Manag.* **2020**, *101*, 102306. [CrossRef]
65. Song, C.-X.; Liu, R.-F.; Oxley, L. Do farmers care about climate change? Evidence from five major grain producing areas of China. *J. Integr. Agric.* **2019**, *18*, 1402–1414. [CrossRef]
66. Dorward, A.; Chirwa, E. The Malawi agricultural input subsidy programme: 2005/06 to 2008/09. *Int. J. Agric. Sustain.* **2011**, *9*, 232–247. [CrossRef]
67. Willett, W.; Rockström, J.; Loken, B.; Springmann, M.; Lang, T.; Vermeulen, S.; Garnett, T.; Tilman, D.; DeClerck, F.; Wood, A. Food in the Anthropocene: The EAT–Lancet Commission on healthy diets from sustainable food systems. *Lancet* **2019**, *393*, 447–492. [CrossRef]
68. Feder, G.; Umali, D.L. The adoption of agricultural innovations: A review. *Technol. Forecast. Soc. Chang.* **1993**, *43*, 215–239. [CrossRef]
69. Ortega, D.L.; Waldman, K.B.; Richardson, R.B.; Clay, D.C.; Snapp, S. Sustainable intensification and farmer preferences for crop system attributes: Evidence from Malawi's central and southern regions. *World Dev.* **2016**, *87*, 139–151. [CrossRef]
70. Paulrud, S.; Laitila, T. Farmers' attitudes about growing energy crops: A choice experiment approach. *Biomass Bioenergy* **2010**, *34*, 1770–1779. [CrossRef]
71. Embaye, W.T.; Bergtold, J.S.; Archer, D.; Flora, C.; Andrango, G.C.; Odening, M.; Buysse, J. Examining farmers' willingness to grow and allocate land for oilseed crops for biofuel production. *Energy Econ.* **2018**, *71*, 311–320. [CrossRef]
72. Nkonya, E.; Schroeder, T.; Norman, D. Factors affecting adoption of improved maize seed and fertiliser in northern Tanzania. *J. Agric. Econ.* **1997**, *48*, 1–12. [CrossRef]
73. McCord, P.F.; Cox, M.; Schmitt-Harsh, M.; Evans, T. Crop diversification as a smallholder livelihood strategy within semi-arid agricultural systems near Mount Kenya. *Land Use Policy* **2015**, *42*, 738–750. [CrossRef]
74. Maas, C.J.; Hox, J.J. Sufficient sample sizes for multilevel modeling. *Methodol. Eur. J. Res. Methods Behav. Soc. Sci.* **2005**, *1*, 86.
75. Vallejo, G.; Tuero-Herrero, E.; Núñez, J.C.; Rosário, P. Performance evaluation of recent information criteria for selecting multilevel models in behavioral and social sciences. *Int. J. Clin. Health Psychol.* **2014**, *14*, 48–57. [CrossRef]
76. Pataczek, L.; Zahir, Z.A.; Ahmad, M.; Rani, S.; Nair, R.; Schafleitner, R.; Cadisch, G.; Hilger, T. Beans with benefits—the role of Mungbean (Vigna radiate) in a changing environment. *Am. J. Plant Sci.* **2018**, *9*, 1577. [CrossRef]
77. Poehlman, J.M. *The Mungbean*; Oxford & IBH: Oxford, UK, 1991.
78. Wheeler, T.R.; Craufurd, P.Q.; Ellis, R.H.; Porter, J.R.; Prasad, P.V. Temperature variability and the yield of annual crops. *Agric. Ecosyst. Environ.* **2000**, *82*, 159–167. [CrossRef]
79. Mani, J.; Abdullahi, A.; Ibrahim, L.; Yusuf, M. Supply response analysis of millet in funakaye local government area of Gombe state, Nigeria: A profit function approach. *J. Agric. Environ.* **2019**, *15*, 7–17.

sustainability MDPI

Review

Commercialization Potential of Six Selected Medicinal Plants Commonly Used for Childhood Diseases in South Africa: A Review

Peter Tshepiso Ndhlovu [1,2], Abiodun Olusola Omotayo [3,*], Wilfred Otang-Mbeng [2] and Adeyemi Oladapo Aremu [1,3,*]

1 Indigenous Knowledge Systems (IKS) Centre, Faculty of Natural and Agricultural Sciences, North-West University, Private Bag X2046, Mmabatho 2745, North West Province, South Africa; Tshepiso.Ndhlovu@ump.ac.za
2 School of Biology and Environmental Sciences, Faculty of Agriculture & Natural Sciences, University of Mpumalanga, Private Bag X11283, Mbombela 1200, Mpumalanga Province, South Africa; Wilfred.Mbeng@ump.ac.za
3 Food Security and Safety Niche Area, Faculty of Natural and Agricultural Sciences, North West University, Private Bag X2046, Mmabatho 2745, North West Province, South Africa
* Correspondence: 25301284@nwu.ac.za (A.O.O.); Oladapo.Aremu@nwu.ac.za (A.O.A.)

Abstract: Globally, the potential of medicinal plants is increasingly being recognized due to their relative availability, particularly in rural areas. This review explored the ethnobotanical and economic values of six selected medicinal plants widely used to treat and manage childhood diseases in South Africa. *Acalypha glabrata*, *Aloe maculata*, *Datura stramonium*, *Gomphocarpus fruticosus*, *Rhoicissus tridentata* and *Vachellia karroo* were selected based on their high relative frequency of citations for treating a wide range of diseases. Information was obtained from various scientific databases and ethnobotanical books. In addition to being popular for treating childhood diseases, the selected medicinal plants possess diverse applications in traditional medicine for other age groups, highlighting their general therapeutic values. This translates to extensive harvesting, trading and consumption of these plants in order to meet demands on local levels. Currently, empirical data on the economic value of the selected plants remain poorly reported. Even though South Africa has many laws to conserve and promote indigenous knowledge and medicinal plants, their commercialization remains low. Particularly the cultivation of the selected plants needs to be promoted under a participatory management action plan to stimulate the economy of the disadvantaged. A collaborative research framework for the inclusive transformation of indigenous medicinal plants is recommended to reveal their concealed beneficial properties.

Keywords: child health; commercialization potential; conservation status; livelihood; policy; rural economy

Citation: Ndhlovu, P.T.; Omotayo, A.O.; Otang-Mbeng, W.; Aremu, A.O. Commercialization Potential of Six Selected Medicinal Plants Commonly Used for Childhood Diseases in South Africa: A Review. *Sustainability* **2022**, *14*, 177. https://doi.org/10.3390/su14010177

Academic Editors: Aaron K. Hoshide and Marc A. Rosen

Received: 5 October 2021
Accepted: 19 December 2021
Published: 24 December 2021

1. Introduction

The use of medicinal plants provides a valuable contribution to alternative primary healthcare [1,2]. An estimated 80% of the global population uses traditional medicine to treat and manage various diseases [3]. The use of medicinal plants has increased in the 21st century, particularly with the emergence of severe diseases such as diabetes, cancer and skin-related conditions [4]. Many publications detail the use of traditional medicine among various ethnic communities, particularly for human and animal-related conditions [5–7]. The number of medicinal plant species is estimated to be between 35,000 and 50,000, with approximately 4000–6000 species entering the global medicinal plant market annually, which is worth approximately US $83 billion [8].

In developing countries, the widespread use of medicinal plants has resulted in traditional healthcare becoming a profitable business for multinational corporations, who have identified the active ingredients and found a lucrative business in medicinal plants [9]. Traditional medicine is commercialized and exported to different countries for product development. This implies that medicinal plants contribute both directly or indirectly to the economy of rural households through providing welfare and economic status [10,11]. Some traditional health practitioners and knowledge-holders have chosen to market their knowledge outside of traditional settings. This contributes to the advancement of pharmaceutical treatments by providing information about the active ingredients derived from plants.

In many instances, there has been inadequate documentation of indigenous knowledge about medicinal plants and their utilization. Medicinal plants have been one of the main ingredients in traditional medical systems, being in high demand in both rural and urban livelihoods and an important source of human survival and well-being [12]. Medicinal plants used to treat and manage childhood diseases in particular are recognised in rural areas due to their ability to provide welfare and wellbeing to traditional health practitioners, particularly those who practice and believe in the use of medicinal remedies [3].

In South Africa, approximately 27 million inhabitants rely on traditional medicine, with an estimated 20,000 tons from 771 medicinal plants being harvested from the wild population [13]. The increasing South African population and urban migration has resulted in the establishment of competitive medicinal plant markets, especially in cities [14]. The traditional medicine industry is unregulated, with an estimated contribution of ZAR 2.9 billion (USD 192,948,107) to the economy per annum [13]. In many rural areas, people die of preventable or curable diseases due to a lack of adequate healthcare facilities, road infrastructure and telecommunication network coverage to reach emergency facilities [15]. As a result, many rural communities resort to the use of traditional medicine as an alternative medicine, a trend that has been observed in urban areas where there are many *'Muthi'* markets (traditional markets).

Although many of the medical floras consumed by the South African public remain poorly explored by scientists [16], the use of medicinal plants has remained popular for historical and cultural reasons [5,17–20]. Most studies have overlooked the medicinal plants used for treating and managing childhood diseases, as well as their economic benefits to the livelihoods of individuals who cultivate or harvest them. While there is a financial benefit from the sale of medicinal plants in the informal sector, it remains largely undocumented and forms part of the 'hidden' economy [21].

There are limited studies on the economic importance of high-value medicinal plants used for child-related diseases [22], and this is an indication of the extent of neglect in monitoring and documenting such medicinal plants [22]. A large number of medicinal plants used to treat and manage childhood diseases are regularly sold as crude, unprocessed drugs in local *'Muthi'* markets in several parts of South Africa [23]. Traders of medicinal plants experience numerous challenges including the absence of formal arrangements provided by city officials with good infrastructure and resources.

In South Africa, the need to reduce infant and child morbidity and mortality is one of the greatest challenges in rural areas [24]. In South Africa, the child mortality rate varies across communities, with the highest (56/1000) being for children under the age of five [15], this being above the WHO's desired level of reducing child mortality rate. A previous study indicated inequalities in the provision of healthcare services between and within South African provinces, with urban areas being better resources than rural ones, which is one of the major causes of poor health and premature death among children [25]. The cultivation and trading of medicinal plants that are used to treat childhood diseases remains an alternative measure to reduce the child mortality rate and increase economic activities. As indicated in several studies [26,27], the demand for medicinal plants creates possibilities for their local cultivation. Such endeavours could help raise employment in the rural economy, boost commerce and possibly contribute to the health of millions. In this regard, this review explored the economic potential of six medicinal plants that are used for childhood diseases in South Africa. We previously generated an inventory of 194 medicinal plants used for childhood disease in South Africa [22].

2. Methods

For the selection of articles for this review, we followed a systematic approach as described by Shamseer et al. [28], called the Preferred Reporting Items for Systematic Reviews and Meta-Analyses (PRISMA). To find relevant articles, we used keywords and terminologies such as economic and industrial relevance, high-value medicinal plants, ethnomedicine, folk-paediatric, paediatric, childhood diseases, indigenous medicine, and phytomedicine. The bibliographies from the retrieved papers were also checked and saved using Endnote reference manager. Plants were selected in this research based on their traditional use in the treatment of multiple childhood diseases. The plants were selected for their health and economic potential with respect to childhood diseases. The six medicinal plants were selected based on their high frequency of citations in the literature and high number of uses for managing three to six childhood diseases. Search engines such as Web of Science, Science Direct, Google Scholar, PubMed, Scopus and JSTOR were accessed for relevant studies. Screening of the published literature was guided by the scope predefined for the current study. Aside from the peer-reviewed journals, theses and dissertations, books were obtained from the libraries of North-West University (NWU) and the University of Mpumalanga (UMP), South Africa. A total of 1635 articles were obtained based on the keywords used for the search. Additional data were collected from 20 sources located at NWU and UMP libraries (Figure 1). Following the removal of duplicates, the remaining 1303 were screened with respect to the eligibility criteria and articles related to ethnoveterinary, wild food and those more than 15 years old were excluded from the selection because they did not meet the criteria and the scope of the study. We accessed the full texts of the resultant 369 publications. The research papers (systematic or literature) exclusively dealing with current childhood diseases were retained. Thereafter, 266 articles were removed due to the lack of sufficient information regarding the scope of this review, while 104 articles were included to complete this study.

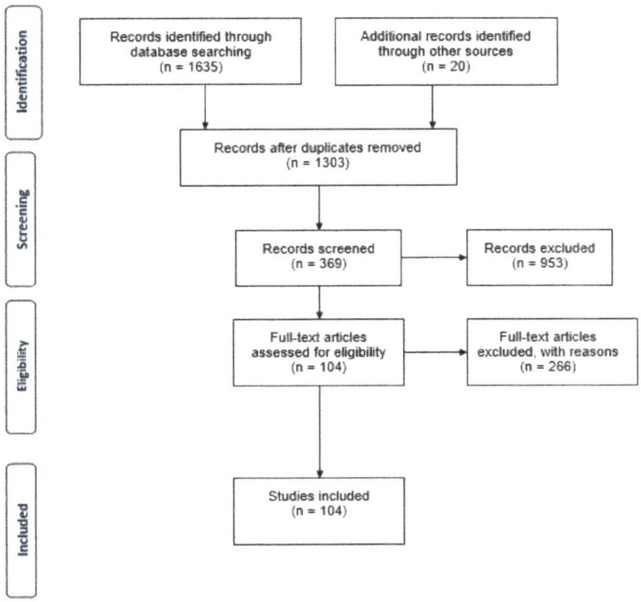

Figure 1. Preferred Reporting Items for Systematic Reviews and Meta-Analyses flow diagram for the exclusion and inclusion of articles for the current review.

3. Distribution and Occurrence of the Selected Medicinal Plants

The six selected medicinal plants identified are (a) *Acalypha glabrata*, (b) *Aloe maculata*, (c) *Datura stramonium*, (d) *Gomphocarpus fruticosus*, (e) *Rhoicissus tridentata* and (f) *Vachellia karroo*, which have diverse distributions globally (Figure 2). For example, the members of the genus *Acalypha* have succulent leaves with sappy stem and are regarded as the fourth largest genus in the Euphorbiaceae [29]. There are approximately 65 *Acalypha* species in tropical Africa and 35 on Indian Ocean islands (Figure 2a). In Asia and Africa, the majority of the species in this family are utilized as medicinal herbs [29], and *A. glabrata* is distributed in four provinces of South Africa [30].

The *Aloe* genus is mainly indigenous to Africa but is widely distributed globally [31,32], being widely distributed in Africa, India and other arid areas (Figure 2b). As one of the most common *Aloe* species, *A. maculata* is well-defined by a suite of characters that includes maculate (spotted) leaves, flat-topped inflorescences and uniformly coloured flowers [33]. Globally, about 548 *Aloe* species are known to exist, with South Africa contributing 140 (26%) species [30]. In South Africa, *A. maculata* occurs in the Western Cape, KwaZulu-Natal, Mpumalanga and Limpopo provinces [30], and is known by different local names and diverse uses [34].

Datura stramonium is a wild-growing plant in the family Solanaceae (Figure 2c). It is widely distributed and easily accessible, particularly in open locations such as grassland, roadsides, waste places, scrub vegetation and open forest [35]. The plant species has wide leaves and large fruit, with few but stout spines [35]. It is indigenous to the Americas and has been introduced into many tropical and subtropical regions [36], having been cultivated in Europe and South America [36,37]. It is a naturalized weed in many developing countries but is probably under-reported.

Gomphocarpus fruticosus (family: Apocynaceae) is a herbaceous, perennial, spindly shrub, often with watery or milky sap with milkweed that is fibrous [38,39]. The plant is a small shrub, approximately 2 m tall, with thin stems that are adorned with large round-shaped seed pods, oppositely connected leaves and yellowish-green blooms in dangling clusters. It is native to South and Tropical Africa (Figure 2d), but is also found in North Africa and the Arabian Peninsula as an exotic plant [40]. Although it has been naturalized and is widely distributed in South Africa, it is an imported weed native to several tropical African countries, such as Botswana, Eswatini, Kenya, Mozambique and Zimbabwe [34]. In South Africa, *G. fruticosus* is found across several provinces, including Mpumalanga, the North West, KwaZulu-Natal, Eastern Cape, Western Cape and Limpopo [41].

Rhoicissus tridentata is a deciduous shrubby creeper of the family Vitaceae, commonly known as wild grape, which has 700 to 800 species spread over 13 to 14 genera [30]. The plant has greenish-yellow flowers that bloom in late summer in tight clusters on a thickly hairy inflorescence, and when young, the trifoliate leaves are bright green, but as they develop, they become grey [42]. Studies have shown that the majority of these genera are thought to be mostly found in tropical and subtropical areas [43] (Figure 2e), ranging from Africa to Asia and the Pacific Islands [43].

Vachellia karroo is distinctly based on several morphological, anatomical and biochemical attributes, and belongs to the *Acacia* genus of the family Leguminosae [44]. *V. karroo* (previously *Acacia karroo*) is indigenous to southern Africa and known as "sweet thorn" [45,46], being found in different habitats from low to highveld areas. The plant has been used as traditional medicine by many local inhabitants of southern Africa and is a common woody species in South Africa. Its distribution ranged from the Western Cape in South Africa to neighbouring countries such as Zambia and Angola (Figure 2f). In tropical Africa, it is replaced by *Acacia seyal*, which has probably been renamed.

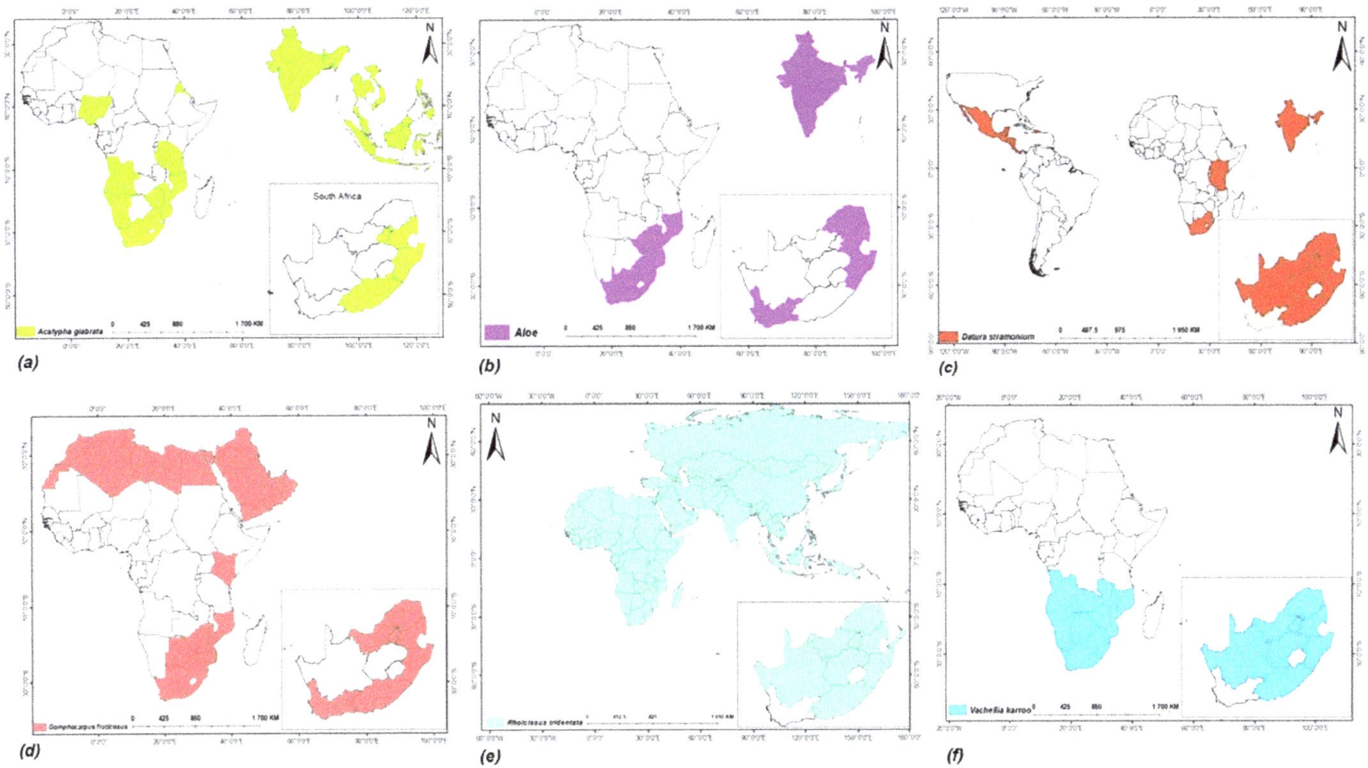

Figure 2. Distribution map of the selected medicinal plants: (**a**) *Acalypha glabrata*, (**b**) *Aloe maculata*, (**c**) *Datura stramonium*, (**d**) *Gomphocarpus fruticosus*, (**e**) *Rhoicissus tridentata* and (**f**) *Vachellia karroo*.

4. Ethnobotanical Uses of the Six Medicinal Plants

The six medicinal plants described in this review have a variety of medical values and are widely used to treat childhood diseases (Figure 3). However, they are characterized by limited exploration in terms of their potential, especially from health and economic perspectives. For instance, *A. glabrata* is known to treat skin-related ailments among children [40,47,48]. As indicated by Van Wyk et al [30], the leaves and bark of the *Acalypha* genus are traditionally used to treat and manage diverse ailments including skin rash [30]. The majority of the *Acalypha* genus species are utilized as medicinal herbs in Asia and Africa [29,30,48]. In South Africa, *Acalypha glabrata* is relatively scarce and traditional health practitioners often find it difficult to cultivate in their home gardens due to limited seed supply [48].

Figure 3. Six South African plants widely used for managing childhood diseases (**a**) *Acalypha glabrata* (**b**) *Aloe maculata*, (**c**) *Datura stramonium*, (**d**) *Gomphocarpus fruticosus*, (**e**) *Rhoicissus tridentata*, (**f**) *Vachellia karroo*.

Aloe maculata is one of 548 accepted *Aloe* species, of which at least one-third are documented as having some indigenous utilitarian value globally [31]. Over 350 multidisciplinary publications surveyed the uses and useful properties of *Aloe* spp [48,49]. Unsurprisingly, *A. maculata* is rich medicinal component and has diverse uses in folk medicine to treat various childhood diseases. *Aloe* flowers are traditionally infused and used as a laxative agent and for pain relief [50], while some species, such as *A. maculata*, are used in ethnoveterinary medicine [50]. Despite the large number of uses, only a few have been commercially exploited, such as 'Forever Living Products,' which are sold globally, with many South Africans being involved in the selling scheme. The commercial trade in *Aloe*-derived natural products is based mainly on two materials (excaudate and the gel) that are obtained from the leaves of certain *Aloe* species. These components are used as laxatives, applied topically for skin ailments or taken internally for digestive complaints and general well-being [50].

Datura stramonium is a commonly known traditional herb, with both poisonous and medicinal properties, which has had great pharmacological potential and a considerable utility in traditional medicine [36,50]. It contains highly toxic tropane alkaloids, including the pharmacologically active compounds atropine and scopolamine [50], the former being a highly toxic substance with a lethal dose in humans in the range of 100 mg [47]. Several studies revealed that the leaves of this plant are used for childhood ailments [5,20,51] and for curing various diseases among adults in Ayurvedic medicine [50]. Furthermore, *D. stramonium* leaves are used to relieve headaches, and the vapours of leaf infusions are used to alleviate the discomfort of rheumatism and gout, while asthma and bronchitis are eased by inhaling the smoke from a burning leaf [36]. In Europe, the plant is used to cure haemorrhoids by steaming the leaf over boiling water, and its juice is applied to the scalp to treat dandruff and hair loss, as well as wounds and sores [37].

The whole plant or leaves of *G. fruticosus* are used to treat various diseases including coughs, diabetes, tumours, skin diseases [4,5,38,48] and impotence [50,52] in humans as well as retained placenta in cattle [53]. It is regarded as a species of ethnobotanical interest in Kenya and Lesotho [54,55]. In tropical Africa, it is used to treat malaria, diabetes, asthma, bronchitis and cardiac palpitations. Furthermore, it is used for managing tumours, skin diseases, scabies and itching in the Arabian Peninsula [40]. In Egypt and South Africa, *G. fruticosus* is regarded as ethnoveterinary medicine [40,53]. Despite being known for its ethnobotanical uses and medicinal properties, the economic value of *G. fruticosus* remains unexplored and unknown (Table 1).

In South Africa, *R. tridentata* has been used to treat various childhood diseases, such as pulsating anterior fontanelle, stomach-ache and to stop vomiting [4,20,56]. In addition, it is a popular remedy for treating and managing diseases, such as broken bones, cuts, epilepsy, menorrhagia, sprained ankles and gastrointestinal disorders [4,30,57]. Evidence of its use as ethnoveterinary medicine has been indicated [58,59], and is listed as one of the 'rare' species (http://redlist.sanbi.org/species; accessed on 15 August 2021). There have been limited economic value chain studies of *R. tridentata* despite it being heavily traded and unsustainably harvested [26].

In humans, *V. karroo* has been used for many conditions such as gastrointestinal disorders, candidiasis and skin infections [4,60–64]. In addition, it is a well-known ethnoveterinary medicine [55,61,65]. Despite being highly utilized, there is limited knowledge regarding its commercial and economic impact, and product development (Table 1). The leaves of *V. karroo* are known to be a meal supplementation for natural pasture, with a fresher appearance and higher protein content. It is economically beneficial for beef farmers as it improves the meat nutritive value and freshness, and therefore increases consumer acceptance of the meat produced from these livestock [66].

Table 1. Most frequently cited plants used for treating and managing childhood diseases in South Africa.

Scientific Name and Family	Childhood Diseases/Conditions	Plant Parts and Preparation	* Conservation Status in South Africa	Reference	Region(s) of Availability	Signature Phytochemicals
Gomphocarpus fruticosus (L.) W.T. AitonApocynaceae	Convulsions, stomach ailments, chest ailments and milk latex used for the treatment of warts	Whole plant, leaves Infusion/poultice	LC	[4,5,48]	North Africa and the Arabian Peninsula [40]; Southern Africa and East Africa [34,41]	Glycosides [40]
Acalypha glabrata Thunb. Euphorbiaceae	Skin-related problems (e.g., skin rashes)	Stem, bark A sufficient amount of fresh stem is cut open and gently heated over an open fire before being applied to the skin.	LC	[67,68]	Indian Ocean islands, Asia and Africa [29,69]	-
Vachellia karroo Hayne Banfi and Galosso Syn: *Acacia karroo* var. transvaalensis (Burtt Davy) Burtt Davy Leguminosae/Fabaceae	Diarrhoea, dysentery, colic and convulsions	Whole plant Infusion and decoction	LC	[4,60,70]	Southern Africa [45,46]	Tannis and flavonoids [30]
Datura stramonium L. Solanaceae	Ailments such as mumps, earache (otitis)	Leaves, Maceration	IA	[5,20,51]	Europe and South America [36,37]	Tropane alkaloids [47,71]
Rhoicissus tridentata (L.f.) Wild and Drummond. Vitaceae	Pulsating anterior fontanelle, stomach-ache and stop vomiting	Roots, leaves Infusion/Maceration	R	[4,20,56,63]	Africa to Asia and the Pacific Islands [43]	Anthocyanidin [30]
Aloe maculata All. Xanthorrhoeaceae	Feverish colds, induce vomiting and weaning	Flower, Infusion	LC	[4,20,38]	Africa, India and other arid areas [31,32]	Acetylated mannans, polymannans, anthroquinones, anthraquinone glycosides, anthrones, lectins [72,73]

* Conservation status: LC = Least concern; IA = Invasive alien species and R = Rare. The botanical names were verified using the world flora online (http://www.worldfloraonline.org/) and conservation status were verified using the South African Red data list (http://redlist.sanbi.org/species, accessed on 15 August 2021).

5. Benefits and Economic Prospects of Selected Medicinal Plants

The African Union (AU) head of states has emphasized that African Traditional Medicine (ATM) is important to local livelihoods, as the mainstream medicines and other medical supplies can be costly, with many developing countries being unable to afford to purchase orthodox medicine to meet people's health needs [74]. Plants such as *A. glabrata*, *A. maculata*, *D. stramonium*, *G. fruticosus*, *R. tridentata* and *V. karoo* are important resources for livelihoods in local communities (Table 1). In particular, *Aloe* species have gained prominence as key natural resources that may contribute to the livelihoods of the poor. According to Idamokoro et al. [75], *V. karroo* together with other *Acacia* species are valuable feed resources that can be utilised by livestock in semi-arid areas. Furthermore, the foliage is a cheap and widespread non-conventional feed resource that can be utilised by low-input communal farmers to boost goat production in semi-arid areas of southern Africa. Feeding goats with *V. karroo* leaf meal can help to improve animal performance, including reducing the worm burden and increasing the growth rate, meat quality and colour. In South Africa, several indirect policy initiatives support the marketing of ATM by attempting to improve the standard of traditional medicine practice. Despite the policies and the legislation developed at the national government level, the promotion and protection of medicinal plants, and construction and regulation of a market for street traders remains low at grassroot levels [13]. Notwithstanding the different policies and initiatives of protecting and promoting indigenous knowledge, there is limited information specifically supporting the marketing of traditional medicinal plants used for childhood diseases at a national and grassroot levels.

Even though the selected plants (*A. glabrata*, *A. maculata*, *D. stramonium*, *G. fruticosus*, *R. tridentata* and *V. karoo*, Figure 3) play important roles in the South African traditional medicine, their potential economic benefit is often underestimated or unknown. Furthermore, medicinal plants may be important tools for addressing poverty in marginalized communities, by contributing to livelihoods, including food security, income, health and sustainable human development [1]. Specifically, *V. karroo*, *R. tridentate* and *G. fruticosus* serve as useful sources for different purposes and could be important income generators, especially for poor people in rural areas [76–78].

6. Challenges to Unlocking Potential of Selected Medicinal Plants

There are a number of challenges regarding unlocking the potential of these selected plants. One of the critical aspects is policy integration. Others include agro-processing and commercialization.

6.1. Policy Integration and Medicinal Plants

A number of acts have been passed in South African to regulate the use and conservation of medicinal plants. The National Environmental Management: Biodiversity Act 10 of 2004, which complies with the Convention on Biological Diversity [79], has spearheaded the Bioprospecting, Access and Benefit-Sharing Notice 329, the National Biodiversity Strategy and Action Plan, and the National Biodiversity Framework to foresee the sustainable use and conservation of biological resources [80]. The National Environmental Management: Protected Areas Act 57 of 2003 provides and facilitates access to natural resources where it has been prohibited before, but with a focus on the principle of sustainability [81]. In addition, the National Forests Act 84 of 1998 protects South African forests and trees through a licensing system.

In 2005, the government of South Africa promulgated the 2004 Traditional Health Practitioners (THP) Act, which recognized the practice of traditional medicine and established an Interim Traditional Health Practitioners Council [82]. According to the policy on African Traditional Medicine Notice 906 of 2008 [83], the South African Department of Health that ratified the AU plan adopted the 2007 Head of State Summit in Lusaka on the Plan of Action of the decade, that African traditional medicine should be institutionalized in the public health systems. The South African Constitution (1996) also allows conventional and indigenous or traditional medicine practices to coexist within the public health system.

In South Africa, the Department of Science and Technology (now referred to as the Department of Science and Innovation) was mandated to harmonize the policy and its implementation to ensure lucidity on issues related to the indigenous knowledge system (IKS) and the "*sui generis*" laws on the intellectual property housed by the Department of Trade and Industry (to facilitate registration of indigenous knowledge holders). In 2004, the policy stance saw the establishment of legislative bodies that include the National Office of IKS (NIKSO), which was tasked with recognizing and promoting indigenous knowledge, known as IKS Act 6 of 2019 [84]. South Africa has several laws, although some lack an integrated legislative framework or an appropriate regulatory body for traditional medicine. There are currently difficulties with the registration of African Traditional Medicine (ATM) and complementary medicines. South Africa still faces many constraints with the indigenous medicinal plant market, such as relevant policy, limited government recognition of the associated opportunities and poor implementation of the existing frameworks.

Most commercial acts in South Africa are silent regarding indigenous and traditional plants, despite acknowledging the use of traditional medicine and the practical constraints associated with low-income plants. The agricultural policy goals are represented in the National Development Plan (NDP) of 2030, which promotes crop commercialization under the prevailing food system [85]. However, it remains mainly silent about the cultivation of medicinal plants. Similarly, an observation was made in the National Food and Nutrition Security Policy that speaks about efficient agricultural output but remains silent on medicinal plants used for childhood or any other diseases [86–88].

Currently, the IKS Act 6 of 2019 reflects a developmental agenda and aims to develop, promote and preserve the local knowledge of different communities in South Africa. The Act aims at reducing the limitations facing local communities regarding their knowledge systems while creating opportunities for the inclusion of indigenous knowledge into local livelihoods. This includes the harmonization of the laws related to the NEMBA Act 10 of 2004, Access and Benefit-sharing Notice 329 of 2007, National Forests Act 84 of 1998, Land act, African Traditional Medicine Notice 906 of 2008 and the THP Act of 2004. However, there are inconsistencies in the policies related to indigenous knowledge and natural recourses as the divergent views held by policy makers and lawmakers on how to address problems of equity, nutrition security and poverty reduction. Medicinal plants have an explorable potential to grow the local economy. However, they are often excluded from the policy and strategic documents, except for the well-known ones.

The recognition of indigenous plants for medicinal purposes could be used to leverage their incorporation into policy documents and implementation processes and to improve local livelihoods [89]. Many South Africans living in rural areas who rely on ATM for healthcare are often caught in a vicious circle of malnutrition, poor access to healthcare and low standards of living, which are exacerbated by poor socio-economic conditions [90]. The effects of low income and poor wellbeing are exacerbated by people's inherent vulnerability to socio-economic factors [91]. Given the evidence presented above, it is pertinent to explore ways and paradigm shift to support vulnerable South African populations, and improve their healthcare status, socio-economics status and well-being through prioritizing indigenous knowledge of medicinal plants.

To create an enabling environment for the use of medicinal plants a different approach needs to occur at the local level, a multi-disciplinary research approach involving medicinal plants, agro-processing and other socio-scientific parameters is required to inform policymakers of the sustainable indigenous plant-based healthcare system that needs to be formalized in South Africa. This will require strengthening the capacity of South Africa to include both complementary and indigenous healthcare systems, in compliance with government commitments to adopt the proposed collaborative framework (Figure 4).

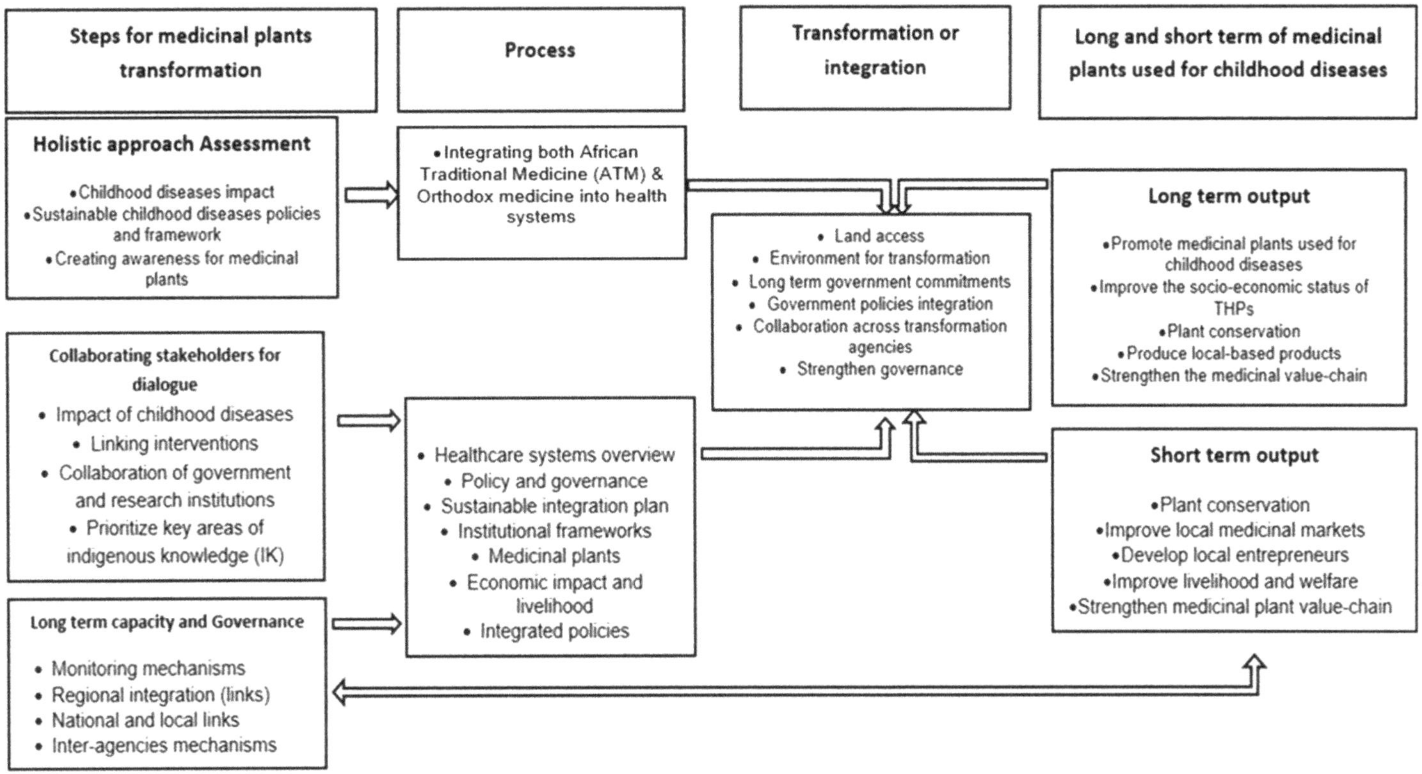

Figure 4. A proposed collaborative framework for medicinal plants used for managing childhood diseases.

Traditional medicine is the total sum of all knowledge, skills and practices based on the beliefs and experiences of the various indigenous cultures [92]. Developing the medicinal plant market could be an important determinant of short- and long-term economic growth in rural areas [8,93]. Creating an awareness of the use of medicinal plants and assessing the impact of different policies on knowledge accumulation is an important strategy for the growth of the local economy. Commercialization policy remains a controversial aspect to both indigenous knowledge and science that needs to be balanced. Policies are needed that (i) allow stronger incentives for economic agents to engage in knowledge-creation activities and (ii) encourage people and organizations to invest in research and development of new medicinal plant products. Research and development is essential for organizational innovation and competition to improve people's quality of life through their participation in small- and medium-sized enterprises, these being the backbone of the local economic sector [94,95].

Integrating both ATM and orthodox medicine into the healthcare systems should be prioritized by the government commitments and policies at all levels. This will be accomplished by engaging stakeholders for discourse, increasing the number of local enterprises to develop indigenous products and strengthening local pharmaceutical markets. Furthermore, to boost the medicinal plant value chain, monitoring methods, ranging from national to regional to local and inter-agency systems should be created to improve the local livelihood of local communities.

6.2. Agro-Processing of Medicinal Plants

In developed countries, agro-processing is regarded as one of the "sunrise sectors" for an economy, with significant development opportunities and socio-economic impacts for jobs and income generation [96]. To avoid spoilage or waste of produce, agro-industry converts agricultural materials into semi-finished goods [97]. Furthermore, agro-processing has a significant potential to prompt development in other sectors of the economy through the multiplier effect (Figure 4). It can create employment away from farms and processing units in sectors, such as transportation, distribution and retailing. Many developing countries continue to depend on importing orthodox medicines while having the potential for substantial production of local indigenous medicinal plants (Table 1). However, the economic value and agro-processing of medicinal plants used for childhood diseases remains poorly studied.

6.3. Commercialization of Medicinal Plants

The demand for plant-based products is increasing at the rate of 15 to 25% annually. Based on the estimate by the World Health Organization (WHO) [3], the demand for medicinal plants is likely to increase by more than USD 5 trillion by 2050 [98]. Despite an increase in the international trade of medicinal plants, their commercial benefits for childhood diseases remain speculative. South Africa exported 6497 tons of medicinal plant materials in 2015, compared to 1143 tons in 2014 [81]. According to the South African government, medicinal plants contributed approximately ZAR 82 million (USD 5,494,763) to the Gross Domestic Product (GDP) in 2014. Although this is currently a small industry, it has potential of making a greater contribution to the GDP, potentially increasing from ZAR 115 million (USD 7,000,874) to ZAR 150 million (USD 10,044,300). Despite the economic impact, the extent of medicinal plants' contribution to the social fabric remains largely unknown in many rural areas. Medicinal plants including *V. karroo* make a range of contributions to the livelihood and profitability of residents in rural areas. Despite the lack of documentation on agro-processing and commercial development, many medicinal plants continue to be traded for rural economic development [99].

Medicinal plants are sold in both urban and rural markets, and provide a substantial income to the small-scale vendors and traditional health practitioners for livelihood substance [100]. However, the adoption of large-scale cultivation of the above-mentioned medicinal plants remains low and unknown due to limited and undocumented economic returns on investments. These results are in line with a previous study, which estimated that only 10% of the South African plants have been completely commercialized, with their scientific evaluation often being unavailable [9]. This is attributed to a lack of organization and networking by the often rural collectors of medicinal plants from the wild and increased transactional costs, a constraint exacerbated by increasingly stringent health and safety requirements in the developed markets. Even basic pricing information on the indigenous plants we studied is scant. For example, a 2005 study surveyed prices of *R. tridentata* at ZAR 0.03/kg (USD 0.002/kg) of dry weight for the tuber and *A. glabrata* at ZAR 0.50/kg (USD 0.031/kg) of dry weight for the bark [78].

7. Prospects of Unlocking the Potential of Selected Medicinal Plants

7.1. Promoting Collaborative Research and Sustainability

The government and private stakeholders have to identify and recognize the importance of *A. glabrata*, *A. maculata*, *D. stramonium*, *G. fruticosus*, *R. tridentata* and *V. karroo* for childhood diseases in rural areas to conserve and protect the associated traditional knowledge for future generations [22]. Therefore, it is necessary to encourage the farmers, pharmaceuticals companies, institutions of higher learning (universities) and other stakeholders along the value chain in managing soil health, producing quality seeds and good marketing to undergo training and provide a policy framework that is beneficial to rural communities [101]. Developing a comprehensive and workable monitoring and evaluation framework for any kind of social system will be beneficial towards building linkages, interactions and resource flows between entities, responding and adapting to new demands and priorities in their environment [102]. Some strategies are not necessarily formal plans but rather abstract visions.

7.2. Technology Application and Product Development

Plant-derived pharmaceuticals are the next major commercial development in biotechnology because of the advantages that they offer in terms of production scale and value chain in both developing and developed countries. The provision of modern technological infrastructural facilities including state-of-the-art storage facilities and processing machinery is required in local communities where the plants originally occur and are harvested [9]. This will provide efficient agro-processing aligned with the drive for a modernized product and the preparation of monographs within the pharmacopeia production systems in Africa [9]. In addition, it will enhance the income and livelihood of households in rural communities and the durability of highly utilized plants such as *A. maculata* and *V. karroo*.

7.3. Promoting a Useful Value Chain

In South Africa, the medicinal plant trade proceeds with little government intervention or documentation. Both traders and traditional health practitioners benefit from this informal market. However, this market works in a grey area and policy makers are often unaware of the importance of medicinal plant value chains, which are essential towards commercialization. For instance, the *Aloe vera* value chain is distinguished from the informal nature of its upstream base, when compared to other *Aloe* species. The consistency of raw material supply as a result of a developing market is critical in marketing the abovementioned plants. Medicinal plants relative to their bio products have different value chains. To promote a successful value chain for the selected medicinal plants, the policymakers and local communities should consider several aspects such as awareness of environmental factors, increased research and opportunities related to promoting active collaboration with local communities, and mainstreaming gender-sensitive approaches [103]. Furthermore, different organizations and policy makers should be consulted and participate in the open process to promote the effectiveness and relevance of medicinal plants (Figure 5).

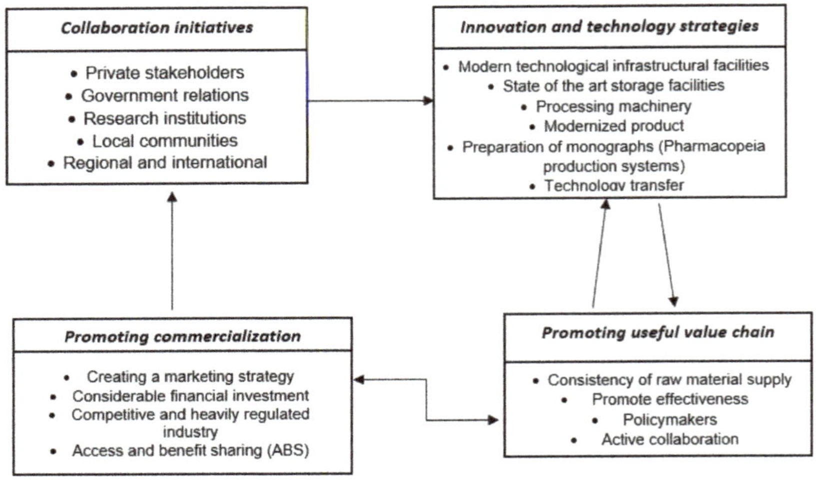

Figure 5. Prospects for unlocking the potentials of the selected medicinal plants widely used for managing childhood diseases in South Africa.

7.4. Potential of the Selected Plants for Commercialization

The need to enhance medicinal plant production and commercialization cannot be overemphasized. These medicines provide healthcare to the communities and can be further explored to generate income for the producers. Plants such as *A. maculata*, *D. stramonium* and *V. karroo* are utilised for primary health for humans and serve as ethnoveterinary medicine. Despite the increased global commerce in *Aloe* species, the benefits to developing countries, particularly growers and producers, remains limited [104]. As a result, in a competitive and heavily regulated industry that requires both specialist knowledge and a significant financial investment to become certified, there is a need to develop marketing, ecological and social strategies that mediate access to benefits from natural resources that are inherently dynamic in character. The realized value of medicinal plants to the rural livelihoods of individuals is also not static. For medicinal plant use, it is often easy to decrease or increase the quantity of *Aloe* used, as household social dynamics allow.

8. Conclusions and Recommendations

The limitations and challenges associated with existing orthodox medicine and healthcare services in rural communities justifies the need for viable alternatives for managing childhood diseases. In addition to being cheaper and more accessible, *A. glabrata*, *A. maculata*, *D. stramonium*, *G. fruticosus*, *R. tridentata* and *V. karroo* can be viable therapeutic options or substitutes if they are properly prepared and standardized. Nevertheless, a multidisciplinary research approach involving traditional health practitioners, paediatricians, chemists, pharmacists, botanists, farmers and policymakers remains crucial. In particular, the cultivation of the selected plants should be promoted under a participatory management action plan to stimulate the economy of the disadvantaged local communities. Furthermore, low-cost technologies are essential to promote sustainable use. Local communities and other stakeholders working in the medicinal plant sector should be properly registered and formally acknowledged so that strong regulations and norms may be adopted to protect community-based traditional knowledge and intellectual property rights, ensuring equitable benefit sharing.

Author Contributions: Conceptualization, P.T.N., A.O.O., W.O.-M. and A.O.A.; Methodology P.T.N.; Resources, W.O.-M. and A.O.A.; Writing—Original Draft Preparation, P.T.N.; Writing—Review and Editing, A.O.O., W.O.-M. and A.O.A.; Supervision, A.O.O., W.O.-M. and A.O.A.; Project Administration, W.O.-M. and A.O.A.; Funding Acquisition, P.T.N.; W.O.-M. and A.O.A. All authors have read and agreed to the published version of the manuscript.

Funding: PTN is grateful for a doctoral bursary from the National Research Foundation, Pretoria, South Africa (Grant UID: 138298). The North-West University and the University of Mpumalanga provided additional financial support. The APC was funded by the Faculty of Natural and Agricultural Sciences, North-West University, South Africa.

Institutional Review Board Statement: This research was approved by the North-West University Health Research Ethics Committee (NWU-HREC), approval number NWU-00485-20-A1.

Informed Consent Statement: Not applicable.

Data Availability Statement: All data used for this study is included in the article.

Acknowledgments: We appreciate the South African National Biodiversity Institute (SANBI) Pretoria, for providing the photos of the selected plants and Boitumelo Selogatwe for assisting with creating the maps.

Conflicts of Interest: The authors declare that there is no conflict of interest. The opinions and conclusions/recommendations in this study are based on the findings of the authors; therefore, the funders (NRF, NWU, UMP) accept no liability whatsoever in this regard. The final version of the manuscript has been read and approved by all authors.

References

1. De Villiers, F.P.R.; Ledwaba, M.J.P. Traditional healers and paediatric care. *S. Afr. Med. J.* **2003**, *2003*, 664–665.
2. Sen, S.; Chakraborty, R. Revival, modernization and integration of Indian traditional herbal medicine in clinical practice: Importance, challenges and future. *J. Tradit. Complement. Med.* **2017**, *7*, 234–244. [CrossRef] [PubMed]
3. World Health Organization (WHO). *World Health Organization Traditional Medicine Strategy 2014–2023*; WHO: Geneva, Switzerland, 2013; pp. 1–78.
4. Hutchings, A.; Scott, A.H.; Lewis, G.; Cunningham, A.B. *Zulu Medicinal Plants: An Inventory*; University of Natal Press: Pietermaritzburg, South Africa, 1996.
5. Hulley, I.M.; Van Wyk, B.-E. Quantitative medicinal ethnobotany of Kannaland (Western Little Karoo, South Africa): Non-homogeneity amongst villages. *S. Afr. J. Bot.* **2019**, *122*, 225–265. [CrossRef]
6. McGaw, L.J.; Famuyide, I.M.; Khunoana, E.T.; Aremu, A.O. Ethnoveterinary botanical medicine in South Africa: A review of research from the last decade (2009 to 2019). *J. Ethnopharmacol.* **2020**, *257*, 112864. [CrossRef] [PubMed]
7. Semenya, S.; Potgieter, M.; Erasmus, L. Ethnobotanical survey of medicinal plants used by Bapedi healers to treat diabetes mellitus in the Limpopo Province, South Africa. *J. Ethnopharmacol.* **2012**, *141*, 440–445. [CrossRef] [PubMed]
8. Street, R.; Prinsloo, G. Commercially important medicinal plants of South Africa: A review. *J. Chem.* **2012**, *2013*, 205048. [CrossRef]
9. Van Wyk, B.E. A review of commercially important African medicinal plants. *J. Ethnopharmacol.* **2015**, *176*, 118–134. [CrossRef]
10. Omotayo, A.O.; Ndhlovu, P.T.; Tshwene, S.C.; Aremu, A.O. Utilization pattern of indigenous and naturalized plants among some selected rural households of North West Province, South Africa. *Plants* **2020**, *9*, 953. [CrossRef]
11. Ajesh, T.P.; Abdulla Naseef, S.A.; Kumuthakalavalli, R. Ethnobotanical documentation of wild edible fruits used by Muthuvan Tribes of Idukki, Kerala- India. *J. Pharm. Biol. Sci.* **2012**, *3*, 479–487.
12. Omotayo, A.O.; Aremu, O.A. Evaluation of factors influencing the inclusion of indigenous plants for food security among rural households in the North West Province of South Africa. *Sustainability* **2020**, *12*, 9562. [CrossRef]
13. Mander, M.; Ntuli, L.; Diederichs, N.; Mavundla, K. Economics of the traditional medicine trade in South Africa care delivery. *S. Afr. Health Rev.* **2007**, *2007*, 189–196.
14. Williams, V.L.; Witkowski, E.T.F.; Balkwill, K. Volume and financial value of species traded in the medicinal plant markets of Gauteng, South Africa. *Int. J. Sustain. Dev. World Ecol.* **2007**, *14*, 584–603. [CrossRef]
15. Stats, S.A. *Living Conditions of Households in South Africa*; Statistics South Africa: Pretoria, South Africa, 2015; pp. 1–232.
16. Masondo, N.A.; Makunga, N.P. Advancement of analytical techniques in some South African commercialized medicinal plants: Current and future perspectives. *S. Afr. J. Bot.* **2019**, *126*, 40–57. [CrossRef]
17. Asong, J.A.; Ndhlovu, P.T.; Khosana, N.S.; Aremu, A.O.; Otang-Mbeng, W. Medicinal plants used for skin-related diseases among the Batswanas in Ngaka Modiri Molema District Municipality, South Africa. *S. Afr. J. Bot.* **2019**, *126*, 11–20. [CrossRef]
18. Magwede, K.; Van Wyk, B.E.; Van Wyk, A.E. An inventory of Vhavenḓa useful plants. *S. Afr. J. Bot.* **2019**, *122*, 57–89. [CrossRef]
19. Moffett, R.O. *Basotho Medicinal Plants–Meriana ya Dimela Tsa Basotho*; Sun Press: Stellenbosch, South Africa, 2016.
20. Mhlongo, L.S.; Van Wyk, B.-E. Zulu medicinal ethnobotany: New records from the Amandawe area of KwaZulu-Natal, South Africa. *S. Afr. J. Bot.* **2019**, *122*, 266–290. [CrossRef]

21. Botha, J.; Witkowski, E.T.F.; Shackleton, C.M. Market profiles and trade in medicinal plants in the Lowveld, South Africa. *Environ. Conserv.* **2004**, *31*, 38–46. [CrossRef]
22. Ndhlovu, P.T.; Omotayo, A.O.; Otang-Mbeng, W.; Aremu, A.O. Ethnobotanical review of plants used for the management and treatment of childhood diseases and well-being in South Africa. *S. Afr. J. Bot.* **2021**, *137*, 197–215. [CrossRef]
23. Van Wyk, B.-E. The potential of South African plants in the development of new food and beverage products. *S. Afr. J. Bot.* **2011**, *77*, 857–868. [CrossRef]
24. UNICEF. *Levels and Trends in Child Mortality: Report 2019. Estimates Developed by the UN Inter-Agency Group for Child Mortality Estimation*; United Nation's Children Fund: New York, NY, USA, 2019.
25. Lake, L.; Shung-King, M.; Hendricks, M.; Heywood, M.; Nannan, N.; Laubscher, R.; Bradshaw, D.; Mathews, C.; Goga, A.; Ramraj, T.; et al. Prioritising child and adolescent health: A human rights imperative. *S. Afr. Child Gauge* **2019**, *2*, 31–62.
26. Dold, A.P.; Cocks, M.L. The trade in medicinal plants in the Eastern Cape Province, South Africa. *S. Afr. J. Sci.* **2002**, *98*, 589–597.
27. Moyo, M.; Aremu, A.O.; Van Staden, J. Medicinal plants: An invaluable, dwindling resource in sub-Saharan Africa. *J. Ethnopharmacol.* **2015**, *174*, 595–606. [CrossRef]
28. Shamseer, L.; Moher, D.; Clarke, V.; Ghersi, D.; Liberati, A.; Petticrew, M.; Shekelle, P.; Stewart, K.M.; the PRISMA-P Group. Preferred reporting items for systematic review and meta-analysis protocols (PRISMA-P) 2015: Elaboration and explanation. *BMJ* **2015**, *349*, g7647. [CrossRef] [PubMed]
29. Seebaluck, R.; Gurib-Fakim, A.; Mahomoodally, F. Medicinal plants from the genus *Acalypha* (Euphorbiaceae)–A review of their ethnopharmacology and phytochemistry. *J. Ethnopharmacol.* **2015**, *159*, 137–157. [CrossRef] [PubMed]
30. Van Wyk, B.-E.; Van Oudtshoorn, B.; Gericke, N. *Medicinal Plants of South Africa*; Briza: Pretotia, South Africa, 1997.
31. Grace, O.M.; Simmonds, M.S.J.; Smith, G.F.; Van Wyk, A.E. Documented utility and biocultural value of *Aloe* L. (Asphodelaceae): A Review. *Econ. Bot.* **2009**, *63*, 167–178. [CrossRef]
32. Corredor-Prado, J.; Conti, D.D.; Sezerino, A.; Guerra, M.; Orth, A. Reproductive biology and absence of fruiting of *Aloe saponaria* (Aiton) Haw.(Xanthorrhoeaceae) outside its place of origin. *Rev. Bras. Plantas Med.* **2015**, *17*, 713–721. [CrossRef]
33. Smith, G.; Figueiredo, E.; Klopper, R.; Crouch, N. *Aloe maculata* All. (Asphodelaceae) in the Free State Province, South Africa, and resurrection of 'var. *ficksburgensis*'. *Bradleya* **2012**, *30*, 13–18. [CrossRef]
34. Van Wyk, B.; Van Wyk, P. *Field Guide to Trees of Southern Africa*; Struik: Cape Town, South Africa, 1997.
35. Gaire, B.P.; Subedi, L. A review on the pharmacological and toxicological aspects of *Datura stramonium* L. *J. Integr. Med.* **2013**, *11*, 73–79. [CrossRef]
36. Soni, P.; Siddiqui, A.A.; Dwivedi, J.; Soni, V. Pharmacological properties of *Datura stramonium* L. as a potential medicinal tree: An overview. *Asian Pac. J. Trop. Biomed.* **2012**, *2*, 1002–1008. [CrossRef]
37. Shagal, M.H.; Modibbo, U.U.; Liman, A.B. Pharmacological justification for the ethnomedical use of *Datura stramonium* stem-bark extract in treatment of diseases caused by some pathogenic bacteria. *Int. Res. Pharm. Pharmacol.* **2012**, *2*, 16–19.
38. Watt, J.M.; Breyer Brandwijk, M. *Medicinal and Poisonous Plants of Southern and Eastern Africa*; E. & S. Livingstone Ltd.: Edinburgh, UK; London, UK, 1962.
39. Van Wyk, B.-E.; Gericke, N. *People's Plants: A Guide to the Useful Plants of Southern Africa*; Briza Publications: Pretoria, South Africa, 2000.
40. Marzouk, A.M.; Osman, S.m.; Gohar, A.A. A newpregnane glycoside from *Gomphocarpus fruticosus* growing in Egypt. *Nat. Prod. Res.* **2016**, *30*, 1060–1067. [CrossRef]
41. Chivandi, E.; Mukonowenzou, N.; Berliner, D. The coastal red-milkwood *(Mimusops caffra)* seed: Proximate, mineral, amino acid and fatty acid composition. *S. Afr. J. Bot.* **2016**, *102*, 137–141. [CrossRef]
42. Urton, N.R.; Olivier, M.C.; Robertson, B.L. The taxonomy of the *Rhoicissus tridentata* (Vitaceae) complex in southern Africa. *S. Afr. J. Bot.* **1986**, *52*, 389–396. [CrossRef]
43. Mukundi, M.J.; Mwaniki, N.E.; Ngugi, M.P.; Njagi, J.M.; Agyirifo, S.D.; Gathumbi, K.P.; Muchugi, N.A. In vivo anti-diabetic effects of aqueous leaf extracts of *Rhoicissus tridentata* in alloxan induced diabetic mice. *J. Dev. Drugs* **2015**, *4*. [CrossRef]
44. Luckow, M.; Miller, J.T.; Murphy, D.J.; Livshultz, T. A phylogenetic analysis of the Mimosoideae (Leguminosae) based on chloroplast DNA sequence data. *Adv. Legume Syst. Part* **2003**, *10*, 197–220.
45. Maroyi, A. Traditional use of medicinal plants in south-central Zimbabwe: Review and perspectives. *J. Ethnobiol. Ethnomed.* **2013**, *9*, 31. [CrossRef] [PubMed]
46. Taylor, C.L.; Barker, N.P. Species limits in *Vachellia (Acacia) karroo* (Mimosoideae: Leguminoseae): Evidence from automated ISSR DNA "fingerprinting". *S. Afr. J. Bot.* **2012**, *83*, 36–43. [CrossRef]
47. Van Wyk, B.-E.; Van Heerden, F.; Van Oudtshoorn, B. *Poisonous Plants of South Africa*; Briza: Pretoria, South Africa, 2005.
48. Amoo, S.O.; Aremu, A.O.; Van Staden, J. Unraveling the medicinal potential of South African *Aloe* species. *J. Ethnopharmacol.* **2014**, *153*, 19–41. [CrossRef] [PubMed]
49. Grace, O. Current perspectives on the economic botany of the genus Aloe L.(Xanthorrhoeaceae). *S. Afr. J. Bot.* **2011**, *77*, 980–987. [CrossRef]
50. Nortje, J.M.; Van Wyk, B.E. Medicinal plants of the Kamiesberg, Namaqualand, South Africa. *J. Ethnopharmacol.* **2015**, *171*, 205–222. [CrossRef]
51. Das, S.; Kumar, P.; Basu, S. Phytoconstituents and therapeutic potentials of *Datura stramonium* Linn. *J. Drug Deliv. Ther.* **2012**, *2*, 4–7. [CrossRef]

52. Philander, A.L. An ethnobotany of Western Cape Rasta bush medicine. *J. Ethnopharmacol.* **2011**, *138*, 578–594. [CrossRef] [PubMed]
53. Semenya, S.S.; Potgieter, M.J. Ethnobotanical survey of medicinal plants used by Bapedi traditional healers to treat erectile dysfunction in the Limpopo Province, South Africa. *J. Med. Plants Res.* **2013**, *7*, 349–357.
54. Moichwanetse, B.I.; Ndhlovu, P.T.; Sedupane, G.; Aremu, A.O. Ethno-veterinary plants used for the treatment of retained placenta and associated diseases in cattle among Dinokana communities, North West Province, South Africa. *S. Afr. J. Bot.* **2020**, *132*, 108–116. [CrossRef]
55. Omino, E.A.; Kokwaro, J.O. Ethnobotany of Apocynaceae species in Kenya. *J. Ethnopharmacol.* **1993**, *40*, 167–180. [CrossRef]
56. Mabaleha, M.B.; Zietsman, P.C.; Wilhelm, A.; Bonnet, S.L. Ethnobotanical survey of medicinal plants used to treat mental illnesses in the Berea, Leribe, and Maseru Districts of Lesotho. *Nat. Prod. Commun.* **2019**, *14*, 1934578X19864215. [CrossRef]
57. Arnold, H.-J.; Gulumian, M. Pharmacopoeia of traditional medicine in Venda. *J. Ethnopharmacol.* **1984**, *12*, 35–74. [CrossRef]
58. Lin, J.; Opoku, A.; Geheeb-Keller, M.; Hutchings, A.; Terblanche, S.; Jäger, A.K.; Van Staden, J. Preliminary screening of some traditional zulu medicinal plants for anti-inflammatory and anti-microbial activities. *J. Ethnopharmacol.* **1999**, *68*, 267–274. [CrossRef]
59. Van der Merwe, D.; Swan, G.E.; Botha, C.J. Use of ethnoveterinary medicinal plants in cattle by Setswana-speaking people in the Madikwe area of the North West Province of South Africa. *J. S. Afr. Vet. Assoc.* **2001**, *72*, 189–196. [CrossRef]
60. Dold, A.P.; Cocks, M.L. Traditional veterinary medicine in the Alice district of the Eastern Cape Province, South Africa: Research in action. *S. Afr. J. Sci.* **2001**, *97*, 375–379.
61. Corrigan, B.M.; VanWyk, B.-E.; Geldenhuys, C.J.; Jardine, J.M. Ethnobotanical plant uses in the KwaNibela Peninsula, St Lucia, South Africa. *S. Afr. J. Bot.* **2011**, *77*, 346–359. [CrossRef]
62. Maroyi, A. Diversity of use and local knowledge of wild and cultivated plants in the Eastern Cape province, South Africa. *J. Ethnobiol. Ethnomed.* **2017**, *13*, 43. [CrossRef] [PubMed]
63. Masevhe, N.A.; McGaw, L.J.; Eloff, J.N. The traditional use of plants to manage candidiasis and related infections in Venda, South Africa. *J. Ethnopharmacol.* **2015**, *168*, 364–372. [CrossRef]
64. Mabogo, D.E.N. *The Ethnobotany of the Vhavenda*; University of Pretoria: Pretoria, South Africa, 1990.
65. Mulaudzi, R.B.; Ndhlala, A.R.; Kulkarni, M.G.; Finnie, J.F.; Staden, J.V. Anti-inflammatory and mutagenic evaluation of medicinal plants used by Venda people against venereal and related diseases. *J. Ethnopharmacol.* **2013**, *146*, 173–179. [CrossRef] [PubMed]
66. Chinsembu, K.C.; Negumbo, J.; Likando, M.; Mbangu, A. An ethnobotanical study of medicinal plants used to treat livestock diseases in Onayena and Katima Mulilo, Namibia. *S. Afr. J. Bot.* **2014**, *94*, 101–107. [CrossRef]
67. Tripathi, R.; Chung, Y.B.; Deering, K.; Saracini, N.; Willoughby, R.; Wills, O.; Mikhail, M.; Warburton, H.; Jayasinghe, D.; Rafanomezana, J. What Works for Women: Proven approaches for empowering women smallholders and achieving food security. *Oxfam Policy Pract. Agric. Food Land* **2012**, *12*, 113–140.
68. Bhat, R.B. Plants of Xhosa people in the Transkei region of Eastern Cape (South Africa) with major pharmacological and therapeutic properties. *J. Med. Plants Res.* **2013**, *7*, 1474–1480.
69. Dlisani, P.B.; Bhat, R.B. Traditional health practices in Transkei with special emphasis on maternal and child health. *Pharm. Biol.* **1999**, *37*, 32–36. [CrossRef]
70. Zahidin, N.S.; Saidin, S.; Zulkifli, R.M.; Muhamad, I.I.; Ya'akob, H.; Nur, H. A review of *Acalypha indica* L. (Euphorbiaceae) as traditional medicinal plant and its therapeutic potential. *J. Ethnopharmacol.* **2017**, *207*, 146–173. [CrossRef] [PubMed]
71. Nzue, M.M.; Pierre, A. *Use and Conservation Status of Medicinal Plants in the Cape Peninsula, Western Cape Province of South Africa*; University of Stellenbosch: Stellenbosch, South Africa, 2009.
72. Sharma, M.C.; Sharma, S. Phytochemical, preliminary pharmacognostical and antimicrobial evaluation of combined crude aqueous extract. *Int. J. Microbiol.* **2010**, *1*, 166–170.
73. Cock, I. *The Genus Aloe: Phytochemistry and Therapeutic Uses Including Treatments for Gastrointestinal Conditions and Chronic Inflammation*; Springer: Basel, Switzerland, 2015; Volume 70, pp. 179–235.
74. Mapiye, C.; Chimonyo, M.; Marufu, M.; Dzama, K. Utility of Acacia karroo for beef production in Southern African smallholder farming systems: A review. *Anim. Feed. Sci. Technol.* **2011**, *164*, 135–146. [CrossRef]
75. Kayombo, E. Traditional methods of protecting the infant and child illness/disease among the Wazigua at Mvomero Ward, Morogoro, Region, Tanzania. *J. Altern. Integr. Med.* **2013**, *2*, 1000103. [CrossRef]
76. Idamokoro, E.M.; Masika, P.J.; Muchenje, V. *Vachellia karroo* leaf meal: A promising non-conventional feed resource for improving goat production in low-input farming systems of Southern Africa. *Afr. J. Range Forage Sci.* **2016**, *33*, 141–153. [CrossRef]
77. Randrianarivony, T.N.; Ramarosandratana, A.V.; Andriamihajarivo, T.H.; Rakotoarivony, F.; Jeannoda, V.H.; Randrianasolo, A.; Bussmann, R.W. The most used medicinal plants by communities in Mahaboboka, Amboronabo, Mikoboka, Southwestern Madagascar. *J. Ethnobiol. Ethnomed.* **2017**, *13*, 19. [CrossRef]
78. Ahmad, M.; Zafar, M.; Shahzadi, N.; Yaseen, G.; Murphey, T.M.; Sultana, S. Ethnobotanical importance of medicinal plants traded in herbal markets of Rawalpindi- Pakistan. *J. Herb. Med.* **2018**, *11*, 78–89. [CrossRef]
79. Keirungi, J.; Fabricius, C. Selecting medicinal plants for cultivation at Nqabara on the Eastern Cape Wild Coast, South Africa. *S. Afr. J. Sci.* **2005**, *101*, 497–501.
80. Potgieter, L.; Irlich, U.M.; Gaertner, M.; Stafford, L. Recommendations for municipalities to become compliant with national legislation on biological invasions. *Bothalia-Afr. Biodivers. Conserv.* **2017**, *47*, 1–11.

81. Van Wyk, A.S.; Prinsloo, G. Medicinal plant harvesting, sustainability and cultivation in South Africa. *Biol. Conserv.* **2018**, *227*, 335–342. [CrossRef]
82. Constitution of Republic of South Africa. National Environmental Management: Protected Areas Amendment Act (Act No. 21 of 2014). *Gov. Gaz. Pretoria S. Afr.* **2014**, *588*, 27274.
83. Constitution of Republic of South Africa. *Republic of South Africa Traditional Health Practitioners Act*; Government Gazzatte South Africa: Cape Town, South Africa, 2007.
84. Constitution of the Republic of South Africa. *Notice: Draft Policy on African Traditional Medicine for South Africa*; South Africa Department of Communications Pretoria: South Africa, 2008; Volume Notice 906 of 2008; p. 49.
85. Constitution of the Republic of South Africa. *Protection Promotion Development and Management of Indigenous Knowledge Act of 2019*; Department of Communications: Cape Town, South Africa, 2019; Volume 42647, pp. 1–28.
86. NPC. *National Development Plan 2030: Our Future-Make It Work*; Sherino Printers: Pretoria, South Africa, 2012.
87. Cotula, L. *Land Grab or Development Opportunity?: Agricultural Investment and International Land Deals in Africa*; Iied: London, UK, 2009.
88. Cooksey, B. The Comprehensive Africa Agriculture Development Programme (CAADP) and Agricultural Policies in Tanzania: Going with or against the Grain. In Proceedings of the Future Agricultures. Available online: http://www.future-agricultures.org/pp-conference-papers/the-political-economy-of-caadp-engagement/1646-caadp-and-agricultural-policies-in-tanzania-going-with-or-against-the-grain (accessed on 1 October 2021).
89. Van Wyk, B.-E.; Smith, G. *Guide to the Aloes of South Africa*; Briza Publications: Pretoria, South Africa, 1996.
90. Herforth, A. Promotion of Traditional African Vegetables in Kenya and Tanzania: A Case Study of an Intervention Representing Emerging Imperatives in Global Nutrition. Ph.D. Thesis, Cornell University, New York, NY, USA, 2010.
91. Khumalo, N.Z. An Assessment of the Contribution of Peri-Urban Agriculture on Household Food Security in Tongaat, eThekwini Municipality. Master's Thesis, University of Zululand, KwaDlangezwa, South Africa, 2018.
92. Mabhaudhi, T.; Chibarabada, T.P.; Chimonyo, V.G.P.; Murugani, V.G.; Pereira, L.M.; Sobratee, N.; Govender, L.; Slotow, R.; Modi, A.T. Mainstreaming underutilized indigenous and traditional crops into food systems: A South African perspective. *Sustainability* **2019**, *11*, 172. [CrossRef]
93. Tesfahuneygn, G.; Gebreegziabher, G. Medicinal plants used in traditional medicine by Ethiopians: A review article. *J. Genet. Genet. Eng.* **2019**, *2*, 18–21.
94. Sofowora, A.; Ogunbodede, E.; Onayade, A. The role and place of medicinal plants in the strategies for disease prevention. *Afr. J. Tradit. Complement. Altern. Med.* **2013**, *10*, 210–229. [CrossRef]
95. Division, N.T.E.P. *Economic Transformation, Inclusive Growth, and Competitiveness: A Contribution towards a Growth Agenda for the South African Economy*; National Treasury: Pretoria, South Africa, 2019.
96. Omotayo, O.A.; Ndhlovu, P.T.; Olagunju, K.O.; Tshwane, S.C.; Aremu, A.O. Determinants of household's income and willingness to pay for indigenous plants in North West Province, South Africa: A Two-stage Heckman Approach. *Sustainability* **2020**, *13*, 5458. [CrossRef]
97. Sharma, K.; Pathania, M.; Lal, H. Value chain analysis and financial viability of agro-processing industries in Himachal Pradesh. *Agric. Econ. Res. Rev.* **2010**, *23*, 515–522.
98. Shaheen, F.A.; Wani, S.A.; Kubrevi, S.S. The landscape of agro-processing industries in Jammu & Kashmir. *Agric. Econ. Res. Rev.* **2019**, *32*, 175–182.
99. Kala, C.P.; Dhyani, P.P.; Sajwan, B.S. Developing the medicinal plants sector in northern India: Challenges and opportunities. *J. Ethnobiol. Ethnomed.* **2006**, *2*, 32. [CrossRef]
100. Paremoer, T. *Regional Value Chains: Exploring Linkages and Opportunities in the Agro-Processing Sector across Five SADC Countries*; Centre for Competition, Regulation & Economic Development (CCRED): New York, NY, USA, 2018.
101. Zhang, L.; Zhuang, H.; Zhang, Y.; Wang, L.; Zhang, Y.; Geng, Y.; Gou, Y.; Pei, S.; Wang, Y. Plants for health: An ethnobotanical 25-year repeat survey of traditional medicine sold in a major marketplace in North-west Yunnan, China. *J. Ethnopharmacol.* **2018**, *224*, 119–125. [CrossRef] [PubMed]
102. Suneetha, M.; Chandrakanth, M. Establishing a multi-stakeholder value index in medicinal plants an economic study on selected plants in Kerala and Tamilnadu States of India. *Ecol. Econ.* **2006**, *60*, 36–48. [CrossRef]
103. Awotedu, B.; Akala, A.; Omolola, T.; Owoeye, E.; Olaoti-Laaro, S.; Ogunsiji, A. Some underutilized indigenous timber trees with high medicinal values: Its challenges and prospects. *World J. Adv. Res. Rev.* **2021**, *10*, 161–168. [CrossRef]
104. Kayombo, E.; Mahunnah, R.; Uiso, F. Prospects and challenges of medicinal plants conservation and traditional medicine in Tanzania. *Anthropol. Educ. Q.* **2013**, *1*, 108. [CrossRef]

sustainability

MDPI

Article

Determinants of Household Income and Willingness to Pay for Indigenous Plants in North West Province, South Africa: A Two-Stage Heckman Approach

Abiodun Olusola Omotayo [1,*], Peter Tshepiso Ndhlovu [2], Seleke Christopher Tshwene [3], Kehinde Oluseyi Olagunju [4] and Adeyemi Oladapo Aremu [1,2]

1 Food Security and Safety Niche Area, Faculty of Natural and Agricultural Sciences, North-West University, Private Bag X2046, Mmabatho 2790, South Africa; Oladapo.Aremu@nwu.ac.za

2 Indigenous Knowledge Systems (IKS) Centre, Faculty of Natural and Agricultural Sciences, North-West University, Private Bag X2046, Mmabatho 2790, South Africa; Tshepiso.Ndhlovu@ump.ac.za

3 Department of Agricultural Economics and Extension, Faculty of Natural and Agricultural Sciences, North-West University, Private Bag X2046, Mmabatho 2790, South Africa; 16969669@nwu.ac.za

4 Economics Research Branch, Agrifood and Biosciences Institute (AFBI), 18a Newforge Lane, Belfast BT9 5PX, UK; kehinde-oluseyi.olagunju@afbini.gov.uk

* Correspondence: 25301284@nwu.ac.za; Tel.: +27-18-389-2573

Abstract: Using a cross-sectional dataset, this study examines the factors influencing the income and willingness to pay for indigenous plants by rural households in the North West Province of South Africa. The method of data analysis was based on a two-stage Heckman model. Based on empirical data, the majority (93%) of the households are willing to pay for the indigenous plants. Furthermore, factors such as the age of the household's head, marital status, size of households, financial returns and economic benefit of indigenous plants significantly influenced income and willingness to pay for indigenous plants by the households. Thus, indigenous plants have the potential to penetrate local and international markets and can be used to improve the economies, livelihood, and food security status of rural households in South Africa. Indigenous plant cultivation can increase agro-food system species diversity while conserving plant species indigenous to this area of South Africa. Encouraging both consumption and production of indigenous plants can also help diversify local economies and communities.

Keywords: consumers' food preference; contingent valuation; purchasing power; market prospects; rural economy; sustainability

Citation: Omotayo, A.O.; Ndhlovu, P.T.; Tshwene, S.C.; Olagunju, K.O.; Aremu, A.O. Determinants of Household Income and Willingness to Pay for Indigenous Plants in North West Province, South Africa: A Two-Stage Heckman Approach. *Sustainability* **2021**, *13*, 5458. https://doi.org/10.3390/su13105458

Academic Editor: Aaron K. Hoshide

Received: 10 March 2021
Accepted: 6 May 2021
Published: 13 May 2021

1. Introduction

Despite the acceleration in the state of technological advancement, the global food system is far behind the exponential increase in population thereby failing to meet the basic food needs and requirements of humans [1,2]. These challenges, as well as the on-going Covid-19 pandemic, have exerted multidimensional shocks on the global food system, thereby making achievements of Sustainable Development targets related to food security and hunger by 2030 doubtful [3,4]. The current global food system remains a diverse mixture of localized and industrialized systems of interconnected food chains [5,6]. However, most of these systems are centered on few exotic crop varieties with little role played by indigenous plants. Food systems supported through indigenous plants are easily sourced within a particular economy and, therefore, are potentially less vulnerable to economic, climatic, as well as the on-going Covid-19 pandemic shocks. The vulnerability related to dependence of a handful of exotic crops is particularly higher in the developing economies which are currently experiencing the effects, as well as of climate change and the unprecedented adverse effects of the Covid-19 pandemic [7–9].

Although South Africa has been embedded in the global agro-industrial food system for decades [1], its rural communities still suffer from poor nutritional status and extreme poverty. This indicates that such interconnectedness with global systems centered on exotic food crops may not solve the fundamental issues of malnutrition and hunger. Even though indigenous plants remain affordable sources of several micronutrients, the knowledge and use of these plants are currently low [10,11]. These plants are regarded as minor crops with little marketing and scientific attention [12]. Despite consolidation of the agro-industrial food system, South African food prices remain too high for a significant portion of the population, who consequently cannot afford to purchase adequate food, thereby leaving 21.3% of the population with low access to food [13].

The concerns regarding environmental degradation, loss of biodiversity, and vulnerability to climate change have prompted a call to rethink current configuration of the South African food system [14]. A focus on re-invigorating economic values and willingness to pay for indigenous plants, and bringing these to the market, has been suggested as an entry point for improving diets and making nutrition more sustainable [14,15]. According to some researchers [16–18], indigenous plants would fetch a higher price at supermarkets than in an informal market. One possible reason is that supermarkets are patronized by more nutritionally aware consumers with higher incomes who are more conscious of their health and are willing to pay higher prices to obtain these nutritional benefits [18,19]. It is also possible that the atmosphere in supermarkets influences prices and willingness to pay a premium price as they use modern retail technology in terms of storage, display, and packaging [16,20,21]. There is a scarcity of studies on the economics of indigenous plants, especially in South Africa. Even though there is no formal market for indigenous plants in the North West province, some anecdotal evidence suggests that there is a rising interest in purchasing such plants among households. The need to identify potential and create awareness for mainstreaming indigenous plants into the South African food system cannot be overemphasized.

Given this backdrop, the present study explored the knowledge of indigenous plants among rural households in the North West province of South Africa and examined factors determining their willingness to pay for indigenous plants. Understanding these determinants is vital in providing important information to promote and direct policy options towards production, adding value to and consumption of indigenous plants. Furthermore, findings from this study may provide potential traders of indigenous plants with important information about the socioeconomic factors influencing household willingness to pay and food consumption decisions, and as such contribute towards a more sustainable food system in South Africa.

2. Willingness to Pay within the Theory of Consumer Choice

The present section provides a brief discussion of competing theories and models of consumers' willingness to pay within the theory of consumer choices on which the analytical framework of this study was based.

2.1. The Key Economic Model

As explained by Ramasubramanian [22], willingness to pay for a particular product is the amount of money an individual or household is willing to pay to purchase a product given a person's income, risk preferences, and other factors. The two determinants of willingness to pay focused on by economic models are income of the individual or household and the use of the good in question. When individuals consider paying for improved environmental quality, such as forest and/or some of its products (e.g., indigenous plants), their choices and responses to valuation questions are constrained by their disposable income [21]. Therefore, a household or an individual's income is expected to correlate to be with the amount of money consumers are willing to spend for better environmental outcomes, such as indigenous plant conservation, by purchasing indigenous rather than non-native plants.

Interestingly, income is mostly included in stated preference surveys and is expected to have a positive effect on consumer willingness to pay. Therefore, there is a direct behavioral link between the use of indigenous plants for consumption and the individual's well-being. This link is expressed by the concept of "use value." If an individual does not use the good (indigenous plants) in question, the only link between the good and the individual's well-being is the "knowledge" the individual has about the good [23]. This link is expressed as "non-use value", which is expected to be weaker than the link based on direct use. It is assumed that the willingness to pay for users is more than for nonusers [24].

2.2. Transactions and Economic Potentials of Indigenous Plants

Evidence of trading indigenous plants and their products varies from place to place, such as between different fruits, fruit juice, seedlings, and other by-products [22,23]. Indigenous fruits are sold in both urban and rural markets and provide a substantial income to small-scale farmers [24,25]. However, existing studies have revealed that the trading of fruits collected from the wild is a profitable enterprise [22]. For example, studies in Malawi, Tanzania, and Zimbabwe found that the percentage of net profit of indigenous fruit products reached 28% with higher profits being obtained in locations that are close to markets [22].

In South Africa, communities collectively harvested about 2000 metric tons of *Sclerocarya birrea* (A. Rich.) Hochst (Marula) and earned $180,000 annually, representing more than 10% of average household income in these communities [26]. In addition, the members of a popular southern African Natural Products Trade Association reported gross revenue of $629,500 from the sale of fruit tree products. The key fruit tree products were obtained from *Sclerocarya birrea* (A. Rich.) Hochst (Marula) and *Adansonia digitata* L (Baobab) that generated $126,420 and $44,120, respectively [22,27]. Based on a recent market projection, the potential market of *Adansonia digitata* was valued at $960 million. Studies in Zimbabwe revealed that improvements in tree yield and earlier fruiting of indigenous fruit trees will create incentives for farmers to cultivate indigenous fruits [27].

Prior studies have found most participants indicating that indigenous plants (grains, vegetables, fruits) have favourable market potential [23,28]. Marketing costs of African leafy vegetables were estimated at between R0.50-R1.50/bundle with an average price paid by consumers of R10/bundle [29,30]. In another study [31], gender, education, marital status, and consumer perceptions influenced the purchasing decision for African leafy vegetables but not the level of expenditure. As highlighted by the authors, the identified factors influenced the purchasing decision but not the level of expenditure; such socio-economic characteristics were gender, educational and marital status, as well as perception factors.

2.3. Theory of Planned Behavior

Although economists rely on the concept of scale of preferences (choices) in order to determine what people value, sociologists have a strong affinity attitude (the desirability of a single action or object) [30,31]. A classical attitude–behavior paradigm would assume that behavior can be predicted by attitudes. This would mean that general attitudes such as environmental concern have a direct and positive effect on willingness to pay (See Figure 1). The basic attitude–behavior model is still part of thinking in social psychology although there are approaches going beyond this simple paradigm [21]. The intention to perform a behavior is the immediate determinant of the behavior in question, including the behavior of paying money for a good [32,33]. Prior research identified three determinants of the behavioral intention, such as attitude towards behavior, subjective norm, and perceived behavioral control [33]. The attitude towards behavior refers to an individual's positive or negative evaluation of performing the behavior. An individual's perception of social pressure from reference group members to enact the behavior is captured by the subjective norm. Perceived behavioral control includes perceived ease or difficulty of performing the behavior. Regarding indigenous plants, willingness to pay is expected to increase with a more favorable attitude towards paying, and with an increasing social pressure towards paying and with an increasing perceived behavioral control regarding paying for the indigenous plants [34–36].

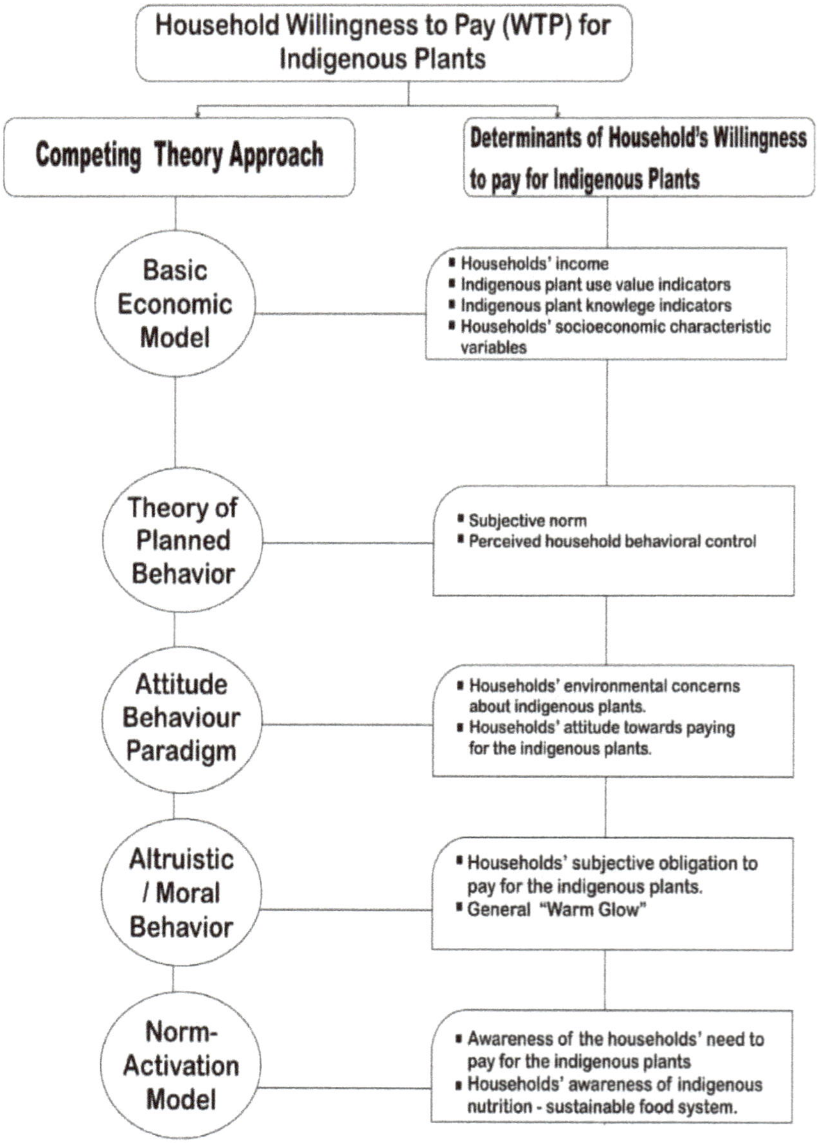

Figure 1. Framework for the theoretical approaches and determinants of households' willingness to pay for indigenous plants.

2.4. Models of Altruistic and Moral Behavior

Models of altruism are based on a broader motivational structure than standard economic models. One way to enable altruistic behavior in economic terms is to use a utility function that incorporates a taste for having other people better off or for having beneficial environmental attributes [23]. Altruistic motivation can lead to perceived obligations to contribute to the preservation of indigenous plants. These concepts can be used to explain what is known as the embedding effect [37], which is the observation that sometimes willingness to pay does not vary with the quantity of the good in question. If people only derive utility from the act of giving, then it does not matter what quantity of the good is provided. However, moral satisfaction may vary with the good as some goods give more satisfaction than others. There is empirical evidence for the purchase of moral satisfaction [37]. Willingness to pay can be positively affected by both a subjective obligation to pay for the specific good and a general warm glow which is independent of the specific good in question [21].

2.5. Norm-Activation Model

Given the awareness of need and awareness of responsibility, a perceived moral obligation can result in specific behavior. Both determinants mediate the effect of a perceived moral obligation on behavior [21,38]. In the context of willingness to pay analysis, the personal norm to pay for the good, the awareness of need with respect to providing the good, and the awareness of responsibility for paying are considered as behavioral determinants. The personal norm equals more or less the subjective obligation to pay which was discussed in the preceding models of altruistic behavior. It is expected that the awareness of need and the awareness of responsibility positively affect willingness to pay when considered as interaction terms with the subjective obligation to pay. Figure 1 presents a summary of the alternative theories and models and related determinants of willingness to pay discussed in previous sections, a combination of which is adopted in this article. Specifically, our study develops two econometric models for empirical analysis: a model of the determinants of a participant's income wherein knowledge and use of indigenous plants are included as explanatory variables along with other socio-demographic factors; and another model of the determinants of a participant's willingness to pay for indigenous plants. Both models were estimated using the two-stage Heckman approach.

3. Methods

3.1. Study Area, Sampling Methods, Procedure, and Sample Size

The study was conducted from all four district municipalities of the North West Province, South Africa. A multistage sampling procedure was used for data collection. Prior to the commencement of the survey, the enumerators were trained on the requirements of the survey and a pre-testing of the questionnaire was undertaken on a few rural households. In each of the selected communities, the traditional leaders provided access to the participants after consultation with them on the purpose and value of the research. Although the questionnaire was designed in English, interviews were conducted in the local languages of the participants, which include Setswana and Northern Sotho.

In the first stage of the sampling, four district municipalities (Ngaka Modiri Molema, Dr. Kenneth Kaunda, Bojanala Platinum, and Dr. Ruth Segomotsi Mompati) were selected. The second stage involved selecting 3 communities from each district municipality to account for the 12 communities studied. The selection of the 12 communities was based on their rural nature, active engagement in smallholder agricultural practices, and poor socioeconomic status of the residents. The last and final stage of sampling was the selection of rural household heads which was simplified by the assistance of resident extension officers. In this study, 133 sample households were utilized after a proper screening of the questionnaires. The allotted questionnaire samples for each selected community were representative, sufficiently robust and satisfactory to give estimates at the local level. Willingness to pay is generally analyzed using the contingent valuation (CV) method which helps to estimate value an individual places on a good. The contingent valuation method is originally designed to value goods and services where the market fails to adequately value these. This is mainly the case for public goods, the environment, and/or healthcare programs [39].

Generally, CV is referred to as a stated preference model, in contrast to a price-based revealed preference model [18]. The survey asks how much money people would be willing to pay (or willing to accept) to use (or be compensated for the loss of) organic food product features. Indeed, CV permits a direct estimation of willingness to pay by means of different elicitation techniques [39,40]. Consumers simply indicate their willingness to pay without purchasing the hypothetical product. As explained, the CV method relies on directly asking individuals about their willingness to pay for a specific commodity. The most important part in applying CV is to choose appropriate survey and elicitation methods to increase survey data accuracy [18]. Various survey methods and questionnaire formats are possible for collection of data. In-person interviews usually produce the highest-quality willingness to pay data, although telephone and mail surveys have been applied in several studies [41,42].

There are various techniques for eliciting consumers' willingness to pay. For instance, in a dichotomous-choice format, the participant is given a question to indicate the ability to pay X amount for the good, or not. Use of open-ended questions about a consumer's willingness to pay is another technique. An alternative method is to present several possible willingness to pay values on a card to the participant, knowns as a "payment card". The participant would choose the nearest quantity to their willingness to pay among others written on the card [43,44]. The chosen amount can be taken as the consumer's willingness to pay [18]. Since a payment card is simple, and it enlightens an unaware participant's picking options by giving them a range of predesigned price premiums, it is an appropriate approach in some studies [40,44].

The data used in this paper were collected through a contingent valuation survey accompanied with a photo album to make plant identification easier for participants. Plants were identified within each category to which they belong. Our study used open-ended questions about the participants' willingness to pay for the indigenous plants. Participants were presented with the following willingness to pay question among others: Suppose your favorite indigenous plant has a price premium, would you pay? How much are you willing to pay? Can you pay X amount for the purchase of this specific plant? These questions were asked in relation to different local price premiums in the study area. Prices vary by indigenous plant category (grain, fruits, vegetable, and beverages). In addition, the participants' willingness to pay parameter (dichotomous) was considered a dependent variable in this study. Households in the North West Province purchase indigenous plants from either farms or informal markets. In this study, 21 indigenous plants from 17 plant families were utilized, which were recorded as staple food, fruits, and beverages by the participants (Appendix A, Tables A1 and A2).

3.2. Determinants of Households' Income and Willingness to Pay for Indigenous Plants

To empirically analyze the determinants of households' income and willingness to pay for indigenous plants in the North West province of South Africa, a two-stage Heckman model was employed as detailed in previous studies [45–47]. The advantage of the application of the Heckman model is that it is able to model factors influencing households' income and their willingness to pay for indigenous plants in a single framework while simultaneously correcting for possible sample selection bias [48]. The first-stage equation of the Heckman model is a probit model, which assumes that the errors are homoscedastic [49].

$$y_{1i} = 0 \text{ if } S_{ni} \leq 0 \quad (1)$$

$$y_{1i} = 1 \text{ if } S_{ni} > 0 \quad (2)$$

where y_1 is the binary response, S_{ni} is the amount of money spent by the household i.

The spending equation can then be written as:

$$y_1^* = \beta_{1I} X_{1I} + \varepsilon_{1I} \quad (3)$$

where y_1^* is a latent variable, which is the utility the household will spend on indigenous plants. The binary model is then stated as:

$$\begin{cases} 1, \text{ if household spends on indigenuos plants} \\ 0, \text{ otherwise} \end{cases} \tag{4}$$

Specifically, the probit model in stage one of estimation is stated as

$$\Pr(y_1) = f(x1,\ x2,\ \ldots\ldots\ldots\ldots xn) \tag{5}$$

where $\Pr(y_1)$ is the probability of a household deciding to spend on indigenous plants or not, $x1,\ x2,\ \ldots\ldots\ldots\ldots xn$ are the variables specified in Table 1 and ε is the normally distributed error term. In the second stage of the Heckman model, ordinary least square (OLS) regression equations are estimated to test the effect of hypothesized factors on the on the extent of spending on indigenous plants. The model is stated as:

$$(S_n) = f(y1,\ y2,\ \ldots\ldots\ldots\ldots yn,\ \varepsilon) \tag{6}$$

where Sn is the amount of money spent on indigenous plants, $y1,\ y2,\ \ldots\ldots yn$ are the variables that were a priori hypothesized to affect the amount of money spent by the households on indigenous plants consumption, while ε is the error term.

Table 1. Explanatory variables used in the models applied in the current study.

Variables	Description
Age of the household's head	Number of years (Continuous)
Gender of the household's head	Dummy; 1 if head is male and 0 if otherwise
Educational status of the household's head	Number of years of academic education (Continuous)
Marital status of the household's head	Dummy; 1 if head is married, 0 otherwise
Household's size	Number of members of the household (Continuous)
Number of working class	Number who work and have income (Continuous)
Involvement in agricultural practices by the household	Dummy; 1 if yes, 0 otherwise
Membership of cooperative society	Dummy; 1 if yes, 0 otherwise
Land ownership	Dummy; 1 if lease, 0 otherwise
Participation in training program	Dummy; 1 if yes, 0 otherwise
Extension officer visit	Dummy; 1 if yes, 0 otherwise
Knowledge of indigenous plants consumption	Dummy; 1 if yes, 0 otherwise
Nutritious benefits of indigenous plants	Dummy; 1 if yes, 0 otherwise
Awareness of the drought-resistant benefits of indigenous plants	Dummy; 1 if yes, 0 otherwise
Awareness of the pest-resistant nature of indigenous plants	Dummy; 1 if yes, 0 otherwise
Economic value benefits of indigenous plants	Dummy; 1 if yes, 0 otherwise
Availability of indigenous plants	Dummy; 1 if yes, 0 otherwise
Access to indigenous plants in local market	Dummy; 1 if yes, 0 otherwise
Medicinal benefits of indigenous plants	Dummy; 1 if yes, 0 otherwise
Household's financial returns	Dummy; 1 if yes, 0 otherwise

3.3. Ethical Approval

The Faculty of Natural and Agricultural Sciences research ethics committee (FNAS-REC) of the North-West University (NWU) approved (ethical clearance no: NWU-01243-19-S9) the study. Permission to access the study area was granted by the North West Provincial Department of Rural, Environment and Agriculture (READ) in South Africa. Participation in the study was voluntary and conducted with full consent of the participants.

4. Results and Discussion

4.1. Demographic Characteristics of the Participants

As shown in Table 2, the average age of household head is 47 years, suggesting that the sampled households are within their active years. This could be economically advantageous given that the active workforce falls within this age category in South Africa [50–52]. In addition, a larger proportion of households are headed by females (56%), signifying the increasing roles of females in decision-making among households in the North West province. A similar finding was reported in previous studies [53,54]. Surprisingly, 44% of the heads of households are single despite their old age. Furthermore, the study indicated low educational attainment by the participants as only 35% of the participants completed a secondary level of education. Education plays a crucial role in the socioeconomic status of households and consciousness about environmental and food nutrition issues [55,56].

Table 2. Socioeconomic characteristics of the participants.

Variables	Frequency	Percentage (%)	Mean (S.D)
Age			
20–30	11	8.30	
31–40	27	20.30	
41–50	38	28.60	47 (11.85)
51–60	44	33.10	
71–80	13	9.80	
Gender			
Male	57	42.90	
Female	75	56.40	
not disclosed	1	0.80	
Marital status			
Married	47	35.30	
Single	58	43.60	
Divorced	14	10.50	
Widow(er)	14	10.50	
Educational attainment			
Standard	38	28.60	
High school	46	34.60	
Diploma	19	14.30	
Degree and Postgraduate	23	17.30	
No formal education	7	5.30	
Race			
Black	132	99.20	
People of color	1	0.80	
White	0	0.00	
Household size members			
1–4	55	41.35	
5–8	71	53.38	5 (1.46)
9–12	7	5.26	

Table 2. *Cont.*

Variables	Frequency	Percentage (%)	Mean (S.D)
Household head's religion			
Christianity	124	93.20	
Islam	2	1.50	
Traditional worshiper	6	4.50	
Others	1	0.80	
Major occupation			
Civil servants	51	38.35	
Entrepreneur	30	22.56	
Farmer	28	21.05	
Others	24	18.05	
Willingness to pay for indigenous plants			
Yes	124	93.23	
No	9	6.77	
Monthly income			
R1000–3000	42	31.58	
R3001–6000	56	42.11	ZAR7499/US$510.82 (11,134.85)
R6001–9000	25	18.79	
R9001–12,000	10	7.52	
No of observations	**133**		

Note: US$1 = ZAR14.68 (https://www.xe.com/currencyconverter/convert, accessed on 5 February 2021).

In the current study, 99% of the participants were black, which reflects the ethnic demography of the North West province, currently dominated by black South Africans [57]. The average household size is five, indicating the potential that rural households in South Africa can supply labor and economically contribute to the South African economy. The household members can also help the head of household with cultivation of indigenous plants for more food secure rural regions in South Africa. Table 2 also reveals that the rural community participants were predominantly civil servants (38% of total), suggesting that a large proportion of households in the North West province depend on wage/salaried employment. Meanwhile, entrepreneur (23%), farmers (21%), and other (18%) categories of occupation practiced in the study area is an indication of job diversification among the participants. The average income was estimated at R7499 (US$510.82) in the study area. This is a relatively low income when evaluated from the household size perspective. Cultivation of indigenous plants could serve as additional source of income as well as a means for better food and nutrition within the marginalized rural poor of South Africa [18,58].

4.2. Participants Willingness to Pay for Indigenous Plants

Most (93.23%) of the participants were willing to pay for indigenous plants for their consumption, while 6.77% were not willing to pay. This observation is consistent with findings from existing studies conducted in many provinces in South Africa [16,18]. This suggests a high potential demand and willingness to pay for indigenous plants by households in the North West Province of South Africa. Thus, there is potential for indigenous plants to gain a larger market share in the North West Province compared to their exotic counterparts.

4.3. Two-Stage Heckman Model Results

The Heckman two-stage approach was used to analyze the determinants of households' willingness to pay for indigenous plants, and subsequently, factors influencing household income. The first stage involves the application of the probit model, where the dependent variable is the household's willingness to pay. This variable is in binary form and takes the value of 1 if the household is willing to pay for indigenous plants and 0 otherwise. In the second stage, the ordinary least square (OLS) regression was applied where the log of a household's income is specified as the dependent variable.

Table 3 further presents the results of the Breusch–Pagen test for heteroscedasticity, with a test statistic of 13.33. When compared to a chi-squared distribution with one degree of freedom, the resulting p-value falls well below the standard 0.05 level. Thus, we have clear evidence to reject the null hypothesis of homoscedasticity and accept the alternative hypothesis that we do in fact have heteroscedasticity in the residual of this regression model.

Table 3. Determinants of income and willingness to pay for indigenous plants by households in North West Province.

Variables	Coefficient	Standard Error	Z	$p > \|z$
Outcome Equation: [Determinants of household's income: ordinary least square regression]				
Age of the households' head (years)	0.05618	0.01617	3.47	0.001 ***
Gender of the household's head (1 = male, 0 = female)	0.05617	0.32896	0.51	0.607
Educational status of the household's head (years of schooling)	−0.27232	0.83488	−0.33	0.744
Marital status of the household's head (married = 1, 0 = otherwise)	0.84260	0.33143	2.54	0.011 **
Size of household (number)	0.24601	0.13214	1.86	0.063 *
Number of working class	−0.01905	0.16168	−0.12	0.906
Farming household (1 = farming, 0 = otherwise)	−0.90351	0.46817	−1.93	0.054 *
Membership of co-operative society (1 = member,0 = otherwise)	−0.14177	0.52930	−0.27	0.789
Land ownership (yes = 1, no = 0)	0.31362	0.33471	0.94	0.349
Participation in training program (yes = 1, no = 0)	−0.03884	0.12217	−0.32	0.751
Visit by extension officers (yes = 1, no = 0)	1.14375	0.60564	1.89	0.059*
Knowledge of indigenous plant consumption (yes = 1, no = 0)	−0.48173	0.64200	−0.75	0.453
Nutritional benefits of indigenous plants (yes = 1, no = 0)	0.45827	0.32879	1.39	0.163
Drought-resistant trait of indigenous plants (yes = 1, no = 0)	0.66097	0.33952	1.95	0.052*
Pest-resistant trait of indigenous plants (yes = 1, no = 0)	0.22832	0.32829	0.70	0.487
Economic benefits of indigenous plants (yes = 1, no = 0)	0.78487	0.43004	1.83	0.068 *

Table 3. *Cont.*

Variables	Coefficient	Standard Error	Z	$p > \|z$
Constant	3.72749	1.94571	1.92	0.055
Breusch–Pagan/Cook–Weisberg test for heteroskedasticity Ho: Constant variance Variables: fitted values of log_Income chi^2 (1) = 13.33 *Prob > chi^2 = 0.0003*				
Selection Equation: Determinants of willingness to pay for indigenous crops: probit regression				
Age of the household's head (years)	0.00282	0.01688	0.17	0.868
Marital status of the household's head (1 = married, 0 = otherwise)	−0.21769	0.39167	−0.56	0.578
Educational status of the household's head (years of schooling)	0.70345	0.57533	1.22	0.221
Farming household (1 = farming, 0 = otherwise)	−0.07060	0.57662	−0.12	0.903
Size of household (number)	0.39314	0.16221	2.42	0.015 **
Accessibility to indigenous plants (yes = 1, no = 0)	−0.55837	0.39234	−1.42	0.155
Extension officers visit (yes = 1, no = 0)	−0.17523	0.69594	−0.25	0.801
Availability of indigenous plants in local market (yes = 1, no = 0)	0.05987	0.34684	0.17	0.863
Economic benefits of indigenous plants (yes = 1, no = 0)	0.39823	0.41248	0.97	0.334
Medicinal benefits of indigenous plants (yes = 1, no = 0)	1.42369	0.89110	1.60	0.110
Financial returns for households (yes = 1, no = 0)	0.78573	0.42039	1.87	0.062 *
Constant	−2.60987	1.39711	−1.87	0.062
Number of observations Lambda Wald chi^2 (20) Prob > chi^2 Rho Sigma	133 0.275024 47.96 0.0000 0.17590 1.5635301			

Note: ***, ** and * means 1%, 5%, and 10% levels of significance, respectively.

4.3.1. Determinants of Household's Income: OLS Regression

About 43% of the coefficients on the variables included in the model have expected signs and are statistically significant at the 10% level or more (Table 3). The age of the household's head is significant and positive ($p < 0.01$) which indicates that the participants' income increases with age, a proxy measure used to capture experience. This might be true as the increase in age of participants might translate into more experience which invariably translates into higher income. This corroborates with existing studies indicating earnings increases with increasing age, specialized skill trainings, and/or education of the household heads [58–60]. In addition, the results also show that marital status of the household head was positively and significantly related to income ($p < 0.05$). This implies that participants who are single are likely to have higher income relative to other categories of marital status. This could be attributed to the fact that they are likely to have more time to do more jobs, hence, have more time to work, thus earning more income than their married counterparts.

The coefficient of the household size is positive and significant ($p < 0.05$), meaning that participants with a large household size will likely have higher income. This is possible as households with mature and working-age members are likely to have higher income than large household with more dependent members [61,62]. The coefficient of involvement of households in agricultural activities is negative and significant ($p < 0.10$), which indicates that farming households in the study area earn relatively less than those who earns mainly from non-agricultural activities. This might be due to the fact that a large number of households are comprised of civil servants (Table 2) with potentially greater incomes than agricultural households. This is in line with a study in Nigeria that reported that non-farm income of households was more than on-farm income and hence, it has a direct effect on food security status [63,64].

It is encouraging that extension visits positively and significantly ($p < 0.10$) influence households' income, which indicates that households with more access to extension officers were able to generate higher income from farming activities compared to their counterparts without such access [65,66]. The coefficient of the drought-resistant benefits of indigenous plants variable was positive and significant ($p < 0.10$), thereby exerting a positive effect on household income. This suggests that participants who are knowledgeable about the drought-resistant attributes of the indigenous plants are more likely to have higher income when cultivating indigenous plants. Similarly, the acknowledgement of economic benefits of indigenous plants had a positive and significant ($p < 0.10$) relation to household income. This is consistent with prior literature on the economic potential of indigenous plants to increase household income, lifting rural communities out of poverty and enhancing food security [18,50,67–69].

4.3.2. Probit Regression

The probit model analysis identified the determinants of the willingness to pay for indigenous plants by the households (Table 3). The coefficient of household size was positive and significant ($p < 0.05$), indicating that participants with a large household size had a higher likelihood of paying for indigenous plants. This might be due to the expected food need for such a large household which will translate into willingness to pay for indigenous plants in the study area. In addition, this might be due to greater expected food needs for larger households, which can increase the demand and thus the willingness to pay for indigenous plants in the study area. However, this is contrary to prior studies which have found that encouraging people to have smaller families is associated with greater willingness and ability to pay for food [16]. The variable representing value of financial return of the households was positive and significant ($p < 0.10$), suggesting that the probability of the willingness to pay for indigenous plants by households increased with greater financial returns. This was expected, given that households with higher financial returns have the ability to pay for indigenous plants in order to ensure that their households are food secure. This observation is in line with basic economic consumer behavior related to income, expenditure, and savings [70,71].

5. Policy Recommendations

Based on the empirical findings of this study, the following policy recommendations remain pertinent since the majority (93%) of the households were willing to pay for indigenous food plants, especially those that are larger in size. However, rural households cultivating indigenous plant species could have some added value and beneficial attributes arising from domestic cultivation of indigenous food plants. In this study, the economic benefits and returns from the indigenous plants were emphasized as significant variables in the model. This indicates that households that are cultivating these indigenous species give some added value or beneficial attributes. Public awareness through different media should be target-promoting the nutritional, and economic benefits of the indigenous plants which can help increase the demand. Moreover, households with lower willingness to pay can be properly oriented on the benefits of this indigenous plants, thereby increasing the overall households' willingness to pay for the indigenous crops. Hence, private markets, nongovernmental organizations and the governments need to be involved in the dissemination of information of the potential economic benefits of indigenous plants in order to enhance households' willingness to pay for them. Furthermore, the promotion of indigenous plants in local markets will promote communities' food-nutrition, socioeconomics, welfare, and health through a more diversified local economy. Particularly, quality education remained a principal legal and regulatory barrier to markets for indigenous plants. On this basis, the need to educate households to promote the production and consumption of indigenous plants which is capable of contributing to economic sustainability cannot be overemphasized. In addition, the average income was estimated at ZAR7499 ($510.82) which was relatively low income when evaluated from the household size perspective. The study recommends that the government should implement policies that will grant lower-income households' permission to access wild indigenous plants on public lands if they are priced out of the local marketplace for these foods, so long as it is assured that they promote "stewardship" over these plants in the wild. This can also help in price control by the farmers of such indigenous plants. It could as well result in greater ecological diversity while improving food access issues, especially among the marginalized rural households. Furthermore, household size was repeatedly significant in the two stages of the model. There should be proper orientation of the rural households on the economic implications of large family sizes, as this can further placing such households in multidimensional poverty. Finally, as a result of the high diversity of indigenous plants in the study area, local communities should take advantage of a variety of plants that are more accessible or abundant to combat hunger and malnutrition. Therefore, government support to the farmers should emphasize target products and assist their production through national markets.

6. Conclusions

This study examined the factors influencing income and willingness to pay for indigenous plants by selected rural households in North West Province of South Africa. The majority (93%) of the households who participated were willing to pay for indigenous food plants which suggests a high potential demand by households and market for indigenous plants in the province. In addition, an average age of 47 years, more (56%) female-headed households and average monthly income of R7499 ($510.82) was recorded in the study area. Empirical analysis revealed that factors such as age, marital status, household size, farming households, and economic benefits of indigenous plants influenced income and willingness to pay for indigenous plants by the households. Particularly, household size and financial returns were the significant determinants of the willingness to pay for indigenous plants by the participants. The study therefore concludes that socioeconomic characteristics are key determinants of households' willingness to pay for indigenous food plants in rural South Africa. This understanding will assist policy makers to implement agricultural and food policies, thereby addressing the food security, nutrition, and health sustainability. Future research on the breeding and value-adding activities to enhance the indigenous plants' cultivation, market accessibility, acceptability, and durability are necessary to encourage household's willingness to pay and consumption of indigenous plants. More so, a more nationally and globally representative dataset for studies on household's willingness to pay for indigenous plants needs to be compiled for future studies to provide further understanding of the subject matter. Such initiatives could have a positive societal value of reducing households' dependence on a handful of crops for nutrition, food security, and sustainability in the developing nations.

Author Contributions: Conceptualization, A.O.O. and A.O.A.; funding acquisition; A.O.A.; investigation; A.O.O., S.C.T., and P.T.N.; project administration, A.O.A.; resources, A.O.A.; supervision, A.O.A.; writing original draft, A.O.O. and K.O.O.; writing review and editing, A.O.O., K.O.O., P.T.N., and S.C.T. All authors have read and agreed to the published version of the manuscript.

Funding: The research was partially funded by the National Research Foundation [NRF UID: 109508], Pretoria, South Africa and the Article Processing Charge was funded by the Faculty of Natural and Agricultural Sciences, North-West University, Mmabatho, South Africa.

Institutional Review Board Statement: The research was revised by the Institutional Review Board with ethic approval number NWU-01243-19-S9.

Informed Consent Statement: Informed consent was obtained from all subjects involved in the study.

Data Availability Statement: Data are available on request from the first author of the manuscript.

Acknowledgments: A.O.O. appreciates the Postdoctoral fellowship from the North-West University, Mmabatho, South Africa.

Conflicts of Interest: We declare no conflict of interest with regards to the current research. The opinions and conclusions/recommendations expressed in this work are that of the authors, and that the National Research Foundation accepts no liability whatsoever in this regard.

Appendix A

Table A1. Indigenous vegetables and grains consumed by the participants in selected rural households in North West Province. Number of mention (NM), Vernacular names: Ts = Setswana and E = English.

Scientific Name and Family	Vernacular Name	Usage	NM
Amaranthus sp **Amaranthaceae**	Amaranthus (E); Thepe (Ts)	Staple food	71
Cajanus cajan (L.) Millsp. **Leguminosae/Fabaceae**	Pigeon bean (E); Dinawa (Ts)	Staple food	26
Cleome gynandra L **Cleomaceae**	Cat's whiskers, African cabbage (E); Lerotho (Ts)	Staple food	4
Colocasia esculenta (L.) Schott **Araceae**	Potato of the tropics (E); Madumbis (Ts)	Staple food	3
Glycine max (L.) Merr. **Leguminosae/Fabaceae**	Soybeans (E); Dinawa (Ts)	Staple food	57
Lagenaria siceraria (Mol.) Standl. **Cucurbitaceae**	Bottle gourd, Calabash (E); Segwana (Ts)	Staple food	9
Manihot esculenta Crantz **Euphorbiaceae**	Cassava (E); Muthupula (Ts)	Staple food	8
Tetragonia decumbens Mill. **Aizoaceae**	Dune spinach (E)	Staple food	32
Sorghum bicolor (L.) Moench **Poaceae**	Sorghum (E); Mabele (Ts)	Beverage and staple food	111
Tylosema esculentum (Burch.) A. Schreib. **Leguminosae/Fabaceae**	Marama bean (E)	Staple food	1

Table A2. Indigenous fruits consumed by the participants in selected rural households in North West Province. Number of mention (NM), Vernacular names: Ts = Setswana and E = English.

Scientific Name and Family	Vernacular Name	Usage	NM
Annona senegalensis Pers. **Annonaceae**	African Custard-apple (E); Mokamanawe (Ts)	Fruit	18
Carissa macrocarpa (Eckl.) A. DC. **Apocynaceae**	Natal plum, big num-num (E)	Fruit	1
Diospyros lycioides Desf. **Ebenaceae**	Monkey plum (E); Lethanyu (Ts)	Fruit	1
Diospyros simii (Kuntze) De Winter. **Ebenaceae**	Climbing Star-apple (E)	Fruit	2
Dovyalis caffra (Hook.f. & Harv.) Sim **Salicaceae**	Kei-apple (E)	Fruit	0
Dovyalis zeyheri (Sond.) Warb. **Salicaceae**	Wild apricot (E)	Fruit	1
Mimusops zeyheri Sond **Sapotaceae**	Transvaal red milkwood (E); Mmupudu (Ts)	Beverage	53
Parinari curatellifolia Planch. ex Benth. **Chrysobalanaceae**	Bosappel (A); Mobola (Ts)	Fruit	4
Sclerocarya birrea (A. Rich.) Hochst. **Anacardiaceae**	Marula (E); Morula (Ts)	Fruit and beverage	71
Strychnos spinosa Lam. **Loganiaceae**	Corky-bark Monkey-orange (E); Morapa (Ts)	Fruit	8
Vangueria infausta Burch. **Rubiaceae**	Chirinda wild-medlar, (E); Mobilo (Ts)	Fruit	15

References

1. Mabhaudhi, T.; Chibarabada, T.; Chimonyo, V.; Murugani, V.; Pereira, L.; Sobratee, N.; Govender, L.; Slotow, R.; Modi, A. Mainstreaming underutilized indigenous and traditional crops into food systems: A South African perspective. *Sustainability* **2019**, *11*, 172. [CrossRef]
2. Borelli, T.; Hunter, D.; Padulosi, S.; Amaya, N.; Meldrum, G.; Beltrame, D.M.d.O.; Samarasinghe, G.; Wasike, V.W.; Güner, B.; Tan, A. Local solutions for sustainable food systems: The contribution of orphan crops and wild edible species. *Agronomy* **2020**, *10*, 231. [CrossRef]
3. Albert, J.R.G.; Abrigo, M.R.M.; Quimba, F.M.A.; Vizmanos, J.F.V. *Poverty, the Middle Class, and Income Distribution Amid COVID-19*; Philippine Institute for Development Studies: Quezon City, Philippine, 2020.
4. Otekunrin, O.A.; Otekunrin, O.A.; Fasina, F.O.; Omotayo, A.O.; Akram, M. Assessing the Zero Hunger Target Readiness in Africa in the Face of COVID-19 Pandemic. *Caraka Tani J. Sustain. Agric.* **2020**, *35*, 213–227. [CrossRef]
5. Ackerman, K.; Conard, M.; Culligan, P.; Plunz, R.; Sutto, M.-P.; Whittinghill, L. Sustainable food systems for future cities: The potential of urban agriculture. *Econ. Soc. Rev.* **2014**, *45*, 189–206.
6. Anderson, C.R.; Bruil, J.; Chappell, M.J.; Kiss, C.; Pimbert, M.P. From transition to domains of transformation: Getting to sustainable and just food systems through agroecology. *Sustainability* **2019**, *11*, 5272. [CrossRef]
7. Van Bavel, J.J.; Baicker, K.; Boggio, P.S.; Capraro, V.; Cichocka, A.; Cikara, M.; Crockett, M.J.; Crum, A.J.; Douglas, K.M.; Druckman, J.N. Using social and behavioural science to support COVID-19 pandemic response. *Nat. Hum. Behav.* **2020**, *4*, 460–471. [CrossRef]
8. Capano, G.; Howlett, M.; Jarvis, D.S.L.; Ramesh, M.; Goyal, N. Mobilizing Policy (In)Capacity to Fight COVID-19: Understanding Variations in State Responses. *Policy Soc.* **2020**, *39*, 285–308. [CrossRef]
9. Silva, P.C.L.; Batista, P.V.C.; Lima, H.S.; Alves, M.A.; Guimarães, F.G.; Silva, R.C.P. COVID-ABS: An agent-based model of COVID-19 epidemic to simulate health and economic effects of social distancing interventions. *ChaosSolitons Fractals* **2020**, *139*, 110088. [CrossRef]
10. Van der Hoeven, M.; Osei, J.; Greeff, M.; Kruger, A.; Faber, M.; Smuts, C.M. Indigenous and traditional plants: South African parents' knowledge, perceptions and uses and their children's sensory acceptance. *J. Ethnobiol. Ethnomed.* **2013**, *9*, 1–12. [CrossRef]

11. Kuhnlein, H.V.; Receveur, O. Dietary change and traditional food systems of indigenous peoples. *Annu. Rev. Nutr.* **1996**, *16*, 417–442. [CrossRef]
12. Weinberger, K.; Lumpkin, T.A. Diversification into horticulture and poverty reduction: A research agenda. *World Dev.* **2007**, *35*, 1464–1480. [CrossRef]
13. Stats, S.A. *Provincial Profile: North West Community Survey 2016*; Statistics South Africa: Pretoria, South Africa, 2018; pp. 1–93.
14. Drimie, S.; Pereira, L. Advances in food security and sustainability in South Africa. In *Advances in Food Security and Sustainability*; Elsevier: Amsterdam, The Netherlands, 2016; Volume 1, pp. 1–31.
15. Pereira, L.M. The future of South Africa's food system: What is research telling us. *SA Food Lab. S. Afr.* **2014**. Available online: https://bio-economy.org.za/wp-content/uploads/2015/02/Pereira-2014-The-Future-of-South-Africas-Food-System-Report.pdf (accessed on 11 May 2021).
16. Chelang'a, P.K.; Obare, G.A.; Kimenju, S.C. Analysis of urban consumers' willingness to pay a premium for African Leafy Vegetables (ALVs) in Kenya: A case of Eldoret Town. *J. Food Secur.* **2013**, *5*, 591–595. [CrossRef]
17. Ngugi, I.K.; Gitau, R.; Nyoro, J. *Access to High Value Markets by Smallholder Farmers of African Indigenous Vegetables in Kenya*; Regoverning Markets Innovative Practice Series; IIED: London, UK, 2007.
18. Senyolo, G.M.; Wale, E.; Ortmann, G.F. Consumers' Willingness-To-Pay for underutilized vegetable crops: The case of African leafy vegetables in South Africa. *J. Hum. Ecol.* **2014**, *47*, 219–227. [CrossRef]
19. Parsa, H.G.; Lord, K.R.; Putrevu, S.; Kreeger, J. Corporate social and environmental responsibility in services: Will consumers pay for it? *J. Retail. Consum. Serv.* **2015**, *22*, 250–260. [CrossRef]
20. Ramasubramanian, J.A. *Willingness to pay for index based crop microinsurance in India. DPhil Candidate (Economics)*; University of Sussex: Brighton, UK, 2012.
21. Liebe, U.; Preisendörfer, P.; Meyerhoff, J. To pay or not to pay: Competing theories to explain individuals' willingness to pay for public environmental goods. *Environ. Behav.* **2011**, *43*, 106–130. [CrossRef]
22. Ham, C.; Akinnifesi, F.K.; Franzel, S.; Jordaan, D.; Hansmann, C.; Ajayi, O.; De Kock, C. Opportunities for commercialization and enterprise development of indigenous fruits in southern Africa. In *Indigenous Fruit Trees in the Tropics: Domestication, Utilization and Commercialization*; World Agroforestry Centre: Nairobi, Kenya; CAB International Publishing: Wallingford, UK, 2007; pp. 254–272.
23. Omotayo, A.O.; Aremu, O.A. Evaluation of factors influencing the inclusion of indigenous plants for food security among rural households in the North West Province of South Africa. *Sustainability* **2020**, *12*, 9562. [CrossRef]
24. Shai, K.N.; Ncama, K.; Ndhlovu, P.T.; Struwig, M.; Aremu, A.O. An exploratory study on the diverse uses and benefits of locally-sourced fruit species in three villages of Mpumalanga Province, South Africa. *Foods* **2020**, *9*, 1581. [CrossRef]
25. Muhanji, G.; Roothaert, R.L.; Webo, C.; Stanley, M. African indigenous vegetable enterprises and market access for small-scale farmers in East Africa. *Int. J. Agric. Sustain.* **2011**, *9*, 194–202. [CrossRef]
26. Ham, C. Plant for food and drink. Southern African Trade Directory of Indigenous Natural Products. Commercial Products from the Wild Group. *Stellenbosch* **2005**, *1*, 17–22.
27. Mithofer, D.; Waibel, H.; Akinnifesi, F.K. The Role of Food from Natural Resources in Reducing Vulnerability to Poverty: A Case Study from Zimbabwe. 2006. Available online: https://ageconsearch.umn.edu/record/25264/ (accessed on 11 May 2021).
28. Senyolo, G.M.; Wale, E.; Ortmann, G.F. A double hurdle analysis of consumers' decisions to purchase African leafy vegetables in Limpopo province. *J. Consum. Sci.* **2019**, *4*, 10–21.
29. Senyolo, G.M.; Wale, E.; Ortmann, G.F. Analysing the value chain for African leafy vegetables in Limpopo Province, South Africa. *Cogent Soc. Sci.* **2018**, *4*, 1509417. [CrossRef]
30. Green, C.; Tunstall, S. *A Psychological Perspective*; Oxford University Press: Oxford, UK, 1999.
31. Kahneman, D.; Ritov, I.; Schkade, D.; Sherman, S.J.; Varian, H.R. Economic preferences or attitude expressions?: An analysis of dollar responses to public issues. In *Elicitation of Preferences*; Springer: New York, NY, USA, 1999; pp. 203–242.
32. Ajzen, I. The theory of planned behavior. *Organ. Behav. Hum. Decis. Process.* **1991**, *50*, 179–211. [CrossRef]
33. Ajzen, I.; Brown, T.C.; Rosenthal, L.H. Information bias in contingent valuation: Effects of personal relevance, quality of information, and motivational orientation. *J. Environ. Econ. Manag.* **1996**, *30*, 43–57. [CrossRef]
34. Meyerhoff, J. Stated willingness to pay as hypothetical behaviour: Can attitudes tell us more? *J. Environ. Plan. Manag.* **2006**, *49*, 209–226. [CrossRef]
35. Moisseinen, E. On behavioural intentions in the case of the Saimaa Seal. Comparing the contingent valuation approach and the attitude-behaviour research. Valuation and the Environment: Theory, Method and Practice. 1999, pp. 183–204. Available online: https://www.econbiz.de/Record/on-behavioural-intentions-in-the-case-of-the-saimaa-seal-comparing-the-contingent-valuation-approach-and-attitude-behaviour-research-moisseinen-eija/10001365087 (accessed on 12 May 2021).
36. Rekola, E.; Mika, P. The theory of planned behavior in predicting willingness to pay for abatement of forest regeneration. *Soc. Nat. Resour.* **2001**, *14*, 93–106. [CrossRef]
37. Kahneman, D.; Knetsch, J.L. Valuing public goods: The purchase of moral satisfaction. *J. Environ. Econ. Manag.* **1992**, *22*, 57–70. [CrossRef]
38. Blamey, R. Contingent valuation and the activation of environmental norms. *Ecol. Econ.* **1998**, *24*, 47–72. [CrossRef]
39. Blumenschein, K.; Johannesson, M.; Yokoyama, K.; Freeman, P. Hypothetical versus real willingness to pay in the health care sector: Results from a field experiment. *Value Health* **2001**, *4*, 79. [CrossRef]

40. Boccaletti, S.; Nardella, M. Consumer willingness to pay for pesticide-free fresh fruit and vegetables in Italy. *Int. Food Agribus. Manag. Rev.* **2000**, *3*, 297–310. [CrossRef]
41. Haghjou, M.; Hayati, B.; Pishbahar, E.; Mohammad, R.R.; Dashti, G.H. Factors affecting consumers'potential willingness to pay for organic food products in iran: Case study of Tabriz. *J. Agric. Sci. Technol.* **2013**, *15*, 191–202.
42. Weinberger, K. *Indigenous Vegetables in Tanzania: Significance and Prospects*; AVRDC-World Vegetable Center: Shanhua, Taiwan, 2004; Volume 600.
43. Fu, T.T.; Liu, J.T.; Hammitt, J.K. Consumer willingness to pay for low-pesticide fresh produce in Taiwan. *J. Agric. Econ.* **1999**, *50*, 220–233. [CrossRef]
44. Gil, J.M.; Gracia, A.; Sanchez, M. Market segmentation and willingness to pay for organic products in Spain. *Int. Food Agribus. Manag. Rev.* **2000**, *3*, 207–226. [CrossRef]
45. Goetz, S.J. A selectivity model of household food marketing behavior in sub-Saharan Africa. *Am. J. Agric. Econ.* **1992**, *74*, 444–452. [CrossRef]
46. Abdullah; Rabbi, F.; Ahamad, R.; Ali, S.; Chandio, A.A.; Ahmad, W.; Ilyas, A.; Din, I.U. Determinants of commercialization and its impact on the welfare of smallholder rice farmers by using Heckman's two-stage approach. *J. Saudi Soc. Agric. Sci.* **2019**, *18*, 224–233. [CrossRef]
47. Deressa, T.T.; Hassan, R.M.; Ringler, C. Perception of and adaptation to climate change by farmers in the Nile basin of Ethiopia. *J. Agric. Sci.* **2011**, *149*, 23–31. [CrossRef]
48. Tilahun, U.; Bedemo, A. Farmers' perception and adaptation to climate change: Heckman's two stage sample selection model. *Ethiop. J. Environ. Stud. Manag.* **2014**, *7*, 832–839. [CrossRef]
49. Winship, C.; Mare, R.D. Models for sample selection bias. *Annu. Rev. Sociol.* **1992**, *18*, 327–350. [CrossRef]
50. Omotayo, A.O.; Ndhlovu, P.T.; Tshwene, S.C.; Aremu, A.O. Utilization pattern of indigenous and naturalized plants among some selected rural households of North West Province, South Africa. *Plants* **2020**, *9*, 953. [CrossRef]
51. Kaona, F.A.D.; Tuba, M. A qualitative study to identify community structures for management of severe malaria: A basis for introducing rectal artesunate in the under five years children in Nakonde District of Zambia. *BMC Public Health* **2005**, *5*, 28. [CrossRef]
52. Mavhura, E.; Manyena, S.B.; Collins, A.E.; Manatsa, D. Indigenous knowledge, coping strategies and resilience to floods in Muzarabani, Zimbabwe. *Int. J. Disaster Risk Reduct.* **2013**, *5*, 38–48. [CrossRef]
53. Flatø, M.; Muttarak, R.; Pelser, A. Women, weather, and woes: The triangular dynamics of female-headed households, economic vulnerability, and climate variability in South Africa. *World Dev.* **2017**, *90*, 41–62. [CrossRef]
54. Tibesigwa, B.; Visser, M. Small-scale subsistence farming, food security, climate change and adaptation in South Africa: Male-female headed households and urban-rural nexus. *Econ. Res. S. Afr. Cape Town* **2015**, *527*, 33.
55. Babar, N.F.; Muzaffar, R.; Khan, M.A.; Imdad, S. Impact of socioeconomic factors on nutritional status in primary school children. *J. Ayub. Med. Coll. Abbottabad* **2010**, *22*, 15–18. [PubMed]
56. Nkonki-Mandleni, B.; Ogunkoya, F.T.; Omotayo, A.O. Socioeconomic factors influencing livestock production among smallholder farmers in the free state province of south Africa. *Int. J. Entrep.* **2019**, *23*, 1–17.
57. Naude, W.; Serumaga-Zake, P. An analysis of the determinants of labour force participation and unemployment in South Africa's North-West province. *Dev. S. Afr.* **2001**, *18*, 261–278. [CrossRef]
58. Sigit, H. Income distribution and household characteristics. *Bull. Indones. Econ. Stud.* **1985**, *21*, 51–68. [CrossRef]
59. Ibrahim, H.; Rahman, S.A.; Envulus, E.E.; Oyewole, S.O. Income and crop diversification among farming households in a rural area of north central Nigeria. *J. Agro-Sci.* **2009**, *8*, 84–89. [CrossRef]
60. Omotayo, A.O. Farming households' environment, nutrition and health interplay in Southwest, Nigeria. *Int. J. Sci. Res. Agric. Sci.* **2016**, *3*, 84–98. [CrossRef]
61. Kamo, Y. Racial and ethnic differences in extended family households. *J. Sociol. Perspect.* **2000**, *43*, 211–229. [CrossRef]
62. Omotoso, A.B.; Daud, A.S.; Adebayo, R.A.; Omotayo, A.O. Socioeconomic determinants of rural households' food crop production in Ogun State, Nigeria. *Appl. Ecol. Environ. Res.* **2018**, *16*. [CrossRef]
63. Adelekan, Y.A.; Omotayo, A.O. Linkage between rural non-farm income and agricultural productivity in Nigeria: A tobit-two-stage least square regression approach. *J. Dev. Areas* **2017**, *51*, 317–333. [CrossRef]
64. Omotayo, A.O. Economic Synergy between Rural Off-farm Income and Households' Poverty in Ekiti State, Nigeria. *J. Hum. Ecol.* **2016**, *56*, 99–106. [CrossRef]
65. Onubuogu, G.C.; Esiobu, N.S.; Nwosu, C.S.; Okereke, C.N. Resource use efficiency of smallholder cassava farmers in Owerri Agricultural zone, Imo State, Nigeria. *Sch. J. Agric. Sci.* **2014**, *4*, 306–318.
66. Olorunfemi, O.D.; Adekunle, O.A.; Oladipo, F.O.; Oladele, T.O.; Oladele, O.I. Training needs of fish farmers on value addition initiatives in Kwara State, Nigeria. *Sarhad J. Agric.* **2017**, *33*, 1–21. [CrossRef]
67. Akinnifesi, F.K.; Ajayi, O.C.; Sileshi, G.; Kadzere, I.; Akinnifesi, A.I. Domesticating and commercializing indigenous fruit and nut tree crops for food security and income generation in Sub-Saharan Africa. In Proceedings of the New Crops International Symposium, Southampton, UK, 1 September 2007.
68. Ndhlovu, P.T.; Omotayo, A.O.; Aremu, A.O.; Otang-Mbeng, W. Herbal-Based Cosmeceuticals and Economic Sustainability among Women in South African Rural Communities. *Economies* **2020**, *8*, 51. [CrossRef]

69. Omotayo, A.O.; Aremu, A.O. Underutilized African indigenous fruit trees and food–nutrition security: Opportunities, challenges, and prospects. *Food Energy Secur.* **2020**, *9*. [CrossRef]
70. Johnstone, D.B.; Marcucci, P.N. *Financing Higher Education Worldwide: Who Pays? Who Should Pay?* The Johns Hopkins Univrsity Press: Baltimore, MD, USA, 2010.
71. Singh, U. Financial literacy and financial stability are two aspects of efficient economy. *J. Financ. Account. Manag.* **2014**, *5*, 59.

sustainability

MDPI

Article

Rural Transformation and Labor Market Outcomes among Rural Youths in Nigeria

Adedoyin Mistura Rufai [1], Adebayo Isaiah Ogunniyi [2], Kabir Kayode Salman [3], Mutiat Bukola Salawu [4] and Abiodun Olusola Omotayo [5,*]

1 World Bank Group, Abuja 900231, Nigeria; dmisty28@yahoo.co.uk
2 International Fund for Agricultural Development (IFAD), Abuja 900101, Nigeria; a.ogunniyi@ifad.org
3 Department of Agricultural Economics, University of Ibadan, Ibadan 200213, Nigeria; kksalmann@gmail.com
4 Federal College of Animal Health and Production Technology, Ibadan 930103, Nigeria; salawumutiat@gmail.com
5 Food Security and Safety Niche Area, Faculty of Natural and Agricultural Sciences, North West University, Private Bag X2046, Mmabatho 2745, South Africa
* Correspondence: 25301284@nwu.ac.za; Tel.: +27-611649252

Abstract: In Nigeria, unemployment in rural areas translates to economic problems, such as high levels of rural–urban migration. Interventions aimed at promoting rural transformation and development are designed to generate employment by promoting the growth of sectors such as manufacturing and services in rural areas. In this study, the General Household Survey (GHS) panel data for the post-planting and post-harvest periods of the 2011/2012 and 2015/2016 cropping seasons for Nigeria was used to investigate developments in rural areas in Nigeria between 2011 and 2015, and identified how these developments influenced labor market outcomes among rural youths. Fixed effect models were employed to control for unobserved heterogeneity that may exist because of the different years in the data used. Key levers of sustainable social and economic development, such as access to finance, health services, markets, and infrastructure such as electricity, were considered. The empirical results from the study revealed that being educated as well as having access to infrastructure and information had positive effects on the number of youths that took up wage/salary employment in the rural areas. The study concluded that the diversification of youths into other sectors would have higher growth effects on the development of rural areas, as they can invest more in agriculture, while also reducing the level of dependence on the sector. The study recommends an increase in budgetary allocations for education and rural development projects, with a special focus on electricity and financial institutions, while increasing access to information on available job opportunities.

Keywords: binary outcome models; off-farm; rural development; principal component analysis; wage employment

Citation: Rufai, A.M.; Ogunniyi, A.I.; Salman, K.K.; Salawu, M.B.; Omotayo, A.O. Rural Transformation and Labor Market Outcomes among Rural Youths in Nigeria. *Sustainability* **2021**, *13*, 13794. https://doi.org/10.3390/su132413794

Academic Editor: Aaron K. Hoshide

Received: 28 September 2021
Accepted: 2 December 2021
Published: 14 December 2021

1. Introduction

Agriculture persists as the main employer of labor in the rural areas in Sub-Saharan Africa (SSA). In Nigeria, about 70% of the youth-dominated labor force lives in rural areas and only very few of them work in the non-farm sector as well as in small and medium-scale enterprises that depend directly or indirectly on agriculture [1–4]. While their employment status may be seasonal or casual, the majority of the rural youths earn very low income/wages while they work under unfavorable and unsafe working conditions [5,6]. This has persisted even though rural development projects which address increasing access to economic resources such as electricity, school infrastructure, health services, employment opportunities, and portable water have been put in place by successive governments to improve the living and working conditions among rural dwellers. Additionally, while there has been some focus on economic development, other key drivers of sustainable development, which promote a balance between economic prosperity, human capital development and a healthy environment, receive little attention [7].

While a rise in non-farm income is an expected outcome of structural transformation, the World Bank and International Fund for Agricultural Development (IFAD) [8] explained that most of the non-farm activities in rural areas are often related to agriculture and the overall food system, hence economies that were successful in rural diversification were those that had strong and continuously growing agricultural sectors. Therefore, to increase the level of rural youth engagement in both farm and non-farm activities in rural areas, there is a need to put in place policies and interventions that would facilitate the need (demand) for rural labor and also improve the quality (supply) of labor in those areas [8–11], while addressing other economic and social issues characteristic of rural areas so people can fulfill their potential in dignity, equality, and a healthy environment (The 2021 Nigeria Sustainability Outlook by PwC) [7]. With the quest to develop the agricultural sector and to reduce the level of unemployment in Nigeria, drivers of sustainable development, such as access to infrastructure including electricity and markets, information, finance, and other welfare indicators, were assessed in terms of their influence on the involvement of youths in both on-farm and non-farm employment over the years remains germane [12,13]. In this regard, the General Household Survey (GHS) panel data covering the post-planting and post-harvest periods of the 2011/2012, 2013/2014 and 2015/2016 cropping seasons for Nigeria was used to assess the relationship between rural transformation and employment choices among youths in this study. In analyzing the data, fixed effect models were employed to control for unobserved heterogeneity that may exist because of the different years.

A growing number of studies have examined the relationship between government investment and policies in rural areas and youth employment using various indicators. This study differentiates itself from studies such as [7,14,15], that have assessed how access to economic resources influenced the engagement of youths in paid off-farm work in Nigeria in several ways. Unlike these studies that used cross-sectional data and focused on specific states within Nigeria, this study used panel data and gave a broader outlook by looking at various types of off-farm employment in rural areas at the national level. Compared to cross-sectional data, the use of panel data allowed the authors to minimize estimation biases, increase efficiency, and account for variability across groups in the analyses. Though Adeoye, et al. [16], used the 2013 GHS data and Van den Broek and Tilip used both 2013 and 2015 GHS data to analyze the extent of off-farm diversification at the national level, this study differentiates itself by using three waves (2011, 2013, and 2015) of the GHS data for Nigeria. Additionally, this study also differentiates itself from both studies by considering youths that might be in dual employment positions and including more indicators of rural transformation, such as access to health facilities, information, and basic needs (housing).

The main objective of the paper was to examine how key levers of sustainable social and economic development, such as access to finance, health services, markets, and infrastructure such as electricity, have influenced labor market outcomes among rural youths. The labor market outcomes among youths were specified by wage/salaried employment, self-employment, and dual employment (i.e., being employed in both wage/salaried employment and self-employment) in the study. The goal of the study is to provide critical policy insights into how interventions aimed at promoting rural transformation and development have influenced employment choices made by rural youths and to provide recommendations on how such interventions can be more effective to promote economic, environment, and community sustainability.

2. Background

2.1. Nigerian Youth Employment and Sustainable Development Policies

In 2017, the ERGP (2017–2020) was designed to restore Nigeria's economic growth in reaction to the negative growth recorded by the country in 2016. The broad objectives of ERGP were to restore growth through macroeconomic stability and economic diversification; to build a globally competitive economy through investment in infrastructure; to improve the business environment and promotion of digital-led growth; and to facilitate investment in the Nigerian people through programs on social inclusion, job creation, youth empowerment, and improved human capital. To achieve these objectives, one of the strategies under the ERGP was to invest in Nigerian citizens by improving the health and education system as well as by addressing severe land degradation and desertification while eliminating gas flaring by 2020. Key components to achieving these objectives were implementing social safety net programs targeted at the vulnerable; boosting job creation and stressing an emphasis on made-in-Nigerian goods; improving the quality of teachers; and expanding the coverage of the NHIS.

Though the country improved in segments, such as in terms of economic diversification and the ease of doing business under the ERGP, the country still needs to improve in terms of job creation and infrastructural development to improve revenue generation by the government. In 2020, the Nigeria Economic Sustainability Plan (NESP) was developed as a successor plan to the ERGP. The NESP is a 12-month NGN 2.3 trillion 'Transit' plan designed to promote local production and the use of local materials; economic stimulation to ensure liquidity and prevent business collapse; and job preservation and both the creation and provision of social protection to vulnerable groups, including women and persons living with disabilities. The NESP includes a plan to reduce annual fuel subsidies and a commitment from the federal government to promote the use of green energy by delivering and maintaining about five million new solar connections under a solar power strategy.

However, several factors influence the decision of youths to engage in off-farm employment and to determine the type of jobs they engage in. References [4,17–19] revealed that due to restricted opportunities in rural areas, desperate youths going into off-farm work choose low-paying, unskilled off-farm wage employment to earn a living [4,18–20]. The low employment opportunities for youths in rural areas have been related to the underdevelopment and poor access to economic resources [21]. References [22,23] explained that the unfavorable conditions in the rural areas encourage youths to migrate to urban areas in search of jobs, thus reducing the availability of labor in rural areas while also widening the informal sectors in urban areas as well as making sub-optimal contributions to economic development.

Bezu and Holden [24] explained that if the non-farm sector in rural areas is strong and viable, most youths would most likely remain in the rural areas and diversify into other sectors while also engaging in activities in the agricultural sector. This is evident as the availability of key infrastructures, such as electricity, markets, and pipe-borne water, have been identified as major drivers of diversification in rural areas in Nigeria [7,16]. Other characteristics such as the level of education [7] and ownership of productive assets such as land [16] have also been identified to influence participation in off-farm jobs. Access to sound education also influences the capacity of youths to take up jobs outside the agricultural sector. Poorly educated youths, without the right skills have slimmer chances of being employed in decent and well-paying jobs. Even when they are engaged in other sectors such as manufacturing and construction work, low-income levels and underemployment remain as common features among them [20,25]. In addressing the challenges of poor development and its economic implications in rural areas, Timmer and Timmer and Akkus [1] opined that effective policies and interventions that promote rural transformation can increase the relative contribution of sectors, such as industries and services, while reducing the contribution from agriculture. Such policies also reduce the level of subsistence farming among agricultural households as it promotes increases in agricultural productivity and raises the quantity of marketable surplus [2]. This is evident as studies such as [7], revealed that off-farm income has positive and significant effects on-farm output and the demand for purchased inputs.

2.2. Theoretical Framework and Econometric Model Specification

Theoretical Perspective

Diversification under the sustainable livelihood framework is considered as one of the options taken by people to reduce poverty and vulnerability, and to cope with and recover from stresses and shocks while maintaining or enhancing their capabilities [26,27]. However, diversification is strongly related to whether or not (and to which extent) people have access to the five forms of capital assets, depicted as human capital, social capital, natural capital, physical capital, and financial capital. Applying this to off-farm employment among youth implies that access to capital assets and the extent of their access influences the choice of youths to diversify into employment outside the agricultural sector. In rural areas, youths struggle in terms of access to natural assets such as land. Rural areas are also disadvantaged in terms of physical infrastructure, such as good roads, potable water, and shelter, while access to health and financial services may be very difficult. The poor level of development in rural areas translates to a low level of investments in rural areas, which keeps job prospects for youths very low. Based on these theoretical perspectives, we construct our first hypothesis in line with the argument that access to capital assets increases the probability of youth participation in off-farm employment opportunities in rural Nigeria.

Hypothesis 1 (H1). *Access to capital assets increases the probability of youth participation in off-farm employment opportunities in rural Nigeria.*

Buchenrieder and Mollers [27], also used the demand-pull/distress-push concept to explain labor shifts from the agricultural sector to the rural non-farm sector in diversification. The demand-pull concept describes situations where agricultural labor engages in more lucrative employment in the rural non-farm economy. Agricultural workers that take up lucrative jobs in rural areas often have characteristics that enable them to do so as demand-pull factors, including education level, skills, and access to social networks. The distress-push concept describes a situation where agricultural workers are pushed into poorly paid jobs because of inadequate agricultural income. Factors that promote distress in push situations include low farm productivity, low financial capacity, small farm size, and low land productivity. Both demand-pull and distress-push factors are constrained by a lack of infrastructure. While distress push factors have benefits of improving the livelihood of the poor by increasing income, they are also constrained by inefficient institutions, high unemployment rates, poor access to markets, and legal and cultural barriers. Thus, this line of argument indicates that when making choices about off-farm engagements, different factors influence decisions regarding which type of off-farm employment to engage in, and this is captured in the second hypothesis.

Hypothesis 2 (H2). *Demand-push factors (level of education) and distress-pull factors (farm productivity) influence the type of off-farm employment opportunities youths engage in, in rural Nigeria.*

In addition to the factor which influences employment choices, the large dependence on agricultural work in rural areas reduces the demand for human capital development particularly in terms of education and skill acquisition [18,23]. Consequently, rural youths are often the most disadvantaged with little experience and low chances of obtaining capital and other production assets [5].

3. Materials and Methods

The General Household Survey (GHS) data for Nigeria was used. The panel data was collected from 5000 agricultural households in the post-planting and post-harvest periods in 2011, 2013, and 2015. In collecting the data, a two-stage sampling procedure was utilized. In the first stage, enumeration areas (EA) were identified and selected using the Probability Proportional to Size (PPS) criteria. The PPS was based on the number of households in each in EA and the number of EAs in each state. In the second stage, ten households were randomly selected from each EA. The data is representative at the national level and has information about the household, women, youths, and children in agricultural households. It contains information about the socio-economic characteristics of each household member, employment type, and sector of employment agricultural activities and productivity, among other factors. The data covers both rural and urban areas in the country. In order to create the study sample, the respondents selected were restricted to youths that lived in rural areas and were within the age group of 18 to 35 years between 2011 and 2015. A balanced panel data set containing 1575 youths across the three years (2011, 2013, and 2015) was used. While there is the probability of rural youths engaging in on-farm work as agricultural labor, this study specifically focuses on the engagement of youths in income earning activities outside the farm. The two labor market outcomes considered in this study were (1) employment in wage/salaried job and (2) self-employment/engagement in a household enterprise.

Model Specification

We estimated the effect of rural transformation on labor market outcomes using the fixed effect (FE) model. This model allows for the consideration of the individuality of time (year) groups available in the model [28]. In this study, the FE was used to consider the differences in labor market outcomes that may occur as a result of transformation in rural areas over time. This model helps to control for unobserved heterogeneity that may exist as a result of the different years. Such time-invariant unobserved factors may correlate with independent variables in the model and, as such, produce biased results. We estimated different models to understand the relationship between rural transformation and labor market outcomes.

First, we modeled the explanatory variables against labor outcomes in the context of the wage/salaried employment model (Models 1–3). Second, we used self-employment in the household model (Models 4–6). Finally, we estimated the role of rural transformation on engaging in dual employment (both wage/salaried employment and self-employment, Models 7–9). For each model, we estimated the likelihood of labor market outcomes, specified as 1 if a youth is engaged in the type of employment and 0 otherwise. The model is specified below:

$$LMO_{it} = \gamma_l SEC_{it} + \gamma_x RTI_{it} + \gamma_x X_{it} + \gamma_x I_{it} + \gamma_x F_{it} + \nu_{it} + \varepsilon_{it} \quad (1)$$

where LMO_{it} denotes the labor outcome indicators (wage/salaried employment, self-employment/employment, and dual employment) for individual youth i at time t. SEC_{it} denotes the socioeconomic characteristics (such as age, sex, and level of education) for individual youth i at time t. RTI_{it} is a vector of indicators for rural transformation, X_{it} denotes the access to infrastructure (such as access to the market, banks, electricity, and health facility), I_i represents the institutional related variables such as access to extension services and credit facilities, F_{it} constitutes the vector of the cluster/district level, and ν_i are household fixed effects.

Following previous studies [29,30] the main drivers of rural transformation are: (1) diversification from complete reliance on agriculture, (2) progressive globalization of agri-food systems, and (3) urbanization of rural areas through increased access to public services. The authors also identified the development of roads and telecommunication services as additional factors essential to the three main factors mentioned, as there can be no diversification or globalization of food systems without adequate infrastructure for transportation. Raising the productivity of the agricultural sector in rural areas through technology, better inputs, and access to credit, amongst other avenues, is also recognized as central to rural development as it promotes the spillover to non-farm and non-agricultural sectors (United Nations-Department of Economic and Social Affairs, New York City, NY, USA, 2021). The rural transformation process generally involves a reorganization process in rural areas, which results in a decline in economic, social, and cultural differences between such areas and urban areas (Wang et al., 2013). These differences are often seen in terms of access to education, health services, information, infrastructure, and productivity, amongst others.

In this study, we focused on the extent of development in rural areas by considering the availability of public services and infrastructure in rural communities among youths in such areas to generate the rural transformation index (RTI) in this study. The indicators for rural transformation are specified as: infrastructure access to markets; access to banks; access to electricity; access to health facilities; access to information, radio, and mobile phones (all dummy variables, yes = 1.0 otherwise); and access to basic needs, e.g., number of rooms in households (continuous).

The RTI was generated using principal component analysis (PCA). Though we acknowledge that dummy variables are problematic for factor analysis (generation of dimensions), the approach is acceptable for our needs in this study. The PCA linearly transforms data into a substantially smaller set of uncorrelated variables called principal components such that it contains most of the information in the original data. The principal components are found by calculating the eigenvectors and eigenvalues of the covariance matrix. The principal components (PC_J) of variable $X_1, \ldots, X_n$ are linear combinations $\alpha_{i1}, \ldots, \alpha_{in}$ such that the dimensions of $a_n < x_n$. The principal component (PC_J) is given as:

$$PC_J = a_j x \tag{2}$$

where $j = 1, \ldots, n$.

The first principal component accounts for the largest variance among the variables and the variance decreases from PC_1 to PC_n. The index is generated as a weighted average of the variable scores with weights equal to the loadings of the first principal component.

$$c_i = \sum_{i=1}^{n} w_1 \, x_1 \tag{3}$$

where C = composite index, w = weight attributed, and n = number of variables.

4. Descriptive Statistics

4.1. Socio-Economic Characteristics

About 51.24% of the respondents were males and 52.32% lived in the northern geopolitical zones of the country. As expected, the average age of youths increased across the years. In 2011, the average age of youths was 23.76 ± 4.30 and this rose to an average of 28.00 ± 4.23 in 2015. The level of education among youths also increased across the years. There was a drop in the proportion of youths with no education and primary education, while there was an increase in the proportion of youths with higher degrees across the years. This is consistent with the findings of [22], which revealed that there has been a continuous increase in school enrollment in both rural and urban areas of Nigeria over the years (see Table 1).

4.2. Labour Market Outcomes

About 10.16% of rural youths were employed in wage/salaried jobs in 2011. This dropped slightly to 9.77% in 2013 and then rose to about 11.68% in 2015. This implies that only a small proportion of youths were engaged in wage/salaried employment over the years. The obvious variation across the years was in the sectors in which youths took up wage/salaried jobs (see Tables A5 and A6). In 2011, about 33.44% of the employed youths earned wages/salaries from jobs in the agricultural sector. However, this dropped to 21.59% in 2013 and 7.64% in 2015. About 11% of youths were employed in construction in 2013 compared to 3.75% in 2011. In 2015, more youths were employed in education (20.14%), financial and personal services (14.58%), health (10.42%), and public administration (17.36%) sectors.

Table 1. Descriptive statistics.

	2011		2013		2015	
	Mean	**SD**	**Mean**	**SD**	**Mean**	**SD**
Age	23.75	4.30	25.65	4.35	28.00	4.22
Female	0.49	0.50	0.49	0.50	0.49	0.50
Education						
None	0.10	0.30	0.09	0.28	0.09	0.29
Primary	0.24	0.43	0.17	0.37	0.16	0.37
Secondary	0.55	0.49	0.59	0.49	0.53	0.49
Higher	0.09	0.29	0.15	0.35	0.20	0.40
Employment						
Wage/salaried job (yes)	0.10	0.30	0.09	0.29	0.11	0.32
Self-employment (yes)	0.26	0.44	0.25	0.43	0.32	0.46
Dual (wage and self-employment)	0.03	0.16	0.01	0.11	0.10	0.10
Access to rural infrastructure						
Electricity (yes)	0.60	0.49	0.61	0.49	0.63	0.48
Market (yes)	0.59	0.49	0.66	0.47	0.76	0.43
Health facility (yes)	0.72	0.45	0.72	0.49	0.85	0.36
Financial services (yes)	0.28	0.45	0.28	0.45	0.31	0.46
Access to information						
Mobile phone (yes)	0.85	0.36	0.91	0.29	0.96	0.19
Radio (yes)	1.00	0.00	1.00	0.00	0.67	0.47
Internet (yes)	0.09	0.29	0.14	0.35	0.33	0.47
Agricultural production						
Value of output	5030.96	20,515.75	7282.47	42,647.86	9013.31	0
Basic needs						
Number of rooms	4.09	2.86	4.18	2.69	4.18	2.74

The proportion of youths engaged in self-employment/household enterprises increased to 31.60% in 2015 compared to 25.17% in 2013 and 25.90% in 2011. The results indicate that more youths were more likely to be self-employed or work in the household enterprise compared to working for a wage/salary in rural areas. The majority of the self-employed youths were employed in three sectors, namely buying and selling, manufacturing, and services, over the years.

4.3. *Indicators of Rural Transformation*

About 60% of youths lived in households that had access to electricity in 2011. The proportion of youths with access to electricity grew narrowly to about 61% in 2013 and 63.11% in 2016. More youths had access to markets in 2013 (66%) and 2015 (75%) compared to 2011 (59%). Regarding health facilities, about 71% of youths had at least a healthcare center in their community in both 2011 and 2013, while in 2015, the proportion of youths with access to health infrastructure increased to about 85%. Less than 15% of youths had communities with banks and other financial institutions in 2011. However, in 2013 (42%) and 2015 (44%), more youths had at least one financial institution in their community.

The majority of the youth could assess information through mobile phones as over 85% had access to mobile phones across the years. Additionally, most of the rural youths also had access to radio as in 2011, about 88% had access to radios. In 2013, the proportion of youths with access to radios rose to about 92%; however, in 2015 it dropped to about

67%. This could be associated with the increased availability of radio applications on most mobile phones. Sambira [31] explained that youths in Africa use their mobile phones for most activities, which include communicating, shopping, listening to the radio, and interacting on social media.

In terms of agricultural productivity, over 80% of youths did not have any agricultural output. However, about a quarter of youths worked as household laborers on plots owned by family members across the years (see Appendix A Table A2). This implies that while a significant proportion of youths in rural areas participated in agricultural/farming activities, few of them had their own personal farms/plots. For youths that had agricultural output, the level of their productivity increased across the years as the average value of output rose from NGN 35,060.92 in 2011 to NGN 47,667.68 in 2015. Regarding basic needs, the average number of rooms remained at four.

5. Results and Discussion

5.1. Distribution of Labor Market Outcomes across Socio-Economic Characteristics of Rural Youths and Zones

A higher proportion of males were employed in wage/salaried employment while more females engaged in self-employment or worked in the household enterprise across the years, as shown in Table 2. Adesugba and Mavrotas [3] revealed that the proportion of females in the youth labor force in Nigeria is on the rise and more are earning jobs in the informal sector than the formal sector. Older youths were more likely to be employed in either type of employment as youths aged between 25 and 35 years were more likely to be employed compared to those aged less than 25 across the years. This is consistent with the findings of [32] which revealed that youths aged between 25 and 35 were more likely to work in non-farm enterprises compared to those who were older. The proportion of employed youths varied across the level of education. The proportion of youths without any education and those with primary education that engaged in wage/salary jobs dropped across the years, while the proportion of those with higher education (26.93%) increased in 2015.

For self-employment/engagement in household enterprises, those without any education had the highest proportion of youths engaged across the years. However, while the proportion of those with primary and secondary education engaged in self-employment/household enterprises increased over the years, the proportion of those with higher education dropped in 2015. In the northern zones, the proportion of youths employed in wage/salary employment positions dropped from 11.65% in 2011 to 10.57% in 2013. However, in the southern zones, it increased from 8.52% in 2011 to 12.88% in 2013. For self-employment/employment in household enterprises, the proportion of youths engaged in the northern zones dropped from 29.49% in 2011 to 25.51% in 2013 and rose to 33.78% in 2015. In the southern zones, there was an increase from 21.97% in 2011 to 29.22% in 2015. The drop in the engagement of youths in the northern zones in off-farm work in 2013 could be associated with the occurrence of insurgencies, particularly in the north-east zone. According to Awojobi [33], the peak of the fatalities of the dreaded Boko Haram attacks occurred between 2012 and 2014 after the initial attack in 2009. Commercial activities in the north-east reduced as financial institutions and businesses did not operate regularly, and investors and human capital continued to leave the area due to unprecedented attacks [34]. The F-test shows that there was a significant difference in the proportion of rural youths employed in wage/salaried employment positions and those that were self-employed/employed in household enterprises across sex, age, level of education, and zone of residence across the years.

Table 2. Distribution of labor market outcomes across socio-economic characteristics and zones.

	Employment in Wage/Salaried Jobs			F-Rest	Self-Employment/Employment in Household Enterprises			F-Test
	2011	2013	2015		2011	2013	2015	
	Yes (%)	Yes (%)	Yes (%)		Yes (%)	Yes (%)	Yes (%)	
Male	12.14	10.85	13.04	11.07 *	18.71	20.35	25.95	88.27 *
Female	8.07	8.62	10.22		33.46	30.29	37.61	
Age								
18–24	6.06	5.56	5.40	81.28 *	15.60	14.59	18.51	238.89 *
25–35	15.50	12.83	13.73		39.33	32.89	35.89	
Education								
None	6.67	4.55	1.30	77.86 *	37.58	30.30	46.10	43.08 *
Primary	5.71	4.40	3.89		32.99	33.60	42.02	
Secondary	10.23	8.72	10.10		22.87	24.38	32.66	
Higher	24.52	25.00	26.93		12.90	14.09	13.62	
Zones								
Northern zones	11.65	9.07	10.57	3.77 *	29.49	25.51	33.78	21.58 *
Southern zones	8.52	10.53	12.88		21.97	24.80	29.22	

Source: Author's computation of GHS 2011, 2013, and 2015; level of significance at * 1%.

5.2. Empirical Results

The effect of rural transformation on labor market outcomes among youths was examined using three different models. For each outcome, the first model examined only the effects of rural transformation (using the transformation index-Appendix A) on the type of employment. The second model included controls for the socio-economic characteristics of youths and their region of residence, while in the third model, the indicators of rural transformation were controlled for individually with the socio-economic and regional variables (see Table 3).

Table 3. Control variables in each model for labor market outcomes.

Variables Controlled for in Each Model	Employment in Wage/Salaried Job Models	Self-Employment/Employment in Household Enterprise Models	Dual Employment Models
Index only	1	4	7
Socio-economic characteristics	2	5	8
Individual indicators of rural transformation with controls for socio-economic characteristics	3	6	9

The selected independent variables were subjected to a variance inflation factor (VIF) test to check for multicollinearity among the variables. For the employment in the wage/salary employment model, the mean VIF was 1.09, while in the self-employment model, the mean VIF was 1.10. This indicates the absence of any significant multicollinearity in any of the models, as it was less than 10 (see Appendix A Tables A3 and A4).

5.2.1. Socio-Economic Characteristics

For the socio-economic characteristics of youths, despite the controls for rural transformation in Models 2 and 3 for self-employment and Models 5 and 6 for wage employment, the results revealed that compared to females, males were more likely to be engaged in wage/salaried employment and less likely to engage in self-employment across time. This is consistent with the findings of [35], which revealed that more men participate in wage and self-employment opportunities compared to women, though the participation of women in off-farm employment opportunities was more common in Nigeria. Age had a positive significant relationship with the employment of youths in wage/salaried employment and self-employment positions. This implies that over time, older youths in rural areas were more likely to engage in diversification strategies, such as engaging in off-farm employment positions in rural areas. This also confirms the distribution of employment across age in Table 4, wherein those aged between 25 and 35 years were found to have a higher proportion of those employed compared to those who were younger. These results are consistent with the findings of [36], which revealed that unemployment is particularly common among younger youths (aged between 15 and 24 years) and young females. Adesugba and Mavrotas [3] associated the high level of underemployment among youths aged between 15 and 24 years to the fact that such youths are usually still in school or receiving some form of education.

The coefficients for the levels of education show that secondary and higher levels of education have a positive and significant relationship with wage/salaried employment over time. The results indicate that educated youths are more likely to engage in off-farm employment compared to those with little or no form of education over time. This result is consistent with the findings of [37], and [38] which found that the length of formal education is positively associated with participation in off-farm employment positions. In assessing the pattern of youth employment in Namibia, the Namibia Statistics Agency [39] revealed that having high school education or higher, being married, or being between the ages of 30 and 34 increases the probability of youth employment.

The results, however, also indicate that having tertiary education has a significant negative relationship with being self-employed among youths. This implies that youths with higher levels of education are less likely to start a business of their own. The result indicates that highly educated youths are more likely to depend on blue collared jobs rather than engage in self-employment. Yeboah, Jayne, Muyanga and Chamberlin [38], explained that increased educational attainment enhances the prospects of youths to secure off-farm employment opportunities and raises their career aspirations beyond agriculture; as farming is associated with lower social status, young people are socialized to have career aspirations beyond farming. These result leads us to accept the hypothesis that demand pull factors, such as being highly educated and skilled, allows youths to engage in well paying off-farm jobs compared to those that are less educated.

The coefficient of the northern zones had significant and negative effects on the engagement of youths in wage/salaried employment over time. This implies that compared to the southern zones, less youths are likely to be employed in wage/salaried employment positions in the northern zones. The low likelihood of northern youths in wage/salary employment could be associated with the poor acquisition of skills and low education level attainment in the region. According to the World Bank [40], there is a large geographic divide in Nigeria as the northern part of the country has a higher proportion of uneducated and underemployed youths compared to the southern part.

Only the coefficient of the south-south region had a positive and significant relationship with dual employment over time. However, the coefficients of all the zones had a positive and significant relationship with self-employment over time. This implies that there has been growth in the proportion of youths that engage in self-employment all over Nigeria, irrespective of their location. This may be associated with the low availability of formal jobs, which has forced many youths to start their own business. Self-employment remains the main form of employment in Nigeria as, according to the World Bank Development indicators (WDI), self-employment as a proportion of the total employment in Nigeria is about 80% compared to the Sub-Saharan African average of 75% and world average of 46% in 2019.

Table 4. Rural transformation and labor market outcomes.

	Labor Market Outcomes								
	Wage Employment			Self-Employment			Dual Employment		
	−1	−2	−3	−4	−5	−6	−7	−8	−9
VARIABLES									
RTI	0.0217 **	0.0112 *		0.0292 ***	0.0125 **		0.144 ***	0.111 ***	
	−0.00372	−0.00321		−0.00502	−0.00128		−0.00201	−0.00146	
Sex (male)		0.0339 *	0.0362 **		−0.0698 **	−0.0607 *		0.000607	0.000548
		−0.00825	−0.0082		−0.0137	−0.0151		−0.00323	−0.00285
Age (years)		0.0136	0.0121		0.0543 **	0.0580 ***		0.00412	0.00397
		−0.0126	−0.0126		−0.00776	−0.00434		−0.00524	−0.00522
Age (squared)		-5.04×10^{-5}	-3.44×10^{-5}		−0.000541 *	−0.000641 **		-4.71×10^{-5}	-4.52×10^{-5}
		−0.000248	−0.000243		−0.000179	−0.000122		9.74×10^{-5}	9.58×10^{-5}
Primary		−0.00254	−0.00615		0.0102	0.000899		0.00949	0.0086
		−0.00628	−0.00636		−0.0215	−0.0239		−0.00445	−0.00344
Secondary		0.0492 *	0.0376		0.0496 *	0.0621 **		0.0127	0.0105
		−0.0157	−0.0162		−0.012	−0.0119		−0.00689	−0.00549
Tertiary		0.186 **	0.169 **		−0.243 **	−0.243 **		0.0223	0.0191
		−0.0258	−0.0272		−0.0361	−0.0382		−0.00835	−0.011
Radio access			0.131 ***			0.0433 **			0.0145 ***
			−0.00469			−0.00476			−0.00296
Phone access			0.283 ***			0.0522			0.108 ***
			−0.0143			−0.0324			−0.00544
Internet access			0.0685 **			0.0874 **			0.0308 ***
			−0.0111			−0.0129			−0.002
Bank			0.196 ***			0.000532			0.0142 ***
			−0.0189			−0.0256			−0.00384
Health			0.0114			0.0217 **			0.0217 ***
			−0.0148			−0.00348			−0.00184
Market			0.006			0.672 ***			0.0329 ***
			−0.0142			−0.0177			−0.00769
Electricity			0.0446 **			0.0913 ***			0.0618 ***
			−0.00573			−0.00723			−0.00444
Rooms			−0.00380 **			−0.00584			7.16×10^{-5}
			−0.000427			−0.00325			−0.00112
ln_ag_pdvty		−0.00733	−0.449 ***		0.0137	0.0910 ***		0.000456	0.00104
		−0.00474	−0.00505		−0.00933	−0.00917		−0.000601	−0.000604

Table 4. *Cont.*

	Labor Market Outcomes								
	Wage Employment			Self-Employment			Dual Employment		
	−1	−2	−3	−4	−5	−6	−7	−8	−9
VARIABLES									
North-central		−0.0258 **	−0.0383 **		0.142 **	0.118 ***		−0.00351	−0.00469
		−0.0051	−0.00596		−0.0165	−0.0116		−0.00593	−0.00523
North-east		−0.00306	−0.0117		0.0197 **	0.0368 ***		−0.00989	−0.0112
		−0.00912	−0.0104		−0.0021	−0.00306		−0.00622	−0.00679
North-west		−0.0296 **	−0.0357 **		0.0730 *	0.0782 *		−0.00834	−0.00840
		−0.0059	−0.00465		−0.0176	−0.0218		−0.00551	−0.00546
South-east		−0.0150	−0.0175		0.0712 *	0.0588		0.0228	0.0227
		−0.0275	−0.0307		−0.0192	−0.0296		−0.0124	−0.0132
South-south		0.0011	0.00562		0.0972 *	0.109 **		0.00967 **	0.0107 **
		−0.0274	−0.024		−0.0286	−0.0254		−0.00183	−0.00221
Constant	0.105 ***	0.267 ***	0.277 ***	0.276 ***	−0.676 **	−0.696 **	0.0154 ***	−0.0733	−0.0825
	0	−0.054	−0.072	−0.009	−0.099	−0.0911	0	−0.0731	−0.0774
Control	No	Yes	Yes	No	Yes	Yes	No	Yes	Yes
Observations	4728	4728	4728	4728	4728	4728	4728	4728	4728
R-squared	0.009	0.374	0.479	0.04	0.124	0.438	0.04	0.112	0.414
Number of waves	3	3	3	3	3	3	3	3	3

Robust standard errors in parentheses. *** $p < 0.01$, ** $p < 0.05$, and * $p < 0.1$.

5.2.2. Agricultural Productivity

The coefficient of the value of output from agriculture had positive growth effects on the engagement of youths in self-employment over time. This implies that youths that produce more agricultural output are more likely to own their own business outside the farm in rural areas of Nigeria. This is consistent with the findings of [24], which found that agricultural productivity growth had a significant positive effect on the growth of informal (small-scale) manufacturing.

The results also revealed that agricultural productivity had a negative effect on wage/salary employment in the model across time. This indicates that youths that have high output from agriculture are less likely to leave the sector to take up jobs in order to earn wage/salary. Adesugba and Mavrotas [3,41,42] explained that the increased use of mechanization and adoption of technology can increase agricultural productivity and make the sector look more attractive to youths. Shilpi and Emran [21] explained that agricultural productivity growth often leads to structural transformation within the services sector. These results lead us to accept the hypothesis that distress push factors, such as low agricultural productivity and income, can force youths to engage in off-farm jobs to raise additional income.

5.2.3. Rural Transformation and Employment among Youths

The coefficient of the transformation index revealed that having access to infrastructure in rural areas had positive growth effects on the engagement of youths in all forms of employment in rural areas. The coefficient of the transformation index was highest in the dual employment model. This indicates that more youths were likely to engage in both wage/salary employment and self-employment with increased access to economic resources. The results of each indicator of rural transformation in Models 3, 6, and 9 provides more insight on the positive influence of increased access to economic resources on labor market outcomes among youths.

Having access to information through mobile phones, radio, and the internet had a significant positive relationship with youth employment. This indicates that youths with access to information are more likely to be engaged in the labor market. According to the World Bank [43], the difficulty youths face in terms of finding employment and gaining access to inputs such as capital and land is usually augmented by the fact that they have less access to information compared to adults. Sibisi [43] explained that even though more youths are aspiring to become entrepreneurs, the challenges regarding a lack of access to information, networking, mentoring, and finances remain barriers to their efforts of creating employment opportunities for themselves.

Increased access to banks, health institutions, and markets had positive effects on the engagement of youths in self and dual employment. Access to electricity had significant and increasing growth effects on the number of youths engaged in all three forms of employment considered in the study. This implies that improving access to electricity in rural areas would also translate into an increased engagement of youths in the off-farm labor market. Klutsey and Ankomah [36] recommended that the development of infrastructure, particularly electricity, will promote growth in the industrial sector and consequently create employment for millions of job seekers, especially Nigerian youths [44,45]. These results lead us to accept the hypothesis that access to capital assets increases the probability of youth participation in off-farm employment opportunities in rural Nigeria.

6. Conclusions

In this study, the relationship between different indicators of rural transformation and labor market outcomes among rural youths was examined. Panel data covering three periods, including 2011, 2013, and 2015, for 1575 youths was examined using a binary outcome model (fixed effect regression) for panel data. Rural transformation was captured using access to infrastructure, access to information, agricultural productivity, and access to basic needs.

The results from the study revealed that despite the growth in the number of educated rural youths across the years, the majority of youths that were educated had only secondary education. Female youths and the uneducated were more likely to start their own businesses or be engaged in the household enterprise. However, more males and the educated were more likely to work for wages/salary. Youth employment in off- farm work positions is influenced by the level of development in rural areas and, across the years, more rural youths took up off-farm jobs outside the agricultural sector. Youths in the rural areas of the northern zones are less likely to work in wage/salary employment positions compared to their counterparts in the southern zones. Most of the youths that were self-employed/employed in household enterprises were engaged in buying and selling, service delivery, and manufacturing activities. There has been very little improvement in the access to electricity among rural households and both access to financial institutions and services also remains low. Having access to information plays a significant role in the engagement of youths in off-farm work.

From the results, it can be concluded that more rural youths are likely to take up off-farm jobs and diversify into other sectors if rural areas are continuously developed. The diversification of youths into other sectors would subsequently have a higher growth effect on the development of rural areas as they can invest more in agriculture while also reducing the level of dependence on the sector.

7. Recommendations

The findings from this study offer important insights for policy direction.

- Firstly, more educated youths are likely to seek wage employment and are less likely to be self-employed. The Federal Government of Nigeria needs to provide more incentives to encourage investment and business creation among young Nigerians. To increase economic growth, rural youths should be encouraged to acquire skills by strengthening policies and interventions such as the Nigerian Industrial Skills Development Program (NISDP) and the Nigerian Enterprise Development Program (NEDP) to ensure they have the necessary skills to start their own businesses. Rural youth empowerment can also translate to environment and community sustainability.
- Secondly, a significant proportion of youths drop out of school after secondary education. The Federal Government should increase budgetary allocations to the education sector to improve the quality of learning and to establish more tertiary institutions. More rural youths should be encouraged to further their education after attending secondary school through scholarships and bursaries. Such policies will promote economic prosperity and address social differences that exist in communities.
- Thirdly, there has been very little improvement in the access to electricity among rural households and both access to financial institutions and health services remains low. The Federal Ministry of Works and Housing needs to ensure that rural development projects, with a special focus on electricity, financial institutions, and basic needs such as portable water, are initiated. The rural electrification project should be strengthened to ensure that more households in rural areas gain access to electricity. Incentives and policies that would encourage the establishment of financial institutions in rural areas should be designed by the ministry of finance. The solar power strategy under the NESP should be expanded to reduce carbon emissions, contribute to a healthy environment, and ultimately promote environmental sustainability.
- The Federal Ministry of Health needs to ensure that more healthcare facilities are made available and accessible in rural areas. The Federal Government should also ensure that infrastructure such as roads and markets are available in rural areas. These policies will facilitate access to the key levers of sustainable social and economic development, and encourage business startups and investments that would improve job prospects for youths.
- Region-specific interventions should be designed and put in place particularly in the northern zones to increase employment among rural youths and address the

regional differences in terms of youth employment. Such interventions will bridge the developmental gap between regions and facilitate community sustainability especially in less developed areas.
- The Central Bank of Nigeria should also implement policies and incentives that would encourage the establishment of financial institutions in rural areas should be designed by the Ministry of Finance. Increased access to finance resources among rural youths will also translate into increased investments and job creation in rural areas, ultimately promoting economic sustainability.

Author Contributions: Conceptualization, A.M.R., A.I.O., K.K.S., M.B.S. and A.O.O.; methodology, A.M.R., A.I.O. and K.K.S.; software, A.M.R. and A.I.O.; formal analysis, A.M.R. and A.I.O.; data curation, A.M.R. and A.I.O.; writing, A.M.R., A.I.O., K.K.S., M.B.S. and A.O.O.; supervision, K.K.S. All authors have read and agreed to the published version of the manuscript.

Funding: Financial support from North-West University is appreciated. The article processing charge (APC) was paid for by the North-West University, Mmabatho, South Africa.

Institutional Review Board Statement: The research relied on a secondary data source and as such was not revised by the Institutional Review Board.

Informed Consent Statement: Informed consent was obtained from all subjects involved in the study.

Data Availability Statement: Data are available upon request from the first author.

Acknowledgments: The authors appreciate their respective institutions for providing an environment to conduct this research study.

Conflicts of Interest: The authors declare no conflict of interest.

Appendix A

Principal Component Analysis

The eigenvalues of the PCA for the healthcare indicators show that the first component has a variance of 1.79 and the second component has a variance of 1.02. The variance of the first and second component represents 44.82% and 25.44% of the total variance regarding the access to infrastructure among youths. Electricity had a negative and reducing effect on the infrastructure index in the first component, while access to banks, markets, and healthcare had reducing effects in the second, third, and fourth components, respectively.

Table A1. Variance of principal components.

Component	Eigenvalue	Difference	Proportion	Cumulative
Comp1	1.79276	0.775214	0.4482	0.4482
Comp2	1.01755	0.138234	0.2544	0.7026
Comp3	0.63519	0.080706	0.1588	0.8614
Comp4	0.55449		0.1386	1.0000

Table A2. Principal components (eigenvectors).

Variable	Comp1	Comp2	Comp3	Comp4
Electricity	−0.0003	0.9777	0.1894	0.0910
Bank	0.5732	−0.1631	0.5713	0.5643
Health	0.5976	0.0376	0.1816	−0.7801
Market	0.5607	0.1271	−0.7776	0.2546

Table A3. Youth participation in the agricultural sector.

	2011		2013		2015	
	Freq.	%	**Freq.**	%	**Freq.**	%
Household agricultural labor						
Yes	472	29.97	454	28.82	364	23.11
No	1103	70.03	1121	71.18	1211	76.89

Source: Authors computation of GHS 2011, 2013, and 2015.

Table A4. Variance inflation factor (VIF) test.

	Employment in Wage/Salaried Job Model			**Self-Employment/Employment in Household Enterprise Model**		
	VIF	**Tolerance**	**R-Squared**	**VIF**	**Tolerance**	**R-Squared**
Employment variable	1.07	0.94	0.06	1.12	0.89	0.10
North	1.09	0.92	0.08	1.09	0.92	0.08
Sex	1.08	0.93	0.07	1.08	0.92	0.07
Age	1.15	0.87	0.13	1.18	0.85	0.15
Infrastructure index	1.12	0.89	0.11	1.12	0.89	0.11
Radio	1.05	0.95	0.05	1.05	0.94	0.05
Phone	1.07	0.94	0.06	1.07	0.93	0.07
Education	1.24	0.81	0.19	1,23	0.81	0.19
Value of output	1.09	0.92	0.08	1.09	0.92	0.08
Land size	1.11	0.91	0.09	1.11	0.90	0.09
No. of rooms	1.04	0.96	0.04	1.04	0.96	0.04
Distance to water source	1.01	0.99	0.01	1.01	0.99	0.01
Mean VIF	1.09			1.10		

Table A5. Sector of youth engagement in wage/salaried employment positions.

Sectors	**2011**	**2013**	**2015**
	%	%	%
Education	12.64	18.06	20.14
Finance and personal services	6.48	9.69	14.58
Health	6.48	2.64	10.42
Public administration	12.97	14.53	17.36
Construction	3.75	10.57	7.64
Manufacturing	2.04	6.60	2.78
Transportation	4.45	5.72	6.94
Buying and selling	4.79	3.52	3.47
Agriculture	33.44	21.59	7.64
Professional/scientific/technical	4.09	1.76	4.17
Others	8.87	5.32	4.86

Table A6. Sectors of engagement for youths in self-employment/employment positions in household enterprises.

Sectors	2011	2013	2015
	%	%	%
Agriculture	1.96	1.76	1.61
Services	11.27	13.85	11.45
Buying and selling	61.27	52.14	53.21
Construction	1.96	0.76	4.02
Manufacturing	13.48	19.46	19.08
Transportation	6.13	6.80	7.83
Professional/scientific/technical	0.74	1.51	0.80
Others	3.19	3.54	2.01

References

1. Timmer, C.P.; Akkus, S. The structural transformation as a pathway out of poverty: Analytics, empirics and politics. *Cent. Glob. Dev. Work. Pap.* 2008, pp. 1–60. Available online: https://www.files.ethz.ch/isn/91306/wp150.pdf (accessed on 18 October 2021).
2. Robinson, S.; Petrick, M. *Trade-Offs among Sustainability Goals in the Central Asian Livestock Sector: A Research Review*; Justus-Liebig Universitat: Giessen, Germany, 2021; pp. 1–149.
3. Adesugba, M.; Mavrotas, G. *Youth Employment, Agricultural Transformation and Rural Labour Dynamics*; International Food Policy Research Institute (IFPRI): Washington, DC, USA, 2016; Available online: https://www.ifpri.org/publication/youth-employment-agricultural-transformation-and-rural-labor-dynamics-nigeria (accessed on 15 July 2021).
4. Ogunniyi, A.; Babu, S.C.; Balana, B.; Andam, K.S. *National Extension Policy and State-Level Implementation: The Case of Cross River State, Nigeria*; International Food Policy Research Institute: Washington, DC, USA, 2020; Volume 1951, Available online: https://www.ifpri.org/publication/national-extension-policy-and-state-level-implementation-case-cross-river-state-nigeria (accessed on 5 July 2021).
5. Food and Agriculture Organization (FAO); UNICEF. *Impact Evaluation of Africa's Child Grants Programme (CGP) and Sustainable Poverty Reduction through Income, Nutrition and access to Government Services (SPRINGS)*; FAO: Rome, Italy, 2019.
6. Omotayo, A.O. Farming households' environment, nutrition and health interplay in Southwest, Nigeria. *Int. J. Sci. Res. Agric. Sci.* **2016**, *3*, 84–98. [CrossRef]
7. Adelekan, Y.A.; Omotayo, A.O. Linkage between rural non-farm income and agricultural productivity in Nigeria: A tobit-two-stage least square regression approach. *J. Dev. Areas* **2017**, *51*, 317–333. [CrossRef]
8. Ogunniyi, A.; Oluseyi, O.K.; Adeyemi, O.; Kabir, S.K.; Philips, F. Scaling Up Agricultural Innovation for Inclusive Livelihood and Productivity Outcomes in Sub-Saharan Africa: The Case of Nigeria. *Afr. Dev. Rev.* **2017**, *29*, 121–134. [CrossRef]
9. King'au, J.M.; Ngâ, C.; Chacha, B.K. Self-Reliance Projects; The Future Core Driver of Rural Youth Livelihood Development Outcomes. *Kabarak J. Res. Innov.* **2017**, *5*, 39–47. Available online: https://africaneditors.org/journal/KJRI/abstract/05889-58558 (accessed on 5 July 2021).
10. Adeniyi, A.B.; Daud, A.S.; Amao, O.; Omotayo, A.O. Determinants of rural women's livelihood in Ibarapa North local government area of Oyo State, Nigeria. *J. Hum. Ecol.* **2016**, *56*, 84–90. [CrossRef]
11. Daud, S.A.; Omotayo, A.O.; Aremu, A.O.; Omotoso, A.B. Rural infrastructure and profitability of food crop production in Oyo state, Nigeria. *Appl. Ecol. Environ. Res.* **2018**, *16*, 4655–4665. [CrossRef]
12. Omotayo, A.O.; Omotoso, A.B.; Daud, S.A.; Tshwene, S.; Nkonki-Mandleni, B. Analysis of Household's Environment and Willingness To Pay For Waste Disposal In South Africa. *Int. J. Sci. Res.* **2020**, *76*, 132–140.
13. Ciburiene, J. Youth employment changes in rural areas of Lithuania. In Proceedings of the International Scientific Conference "Economic Science for Rural Development", Jelgava, Latvia, 21–22 April 2016; pp. 38–44. Available online: https://agris.fao.org/agris-search/search.do?recordID=LV2016000654 (accessed on 5 July 2021).
14. Babatunde, R.O.; Olagunju, F.I.; Fakayode, S.B.; Adejobi, A.O. Determinants of participation in off-farm employment among small-holder farming households in Kwara State, Nigeria. *Prod. Agric. Technol.* **2010**, *6*, 1–14.
15. Bingen, J.; Serrano, A.; Howard, J. Linking farmers to markets: Different approaches to human capital development. *Food Policy* **2003**, *28*, 405–419. [CrossRef]
16. Adeoye, I.D.; Seini, W.; Sarpong, D.B.; Amegashie, D. Off-farm income diversification among rural farm households in Nigeria. *Agric. Trop. ET Subtrop.* **2019**, *52*, 149–156. Available online: https://sciendo.com/pdf/10.2478/ats-2019-0017 (accessed on 5 July 2021). [CrossRef]
17. Mece, M. Youth Employment in Rural Albania: A Theoretical Approach. *Eur. Sci. J.* **2016**, *12*, 340–351. [CrossRef]

18. Pels, J.; Sheth, J.N. Designing Solutions for the Low Income Consumer Markets: Four Schools of Thought. In *Bottom of the Pyramid Marketing: Making, Shaping and Developing BoP Markets*; Emerald Publishing Limited: Bingley, UK, 2018; pp. 89–109.
19. Ghebru, H.; Amare, M.; Mavrotas, G.; Ogunniyi, A. *Role of Land Access in Youth Migration and Youth Employment Decisions: Empirical Evidence from Rural Nigeria*; International Food Policy Research Institute: Washington, DC, USA, 2018; Volume 58.
20. Olagunju, K.O.; Ogunniyi, A.; Oguntegbe, K.F.; Oyetunde-Usman, Z.A.; Adenuga, A.H.; Andam, K.S. *The Impact of Quality Foundational Skills on Youth Employment in Africa: Does Institutional Quality Matter?* International Food Policy Research Institute: Washington, DC, USA, 2020; Volume 1963.
21. Shilpi, F.; Emran, S. *Agricultural Productivity and Non-Farm Employment: Evidence from Bangladesh*; Policy Research Working Paper; No. 7685; World Bank: Washington, DC, USA, 2016.
22. Olaniyan, O. The determinants of child schooling in Nigeria. *Afr. Econ. Res. Consort.* **2011**, *217*, 1–49.
23. Omotayo, A.O.; Ogunniyi, A.I.; Tchereni, B.H.; Nkonki-Mandleni, B. Understanding the link between households' poverty and food security in South West Nigeria. *J. Dev. Areas* **2018**, *52*, 27–38. [CrossRef]
24. Bezu, S.; Holden, S. Are rural youth in Ethiopia abandoning agriculture? *World Dev.* **2014**, *64*, 259–272. [CrossRef]
25. Osadebe, N. *Socio-Cultural Factors Affecting the Performance of Micro and Small Scale Enterprises in Anambra and Abia States, Nigeria*; University of Nigeria: Nigeria, Nsukka, 2017; Available online: https://freeresearchproject.com.ng/research/socio-cultural-factors-affecting-the-performance-of-micro-and-small-scale-enterprises-in-anambra-and-abia-states-nigeria/ (accessed on 5 July 2021).
26. Serrat, O. The sustainable livelihoods approach. In *Knowledge Solutions*; Springer: Singapore, 2017; pp. 21–26.
27. Buchenrieder, G.; Mollers, J. A synthesis of theoretical concepts for analysing non-farm rural employment. In Proceedings of the International Association of Agricultural Economists (IAAE)—2006 Annual Meeting, Queensland, Australia, 12–18 August 2006; pp. 1–12.
28. Ogunniyi, A.I.; Mavrotas, G.; Olagunju, K.O.; Fadare, O.; Adedoyin, R. Governance quality, remittances and their implications for food and nutrition security in Sub-Saharan Africa. *World Dev.* **2020**, *127*, 104752. [CrossRef]
29. Currie-Alder, B.; Kanbur, R.; Malone, D.M.; Medhora, R. *International Development: Ideas, Experience, and Prospects*; OUP Oxford: Oxford, UK, 2014.
30. Wang, L.; Khan, Q.U.; Zhang, D. Rural transformation index: Measuring rural–urban disparities. In *Skills Development for Inclusive and Sustainable Growth in Developing Asia-Pacific*; Springer: Dordrecht, The Netherlands, 2013; pp. 213–240.
31. Sambira, J. Africa's mobile youth drive change. *Afr. Renew.* **2013**, *27*, 19. [CrossRef]
32. Schmidt, E.; Bekele, F. *Rural Youth Employment in Rural Ethiopia*; Ethiopian Development Research Institute (EDRI); International Food Policy Research Institute Ethiopia: Addis Ababa, Ethiopia, 2016.
33. Awojobi, O.N. The socio-economic implications of Boko Haram insurgency in the north-east of Nigeria. *Int. J. Innov. Sci. Res.* **2014**, *11*, 144–150.
34. Okereocha, C. Heartache for the Economy. *Tell* **2012**, *14*, 46–47.
35. Van den Broeck, G.; Kilic, T. Dynamics of Off-Farm Employment in Sub-Saharan Africa. *Dev. Econ.* **2018**, 1–38. Available online: https://openknowledge.worldbank.org/handle/10986/30230 (accessed on 18 October 2021).
36. Klutsey, E.E.; Ankomah, A. Factors associated with induced abortion at selected hospitals in the Volta Region, Ghana. *Int. J. Women's Health* **2014**, *6*, 809. Available online: https://pubmed.ncbi.nlm.nih.gov/25187740/ (accessed on 5 July 2021).
37. Anang, B.T. *Effect of Non-Farm Work on Agricultural Productivity: Empirical Evidence from Northern Ghana*; 9292562622; The United Nations University World Institute for Development Economics Research (UNU-WIDER): Helsinki, Finland, 2017.
38. Yeboah, F.K.; Jayne, T.; Muyanga, M.; Chamberlin, J. Youth access to land, migration and employment opportunities: Evidence from sub-Saharan Africa. *IFAD Res. Ser.* **2019**, *53*, 50.
39. Namibia Statistics Agency. *Analysis of Youth Employment and Unemployment in Namibia 2012–2013*; Namibia Statistics Agency: Windhoeck, Namibia, 2015.
40. World Bank. Nigeria's Booming Population Requires More and Better Jobs. Available online: https://www.worldbank.org/en/news/press-release/2016/03/15/nigerias-booming-population-requires-more-and-better-jobs (accessed on 15 July 2021).
41. Miller, D.C.; Muñoz-Mora, J.C.; Christiaensen, L. Prevalence, economic contribution, and determinants of trees on farms across Sub-Saharan Africa. *For. Policy Econ.* **2017**, *84*, 47–61. [CrossRef]
42. Christiaensen, L.; Demery, L. Agriculture in Africa: Telling myths from facts. *Food Policy* **2017**, *67*, 1–11. [CrossRef] [PubMed]
43. Sibisi, S. Africa: Youth Employment Critical for Africa's Economic Growth. Available online: https://allafrica.com/stories/201810180532.html (accessed on 18 October 2021).
44. Ogunniyi, A.I.; Omotoso, S.O.; Salman, K.K.; Omotayo, A.O.; Olagunju, K.O.; Aremu, A.O. Socio-economic Drivers of Food Security among Rural Households in Nigeria: Evidence from Smallholder Maize Farmers. *Soc. Indic. Res.* **2021**, *155*, 583–599. Available online: https://link.springer.com/article/10.1007/s11205-020-02590-7 (accessed on 18 October 2021). [CrossRef]
45. Omotayo, A.O.; Ijatuyi, E.J.; Olorunfemi, O.D.; Agboola, P.T. Food security situation among South African urban agricultural households: Evidence from Limpopo Province. *Acta Univ. Danubius. Acon.* **2017**, *14*, 60–74.

sustainability

Article

Organic Fertilization with Biofertilizer Alters the Physical and Chemical Characteristics of Young Cladodes of *Opuntia stricta* (Haw.) Haw.

Márcio S. Silva [1,*], Jackson S. Nóbrega [1], Cleberton C. Santos [2], Franciscleudo B. Costa [1], Daniel C. Abreu [3], Wininton M. Silva [3,4], Aaron Kinyu Hoshide [3,5], Fernando A. L. Gomes [6], Ulisses S. Pereira [1], Jéssica A. Linné [2] and Silvana P. Q. Scalon [2]

[1] Center of Science and Agri-Food Technology, Federal University of Campina Grande, Street Jario Vieira Feitosa, 1770, Pombal 58840-000, PB, Brazil
[2] Faculty of Agricultural Science, Federal University of Grande Dourados, Highway Dourados, Itahum/km 1, Dourados 79804-970, MS, Brazil
[3] AgriSciences, Universidade Federal do Mato Grosso, Caixa Postal 729, Sinop 78550-970, MT, Brazil
[4] Empresa Mato-Grossense de Pesquisa, Assistência e Extensão Rural (EMPAER-MT), Centro Político Administrativo, Cuiabá 78049-903, MT, Brazil
[5] College of Natural Sciences, Forestry and Agriculture, The University of Maine, Orono, ME 04469, USA
[6] Department of Plant Science and Environmental Science, Federal University of Paraíba, Unit II, Areia 58397-000, PB, Brazil
* Correspondence: marcyyo@outlook.com; Tel.: +55-(067)-999447958

Abstract: Cactus cladodes are consumed by humans in arid and semiarid regions of the world. The use of biofertilizers when cultivating cacti can improve the physical and chemical characteristics of the soil, as well as the cladodes' productivity and physical-chemical quality. We evaluated the physical and physical-chemical qualities of different lengths of *Opuntia stricta* (Haw.) Haw. Cladodes were grown with different biofertilizer doses. The 3 × 5 factorial design employed corresponded to three cladode sizes (8–12, 12–16, and 16–20 cm) and five doses of biofertilizer (0, 5, 10, 15, and 20%) with three repetitions in a completely randomized design. Cladode characteristics were evaluated 40 days after emergence: diameter, fresh mass, soluble solids, pH, titratable acidity, soluble solid and titratable acidity ratio (SS/TA), ascorbic acid, phenolic compounds, total soluble sugars, chlorophyll *a*, *b*, and total, carotenoids, and respiration. The *Opuntia stricta* cladodes sized 16–20 cm exhibited better physical and physical-chemical qualities as well as better respiratory rates. The biofertilizer improved the cladodes' physical and physical-chemical qualities, regardless of the cladode's size. *Opuntia stricta* (Haw.) Haw. cladodes had levels of antioxidant compounds similar to those of some conventional vegetables, making them suitable for improving human health in arid environments.

Keywords: human food; *Opuntia stricta*; organic production; non-conventional vegetable; soluble solids

Citation: Silva, M.S.; Nóbrega, J.S.; Santos, C.C.; Costa, F.B.; Abreu, D.C.; Silva, W.M.; Hoshide, A.K.; Gomes, F.A.L.; Pereira, U.S.; Linné, J.A.; et al. Organic Fertilization with Biofertilizer Alters the Physical and Chemical Characteristics of Young Cladodes of *Opuntia stricta* (Haw.) Haw.. *Sustainability* **2023**, *15*, 3841. https://doi.org/10.3390/su15043841

Academic Editor: George Mavrotas

Received: 20 December 2022
Revised: 9 February 2023
Accepted: 17 February 2023
Published: 20 February 2023

1. Introduction

Cacti (*Opuntia* spp.) are species that belong to the Cactaceae family. They are abundantly grown in Brazil's northeastern semiarid region and employed as a source of human and animal nutrition [1,2]. Within this cactus genus, *Opuntia tuna* (L.) Mill is highlighted for presenting resistance against the prickly pear cochineal and for yielding satisfactory productions in low-fertility soils [3]. Another species, *Opuntia stricta* (Haw.) Haw. is used in human nutrition due to its nutraceutical properties. Consequently, young cladodes are consumed in juices or dehydrated powders with high fiber contents [4]. Its use has become viable due to its low acquisition cost and because its constitution presents mineral elements, phenolic compounds, vitamins, proteins, and antioxidant carbohydrate compounds [5–7].

Besides these characteristics, cladodes are employed for phytotherapeutic purposes to treat gastritis, hyperglycemia, diabetes, arteriosclerosis, and prostatic hypertrophy [8].

To obtain quality cladodes, alternative techniques can be adopted to reduce production costs. One such technique is to use biofertilizers, which can increase yield by acting on the accumulation of solutes and the adjustment of vegetable metabolism, which increases biomass production [9]. Furthermore, biofertilizers have organic substances that can improve the soil's properties and increase plant growth by satisfying the cultivar's needs for both macro- and micro-nutrients as well as organic matter [10,11]. Therefore, the use of biofertilizers can increase the production of cactus cladodes, improve the chemical, physical, and biological characteristics of the soil, control pests and diseases, and improve the physical-chemical qualities of different vegetables [12].

Currently, the Northeast region of Brazil has the largest cultivated area of forage palm in the world, reaching around 550,000 hectares, with the planted area in Brazil being approximately 600,000 hectares [12,13]. In arid and semi-arid regions, such as those found in Brazil, cactus pear is clearly important as a source of animal feed and is used as a strategy for coping with drought. The main factor(s) that encourage or discourage prominent cultivation of *Opuntia stricta* (Haw.) Haw. are the supply of food alternatives to these cacti's cladodes and fruits for (1) foraging for animals with high nutritional quality and (2) human food that can guarantee both food security and nutritional quality [14].

Cactus use in human food is widespread in some parts of the world, such as Mexico [4]. In Brazil, the use of young *Opuntia* spp. cladodes is still relatively new, requiring the development of technologies for the production of cladodes with higher nutritional quality. In general, the management of this cactus is low-intensity, with organic fertilization being a viable alternative for producers in arid climates. Efficient use of organic fertilizers such as animal manures can reduce production costs and improve the nutritional qualities of cactus cladodes intended for human consumption [1].

In our research, we hypothesize that increased biofertilizer availability alters the physical and physicochemical quality of *Opuntia stricta* at different growth stages. We expect that there is an optimal amount of biofertilizer that should be added to maximize the size and quality of *Opuntia stricta* (Haw.) Haw. cactus cladodes for human consumption. In this context, the objective of our research was to evaluate the physical and physical-chemical qualities of different lengths of *Opuntia stricta* cladodes grown with different doses of biofertilizer.

2. Materials and Methods

2.1. Plant Material and Cultivation Conditions

The mature *Opuntia stricta* plants used to propagate the plants for our study were obtained from an experimental area of the Center of Agri-food Science and Technology (CCTA) of the Federal University of Campina Grande (UFCG), Pombal, Paraíba, Brazil, located at 6°48′16″ of south latitude and 37°49′15″ of west longitude, at an altitude of 175 m. The mature *Opuntia stricta* cladodes were planted in the CCTA's Experimental Farm Professor Rolando Enrique Rivas Castellón, which belongs to the UFCG, in São Domingos, Paraíba, Brazil, located at 6°48′45″ of south latitude and 37°55′43″ of west longitude, at an altitude of 190 m, around 38 km from Pombal, Paraíba. The predominant climate of the region is of type BSh (Köppen), a hot semiarid with an average annual precipitation of 750 mm and concentrated rainfalls from the months of December to April [15]. The temperature and relative humidity are 36 °C and 20%, respectively.

The experiment was installed in June 2018, using 400 cladodes of *O. stricta*, harvested completely mature at 19 months of age, as propagation material. These cuttings were dried in the shade for six days for healing using an application of Bordeaux mixture to prevent fungal diseases and control pests. The experimental design used at the research site and during laboratory analyses was the Completely Randomized Design (CRD), in a factorial design of 3 × 5, with 3 repetitions for each treatment, that is, three cladode size groups: 8–12, 12–16, and 16–20 cm (Figure 1), and five biofertilizer doses (0, 5, 10, 15, and

20%). The cladodes were planted in blocks with an area of 342 m^2 in the vertical position and in the east–west direction. The CRD consisted of four blocks, with each block being 19 m × 9 m = 171 m^2, containing 5 experimental patches. The patches were composed of four plantation lines with five plants spaced 0.5 m apart, totaling 20 experimental patches with 4 repetitions. The space between lines, patches, and blocks was 1.6, 1.6, and 2.0 m, respectively.

Figure 1. (**a**) *Opuntia stricta* (Haw.) Haw. plant evaluated for different biofertilizer applications; (**b**) cladodes with 8 to 12, 12 to 16, and 16 to 20 cm; and (**c**) *Opuntia stricta* minimally processed for human consumption. Source: the authors.

When planting, about two-thirds of each cladode was buried in the soil. Irrigation was performed twice a week for 30 min at the end of the day through a drip system with a flow of 2.2 L per hour, with drip lines spaced 0.5 m apart. The cactus plantings were thinned three times within 107 days after planting (22 September2018, 3 November 2018, and 15 December 2018), after which only the main cladodes were left for the evaluation of the cladodes' growth. After the second thinning, biofertilizer doses of 0% (just water), 5%, 10%, 15%, and 20% were applied to the mature cladodes. The biofertilizer used had a macronutrient analysis for nitrogen (N), phosphorus (P), and potassium (K) shown in Table 1. The applications were performed every 20 days with a volume of 312.5 mL. The control of spontaneous vegetation was performed through weeding.

Table 1. Physical and chemical characteristics of the biofertilizer used for the cultivation of *Opuntia stricta* (Haw.) Haw.

Sample	Chemical and Physical Characteristics *								
	N	P	K^+	Ca^{2+}	Mg^{2+}	Na	OM	EC	pH
	%	mg/dm^3		mg/dm^3		mg/dm^3	g/dm^3	dS/m	(H_2O)
Biofertilizer	0.2	546.6	8844.1	3.3	1.1	5.54	79.4	7.9	9.7

* Measured characteristics included percent (%) nitrogen (N), phosphorus (P), potassium (K^+), calcium (Ca^{2+}), magnesium (Mg^{2+}), sodium (Na), and organic matter (OM) measured in milligrams (mg) per cubic decimeter (dm^3). Electric conductivity (EC) was measured in deciSiemens (dS) per meter (m). The pH was measured in aqueous extract (1:2.5).

The material harvested after the second thinning and after the application of the treatments had its physical and physical-chemical features analyzed. The plants were evaluated in the Laboratory of Food Analysis, Chemistry and Biochemistry of Foods at the Universidade Federal de Campina Grande's Pombal campus, Paraiba state, Brazil. Laboratory samples were taken from plants in the central six of each patch's middle line; plants along the borders were not sampled.

The biofertilizer used was prepared as follows: Ten kilograms of cattle manure were added to 40 L of water in a plastic container with a volume of 100 L, after which it was fermented for 30 days and agitated once a week. The first and second application of the biofertilizer happened on 3 November 2018 and 23 November 2018, 149 and 169 days after planting, respectively. The physical and chemical characteristics of the biofertilizer are presented in Table 1.

The characteristics of the soil used in our study is of the Eutrophic Fluvic Neosol (RYve) type (Table 2), sampled from the CCTA/UFCG experimental farm [16]. Fluvial Neosols (RY) are non-hydromorphic mineral soils, originating from recent sediments from the Quaternary period. They are formed by overlapping layers of recent alluvial sediments without pedogenetic relationships between them, due to their low pedogenetic development. Generally, they present very diversified thickness and granulometry, along the soil profile, due to the diversity and deposition forms of the original material: Fluvisols - BRAZILIAN SYSTEM Fluvisols - WRB/FAO Entisols (Fluvents) - Soil Taxonomy [17]

Table 2. Physical and chemical characteristics of the Eutrophic Fluvic Neosol (RYve) soil used in the experimental area for the production of *Opuntia stricta* (Haw.) Haw.

Soil Type of Eutrophic Fluvic Neosol (RYve)	Chemical and Physical Characteristics *								
	P	K^+	Ca^{2+}	Mg^{2+}	Na	SB	CEC	OM	pH
	mg/dm^3		cmolc/dm^3					g/kg	(H_2O)
0–0.2 m depth	148.9	263.7	3.0	1.34	0.07	5.09	6.42	7.1	7.6

* Measured characteristics included percent (%) nitrogen (N), phosphorus (P), potassium (K^+), calcium (Ca^{2+}), magnesium (Mg^{2+}), sodium (Na), SB (sum of bases), CEC (cation exchange capacity), and organic matter (OM), measured in milligrams (mg) per cubic decimeter (dm^3). Electric conductivity (EC) was measured in deciSiemens (dS) per meter (m). The pH was measured in aqueous extract (1:2.5). Soil sampled at a depth of 0 to 0.2 m.

2.2. Physical and Physical-Chemical Analyses

The longitudinal and transversal diameters of the cladode were calculated with a digital caliper. Cladode dimensions were measured in centimeters (cm). The fresh mass measured in grams (g) was determined by weighting the young cladodes on a semi-analytical scale with a precision of 0.01 g.

Physico-chemical variables of soluable solids, pH, titratable acidity, ascorbic acid, phenolic compounds, total soluble sugars, chlorophyll and carotenoids, and respiration rate of palm cladodes were evaluated. Chemical characteristics such as soluble solids of the young cladode extract were determined in a digital refractometer (ITREFD65) with automatic temperature compensation. The results were expressed in percentages. Power of Hydrogen (pH) of the young cladode extracts was determined with a benchtop digital potentiometer (Digimed, model DM-22) [18]. The H^+ ion concentration, measured in micromolar (μM), required a direct reading of the pulp in digital potentiometer and calculated according to the equation: pH = log [H^+]. To measure titratable acidity, one gram of the young cactus cladode extract was added to 50 mL of distilled water. The solution was titrated with sodium hydroxide (NaOH) at 0.1 molar (M) until the endpoint of the phenolphthalein indicator, as confirmed by the pH range of 8.2 [18]. The total titratable acidity was expressed as percentage of malic acid. The soluble solid and titratable acidity ratio (SS/TA) was calculated by dividing the soluble solid values by the titratable acidity values.

2.3. Nutritional Analyses

Nutritional characteristics of cactus cladodes included ascorbic acid, which was measured by taking one gram of the young cactus cladode extract and adding to 49 mL of oxalic acid at 0.5% concentration. This was then titrated with Tillmans' solution until a pink color was obtained, according to the method (365/IV) described by Institute [18]. Soluble phenolic compounds were estimated through the Folin-Ciocalteu method [19]. Water and the Folin-Ciocalteu reagent were added to the sample, followed by agitation and rest for 5 min. After the reaction time, 250 microliter (μL) of sodium carbonate was added, followed by more agitation, and then rest in a water bath at 40 °C for 30 min. The sample was cooled, and the reading was performed at 765 nm in a spectrophotometer.

Other nutritional characteristics included total soluble sugars (%) determined through the Anthrone method [20]. The extract was obtained by the dilution of 1 g of cactus pulp in 10 mL of distilled water. The samples were prepared in ice bath through the addition, in a tube, of 0.5 mL of the extract, 0.5 mL of distilled water, and 2 mL of the anthrone solution (0.2%), followed by agitation and rest in a thermostatic bath at 100 °C for 3 min. The samples' reading was performed in a spectrophotometer at 620 nm, employing glucose as reference to obtain the standard curve.

Chlorophyll and carotenoid levels were determined as described by Lichtenthaler [21]. Two grams (2.0 g) of the cladodes' cellular extract were weighed and ground in a mortar, after which 0.2 g of calcium carbonate ($CaCO_3$) and about 3 mL of acetone at 80% were added to the mortar. After the extract's transition, it was transferred into a centrifuge tube. Then, the leftover in the mortar was washed with 2 mL of acetone at 80% and added to the tubes. They were centrifuged for 10 min at 10 °C and 3000 revolutions per minute (rpm). An aliquot of the supernatant was poured into a cuvette. The samples' readings were performed in a spectrophotometer at the wavelengths of 470.646, and 663 nm.

2.4. Respiratory Rate

The respiratory rate of cactus cladodes was measured in milligrams of carbon dioxide (CO_2)/kg/hour was determined according to [22] and adaptations described by [23]. The *Opuntia stricta* cladodes were stored in 750 mL plastic pots with covers for 6 h, remaining over a bench at a controlled room temperature of 24 ± 1 °C and 32 ± 2% relative humidity (RH). Another container holding NaOH 0.5 N was placed in the aforementioned containers. They worked as fixers of the CO_2 generated in the respiration process. In the experiment, 0.5 mL of NaOH 0.5 N was employed, containing five repetitions, one of which was the sample called white test (repetition prepared without the cladode). To avoid gas exchange with the environment, the containers' covers were wrapped with a silicon layer. After the 6-h period, the NaOH solution was removed from the container and received three drops of the phenolphthalein indicator and 10 mL of $BaCl_2$ 0.2 N in an Erlenmeyer flask before being submitted to titration with hydrochloric acid at 0.1 N. The final calculation of the respiratory rate, at each analysis time, was based on the repetitions, whose result was expressed in mg CO_2/gram of fresh mass.

2.5. Statistical Analyses

Statistical analyses comparing all physical-chemical contrasts were performed using the Sisvar® program [24]. The final data were submitted for analysis of variance at 5% probability through the F-test. In cases of significant effects, Tukey's test at 5% ($p > 0.05$) was used when comparing cladodes of different sizes. Non-linear regression analysis was used to model functional relationships between each physical-chemical characteristic and biofertilizer doses (0—just water, 5, 10, 15, and 20%).

3. Results

The interaction between the factors was significant for the cladodes' thickness and fresh mass. The isolated effects were significant for the biofertilizer doses and cladode sizes (Table 3). Organic fertilization via biofertilizer promotes improvements in the physical attributes of the palm cladodes; it was observed that the thickness of the cladodes at 16–20 cm (cm) was greater (0.80 cm) at the biofertilizer level of 10%. As for the cladodes of 8–12 and 12–16 cm, they presented their maximum increases (0.60 and 0.73 cm) at the doses of 10 and 20%, respectively (Figure 2a). The greatest fresh mass of the cladodes of 16–20, 8–12, and 12–16 cm was seen at the biofertilizer dose of 20%, with averages of 143.55, 29.52, and 121.99 g, respectively (Figure 2b).

Table 3. Summary of analysis of variance for longitudinal diameter, transverse diameter, thickness, and fresh mass of palm cladodes *Opuntia stricta* (Haw.) Haw. for different biofertilizer doses and cactus cladode sizes.

Variation Sources	Degrees of Freedom	Medium squares			
		Longitudinal Diameter	Transverse Diameter	Thickness	Fresh Mass
Doses (D)	4	262.9 **	106.02 **	0.013 **	32,118.6 **
Length (C)	2	1.24 *	10.76 **	0.12 **	1907.9 **
D × C	8	0.77 ^{ns}	2.08 ^{ns}	0.0059 *	608.3 *
Residue	28	0.38	2.17	0.0022	219.6
Average		14.94	10.85	0.63	75.6
Coefficient of Variation (%)		4.14	13.6	7.53	19.6

The symbol ^{ns} denotes differences between factors are not significant, while ** and * indicate significant differences at 1% and 5% probability, respectively, using the F-test.

The longitudinal (19.10 cm) and transversal (13.28 cm) diameters of the cladodes with 16–20 cm were greater in comparison with the cladodes of 12–16 and 8–12 cm, whose values were (14.80 and 11.28 cm) and (10.70 and 8.01 cm), respectively (Figure 3a,b). Biofertilization also increased the growth of cladodes, with gains in longitudinal diameter (15.4 cm) and transversal diameter (12.2 cm) that were superior in the cladodes that received a biofertilizer dose of 20% (Figure 4a,b).

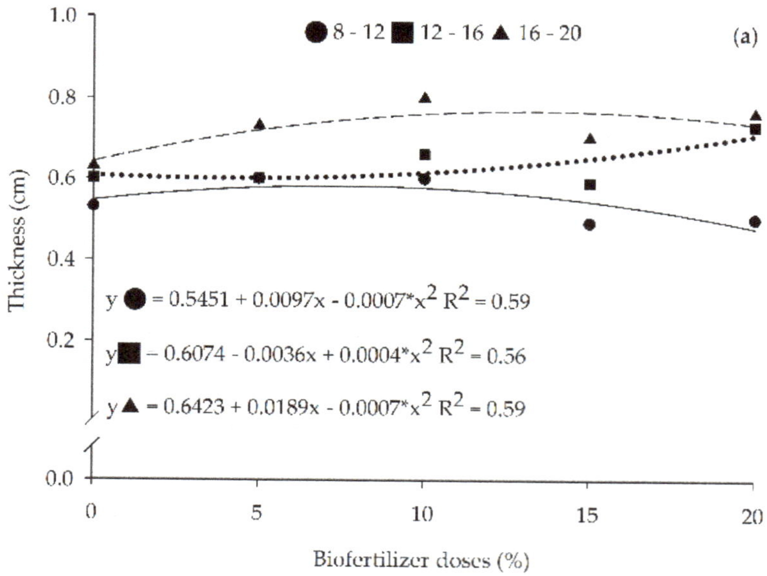

Figure 2. *Cont.*

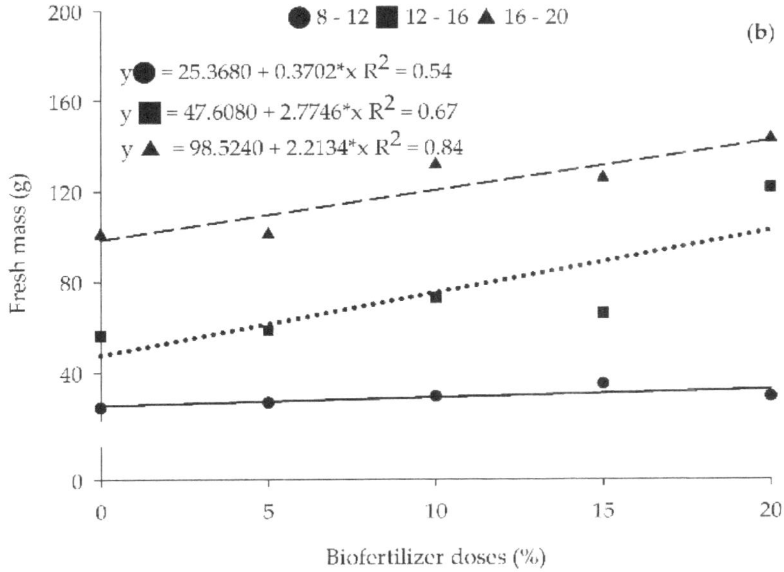

Figure 2. (**a**) Thickness in centimeters (cm) and (**b**) fresh mass in grams (g) for cladodes of *Opuntia stricta* (Haw.) Haw. cultivated with biofertilizer. * indicate significant differences at 5% probability.

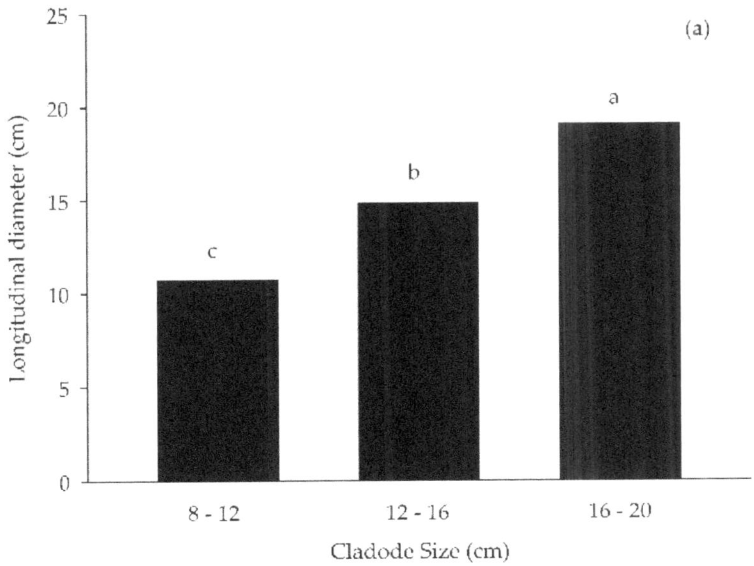

Figure 3. *Cont.*

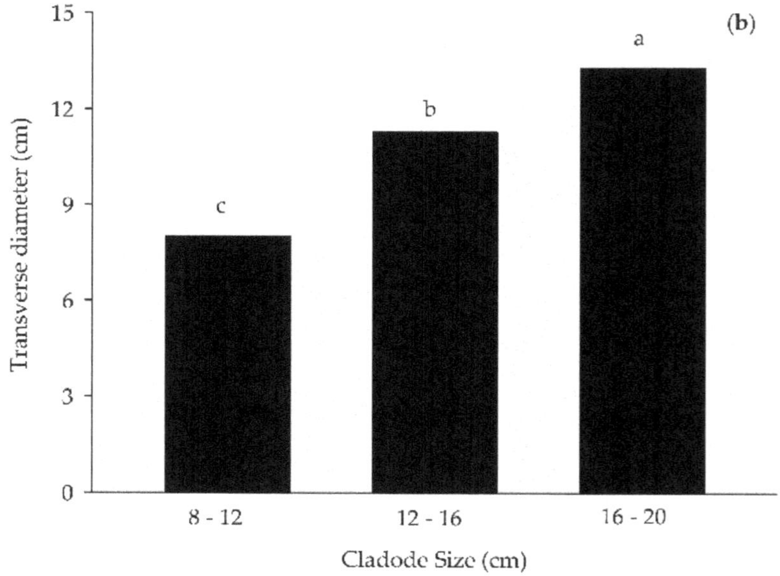

Figure 3. (**a**) Longitudinal diameter and (**b**) transverse diameter of *Opuntia stricta* (Haw.) Haw. cactus cladodes in centimeters (cm). Means followed by same letter are not statistically different using Tukey test at 5% ($p > 0.05$).

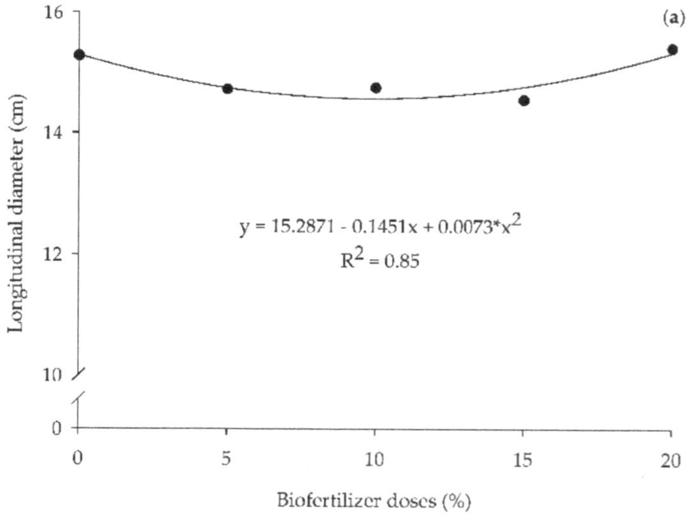

Figure 4. *Cont.*

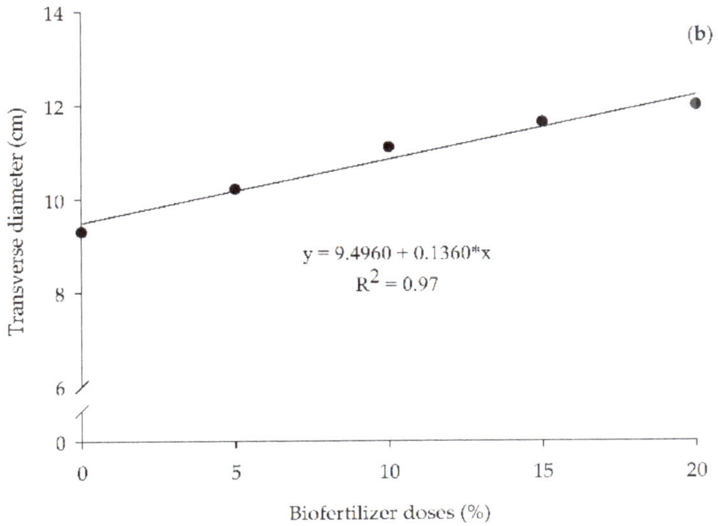

Figure 4. (**a**) Longitudinal diameter and (**b**) transverse diameter of *Opuntia stricta* (Haw.) Haw. cactus cladodes in centimeters (cm) with increasing biofertilizer doses (0% to 20%). * indicate significant differences at 5% probability.

According to the analysis of variance for the physical-chemical parameters, there was an interaction between the doses of biofertilizer and the cladode length for pH, H+ ions, soluble solids (SS), soluble solids/titratable acidity ratio (SS/TA), and ascorbic acid (AA) (Table 4). The pH of palm cladodes increases as a function of the length and dose of the biofertilizer. Observing the pH, the greatest values were 4.38, 4.33, and 4.05 for the cladode lengths of 8–12, 12–16, and 16–20 cm, respectively, reached at the biofertilizer doses of 10, 15, and 20%. It can be seen in Figure 5a. For the concentration of hydrogen (H^+) ions, the cladodes 12–16 cm long were greater in comparison with the others, with the greatest value at the dose of 0% biofertilizer, with an average of 227.43 μM, followed by cactus cladodes with a length of 8–12 cm, with an average of 217.19 μM at the same dose of biofertilizer. As for cladodes that were 16–20 cm long, the greatest value was at the dose of 5% biofertilizer, with an average of 157.33 μM. Thus, pH was inversely proportional to the increase in biofertilizer doses, indicating that increasing biofertilizer application increases the concentration of H^+ ions (Figure 5b).

Table 4. Summary of analysis of variance for pH, H+ ions, soluble solids (SS), titratable acidity (AT), soluble solid ratio and titratable acidity (SS/TA), and ascorbic acid (AA) of palm cladodes *Opuntia stricta* (Haw.) Haw. fertilized with different doses of biofertilizer.

Variation Sources	Degrees of Freedom	Medium Squares					
		pH	H^+	AT	SS	SS/TA	AA
Doses (D)	4	0.10 **	4951.2 **	0.0099 ns	1.02 **	0.98 **	28.25 **
Length (C)	2	0.28 **	22,189.3 **	0.054 ns	7.65 **	6.65 **	17.01 **
D × C	8	0.11 **	5776.7 **	0.058 ns	1.18 **	1.22 **	9.36 **
Residue (error)	28	0.0012	125.79	0.035	0.086	0.099	0.62
Average		2.93	129.98	1.73	4.18	2.45	19.47
Coefficient of Variation (%)		0.59	8.63	10.1	7.02	12.84	4.06

The symbol ns denotes differences between factors are not significant, while ** indicate significant differences at 1% probability, respectively, using the F-test.

The soluble solid content was greater in the cactus cladodes of 12–16 cm (6.69%) at the dose of 15% in comparison with the cladodes of 8–12 (5.53%) and 16–20 cm (4.13%) for the doses of 10 and 15%, respectively (Figure 5c). The SS/TA ratio of the cladodes of 12–16 cm at the biofertilizer dose of 15% (4.42) surpassed those of the cladodes of 8–12 and 16–20 cm, with SS/TA values of 2.90 and 2.57 at the biofertilizer doses of 0 and 15%, respectively (Figure 5d). The content of ascorbic acid was greater for cactus cladode lengths of 8–12 cm and 12–16 cm (22.36 mg/100 g) at biofertilizer doses of 10 and 5% when compared with the cladodes 16–20 cm long (20.30 mg/100 g) at the biofertilizer dose of 20% (Figure 5e).

Also significant were the effects of the interaction between the doses of biofertilizer and the length of the cladode on the contents of chlorophyll, carotenoids, phenolic compounds, and total sugars (Table 5). For the chlorophyll a and total contents, there was an increase in values depending on the size of the cladodes, with the highest values being observed for (0.69, 0.64, and 0.75 mg/100 g) and (1238.2, 1249.9, and 1446.0 mg/100 g) for cladode lengths of 8–12, 12–16, and 16–20 cm, respectively, both for the 0% biofertilizer dose (Figure 6a,b). The chlorophyll *b* contents were 0.53, 0.70, and 0.69 mg/100 g for the cladodes of 8–12, 12–16, and 16–20 cm at the doses of 0, 15, and 0% (Figure 6c). The greater carotenoid contents were 78.81, 80.81, and 96.74 μg/100 g in the cladode lengths of 8–12, 12–16, and 16–20 cm at the 0% and 20% doses of biofertilizer, respectively (Figure 6d).

The greatest total sugar contents of 0.50%, 0.23%, and 0.25% were at biofertilizer doses of 0%, 5%, and 15% and at cactus cladode lengths of 8–12 cm, 12–16 cm, and 16–20 cm, respectively. Therefore, the cladode length of 8–12 cm generated increased sugar contents (Figure 6e). The phenolic compound contents presented an increase at the biofertilizer level of 20% for both length groups, with values of 8.53, 8.94, and 9.44 mg/100 g in the cladodes with lengths of 8 to 12, 12 to 16, and 16 to 20 cm for each unitary increase in the biofertilizer doses, respectively (Figure 6f).

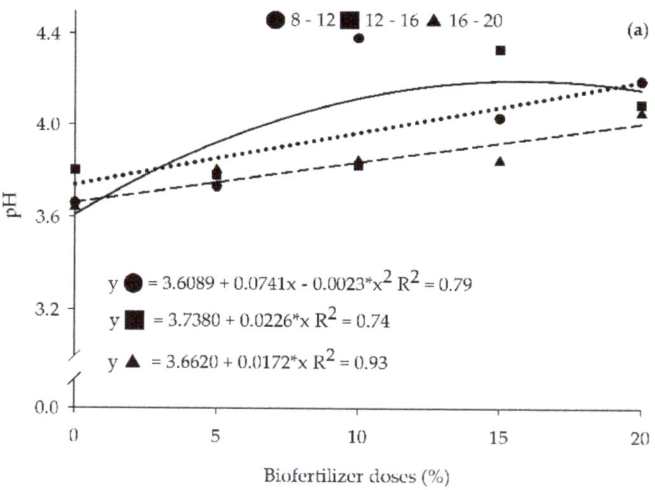

Figure 5. *Cont.*

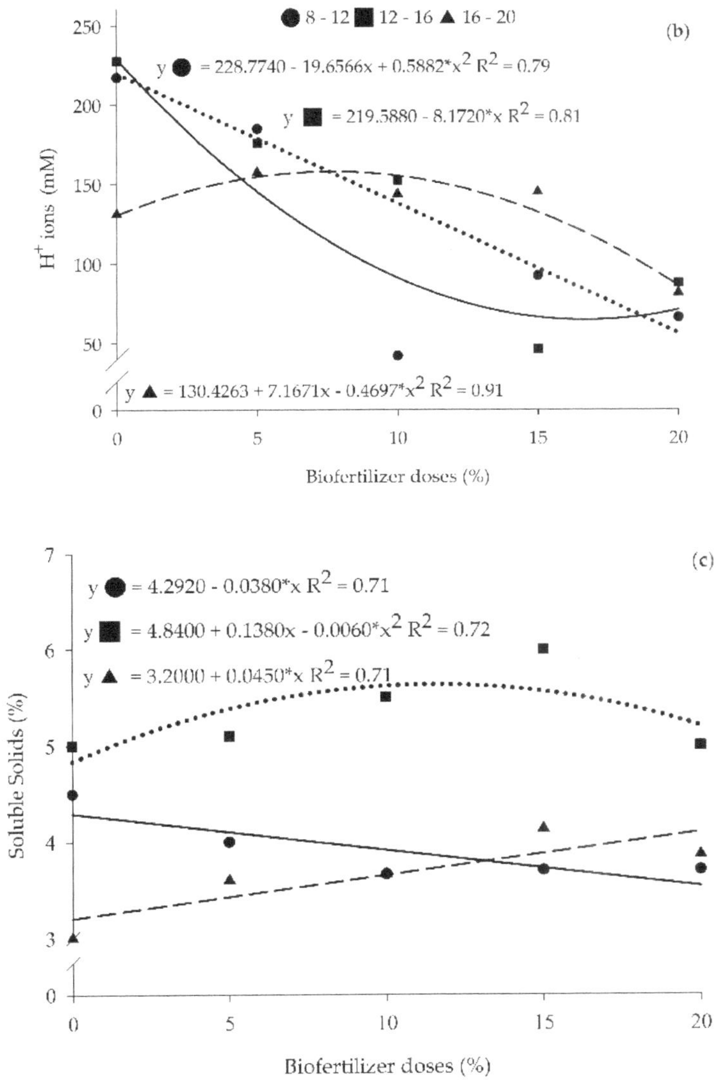

Figure 5. *Cont.*

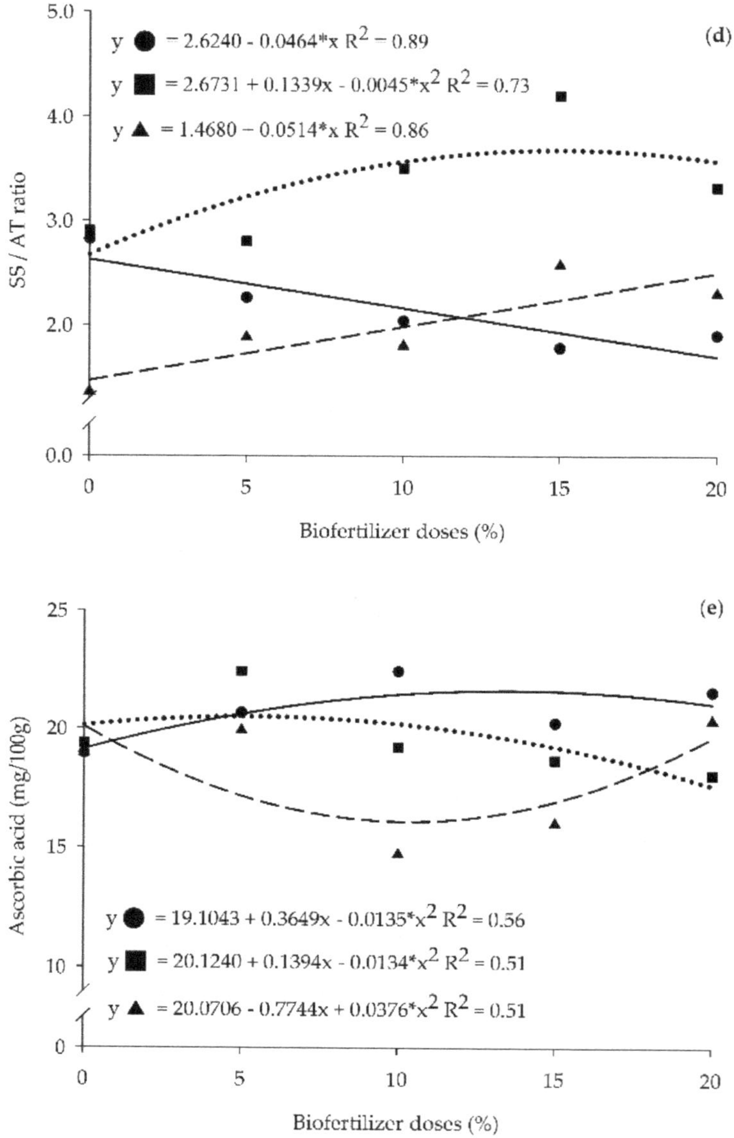

Figure 5. (**a**) pH, (**b**) H + ions, (**c**) soluble solids (%), (**d**) soluble solids/titratable acidity ratio (SS/TA), and (**e**) ascorbic acid (milligrams per 100 g) of *Opuntia stricta* (Haw.) Haw. cactus cladodes of different lengths and applied doses (0% to 20%) biofertilizer. * indicate significant differences at 5% probability.

Table 5. Summary of analysis of variance of *Opuntia stricta* (Haw.) Haw. cactus cladode chlorophyll *a* (Clo *a*), chlorophyll *b* (Clo *b*), total chlorophyll (total clo), carotenoids (Car), phenolic compounds (Phnl.comp.), and total sugars for different cladode lengths and biofertilizer doses.

Variation Sources	Degrees of Freedom	Medium Squares					
		Clo *a*	Clo *b*	Clo Total	Car	Phnl.comp.	Total Sugars
Doses (D)	4	0.014 **	0.021 **	18,863.1 **	2797.2 **	1.85 **	0.016 **
Length (C)	2	0.10 **	0.058 **	286,522.4 **	2357.1 **	2.65 **	0.013 **
D × C	8	0.012 **	0.027 **	67,914.7 **	258.27 **	0.60 **	0.026 **
Residue (error)	28	0.00015	0.00020	367.45	6.11	0.0071	0.000010
Average		0.53	0.52	1057.5	60.1	8.10	0.21
Coefficient of Variation (%)		2.33	2.76	1.81	4.12	1.04	1.48

The symbol ** indicates significance at the 1% probability using the F-test.

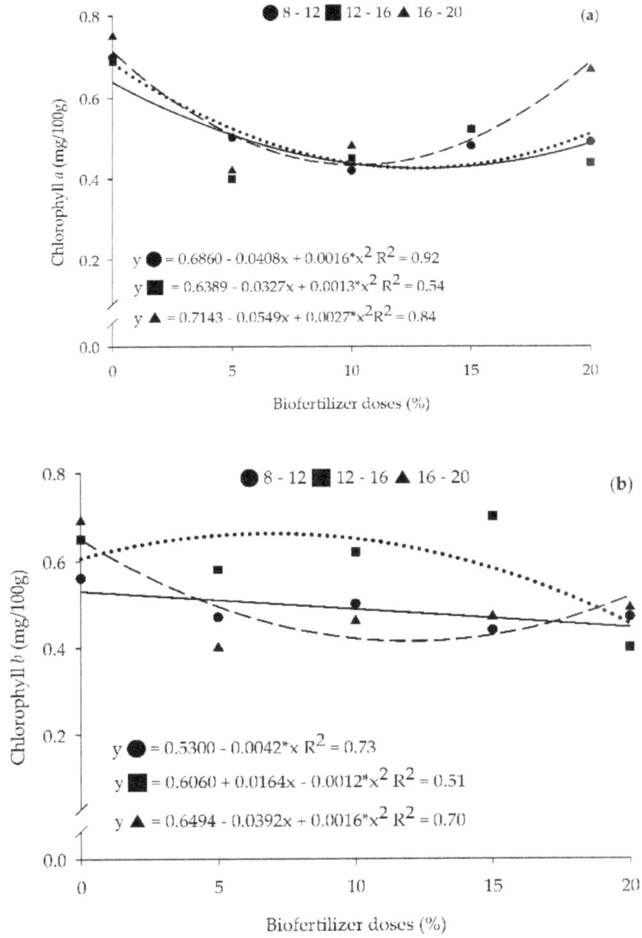

Figure 6. *Cont.*

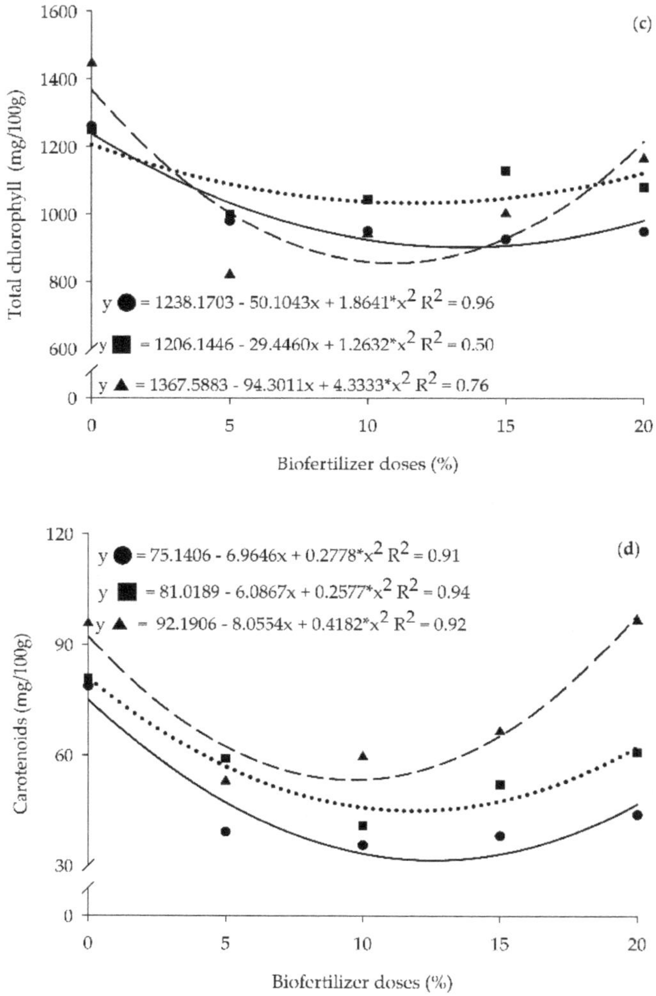

Figure 6. *Cont.*

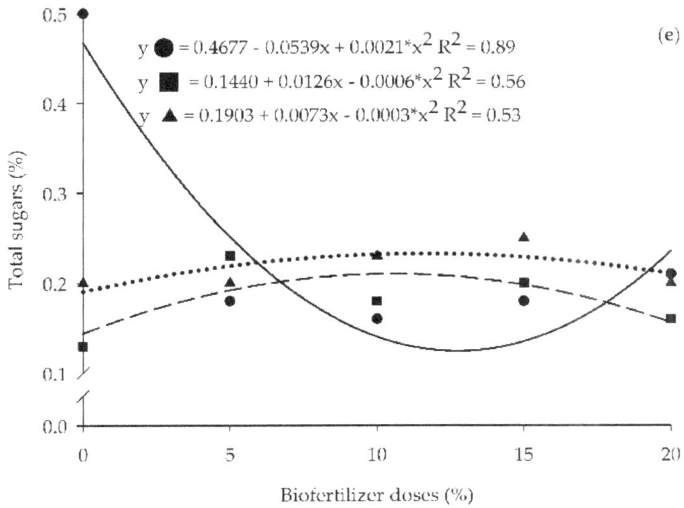

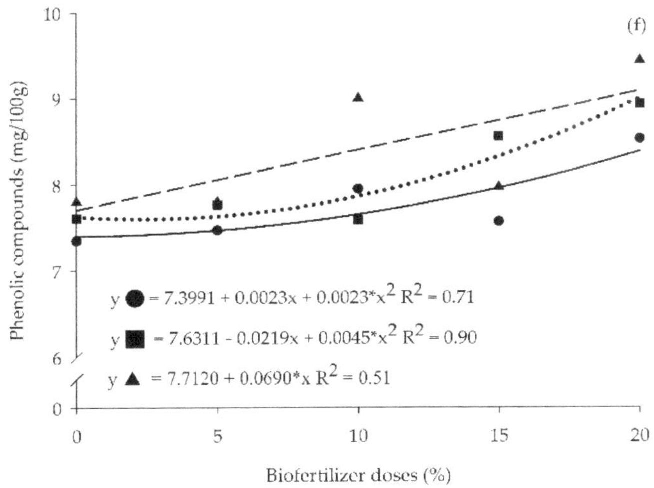

Figure 6. Indices of (**a**) chlorophyll a, (**b**) chlorophyll b, (**c**) total chlorophyll, (**d**) carotenoids, (**e**) total sugars, and (**f**) phenolic compounds of *Opuntia stricta* (Haw.) Haw. cactus cladodes of different lengths and applied doses of biofertilizer. Except for total sugars (%), all indices measured in milligrams (mg) per 100 g (g) of cactus cladode. * indicate significant differences at 5% probability.

As shown in Figure 7, young cactus cladodes with biofertilizer had great similarity in their respiratory rates. The cactus cladode size of 4–8 cm, for both treatments, presented higher respiratory rate values of 344.10 mg of CO_2/kg/h for those grown without biofertilizer and 278.85 mg of CO_2/kg/h for those grown with biofertilizer. The respiratory rate of the cladodes 4 to 8 cm long grown without biofertilizer was superior to the other sizes with an average of 344.10 milligrams (mg) of CO_2/kg/h, which was superior to the cactus cladodes grown with biofertilizer (278.85 mg CO_2/kg/h) in the same size group.

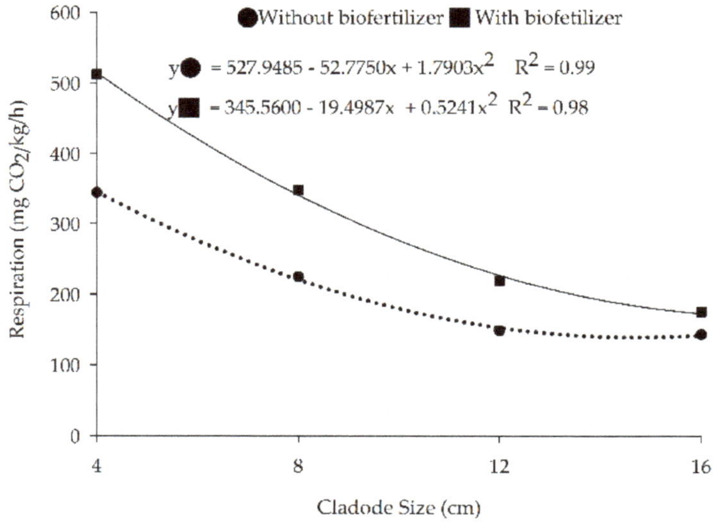

Figure 7. Respiratory rates of *Opuntia stricta* (Haw.) Haw. cactus cladodes of different lengths, both with and without biofertilizer, measured in milligrams (mg) of CO_2 per kilogram (kg) of cactus cladode per hour.

4. Discussion

4.1. Underlying Drivers of Cactus Responses to Biofertilization

Foliar fertilization via biofertilizer improves the physical attributes of palm cladodes, enabling greater productivity and yield. These results suggest that liquid biofertilizers increase cactus development, providing a better nutritional profile based on the availability of nutrients. Biofertilization of *Opuntia stricta* (Haw.) Haw. favored an increase in physical attributes and biomass production. A greater supply of organic matter in the soil enables an increased nutritional and physiological balance, consequently favoring the production and accumulation of photoassimilates [24]. The provision of organic matter to the soil through biofertilizers promotes improvements to the plant's physical attributes. Since this material is in a liquid state, which is easily absorbed, it generates better conditions for the plants' development, improving their physical and nutritional performance. These results are similar to those obtained by [1] when working with organic fertilization, where increases in physical attributes were observed for plants that were fertilized with cattle manure, with values ranging from 10.9 to 18.3 cm. The transversal and longitudinal diameters were superior for cladodes grown at a biofertilization level of 20% at 12.20 cm and 15.40 cm, respectively.

The greatest longitudinal and transversal diameters of the cladodes with lengths of 16–20 cm occurred because they were in a development stage that was superior to the others, since a larger diameter is an indicator of cactus cladodes in more advanced development stages. Prior research [25], working with different length groups of giant and round *Opuntia* species, also noticed that the larger the cladode size, the greater the values of cladode diameters, and with 74.02% and 96.01% greater yields, while comparing the results from bigger and smaller cladodes, respectively.

As the biofertilizer dose increased, the pH value tended to increase, regardless of the cladode lengths, because the pH of the biofertilizer used in the research was alkaline. Another explanation for this difference in the cladodes' pH values can be related to the Crassulacean acid photosynthetic metabolism (CAM) pathway, which has the tendency to present acidity variation [26]. The variation in the H^+ ion concentration in cladodes has been reported in other research, such as [27], which observed variation between 129.7 and 344.4 μM in cladodes of the erect prickly pear species of different sizes. This variation can be driven by the cladodes' development [26].

It has been reported that the increase in the cladodes' length results in greater soluble solid contents. The high soluble solid content in the cladodes 12–16 cm long can be related to environmental conditions, directly affecting the plants' metabolism during harvest, as reported by [28]. The greatest soluble solid (SS) to titratable acidity (TA) or SS/TA ratio in the cladodes 12–16 cm long at 15% biofertilizer is an indication that these cladodes present pleasant characteristics to the palate, since the greater the SS/TA ratio, the greater the concentration of soluble solids, and the more pleasant the taste, as reported by [27].

These values are superior to those reported by [1], where there was a reduction in ascorbic acid as the organic fertilization doses were increased, presenting maximum values of 7.54 milligrams (mg) per 100 g (g). Prior research [5], while working with different cactus species, reported values of 5.7, 4.3, and 1.6 mg of ascorbic acid/100 g of cactus in the cultivars *Opuntia strica*, *Opuntia ficus-indica*, and *Nopalea cochenillifera*, respectively. Similar results were reported by [29], where increasing amounts of biofertilizer were applied to beets, which also resulted in an increase in ascorbic acid content.

For both types of chlorophyll, there are variations in the chlorophyll contents with increased biofertilizer doses. These variations may have been caused because both chlorophylls, especially chlorophyll *a*, and carotenoids have the tendency to be inversely proportional to biofertilizer dose. Nonetheless, the species seems to adjust itself to the increased supply of nutrients, where there is a tendency for elevated chlorophyll contents associated with biofertilizer doses greater than 10%. The greatest chlorophyll values were associated with a cactus cladode length of 12–16 cm at the biofertilizer dose of 15%. The successive application of organic fertilizer over a longer period can increase the nutrient source in the soil, promoting the supply of nitrogen (N) and magnesium (Mg), in addition to other nutrients that take part in the synthesis of this pigment. These chlorophyll *b* variations can be related to several factors, such as the species and cultivar, as well as environmental conditions, such as luminosity and temperature, which can promote its degradation, as reported by [1] for *Opuntia stricta*.

We reported a greater carotenoid content for cactus cladodes with lengths of 16–20 cm at the biofertilizer level of 20%. The more developed cladodes had greater carotenoid contents because they already had a darker color in comparison with the younger cladodes. This is also observed in vegetables with more intense colors and more advanced development. These pigments' evaluation is important because they have antioxidant and anticancer effects [28].

The increase in the size of the cladodes induces a greater production of phenolic compounds. As the cladode expands and grows older, there is greater accumulation and variation of photoassimilates and reserves that are constantly changed into other compounds, such as phenolics. Farias [30] reported similar behavior in 'Giant' (*Opuntia ficus indica* L. Miller) and 'Round' (*Opuntia* spp.) cactus cladodes.

We observed an inverse relationship between cladode maturity and total sugar content. During their development, cactus cladodes reach greater physiological maturity, which reduces total sugar content. This fact was verified for giant and round cactus cladodes, where sugar contents ranging from 0.59% to 0.72% and from 0.52% to 0.72% were obtained, respectively [30]. These low levels may be related to the presence of other substances dissolved in the aqueous medium of the plant, such as organic acids, pectins, mucilages, and phenolics.

Increasing biofertilizer doses promote greater cactus cladode quality, regardless of cladode length. This effect is likely driven by the greater availability of nutrients from biofertilization, which results in more nutrients being absorbed by the plants and resulting in more plant development. This happens because these compounds have nitrogen in their composition, which is accumulated to act especially in the plant's defense [31]. Therefore, the provision of organic matter to the soil via biofertilizer promotes improvements to the soil's characteristics, consequently improving plant development [32].

Regardless of the production method, respiration tends to decrease as the cladodes' maturation stage advances due to greater energy consumption, just as the respiratory rate increases in younger cladodes when they are removed from the plant. The cladodes grown with biofertilizer presented lower respiratory rates. This condition can be related to an increased supply of nutrients from biofertilization, which enables the construction of several molecules that assist both in the protection of and in the transportation of nutrients in the plants. When the biofertilizer is applied foliarly, the absorption of nutrients is stimulated, especially nitrogen, which has a role in the assimilation of countless amino acids, which are then incorporated into proteins and nucleic acids, which mold chloroplasts, mitochondria, and other structures that hold most of the biochemical reactions [33].

According to [34], plant respiration involves the oxidative decomposition of complex substances present in the cells (such as starch, sugars, and organic acids) into simple molecules (CO_2 and H_2O) for energy production. Generally, the respiratory rate is directly linked to the deterioration rate of a harvested product, and the temperature to which it is exposed directly affects the respiration. In other words, the increase in temperature raises the respiratory rate, therefore decreasing its post-harvest life. The results showcased in our research confirmed our hypothesis that biofertilizers increase the physical and physical-chemical qualities of cladodes. Nonetheless, bigger cladodes are more responsive to the use of biofertilizer.

4.2. Human Nutrition

Opuntia species have had a long history of use for both human food and animal feed in the Americas. These cacti have been used both pre- and post-colonization by both indigenous peoples and colonial settlers, particularly in Mexico [35]. The use of such cactus species has been and can continue to be relevant in adapting to drought given the limits and challenges of fighting against drought by increasing water sources (e.g., reservoirs, wells) [36].

Opuntia is a genus that plays a strategic role in agriculture in regions with arid and semi-arid climates. In addition to the potential for adapting to edaphoclimatic conditions, *Opuntia* spp. have nutraceutical compounds that promote improvements in human nutrition and health, establishing food security, in addition to being a species with applications in the food industry [37,38]. According to [38], this cactus genus has high levels of phosphorus (62.2 mg/100 g), sodium (9.1 mg/100 g), calcium (1.18 to 1.28 mg/100 g), iron (3.0 to 3.8 mg/100 g), potassium (45–47 mg/100 g), and flavonoids (196.7 mg/100 g). This demonstrates this cactus' social and cultural importance in rural communities, especially in the Brazilian Northeast. In our study, the increase in phenolic compounds with the application of the biofertilizer was desirable, as it increased the benefits of inserting *Opuntia* spp. in human food without worrying about toxicity [39].

In addition, the benefit of consuming *Opuntia* spp. is also due to the cladodes, pulp, fruits, seeds, and bark having high levels of antioxidants. This type of cactus is also anti-diabetic, anti-bacterial, anti-viral, anti-inflammatory, and analgesic, and it is used in neopathy and folk medicine [39,40]. According to [41], *Opuntia* spp. can be used in the preparation of cosmetics and dyes, adding value and becoming a raw material of economic interest. Although *Opuntia* is a genus with diverse nutraceutical characteristics, few people adopt its use in human food in Brazil, although this can vary according to cactus distribution, cultural aspects, and even differences in family financial backgrounds. Thus, our study describing its physicochemical characteristics makes its use for human food and supplementation feasible. According to [42], these cacti have a lot of potential to combat hunger. In addition, products made with *Opuntia* have good sensory acceptance [43] and can be a viable option to encourage sustainable agricultural development in Brazil.

4.3. Management and Ideal Size of Cactus Cladodes for Human Consumption

Generally in Brazil, cactus is used for animal feed, making its management carried out with low intensity and few technologies being used by producers. This is true not only during planting but also later during the growth and development phases of the crop. Normally, cactus producers in the northeast region of Brazil make use of organic fertilizers such as bovine, goat, and poultry manure, which are added in furrows or holes during planting. The use of organic amendments in agricultural systems for cacti has been widespread, given their important economic and environmental contributions [44].

From an economic point of view, cactus producers can realize increased income due to the decrease in the use of industrial fertilizers. This makes cactus cultivation possible for less capitalized farmers, which can generate promising results from such mineral fertilization [44].

Other studies verify the benefits of using organic fertilizers during cactus production. For example, researchers have found that cacti respond well to organic fertilization since soils in semi-arid regions tend to have low organic matter content. Therefore, organic fertilization with livestock manure can increase organic matter in these soils and be essential to crop production [45]. It is recommended to fertilize cacti with bovine and goat manure at a rate of 10 to 30 metric tons /hectare during planting and then every two years during the beginning of the rainy season [46]. The presence of animal manure, mainly from cattle, on most agricultural properties can help maintain soil fertility in Brazil's semi-arid regions [47].

Few studies have focused on the cultural management of *Opuntia* spp. cacti in Brazil. Most research has focused on the use of cactus cladodes in human food as well as the manufacture of differentiated, value-added products [30]. Research is needed to determine the ideal size of cactus cladodes for human consumption.

In general, in Brazilian cuisine, there is no consumption of cactus cladodes. In some states of Brazil, there is a small trade in such cladodes, mainly at small urban fairs. Thus, this work is a first step in characterizing the post-harvest quality and size of *Opuntia* spp. cladodes intended for human consumption using organic fertilizer. Our research results suggest that cactus production is a viable commercial alternative for small- and medium-sized cactus producers in Brazil's semi-arid regions. It is important for cactus cladodes to be balanced in terms of palatability, measured as the soluble solid to titratable acidity ratio (SS/TA), for the more mature developmental stages of the crop.

Likewise, it was observed for the 'Redonda' and 'Gigante' cacti cultivars that yield, physical attributes, and bioactive compounds of the cactus buds increased with developmental stage advancement up to 20 cm, which is the ideal for consumption [30]. This same study also reported that yield varied according to the size of the cactus cladodes and the number of thorns [30]. Cactus yield for food can play an important role when selecting the best-sized shoots to optimize propagation for both consumption and agro-industrial processing.

In recent years in Brazil, some authors have determined that the cactus cladode size of 15 to 20 cm is ideal for human consumption. Cladodes should be harvested 30 to 60 days after sprouting and weigh 80 to 120 g. They must also be tender, young, fine, fresh in appearance, turgid, and bright green in color so that they can also be used in various culinary recipes from around the world [30]. In Mexico, researchers have determined that the ideal size of cactus cladodes for human food is 10 to 12 cm in length and that the cladodes need to be tender [48]. Another study concluded that ideal cladodes were 20 cm long and weighed 90 to 100 g [49]. Cladodes to be exported should be between 17 and 21 cm in length, and those selected for national consumption should be between 21 and 25 cm [50], while [51,52] state that cladodes are ready for consumption when they are between 7 and 30 cm long.

Mexico is one of the origin centers of *Opuntia* spp., where consumption of cladodes is ancestral [53]. Cladodes are consumed as vegetables. Harvested area has more than doubled from 5269 to 12,105 ha between 1990 and 2012 [50]. Cactus plants, due to their adaptation to moisture deficits and semi-desert and desert climates, are a food source with great potential for the development of plantations [54]. Mexico is the main producer of cactus cladodes in the world, accounting for 74% of global production, and is the largest consumer of both fresh and processed cactus. However, other markets, such as those in the United States and Canada, present a growing opportunity for export [50]. Due to their nutraceutical characteristics, cactus cladodes have shown great interest in the world market [53], as they are beneficial for various treatments [50]. Furthermore, they contain vitamin C, minerals, as well as soluble and insoluble fiber [50,53].

For the commercial use of cactus cladodes in Mexico, producers already use commercial size standards. However, it is important to follow similar commercial standards for export and national trade. These standards do not yet exist in Brazil since consumption of cactus cladodes in Brazil is still regional and limited. Further research and incentives from governments and public entities are needed in order to expand the production and consumption of cactus as a vegetable. Brazil has great potential for developing cactus as an export crop since it is tied with Tunisia in global production (600,000 hectares for each country), with Mexico at third with 230,000 hectares of cultivated cactus [55]. This holds great importance not only for Mexico but also for other countries and markets around the world.

5. Conclusions

The *Opuntia stricta* cladodes with a length of 16 to 20 cm present greater physical and physical-chemical attributes when submitted to a range of biofertilizer doses. The biofertilizer dose of 20% promotes an increase in the contents of soluble solids, phenolic compounds, and total sugars. Nevertheless, it did not affect the content of pigments. The cladodes (4–8 cm long) grown with and without biofertilizers presented greater respiratory rates. The use of biofertilizer is a promising practice for the cultivation of *Opuntia stricta* in agroecological or organic systems. The physical-chemical characteristics presented in our study provided information that enables the use of this cactus in human food and other alternatives for economic income.

Author Contributions: Conceptualization, M.S.S.; methodology, M.S.S. and J.S.N.; software and validation, C.C.S.; formal analysis, F.B.C.; investigation, D.C.A.; resources, A.K.H. and W.M.S.; data curation, M.S.S.; writing—original draft preparation, writing—review and editing, A.K.H., C.C.S., F.B.C., J.A.L. and M.S.S.; visualization, supervision, U.S.P. and F.A.L.G.; project administration, S.P.Q.S.; funding acquisition, M.S.S. All authors have read and agreed to the published version of the manuscript.

Funding: This research was funded by CNPq-Conselho Nacional de Desenvolvimento Científico e Tecnológico, grant number 456518/2014-2.

Institutional Review Board Statement: Not applicable.

Informed Consent Statement: Not applicable.

Data Availability Statement: Not applicable.

Acknowledgments: I would like to thank the CAPES (Coordenação de Aperfeiçoamento de Pessoal de Nível Superior) by financial assistance (001), Food Science, Technology and Engineering Research Group (GPCTEA), AgriSciences, the collaborators of Revista Sustentabilidade (MDPI), and all the partners who contributed to the planning, execution, writing, and publication of the work. We also thank two anonymous reviewers whose comments and edits improved the quality of our work.

Conflicts of Interest: The authors declare no conflict of interest.

References

1. Silva, M.S.; Costa, F.B.; Formiga, A.S.; Nóbrega, J.S.; Brasil, Y.L.; Lira, R.P.; Silva, J.L.; Silva, K.G.; Sales, G.N.B.; Pereira, U.S.; et al. Qualitative assessment of early palm cladodes of *Opuntia tuna* L. (Mill) grown with organic manures. *J. Exp. Agri. Int.* **2018**, *26*, 1–14. [CrossRef]
2. Almeida, H.A.; Soares, E.R.A.; Santos Neto, J.A.; Pinto, I.O. Social and productive indicators of forage palm and the survival of livestock activity in the semi-arid region of Northeastern Brazil. *Asian J. Adv. Agri. Res.* **2019**, *10*, AJAAR.49011. [CrossRef]
3. Vasconcelos, A.G.V.; Lira, M.A.; Calvalcanti, V.L.B.; Santos, M.V.F.; Willadino, L. Seleção de clones de palma forrageira resistentes à cochonilha do carmim (*Dactylopius ceylonicus*). *Rev. Bras. Zoot.* **2009**, *38*, 827–831. [CrossRef]
4. Food and Agriculture Organization of the United Nations (FAO). *Ecologia del Cultivo, Manejo y Usos del Nopal*; Inglese, P., Mondragon, C., Nefzaoui, A., Sáenz, C., Eds.; FAO: Rome, Italy, 2018; pp. 1–248. ISBN 978-92-5-130494-5.
5. Lima, D.P.; Pereira, E.M.; Silva, N.J.P.; Magalhães, W.B.; Henrique, R.S. Qualidade pós-colheita em brotos de palma de diferentes cultivares. *Rev. Ve. Agro. Des. Sust.* **2017**, *12*, 280–283. [CrossRef]
6. Vila Nova, S.R.M.; Barros, J.G.; Paixão, A.E.A.; Tonholo, J.; Uchoa, S.B.B. Palma forrageira: Seu uso em alimentação e novas oportunidades de negoçioes de elevado valor agregado. *Cader. Pros.* **2017**, *10*, 738–753. Available online: https://periodicos.ufba.br/index.php/nit/article/view/22939 (accessed on 4 September 2022). [CrossRef]
7. Andreu, L.; Nuncio-Jáuregui, N.; Carbonell-Barrachina, Á.A.; Legua, P.; Hernández, F. Antioxidant properties and chemical characterization of Spanish *Opuntia ficus-indica* Mill. cladodes and fruits. *J. Sci. Food Agric.* **2018**, *98*, 1566–1573. [CrossRef]
8. Diaz, M.S.S.; Rosa, A.-P.B.; Toussaint, C.H.; Guéraud, F.; Nègre-Salvayre, A. *Opuntia* spp: Characterization and benefits in chronic diseases. *Oxi. Med. Cel. Long.* **2017**, *2017*, 8634249. [CrossRef]
9. de Figueredo, L.F.; Júnior, S.D.O.M.; de Figueredo, J.P.; da Silva, J.N.; de Sousa Ferreira, R.; Andrade, R. Desempenho agronômico de amendoim sob diferentes fontes e doses de biofertilizantes. *Acta Iguazu* **2018**, *7*, 17–26. [CrossRef]
10. Assouline, S.; Narkis, K.; Gherabli, R.; Sposito, G. Combined effect of sodicity and organic matter on soil properties under long-term irrigation with treated wastewater. *Vad. Zon. J.* **2016**, *15*, 1–10. [CrossRef]
11. Alves, L.S.; Véras, M.L.M.; Melo Filho, J.S.; Irineu, T.H.S.; Dias, T.J. Salinidade na água de irrigação e aplicação de biofertilizante bovino no crescimento e qualidade de mudas de tamarindo. *Irriga* **2019**, *24*, 254–273. [CrossRef]
12. Xavier Junior, O.S.; Jardim, A.M.R.F.; Souza, L.S.B.; Souza, C.A.A.; Cruz Neto, J.F.; Araújo Júnior, G.N.; Leite, R.M.C.; Silva, T.G.F. Utilização de biofertilizantes alternativos no cultivo de palma forrageira: Uma revisão. *Res. Soc. Dev.* **2021**, *10*, e46110616024. [CrossRef]
13. Garcete-Gómez, J.M.; Conforto, C.; Domínguez-Monge, S.; Flores-Sánchez, J.L.; Mora-Aguilera, G.; Michereff, S.J. Sample size for assessment of cladode brown spot in prickly pear cactus. *Eur. J. Plant. Pathol.* **2017**, *149*, 759–763. [CrossRef]
14. Alves, H.K.M.N.; Jardim, A.M.R.F.; Souza, L.S.B.; Silva, T.G.F. The application of agrometeorological techniques contributes to the agricultural resilience of forage cactus: A review. Amazon. *J. Plant Res.* **2018**, 2, 207–220. [CrossRef]
15. Souza, F.M.; Lima, E.C.S.; Sá, F.V.S.; Souto, L.S.; Araújo, J.E.S.; Paiva, E.P. Doses de esterco de galinha e água disponível sob o desenvolvimento inicial de milho. *Rev. Verde De Agroecol. E Desenvolv. Sustentável* **2016**, *11*, 64–69. [CrossRef]
16. Souza, F.M.; Pereira, W.E.; Dantas, J.S.; Nóbrega, J.S.; Lima, E.C.S.; Sá, F.V.S. Initial growth of *Moringa oleifera* Lam. as a function of poultry litter doses and granulometry. *Pesqui. Agropecuária Trop.* **2018**, *48*, 399–404. [CrossRef]
17. Santos, H.G.; Jacomine, P.K.T.; Anjos, L.H.C.; Oliveira, V.A.; Lumbreras, J.F.; Coelho, M.R.; Almeida, J.A.; Araújo Filho, J.C.; Oliveira, J.B.; Cunha, T.J.F. *Brazilian System of Soil Classification*, 5th ed.; Embrapa: Brasília, Brazil, 2018; pp. 1–356. Available online: https://www.embrapa.br/solos/busca-de-publicacoes/-/publicacao/1094003/sistema-brasileiro-de-classificacao-de-solos (accessed on 15 September 2022).
18. Institute, A.L. *Métodos Físico-Químicos para Análises de Alimentos*, 4th ed.; Instituto Adolf Luther: São Paulo, Brazil, 2008; pp. 1–1020. Available online: http://www.ial.sp.gov.br/ial/publicacoes/livros/metodos-fisico-quimicos-para-analise-de-alimentos (accessed on 15 August 2022).
19. Waterhouse, A. Folin-Ciocalteu Micro Method for Total Phenol in Wine. 2018. Available online: http://www.waterhouse.ucdavis.edu/phenol/folinmicro.htm (accessed on 10 September 2022).
20. Yemm, E.W.; Willis, A.J. The estimation of carbohydrates in plant extracts by anthrone. *Bioch. J.* **1954**, *57*, 508–515. [CrossRef]
21. Lichtenthaler, H.K. Chloropkylls and carotenoids: Pigments of photosynthetic biomembranes. In *Methods in Enzymology*; Packer, L., Souce, R., Eds.; Academic Tress: London, UK, 1987; Volume 148, pp. 350–382.
22. Crispim, J.E.; Martins, J.C.; Pires, J.C.; Rosolem, C.; Cavarani, C. Determinação da taxa de respiração em sementes de soja pelo método da titulação. *Pesquisa Agropecuária Brasileira* **1994**, *29*, 1517–1521. Available online: https://seer.sct.embrapa.br/index.php/pab/article/view/4203 (accessed on 12 September 2022).
23. Silva, J.L.; Costa, F.B.; Nascimento, A.M.; Sousa, F.F.; Santos, K.P. Taxa respiratória de frutos de juazeiro (*Ziziphus joazeiro* Mart.) armazenado sob temperatura ambiente. *Revista Verde de Agroecologia e Desenvolvimento Sustentável* **2017**, *12*, 343–347. [CrossRef]
24. Ferreira, D.F. Sisvar: A Guide for its Bootstrap procedures in multiple comparisons. *Ciên. Agrotec.* **2014**, *38*, 109–112. [CrossRef]
25. Pinto, S.M.; Farias, V.F.S.; Costa, F.B. Avaliação dos aspectos físico e químicos de brotos de palma (*Opuntia* sp.) para o consumo humano. *J. Bio. Pha. Agri. Man.* **2019**, *15*, 1–8. Available online: https://www.semanticscholar.org/author/Stef%C3%A2nia-Morais-Pinto/2058979915 (accessed on 10 September 2022).

26. Sáenz, C.; Berger, H.; Garcia, J.C.; Galletti, L.; Cortázar, V.G.; Higuera, I.; Mondragón, C.; Rodriguez-Félix, A.; Sepúlveda, E.; Varnero, M.T. Características y composición química de los nopales. Utilización agroindustrial del nopal. *Roma: Fa.* **2006**, *162*, 7–22. Available online: https://www.fao.org/3/a0534s/a0534s00.pdf (accessed on 15 September 2022).
27. Galvão, T.H.B.; Costa, F.B.; Nascimento, A.M.; Brasil, Y.L.; Silva, Á.G.F.; Silva, K.G.; Santos, K.P.; Silva, J.L.; Silva, M.S.; Sales, G.N.B.; et al. Physico-chemical quality changes of young cladodes of "mexican elephant ear" minimally processed during refrigerated storage. *J. Exp. Agri. Inter.* **2018**, *27*, JEAI.44273. [CrossRef]
28. Pereira, E.M.; Costa, F.B.; Formiga, A.S.; Sales, G.N.B.; Albuquerque, J.R.T.; Rocha, T.C. Physiology and postharvest quality of palm sprouts (*Opuntia ficus-indica*) harvested at different times. *J. Exp. Agri. Int.* **2018**, *26*, JEAI.44248. [CrossRef]
29. Albuquerque, J.R.T.; Formiga, A.S.; Rocha, T.C.; Costa, F.B.; Gondim, A.R.O. Qualidade pós-colheita de beterraba submetida à adubação com biofertilizante fermentado. *Revista Verde de Agroecologia e Desenvolvimento Sustentável* **2015**, *10*, 41–46. [CrossRef]
30. Farias, V.F.S. Evaluation of Development, Quality and Antioxidant Capacity in Palm Shoots (*Opuntia* sp.) for Human Consumption. Ph.D Thesis, Federal University of Campina Grande, Campina Grande, Brazil, 2013. Available online: http://dspace.sti.ufcg.edu.br:8080/xmlui/bitstream/handle/riufcg/709/VERL%C3%82NIA%20FAB%C3%8DOLA%20DE%20SOUSA%20FARIAS%20-%20DISS (accessed on 15 September 2022).
31. Alves, F.A.L.; Andrare, A.P.; Bruno, R.L.; Silva, M.G.V.; Souza, M.F.V.; Santos, O.C. Seasonal variability of phenolic compounds and antioxidant activity in prickly pear cladodes of *Opuntia* and *Nopalea* genres. *Food Sci. Technol.* **2017**, *36*, 536–543. [CrossRef]
32. Nóbrega, J.S.; Silva, I.C.M.; Silva, I.A.; Nobre, R.G.; Figueredo, F.R.A.; Souza, F.M.; Fátima, R.T.; Ferreira, J.T.A.; Nascimento, R.G.S. *Malpighia emaginata* D.C. Growth in several substrates and salt waters. *J. Agri. Sci.* **2018**, *10*, 352–361. [CrossRef]
33. Awasthi, R.P.; Godara, R.K.; Kaith, N.S. Interaction effect of VA-mycorrhizae and *Azotobacter* inoculation on micronutrient uptake by peach seedlings. *J. Hort.* **1998**, *11*, 1–5. Available online: https://eurekamag.com/research/003/180/003180207.php (accessed on 15 September 2022).
34. Kluge, R.A.; Nachegal, J.C.; Fachinello, J.C.; Bilhalva, A.B. *Fisiologia e Manejo de Frutas de Clima Temperado*, 2nd ed.; Livraria e Editora Rural: Campinas, Brazil, 2002; pp. 1–214.
35. Anaya-Pérez, M.A.; History of the Use of Opuntia as Forage in Mexico. Food and Agriculture Organization. Available online: https://www.fao.org/3/y2808e/y2808e05.htm (accessed on 2 September 2022).
36. Cavalcante, L.; Dewulf, A.; van Oel, P. Fighting against, and coping with, drought in Brazil: Two policy paradigms intertwined. *Reg. Environ. Chang.* **2022**, *22*, 111. [CrossRef]
37. Ncebere, J.M.; Mbuthia, P.G.; Waruiru, R.M.; Gathumbi, P.K. Gross and Histopathology of Goats Feeding on *Opuntia stricta* in Laikipia County, Kenya. *Vet. Med. Int.* **2021**, 8831996. [CrossRef]
38. Kunyanga, C.N.; Vellingiri, V.; Imungi, K.J. Nutritional quality, phytochemical composition and health protective effects of an under-utilized prickly cactus fruit (*Opuntia stricta* Haw.) collected from Kenya. *Afr. J. Fd Agric. Nutr. Develop.* **2014**, *14*, 9561–9577. [CrossRef]
39. Kampamba, M.; Ezeala, C.C.; Hikaambo, C.N.; Bwalya, A.G. *Opuntia stricta* (Haw): A review on its chemical composition, putative in-vitro antidiabetic mechanism of action and potential pharamacological uses in chronic disease. *J. Pharm. Phytochem.* **2022**, *11*, 1–8. [CrossRef]
40. Izuegbuna, O.; Otunola, G.; Bradley, G. Chemical composition, antioxidant, anti-inflammatory, and cytotoxic activities of *Opuntia stricta* cladodes. *PLoS ONE* **2019**, *14*, e0209682. [CrossRef]
41. Nova, S.; de, R.M.V.; Barros, J.G.; Paixão, A.E.A.; Tonholo, J.; Uchoa, S.B.B. Palma forrageira: Evidências de sua utilização econômica. *Cad. Prospec.* **2017**, *10*, 738–753. [CrossRef]
42. FAO. Food and Agriculture Organization of the United Nations. *Global Report on Food Crises 2017*. 2017. Available online: https://www.wfp.org/publications/global-report-food-crisis-2017 (accessed on 10 September 2022).
43. Amaral, S.M.B.; Almeida, A.P.F.; Marinho, R.M.O.; Silva, Y.Y.V.; Frota, M.M.; Damaceno, M.N. Utilização da palma forrageira na elaboração de produtos alimentícios: Uma revisão. *Holos* **2022**, *1*, 1–9. [CrossRef]
44. Pessoa, R.M.S.; Pessoa, A.M.S.; Costa, D.C.C.C.; Silva, P.C.A.; Gois, G.C.; Campos, F.S.; Vicente, S.L.A.; Fereira, J.M.S.; Araújo, C.A.; Lima, D.O. Cactus pear: Organic and mineral fertilization. *Res. Soc. Dev.* **2022**, *11*, e12111334257. [CrossRef]
45. Sales, A.T.; Alves, A.Q.; Ramos, J.P.F.; Nascimento, J.P.; Leite, M.L.M.V. Eficiência de Utilização da Adubação Orgânica pela Palma Forrageira em Função da Densidade Populacional. *Rev. Científica de Produção Animal* **2012**, *14*, 32–35. [CrossRef]
46. Lédo, A.A.; Donato, S.L.R.; Aspiazu, I.; Silva, J.A.; Donato, P.E.R.; Carvalho, A.J. Nutrient concentration and nutritional efficiency in 'Gigante' cactus pear submitted to different spacings and fertilizations. *Rev. Bras. de En. Agrícola e Ambient.* **2020**, *24*, 154–161. [CrossRef]
47. Soares, F.M.F.; Albuquerque, A.L.S.; Silva, W.R.T. Avaliação do crescimento vegetativo de palma forrageira no Semiárido Alagoano. *Diversitas J.* **2021**, *6*, 1777–1785. [CrossRef]
48. Santos, A.I.V. Propriedades Biológicas e Aplicações dos Cladódios de *Opuntia ficus-indica* (L.) Mill. Master's Thesis, Pharmaceutical Sciences—Universidade do Algarve Faculdade de Ciências e tecnologia, Algarve, México, 21 December 2016; pp. 1–76. Available online: https://core.ac.uk/reader/92441160 (accessed on 2 February 2023).
49. Apodaca-Pérez, J.M.; Martínez-Miranda, M.L.; Robles-Burgueño, M.D.; Féliz, A.R. Polifenoloxidasa, fenoles totales y oscurecimiento de nopal verdura. *Rev. Mex. Cienc. Agríc.* **2016**, *7*, 531–543. [CrossRef]

50. Maki-Díaz, G.; Péna-Valdivia, C.B.; García-Nava, R.; Arévalo-Galarza, M.L.; Calderón-Zavala, G.; Anaya-Rosales, S. Caracteristicas físicas y químicas de nopal verdura (Opuntia ficus-indica) para exportacion y consumo nacional. *Agrociencia* **2015**, *49*, 31–51. Available online: https://www.redalyc.org/pdf/302/30236850003.pdf (accessed on 4 September 2022).
51. Toivonen, P.M.A.; Brummell, D.A. Biochemical bases of appearance and texture changes in fresh-cut fruit and vegetables. *Postharvest Biol. Technol.* **2008**, *48*, 1–14. [CrossRef]
52. Calvo-Arriaga, A.O.; Hernández-Montes, A.; PeñaValdivia, C.B.; Corrales-García, J.; Aguirre-Mandujano, E. Preference mapping and rheological properties of four nopal (Opuntia spp.) cultivars. *J. Prof. Assoc. Cactus Develop.* **2010**, *12*, 127–142. [CrossRef]
53. Peña-Valdivia, C.B.; Trejo, C.; Arroyo-Peña, V.B.; Sánchez Urdaneta, A.B.; Morales, A.B. Diversity of unavailable polysaccharides and dietary fiber in domesticated nopalito and cactus pear fruit (Opuntia spp.). *Chem. Biodivers.* **2012**, *9*, 1599–1610. [CrossRef] [PubMed]
54. Reyes-Agüero, J.A.; Aguirre, J.R.; Hernández, H.M. Systematic notes and a detailed description of Opuntia *ficus-indica* (L.) Mill. (Cactaceae). *Agrociencia* **2005**, *39*, 395–408.
55. Kumar, K.; Singh, D.; Singh, R.S. *Cactus Pear: Cultivation and Uses*; Technical Bulletin No. 73; ICAR-Central Institute for Arid Horticulture: Rajasthan, India, 2018; pp. 1–38. Available online: https://krishi.icar.gov.in/jspui/bitstream/123456789/17886/1/Cactus%20pear%20Cultivation%20and%20uses.pdf (accessed on 5 February 2023).

sustainability

MDPI

Review

Agricultural Support and Public Policies Improving Sustainability in Brazil's Beef Industry

Luana Molossi [1,*], Aaron Kinyu Hoshide [1,2], Daniel Carneiro de Abreu [1,3] and Ronaldo Alves de Oliveira [1]

1 AgriSciences, Universidade Federal de Mato Grosso, Caixa Postal 729, Sinop 78550-970, MT, Brazil
2 College of Natural Sciences, Forestry and Agriculture, The University of Maine, Orono, ME 04469, USA
3 Instituto de Ciências Agrárias e Ambientais (ICAA), Universidade Federal do Mato Grosso, Campus Universitário de Sinop, Avenida Alexandre Ferronato, 1200, Sinop 78550-728, MT, Brazil
* Correspondence: luana_molossi@hotmail.com; Tel.: +55-66-99698-1573

Abstract: Since the dawn of Brazilian trade, extensive cattle farming has predominated. Brazil's extensive pasture-based system uses pasture plants adapted to climate and soil conditions with limited use of purchased inputs. However, new technologies such as integrated crop and livestock systems have recently been adopted, with government support and public policies that are intended to encourage increased agricultural production in Brazil. Domestic and international stakeholders have prioritized sustainable agricultural development in Brazil's beef sector to reduce deforestation and other natural-habitat conversions. This review provides an overview of beef production in Brazil, focusing particularly on (1) historical factors that have encouraged an extensive, low-intensity style of production and (2) how national public policies supporting agriculture have improved sustainability in Brazil's beef industry. Since the beginning of the twenty-first century, specific public policies for rural areas began to implement changes that addressed environmental concerns. Programs aimed at protecting secondary forests and increasing their areas are needed to offset the 42% of Brazil's greenhouse gas emissions that come from land-use change. To produce more beef with less environmental impact, cattle ranchers need to use their land more productively. Thus, public policy initiatives need to combat deforestation and preserve the environment and local communities, while sustainably intensifying Brazil's beef production.

Keywords: Amazon; beef; Brazil; deforestation; environmental impacts; greenhouse gases; livestock intensification strategies

Citation: Molossi, L.; Hoshide, A.K.; de Abreu, D.C.; de Oliveira, R.A. Agricultural Support and Public Policies Improving Sustainability in Brazil's Beef Industry. *Sustainability* **2023**, *15*, 4801. https://doi.org/10.3390/su15064801

Academic Editor: Christopher Brewster

Received: 30 January 2023
Revised: 3 March 2023
Accepted: 5 March 2023
Published: 8 March 2023

1. Introduction: History of Cattle Breeding and Production Systems in Brazil

The growth of beef cattle in Brazil has solidified the country in international markets as one of the largest exporters of beef. In 2021, the Brazilian herd was estimated at 196.47 million head, with 39.14 million head slaughtered. The volume of meat produced was 9.71 million metric tons of carcass-equivalent weight. Of this total volume, 25.51% of Brazil's beef production—2.48 million metric tons—was exported, while 7.24 million metric tons—equivalent to 74.49% of Brazil's beef production—were destined for the domestic market [1]. Brazil's beef production has historically and currently been dominated by an extensive pasture-based system, in which animals typically take two to four years to reach slaughter weight [2].

Brazilian cattle are predominantly tropic breeds (*Bos indicus*, such as the Nelore breed), with temperate breeds (*Bos taurus*) more prevalent in southern Brazil. During their evolution, *Bos indicus* cattle acquired genes that confer a greater thermotolerance in response to heat stress than that of European breeds. This is one of several reasons that *Bos indicus* (e.g., Nelore) are the predominant cattle in central Brazil, which has high temperatures and a dry climate throughout the year [3]. *Bos taurus* cattle are generally more adapted to environments with milder and more humid temperatures, such as those as found in the southern region of Brazil [4]. Cattle gain weight during the wet season (October through March) but lose weight during the dry season (April through September) as pasture productivity diminishes (Figure 1). Brazil's pastures comprise approximately 151 million hectares, including areas that are both natural and cultivated (Figure 2).

(a) (b)

Figure 1. Beef cattle (*Bos indicus*, such as the Nelore breed) grazing on extensive pasture (*Brachiaria* spp.) during the (**a**) wet season and (**b**) dry season in midwest region of Brazil (Source: corresponding author).

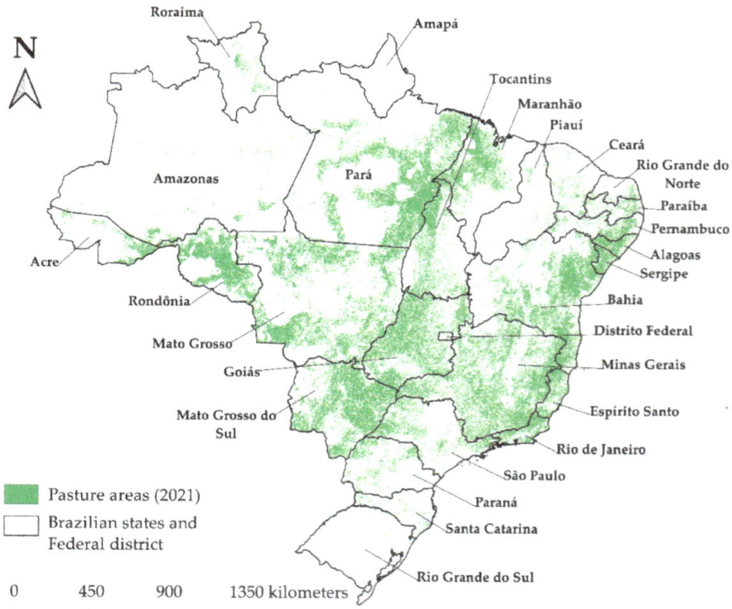

Figure 2. Spatial distribution of pasture areas in Brazil in states and the federal district in 2021 [5].

However, Brazil's current beef production, trade, and marketing are completely different from those practices in Brazil's beef industry 40 years ago. Then, the total beef herd was less than half of the current total, and beef production did not completely meet the Brazilian population's consumer demand [6]. In 2021, the beef production of 9.71 million metric tons of carcass-equivalent weight [1] was enough to meet the domestic demand for beef, which was 36.4 kg per person per year [7]. Even with more recent increases in beef production, the total pasture area associated with beef cattle has declined due to intensification strategies used in Brazil's beef production systems (Figure 3).

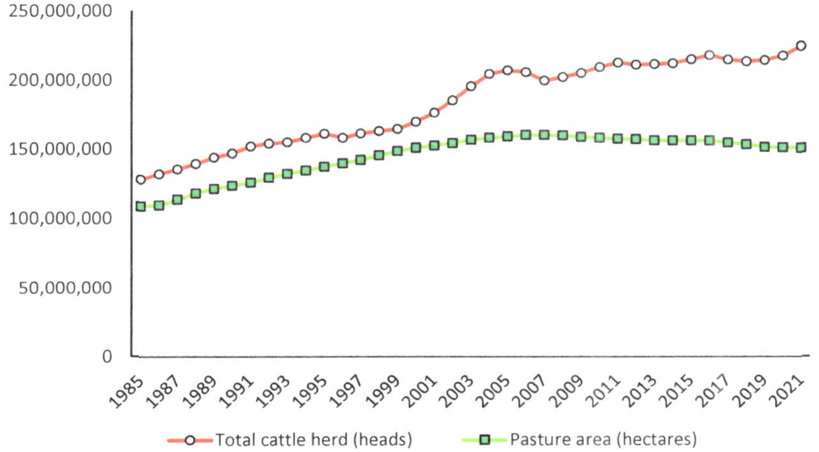

Figure 3. Total Brazilian cattle herds and pasture areas from 1985 to 2021 [8].

The first cattle arrived in Brazil in 1533, during the establishment of the first Portuguese colony on the island of São Vicente in the state of São Paulo [9]. In the middle of the 16th century, the Portuguese royal court encouraged the export of cattle to the Bahian Recôncavo region in northeastern Brazil. Gradually, with the growth of the economy in coastal areas, cattle raising expanded into the country's interior [10]. Since these commercial beginnings, Brazil's beef production has relied on an extensive pasture-based system, using plants adapted to local climate and soil conditions, with limited use of inputs [11].

With the opening of the Brazilian economy and the greater financial support of the agricultural sector in the 1990s, profound changes in Brazil's beef industry took place thereafter [12]. The development of practices aimed at increasing productivity has led to increases in intensive production systems in some regions. These technologies involve the genetic improvement of the animals, control of the economic management of the property, and a supply of concentrated feed for the animals, using feedlots or semi-feedlots to reduce the time to slaughter and increase profitability [13]. Thus, there has been a recent increase in cattle herd size (Figure 4), together with increases in cattle stocking density (head/hectare) (Figure 5), in particular regions in Brazil. Regions with such increases include the northern and central parts of Brazil, such as Brazil's center-west and north regions. While cattle herd numbers have stayed relatively stable from 1985 to 2021 in Brazil's northeastern, southeastern, and southern regions, the numbers have expanded in the north (from 5.3 to 55.7 million) and center-west (from 41.1 to 75.4 million) over these 36 years (Figure 4). Stocking densities of cattle are relatively high (>1 head/hectare), except for states along Brazil's southeastern coast. The adoption of the new technologies that allow for increased productivity and profitability in the beef cattle industry was possible due to the support and public policies developed for agricultural production in Brazil, with a focus on these developments occurring in a sustainable way [14].

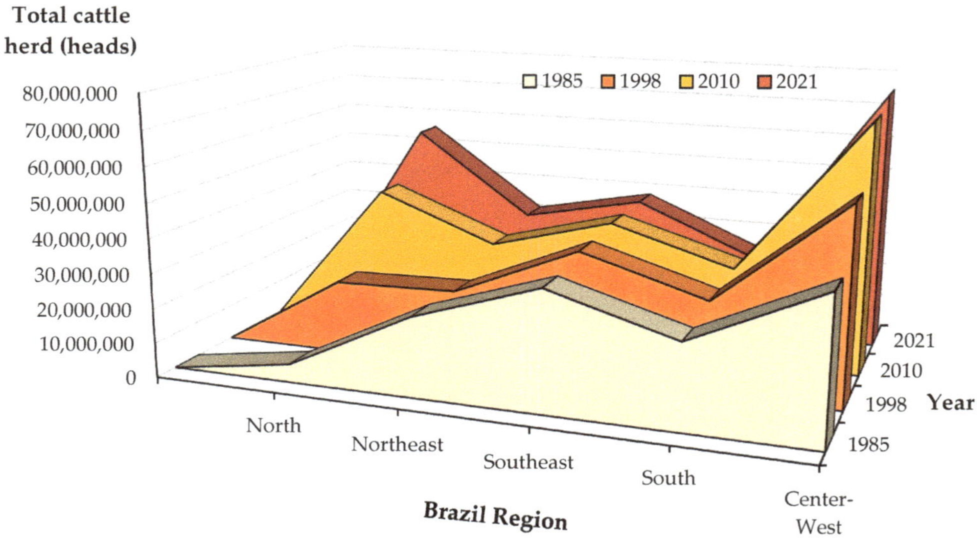

Figure 4. Cattle herd sizes (heads) in the northern, northeastern, southeastern, southern, and center-west regions of Brazil in the years 1985, 1998, 2010, and 2021 [8].

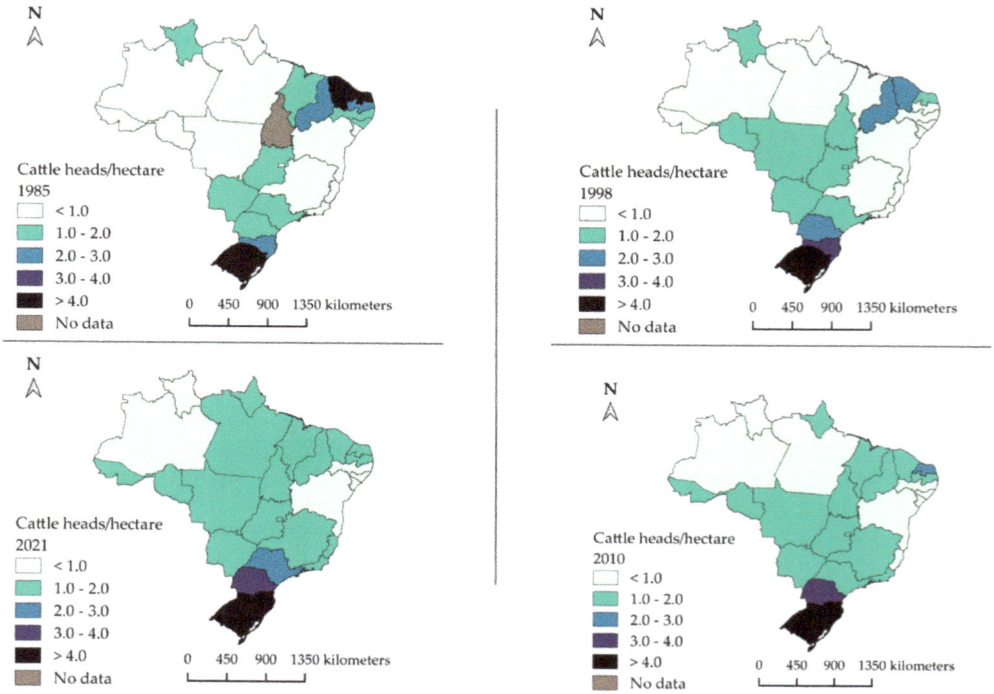

Figure 5. Cattle stocking densities (heads/hectares) by Brazilian states in 1985, 1998, 2010, and 2021 [15].

Such incentives have occurred because extensive systems have low efficiency, as they are based on cattle production with low technological intensity and management standards [13]. Thus, extensive systems for producing Brazilian beef have low productivity, requiring large amounts of land for grazing on seeded pastures that, over time, can become over-grazed and degraded [16]. Amazon rainforest deforestation, which precedes the establishment of pastures, has focused international attention on reducing deforestation [7]. This attention has encouraged three types of intensification in Brazil's beef-production systems to produce more beef on already existing pastureland: (1) re-seeding degraded pastures, (2) feeding grains in pastures, and (3) semi-intensive feedlots [17]. Pasture re-seeding involves the new establishment of pastures through forage sowing and fertilization to meet the plant requirements that are needed to optimize production [17]. Beef cattle can be fed grain at feeding stations in pastures that typically have low nutritional value [18]. Feedlots typically supply concentrated feeds for finishing animals, where the feeds normally consist of an energy and protein source [19]. Management-intensive rotational grazing—where cattle are rotated daily between paddocks that have been created by portable, electrified poly-wire—is much less popular [20], reinforcing producers' preferences for management systems that take less time.

Thus, the goal of this review is to provide an overview of beef production in Brazil, focusing particularly on (1) historical factors that have encouraged an extensive, low-intensity style of production and (2) how national agricultural support public policies have improved sustainability in Brazil's beef industry in order to better balance agricultural economic growth while reducing the environmental impacts of beef cattle. Such public policies have directly targeted the beef sector or indirectly targeted beef sustainability by focusing on reducing deforestation. We also highlight prospects for Brazil's beef production chain and compliance with market requirements. Current and future strategies for the sustainable intensification of Brazilian beef production systems are also explored. These strategies can reduce Brazilian beef's long shadow over the Amazon rainforest by sustainably intensifying land that has already been cleared, without further deforestation [21,22].

2. Historical Factors Encouraging Extensive-Pastured Beef

The agricultural sector has been important for the Brazilian economy since the beginning of colonization [21]. However, Brazilian beef cattle producers are historically characterized by resistance to technological innovations and by more primitive management, which negatively characterized the activity over the past several decades [22]. At the beginning of Brazil's colonization, the coastal lands were used to produce sugarcane, which was the main economic activity. Therefore, livestock was relegated to the interior of the country. In Brazil's northeastern region, livestock was concentrated in the back country (*sertão*) that supplied the northeastern coast from Maranhão state to Bahia state. In the south of the country, livestock farming in Brazil's pampas grassland biome is distinct from that in the rest of the country, since beef production in this region is based on the use of highly diversified native pastures (in contrast to exotic introduced pastures). If animals are not over-stocked. then livestock can contribute to pasture conservation via more sustainable management of these grassland biome agro-ecosystems [23].

However, despite developing more autonomously in the southern region than in the northeast, the growth in livestock industries was not continuous nor consistent [10]. The stagnation of livestock growth was also attributed to the fact that cattle served as a capital reserve during Brazil's inflationary periods. In addition, due to the extensive-exploration model and the availability of large areas of land for exploration, beef cattle were historically used as a land claim to open new areas on the agricultural frontier, while already-established areas were converted to agriculture uses [24]. This extensive-grazing strategy was characterized by low livestock productivity. Generally, meat producers occupy frontiers and use the land for extracting nutrients from the soil without replacing them. Instead of maintaining pasture quality and re-seeding degraded pastures, new areas are deforested, either on farms themselves or in other areas, such as new frontiers where beef production is being established. Such factors are predominant in Brazil's extensive pasture-based system, leading to pasture degradation as well as soil degradation and compaction [13,25].

Additionally, the state and federal governments have implemented programs to regulate illegal possessions of land through donations or sales of lands at below-market prices, encouraging speculative land occupations. These land occupations favor the formation of small unproductive properties [25], which results in negative consequences for the sustainability of livestock, such as low zootechnical indices, environmental impacts, and reduced economic returns [26]. Livestock activities continue to be practiced, to a large extent, within the traditional, extensive system [27]. However, with increases in the demand for food and the technological advancements in agriculture, new production techniques have been introduced for raising cattle, such as integrated systems [28].

Integrated crop and livestock systems can be used in degraded pastures, reducing the need for agricultural expansion. In addition to promoting the increase and diversification of production, integrated systems can enhance carbon stock and soil fertility [29]. However, integrated crop and livestock systems are limited in Brazil [30]. Like crop-livestock integration, rural development is a multidimensional process involving a wide range of actors, institutions, and institutional infrastructure. This can range from communities and farmers to public-policy makers, passing through organizations representing broad social and productive sectors, as well as science and technology institutions across different levels [31]. Public policies are actions and decisions formulated in different spheres of legislative and executive power for the purpose of solving public problems. The process of formulating a public policy begins with the detection of an element with respect to which the government must act. The process must be configured as identifying a problem to be solved, understanding why it is important to provide solutions to such a problem, and anticipating the expected results of adopted solutions [32].

Changes in public policies prioritizing environmental concerns began to be implemented in Brazil, especially for family farming, starting in the early 2000s [33]. Programs that aim to protect secondary forests and increase their areas are necessary to offset Brazil's greenhouse gas emissions, approximately 42% of which come from global land-use change [34]. It is expected that increased beef cattle production will occur via increases in productivity rather than by expansion of pasture areas, transforming current extensive systems into systems with greater livestock intensification [35]. Such strategies will also result in a significant net reduction in greenhouse gas emissions [36]. Thus, it is essential to implement public policy mechanisms that counter environmental degradation and encourage the conservation of natural biomes and the sustainable use of natural resources, while allowing for gains in productivity [37]. This approach can counter illegal land occupation practices, as well as speculative and destructive actions in forest preservation areas [38].

3. Recent Public Policies for More Sustainable Livestock in Brazil

In this section, we discuss two recent types of sustainable agricultural development policies in Brazil. In Section 3.1, we cover agricultural policies that have had a direct impact on improving sustainable intensification (SI) in Brazil's beef-production industry. In Section 3.2, we highlight environmental public policies that have had an indirect impact on the SI of Brazil's beef cattle herd. These environmental policies have typically preceded the more direct policies discussed in Section 3.1 and have involved a reduction of Amazon deforestation and a conversion of land use in other natural Brazilian habitats for cattle pastures.

3.1. *Agricultural Policies Directly Supporting Sustainable Livestock*

We ordered recently enacted public policies in Brazil from the least challenging to the most challenging for producers to adopt. Most of these policies, with the exception of the 1965 Brazilian Forest Code, have been adopted over the past two decades. It is important to note how involved beef producers have been with these programs and to what extent these policies encourage sustainable intensification strategies for Brazil beef production. These strategies include good agricultural practices, low-carbon production, integrated crop–livestock–forest systems, pasture-based grain supplementation, pasture rehabilitation, and semi-intensive feedlots.

3.1.1. Agriculture and Livestock Plan (Plano Agrícola e Pecuário)

Brazil's Agriculture and Livestock Plan is the main instrument for directing public policies aimed at the agricultural sector. The plan includes measures to encourage the production of certain products, while providing resources for agricultural producers, including credit at favorable interest rates that is made available throughout the harvest year. Brazil's harvest year runs from July to June. The amount of money devoted to the Agriculture and Livestock Plan depends on the budget of the National Treasury and the amount allocated to financial subsidies for the agricultural sector [39].

The Agriculture and Livestock Plan embodies the main measures to support commercialization, rural risk management, and credit support. Thus, government actions are necessary to ensure the continuity of the advances that have been already achieved in increasing agricultural productivity. The Agriculture and Livestock Plan also can sustain the income of rural producers and ensure the flow of food, fuel, and fiber to both domestic and international markets. Brazil has had favorable conditions related to production costs, which have increased the competitiveness of its agricultural exports [40].

The annual publication of the previous Crop Plan became a tradition, dealing only with questions related to crops and marginalizing the livestock sector. In 2000, thanks to requests from entities representing milk producers and their cooperatives and to the sensitivity of the Brazilian government, the famous Crop Plan proposed measures related to dairy production for the first time. In the same year, the previous Crop Plan was renamed the Agriculture and Livestock Plan to definitively address livestock via announced measures [41].

The Brazilian Agricultural Research Corporation (Embrapa), which is considered to be an important developer of technologies in the Brazilian agricultural sector, is mentioned in the Agriculture and Livestock Plan 2012–2013. Among the highlights of Embrapa's programs is the Good Agricultural Practices Program (GAPP) for beef cattle. The GAPP is not an agricultural credit measure, but rather a mechanism within the plans that can differentiate access to credit by rural producers. Created in 2005, this program encompasses a set of norms and procedures that must be observed by rural producers in order to make their properties more sustainable. Various factors, such as the management and the social function of rural properties, human resources management, environmental management, rural facilities, pre-slaughter management, animal welfare, pastures, food supplementation, animal identification, sanitary control, and reproductive management, are crucial for the effectiveness of GAPP's adoption on farms [42].

3.1.2. ABC Plan, or Low-Carbon Agriculture Plan (Agricultura de Baixa Emissão de Carbono)

The Sectorial Plan for Mitigation and Adaptation to Climate Change for the Consolidation of a Low-Carbon Economy in Agriculture, also known as the ABC plan, is one of the sectoral plans prepared in accordance with Article 3 of Decree No. 7.390/2010. Its purpose is to organize and plan actions to be carried out for the adoption of sustainable production technologies. These sustainable technologies are selected with the objective of responding to Brazil's commitments to reduce greenhouse gas (GHG) emissions in the agricultural sector [43]. The ABC plan addresses climate change, ecosystem and biodiversity management, resource use efficiency, and sustainable consumption/production, in addition to presenting guidelines for environmental governance, thereby contributing to the exchange of information and experiences among the public, private, and academic sectors [44].

The ABC plan is composed of seven programs, six of which refer to mitigation technologies, and another that includes actions needed to adapt to climate change. The first program involves rehabilitating degraded pastures. The second program is the integration of crop–livestock–forest (ICLF) systems and agroforestry systems (AFSs), which involves integrating commercial forestry species (e.g., *Eucalyptus* sp.) with commodity crops, such as soybeans (*Glycine max* L.), corn (*Zea mays* L.), and cotton (*Gossypium* sp.), and livestock, such as beef (*Bos indicus*, such as Nelore cattle). The third and fourth programs comprise the direct planting system (DPS) and biological nitrogen fixation (BNF) programs, which involve soil mobilization only in a sowing line or planting hole, the permanent maintenance of soil cover, species diversification, and increased fertilization efficiency. The fifth program involves re-forestation, while the sixth program focuses on the treatment of animal waste. Finally, the seventh program addresses climate-change adaptation [45].

The ABC program provides agricultural producers with opportunities to incorporate sustainable technologies into their production processes for more efficient production. This can increase income through increased productivity and product diversification. It also can mitigate environmental liabilities, reduce pressure on native forests, and lower GHG emissions, thereby enhancing sustainable agricultural production of food for local Brazilian and export markets. This new sustainable agricultural program involves government incentives that provide attractive alternatives to existing financing instruments in the marketplace [46]. With the adoption of more sustainable techniques and production systems, it is possible to increase productivity, reduce deforestation, reconcile soil and water conservation, adapt rural properties to environmental legislation, expand the area of cultivated forests, and encourage the recovery of degraded areas [47].

3.1.3. National Integrated Crop–Livestock–Forest Policy (Política Nacional de ICLF)

The silvopastoral system is a technological option for integrated crop–livestock–forest integration that consists of an intentional combination of trees, pastures, and cattle in the same area at the same time. The approval of Law 708/07 on 4 February 2013 established the National Integrated Crop–Livestock–Forest Integration (ICLF) policy in Brazil. The ICLF policy reinforces the growing interest in the use of sustainable production systems. This law integrates agricultural and forestry activities carried out in the same area, in consortium, in succession, or in rotation. It seeks synergistic effects between the components of the agroecosystem, with objectives of recovering degraded areas, and enhancing economic viability, and supporting environmental sustainability [48].

The ICLF policy is a production strategy that includes the economic, social, and environmental aspects of sustainability. With the growing concern about the relationship between the environment and livestock, the challenge of establishing sustainable production systems is paramount. Silvopastoral systems are capable of meeting this challenge [49].

The ICLF policy's core principles involve the preservation and improvement of the soil's physical, chemical, and biological conditions and compliance with environmental protection laws. Cooperation between the public and private sectors and non-governmental organizations is recommended to foster the diversification of economic activities. Another policy guideline is the encouragement of direct planting in crop residue from the preceding crop as a soil-conservation management practice [50]. The ICLF policy also aims to mitigate deforestation caused by land-use conversion of native vegetation into pastures and/or crops and contributes to the maintenance of permanent preservation areas and legal reserves. The recovery of degraded pasture areas is also encouraged via sustainable production systems, such as the adoption of conservation practices and agricultural systems that maintain higher levels of organic matter in the soil and that reduce greenhouse gas emissions [51].

3.1.4. Agriculture Modernization and Natural Resources Conservation Program (Programa de Modernização da Agricultura e Conservação de Recursos Naturais—Moderagro)

The Moderagro (the Program for the Modernization of Agriculture and Conservation of Natural Resources) aims to support and encourage the production, processing, industrialization, packaging, and storage of agricultural products. This program is a Banco Nacional do Desenvolvimento (BNDES) project that enables rural producers to finance actions to recover soils, defend animals, acquire and apply agricultural fertilizers, and build facilities for the storage of agricultural machinery and implements, as well as for the storage of inputs [52]. The program supports and encourages the sectors of production, processing, industrialization, packaging, and storage of animal products from the beekeeping, aquaculture, poultry, chinchilla, rabbit, sheep and goat, frog, pig, and dairy farming industries. The agricultural production of floriculture, fruits, olives, horticulture, palm trees, yerba mate, nuts, and fishing are also encouraged [53].

3.2. Agricultural Policies Indirectly Supporting Sustainable Livestock

Other recently enacted environmental policies in Brazil have had more of an indirect impact of improving the sustainability of Brazil's livestock. These public policies have reduced Amazon deforestation and Cerrado habitat conversion. In general, these public policies were enacted earlier than policies that directly focus on livestock (Figure 6). We discuss whether beef producers were engaged and involved in the writing and implementation of these public policies. We also highlight how influential the limitation of grazing areas for cattle by preserving native habitat has been in encouraging beef producers to sustainably intensify their production systems.

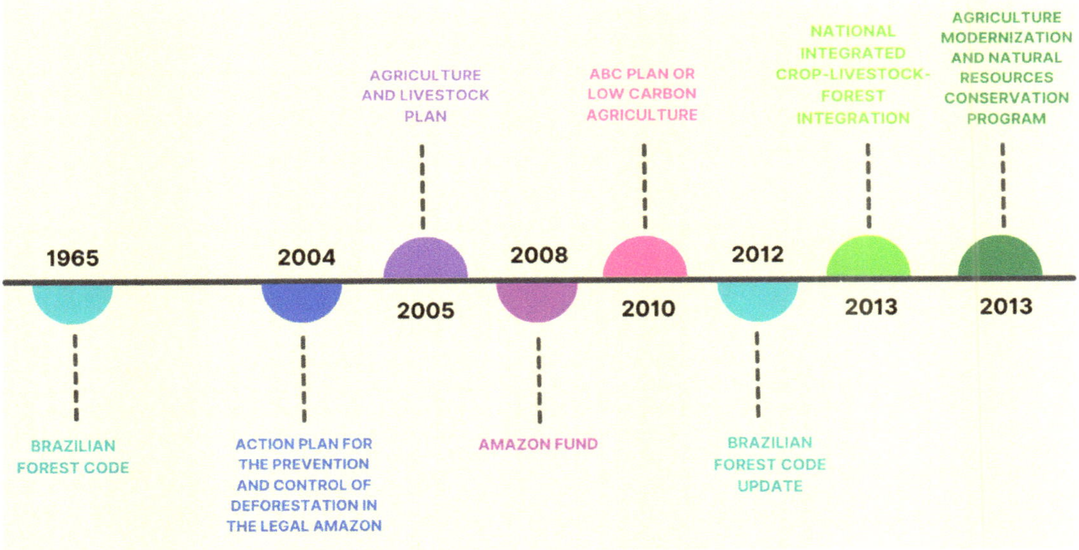

Figure 6. Historical timeline of direct and indirect public policies impacting sustainability in Brazil's beef cattle industry from 1965 to 2013.

3.2.1. The 1965 Brazilian Forest Code & 2012 Update (1965 & 2012 Código Florestal Brasileiro)

The Brazilian Forest Code is an indispensable political instrument for managing the country's economic development. Focusing on the different historical periods in Brazil, the evolution of the Brazilian Forest Code (BFC) reflects past political, economic, and environmental events, as well as the development intentions articulated through the law [54]. The BFC was created in 1965 with the aim of preserving forests and streamlining their management. At the time the code was created, the main agricultural activities were coffee (*Coffea arabica* and *Coffea robusta*) and sugar cane (*Saccharum* spp.). The code also contained several provisions, such as the prohibition of occupying steep slopes and a determination for rural landowners to maintain a reserve of native vegetation on their farms to contribute to preserving existing forests [55].

Although instituted in 1965, it was only in the 1980s that the Legal Reserve and Permanent Preservation Areas were effectively introduced into law, via a provisional measure. Another important aspect of the 1965 code was the creation of the National System of Conservation Units, which covered the different types and categories of protected areas within a single management system [56]. Law 12,651, which dates from 25 May 2012, introduced a series of new forest regulations. The 2012 Forest Code, currently on the books, consolidated protected areas from the previous code and included conceptualizations/specifications for delimitation of each area provided for in the law [54].

The BFC laws' innovations resulted in the creation of the Rural Environmental Registry (RER), as well as implementation of the Environmental Regularization Program. Under the RER, it is possible for the federal government and state environmental agencies to determine the location of each rural property and the status of its adherence to environmental standards for the preservation of native vegetation on the property. Additionally, the new law authorizes a series of benefits for family farmers or owners of smaller properties by including their properties in the RER [57].

The RER is a nationwide electronic registration system for gathering data on rural properties/possessions that are used for environmental and economic planning and for combating deforestation. The RER is the first step enabling rural producers to show that their rural property complies with the Forestry Code. If an owner does not register in the system, the owner is prevented from having access to agricultural credit from financial institutions. In addition, it can be difficult for rural producers to sell their agricultural products, as some companies require RER documentation from producers in order to buy their products [58].

3.2.2. Plan to Prevent and Control Deforestation in the Brazilian Amazon

The Action Plan for the Prevention and Control of Deforestation in the Legal Amazon (APCDAm) was created in 2004 with the aim of continuously reducing deforestation and creating conditions for the transition to a sustainable development model in the Legal Amazon. One of the main initial challenges was to integrate the fight against deforestation into Brazilian State policies [40]. Thus, the APCDAm became a strategic initiative of the Brazilian government that was included in the guidelines and priorities of this sustainable development plan for the Amazon. Therefore, the problem of the Amazon became part of the political agendas at the highest levels of the federal government and ministries [32].

Because the fight against the causes of deforestation could no longer be conducted in isolation by environmental agencies, the complexity of the challenge required coordinated efforts from different sectors of the federal government [40]. The APCDAm is implemented by more than a dozen government ministries; it was coordinated by the Civil House until March 2013, and thereafter by the Ministry of the Environment. The APCDAm is structured to address the causes of deforestation in a comprehensive, integrated, and intensive way, with actions articulated around three themes: land and territorial organization, environmental monitoring and control, and the promotion of sustainable production [59].

3.2.3. Amazon Fund (Fundo Amazon)

The Amazon Fund aims to encourage Brazil and other developing countries that have tropical forests to maintain and increase voluntary reductions in the emission of greenhouse gases caused by deforestation and land degradation [60]. The Amazon Fund was created by Decree No. 6527 on 1 August 2008. This fund raises donations for non-reimbursable investments to prevent, monitor, and combat deforestation and to promote the conservation and sustainable use of forests in the Amazon biome. Its creation was a consequence of the success achieved by the APCDAm in reducing deforestation in the Amazon since its implementation in 2004. The creation and raising of resources by the Amazon Fund have led to funding Brazilian efforts to reduce the loss of forests via projects that work on this theme, in synergy with government agencies [59].

4. Comparison of Agricultural Public Policies

In Table 1, we summarize and compare the program impacts and costs of both the direct and indirect public policies that affect Brazil's beef industry. Direct public policies share three common themes. The first is to provide financial credit to agricultural producers in order to intensify their production in a sustainable way. The second is to increase production without the need to expand into new areas (e.g., without Amazon deforestation). The third shared theme is to reduce the environmental impacts of agricultural production. Indirect public policies support direct policies with the main objective of preserving the environment. These indirect policies can help reduce the adverse environmental impacts of meat production in Brazil. For example, the Amazon Fund finances programs that reduce Amazon rainforest deforestation and greenhouse gas emissions. The Amazon Fund's cost is high at USD 337.3 million in 2021 (see Table 1).

The costs, in 2021, of each program to finance farmers and ranchers varied. The Agriculture and Livestock Plan is the main public policy for providing rural credit in Brazil. Therefore, it is the policy that bears the greatest cost for the Brazilian government, at USD 48.3 billion in 2021 (see Table 1). On the other hand, the National Integrated Crop-Livestock-Forest Integration public policy had the lowest cost in 2021 (USD 13.1 million) as this policy specifically focuses on encouraging integrated crop–livestock–forestry production. This policy is unlike other policies, such as the Agriculture Modernization and Natural Resources Conservation Program, which can cover different forms of modernization within rural properties, thus requiring greater financial resources (USD 68.1 million) annually (see Table 1).

Sustainability challenges in the livestock industry require simultaneous progress in production and environmental performance [61]. The public rural credit policy in Brazil incentivizes rural producers to recover fragile areas and pastures, to reduce production in unproductive soils and degraded areas, to plant forests, and to preserve natural resources. This credit policy also encourages the implementation and improvement of agricultural production systems, such as organic livestock systems and direct-planting systems [62].

The Agriculture and Livestock Plan consolidates the main actions and public policies aimed at the agricultural sector, with an emphasis on rural credit [59]. Some credit programs, such as the Agriculture Modernization and Natural Resources Conservation Program, are aimed at innovation, such as the implementation of an animal traceability system for human consumption. However, credit programs can also support other on-farm investments, such as recovering soils by financing the acquisition, transport, application, and incorporation of agricultural fertilizers [63]. The ABC program is another important program for modernizing sustainable production systems and mitigating emissions through low-carbon agriculture [62]. The same is true for the National Integrated Crop–Livestock–Forest Integration program, which optimizes land use, raises productivity levels, diversifies production, and generates quality products via integrated systems [41].

Other policies provide security and support for direct credit to rural producers, such as policies that establish environmental criteria and inspection actions to encourage environmental improvements. Such policies must be evaluated against specific outcomes, such as targets related to deforestation and greenhouse gas emissions. These assessments should serve as bases for improving the formulation of environmentally conditional policies and specific programs, such as initiatives against deforestation and the preservation of the Amazon biome [62].

Table 1. Comparison of direct and indirect public policies for the beef industry, the impacts generated, and the costs of these programs for the Brazilian government.

Public Policy Type	Brazilian Sustainable Agricultural Public Policy	Year Enacted	Program Impacts [Reference]	Program Cost in 2021 USD
Direct	Agriculture and Livestock Plan	2005	Financing with reduced interest rates [64]. Priority given to technological innovation, storage, irrigation, and low-carbon agriculture [64]. Priority of financing for small- and medium-sized producers [64].	48,311,538,461
	ABC Plan or Low Carbon Agriculture	2010	Reduce GHG emissions in agriculture [59]. Improve efficiency of natural resource use [59]. Encourage the adoption of Sustainable Production Systems [59].	961,000,000
	National Integrated Crop–Livestock–Forest Integration	2013	Sustainably improve productivity, product quality, and income from agricultural activities through application of integrated systems [65]. Stimulate research, development, and technological innovation activities [65]. Promote recovery of degraded pasture areas through sustainable production systems [65].	13,076,923
	Agriculture Modernization and Natural Resources Conservation Program	2013	Soil recovery [52]. Build facilities for the storage of agricultural machinery and implements [52]. Aimed at medium and large rural producers who wish to invest in diverse production [52].	68,076,923
In-direct	Brazil Forest Code	1965	Declares existing forests as assets of common interest to the entire population and limits the use of rural property by its owners [66]. Maintains and protects permanent preservation areas and on-farm reserves [66].	-
	Action Plan for the Prevention and Control of Deforestation in the Legal Amazon	2004	Reduce rate of deforestation in the Amazon [40]. Environmental monitoring and control [40]. Promote sustainable productive activities [40].	3,990,384
	Amazon Fund	2008	Deforestation reduction with sustainable development in the Amazon [60]. Financing actions for prevention, monitoring, and conservation of Amazon biome [60].	337,307,692
	Brazil Forest Code Update	2012	Conditions subject to use or management of native vegetation on rural properties [54]. Incentives for technology adoption/good practices reconciling agricultural/forestry productivity with reduced environmental impacts [54]. Recognition of positive impacts in the field in search for sustainable production [54].	-

5. Future Directions in Agricultural Public Policies for Brazil Beef Production and Sustainability

In the long term, the likely impacts arising from climate change could significantly compromise agricultural activities such as beef cattle production. Some models point to negative scenarios for Brazilian climatic conditions, indicating possible reductions in the availability of water in certain regions and an increase in the availability of water in other regions. In addition to water insecurity, Brazilian agriculture could be impacted by an increase in atmospheric temperatures, which could jeopardize food production and security. These changes may also reduce the profitability of dual cropping systems in Brazil, which are mainly due to shorter rainy seasons, leading to a future shift back to growing only one crop per year instead of two or more [67]. Climate change may also favor nitrous oxide (N_2O) emissions if cropping and soil management systems are maintained at their current status [68]. These potential impacts from climate change could result in a negative balance of payments with reductions in products that are destined for export [59].

For the implementation of sustainable production systems, it is first necessary to adopt good agricultural production practices in order to preserve natural resources (e.g., soil, water, biodiversity, and natural forests) that will ensure future production and ecosystem integrity. Combating erosion, recovering degraded soils, and maintaining water sources, natural forests, and biodiversity are priorities that should guide the actions of rural producers and frame public policy [21]. Sustainable agricultural systems are more likely to be adopted if they are technically efficient, environmentally suitable, economically viable, and socially accepted [28]. Adequate planning and cost management for the best use of available resources and production factors, with a focus on greater productivity, will become essential for more widespread adoption of economically, socially, and financially sustainable beef cattle production [13].

Over the past decade, new production technologies have been disseminated within Brazil's beef production systems. Technological processes, such as strategic supplementation, semi-confinement, the use of multiple mixtures, genetic crossings, and new forage varieties have enabled Brazil's beef producers to shorten beef production cycles. In addition, technological management methods were incorporated and integrated to reduce production costs and increase economic margins, allowing the beef cattle industry to be one of the more prominent agribusinesses in Brazil [24]. However, the cumulative negative environmental impacts of this increased production of beef have increasingly forced public authorities to question and reconfigure the main notions about food production, which are linked in some ways to forms of development [68].

The role of agriculture in the future could substantially exceed the current traditional systems, requiring joint efforts by the public and private sectors [21]. In order to produce more beef with less environmental impact, Brazil's beef industry must use land more productively. Therefore, it may be necessary to discourage the expansion of speculative, inefficient, and/or riskier agricultural frontiers and to provide services and infrastructure that facilitate investments in areas that have already been deforested [25]. For example, Brazil's final agricultural frontier of Matopiba in northeastern Brazil is drier and more likely not to have enough rainfall for double-cropping (e.g., soybeans followed by corn in the same production year) and not to have adequate groundwater for irrigation [69]. By contrast, agricultural production in Brazil's Amazon and Cerrado (i.e., savannah) biomes can potentially be doubled on existing cleared land without additional deforestation [70].

Assuming the continued implementation of the Brazilian Forest Code and active efforts to re-forest Brazil, beef cattle are projected to increase by 57% on the same amount of pasture, while cropland is expected to increase by 85% by 2050 [71]. Increasing such agricultural production and productivity on land that has already been cleared of forest and native habitat involves sustainable intensification of beef, pasture, and both annual and perennial crops. For Brazil's beef, this can involve reducing greenhouse gas emissions by reducing time to slaughter [72] and by using grain supplementation [18,73]. Pasture productivity in Brazil can be maintained to avoid degradation by re-seeding [18] or through on-farm integration of pasture and commodity crops [74]. Management-intensive rotational grazing (MIRG) was estimated to have about double the carbon removal potential, compared to confined feeding for Brazilian cattle, and MIRG can reduce the pasture area [75] that is required to boost beef productivity. However, MIRG can be more labor-intensive for farmers and farm workers, compared to crop–livestock integration, as cattle have to be frequently rotated between paddocks. Even with public-policy support, farmer adoption may be challenging unless farmers have sufficient resources to commit to the additional labor that is required [74].

Public policies encouraging on-farm integration may be more successful, as producers do not need to coordinate with other farmers. Integration between specialized-crop and livestock farms can be challenging, due to the needs for added coordination and to be close enough to integrate livestock with crops [76,77]. Unlike France, Brazil had limited experience with supra-regional transport of manure and feed over long distances (100 to 500 km) to facilitate crop–livestock integration [78]. Although Brazil has recently increased the semi-confined feeding of beef cattle [79], dairy farms, poultry, and hog farms are typically smaller [80], so manure production does not exceed the capacity of farms to absorb manure nutrients [81]. Additionally, there are no current public policies regulating manure production and application [82]. Despite these challenges, past research suggests that rural credit can be successfully used to incentivize these more complex forms of integration, such as integrated crop–livestock–forest (ICLF) systems in the Brazilian state of São Paulo [83]. Therefore, the Brazilian government could continue to use the availability of rural credit that is conditional on producers adopting more sustainable agricultural practices. For Brazil's beef producers, the use of rural credit has been associated with reduced deforestation [84].

Such government policies can be easier for agricultural producers to accept if there are clear potential economic benefits, by adopting systems-based approaches such as the ICLF policy or less management-intensive strategies such as sustainable intensification. When designing such recommendations, it is necessary to redirect Brazilian government subsidies to livestock. This guideline is essential in a scenario of budget constraints and current and future climate change [25]. Thus, the attention of national and international organizations and public opinion on illegal deforestation demands intelligence, articulation, and communication in order to guarantee the preservation of natural resources and not to compromise future agricultural exports. Technologies and knowledge are essential factors in promoting sustainable development, mainly by encouraging more sustainable use of resources in the regions where agricultural activities are already concentrated [21]. It is important for public policies to be accessible to rural producers so that Brazil can produce enough food, bioenergy, and commodities for national consumption, while maintaining competitive advantages within the global economy [85].

Although Brazil has abundant water resources, specific commodity-cropping regions, such as Matopiba, face water shortages [69]. Sustainable intensification of crops can involve traditional breeding, especially for maize [86], unlike soybeans. Dry-season irrigation can also increase agricultural output for third crops on the same land base following soybeans and corn [87]. However, Brazilian commodities, such as soybeans and beef cattle, could be sensitive to future drought caused by climate change [88]. Therefore, future agricultural public policies in Brazil can encourage producers to use less water. Due to increasing water scarcity and rising irrigation costs, there has been a growing interest in improving the productivity of water use in agricultural production, with the need to understand the effects of combining water-irrigation management with other agronomic practices for efficient water management and satisfactory yields [89]. Such sustainable intensification of crops can increase environmental sustainability, improve soil conditions, and reduce water pollution. These improvements can potentially benefit the environment, while increasing agricultural productivity [90].

6. Conclusions and Implications

The extensive cattle-production system in Brazil is characterized by producers who have been resistant to technological innovations and by the adoption of more intensified management practices. However, in recent years, new technologies have been disseminated in beef production systems in Brazil, such as strategic supplementation, semi-confinement, the use of multiple mixtures of concentrated feeds, genetic crossings, and new forage varieties, which have led to increases in productivity and economic returns. However, the increase in production has generated cumulative environmental impacts, such as deforestation, pasture degradation, and greenhouse gas emissions. Public policies have played an important role in encouraging the adoption of new technologies to mitigate these environmental impacts. However, the impacts caused by beef production in Brazil have forced authorities to develop public policies that allow for increased production but prioritize sustainability, with a focus on the preservation of biomes and natural resources. In order to produce more beef with less environmental impact, it is necessary to encourage livestock farmers to use their land more productively. It is also important to discourage the expansion of speculative and inefficient agricultural frontiers and to provide services and infrastructure that facilitate sustainable agricultural development investments in the regions in Brazil that produce beef cattle. Thus, public policies must continue to be evaluated for their ability to balance agricultural production with resource conservation and environmental preservation.

Author Contributions: Conceptualization, L.M., A.K.H. and D.C.d.A.; methodology, A.K.H. and L.M.; validation, L.M. and A.K.H.; formal analysis, L.M., R.A.d.O. and A.K.H.; investigation, L.M.; resources, D.C.d.A., A.K.H. and L.M.; writing—original draft preparation, L.M., A.K.H. and D.C.d.A.; writing—review and editing, all authors; visualization, A.K.H. and R.A.d.O.; supervision, D.C.d.A. and A.K.H.; project administration, D.C.d.A.; funding acquisition, D.C.d.A. and A.K.H. All authors have read and agreed to the published version of the manuscript.

Funding: This research was funded by Grupo Osvaldo Sobrinho (Fazenda Boa Vista e Taguá Agropecuária), Serviço Nacional de Aprendizagem Rural de Mato Grosso (SENAR-MT), Programa Global REDD Early Movers de Mato Grosso (REM-MT), and Projeto Rural Sustentável—Cerrado (PRS—Cerrado) e Coordenação de Aperfeiçoamento de Pessoal de Nível Superior (CAPES) for granting a scholarship.

Institutional Review Board Statement: Not applicable.

Informed Consent Statement: Not applicable.

Data Availability Statement: Not applicable.

Acknowledgments: We wish to give special thanks to the AgriSciences program at the Federal University of Mato Grosso and Osvaldo Sobrinho for helping with the necessary structure to carry out our research. We also thank the ICAA of the Federal University of Mato Grosso, SENAR-MT, PRS—Cerrado e REM—MT for supporting and encouraging the publication of our study. Finally, we thank three anonymous reviewers whose comments and edits substantially improved the quality of this work.

Conflicts of Interest: The authors declare no conflict of interest. Supporting entities had no role in the design of the study; in the collection, analyses, or interpretation of data; in the writing of the manuscript; or in the decision to publish the results.

References

1. Abiec Associação Brasileira das Indústrias Exportadoras de Carnes. Beef Report. *Perfil da Pecuária No Brasil, Disponíve*. 2022. Available online: https://www.abiec.com.br/publicacoes/beef-report-2022/ (accessed on 15 December 2022).
2. Ferraz, J.B.S.; de Felício, P.E. Production systems-An example from Brazil. *Meat Sci.* **2010**, *84*, 238–243. [CrossRef] [PubMed]
3. Hansen, P.J. Physiological and cellular adaptations of zebu cattle to thermal stress. *Anim. Reprod. Sci.* **2004**, *82–83*, 349–360. [CrossRef] [PubMed]
4. Utsunomiya, Y.T.; Fortunato, A.A.A.D.; Milanesi, M.; Trigo, B.B.; Alves, N.F.; Sonstegard, T.S.; Garcia, J.F. *Bos taurus* haplotypes segregating in Nellore (*Bos Indicus*) cattle. *Anim. Genet.* **2021**, *53*, 58–67. [CrossRef] [PubMed]
5. Atlas das Pastagens Brasileiras. 2013. Available online: https://atlasdaspastagens.ufg.br/map (accessed on 20 January 2023).
6. Vale, R.; Vale, P.; Gibbs, H.; Pedrón, D.; Engelmann, J.; Pereira, R.; Barreto, P. Regional expansion of the beef industry in Brazil: From the coast to the Amazon, 1966–2017. Regional Studies. *Reg. Sci.* **2002**, *9*, 641–664. [CrossRef]
7. Magalhaes, D.R.; Maza, M.T.; Prado, I.N.D.; Fiorentini, G.; Kirinus, J.K.; Campo, M.D.M. An Exploratory Study of the Purchase and Consumption of Beef: Geographical and Cultural Differences between Spain and Brazil. *Foods* **2022**, *11*, 129. [CrossRef] [PubMed]
8. Instituto Brasileiro de Geografia e Estatistica (IBGE). 2023. Available online: https://www.ibge.gov.br/estatisticas/economicas/agricultura-e-pecuaria/9107-producao-da-pecuaria-municipal.html?=&t=series-historicas (accessed on 18 January 2023).
9. Amaral, G.; Carvalho, F.; Capanema, L.; Carvalho, C.A. Panorama da Pecuária Sustentável. BNDES Setorial 36. 2012; pp. 249–288. Available online: https://web.bndes.gov.br/bib/jspui/handle/1408/1491 (accessed on 24 October 2022).
10. Miranda, R.A. Breve história da agropecuária brasileira. In *Dinâmica da Produção Agropecuária e da Paisagem Natural No Brazil Nas Últimas Décadas: Sistemas Agrícolas, Paisagem Natural e Análise Integrada Do Espaço Rural*; Landau, E.C., Silva, G.A.D., Moura, L., Hirsch, A., Guimaraes, D.P., Eds.; Cenário Histórico, Divisão Política, Características Demográficas, Socioeconômicas e Ambientais; Embrapa: Brasília, Brazil, 2020; pp. 31–57.
11. Silva, M.C.; Boaventura, V.M.; Fioravanti, M.C.S.; História do Povoamento Bovino No Brasil Central. Revista UFG. *Ano XIII nº 13*. 2012. Available online: https://revistas.ufg.br/revistaufg/article/view/48451 (accessed on 9 December 2022).
12. Romanzini, E.P.; Barbero, R.P.; Reis, R.A.; Hadley, D.; Malheiros, E.B. Economic evaluation from beef cattle production industry with intensification in Brazil's tropical pastures. *Trop. Anim. Health Prod.* **2020**, *52*, 2659–2666. [CrossRef]
13. Daniel, J.L.; Bernardes, T.F.; Jobim, C.C.; Schmidt, P.; Nussio, L.G. Production and utilization of silages in tropical areas with focus on Brazil. *Grass Forage Sci.* **2019**, *74*, 188–200. [CrossRef]
14. Lamas, F.M. *Artigo: A Tecnologia Na Agricultura*; Embrapa: Brasília, Brazil, 2017. Available online: https://www.embrapa.br/busca-de-noticias/-/noticia/30015917/artigo-a-tecnologia-na-agricultura (accessed on 15 December 2022).

15. MapBiomas. 2023. Available online: https://plataforma.brasil.mapbiomas.org/cobertura?activeBaseMap=9&layersOpacity=100&activeModule=coverage&activeModuleContent=coverage%3Acoverage_main&activeYear=2021&mapPosition=-15.114553%2C-51.416016%2C4&timelineLimitsRange=1985%2C2021&baseParams[territoryType]=1&baseParams[territories]=1%3BBrasil%3B1%3BPa%C3%ADs%3B-33.751177993999946%3B-73.9904499689999%3B5.271841077000019%3B-28.847639913999956&baseParams[activeClassesLevelsListItems]=1%2C7%2C8%2C9%2C10%2C2%2C11%2C12%2C13%2C14%2C15%2C16%2C3%2C17%2C18%2C27%2C37%2C38%2C39%2C40%2C41%2C28%2C42%2C43%2C44%2C19%2C20%2C4%2C21%2C22%2C23%2C24%2C5%2C25%2C26%2C6 (accessed on 20 January 2023).
16. Figueiredo, E.B.; Jayasundara, S.; Bordonal, R.O.; Berchielli, T.T.; Reis, R.A.; Wagner-Riddle, C.; La Scala, N. Greenhouse gas balance and carbon footprint of beef cattle in three contrasting pasture-management systems in Brazil. *J. Clean. Prod.* **2017**, *142*, 420–431. [CrossRef]
17. Vale, P.; Gibbs, H.; Vale, R.; Christie, M.; Florence, E.; Munger, J.; Sabaini, D. The Expansion of Intensive Beef Farming to the Brazilian Amazon. *Glob. Environ. Change* **2019**, *57*, 101922. [CrossRef]
18. Molossi, L.; Hoshide, A.K.; Pedrosa, L.M.; Oliveira, A.S.D.; Abreu, D.C. Improve pasture or feed grain? Greenhouse gas emissions, profitability, and resource use for Nelore beef cattle in Brazil's Cerrado and Amazon biomes. *Animals* **2020**, *10*, 1386. [CrossRef]
19. Da Rosa e Silva, P.I.J.L.; Vilas Boas e Silva, Y.R.; Paulino, P.V.R.; de Paula Sousa, D.; Possamai, A.J.; da Freiria, L.B.; Rolim, H.C.L.; de Castro Dias Júnior, W.; da Silva Cabral, L. Dried distiller's grains for feedlot Nellore cattle fed non-forage-based diets. *Trop. Anim. Health Prod.* **2022**, *54*, 230. [CrossRef]
20. Savian, J.V.; Schons, R.M.T.; Mezzalira, J.C.; Barth Neto, A.; Silva Neto, G.F.; Benvenutti, M.A.; Carvalho, P.C.F. A comparison of two rotational stocking strategies on the foraging behaviour and herbage intake by grazing sheep. *Animal* **2020**, *14*, 2503–2510. [CrossRef]
21. Embrapa. *Plano Agrícola e Pecuário*; Embrapa: Brasília, Brazil, 2020. Available online: https://www.embrapa.br/en/agencia-de-informacao-tecnologica/criacoes/gado_de_leite/pre-producao/politicas/conquistas/plano-agricola-e-pecuario (accessed on 16 December 2022).
22. Costa, F.B.; Dias, F.R.T.; Gomes, R.D.C.; Pereira, M.A. Indicadores de Desempenho Na Pecuária de Corte: Uma Revisão No Contexto da Plataforma + Precoce. 2018, pp. 1–28, (Documento/Embrapa Gado de Corte). Available online: https://www.embrapa.br/en/busca-de-publicacoes/-/publicacao/1090951/indicadores-de-desempenho-na-pecuaria-de-corte-uma-revisao-no-contexto-da-plataforma--precoce (accessed on 15 November 2022).
23. Severo, C.M.; Matte, A. Public policies for livestock in the Pampa biome: Analyze for Brazil and Uruguay. *Rev. Agric. Famil. Pesq. Form. Desenvol.* **2020**, *14*, 14–40. Available online: https://periodicos.ufpa.br/index.php/agriculturafamiliar/article/view/7732 (accessed on 28 January 2023).
24. Barcellos, J.O.J.; Suñe, Y.B.P.; Semmelmann, C.E.N.; Grecellé, R.A.; Costa, E.C.D.; Montanholi, Y.R.; Christofari, L. Bovinocultura de Corte Frente a Agriculturização no Sul do Brasil. In Proceedings of the Conference Presentation, XI Ciclo de Atualização em Medicina Veterinária, Centro Agroveterinário de Lages, CAV/UDESC, Lages, Brazil, 14–16 April 2004. Available online: https://vdocuments.pub/a-bovinocultura-de-corte-frente-a-a-bovinocultura-de-corte-frente-a-agriculturizacao.html?page=1 (accessed on 15 November 2022).
25. Baretto, P. *Políticas Para Desenvolver a Pecuária na Amazônia Sem Desmatamento*; Amazonia 2030; Technical Report, 2020; Amazonia: Seattle, WA, USA, 2020. [CrossRef]
26. Broom, D.M. A method for assessing sustainability, with beef production as an example. *Biol. Rev.* **2021**, *96*, 1836–1853. [CrossRef]
27. Berndt, A.; Tomkins, N.W. Measurement and mitigation of methane emissions from beef cattle in tropical grazing systems: A perspective from Australia and Brazil. *Animal* **2013**, *7*, 363–372. [CrossRef]
28. Cardoso, A.S.; Berndt, A.; Leytem, A.; Alves, B.J.R.; Carvalho, I.N.O.; Soares, L.H.B.; Urquiaga, S.; Boddey, R.M. Impact of the intensification of beef production in Brazil on greenhouse gas emissions and land use. *Agric. Syst.* **2016**, *143*, 86–96. [CrossRef]
29. Rodrigues, R.A.R.; Conceição, M.C.G.; Bidone, E.D.; Matos, E.S.; Cordeiro, R.C.; Selva, G.V. The actions of the Brazilian agricultural sector in the context of climate change negotiations. *Sustain. Debate* **2019**, *10*, 28–37. [CrossRef]
30. Kruchelski, S.; Trautenmuller, J.W.; Orso, G.A.; Triches, G.P.; Porfírio-Da-Silva, V.; Moraes, A. Growth and productivity of Eucalyptus benthamii in integrated crop–livestock systems in southern Brazil. *Agroforest. Syst.* **2023**, *97*, 45–57. [CrossRef]
31. Stumpf Junior, W.; Balsadi, O.V. Políticas públicas e pesquisa para o desenvolvimento rural no Brasil. In *Políticas públicas de desenvolvimento rural no Brasil*; Grisa, G., Schneider, S., Eds.; Org. Editora da UFRGS: Porto Alegre, Brazil, 2015; pp. 511–529.
32. Yalmanov, N. Public Policy and Policy-Making. In *XXIII International Conference Culture, Personality, Society in the Conditions of Digitalization: Methodology and Experience of Empirical Research Conference*; KnE Social Sciences: Dubai, United Arab Emirates, 2020; pp. 558–564. [CrossRef]
33. Omotehinse, A.O.; De Tomi, G. Mining and the sustainable development goals: Prioritizing SDG targets for proper environmental Governance. *Ambio* **2022**, *52*, 229–241. [CrossRef]
34. Aragão, L.E.O.C.; Poulter, B.; Barlow, J.B.; Anderson, L.O.; Malhi, Y.; Saatchi, S.; Phillips, O.L.; Gloor, E. Environmental change and the carbon balance of Amazonian forests. *Biol. Rev.* **2014**, *89*, 913–931. [CrossRef]
35. Merry, F.; Soares-Filho, B. Will intensification of beef production deliver conservation outcomes in the Brazilian Amazon? *Elem. Sci. Anth.* **2017**, *5*, 24. [CrossRef]
36. Eckard, R.J.; Grainger, C.; Klein, C.A.M.D. Options for the abatement of methane and nitrous oxide from ruminant production: A review. *Livest. Sci.* **2010**, *130*, 47–56. [CrossRef]

37. Vasconcelos, P.G.A.; Angelo, H.; Almeida, A.N.D.; Matricardi, E.A.T.; Miguel, E.P.; Souza, A.N.D.; Paula, M.F.; Goncalez, J.C.; Joaquim, A.S. Determinants of the Brazilian Amazon deforestation. *Afr. J. Agric. Res.* **2017**, *12*, 169–176.
38. Bustamante, M.M.C.; Nobre, C.A.; Smeraldi, R.; Aguiar, A.P.D.; Barioni, L.G.; Ferreira, L.G.; Longo, K.; May, P.; Pinto, A.S.; Ometto, J.P.H.B. Estimating greenhouse gas emissions from cattle raising in Brazil. *Clim. Change* **2012**, *115*, 559–577. [CrossRef]
39. Yanai, A.M.; Graça, P.M.L.A.; Ziccardi, L.G.; Escada, M.I.S.; Fearnside, P.M. Brazil's Amazonian deforestation: The role of landholdings in undesignated public lands. *Reg. Environ. Change* **2022**, *22*, 30. [CrossRef]
40. Ministério do Meio Ambiente (MMA). Acompanhamento e a Análise de Impacto das Políticas Públicas—PPCDAm. 2017. Available online: http://redd.mma.gov.br/pt/acompanhamento-e-a-analise-de-impacto-das-politicas-publicas/ppcdam (accessed on 10 November 2022).
41. Embrapa. What is "ICLFS"? 2021. Available online: https://www.embrapa.br/en/tema-integracao-lavoura-pecuaria-floresta-ilpf/nota-tecnica (accessed on 14 December 2022).
42. Aguiar, D. Boas Práticas Agropecuárias Impulsionam Ações em Plano Agrícola e Pecuário. *Boas Práticas Agropecuárias Impulsionam Ações em Plano Agrícola e Pecuário—Portal Embrapa*. 2012. Available online: https://www.embrapa.br/en/busca-de-noticias/-/noticia/1481822/boas-praticas-agropecuarias-impulsionam-acoes-em-plano-agricola-e-pecuario---#:~:text=O%20Plano%20indica%20que%20os%20ganhos%2C%20sist%C3%AAmicos%2C%20com,de%20sua%20inclus%C3%A3o%20no%20Plano%20Agr%C3%ADcola%20e%20Pecu%C3%A1rio (accessed on 28 November 2022).
43. Ministério da Agricultura, Pecuária e Abastecimento (MAPA). Plano Agrícola e Pecuário 2010–2011. *Agricultura é Sustentabilidade e Crescimento*. 2010. Available online: https://www.bb.com.br/docs/pub/siteEsp/agro/dwn/PAP20102011.pdf (accessed on 28 November 2022).
44. Reis, J.C.; Rodrigues, R.A.R.; Marcela, C.G.C.; Martins, C.M.S. Integração lavoura-pecuária-floresta no Brasil: Uma estratégia de agricultura sustentável baseada nos conceitos da green economy initiative. *Sustentabilidade Em Debate Embrapa Soils Brasiília* **2016**, *7*, 58–73. [CrossRef]
45. Governo do Brasil (Gov.br). Ações do Plano. 2016. Available online: https://www.gov.br/agricultura/pt-br/assuntos/sustentabilidade/plano-abc/acoes-do-plano (accessed on 30 November 2022).
46. Finger, A.; Ely, C.; Bombardelli, L.; Dos Santos, W.; Schneider, I. *Plano ABC—Agricultura de Baixo Carbono*; Encitech: Halmstad, Sweden, 2016. Available online: https://www2.fag.edu.br/coopex/inscricao/arquivos/encitec/20161023-205240_arquivo.pdf (accessed on 16 December 2022).
47. Maio, A. Práticas Agropecuárias Sustentáveis Começam a Influenciar o Consumo de Carne No Brasil. 2019. Available online: https://www.embrapa.br/en/busca-de-noticias/-/noticia/46109238/praticas-agropecuarias-sustentaveis-comecam-a-influenciar-o-consumo-de-carne-no-brasil (accessed on 29 December 2022).
48. Sekaran, U.; Lai, L.; Ussiri, D.A.N.; Kumar, S.; Clay, S. Role of integrated crop-livestock systems in improving agriculture production and addressing food security—A review. *J. Agric. Food Res.* **2021**, *5*, 100190. [CrossRef]
49. Garrett, R.D.; Niles, M.T.; Gil, J.D.B.; Gaudin, A.; Chaplin-Kramer, R.; Assmann, A.; Assmann, T.S.; Brewer, K.; Faccio Carvalho, P.C.D.; Cortner, O.; et al. Social and ecological analysis of commercial integrated crop livestock systems: Current knowledge and remaining uncertainty. *Agric. Syst.* **2017**, *155*, 136–146. [CrossRef]
50. Nepstad, D.; McGrath, D.; Sticker, C.; Alencar, A.; Azevedo, A.; Swette, B.; Bezerra, T.; Digiano, M.; Shimada, J.; Motta, R.S.; et al. Slowing amazon deforestation through public policy and interventions in beef an soy supply chains. *Science* **2014**, *344*, 1118–1123. [CrossRef]
51. Bungenstab, D.J. *Brazilian Beef Cattle: Reducing Global Warming by the Production Systems Efficiency*; Documents/Embrapa Gado de Corte; Embrapa: Brasília, Brazil, 2012; pp. 1–38. ISSN 1983-974X.
52. Grandchamp, L. Principais Tipos de Crédito Rural e Suas Taxas de Juros. *Dia Rural—Ao Lado do Agronegócio Brasileiro*. 2021. Available online: https://diarural.com.br/principais-tipos-de-credito-rural-e-suas-taxas-de-juros/ (accessed on 6 November 2022).
53. Banco do Brasil. Moderagro. Crédito Para o Desenvolvimento do Seu Agronegócio. 2023. Available online: https://www.bb.com.br/pbb/pagina-inicial/agronegocios/agronegocio---produtos-e-servicos/grande-produtor/investir-em-sua-atividade/moderagro#/ (accessed on 10 November 2022).
54. Lins, C.F.; Moreira, S.N.; Leitão, T.F.S.; Almeida, A.N. Brazilian Forest Code: 1965–2012, from dictatorship to democracy. *Rev. Foco.* **2022**, *15*, 1–27. [CrossRef]
55. Valle, R.S.T.D. Código Florestal: Mudar é preciso. Mas para onde? In *Código Florestal- Desafios e Perspectivas*; Silva, S.T.D., Cureau, S., Leuzinger, M.D., Eds.; Fiuza: São Paulo, Brazil, 2010; pp. 1–360. ISBN 8562354082.
56. Silva Júnior, J.P.; Aparecida, R.; Assis, R.B.; Santos, C.C.J. Histórico e Conceitos do Código Florestal de 1965. *JUS*. 2017. Available online: https://jus.com.br/artigos/58371/historico-e-conceitos-do-codigo-florestal-de-1965 (accessed on 23 October 2022).
57. Embrapa. Entenda a Lei 12.651 de 25 de Maio de 2012. 2013. Available online: https://www.embrapa.br/codigo-florestal/entenda-o-codigo-florestal (accessed on 12 December 2022).
58. Governo do Brazil (Gov.br). Inscrever Imóvel Rural no Cadastro Ambiental Rural (CAR). 2023. Available online: https://www.gov.br/pt-br/servicos/inscrever-imovel-rural-no-cadastro-ambiental-rural-car (accessed on 15 January 2023).
59. Ministerio do Meio Ambiente (MMA). Plano Setorial de Mitigação e de Adaptação às Mudanças Climáticas Para a Consolidação de Uma Economia de Baixa Emissão de Carbono Na Agricultura: Plano ABC (Agricultura de Baixa Emissão de Carbono). 2013. Available online: https://www.gov.br/agricultura/pt-br/assuntos/sustentabilidade/plano-abc/plano-abc-agricultura-de-baixa-emissao-de-carbono (accessed on 5 December 2022).

60. O Banco Nacional do Desenvolvimento (BNDES). Fundo Amazonia. In *Documento do Projeto*; 2013. Available online: https://www.fundoamazonia.gov.br/pt/fundo-amazonia/ (accessed on 6 December 2022).
61. Bonaudo, T.; Bendahan, A.B.; Sabatier, R.; Ryschawy, J.; Bellon, S.; Leger, F.; Magda, D.; Tichit, M. Agroecological principles for the redesign of integrated crop–livestock systems. *Eur. J. Agron.* **2014**, *57*, 43–51. [CrossRef]
62. OECD. *Agricultural Policy Monitoring and Evaluation 2021: Addressing the Challenges Facing Food Systems*; OECD Publishing: Paris, France, 2021. [CrossRef]
63. Ministério da Agricultura, Pecuária e Abastecimento (MAPA). Plano Agrícola e Pecuário 2017–2018. 2017. Available online: https://www.gov.br/agricultura/pt-br/assuntos/politica-agricola/todas-publicacoes-de-politica-agricola/plano-agricola-pecuario/PAP1718.pdf/view (accessed on 14 December 2022).
64. Governo do Brasil (Gov.br). Plano Safra 21/22 Aumenta Recursos Para Técnicas Agrícolas Sustentáveis. 2021. Available online: https://www.gov.br/pt-br/noticias/agricultura-e-pecuaria/2021/06/plano-safra-21-22-aumenta-recursos-para-tecnicas-agricolas-sustentaveis#:~:text=O%20Governo%20Federal%20lan%C3%A7ou%2C%20nessa%20ter%C3%A7a-feira%20%2822%29%2C%20o,de%202021%20a%2030%20de%20junho%20de%202022. (accessed on 15 January 2023).
65. Nunes, S.; Barlow, J.; Gardner, T.; Sales, M.; Monteiro, D.; Souza Jr., C. Uncertainties in assessing the extent and legal compliance status of riparian forests in the eastern Brazilian Amazon. *Land Use Policy* **2019**, *82*, 37–47. [CrossRef]
66. Gomes, L.C.; Bianchi, F.J.J.A.; Cardoso, I.M.; Fernandes, R.B.A.; Fernandes Filho, E.I.; Shulte, R.P.O. Agroforestry systems can mitigate the impacts of climate change on coffee production: A spatially explicit assessment in Brazil. *Agric. Ecosyst. Environ.* **2020**, *294*, 106858. [CrossRef]
67. Carauta, M.; Parussis, J.; Hampf, A.; Libera, A.; Berger, T. No more double cropping in Mato Grosso, Brazil? Evaluating the potential impact of climate change on the profitability of farm systems. *Agric. Syst.* **2021**, *190*, 103104. [CrossRef]
68. Weiler, D.A.; Tornquist, C.G.; Parton, W.; Santos, H.P.; Santi, A.; Bayer, C. Crop Biomass, Soil Carbon, and Nitrous Oxide as Affected by Management and Climate: A DayCent Application in Brazil. *Soil Sci. Soc. Am. J.* **2017**, *81*, 945–955. [CrossRef]
69. Lopes, G.R.; Lima, M.G.B.; doe Reis, T.N.P. Maldevelopment revisited: Inclusiveness and social impacts of soy expansion over Brazil's Cerrado in Matopiba. *World Dev.* **2021**, *139*, 105316. [CrossRef]
70. Brandão, A., Jr.; Rausch, L.; Durán, A.P.; Costa, C., Jr.; Spawn, S.A.; Gibbs, H.K. Estimating the Potential for Conservation and Farming in the Amazon and Cerrado under Four Policy Scenarios. *Sustainability* **2020**, *12*, 1277. [CrossRef]
71. Soterroni, A.C.; Mosnier, A.; Carvalho, A.X.Y.; Câmara, G.; Obersteiner, M.; Andrade, P.R.; Souza, R.C.; Brock, R.; Pirker, J.; Kraxner, F.; et al. Future environmental and agricultural impacts of Brazil's Forest Code. *Environ. Res. Lett.* **2018**, *13*, 074021. [CrossRef]
72. Skidmore, M.E.; Sims, K.M.; Rausch, L.L.; Gibbs, H.K. Sustainable intensification in the Brazilian cattle industry: The role for reduced slaughter age. *Environ. Res. Lett.* **2022**, *17*, 064026. [CrossRef]
73. Pereira, C.H.; Patino, H.O.; Hoshide, A.K.; Abreu, D.C.D.; Rotz, C.A.; Nabinger, C. Grazing supplementation and crop diversification benefits for southern Brazil beef: A case study. *Agric. Syst.* **2018**, *162*, 1–9. [CrossRef]
74. Pedrosa, L.M.; Hoshide, A.K.; Abreu, D.C.; Molossi, L.; Couto, E.G. Financial transition and costs of sustainable agricultural intensification practices on a beef cattle and crop farm in Brazil's Amazon. *Renew. Agric. Food Syst.* **2019**, *36*, 26–37. [CrossRef]
75. Palermo, G.C.; Almeida d'Avignon, A.L.D.; Freitas, M.A.V. Reduction of emissions from Brazilian cattle raising and the generation of energy: Intensification and confinement potentials. *Energy Policy* **2014**, *68*, 28–38. [CrossRef]
76. Hoshide, A.K.; Dalton, T.J.; Smith, S.N. Profitability of coupled potato and dairy farms in Maine. *Renew. Agric. Food Syst.* **2006**, *21*, 261–272. [CrossRef]
77. Gil, J.; Siebold, M.; Berger, T. Adoption and development of integrated crop-livestock-forestry systems in Mato Grosso, Brazil. *Agric. Ecosyst. Environ.* **2015**, *199*, 394–406. [CrossRef]
78. Asai, M.; Moraine, M.; Ryschawy, J.; de Wit, J.; Hoshide, A.K.; Martin, G. Critical factors to crop-livestock integration beyond the farm level: A cross-analysis of worldwide case studies. *Land Use Policy* **2018**, *72*, 184–194. [CrossRef]
79. Rocha, T.C.; Fontes, C.A.A.; Silva, R.T.S.; Processi, E.F.; Valle, F.R.A.F.; Lombardi, C.T.; Oliveira, R.L.; Bezerra, L.R. Performance, nitrogen balance and microbial efficiency of beef cattle under concentrate supplementation strategies in intensive management of a tropical pasture. *Trop. Anim. Health Prod.* **2016**, *48*, 673–681. [CrossRef]
80. Schmidt, N.S.; Silva, C.L. Pesquisa e Desenvolvimento na Cadeia Produtiva de Frangos de Corte no Brasil. *RESR* **2018**, *56*, 467–482. [CrossRef]
81. Koehn, A.C.; Bjorneberg, D.L.; Malone, R.W.; Leytem, A.B.; Moore, A.; Ma, L.; Bartling, P.N.S. Simulating soil nitrogen fate in irrigated crop production with manure applications. *Sci. Total Environ.* **2021**, *793*, 148510. [CrossRef]
82. Souza, C. Políticas Públicas: Uma revisão da literatura. *Sociologias* **2006**, *16*, 20–45. [CrossRef]
83. Carrer, M.J.; Maia, A.G.; Vinholis, M.D.M.B.; de Souza Filho, H.M. Assessing the effectiveness of rural credit policy on the adoption of integrated crop-livestock systems in Brazil. *Land Use Policy* **2020**, *92*, 104468. [CrossRef]
84. Assunção, J.; Gandour, C.; Rocha, R.; Rocha, R. The Effect of Rural Credit on Deforestation: Evidence from the Brazilian Amazon. *Econ. J.* **2019**, *130*, 290–330. [CrossRef]
85. Romero, J.P.; Jayme Jr, F.G. Financial System, Innovation and Regional Development: The Relationship between Liquidity Preference and Innovation in Brazil. *Rev. Polit. Econ.* **2012**, *24*, 623–642. [CrossRef]

86. Pinheiro, D.T.; Santos, D.M.S.; Martins, A.R.R.; Silva, W.M.D.; Araújo, C.V.D.; Abreu, D.C.D.; Hoshide, A.K.; Molossi, L.; Oliveira, R.A.D. Closing the gap: Sustainable intensification implications of increased corn yields and quality for second-crop (*safrinha*) in Mato Grosso, Brazil. *Sustainability* **2021**, *13*, 13325. [CrossRef]
87. Silva, W.M.D.; Bianchini, A.; Amorim, R.S.S.; Couto, E.G.; Santos Weber, O.L.D.; Hoshide, A.K.; Pereira, P.S.X.; Cremon, C.; Abreu, D.C.D. Soil efflux of carbon dioxide in Brazilian Cerrado wheat under variable soil preparation and irrigation. *Agriculture* **2022**, *12*, 163. [CrossRef]
88. Gil, J.D.B.; Garrett, R.D.; Rotz, A.; Daioglou, V.; Valentim, J.; Pires, G.F.; Costa, M.H.; Lopes, L.; Reis, J.C. Tradeoffs in the quest for climate smart agricultural intensification in Mato Grosso, Brazil. *Environ. Res. Lett.* **2018**, *13*, 064025. [CrossRef]
89. Ierna, A. Towards Sustainable Intensification of Crop Production—Yield Gaps and Water Use Efficiency in Farming Systems. *Agronomy* **2021**, *11*, 710. [CrossRef]
90. Mouratiadou, I.; Stella, T.; Gaiser, T.; Birka Wicke, B.; Nendel, C.; Ewert, F.; Hilst, F.V.D. Sustainable intensification of crop residue exploitation for bioenergy: Opportunities and challenges. *Glob. Change Biol. Bioenergy* **2020**, *12*, 71–89. [CrossRef]

sustainability MDPI

Review

Sustaining Forest Plantations for the United Nations' 2030 Agenda for Sustainable Development

Lydie-Stella Koutika [1,*], Rosalie Matondo [2], André Mabiala-Ngoma [1], Viviane Sogni Tchichelle [1,3], Mélanie Toto [1], Jean-Claude Madzoumbou [1], Juste Armand Akana [1], Hugues Y. Gomat [1,4], François Mankessi [2], Armel Thongo Mbou [5], Tiburce Matsoumbou [1], Alpiche Diamesso [1], Aubin Rachel Saya [1] and Jean de Dieu Nzila [4]

1 Research Centre on the Durability and the Productivity of Industrial Plantations (CRDPI), Pointe-Noire P.O. Box 1291, Congo
2 Ecole Nationale Supérieure d'Agronomie et de Foresterie (ENSAF), Université Marien N'gouabi, Brazzaville, Congo
3 Horizon International Sarl, Pointe-Noire, Congo
4 Ecole Normale Supérieure (ENS), Université Marien N'gouabi, Brazzaville, Congo
5 Institut de Formation et de Conseil Agricole (IFCA), 97470 Saint Benoit, France
* Correspondence: ls_koutika@yahoo.com; Tel.: +242-068-133-452

Abstract: Located in the hearth of Africa, the Congo basin is the world's second largest rainforest ecosystem, spanning over nine countries including the Republic of the Congo. Nature-based solutions, i.e., afforestation, reforestation or agroforestry supplying wood energy, halting food insecurity, restoring land desertification and fostering mitigation and adaptation to climate warming, have been increasingly used in the past decades. Within this framework, Congolese coastal plains have been afforested using fast growing trees since the early 1950s. Due to the low forest productivity and soil fertility, sustainable management of these forest ecosystems (trees, soils and environment) have been performed. Improved germplasms, increased stand wood biomass and healthier soils have the potential to enhance wood and fuel wood energy supply, mitigation and adaptation to climate change, food security, restoration of land and ecosystem biodiversity. This meets ten out of the seventeen sustainable development goals (SDG #), specifically goals related to alleviating poverty (1) and hunger (2), improving health (3), education (4), sanitation and access to clean water (6). Other goals include providing affordable clean energy (7), sustainable production and consumption (12), action on climate change (13), life on land (15), and partnerships for goals (17). Nature-based solutions help to face important societal challenges meeting more than half of SDGs of the United Nations.

Keywords: fast growing tree plantations; inherently nutrient-poor soil; improved germplasms; increased stand wood biomass; healthier soils; sustainable development goals

Citation: Koutika, L.-S.; Matondo, R.; Mabiala-Ngoma, A.; Tchichelle, V.S.; Toto, M.; Madzoumbou, J.-C.; Akana, J.A.; Gomat, H.Y.; Mankessi, F.; Mbou, A.T.; et al. Sustaining Forest Plantations for the United Nations' 2030 Agenda for Sustainable Development. *Sustainability* **2022**, *14*, 14624. https://doi.org/10.3390/su142114624

Academic Editors: Marina Moura Morales, Aaron K. Hoshide and Jim Lynch

Received: 19 August 2022
Accepted: 3 November 2022
Published: 7 November 2022

Publisher's Note: MDPI stays neutral with regard to jurisdictional claims in published maps and institutional affiliations.

1. Introduction

Afforestation and/or reforestation involve carbon (C) sequestration in both above and below ground [1,2], fostering soil health improvement and wood pulp and fuel energy production for industry and population [3,4]. This also enhances ecosystem biodiversity, furnishes other products (non-timber forest, food such as honey, insects, and game) and enables payment of other environmental services [5,6] provided by forest ecosystems. In addition to the above-mentioned ecosystem services, practices using trees in nature-based approaches, such as agroecology including agroforestry, also contribute to food security, climate warming mitigation and restoration of degraded lands, as well as sustainable production and human well-being, especially in southern regions [7,8]. These aforementioned principles, concepts, and practices address important societal challenges at the beginning of the 21st century.

This contributes directly and/or indirectly to more than half of the sustainable development goals (SDG) of the Agenda 2030 of the United Nations specifically, SDG 1 "No

Poverty", SDG 2 "Zero Hunger", SDG 3 "Good Health and Well-Being", SDG 4 "Quality Education", SDG 6 "Clean Water and Sanitation", SDG 7 "Affordable and Clean Energy", SDG 12 "Sustainable Production and Consumption", SDG 13 "Climate Action", SDG 15 "Life on Land" and SDG 17 "Partnership for the Goals". Nature-based approaches and practices enabling sustainable management of forest ecosystems, e.g., boosting C sequestration in both soil and biomass, are crucial to tackle food insecurity, climate warming, degraded lands and to enhance other ecosystem services they provide [5,9–11]. Healthier soils rich in carbon (C) harbor improved physical, chemical and biological properties and have the ability to boost the resilience of human societies to pandemics and other crises [12]. The carbon cycle in tropical forest ecosystems, both natural and planted, includes aboveground C, plant respiration, photosynthesis, atmospheric deposition, C mineralisation/decomposition, litterfall, rood exudation, soil C stock and drainage [10], stores about 471 ± 93 Gt C of living, soil and necromass [1].

Sustainability of forest ecosystems therefore relies on C sequestration (soil, biomass) which improves soil fertility, mitigates climate change, restores land and ecosystem biodiversity, and fosters the well-being of surrounding population [1,2,6,10]. Native tropical savannas on sandy soils improper to agriculture in the Congolese coastal plains started to be afforested using fast growing trees (eucalyptus (*Eucalyptus* spp.) and tropical pines) in the beginning of the 1950s [13]. Production of wood pulp and fuel energy for industry and rural population was the primary goal. Nowadays, this nature-based solution greatly contributes to SDG 7 (Affordable Clean Energy for all) [14,15]. In fact, forest ecosystems help around 94% of Congolese households to fulfil their needs in fuel wood energy, e.g., wood and charcoal [6]. In addition, these forest plantations also currently contribute to climate change mitigation through C storage in both soil and woody biomass meeting SDG 12 (Sustainable Production and Consumption), SDG 13 (Climate Action) [16–20], as well as conserving natural forests for SDG 15 (Life on Land) [21].

Research conducted in the Congolese coastal plains on forest planted on previous native tropical savannas has regional importance. This savanna ecosystem extends to around 6 million hectares from the Democratic Republic of the Congo, Gabon and the Republic of the Congo [22]. Research priorities for the sandy soils beneath both ecosystems' savanna and forest have been set up for their sustainable use [11]. The authors set as primary priority building and maintaining health of these sandy soils, strongly related to two other priorities [11], namely improving crop and/or forest production (SDGs 2, 12 and 15), and fulfilling the need for fuelwood (SDG 7). Our research area is located in the world's second largest rainforest ecosystem, the Congo basin. Most countries in the Congo basin rely on trees for more than 90% of their fuel energy [6].

The forest plantations established in the Congolese coastal plains have been largely studied at Centre de Recherche sur la Durabilité et la Productivité des plantations Industrielles (CRDPI). The Centre has the primary mission of managing research projects devoted to planted forest productivity and sustainability. It was created in 1994 as an Association of CIRAD (France), Industrial and Congolese Research Department. CRDPI is a national institution since 2017 after the dissolution of this Association. It is composed of three units: (1) Genetics, Breeding and Diversity (GBD); (2) Plant and Environment Interactions (PEI) and (3) Socio-Environmental Management (SEM) units (Figure 1).

The Centre is one of the most important forest research institutions in Central Africa. It has a considerable genetic material heritage (20 certified clones, more than 1300 clones to be certified), impressive results on biogeochemical processes (soils, trees and water), and socio-economic and environmental contexts of forests with local population and environment. Since 1994, more than 120 peer-reviewed papers have been published, while over 15 PhD and 50 Masters I and II have been supervised in genetics, plants and environment relationship, and environmental and social management research units. Overall performed research greatly contributes to meet 10 SDGs (1, 2, 3, 4, 6, 7, 12, 13, 15 and 17) out of the 17 SDGs of the United Nations of Agenda 2030.

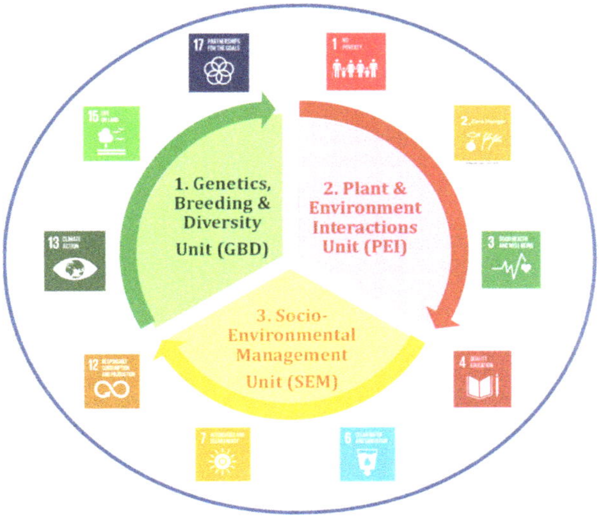

Figure 1. The 3 units of CRDPI related to ten out of the seventeen sustainable development goals (SDG) of the United Nations, Agenda 2030 i.e., SDG 1 "No Poverty", SDG 2 "Zero Hunger", SDG 3 "Good Health and Well-Being", SDG 4 "Quality Education", SDG 6 "Clean Water and Sanitation", SDG 7 "Affordable and Clean Energy", SDG 12 "Sustainable Production and Consumption", SDG 13 "Climate Action", SDG 15 "Life on Land" and SDG 17 "Partnerships for the goals".

From a natural hybrid derived from an unknown antecedent, the productivity of the clonal forest has increased from 7–10 metric tons/hectare (t/ha) to more than 30 t/ha nowadays [14,23–30]. Studies on plants and environment including soils have been also carried out [30–32]. For instance, optimal density of planting (800 stems/ha), the effects of density on the mortality and growth of plants to evaluate the maximum productivity of clones [33,34], and appropriate fertilization (starter dose and no additional fertilization beyond 2 years) [35] have been established.

The richness variability in mineral elements of soils and trees along with the variability of tree growth [36], the use of coppice (no more than two stems), planting spacing (2.65 m × 4.70 m) and other silvicultural studies [37] have been also determined. Soil organic matter, water-balance and nutrient (C, N, P) dynamics have been also studied [19,38–46]. Soils beneath these forest plantations previously under savannas are mainly Arenosols, the most represented soil type in Africa [8]. These soils are inherently nutrient-poor, especially in nitrogen (N) (less than 0.08% in the top 0.5 cm) [22,47]. To both sustain the productivity of eucalyptus plantations and improve soil attributes, two main practices have been experimented with: the management of organic residues [18,38,43,44] and the introduction of nitrogen-fixing trees (NFTs) such as acacia (*Acacia mangium* Willd) in eucalyptus plantation [8,45,48,49].

The management of organic residues evidenced the best tree growth in stands containing the higher quantity of residues [38]. The introduction of acacias in eucalyptus plantations showed higher stand wood biomass in mixed-species (half eucalyptus and half acacia) relative to monocultures of acacia and eucalyptus [18]. Mixed-species plantations of acacias and eucalyptus enhance forest productivity through the transfer of N fixed by acacia to non-NFTs, such as eucalyptus [18,20], improve soil fertility mainly nitrogen status at the end of the 7-year rotation [19,45,46]. Higher N concentration was found in forest floor litter of acacia at 1.66% ± 0.16% compared to eucalyptus stands (0.44% ± 0.06%), with N stocks of 254 ± 36 kilograms (kg)/hectare (ha) (acacia) and 72 ± 6 kg/ha (eucalyptus) [45].

Carbon is sequestered in both soil and biomass [18–20,45,50]. For instance, C stocks increased in the mixed-species (half acacia and half eucalyptus) stands (17.8 ± 0.7 t/ha) relative to pure acacia (16.7 ± 0.4 t/ha) and eucalyptus stands (15.9 ± 0.4 t/ha) in the first 25 cm of the soil profile at the end of the first 7-year rotation [19]. Available phosphorus (P) was higher in coarse particulate organic matter (cPOM, 4000–250 μm) of the afforested stands (mixed-species and monoculture of acacia and eucalyptus) (>60 milligrams/kilograms (mg/kg)) relative to savanna (11 mg/kg) [51]. Higher immobilized P in organic forms was reported in stands containing acacias [52] and higher P concentration in acacia wood (0.61 g of P/kg of dry mass) relative to eucalyptus (0.57 g of P/kg of dry mass) in mixed-species stands sampled 2 years into the second rotation [52,53].

Partial comparative analyses can be made comparing eucalyptus plantation ecosystems in the Congolese plains to those in Brazil, the world's largest eucalyptus agro-forestry producer [18,35,54]. Acacias have been introduced in the *Eucalyptus urophilla x E. grandis* plantations at Tchissoko (Congo), and *Eucalyptus grandis* at Itatinga (State of Sao Paulo, Brazil) [18,35] to sustain forest productivity and improve soil fertility. Main characteristics of the two experimental sites such as climate, altitude, and relief are available in [18]. There is a positive effect of growing eucalyptus with acacias on stand wood production in Congo (Tchissoko) but not in Brazil (Itatinga). Soil type is Ferralic Arenosol (with a ratio of clay: silk: sand of 3:6:91) in Congo, and Ferralsol (with a ratio of clay: silk: sand of 13:3:84) in Brazil. The mean temperature is of 25.7 °C in Congo compared to 19 °C in Brazil.

Studies showed no changes in soil C and N stocks in the upper 0–15 cm layer in the mixed-species (half acacia and half eucalyptus) stands, acacia and eucalyptus monocultures after 6 years of plantation in Brazil [55]. However, soil C and N stocks increased in the 0–25 cm in the mixed-species stands relative to eucalyptus monoculture after 7 years of plantation in the Congo [19]. Higher C stocks in mixed-species plantations have been reported in other studies showing higher decomposition rates in mixed eucalyptus–acacia litter in Australia [56,57]. This is due to the accelerated decomposition of litter and enhanced soil microbial activities enabling more effective nutrient cycling [58]. An increase in microbial activity has been also reported in acacia and eucalyptus stands relative to monoculture or fertilized stands at 27 months after planting in Brazil [54,58–61]. Nevertheless, a similarity has been reported between the Congolese and Brazilian sites regarding the type of soil phosphorus (P) form, specifically 70% of soil P in both locations is in the inorganic form [52,62].

Up to now, an overview of most important findings to sustain forest plantations established in the Congolese coastal plains has never been made. The aim of this paper is to highlight the importance of research conducted for almost three decades for sustainable development. This research has been made to elaborate improved germplasms (resistant to diseases, climate warming, e.g., drought), build healthy soils (rich in C, N and available P), and manage the impact of forest plantations on the environment and the welfare of rural populations. Sustainable management of forest plantations will be presented in three main steps or hypotheses in this paper: (i) Improved germplasms will perform better on nutrient-poor soils in the Congolese coastal plains than the natural hybrids; (ii) Appropriate practices (management of organic residues and introduction of nitrogen fixing species) will enable the creation of healthier soils and environment to foster climate change mitigation, land restoration and conservation of ecosystem biodiversity; (iii) Sustainable management of these forest plantation ecosystems facilitated by efficient social and environment management will contribute directly and/or indirectly to ten out of seventeen SDGs of the Agenda 2030 of the United Nations. Our research aims to boost sustainable development through nature-based solutions and appropriate practices in the Republic of the Congo as well as neighboring countries.

2. Research Conducted at CRDPI and the Agenda 2030 of the United Nations

The studied area is the previous savanna ecosystems on Arenosols common in other surrounding countries such as Gabon and Democratic Republic of Congo [11,22]. To sustain forest plantations established on Arenosols and improve livelihood in the Congolese coastal plains, several studies have been conducted in the Genetics, Breeding and Diversity (GBD), Plant and Environment Interactions (PEI) and Socio-Environmental Management (SEM) units of CRDPI. These studies may be applicable to other regions with similar pedoclimatic conditions in the Central African region [11], or to the entire continent as Arenosols are the most common soil type in Africa [8].

Sustaining forest plantations and improving human well-being, research conducted in the Congolese coastal plains contributes directly and/or indirectly to ten SDGs of United Nations (Figure 1). In the following sections, we will show how conducted research in the Congolese coastal plains may be linked to Agenda 2030. This can enhance sustainable development in other areas and countries.

2.1. Sustainable Development Goal 1 of 'No Poverty'

Since its creation in 1994, Centre de Recherche sur la Durabilité et la Productivité des plantations Industrielles (CRDPI) has been contributing to Sustainable Development Goal (SDG) 1 'No Poverty' through research conducted on fast growing tree plantations installed on soils commonly identified as unsuitable for agriculture in the Congolese coastal plains. These forest plantations were mostly eucalyptus and tropical pines [14] at the beginning. From 1953 to 1981, seeds of 62 eucalyptus species that originated from Brazil, South Africa, and Australia have been introduced in Republic of the Congo ([63], Table 1).

Eucalyptus species names and characteristic features for these introduced species are indicated as well as their potential value for commercialization (suitable vs. unsuitable based on height). However, nowadays, only hybrids are mostly used in forest plantations in the Republic of the Congo (Table 1). In fact, most species introduced had very low growth and productivity (less than 7 cubic meters (m^3)/hectare (ha)/year) and were unsuitable for the environment [15], (Table 2), while some specific diseases were reported [64]. Productivity of forest plantations slightly increased to 10 m^3/ha/year in 1960s [14], (Table 2). However, through extensive research conducted on genetics, the productivity of hybrid clones could reach 40 m^3/ha/year in the 1990s [24,27–29,65–68], (Table 2). Nowadays, *E. urophilla x grandis* is the mostly used hybrid clone in plantations on the Congolese coastal plains.

Studies have also been conducted to investigate the relation between improved germplasms and the environment [30,31,37,39–41]. Other research has focused on socio-economic aspects, namely the relation between forest plantations and local population [5,6,21,69]. It has been reported that more than 90% of households used fuel-wood in most countries of Central Africa [6] and rural population relies strongly on non-timber forest products and payment for other environment services [5].

Overall, studies aim to contribute to SDG 1 through two main pathways: (i) increase in incomes (creation of healthier soils to boost crop production and supply of forest products such as wood, non-timber forest products etc.) [5,8] and (ii) secure fuel-wood energy supply and biodiversity conservation [6,69]. This shows how directly or indirectly sustainable management of forest ecosystems alleviates poverty (increased incomes, improved access to forest products and education) of rural population and meets the SDG 1 of the United Nations.

Table 1. List of the sixty-two eucalyptus species introduced in the Congo from 1953 to 1981 [63] [1] (Species were mostly unsuitable and research has been performed on hybrid clones. Currently mostly clones of *E. urophylla x grandis* are used in forest plantations in the Republic of the Congo).

Eucalyptus Species Name	Characteristic Features	Height (m)	Suitable/Unsuitable
Acmenioïdes	Clear branch stem, two-tone leaves	30–40	Suitable
Alba	Forked and flexuous tree, broad, deltoid to broadly lanceolate leaves	<15	Unsuitable
Andrewsii	Persistent sub fibrous bark on large branches	up to 40	Suitable
Apodophylla	Forked and flexuous tree	<15	Unsuitable
Argophloia	Grows on slopes and creeds	<15	Unsuitable
Bigalerita	Similar to E. alba	<15	Unsuitable
Botryoïdes	Evergreen bark, thick and bicolored leaves	30–40	Suitable
Brassiana	Rough trunk, pedunculate and lanceolate leaves	30	Suitable
Brassii	Light cutlery and defective shape	15	Unsuitable
Camaldulensis	Smooth bark of light color with buff and bluish hues, elongated leaves	40	Suitable
Citriodora	Smooth, white or bluish bark, lemon-smelling leaves	20–40	Suitable
Cloeziana	Poor shape, leaves and pruning	>20	Unsuitable
Contracta	Vigor and average shape (only 7.1 m at 52 months)	<15	Unsuitable
Crebra	Fairly narrow crown, quite narrow, green or grey-green leaves	20–30	Suitable
Cullenii	Similar to E. creba (but only 13 m height at 55 months)	<15	Unsuitable
Deanei	Gum-like bark, broad leaves and also lanceolate leaves (adults)	45–55	Suitable
Deglupta	Thin, variegated bark, conical crown with horizontal branches	30–60	Suitable
Drepanophylla	Very dark ironbark bark type, lanceolate leaves	25–30	Suitable
Exserta	Sub fibrous bark, very narrow linear, lanceolate leaves	up to 25	Unsuitable
Ferruginea	No information available	-	-
Grandis	Strip bark at base, pedunculate, lanceolate and wavy leaves	40–60	Suitable
Hendersonii	Vigor and average shape (8.2 m at 52 months)	<15	Unsuitable
Houseana	Similar to E. alba (13.1 m at 70 months)	<15	Unsuitable
Huberiana	No information available	-	-
Kirtoniana	Stable hybrid tereticornis X robusta	-	-
Leptophleba	Unsuitable species with a height of 11.5 m at 55 months	-	-
Maculata	Wide crown, thick and smooth bark, pedunculate and lanceolate leaves	35–45	Suitable
Maïdenii	Barrel without branches on more than half the height	60–75	Suitable
Melanophloïa	Stem often flexuous, glaucous, amplexicaule and lanceolate leaves	5–30	Unsuitable
Microcorys	Persistent sub fibrous bark, bicolored and acuminate leaves	30–55 m	Suitable
Microtheca	Short-trunked and often multiple-trunked tree	15–20	Unsuitable
Miniata	Straight barrel, grey or rusty bark, oblong or lanceolate leaves	15–30	Suitable

Table 1. *Cont.*

Eucalyptus Species Name	Characteristic Features	Height (m)	Suitable/Unsuitable
Moluccana	Straight barrel, evergreen bark grey, glossy green and lanceolate leaves	20–30	Suitable
Nesophila	Balanced barrel and crown, bark in small scales, pedunculate leaves,	25–30	Suitable
Normantonensis	No information available	-	-
Oligantha	Unsuitable species (6.0 m ate 62 months)	<15	Unsuitable
Pachycalyx	Unsuitable species (7.9 m at 55 months)	<15	Unsuitable
Papuana	Good shaped barrel, coarse bark, lanceolate and bicolored leaves	25–30	Suitable
Phaeotricha	Fibrous bark, pedunculate, lanceolate and somewhat falsified leaves	24–30	Suitable
Paniculata	Pedunculate, lanceolate, bicolored with a green top leaf	25–30	Suitable
Pellita	Strongly framed crown, thick, pedunculate, and a little falsified leaf	25–45	Suitable
Peltata	Peeled leaves (11.8 m tall at 52 months)	<15	Unsuitable
Pilularis	Rough bark, pedunculate, dark green, lanceolate, varnished leaves	35–60	Suitable
Polycarpa	Bark in strip or rough scales, bicolored and lanceolate leaves	20	Unsuitable
Propinqua	Plate bark and rough surface, thick, pedunculate and bicolored leaves	35–40	Suitable
Punctata	Bark in irregular plates, pedunculate, lanceolate and bicolored leaves	35	Suitable
Raveretiana	Scaled bark, pedunculate, lanceolate, bicolored, firm leaves	15–25	Suitable
Regnans	Shaft straight and net of branches	90	Suitable
Resinifera	Red-brown fibrous bark, pedunculate, lanceolate, glossy green leaves	30–40	Suitable
Robusta	Subfibrous bark, glossy green bicolored on top, thick, lanceolate leaves	25–30	Suitable
Rudis	Bad shape and unsuitable species for the region	10–20	Unsuitable
Saligna	Persistent bark at base, pedunculate, lanceolate and bicolored leaves	40–55	Suitable
Scabra	No information available	-	-
Sideroxylon	Dark brown coarse bark, concolored, pedunculate and lanceolate leaves	20–30	Suitable
Tereticornis	Gum-like bark, whitish gray surface, pedunculate and lanceolate leaves	>45	Suitable
Tesselaris	Short trunk, bark in scales, pedunculate and lanceolate leaves	>30	Suitable
Tetrodonta	Bark in long compact fibres, lanceolate, curved and concoloured leaves	25–30	Suitable
Thozetiana	Bark leaving in plates, alternate and lanceolate leaves	20–25	Suitable
Torelliana	Subfibrous then scaly bark, lanceolate and two-colored leaves	25_30	Suitable
Umbra	Dense crown and fairly straight trunk (height < 10 m to 60 months)	<10 m	-
Urophylla	Fibrous bark and smooth (top), pedunculate, lanceolate, pointed leaves	up to 50	Suitable
Viminaliso	Rough bark, alternate pedunculate and lanceolate leaves	30–55	Suitable

[1] Seeds originated from Brazil, South Africa, and Australia.

2.2. Sustainable Development Goal 2 of 'Zero Hunger'

On a regular basis and according to the United Nations World Food Programme, there are around 957 million people worldwide who do not have enough food. There are several ways to address the SDG 2 of the United Nations, either directly (increase crop production and food supply through agroforestry and forestry systems) or indirectly (sustain food production systems by creating healthy soils, plants and environment).

The population of the most populous country in the Congo basin, the Democratic Republic of the Congo (DRC), is estimated at nearly 80 million, while that of the Republic of the Congo (RoC) is around 16 times lower [50]. However, the Republic of the Congo is still struggling to feed its population due to the poorly developed agriculture. Several studies conducted in the Congolese coastal plains lead to reduce hunger directly and/or indirectly, especially through agroforestry systems but also via forestry systems (non-timber forest products, game, insects, etc.) [8], (Table 3). Table 3 highlights some important findings of sustainable management of forest plantations. These findings took into account the specificity of the region, e.g., hybrids instead of introduced species, planting characteristics (density, spacing, length of rotation etc.) and practices adapted to environment and needs of local population.

Nature-based solutions (afforestation, agroforestry) and practices (organic residues management, introduction nitrogen-fixing species) lead to improve soil health and nutrient availability or dynamics (C, N, P and S). They have beneficial impact on crop production and food production systems. Research conducted in the Congolese coastal plains (Republic of the Congo, RoC) and surrounding countries is therefore indirectly linked to SDG 2 'Zero Hunger' (Figure 2). For instance, a four-fold and two-fold increase in yields of two major crops, cassava and maize, respectively, were reported after eight years, as result of the improved soil N status in agroforestry relative to savanna ecosystems in the Democratic Republic of the Congo (DRC) [50,70].

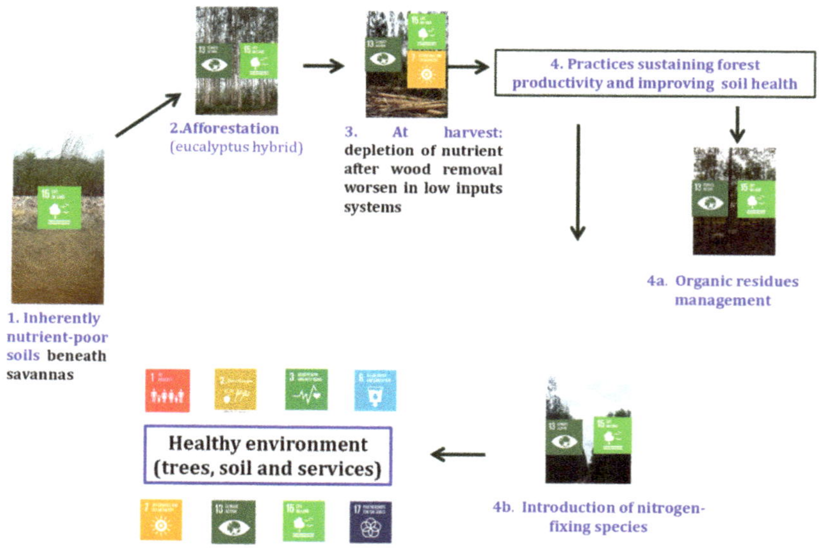

Figure 2. An example of nature-based solutions to improve environment (sustain forest productivity and improved soil health and services) and contribute to eight out of the seventeen Sustainable Development Goals of the United Nations Agenda 2030.

In the same country (Mampu, Batéké plateau), agroforestry using *Acacia auriculiformis* (1987–1993) produced 10,000 metric tons (t)/hectare (ha)/year of cassava, 1200 ha/year of maize, and 6 t/year of honey [9]. An increased yields of both cassava and maize by 6 t/ha

(9 to 15 t/ha) and 1 t/ha (0.5 to 1.5 t/ha) were also noticed [71]. This has been presented in a review paper on C sequestration in mixed-species plantations using nitrogen-fixing species and their benefits in forestry (Republic of the Congo) and agroforestry systems (Democratic Republic of the Congo) [8,9,50,72]. Another study [73] reported that sustainable management of trees in agroforestry and multifunctional agriculture systems greatly benefit to local population with regards to aspects including food security in Cameroon.

Both forest and agroforestry systems can alleviate hunger. This can happen through investigations into soil and plant functions and processes, C sequestration, nutrient dynamics and environment interaction in forestry systems (RoC), and how it may be tested in agroforestry (DRC). Sequestering C in agroforestry or forestry systems creates healthy soils and co-benefits such as securing crop production, food availability and wood energy supply [8–10,72]. Therefore, through sustainable management of forest and agroforestry systems, an inherently nutrient-poor soil may become healthier with enhanced physical, chemical and biological properties [72]. Healthy soil has a great potential to increase crop production (secure food quantity and quality, halt hunger) and meet the SDG 2 of Agenda 2030 for sustainable development.

2.3. Good Health and Well-Being

Some practices linked to nature-based solutions have been adopted to sustain the forest plantations on inherently nutrient-poor soil previously beneath savannas and foster healthy environment and healthy people [27,28,35,48]. Organic residues management in forest plantations induce a decrease in wood productivity in the stands where organic residues were removed twelve months after planting [38]. However, no difference was found in soil C concentration and forest productivity after 21 years (3 rotations of 7 years each) [44]. Introduction of nitrogen-fixing trees (NFT) such as acacia and eucalyptus increased stand wood biomass, soil C and N stocks, and led to build healthier soils with the potential to enable a healthy environment and improve human welfare and health [44,72].

To be healthy, one must eat healthy food and live in a healthy environment [12]. Studies of major processes occurring between plant [38,55,66,74,75], soil and root systems [30,38–41,76,77] have been investigated to ensure healthy soils and environment. The two previous sections supported the positive effects of forest plantations on SDG 1 and 2 which further foster SDG3. Increased incomes will enable access to education, health and food. Improved soil health, has the potential to boost food production and security, improve social, environmental and economic situations, foster healthy populations [6,21,69,71,72] and therefore contribute to SDG3. Sustainable management of forest plantations established in the Congolese coastal plains meets all of the first three Sustainable Development Goals (Figure 1, Table 3).

Table 2. History of elaboration of improved germplasm through genetic performance since the introduction of the species in the Congolese coastal plains.

Period	Activities	Forest Productivity (m^{-3} ha^{-1} y^{-1})	Remarks	References
The 1950s	Introduction of 62 species [1]	7	Most species unsuitable for the agro ecological area	[14]
The 1960s	Species sorting	10	Best species (*E. torelliana, saligna, tereticornis, urophylla, alba*)	[14]
The 1960s	Appearance of natural hybrids (unknown fathers)	<12	*E.* PF1 *E.* 12ABL x *E. saligna*	[14]

Table 2. *Cont.*

Period	Activities	Forest Productivity (m^{-3} ha^{-1} y^{-1})	Remarks	References
The 1980s	Herbaceous cuttings Implementation of clonal tests	Production cap: 18 Mean production: 12	Possibility of multiplication of clones	[15]
The 1980s	First controlled crossings	Production cap: 20 Best clones: 30	without precise scheme (SP)	[24,65,66]
The 1990s	Reciprocal recurrent production scheme (SRR)	Production cap: 25 Best clones: 40	*E. urophylla x E. grandis* and *E. urophylla x E. pellita*	[27–29,67,68,75]
The 1990s	Reciprocal recurrent production scheme (SRR)	Production cap: 25 Best clones: 40	*E. urophylla x E. grandis* and *E. urophylla x E. pellita*	[27–29,67,68,75]

[1] Seeds originated from Brazil, South Africa, and Australia.

Table 3. Forest research conducted in the three units of Centre de Recherche sur la Durabilité et la Productivité des plantations Industrielles (CRDPI) contribute directly and/or indirectly to ten Sustainable Development Goals (SDG) of Agenda 2030 of the United Nations.

SDG and CRDPI's Units	Some Important Research Findings	Direct SDGs	Indirect SDGs
Genetics, Breeding and Diversity (GBD)	- Improved germplasms [24,65,67,68,75] - Increased productivity (from 7 to 40 m^3/ha/year) [14,24,67,68,75].	SDG 15 SDG 13 SDG 17	SDG 1 SDG 2 SDG 3 SDG 7 SDG 12
Plant and Environment Interactions (PEI)	- Planting spacing, density, use of starter dose, optimal growth), [30,34]. - Nitrogen transfer from NFTs to non-NFTs [18,35]. - Increased stand wood biomass in mixtures (NFTs and non-NFTs) [18,20]. - Improved soil health in mixed-species plantations [19,45,46]	SDG 12 SDG 13 SDG 15 SDG 17	SDG 1 SDG 2 SDG 3 SDG 6 SDG 7
Socio-environmental management (SEM)	- High reliance and consumption of fuel wood energy in most countries located in the Congo basin [5,6] - Benefits of mixed-species plantations (incomes, crop and fuel wood supply and other forest products) [5,6,8] - Transfer of knowledge and technologies	SDG 1 SDG 2 SDG 3 SDG 4 SDG 7 SDG 13 SDG 12 SDG 15 SDG 17	SDG 6

2.4. Sustainable Development Goal 4 of 'Quality Education'

As for the challenge of elaborating the first eucalyptus clone [78], studies conducted in the Congolese coastal plains contribute directly and indirectly to quality education through training, namely PhD and Master Programs (Figure 1, Table 3). In fact, since its creation, more than fifteen PhD programs have been fully or partially supervised and conducted at

CRDPI. These studies were related to forest plantation sustainability with regard to genetic studies in improving germplasms [1,27–29,65–68,75].

Applied research has also improved field issues such as density (800 stems/ha), spacing of plantation (2.65 m × 4.70 m), fertilization planning (starter dose and no fertilization beyond 2 years) and rotations length (e.g., 7 years). These studies also help to figure out if genetics and/or environmental factors impact tree growth [34]. Other research has investigated management and improvement of nutrients [30,37], soil attributes through rooting [76,77], organic residues management [43,44,79] or intensive sylviculture (introduction of nitrogen-fixing trees) to sustain forest productivity and improve soil health [8,19,20,45,46,50].

This shows how sustainable forest management (improved germplasms, soil health and socio-economic and environmental contexts) allow the forest productivity to reach 40 metric tons (t)/hectare (ha)/year on inherently nutrient-poor soils and sandy soils of the Congolese coastal plains. Sustainable management of the forest plantations established in the Congolese coastal plains reported increased C stocks (an additional 0.8 t/ha/year for acacia monoculture) and 1.9 t/ha/year (half acacia and half eucalyptus relative to eucalyptus) in an FAO manual on re-carbonizing global soils [80]. Finally, research conducted at CRDPI through PhD and Master I and II programs, contributes to the SDG 4 'Quality Education' of the Agenda 2030 of the United Nations. This currently enables education of several people with extensive responsibilities for the Republic of the Congo as well as other countries.

2.5. Sustainable Development Goal 6 of 'Clean Water and Sanitation'

This is one of the more indirectly linked SDGs to support research on forestry in the Congolese coastal plains (Table 3). For instance, findings highlighted an improvement in soil health, especially regarding its physical, chemical and biological soil fertility [18–20,45,46]. Improved soil health may be positively linked to clean water. In fact, it well knows that enhanced soil health led to clean water. A healthy soil rich in C (organic residues) harbors higher ability not only to retain water in aggregates favoring plant growth, but also to enable the availability of clean ground water [41,45,46].

As result of organic residue management to sustain productivity of monoculture plantations and improve soil fertility, nutrient cycling did improve soil services especially the potential to create clean ground water [38,43,44]. At the same time, healthier soil rich not only in C and N stocks but also in available P reported in acacia and eucalyptus plantations established in the Congolese coastal plains [10,11,45,46,72], increased their potential to enhance water holding capacity and cycling. In fact, regarding water quality, Laclau et al. [41] reported extremely low nutrient concentrations in solutions collected by tension lysimeters at all depths measured. This is probably due to high nutrient requirement of low-input forest plantations and low nutrient availability of weather soils commonly found in the Congolese coastal plains.

With regard to water use, the productivity of eucalyptus plantations established nearby Tchissoko village located in the Congolese coastal plains was not limited by water demand [81]. With annual rainfall of 1125 mm, evapotranspiration of forest plantations was only 747 mm (66%) at Tchissoko (Congo), while it reached 1288 mm which was 95% of the annual rainfall of 1360 mm in Itatinga in Sao Paulo state, Brazil [81]. Therefore, studies conducted in the Congolese plains indirectly contribute to the SDG 6 of Agenda 2030 for sustainable development (Figure 1, Table 3).

2.6. Sustainable Development Goal 7 of 'Affordable and Clean Energy'

Forest plantations established on the Congolese coastal plains enable (i) pulp wood for industry, fuel-wood energy and other forest products (non-timber forest products, etc.) for local populations [5,8]. The region as the entire country harbors high consumption of forest products. The primary goals of forest plantation establishment in the Congolese coastal plains are to insure adequate supply of industrial pulp wood as well as local wood energy [6]. Over 96% of households depended on wood fuel (charcoal and fuelwood) as their energy source in the Republic of the Congo [6].

In 2009, fuelwood consumption was evaluated at 35%, with 75% and 25% of fuelwood from planted and natural forests, respectively [6]. Forest plantations therefore have a major role in reducing pressure of the local population on natural forests which may have diminished by over 1000 hectares/year otherwise [6]. Forest plantations help to preserve natural forests and biodiversity and contribute to supply fuel wood energy for local populations. Research conducted to sustain forest plantations at the Centre de Recherche sur la Durabilité et la Productivité des Plantations Industrielles improves population livelihoods by providing access to fuel energy to satisfy SDG 7.

2.7. Sustainable Development Goal 12 of 'Sustainable Production and Consumption'

Studies conducted to sustain forest plantations of the Congolese coastal plains are directly and/or indirectly linked to SDG 12 (Figure 1, Table 3). Production and consumption of wood have been estimated in most countries of the Congo Basin including the Republic of the Congo in 2010 [6]. The authors estimated that more than 90% of households rely on fuel wood, charcoal and wood waste for their fuel energy with around 213,880 metric tons (t) (fuel wood) 25,461 t (charcoal). Overall consumption was estimated at 1,029,856 t equivalent of wood with the higher consumption noticed in the four more populated towns of the country [6]. Therefore, producing sustainable fuel wood in the Republic of the Congo is paramount.

Past research contrasted wood production and water use for forest plantations in Tchissoko, Republic of the Congo versus Itatinga, Sao Paulo state, Brazil [81]. The authors [81] estimated rainfall and evaporation were 1360 mm and 1288 mm, respectively, at these forest plantations in Brazil which produced 18 metric tons (t) of wood/hectare (ha)/year. Wood production was only 16.6 t/ha/year in the Republic of the Congo with rainfall and evapotranspiration of 1125 mm and 747 mm, respectively. It was concluded that wood production in the Congolese coastal plains mostly depends on soil fertility and not on water needs [81].

Practices leading to improved soil health such organic residues management and introduction of nitrogen-fixing species have been thereafter adopted to intensify forest production and improve soil fertility and health [18,20,35,38,43,48]. In addition, studies on biogeochemical processes in forest ecosystems conducted in the Congolese coastal plains may improve agriculture which is poorly developed in the Republic of the Congo. This can foster sustainable food production and consumption for both agriculture and forest products, as it has been carried out in the neighboring country of the Democratic Republic of the Congo [8,9,70,74,82]. Research conducted in the Congolese coastal plains to sustain forest plantations may fully contribute to the SDG 12 through enhanced socio-economic and environmental management (participatory research) (Table 3).

2.8. Sustainable Development Goal 13 of 'Climate Action'

Mitigating anthropogenic global warming was not the primary goal of establishing forest plantations in the Congolese coastal plains at the very beginning [14]. Nevertheless, it has become a crucial goal currently, along with the fulfilment of needs for wood (pulp and fuel energy) [6,69], other forest products, and the conservation of natural forests [32,72,83]. Nature-based solutions aiming to manage organic residues and introduce nitrogen-fixing trees can increase forest productivity (stand wood biomass) and mitigate climate change (C sequestration in biomass and soil) [18,44,84]. C sequestration and reduction of greenhouse gas emissions have become crucial [32,83,84] due to their impact not only in addressing climate change but also regarding other co-benefits such as soil and environment health and services, which can increase adaptation and resilience to climate change and pandemics [8,11,12,50,72].

Soil organic matter status or characterization is vital to evaluate soil and/or ecosystem health [72,85,86]. Ecosystem health depends on both soil attributes and biomass characteristics. Transferred nitrogen (N) from acacia to eucalyptus boosts forest productivity (stand wood biomass) [18,35] and improves soil attributes through litterfall and roots [76,77,87]. This further enables C sequestration (soil and biomass) [18–20,80], and enhanced availability of nutrients such as P in forest litter of mixed-species stands [52]. Increased carbon stocks in the first 25 centimeters (cm) of the soil profile contributed to higher biomass yields for mixed-species (half acacia and half eucalyptus) stands of 1.9 metric tons (t)/hectare (ha) compared to acacia monoculture stands (0.8 t/ha) [80].

Due the strong correlation between C and N but also with P [88,89], higher values of the cumulative net N stocks were reported in acacia (343 ± 21 kilograms (kg)/hectare (ha) and acacia–eucalyptus mixed-species (287 ± 17 kg/ha) stands relative to eucalyptus (189 ± 12 kg/ha) [45]. Sustainable management of forest plantations through practices, such as managing organic residues or introducing nitrogen-fixing species, fostered healthier soils and agro-forestry species with the potential to both mitigate and adapt to climate change. Carbon storage occurring in both trees with greater woody biomass and in soil with higher carbon stocks, along with other soil attributes have the potential to mitigate climate change and contribute to Sustainable Development Goal 13.

2.9. Sustainable Development Goal 15 of 'Life on Land'

Overall studies conducted in the Congolese coastal plains are strongly linked to SDG 15. Soil and especially soil carbon (C) sequestration has been at the center of several studies and concepts over the last decades [90]. It has been reported how sustainable soil C management may prevent future pandemics [12]. In fact, soil C plays a vital role in enhancing environmental and human health through its positive impact on soil health on which rely food quantity and quality (nutritious value) [12]. Similar to previous SDG 3 related to human health and well-being, carbon sequestration and co-benefits such as healthy soils can also address Life on Land (SDG 15) (Figure 2).

To foster C sequestration and enable healthy soils and other services terrestrial ecosystems provide including life on land, one must consider soil biota at the very beginning. Soil biota drive interaction between soil function, environment, and human activities to further boost C sequestration and build healthy soils and environment [50,85]. Sustaining forest ecosystems through introduction nitrogen-fixing species such as acacias [20,35,48] or managing organic residues [43,44] had a tremendously positive impact on both forest productivity and soil attributes [18,19,42,44,91].

Using ^{15}N dilution method to estimate symbiotic nitrogen fixation and its accumulation in acacia monoculture and mixture, a prior study reported beneficial effects on physiological trends. Here, eucalyptus trees were taller when acacias were in greater proportion in mixed-species stands relative to eucalyptus monocultures and nitrogen (N) mineral mass was greater for mixed-species stands compared to monocultures [46]. In addition, N derived from the atmosphere in the mixture was 60% higher than that expected based on the amount found in acacia monoculture. It was also found that 16% of N accumulated in eucalyptus trees and aboveground eucalyptus litterfall derived from the atmosphere. Reduced competition for light and soil water also contributed to the increased growth of acacias in the mixture, showing that both species benefit from growing in a mixed stand [46].

Establishment of forest ecosystems on former savannas can positively impact the livelihoods of local people. Storage of C in woody biomass is critical to meeting local needs for fuel wood. Soil C in soils can benefit other agro-forestry components such as non-timber forest products, food crops, and ecosystem services [5,6,18,71,73], meeting the Sustainable Development Goals (SDGs) we have reviewed. Management of forest ecosystems through C sequestration in both soil and biomass [18,19] also has great potential to mitigate climate change and increase crop production by improving soil fertility and soil health. This also further restores land and ecosystem biodiversity (enhanced soil fauna, animals, insects) [9,50,71,73], consistent with SDGs 2 and 13. Overall benefits of forest ecosystems enable life on land including humanity.

Forest and agro-forestry nature-based solutions that introduce nitrogen-fixing trees (NFTs) such as acacias may be beneficial in areas where soils are nutrient-poor (especially N), unsuitable for agriculture, and where fuel wood consumption is high. However, given the fact that phosphorus (P) is needed for symbiotic N fixation, one must be aware of the potential risk of depletion in soil available P [89]. In addition, a preliminary study on invasiveness risk of *Acacia mangium* must be conducted prior to any establishment, since this species is invasive in some countries [92]. Preference must be given to local NFTs species if available to fully meet the United Nations' SDG 15.

2.10. Sustainable Development Goal 17 of 'Partnership for the Goals'

Research conducted on forest plantations established in the Congolese coastal plains have been always strongly linked to several partners at national, regional and global levels. Overall research such as published papers and theses (PhDs and Masters) are mostly the results of different projects involving European, Asian and South American countries. In the last two decades, there have been at least six showcase projects.

First, the 'ABIOGEN' project aimed to describe the mechanisms for adaptation to drought for two tree species used on plantations, namely pine and eucalyptus. Second, the 'SEPANG' project defined the conditions for applying the genomic selection in the context of plant breeding in the tropics. The third project called 'ClimaAfrica' evaluated the impact of climatic changes on water availability in agriculture and forestry. The fourth 'WUETREE' project sought to understand the genetic and environmental determinants of water use efficiency to improve sustainability on tree plantations. The fifth project titled 'Site Management and productivity in Tropical Plantation forests' estimated the effects of litter management practices on soil fertility and forest productivity. This was set up similar to the structure of CIFOR. Finally, the sixth project called 'Intens&Fix' promoted ecological intensification of planted forest ecosystems through biophysical modelling and socioeconomically assessment of associated nitrogen fixing species.

The last two projects involved several countries. The project run by CIFOR involved Australia, Brazil, Congo, China, India, Indonesia, South Africa and Vietnam. The last project called 'Intens&Fix' included Brazil, Congo and France. This project shed light on the positive impact of sustainable forest management, including organic residue management and introduction of nitrogen-fixing trees (NFTs), on forest productivity and soil attributes [38–41,43,44]. They also help to understand how N may be transferred from NFTs to non-N-fixing species enabling an increase in stand wood biomass [18,20,35] and healthier soils [19,20,45,46,52,88,93,94]. A project funded by UNESCO (International Geoscience Programme) aiming to evaluate the capacity networks mapping and monitoring C stocks involving Mexico, Chile, United States of America and Republic of the Congo (CRDPI), is currently in progress and will contribute to the United Nations' SDGs 13, 15, and 17.

Recent studies were conducted in frame of a program involving The World Academy of Science (TWAS) and Agenzia Nazionale per le Nuove tecnologie, l'Energia e lo Sviluppo economico sostenibileand (ENEA, Italy). They highlighted the potential impact of human activities (afforestation and oil exploitation) on bacterial community composition [93]. Oil exploitation occurring in the Congolese coastal plains since the end of the 1960s may has boosted the prevalence of *Actinobacteria* in the bacterial community composition of forest soil [93]. This phylum is dominant in the areas with high concentrations of H_2S [95]. *Actinobacteria* biomass growth is enhanced by sulfur (S) amendments [96], and the phylum has the ability to decompose organic residues [97]. This may explain the increased C stocks reported in the mixed-species plantations at the end of 7-year rotation [19,72,80] (Figure 3). Overall research conducted through different projects in strong partnership contributes to the SDG 17 'Partnerships for All Goals' as it also meets previously mentioned SDGs (Figures 1 and 3; Table 3).

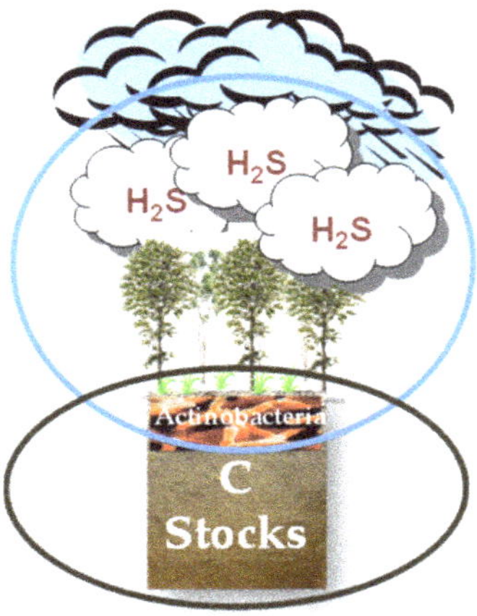

Figure 3. Link between human activities (afforestation and oil exploitation) and soil attributes driven by soil microorganisms (e.g., *Actinobacteria*) contributing to Sustainable Development Goals from Agenda 2030 by the United Nations.

Our research summarizes a large set of findings in genetics, plant and environment interactions and socio-economic and environmental management during almost three decades of forest plantation research on nutrient-poor soils in the Congolese coastal plains. However, there are some limitations. More details on other methods are not provided in this review because of the large set of results investigated. This may make reproduction of our results more challenging. Nevertheless, all cited materials are available for further investigation. The aim of this review is to link conducted research to Sustainable Development Goals (SDGs) of Agenda 2030 of the United Nations. This will benefit targeted countries in the Congo basin as well as other countries with similar conditions for sustainable development.

3. Recommendations

To sustain forest plantations especially when they are established on inherently nutrient-poor soils such as those of the Congolese coastal plains, not only improved germplasms must be elaborated but also sustainable management of both soils and environment through appropriate practices in close link with rural population. Improved germplasms through genetics must be adapted to the specificity of the area of the implementation. Preference must be given to improved germplasms which are more adapted to local conditions than those that are introduced (e.g., the 62 species introduced from Brazil, South Africa and Australia in this case).

This is supported by the increase in productivity from 7 to 40 metric tons (t)/hectare (ha)/year. Technical parameters of plantation and appropriate practices such as organic residues management and introduction of nitrogen-fixing trees, must also be adopted according the specificity of area of establishment taking into account planting materials, soils, and socio-economic and environmental contexts. For instance, the optimal length of rotation is 7 years with a density of 800 planted tree stems/ha. No fertilization is carried out beyond 2 years in the low-input forest plantations established on Arenosols in the Congolese coastal plains. At Itatinga in Sao Paulo state in Brazil, the length of rotation is 6 years, the density of plantation is 1111 planted tree stems/ha, and unlike Congolese forest plantations, fertilization is carried out beyond 2 years in the better managed forest plantations installed on Ferralsols, richer in soil nutrients.

Finally, adoption of nature-based solutions and practices for sustainable development may be successful when they are easily endorsed by the local population. People living around these forest plantations need to obtain financial/economic and environmental benefits. Financial/economic benefits include income which enables access purchasing food, education, health benefits, welfare, fuel wood as an energy source, non-timber forest products and food (e.g., wild game, insects). An example of a potential, long-term, spillover environmental benefit for those living around more sustainable forest plantations is increase in biodiversity.

4. Conclusions

For the first eucalyptus clones planted on the Congolese coastal plains [78], overall research conducted in the Congolese coastal plains from the creation of CRDPI (1994) to date (2022) evidences a challenge: sustaining forest plantations on inherently nutrient-poor and sandy soils strengthened by socio-environmental management. Important findings such as improved germplasms leading to increased forest productivity, optimal technical planting criteria (spacing, density, fertilization, etc.) associated with the adoption of sustainable practices such as increasing organic residues and integrating nitrogen-fixing trees and plants. Additionally, there are also socio-economic benefits such as increasing family incomes, supply of wood-based fuel, and education, as well as environmental benefits.

Sustainable forest management in the Congolese coastal plains is crucial for the Republic of the Congo, which is strongly committed to sustainable development by promoting afforestation, reforestation and conservation of the natural forests. The country is located in the world's second largest rainforest ecosystem. However, high consumption of wood fuel energy, where more than 90% of households consume this type of energy, may threaten natural forests but also forest plantations, and thus compromise the potential to contribute to the United Nations' Agenda 2030 for sustainable development.

Nevertheless, the Republic of the Congo has a good afforestation program through national institutions or services such as ProNAR (Programme National d'Afforestation et de Reboisement) and SNR (Service National de Reboisement) using national and/or private funds. Other regions of the Republic of the Congo such as Pool, Plateaux, Niari, Lékoumou, Cuvette, are being afforested. These regions may therefore benefit from the studies conducted in the Congolese coastal plains over the last three decades.

Author Contributions: Conceptualization, L.-S.K., M.T., A.M.-N. and J.-C.M.; writing—original draft preparation, L.-S.K.; writing—review and editing, L.-S.K., R.M., A.M.-N., V.S.T., M.T., J.-C.M., J.A.A., H.Y.G., F.M., A.T.M., T.M., A.D., A.R.S. and J.d.D.N.; visualization, L.-S.K., R.M., A.M.-N., V.S.T., M.T., J.-C.M., J.A.A., H.Y.G., F.M., A.T.M., T.M., A.D., A.R.S. and J.d.D.N.; supervision, L.-S.K. All authors have read and agreed to the published version of the manuscript.

Funding: This research received no external funding.

Institutional Review Board Statement: Not applicable.

Informed Consent Statement: Not applicable.

Data Availability Statement: Not applicable.

Acknowledgments: The authors thank the Centre de Coopération Internationale en Recherche Agronomique pour le Développement (CIRAD) for his financial and scientific commitment (1994–2017), Congolese government for his continual financial support (1994-to date), the reviewers and editors for their useful inputs and outstanding commitment to improve this paper. This paper is dedicated to Aubin Rachel Saya (†), who worked for more than 25 years at CRDPI and contributed highly to research achievements of center, and passed away 13 March 2022, while retired for less than a year.

Conflicts of Interest: The authors declare no conflict of interest.

References

1. Pan, Y.; Birdsey, R.A.; Fang, J.; Houghton, R.; Kauppi, P.E.; Kurz, W.A.; Phillips, O.L.; Shvidenko, A.; Lewis, S.L.; Canadell, J.G.; et al. A large and persistent carbon sink in the world's forests. *Science* **2011**, *333*, 988–993. [CrossRef] [PubMed]
2. Jandl, R.; Rasmussen, K.M.; Tomé, M.; Johnson, D.W. The Role of Forests in Carbon Cycles, Sequestration, and Storage 4; Forest Management and Carbon Sequestration. 2006. Available online: http://www.iufro.org/science/taskforces/carbonsequestration/ (accessed on 19 February 2022).
3. Lal, R. Climate change and soil degradation mitigation by sustainable management of soils and other natural resources. *Agric. Res.* **2012**, *1*, 199–212. [CrossRef]
4. Mayer, M.; Prescott, C.E.; Abaker, W.E.A.; Augusto, L.; Cecillon, L.; Ferreira, G.W.D.; James, J.; Jandl, R.; Katzensteiner, K.; Laclau, J.-P.; et al. Tamm Review: Influence of forest management activities on soil organic carbon stocks: A knowledge synthesis. *For. Ecol. Manag.* **2020**, *466*, 118127. [CrossRef]
5. Lescuyer, G.; Karsenty, A.; Eba'a Atyi, R. A new tool for sustainable forest management in Central Africa: Payments for environmental services. In *The Forests of the CongoBasin: State of the Forest 2008?* de Wasseige, C., Devers, D., de Marcken, P., Eba'a Atyi, R., Nasi, R., Mayaux, P., Eds.; Publications Office of the European Union: Luxembourg; CIFOR: Kinshasa, Democratic Republic of the Congo, 2009; pp. 131–143.
6. Shure, J.; Marien, J.N.; de Wasseige, C.; Drigo, R.; Salbitano, F.; Dirou, S.; Nkoua, M. Contribution du bois énergie à La Satisfaction des besoins énergétiques des populations d'afrique centrale: Perspectives pour une gestion durable des ressources disponibles. In *Les Forêts du Bassin du Cong: Etat des Forêts 2010*; Office des Publications de l'Union Européenne: Luxembourg, 2012; Volume 5, pp. 109–122, ISBN 978-92-79-22717-2. [CrossRef]
7. Mbow, C.; Smith, P.; Skole, D.; Duguma, L.; Bustamante, M. Achieving mitigation and adaptation to climate change through sustainable agroforestry practices in Africa. *Curr. Opin. Environ. Sustain.* **2014**, *6*, 8–14. [CrossRef]
8. Koutika, L.-S.; Zagatto, M.R.G.; Pereira, A.P.d.A.; Miyittah, M.; Tabacchioni, S.; Bevivino, A.; Rumpel, C. Does the introduction of N_2-fixing trees in forest plantations on tropical soils ameliorate low Fertility and enhance carbon sequestration via interactions between biota and nutrient availability? Case studies from Central Africa and South America. *Front. Soil Sci.* **2021**, *1*, 752747. [CrossRef]
9. Bisiaux, F.; Peltier, R.; Muliele, J.C. Plantations industrielles et agroforesterie au service des populations des plateaux Batéké, Mampu, en République démocratique du Congo. *Bois For. Trop.* **2009**, *301*, 21–32. [CrossRef]
10. Koutika, L.-S. Boosting C Sequestration and Land Restoration through Forest Management in Tropical Ecosystems: A Mini-Review. *Ecologies* **2022**, *3*, 13–29. [CrossRef]
11. Koutika, L.S.; Obame, R.M.; Nkouamoussou, C.K.; Musadji, N.Y. Research priorities for sandy soils in Central Africa. *Geoderma Reg.* **2022**, *29*, e00519. [CrossRef]
12. Rumpel, C.; Amiraslani, F.; Bossio, D.; Chenu, C.; Henry, B.; Fuentes Espinoza, A.; Koutika, L.-S.; Ladha, J.; Madari, B.; Minasny, B.; et al. The Role of Soil Carbon Sequestration in Enhancing Human Resilience in Tackling Global Crises including Pandemics. *Soil Secur.* **2022**, *8*, 100069. [CrossRef]
13. Makany, L. The Atlantic coast of the Congo geographical and geological frameworks, their influence on the distribution of vegetation and on the agricultural possibilities of the territory. In *The Peking Science Symposium*; Institute of Current World Affairs: Washington, DC, USA, 1964; pp. 891–907.
14. Delwaulle, J.C.; Garbaye, J.; Laplace, Y. Ligniculture en milieu tropical: Les reboisements en eucalyptus hybrides de la savane côtière Congolaise. *Rev. For. Fr.* **1981**, *3*, 248–255. [CrossRef]
15. Louppe, D.; Depommier, D. Expansion, Research and Development of the Eucalyptus in Africa Wood Production, Livelihoods and Environmental Issues: An Unlikely Reconciliation? In Proceedings of the FAO/MEEATU Workshop "Eucalyptus in East Africa, the Socio-Economic and Environmental Issues", Bujumbura, Burundi, 31 March–1 April 2010. Available online: https://www.google.com/url?sa=t&rct=j&q=&esrc=s&source=web&cd=&cad=rja&uact=8&ved=2ahUKEwjonLXHkcb5AhVkolwKHeTGCRMQFnoECCIQAQ&url=https%3A%2F%2Fagritrop.cirad.fr%2F554151%2F1%2Fdocument_554151.pdf&usg=AOvVaw36q9uLK1hcRJ2aqrLZFvPg (accessed on 22 February 2022).
16. Marsden, C.; Nouvellon, Y.; Thongo M'Bou, A.; Saint-Andre, L.; Jourdan, C.; Kinana, A.; Epron, D. Two independent estimations of stand-level root respiration on clonal Eucalyptus stands in Congo: Up scaling of direct measurements on roots versus the trenched-plot technique. *New Phytol.* **2008**, *177*, 676–687. [CrossRef]
17. Nouvellon, Y.; Epron, D.; Kinana, A.; Hamel, O.; Mabiala, A.; D'Annunzio, R.; Deleporte, P.; Saint-Andre, L.; Marsden, C.; Roupsard, O.; et al. Soil CO_2 effluxes, soil carbon balance, and early tree growth following savannah afforestation in Congo: Comparison of two site preparation treatments. *For. Ecol. Manag.* **2008**, *255*, 1926–1936. [CrossRef]
18. Epron, D.; Nouvellon, Y.; Mareschal, L.; Moreira, R.M.E.; Koutika, L.-S.; Geneste, B.; Delgado-Rojas, J.S.; Laclau, J.-P.; Sola, G.; Gonçalves, J.L.M.; et al. Partitioning of Net Primary Production in Eucalyptus and Acacia Stands and in Mixed-Species Plantations: Two Case-Studies in Contrasting Tropical Environments. *For. Ecol. Manag.* **2013**, *301*, 102–111. [CrossRef]
19. Koutika, L.S.; Epron, D.; Bouillet, J.P.; Mareschal, L. Changes in N and C Concentrations, Soil Acidity and P Availability in Tropical Mixed Acacia and Eucalypt Plantations on a Nutrient-Poor Sandy Soil. *Plant Soil* **2014**, *379*, 205–216. [CrossRef]

20. Tchichelle, V.S. Production de biomasse et quantification des flux d'azote dans une plantation mixte *d'Eucalyptus urophylla x grandis* et *d'Acacia mangium* au Congo. Ph.D. Thesis, Université de Lorraine, Nancy, France, 2016. Available online: https://scanr.enseignementsup-recherche.gouv.fr/publication/these2016LORR0115 (accessed on 22 February 2022).
21. Missamba-Lola, A.P.; Matondo, R.; Marien, J.N.; Samba-Kimbata, M.; Gillet, J.F. Dynamique spatio-temporelle des recrûs forestiers au bord des pistes secondaires: Cas des UFA-CIB dans la cuvette congolaise. *Rev. Sci. Tech. Environ. Bassin Congo* **2015**, *4*, 34–45.
22. Schwartz, D.; Namri, M. Mapping the total organic carbon in the soils of the Congo. *Glob. Planet Change* **2002**, *33*, 77–93. [CrossRef]
23. Bouvet, J.-M.; Vigneron, P.; Gouma, R.; Saya, A.R. Trends in variances and heritabilities with age for growth traits in eucalyptus spacing experiments. *Silvae Genet.* **2003**, *52*, 121–133.
24. Saya, A.R.; Mankessi, F.; Toto, M.; Marien, J.N.; Monteuuis, O. Advances in mass clonal propagation of *Eucalyptus urophylla x E. grandis* in Congo. *Bois For. Trop.* **2008**, *297*, 15–26.
25. Mankessi, F.; Saya, A.; Toto, M.; Monteuuis, O. Propagation of Eucalyptus urophylla × Eucalyptus grandis clones by rooted cuttings: Influence of 171 genotype and cutting type on rooting ability. *Propag. Ornam. Plants* **2010**, *10*, 42–49.
26. Mankessi, F.; Saya, A.; Boudon, F.; Guedon, Y.; Montes, F.; Lartaud, M.; Verdeil, J.L.; Monteuuis, O. Phase change-related variations of dome shape in Eucalyptus urophylla x Eucalyptus grandis shoot apical meristems. *Trees* **2010**, *24*, 743–752. [CrossRef]
27. Mankessi, F.; Saya, A.; Toto, M.; Monteuuis, O. Cloning field growing Eucalyptus urophylla × Eucalyptus grandis by rooted cuttings: Age, within-shoot position and season effects. *Propag. Ornam. Plants* **2011**, *11*, 3–9.
28. Mankessi, F.; Saya, A.; Favreau, B.E.; Doulbeau, S.; Conejero, G.; Lartaud, M.; Verdeile, J.L.; Monteuuis, O. Variations of DNA methylation in Eucalyptus urophylla × Eucalyptus grandis shoot tips and apical meristems of different physiological ages. *Physiol. Plant.* **2011**, *143*, 178–187. [CrossRef] [PubMed]
29. Mankessi, F.; Saya, A.; Montes, F.; Boudon, F.; Lartaud, M.; Verdeil, J.L.; Monteuuis, O. Histocytological characteristics of *Eucalyptus urophylla* X *Eucalyptus grandis* shoot apical meristems of different physiological. *Trees* **2011**, *25*, 415–424. [CrossRef]
30. Safou-Matondo, R.; Deleporte, P.; Laclau, J.-P.; Bouillet, J.-P. Hybrid and clonal variability of nutrient content and nutrient use efficiency in Eucalyptus stands in Congo. *For. Ecol. Manag.* **2005**, *210*, 193–204. [CrossRef]
31. Nouvellon, Y.; Laclau, J.-P.; Epron, D.; Kinana, A.; Mabiala, A.; Roupsard, O.; Bonnefond, J.-M.; le Maire, G.; Marsden, C.; Bontemps, J.-D.; et al. Within-stand and seasonal variations of specific leaf area in a clonal Eucalyptus plantation in the Republic of Congo. *For. Ecol. Manag.* **2010**, *259*, 1796–1807. [CrossRef]
32. Nouvellon, Y.; Epron, D.; Marsden, C.; Kinana, A.; Maire, G.L.; Deleporte, P.; Saint-André, L.; Bouillet, J.-P.; Laclau, J.-P. Age-related changes in litter inputs explain annual trends in soil CO_2 effluxes over a full Eucalyptus rotation after afforestation of a tropical savannah. *Biogeochemistry* **2012**, *111*, 515–533. [CrossRef]
33. Gomat, H.; Deleporte, P.; Moukini, R.; Mialounguila, G.; Ognouabi, N.; Saya, A.; Vigneron, P.; Saint-Andre, L. What factors influence the stem taper of Eucalyptus growth, environmental conditions, or genetics? *Ann. For. Sci.* **2011**, *68*, 109–120. [CrossRef]
34. Gomat, H.Y. La production des plantations clonales d'eucalyptus dans la plaine côtière du Congo/Brazzaville: Effet des facteurs densité de plantation, fertilisation et régimes d'éclaircies. Ph.D. Thesis, l'Université Marien Ngouabi, Brazzaville, Congo, 2013.
35. Bouillet, J.P.; Laclau, J.P.; Gonçalves, J.L.M.; Voigtlaender, M.; Gava, J.L.; Leite, F.P.; Hakamada, R.; Mareschal, L.; Mabiala, A.; Tardy, F.; et al. Eucalyptus and Acacia tree growth over entire rotation in single- and mixed-species plantations across five sites in Brazil and Congo. *For. Ecol. Manag.* **2013**, *301*, 89–101. [CrossRef]
36. Bikindou, F.D.A.; Gomat, H.Y.; Deleporte, P.; Bouillet, J.-P.; Moukini, R.; Mbedi, Y.; Ngouaka, E.; Brunet, D.; Sita, S.; Diazenza, J.-B.; et al. Are NIR Spectra Useful for Predicting Site Indices in Sandy Soils under Eucalyptus Stands in Republic of Congo? *For. Ecol. Manag.* **2012**, *266*, 126–137. [CrossRef]
37. Laclau, J.P.; Ranger, J.; Deleporte, P.; Nouvellon, Y.; Saint André, L.; Marlet, S.; Bouillet, J.P. Nutrient cycling in a clonal stand of eucalyptus and an adjacent savanna ecosystem in Congo. 3. Input-output budget and consequences for the sustainability of the plantations. *For. Ecol. Manag.* **2005**, *210*, 375–391. [CrossRef]
38. Nzila, J.D.; Bouillet, J.-P.; Laclau, J.-P.; Ranger, J. The effects of slash management on nutrient cycling and tree growth in Eucalyptus plantations in the Congo. *For. Ecol. Manag.* **2002**, *171*, 209–221. [CrossRef]
39. Laclau, J.P.; Deleporte, P.; Ranger, J.; Bouillet, J.P.; Kazotti, G. Nutrient Dynamics throughout the Rotation of Eucalyptus Clonal Stands in Congo. *Ann. Bot.* **2003**, *91*, 879–892. [CrossRef] [PubMed]
40. Laclau, J.-P.; Ranger, J.; Bouillet, J.-P.; Nzila, J.D.; Deleporte, P. Nutrient cycling in a clonal stand of Eucalyptus and an adjacent savanna ecosystem in Congo: 1. Chemical composition of rainfall, throughfall and stemflow solutions. *For. Ecol. Manag.* **2003**, *176*, 105–119. [CrossRef]
41. Laclau, J.-P.; Ranger, J.; Nzila, J.d.D.; Bouillet, J.-P.; Deleporte, P. Nutrient cycling in a clonal stand of Eucalyptus and an adjacent savanna ecosystem in Congo: 2. Chemical composition of soil solutions. *For. Ecol. Manag.* **2003**, *180*, 527–544. [CrossRef]
42. D'Annunzio, R.; Conche, S.; Landais, D.; Saint-Andre, L.; Joffre, R.; Barthes, B.G. Pairwise comparison of soil organic particle-size distributions in native savannas and Eucalyptus plantations in Congo. *For. Ecol. Manag.* **2008**, *255*, 1050–1056. [CrossRef]
43. Versini, A.; Zeller, B.; Derrien, D.; Mazoumbou, J.C.; Mareschal, L.; Saint André, L.; Ranger, J.; Laclau, J.P. The role of harvest residues to sustain tree growth and soil nitrogen stocks in a tropical *Eucalyptus* plantation. *Plant Soil* **2014**, *376*, 245–260. [CrossRef]
44. Epron, D.; Mouanda, C.; Mareschal, L.; Koutika, L.-S. Impacts of organic residue management on the soil C dynamics in a tropical eucalypt plantation on a nutrient-poor sandy soil after three rotations. *Soil Biol. Biochem.* **2015**, *85*, 183–189. [CrossRef]

45. Tchichelle, S.V.; Epron, D.; Mialoundama, F.; Koutika, L.S.; Harmand, J.M.; Bouillet, J.P.; Mareschal, L. Differences in nitrogen cycling and soil mineralization between a eucalypt plantation and a mixed eucalypt and *Acacia mangium* plantation on a sandy tropical soil. *South. For.* **2017**, *79*, 1–8. [CrossRef]
46. Tchichelle, S.; Mareschal, L.; Koutika, L.-S.; Epron, D. Biomass production, nitrogen accumulation and symbiotic 1 nitrogen fixation in a mixed species plantation of eucalypt and acacia on a nutrient-poor tropical soil. *For. Ecol. Manag.* **2017**, *403*, 103–111. [CrossRef]
47. Mareschal, L.; Nzila, J.D.D.; Turpault, M.P.; M'Bou, A.T.; Mazoumbou, J.C.; Bouillet, J.P.; Ranger, J.; Laclau, J.P. Mineralogical and physico-chemical properties of Ferralic Arenosols derived from unconsolidated Plio-Pleistocenic deposits in the coastal plains of Congo. *Geoderma* **2011**, *162*, 159–170. [CrossRef]
48. Bernhard-Reversat, F. Dynamics of Litter and Organic Matter at the Soil-Litter Interface in Fast-Growing Tree Plantations on Sandy Ferrallitic Soils (Congo). *Acta Oecol.* **1993**, *14*, 179–195.
49. Bouillet, J.P.; Safou Matondo, R.; Laclau, J.-P.; Nzila, J.D.; Ranger, J.; Deleporte, P. Pour une production durable des plantations d'eucalyptus au Congo: La fertilisation. *Bois Forêts Trop.* **2004**, *279*, 23–35. [CrossRef]
50. Koutika, L.-S.; Taba, K.; Ndongo, M.; Kaonga, M. Nitrogen-fixing trees increase organic carbon sequestration in forest and agroforestry ecosystems in the Congo basin. *Reg. Environ. Chang.* **2021**, *21*, 109. [CrossRef]
51. Koutika, L.-S.; Mareschal, L. Planting acacia and eucalypt change P, N and C concentrations in POM of arenosols in the Congolese coastal plains. *Geoderma Reg.* **2017**, *11*, 37–43. [CrossRef]
52. Koutika, L.S.; Cafiero, L.; Bevivino, A.; Merino, A. Organic Matter Quality of Forest Floor as a Driver of C and P Dynamics in Acacia and Eucalypt Plantations Established on a Ferralic Arenosols, Congo. *For. Ecosyst.* **2020**, *7*, 40. [CrossRef]
53. Koutika, L.-S.; Ngoyi, S.; Cafiero, L.; Bevivino, A. Soil organic matter dynamics along rotations in acacia and eucalyptus plantations in the Congolese coastal plains. *For. Ecosyst.* **2019**, *6*, 69. [CrossRef]
54. Pereira, A.P.A.; De Andrade, P.A.M.; Bini, D.; Durrer, A.; Robin, A.; Bouillet, J.P.; Andreote, F.D.; Cardoso, E.J.B.N. Shifts in the Bacterial Community Composition along Deep Soil Profiles in Monospecific and Mixed Stands of Eucalyptus Grandis and Acacia Mangium. *PLoS ONE* **2017**, *12*, e018037. [CrossRef]
55. Voigtlaender, M.; Laclau, J.P.; de Gonçalves, J.L.M.; de Piccolo, M.C.; Moreira, M.Z.; Nouvellon, Y.; Ranger, J.; Bouillet, J.P. Introducing Acacia Mangium Trees in Eucalyptus Grandis Plantations: Consequences for Soil Organic Matter Stocks and Nitrogen Mineralization. *Plant Soil* **2012**, *352*, 99–111. [CrossRef]
56. Forrester, D.I.; Bauhus, J.; Cowie, A.L. Nutrient cycling in a mixed-species plantation of Eucalyptus globulus and Acacia mearnsii. *Can. J. For. Res.* **2005**, *35*, 2942–2950. [CrossRef]
57. Forrester, D.I.; Pares, A.; O'Hara, C.; Khanna, P.K.; Bauhus, J. Soil organic carbon is increased in mixed-species plantations of Eucalyptus and nitrogen-fixing Acacia. *Ecosystems* **2013**, *16*, 123–132. [CrossRef]
58. Pereira, A.P.A.; Zagatto, M.R.G.; Brandani, C.B.; Mescolotti, D.D.L.; Cotta, S.R.; Gonçalves, J.L.M.; Cardoso, E.J.B.N. Acacia Changes Microbial Indicators and Increases C and N in Soil Organic Fractions in Intercropped Eucalyptus Plantations. *Front. Microbiol.* **2018**, *9*, 655. [CrossRef] [PubMed]
59. Bini, D.; Figueiredo, A.F.; Silva, M.C.P.; Vasconcellos, R.L.F.; Cardoso, E.J.B.N. Microbial Biomass and Activity in Litter during the Initial Development of Pure and Mixed Plantations of Eucalyptus Grandis and Acacia Mangium. *Rev. Bras. Ciência Solo* **2013**, *37*, 76–85. [CrossRef]
60. Bini, D.; Dos Santos, C.A.; Bouillet, J.-P.; Gonçalves, J.L.M.; Cardoso, E.J.B.N. Eucalyptus Grandis and Acacia Mangium in Monoculture and Intercropped Plantations: Evolution of Soil and Litter Microbial and Chemical Attributes during Early Stages of Plant Development. *Appl. Soil Ecol.* **2013**, *63*, 57–66. [CrossRef]
61. Bini, D.; dos Santos, C.A.; da Silva, M.C.P.; Bonfim, J.A.; Cardoso, E.J.B.N. Intercropping Acacia Mangium Stimulates AMF Colonization and Soil Phosphatase Activity in Eucalyptus Grandis. *Sci. Agric.* **2018**, *75*, 102–110. [CrossRef]
62. Rocha, J.H.T.; Menegale, M.L.C.; Rodrigues, M.; Gonçalves, J.L.M.; Pavinato, P.S.; Foltran, E.C.; Harrison, R.; James, J.N. Impacts of timber harvest intensity and P fertilizer application on soil P Fractions. *For. Ecol. Manag.* **2019**, *437*, 295–303. [CrossRef]
63. Brezard, J.M. *Les Eucalyptus Introduits au Congo 1953–1981*; GERDAT-CTFT: Nogent-sur-Marne, France, 1983; 148p. Available online: https://agritrop.cirad.fr/361610. (accessed on 1 August 2022).
64. Roux, J.; Coutinho, T.A.; Wingfield, M.J.; Bouillet, J.P. Diseases of plantation *Eucalyptus* in the Republic of Congo. *S. Afr. J. Sci.* **2000**, *96*, 454–456.
65. Bouvet, J.M.; Vigneron, P.; Saya, A.; Gouma, R. Early Selection of Eucalyptus clones in retrospective nursery test using growth, morphological and dry matter criteria, in Congo. *South. Afr. For. J.* **2004**, *200*, 5–17. [CrossRef]
66. Bouvet, J.M.; Vigneron, P.; Saya, A. Phenotypic Plasticity of Growth Trajectory and Ontogenic Allometry in Response to Density for Eucalyptus hybrid clones and familles. *Ann. Bot.* **2005**, *96*, 811–821. [CrossRef]
67. Makouanzi, G.; Bouvet, J.-M.; Denis, M.; Saya, A.; Mankessi, F.; Vigneron, P. Assessing the additive and dominance genetic effects of vegetative propagation ability in Eucalyptus -influence of modeling on genetic gain. *Tree Genet. Genomes* **2014**, *10*, 1243–1256. [CrossRef]
68. Makouanzi, G.; Chaix, G.; Nourissier, S.; Vigneron, P. Genetic variability of growth and wood chemical properties in a clonal population of *Eucalyptus urophylla* × *Eucalyptus grandis* in the Congo. *South. For. South. For. J. For. Sci.* **2017**, *80*, 151–158. [CrossRef]
69. Tassin, J.; Missamba-Lola, A.P.; Marien, J.N. Biodiversité des plantations d'eucalyptus. *Bois For. Trop.* **2011**, *309*, 27–35. [CrossRef]

70. Lejolly, J. Modèle agroforestier pour le développement à Ibi/Plateau de Batéké, Université Libre de Bruxelles, Belgique, 2018. p. 12. Available online: www.cairn.info/revue-mondes-en-developpement-2019-3-page113.html (accessed on 27 April 2022).
71. Nsombo, B.M. Evaluation des nutriments et du carbone organique du sol dans le système agroforestier du plateau des Batéké en République Démocratique du Congo. Ph.D. Thesis, Ecole Régionale Post-Universitaire d'Aménagement et de Gestion Intégrés des Forêts et Territoires Tropicaux (ERAIFT), Kinshasa, Democratic Republic of the Congo, 2016; pp. 79–80.
72. Koutika, L.-S. How hydrogen sulphide deposition from oil exploitation may affect bacterial communities and the health of forest soils in Congolese coastal plains? *Front. Soil Sci.* **2022**, *2*, 920142. [CrossRef]
73. Asaah, E.K.; Tchoundjeu, Z.; Leakey, R.R.B.; Takousting, B.; Njong, J.; Edang, I. Trees, agroforestry and multifunctional agriculture in Cameroon. *Int. J. Agric. Sustain.* **2011**, *9*, 110–119. [CrossRef]
74. Dubiez, E.; Freycon, V.; Marien, J.M.; Peltier, R.; Harmand, J.M. Long term impact of *Acacia auriculiformis* woodlots growing in rotation with cassava and maize on the carbon and nutrient contents of savannah sandy soils in the humid tropics (Democratic Republic of Congo). *Agrofor. Syst.* **2019**, *93*, 1167–1178. [CrossRef]
75. Mankessi, F.; Saya, A.; Baptiste, C.; Nourissier, S.; Monteuis, O. In vitro rooting of genetically related *Eucalyptus urophylla* × *Eucalyptus grandis* clones in relation to the time spent in culture. *Trees* **2009**, *23*, 931–940. [CrossRef]
76. Thongo M'bou, A.; Jourdan, C.; Deleporte, P.; Nouvellon, Y.; Saint-André, L.; Bouillet, J.-P.; Mialoundama, F.; Mabiala, A.; Epron, D. Root elongation in tropical Eucalyptus plantations: Effect of soil water content. *Annals For. Sci.* **2008**, *65*, 609–615. [CrossRef]
77. Thongo M'Bou, A.; Saint-André, L.; de Grandcourt, A.; Nouvellon, Y.; Jourdan, C.; Mialoundama, F.; Epron, D. Growth and maintenance respiration of roots of clonal Eucalyptus cuttings: Scaling to stand-level. *Plant Soil* **2010**, *332*, 41–53. [CrossRef]
78. Martin, B.; Quillet, G. Bouturage des arbres forestiers au Congo. 1e partie: Résultats des essais effectués à Pointe-Noire de 1969 à 1973. *Bois For. Trop.* **1974**, *154*, 41–57. [CrossRef]
79. Versini, A.; Mareschal, L.; Matsoumbou, T.; Zeller, B.; Ranger, J.; Laclau, J.-P. Effects of litter manipulation in a tropical Eucalyptus plantation on leaching of mineral nutrients, dissolved organic nitrogen and dissolved organic carbon. *Geoderma* **2014**, *232*, 426–436. [CrossRef]
80. Koutika, L.-S. Soil fertility improvement of nutrient-poor and sandy soils in the Congolese coastal plains. Recarbonizing global soils: A technical manual of recommended management practices. In *Forestry, Wetlands, Urban—Case Studies*; FAO & ITPS: Rome, Italy, 2021; Volume 6, pp. 4–13. [CrossRef]
81. Epron, D.; Mareschal, L.; Marron, N.; Laclau, J.P.; Marsden, C.; Ranger, J.; Nouvellon, Y.; Saint-André, L.; Bouillet, J.-P. Les enjeux environnementaux des plantations forestières intensives. *Forêt Privée* **2011**, *320*, 66–74.
82. Kasongo, R.K.; Van Ranst, E.; Verdoodt, A.; Kanyankagote, P.; Baert, G. Impact of *Acacia auriculiformis* on the chemical fertility of sandy soils on the Batéké plateau, D.R. Congo. *Soil Use Manag.* **2009**, *25*, 21–27. [CrossRef]
83. Epron, D.; Marsden, C.; Thongo M'Bou, A.; Saint-André, L.; d'Annunzio, R.; Nouvellon, Y. Soil carbon dynamics following afforestation of a tropical savannah with Eucalyptus in Congo. *Plant Soil* **2009**, *323*, 309–322. [CrossRef]
84. Epron, D.; Nouvellon, Y.; Deleporte, P.; Ifo, S.; Kazotti, G.; Thongo M'Bou, A.; Mouvondy, W.; Saint-André, L.; Roupsard, O.; Jourdan, C.; et al. Soil carbon balance in a clonal Eucalyptus plantation in Congo: Effects of logging on carbon inputs and soil CO_2 efflux. *Glob. Change Biol.* **2006**, *12*, 1021–1031. [CrossRef]
85. Hassink, J.; Bouwman, L.A.; Zwart, K.B.; Bloem, J.; Brussaard, L. Relationships between soil texture, physical protection of organic matter, soil biota, and C and N mineralization in grassland soils. *Geoderma* **1993**, *57*, 105–128. [CrossRef]
86. Koutika, L.-S.; Andreux, F.; Hassink, J.; Choné, T.; Cerri, C.C. Characterisation of soil organic matter in the topsoils under rain forest and pastures in the eastern Brazilian Amazon Basin. *Biol. Fertil. Soils* **1999**, *29*, 309–313. [CrossRef]
87. Laclau, J.P.; Toutain, F.; Thongo M'Bou, A.; Arnaud, M.; Joffre, R.; Ranger, J. The Function of the Superficial Root Mat in the Biogeochemical Cycles of Nutrients in Congolese Eucalyptus plantations. *Ann. Bot.* **2004**, *93*, 249–261. [CrossRef]
88. Koutika, L.-S.; Tchichelle, S.V.; Mareschal, L.; Epron, D. Nitrogen dynamics in a nutrient-poor soil under mixed-species plantations of eucalypts and acacias. *Soil Biol. Biochem.* **2017**, *108*, 84–90. [CrossRef]
89. Koutika, L.S.; Mareschal, L.; Epron, D. Soil P Availability under Eucalypt and Acacia on Ferralic Arenosols, Republic of the Congo. *Geoderma Reg.* **2016**, *7*, 153–158. [CrossRef]
90. Hartemink, A.E.; Lal, R.; Gerzabek, M.H.; Jama, B.; McBratney, A.B.; Six, J.; Tornquist, C.G. Soil carbon research and global environmental challenges. *PeerJ PrePrints* **2014**, *2*, e366v1. [CrossRef]
91. Tarus, G.K.; Nadir, S.W. Effect of Forest management types on soil carbon stocks in montane forest: A case study of Eastern M au Forest in Kenya. *Inter. J. For. Res.* **2020**, *2020*, 8862813. [CrossRef]
92. Koutika, L.-S.; Richardson, D.M. *Acacia mangium* Willd: Benefits and threats associated with its increasing use around the world. *For. Ecosyst.* **2019**, *6*, 2. [CrossRef]
93. Koutika, L.-S.; Fiore, A.; Tabacchioni, S.; Aprea, G.; Pereira, A.P.A.; Bevivino, A. Influence of *Acacia mangium* on soil fertility and bacterial community in *Eucalyptus* Plantations in the Congolese Coastal Plains. *Sustainability* **2020**, *12*, 8763. [CrossRef]
94. Koutika, L.-S.; Mareschal, L.; Rudowski, S. Fate of *Acacia mangium* in eucalypt mixed-species plantation during drought conditions in the Congolese coastal plains. *Bosque* **2018**, *39*, 131–136. [CrossRef]
95. Dong, Q.; Shi, H.; Liu, Y. Microbial Character Related Sulfur Cycle under Dynamic Environmental Factors Based on the Microbial Population Analysis in Sewerage System. *Front. Microbiol.* **2017**, *8*, 64. [CrossRef]

96. Xu, Y.; Fan, J.; Ding, W.; Bol, R.; Chen, Z.; Luo, J.; Bolan, N. Stage-specific response of litter decomposition to N and S amendments in a subtropical forest soil. *Biol. Fertil. Soils* **2016**, *52*, 711–724. [CrossRef]
97. Anandan, R.; Dharumadurai, D.; Manogaran, G.P. An introduction to actinobacteria. In *Actinobacteria—Basics and Biotechnological Applications*; Intech Open Sciences and Open Minds: London, UK, 2016; pp. 3–37. [CrossRef]

sustainability

MDPI

Article

Fuzzy Set Qualitative Comparative Analysis of the Factors Affecting Satisfaction with the Policy of Ecological Forest Rangers

Yonghua Zhang, Xue Wang, Shenwei Wan and Hongge Zhu *

College of Economics and Management, Northeast Forestry University, Harbin 150040, China
* Correspondence: honggebill@163.com; Tel.: +86-139-4518-6371

Abstract: The policy of ecological forest rangers (EFRs) is one of the important policy tools to consolidate the achievements of poverty alleviation in China. An in-depth analysis of the factors affecting the satisfaction of EFRs, and targeted improvement of related issues are conducive to the promotion of the policy implementation effect, which is of great significance to further consolidate the achievements of ecological poverty alleviation, promoting rural revitalization. Based on the field survey data of 412 ecological forest rangers in Qianshan City, Anhui Province, China, this paper uses the multivariable interaction fuzzy set qualitative comparative analysis method to explore the level of ecological forest rangers' policy satisfaction and the associated influencing factors. The results showed that (1) the overall evaluation of the ecological ranger groups' satisfaction with EFRs was between "general" and "satisfied"; (2) the lack of policy identity and information mastery are the necessary conditions for low and high satisfaction of EFRs, respectively; (3) perception of implementation played a core role in high policy satisfaction, while a lack of information mastery and perception of implementation were the core variables that caused low policy satisfaction. Through comprehensive comparison, it was found that the conditional variables of policy cognition had an important impact on both high and low policy satisfaction. Enhancing the information grasp degree of ecological forest rangers and improving the perception level of policy implementation was the best strategy to effectively improve the satisfaction levels of ecological forest rangers with the policy. By exploring the influencing factors of the satisfaction with the current ecological forest ranger policy and analyzing the comprehensive effect of the configuration of each factor, this paper provides a reference for further improving the ecological forest ranger policy and consolidating the ecological poverty alleviation results in the future.

Keywords: ecological forest ranger; policy satisfaction; fuzzy set; qualitative comparative analysis; influencing factors

Citation: Zhang, Y.; Wang, X.; Wan, S.; Zhu, H. Fuzzy Set Qualitative Comparative Analysis of the Factors Affecting Satisfaction with the Policy of Ecological Forest Rangers. *Sustainability* **2023**, *15*, 6743. https://doi.org/10.3390/su15086743

Academic Editor: Aaron K. Hoshide

Received: 19 January 2023
Revised: 9 April 2023
Accepted: 11 April 2023
Published: 17 April 2023

1. Introduction

Poverty is a serious challenge and a major problem that needs to be solved for all humankind [1], and the 2030 Agenda for Sustainable Development set by the United Nations has made "the eradication of poverty in all its forms worldwide" the primary goal of human sustainable development [2]. In the journey of poverty eradication, countries around the world have made many explorations in financial poverty alleviation, industrial poverty alleviation, ecological poverty alleviation, and scientific and technological poverty alleviation [3,4]. Ecological poverty alleviation, which works by changing the ecological environment of poor areas to escape from poverty [5], combines both ecological conservation and poverty reduction [6] and has received much attention in the international community. The ecological poverty alleviation paths are multiple and varied, mainly including the construction of major ecological projects, implementation of ecological compensation policies, development of ecological characteristic industries, introduction of ecological public welfare jobs, and implementation of relocation to alleviate poverty [7,8]. The policy of ecological forest rangers is an exploration of ecological poverty alleviation through setting ecological public welfare positions in China, which started in 2016. This refers to the policy of selecting and hiring people involved in forest resource management services by arranging subsidized funds from the central government or provincial governments to purchase labor services within the scope of the documented poor population [9]. This is a product of combining ecological compensation policies with precise poverty alleviation policies in China [10].

As an ecological poverty alleviation realization mechanism, the implementation process of the policy of ecological forest rangers involves the selection of one person from an eligible poor family (mostly poor households with established cards) by the local government to serve as an ecological ranger and engage in the management of forest resources, and he or she can earn 1000 to 10,000 CNY of wage income per year, helping to drive poor families out of poverty [11]. The implementation of the policy of ecological forest rangers in the vast ecologically vulnerable areas has not only driven the local poor out of poverty and increased the livelihood capital stock of poor households [12], but also has improved the local ecological environment and achieved a balance between economic, ecological, and social benefits. However, after the implementation of the ecological forest ranger policy, the protection of local ecological resources has increased, and measures such as logging and hunting bans have caused some poor households to lose their original sources of livelihood. The salary income of ecological forest rangers cannot fully compensate for the losses caused by logging and hunting bans. Therefore, there may be a phenomenon of low satisfaction during the implementation of policies, which directly affects the effectiveness of policy implementation [13].

Satisfaction is a crucial aspect of people's perception and judgment of pro-poor policies, which reflects their subjective evaluation of policy formulation and implementation [14]. Understanding satisfaction with poverty alleviation policies is important as it can serve as an indicator of the performance level of pro-poor policies [15]. Recent research has focused on evaluating satisfaction with poverty alleviation policies and identifying the factors that influence this satisfaction. A quantitative evaluation of satisfaction with poverty alleviation policies involves constructing an evaluation index system, and exploring the factors that affect satisfaction with policies. These factors are found to be complex and diverse, including individual characteristics such as gender, age, and income of the study subjects. For instance, older poor people tend to have limited fixed income, and poverty alleviation policies may bring greater marginal income, leading to higher satisfaction with policies [16]. The research subjects' cognition of the policy is also an essential factor affecting satisfaction with a poverty alleviation policy. The wider the channels for research subjects to obtain information and the clearer their understanding of poverty alleviation policies, the higher their satisfaction with the policies [17]. External factors such as the timeliness of poverty alleviation funds and the democratic atmosphere of the collective where the research subjects belong also have an important impact on satisfaction with poverty alleviation policies [18]. In summary, the existing studies on satisfaction with poverty alleviation policies indicate that the livelihood situation, individual characteristics, knowledge of policies, policy cognition, and information availability of poor people are significant factors affecting policy satisfaction [19,20].

Fuzzy set qualitative comparative analysis (fsQCA) is a method that can analyze complex causality and interactions among multiple factors by identifying causal configurations that lead to a specific outcome [21,22]. It is based on the principles of fuzzy logic and Boolean algebra, and it allows for the consideration of multiple combinations of variables that may contribute to a specific outcome [23]. By applying this method, researchers can identify necessary and sufficient conditions that lead to satisfaction with the ecological forest rangers policy, as well as the different combinations of factors that may lead to this outcome. Using a histological perspective means that the analysis will take into account the historical context and the different factors that have shaped the policy of ecological forest rangers over time. This approach can provide a more nuanced understanding of the factors that influence policy satisfaction and can help to identify potential areas for policy improvement. Overall, fsQCA provides a useful alternative to traditional regression analysis by allowing for a more comprehensive and nuanced analysis of complex causal relationships among multiple factors [24].

The research presents a novel approach to analyzing the influencing factors of policy satisfaction of ecological forest rangers using the qualitative comparative analysis method of fuzzy sets. The marginal contribution of the paper is two-fold. Firstly, it explores the influencing factors and interaction relationships of ecological forest rangers' policy satisfaction from a configuration perspective, which enriches the existing research methods of ecological poverty alleviation policy satisfaction. Secondly, the study evaluates the implementation effect of ecological forest ranger policies from a subjective evaluation perspective, which is relatively novel in the current research on ecological poverty alleviation policies.

Research in this area is particularly relevant given that China has recently eliminated absolute poverty, and preventing the return of absolute poverty has become a critical focus of the country's ecological poverty alleviation policies. The findings of the study can provide valuable insights into the optimization of ecological forest ranger policies, thereby contributing to poverty reduction and governance efforts. This investigation makes a significant contribution to the research on ecological poverty alleviation policies, and offers important practical implications for policymakers and stakeholders involved in poverty reduction efforts.

2. Research Framework

Based on the analysis of ecological poverty alleviation policy satisfaction studies, and considering the representativeness of indicators, the research framework (Figure 1) was constructed from the perspective of ecological forest rangers. Independent variables included "livelihood capital", "livelihood outcome", "policy cognition", "policy identity", "information mastery", and "implementation perception". The satisfaction of ecological forest rangers with the policy was an dependent variable. Group matching was used from an analysis perspective in order to analyze factors influencing ecological forest rangers' policy satisfaction.

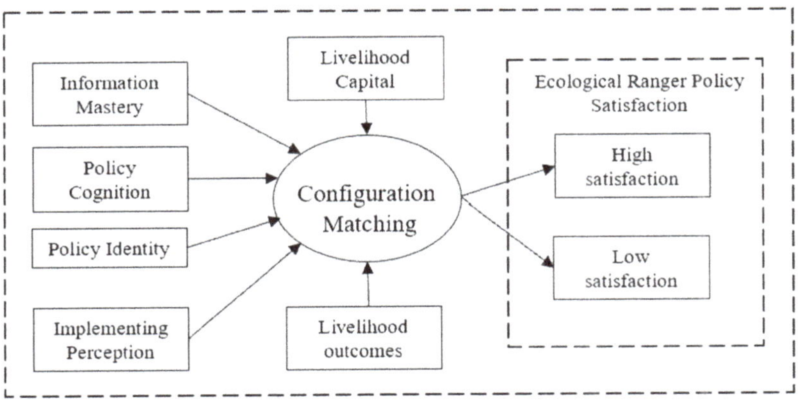

Figure 1. Research framework for qualitative comparative analysis of factors influencing the policy of ecological forest rangers.

(1) Livelihood capital. Using the sustainable livelihoods analysis framework developed by the UK's Department for International Development (DFID), the human capital (H), natural capital (N), social capital (S), physical capital (P), and financial capital (F) possessed by ecological forest rangers were assessed to explore their impact on the satisfaction with the policy of ecological forest rangers. Among them, natural capital is characterized by the area of forest land owned by ecological forest rangers, the area of farmland, and the quality of farmland, and natural capital endowment has a significant effect on the satisfaction with the forest land subsidy policy, farmland subsidy policy, and land transfer policy [25]; therefore, it is inferred that natural capital also has an effect on the satisfaction with the ecological ranger policy. Financial capital is characterized by the share of forestry subsidy income, the share of agricultural income, and the amount of annual income. Xie and Cai [26], in 2021, argued that physical and financial capital positively influenced farmers' satisfaction with the implementation of farmland protection compensation policies, and therefore inferred that physical and financial capital also have important effects on ecological forest rangers' policy satisfaction. Human capital was characterized by the number of years of education and the proportion of labor force of the users. Social capital is characterized by the number of governmental institution personnel owned by the ecological ranger's household, the number of friends and relatives of long-term urban residents, and whether the neighborhood is cordial. Social resources have a significant positive effect on farmers' income, while human capital has an income enhancing effect [27], and ecological forest rangers' satisfaction with the policy is closely related to their income level. So, human capital and social capital are considered as important factors affecting ecological forest rangers' policy satisfaction. Combining the studies of existing scholars, this paper uses livelihood capital as a comprehensive indicator to analyze its impact on the satisfaction with the policy of ecological forest rangers.

(2) Livelihood outcomes. Livelihood outcomes are the standard of living achieved by ecological forest rangers, using different livelihood strategies to integrate and utilize the livelihood capital they possess [28]. Determining whether the living standard of ecological forest rangers has improved is an important criterion to test the effectiveness of the implementation of the ecological ranger policy. After the implementation of the ecological ranger policy, eligible poor households received an important source of income and their living standards were improved. In addition to this, each forestry station (a subordinate unit of the Forestry Bureau) provides additional training to ecological forest rangers in planting techniques and farming techniques to enhance their ability to withstand livelihood risks. However, in the interviews with ecological forest rangers, it was also found that since the implementation of the ecological ranger policy, the workload of ecological forest rangers has been relatively heavy, especially during the forest fire prevention period. During this period, forest management work takes up almost a whole day of their time, and there are situations where the ecological forest rangers' own work conflicts with other family business activities, making it necessary for them to give up other sources of income for a short time. Thus, the ecological ranger policy as a new livelihood strategy leads to changes in livelihood outcomes, and different livelihood outcomes will inevitably have an impact on the satisfaction with the policy of ecological forest rangers.

(3) Policy cognition. This variable is interpreted as the ecological forest rangers' perceptions of ecological ranger policies, including their perceptions and evaluations of the effects of policy formulation, implementation, and enforcement [29]. The satisfaction of ecological forest rangers with policies depends on the deviation between policy expectations and actual implementation effects. There is a positive bias and higher satisfaction when the effect of policy implementation is higher than the effect of policy expectation [30]. Therefore, policy cognition is also one of the factors that influences ecological forest rangers' policy satisfaction.

(4) Policy identity. Policy implementation theory suggests that the target group's identification with the policy is the basis for policy implementation [31] and also influences the effectiveness of policy implementation. In this paper, policy identity refers to the degree of ecological forest rangers' support for EFRs implementation. Analyzing the degree of ecological forest rangers' identification with policy implementation is an important step in evaluating EFRs' satisfaction [32].

(5) Information Mastery. Information mastery is explained as the degree of an ecological ranger's understanding of the EFRs, and the degree of an ecological ranger's mastery of the policy reflects the effectiveness of policy promotion. The more the ecological forest rangers understand the policy, and the deeper their knowledge of the ecological poverty alleviation policy, the more beneficial it is for them to actually work harder and enhance their work identity [33], and, accordingly, their policy satisfaction increases.

(6) Implementation Perception. Implementation perception is the degree to which ecological forest rangers perceive the effect of EFRs implementation, including willingness to participate, demand perception, process perception, service perception, and policy demands and trust [34], and the results can be divided into five grades: "very insignificant, less insignificant, average, more insignificant, and very insignificant". The better the ecological forest rangers' perceptions of policy implementation are, the more willing they will be to participate in public policies, which in turn enhances the effectiveness of policy implementation.

In summary, livelihood capital and livelihood outcomes are important outcome manifestations of ecological forest ranger policies; policy cognition and policy identity are the ecological forest rangers' awareness and evaluation of the relevant policies and the degree of support they provide; information mastery reflects the degree of openness and transparency of the EFRs implementation process; implementation perception is an important reflection of the effectiveness of EFRs implementation. In the group perspective, livelihood capital, livelihood outcomes, policy cognition, policy identity, information mastery, and implementation perception are not independent; the six variables work synergistically through a linkage matching approach to jointly influence satisfaction with ecological ranger policies (Supplementary Materials).

3. Materials and Methods

3.1. Research Area

The research site is Qianshan City, Anhui Province (Figure 2), located in the southwestern part of Anhui Province, on the north bank of the lower reaches of the Yangtze River and the southeastern foothills of the Dabie Mountains, with a complex topography dominated by mountains and hills. Qianshan City was once a focus area of China's national poverty alleviation and development scheme. It is rich in forest resources. The economic activities in mountainous areas are dominated by forestry production. Forestry plays an important role in its economic development. The implementation effect of the local ecological forest ranger policy is relatively significant. Based on the principles of representativeness and convenience, ecological forest rangers in the town of Qianshan City were selected for this study.

3.2. Data Sources

From June to August 2021, the research team went to Qianshan City, Anhui Province, to conduct field research, following the principle of randomness and adopting a multi-stage sampling method to conduct the survey. A simple random sampling method was used to randomly select 10 towns, including Huangbai Town, Doumu Town, and Shuihou Town, among 16 townships in Qianshan City. A total of 512 ecological forest rangers were selected as a sample for the questionnaire survey, according to the amount of ecological forest rangers in each town in proportion to their total number (Table 1). After eliminating the questionnaires containing missing values and outliers, 412 valid questionnaires were obtained, with a valid return rate of 80.6%. In addition to the questionnaire survey, the research team also conducted in-depth interviews with 2 to 4 ecological forest rangers randomly selected from each township to understand the implementation of the ecological forest ranger policy.

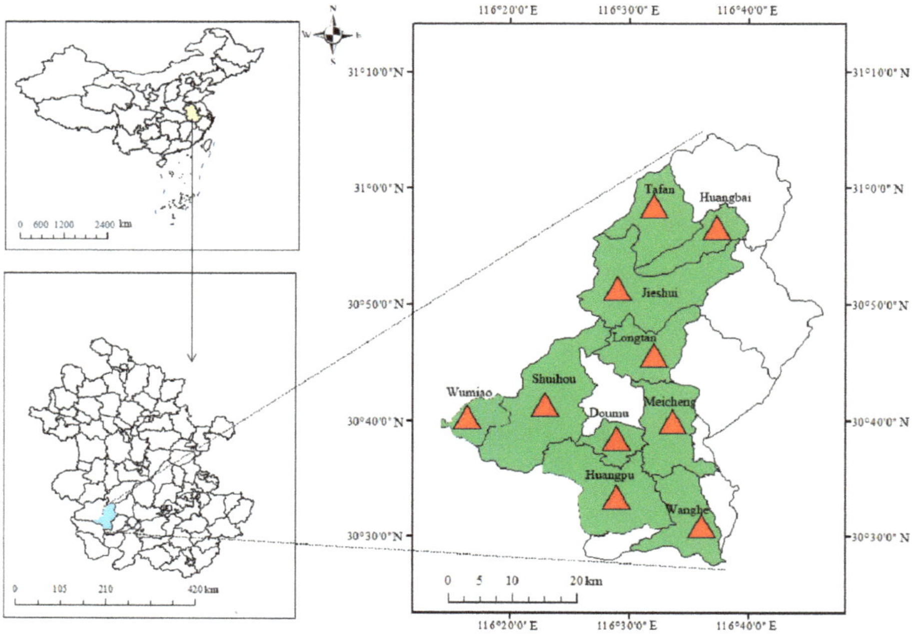

Figure 2. Location of the studied towns in Qianshan City, Anhui Province, China.

Table 1. Township distribution of the sample of ecological forest rangers.

Sample Area	Sample Size (Person)	Sample Share (%)
Huangbai	50	12.14
Doumu	33	8.01
Shuihou	51	12.38
Wumiao	44	10.68
Chashui	72	17.48
Tafan	30	7.28
Wanghe	11	2.67
Huangpu	52	12.62
Longtan	39	9.46
Meicheng	30	7.28
Total	412	100.00

From the returned questionnaires, the age of the respondents was concentrated in the age range of 46–66. The number of years of education was generally low, concentrated in the range of 6 years and below. The overall health status of the respondents was concentrated in "average" and "better" categories. The basic characteristics of the survey respondents are shown in Table 2.

Table 2. Basic characteristics of the sample.

Variables	Variable Properties	Quantity/Person	Percentage/%
Age (years)	≤35	8	1.94
	36~45	37	8.98
	46~55	225	54.61
	≥56	142	34.47
Years of education (years)	≤6	238	57.77
	7~9	155	37.62
	10~12	16	3.88
	≥13	3	0.73
Self-assessment of health status	Very poor	10	2.43
	Poor	42	10.19
	Fair	185	44.90
	Better	116	28.16
	Very good	59	14.32

3.3. Research Methods

Qualitative comparative analysis (QCA) is a research method that combines qualitative and quantitative approaches to analyze complex problems with multiple causal factors [35]. QCA includes three types: clear sets (cs), fuzzy sets (fs), and multi-valued sets (mv). Fuzzy set QCA is based on fuzzy sets and uses data elements to explore multiple combinations of conditions that lead to outcomes, using fuzzy scores to represent conditional variables and outcomes between 0 and 1 [36]. This helps to avoid data transformation losses and effectively describes the interactions between different variables [37]. Fuzzy set QCA is particularly useful for exploring factors that affect satisfaction, as it can handle issues related to degree variation and partial membership. Compared to cs QCA and mv QCA, which are only suitable for categorical issues, fuzzy set QCA has the advantage of dealing with more complex and nuanced issues [38].

This paper uses the qualitative comparative analysis method of fuzzy sets to analyze the influencing factors of policy satisfaction of ecological forest rangers, mainly including the following three steps. Firstly, the original data of various factors and results that affected the satisfaction of ecological forest rangers were calibrated to the corresponding fuzzy set membership scores, and the number of cases and membership scores corresponding to all possible combinations of conditions were listed. Secondly, the corresponding judging criteria and thresholds were determined, the combinations of conditions relevant to the results were screened, and the case distributions of the various combinations of conditions were evaluated. Finally, the consistency and coverage of the fuzzy subsets of condition combinations constituting the results were evaluated to analyze the stability of the paths of influence of condition combinations on the results, which were mainly measured by two indicators, consistency and coverage, calculated as follows.

$$\text{Consistency } (x_i \le y_i) = \sum [\min (x_i, y_i)] / \sum x_i \quad (1)$$

Consistency reflects the degree of consistency of the influence of the condition variables, or the path of the combination of condition variables on the outcome. x_i is the affiliation score in the combination of conditions and y_i is the affiliation score in the outcome, and the consistency takes values between 0 and 1. The larger the value, the stronger and more convincing the significance of the condition variables' influence on the outcome variables. When the consistency reaches 0.9, it means that the condition is necessary to form the result, where coverage is the assessment of the degree of explanation of the conditional variables for the results after the consistency operation. x_i and y_i have the same meaning as Equation (1), and coverage takes values between 0 and 1. The higher the coverage, the stronger the explanatory power of the cause and effect.

$$\text{Coverage } (x_i \leq y_i) = \sum [\min (x_i, y_i)] \ / \ \sum y_i \quad (2)$$

3.4. Data Calibration

In fuzzy set qualitative comparative analysis, each condition variable and outcome variable are an ensemble, respectively. Prior to data analysis, the condition and outcome variables are calibrated to a fuzzy number between 0 and 1, which represents the degree to which different cases belong to a certain set. A threshold value closer to 1 indicates a higher affiliation; when the threshold value is zero, it means no affiliation at all; when the threshold value is between 0 and 0.5, it means weak affiliation to the set; when the threshold value is between 0.5 and 1, it means strong affiliation. The direct calibration method and the indirect calibration method are used to assign values to the data according to the differences in the data types of each variable. The livelihood capital evaluation index system is established and weighted, and the continuous variable livelihood capital is calibrated using the direct calibration method. Livelihood outcomes, policy cognition, policy identity, information mastery, implementation perception, and ecological forest rangers' policy satisfaction are five-scale variables that are calibrated using an indirect calibration method for fuzzy set affiliation scores.

3.4.1. Result Variables

The research focus of this paper is to analyze the satisfaction of ecological forest rangers with the policy of ecological forest rangers. The result variable is satisfaction. The indirect calibration method is used to calibrate the membership score of fuzzy sets. According to the differences in the ecological forest rangers' satisfaction with the policy, the policy satisfaction was divided into five levels from low to high: "very dissatisfied", "relatively dissatisfied", "basically satisfied", "relatively satisfied", and "very satisfied", and the values were 0, 0.25, 0.5, 0.75, and 1, respectively.

3.4.2. Condition Variables

The livelihood capital composite results of ecological forest rangers were obtained based on the livelihood capital evaluation index system, and the variables were calibrated using the direct calibration method with the 95% (fully affiliated), 50% (qualitative intersection), and 5% (fully unaffiliated) quartiles of the sample data taken as calibration points. A livelihood capital evaluation index system was established to comprehensively evaluate livelihood capital. In order to avoid the influence of human factors on the evaluation results, and the problems of information crossover and superposition among the composite indicators in the indicator system, the entropy weighting method was used to determine the weights of each indicator [39]. Each measurement indicator, the meaning of the indicator, and the weight are shown in Table 3.

Table 3. Establishment of livelihood capital evaluation index system and weight setting.

Indicator Name	Measurement Standard	Indicator Meaning	Specific Gravity
Natural capital (N)	Cultivated area (N1)	Having arable land area (mu)	0.092
	Quality of cultivated land (N2)	The quality of cultivated land, "very poor", "worse", "general", "better", "very good", respectively. 1, 2, 3, 4, 5	0.011
	Forest area (N3)	Having forest land area (mu)	0.078

Table 3. *Cont.*

Indicator Name	Measurement Standard	Indicator Meaning	Specific Gravity
Material capital (P)	House material (P1)	Material of the owned house. Four dummy variables set for "civil house", "brick house", "masonry house", "concrete house".	0.007
	The value of durable goods and other agricultural production equipment (P2)	Paid value of durable goods and other agricultural production equipment	0.010
	Number of houses (P3)	The number of houses (rooms) that ecological forest guards have	0.045
Human capital (H)	Householder education period (H1)	The year of the householder receiving education (year)	0.017
	Proportion of labor (H2)	Number of family members undergoing labor for family Labor/number of families (%)	0.026
Financial capital (F)	Forestry subsidy income proportion (F1)	Forestry subsidy income/total family income (%)	0.036
	Agricultural income ratio (F2)	Agricultural operating income/total household income (%)	0.131
	Annual savings (F3)	The value of the annual savings amount	0.138
Social capital (S)	Whether the neighborhood relationship is harmonious (S1)	0 = No; 1 = Yes	0.003
	Number of staff members of government institutions (S2)	Number of people involved employed in government institutions (people)	0.206
	The number of relatives and friends of the long-term residents of the town (S3)	Number of long-term urban residents' relatives and friends (people)	0.200

After obtaining the composite evaluation score of ecological forest rangers' livelihood capital, the data were calibrated using the direct calibration method, using the 95% quantile value of the composite score (4.437) as the full affiliation threshold, the 50% quantile value (1.362) as the crossover point, and the 5% quantile value (0.446) as the full non-affiliation threshold. For livelihood outcomes, policy cognition, policy identity, information mastery, and implementation perception, the indirect calibration method was used, and the data results are shown in Table 4.

Table 4. Conditional variable data calibration.

Variable Name	Indicator Meaning	Calibration Result
Livelihood capital	Life Capital Comprehensive Score	Fully belong to (4.437); intersection (1.362); not affiliated at all (0.446)
Livelihood outcomes	"How does your family living level change after the implementation of ecological forest protection policies?"	"Reduce a lot" (0); "Lower some" (0.25); "Unchanged" (0.5); "Improve some" (0.75); "Improve a lot" (1)
Policy cognition	"Your support for the continuation of ecological forest guard policies"	"Very unsuitable" (0); "less supportive" (0.25); "Indifferent" (0.5); "More supportive" (0.75); "Very supportive" (1)
Policy identity	"Do you think the degree of implementation of ecological forest protection policies?"	"Very unnecessary" (0); "more unnecessary" (0.25); "Indifferent" (0.5); "necessary" (0.75); "very necessary" (1)

Table 4. *Cont.*

Variable Name	Indicator Meaning	Calibration Result
Information mastery	"Do you understand the policy of ecological forest guard?"	"don't understand" (0); "don't know more about" (0.25); "General understanding" (0.5); "More understanding" (0.75); "Know well" (1)
Implement perception	"Your perception of the implementation of ecological forest guard policies"	"Very unknown" (0); "Not obvious" (0.25); "Generally obvious" (0.5); "More obvious" (0.75); "very obvious" (1)

4. Results

4.1. Descriptive Analysis

According to the statistical analysis, the average satisfaction degree of ecological forest rangers with policies is 3.597. This satisfaction score is between "average" and "satisfied", indicating that most ecological forest rangers have a positive attitude towards policies. It is believed that the policy of ecological forest rangers has a positive effect on their livelihood improvement.

4.2. Necessary Conditions Analysis

The necessity of each condition variable was analyzed separately using fsQCA 3.0 software and the results are shown in Table 5. In the analysis of the necessity of low satisfaction, the consistency test for the absence of policy identity was 0.900 and the test results for the remaining conditional variables were below 0.900, indicating that the absence of policy identity may be a necessary condition for low satisfaction. In the analysis of the necessity of high satisfaction, the consistency test for information mastery was 0.912, and the consistency of the remaining conditional variables was below 0.900, suggesting that information mastery may be a necessary condition for high satisfaction with the policy of ecological forest rangers. The degree of policy satisfaction of ecological forest rangers is the result of the comprehensive effect of each group. Therefore, in the subsequent analysis of the comprehensive effect of each group, the two conditional variables of policy identification and information mastery are eliminated.

Table 5. Test of necessity for conditional variables.

Condition Variable	Low Satisfaction		High Satisfaction	
	Consistency	Coverage	Consistency	Coverage
Livelihood capital	0.715	0.502	0.676	0.745
~Livelihood capital	0.636	0.556	0.548	0.751
Livelihood outcome	0.664	0.504	0.713	0.849
~Livelihood outcome	0.801	0.640	0.583	0.731
Policy cognition	0.583	0.493	0.658	0.872
~Policy cognition	0.849	0.613	0.617	0.699
Policy identity	0.491	0.489	0.577	0.901
~Policy identity	0.900	0.576	0.673	0.674
Information mastery	0.665	0.414	0.912	0.889
~Information mastery	0.821	0.856	0.399	0.651
Implementation perception	0.578	0.456	0.688	0.850
~Implementation perception	0.810	0.624	0.560	0.675

Note: "~" means that the condition is missing.

4.3. Condition Group Results

4.3.1. Group Analysis of High Satisfaction Conditions

As can be seen from Table 6, the consistency of the solution is 0.969, and the consistency of each group is higher than 0.9. This indicates that 96.9% of the cases meet the high satisfaction level among all the research subjects who meet these six condition groups. The coverage of the solution was 0.568, indicating that these six groupings were able to explain 56.8% of the highly satisfied cases.

Table 6. High satisfaction group analysis.

Variables	Group 1	Group 2	Group 3	Group 4	Group 5	Group 6
Livelihood capital	⊗	⊗	⊗		•	•
Livelihood outcome	•	•		•	⊗	⊗
Policy cognition		•	•	•	•	⊗
Policy identity	⊗	•	•	•	⊗	•
Implementation perception	•		•	•	•	•
Consistency	0.974	0.988	0.991	0.999	0.982	0.990
Coverage	0.301	0.254	0.240	0.322	0.279	0.226
Unique coverage	0.073	0.036	0.009	0.050	0.037	0.023
Consistency of solution	0.969					
Coverage of solution	0.568					

Note: "⊗" means the condition is missing; "•" means the condition exists; the big circle is the core condition; the small circle is the auxiliary condition.

In group 1, the absence of livelihood outcomes and implementation perceptions and livelihood capital played a central role, with the absence of policy identity as an auxiliary condition. The coverage of group 1 was 0.301, indicating that group 1 can account for 30.1% of the high-satisfaction cases; the unique coverage was 0.073, indicating that 7.3% of the high-satisfaction cases can be accounted for by group 1 only. In group 2, the absence of livelihood capital and livelihood outcomes, policy cognitions, and policy identity were all core conditions. The coverage of group 2 was 0.254, indicating that group 2 can account for 25.4% of the high-satisfaction cases; the unique coverage was 0.036, indicating that 3.6% of the high-satisfaction cases can be accounted for by group 2 only. In group 3, the absence of livelihood capital, policy identity, and implementation perceptions played a central role, with policy cognitions playing an auxiliary role. The coverage of group 3 was 0.240, indicating that 24.0% of the high-satisfaction cases could be accounted for by group 3; the unique coverage was 0.009, indicating that 0.9% of the high-satisfaction cases could be accounted for by group 3 only. In group 4, livelihood outcome, policy cognition, policy identity, and implementation perception were all core condition variables, and the coverage of group 4 was 0.322, indicating that group 4 can account for 32.2% of the high satisfaction cases; the unique coverage was 0.050, indicating that 5% of the high satisfaction cases were uniquely accounted for by group 4. In group 5, livelihood capital, the lack of livelihood outcomes, policy cognition, and implementation perception were the core conditions, and the lack of policy identity played an auxiliary role. The coverage of group 5 was 0.279, indicating that group 5 could account for 27.9% of the high satisfaction cases; the unique coverage was 0.037, indicating that 3.7% of the high satisfaction cases could be accounted for by group 5 only. In group 6, livelihood capital, lack of livelihood outcome, policy identity, and implementation perception played a central role, and a lack of policy cognition played a supporting role. The coverage of group 6 was 0.226, indicating that 22.6% of high satisfaction cases could be accounted for by group 6; the unique coverage was 0.023, indicating that 2.3% of high satisfaction cases could be uniquely accounted for by group 6.

4.3.2. Group Analysis of Low Satisfaction Conditions

In the low satisfaction condition group analysis, the combination of each condition variable is divided into group 1 and group 2. As can be seen from Table 7, the consistency and overall agreement of the groups is above the minimum standard of 0.8. Among them, the consistency of the solution is 0.960, indicating that 96.0% of the research respondents who meet the two conditions have low satisfaction with the policy of ecological forest rangers. The coverage of the solution is 0.638, indicating that group 1 and group 2 can account for 63.8% of the low satisfaction cases.

Table 7. Low satisfaction group analysis.

Variables	Group 1	Group 2
Livelihood capital		•
Livelihood outcome		⊗
Policy cognition	⊗	
Information mastery	⊗	⊗
Implementation perception	⊗	⊗
Consistency	0.965	0.968
Coverage	0.598	0.439
Unique coverage	0.199	0.040
Consistency of solution	0.960	
Coverage of the solution	0.638	

Note: "⊗" means the condition is missing; "•" means the condition exists; the big circle is the core condition; the small circle is the auxiliary condition.

In group 1, the lack of information mastery and the lack of implementation perception played a central role, and the lack of policy cognition played an auxiliary role. The coverage of group 1 was 0.598, indicating that group 1 can explain 59.8% of the low satisfaction cases. The unique coverage of 0.199 indicated that 19.9% of the low satisfaction cases can be explained only by group 1. In group 2, the absence of information mastery and the absence of implementation perception also played a central role, with the absence of livelihood capital and livelihood outcome as auxiliary condition variables. The coverage of group 2 was 0.439, indicating that group 2 can account for 43.9% of the low satisfaction cases. The unique coverage of 0.040 indicates that 4.0% of the low satisfaction cases can be accounted for only by group 2.

4.4. Analysis of the Combined Effects of Groups

In the high satisfaction group analysis, the implementation of the perceived condition variable played a central role in all five groups. This can be explained by the fact that the more pronounced the ecological ranger's perception of the implementation of the relevant policy, the higher the satisfaction with the ecological ranger's policy, given the differences in other condition variables. Namely, implementation perception is a key factor for high satisfaction with the ecological forest ranger policy. In general, the more obvious the ecological ranger's perception of the implementation effect of the policy, the more he or she enjoys the "benefits" yielded by the policy; specifically, the ecological ranger feels the improvement in income and living standard due to the policy. The work of ecological forest rangers also improves the local ecological environment and promotes the development of environmental protection in ecologically fragile areas. The improved living conditions and the effectiveness of the ecological forest rangers' work bring them a great sense of "work identity", and they also "feel that their work is very valuable, and they can make their own contribution to the country while enjoying the national preferential policies", and the pride brought on by their work also leads to higher policy satisfaction.

In the high satisfaction condition grouping, the absence of livelihood capital was the core condition variable for group 1, group 2, and group 3, which had an impact on high satisfaction with the policy of ecological forest rangers. The coverage of group 4 was the highest among the six groups, indicating that livelihood outcomes, policy cognition, policy identity, and implementation perception, which were core condition variables, all had a positive effect on high policy satisfaction. In the comparative analysis of group 5 and group 6, ignoring the effect of missing auxiliary conditions, both policy cognition and policy identity had a positive effect on high policy satisfaction, but in terms of the difference in coverage between the two (27.9% for group 5 and 22.6% for group 6), the effect of policy cognition on high policy satisfaction was more significant. As an important ecological poverty alleviation measure, the policy of ecological forest rangers greatly improved the livelihoods of ecological forest rangers. According to the field interviews with ecological forest rangers, more than 90% of the groups had a high degree of agreement with the policy. "They thought that the implementation of the policy of ecological forest rangers was very necessary". Therefore, the policy identification condition variables had no significant impact on the differences in the satisfaction with the ecological forest rangers policy. In terms of policy cognition, through interviews, it was found that the ecological forest rangers' understandings of the formulation, implementation, and implementation effects of relevant policies varied greatly. Due to the differences in the basic situations of the ecological forest rangers themselves, their expectations of the implementation effect of the policy were also different. Generally, the group of ecological forest rangers who had a better understanding of the policy had a higher awareness of the policy, and the greater the degree of support for the continuation of the policy of ecological forest rangers, the higher the degree of policy satisfaction.

In the low satisfaction group analysis, both the lack of information mastery and the lack of implementation perception played a central role, which corresponded to the high-satisfaction group analysis in the previous section, where more "implementation perception" corresponded to higher policy satisfaction. The lack of information mastery as one of the core conditions leading to low satisfaction was shown by the fact that incomplete information mastery of the policy reduced the satisfaction of ecological forest rangers with the policy even if they had high livelihood capital. Since the implementation of the policy of ecological forest rangers, all regions have been selecting qualified ecological forest rangers in strict accordance with the selection procedure. Generally, the education level of the ecological forest rangers on duty was concentrated in primary school level and below. Due to the limitation of cultural knowledge level, there were obstacles to the understanding and interpretation of relevant policies of ecological forest rangers; this made it easy to reduce recognition of work and difficult to obtain "work happiness". Therefore, the satisfaction with the policy was low.

The lack of policy cognition also became a supporting condition for low satisfaction in group 1. Group 2 showed that even if ecological forest rangers had a high amount of livelihood capital, the lack of information mastery, the lack of implementation perceptions, and the lack of supporting livelihood outcomes all contributed to the low policy satisfaction of ecological forest rangers. Through the comparison between group 1 and group 2, it was found that under the premise that group 2 had a higher amount of livelihood capital, the coverage of configuration 1 was 15.9% higher than that of configuration 2 due to the lack of policy awareness and the lack of auxiliary conditions for livelihood results. This showed that although the lack of policy cognition and the lack of livelihood results had a certain impact on the low satisfaction with the policy, the impact of the lack of policy cognition was significantly higher than the impact of the lack of livelihood results. This also corresponded with the previous research results of high satisfaction configuration analysis.

5. Discussion

Qianshan City of Anhui Province is located in the southeast foot of Dabie Mountain Area, one of the 14 contiguous poverty-stricken areas in China, with backward economic development and poor ecological environment. Qianshan City once developed its economy at the expense of the ecological environment, making environmental problems increasingly prominent, which affected the poverty reduction of the poor population, severely restricted the economic development of Qianshan City, and plunged the local area into a vicious cycle of ecological poverty. With the organic unity of ecological environmental protection and economic and social development as the starting point, the ecological forest ranger policy plays an important role in helping Qianshan City to get rid of the vicious circle of ecological poverty. Additionally, it plays a role in eliminating absolute poverty by constructing a long-term poverty relief mechanism, taking a sustainable ecological poverty alleviation path [40]. With the gradual promotion of the ecological forest ranger policy, the problem of absolute poverty has been solved, and now we have entered the "post-poverty alleviation era" [41], which focuses on consolidating the results of poverty eradication and preventing the return to poverty. It is particularly important to further consolidate and optimize the ecological forest ranger policy for the implementation of the ecological poverty alleviation mechanism. Therefore, it is of great significance to analyze the factors influencing the satisfaction with the implementation of the ecological forest ranger policy, and further improve the sustainable livelihood ability of the ecological forest ranger group in China.

The degree of information mastery is an important factor affecting ecological forest rangers' policy satisfaction, and information asymmetry theory suggests that the degree of information mastery reflects differences in the ability of different economic individuals to obtain information [42] and has an important impact on individual decision making. The degree of information mastery refers, first of all, to the ecological ranger's understanding and cognition of the ecological forest ranger policy. The deeper their cognition of the policy, the better the understanding of the essence of the policy, which helps them to evaluate the services of the relevant departments more rationally, leading to higher policy satisfaction accordingly. For ecological forest rangers, their main income is not only wage income but also income from agricultural production and operation. Agriculture, as a weak industry, has greater natural and market risks [43]. The higher the information mastery of ecological forest rangers, the better it is for them to avoid market and other risks, obtain more agricultural income, and improve their living standards, thus positively influencing policy satisfaction.

Another important factor affecting ecological forest rangers' policy satisfaction is the perception of policy implementation. The degree of ecological forest rangers' perception of policy implementation is closely related to their own needs, expectations, and social comparisons [44], which mainly include perceived needs, perceived processes, and perceived services. Ecological forest rangers, with their single livelihood strategy and generally low standard of living, have a high degree of demand for ecological compensation income. According to marginal utility theory, when the compensation standard has not yet reached its extreme value, the marginal utility of ecological forest rangers increases as the compensation standard increases [45], and policy satisfaction increases accordingly. In addition, if the policy of ecological forest rangers takes into account the actual needs of different types of ecological forest rangers and adopts corresponding strategies, the policy satisfaction will also be improved. Process perception affects the degree of familiarity of the ecological forest rangers with policy-related information, and whether they can participate in the whole process of the implementation of the ecological forest ranger policies has a great impact on the satisfaction with the policies. The more open and transparent the implementation process of ecological forest rangers is, the fairer and more objective the government is perceived to be [46]. Ecological forest rangers can also play a supervisory role and improve policy satisfaction. The perception of service is mainly reflected in the timeliness and convenience of the salary payment of ecological forest rangers, that is, the cash situation of funds. The cashing of funds affects the ecological forest rangers' trust in the policy, and the timely release of compensation funds is conducive to improving policy satisfaction.

In summary, according to the varied importance of the influences of various conditions on policy satisfaction, enhancing the information grasp degree of ecological forest rangers and improving the perception level of policy implementation is the best strategy to effectively improve the policy satisfaction of ecological forest rangers. On the one hand, it is necessary to strengthen the publicity of the ecological forest ranger policy, popularize the knowledge related to the ecological forest ranger policy, and improve the ecological protection awareness and information mastering ability of ecological forest rangers. This will further promote the degree of support for the continuation of the ecological forest ranger policy, and increase ecological forest rangers' enthusiasm regarding participation in the national ecological poverty alleviation cause. On the other hand, relevant national departments should establish and improve the monitoring mechanism of ecological forest rangers' livelihood [47], effectively understand the degree of ecological forest rangers' perception of policy implementation, and establish a sound dynamic adjustment mechanism. Through regular monitoring of people's livelihood, timely feedback of ecological forest rangers on relevant policies should be obtained, and targeted improvements should be made to effectively solve related problems that lead to low satisfaction with ecological forest ranger policies, and the implementation efficiency of ecological forest rangers policies should also be improved.

6. Conclusions

Based on research data from Qianshan City, Anhui Province, we used the fuzzy set qualitative comparative analysis method to investigate the complex relationships between "livelihood capital", "livelihood outcomes", "policy perceptions", and "implementation perceptions" from a histological perspective. In addition, we investigated "policy identity", "information acquisition", and "implementation perception", and analyzed the complex relationship between the six variables and policy satisfaction in the discussion section. The findings are as follows.

In general, the implementation effect of the ecological ranger policy is recognized by most audiences, but there is still much room for improvement. The necessity test analysis showed that ecological rangers with low policy satisfaction usually lacked a sense of identification with the policy. Ecological rangers with high policy satisfaction had a better ability to grasp information.

In the conditional grouping analysis of high policy satisfaction, implementation perception is the core conditional variable, which plays a key role in high policy satisfaction. In the comparative analysis of other important core condition variables, policy perception and policy agreement, it was found that policy perception had a more significant effect on high policy satisfaction. There are two conditional groups that lead to low policy satisfaction among ecological rangers: policy perception*information mastery*implementation perception and livelihood capital*livelihood outcome*information mastery*implementation perception, among which the lack of information mastery and implementation perception play a central role, further validating the importance of implementation perception for policy satisfaction. The absence of policy perceptions and the absence of livelihood outcomes, on the other hand, were the auxiliary condition variables, and the comparative analysis revealed that the effect of the absence of policy perceptions was significantly higher than the effect of the absence of livelihood outcomes in the low satisfaction group analysis, corresponding to the findings in the high satisfaction group analysis.

Undeniably, there are still many shortcomings in the article. The subjective satisfaction of ecological rangers with relevant policies can lead to different perceived outcomes due to individual differences, and the article uses fuzzy set qualitative comparative analysis for the analysis of satisfaction influencing factors, which makes it difficult to ensure the complete objectivity of the evaluations made. In a subsequent study, the macro-level livelihood monitoring data should be combined to corroborate and supplement the results of the implementation of the ecological ranger policy, so as to improve the satisfaction and policy adaptability of ecological rangers in a targeted manner.

Supplementary Materials: The following supporting information can be downloaded at: https://www.mdpi.com/article/10.3390/su15086743/s1.

Author Contributions: Conceptualization, Y.Z. and X.W.; methodology, Y.Z. and S.W.; software, S.W.; validation, Y.Z. and S.W.; formal analysis, Y.Z.; investigation, Y.Z.; resources, H.Z. and Y.Z.; data curation, S.W.; writing—original draft preparation, Y.Z. and X.W.; writing—review and editing, Y.Z., X.W. and H.Z.; visualization, S.W.; supervision, S.W.; project administration, S.W. and H.Z.; funding acquisition, H.Z. All authors have read and agreed to the published version of the manuscript.

Funding: This research was funded by an innovative training program for college students at Northeast Forestry University "Research on the livelihood strategies of ecological forest rangers after poverty alleviation and their impact on poverty alleviation stability—based on the survey conducted in Anqing City, Anhui Province" (No. 202110225153) and Northeast Forestry University's undergraduate innovation project (No. 202210225022).

Institutional Review Board Statement: Not applicable.

Informed Consent Statement: Not applicable.

Data Availability Statement: No new data were created in this study. Data sharing is not applicable to this research.

Conflicts of Interest: The authors declare no conflict of interest.

References

1. Li, K.; Lloyd, B.; Liang, X.-J.; Wei, Y.-M. Energy poor or fuel poor: What are the differences? *Energy Policy* **2014**, *68*, 476–481. [CrossRef]
2. Pradhan, P.; Costa, L.; Rybski, D.; Lucht, W.; Kropp, J. A systematic study of sustainable development goal (SDG) interactions. *Earth's Future* **2017**, *5*, 1169–1179. [CrossRef]
3. Duan, M.; Zhang, C.; Xiao, H.; Liu, J.; Li, Z.; Li, W. Practice and thinking of poverty alleviation mode of ecological fishery poverty reduction work in Enshi city, Hubei province and Liupanshui city, Guizhou province, China. *Bull. Chin. Acad. Sci.* **2019**, *34*, 114–120. Available online: https://www.webofscience.com/wos/alldb/summary/b8402be5-c565-41bb-9bd1-45ae31bc4c34-69ea7df3/relevance/1 (accessed on 11 April 2023).
4. Srivastava, S.K. Making a technological choice for disaster management and poverty alleviation in India. *Disasters* **2009**, *33*, 58–81. [CrossRef] [PubMed]
5. Qin, B.; Yu, Y.; Ge, L.; Yang, L.; Guo, Y. Does eco-compensation alleviate rural poverty? New evidence from national key ecological function areas in China. *Int. J. Environ. Res. Public Health* **2022**, *19*, 10899. [CrossRef] [PubMed]
6. Zhang, K.; Dearing, J.A.; Dawson, T.P.; Dong, X.; Yang, X.; Zhang, W. Poverty alleviation strategies in Eastern China lead to critical ecological dynamics. *Sci. Total Environ.* **2015**, *506*, 164–181. [CrossRef]
7. Bulte, E.H.; Lipper, L.; Stringer, R.; Zilberman, D. Payments for ecosystem services and poverty reduction: Concepts, issues, and empirical perspectives. *Environ. Dev. Econ.* **2008**, *13*, 245–254. [CrossRef]
8. Li, D.; Xu, D.; Wang, Z.; Ding, X.; Song, A. Ecological compensation for desertification control: A review. *J. Geogr. Sci.* **2018**, *28*, 367–384. [CrossRef]
9. Yan, Z.; Wei, F.; Deng, X.; Li, C.; He, Q.; Qi, Y. Does the policy of ecological forest rangers (EFRs) for the impoverished populations reduce forest disasters?—Empirical evidence from China. *Forests* **2022**, *13*, 80. [CrossRef]
10. Liu, C.-L.; Xu, M.; Zhou, K.-Y.; Zeng, F.-C.; Liu, Z.-M. Coupling development mechanism and typical ways of targeted poverty alleviation and eco-compensation in China: Case analysis based on forestry. *J. Nat. Resour.* **2019**, *34*, 989–1002. Available online: https://www.webofscience.com/wos/alldb/summary/99e66f02-ce7f-4627-90c7-b16d7034b150-69eaa3d1/relevance/1 (accessed on 18 March 2023).
11. Wang, Y.; Wang, D.; Zhao, R. The effectiveness of the ecological forest rangers policy in southwest China. *Forests* **2021**, *12*, 746. [CrossRef]
12. Wang, W.; Lan, Y.; Wang, X. Impact of livelihood capital endowment on poverty alleviation of households under rural land consolidation. *Land Use Policy* **2021**, *109*, 105608. [CrossRef]
13. Qiu, S.; Jin, L. Forest ecological benefit compensation, income level and policy satisfaction: Empirical analysis based on the survey data of farmers in ecological protection red line area. *Resour. Environ. Yangtze Basin* **2022**, *31*, 234–243. Available online: https://www.webofscience.com/wos/alldb/summary/4fa7657e-eea8-4df4-b8be-59ce2196b7ff-69eaadad/relevance/1 (accessed on 25 March 2023).
14. Shir, N.; Nikolaev, B.N.; Wincent, J. Entrepreneurship and well-being: The role of psychological autonomy, competence, and relatedness. *J. Bus. Ventur.* **2019**, *34*, 105875. [CrossRef]
15. Ziaei, H.; Katibeh, M.; Eskandari, A.; Mirzadeh, M.; Rabbanikhah, Z.; Javadi, M.A. Determinants of patient satisfaction with ophthalmic services. *BMC Res. Notes* **2011**, *4*, 7. [CrossRef]
16. Li, J.; Wang, Z.; Cheng, X.; Shuai, J.; Shuai, C.; Liu, J. Has solar PV achieved the national poverty alleviation goals? Empirical evidence from the performances of 52 villages in rural China. *Energy* **2020**, *201*, 117631. [CrossRef]

17. Pang, J.; Jin, L.; Yang, Y.; Li, H.; Chu, Z.; Ding, F. Policy cognition, household income and farmers' satisfaction: Evidence from a wetland ecological compensation project in the Poyang lake area at the micro level. *Sustainability* **2022**, *14*, 955. [CrossRef]
18. Yang, D.; Luan, W.; Yang, J.; Xue, B.; Zhang, X.; Wang, H.; Pian, F. The contribution of data-driven poverty alleviation funds in achieving mid-21st-century multidimensional poverty alleviation planning. *Humanit. Soc. Sci. Commun.* **2022**, *9*, 179. [CrossRef]
19. Gakou-Kakeu, J.; di Gregorio, M.; Paavola, J.; Sonwa, D.J. REDD plus policy implementation and institutional interplay: Evidence from three pilot projects in Cameroon. *For. Policy Econ.* **2022**, *135*, 102642. [CrossRef]
20. Mitonga-Monga, J. Ethical climate influences on employee commitment through job satisfaction in a transport sector industry. *J. Psychol. Afr.* **2018**, *28*, 15–20. [CrossRef]
21. Cao, M.; Xu, D.; Xie, F.; Liu, E.; Liu, S. The influence factors analysis of households' poverty vulnerability in southwest ethnic areas of China based on the hierarchical linear model: A case study of liangshan yi autonomous prefecture. *Appl. Geogr.* **2016**, *66*, 144–152. [CrossRef]
22. Li, M.; Zhao, P.; Wu, L.; Chen, K. Effects of value perception, environmental regulation and their interaction on the improvement of herdsmen's grassland ecological policy satisfaction. *Int. J. Environ. Res. Public Health* **2021**, *18*, 3078. [CrossRef] [PubMed]
23. Gabriel, A.S.; Campbell, J.T.; Djurdjevic, E.; Johnson, R.E.; Rosen, C.C. Fuzzy profiles: Comparing and contrasting latent profile analysis and fuzzy set qualitative comparative analysis for person-centered research. *Organ. Res. Methods* **2018**, *21*, 877–904. [CrossRef]
24. Fiss, P.C. Building better causal theories: A fuzzy set approach to typologies in organization research. *Acad. Manag. J.* **2011**, *54*, 393–420. [CrossRef]
25. Barnes, C.; Claus, R.; Driessen, P.; Dos Santos, M.J.F.; George, M.A.; van Laerhoven, F.; Dos Santos, M.F. Uniting forest and livelihood outcomes? Analyzing external actor interventions in sustainable livelihoods in a community forest management context. *Int. J. Commons* **2017**, *11*, 532–571. [CrossRef]
26. Xie, J.; Cai, Y. The impacts of farmer livelihood endowment on participation effectiveness in farmland conservation compensation policy: Chengdu and Suzhou as typical innovational practice areas. *Resour. Sci.* **2016**, *38*, 2082–2094. Available online: https://www.webofscience.com/wos/alldb/summary/1f9d2d4c-041b-4d3a-b899-6ead5ae1cfcd-69eab973/relevance/1 (accessed on 4 March 2023).
27. Victora, C.G.; Adair, L.; Fall, C. Maternal, Maternal and child undernutrition: Consequences for adult health and human capital. *Lancet* **2008**, *371*, 340–357. Available online: https://www.webofscience.com/wos/alldb/summary/88799520-f05c-4f7d-85f9-99450f7cdbf8-69eab472/relevance/1 (accessed on 4 March 2023). [CrossRef]
28. Schure, J.; Ingram, V.; Sakho-Jimbira, M.S.; Levang, P.; Wiersum, K.F. Formalisation of charcoal value chains and livelihood outcomes in central- and west Africa. *Energy Sustain. Dev.* **2013**, *17*, 95–105. [CrossRef]
29. Goldberg, A.C.; Robinson, J.G.; Cromwell, W.C.; Ross, J.L.; Ziajka, P.E. Future issues, public policy, and public awareness of familial hypercholesterolemias: Recommendations from the national lipid association expert panel on familial hypercholesterolemia. *J. Clin. Lipidol.* **2011**, *5*, S46–S51. [CrossRef]
30. Lee, M.J.; Hwang, S.O.; Cha, K.C.; Cho, G.C.; Yang, H.J.; Rho, T.H. Influence of nationwide policy on citizens' awareness and willingness to perform bystander cardiopulmonary resuscitation. *Resuscitation* **2013**, *84*, 889–894. [CrossRef]
31. Ansell, C.; Sørensen, E.; Torfing, J. Improving policy implementation through collaborative policymaking. *Policy Politics* **2017**, *45*, 467–486. [CrossRef]
32. Davis, S.; Murphy, S.A.; Watkins, J. Work changes and employee perceptions of co-worker flexible work policy use: A moderated mediation study. *Empl. Relat. Int. J.* **2022**, *45*, 516–534. [CrossRef]
33. Chen, S.; Ji, Y.-S. The effects of employees' job satisfaction on work performance in social organization: Focusing on psychological capital and organizational identity as a mediator. *Korean J. Bus. Adm.* **2021**, *34*, 1955–1978. Available online: https://www.webofscience.com/wos/alldb/summary/3e1b6243-a7f7-4959-ab92-f0dd6db01f12-69ea6946/relevance/1 (accessed on 25 March 2023). [CrossRef]
34. Pradhan, N.S.; Fu, Y.; Zhang, L.; Yang, Y. Farmers' perception of effective drought policy implementation: A case study of 2009–2010 drought in yunnan province, China. *Land Use Policy* **2017**, *67*, 48–56. [CrossRef]
35. Llopis-Albert, C.; Merigó, J.M.; Xu, Y.; Liao, H. Application of fuzzy set/qualitative comparative analysis to public participation projects in support of the eu water framework directive. *Water Environ. Res.* **2018**, *90*, 74–83. [CrossRef]
36. Hug, S. Qualitative comparative analysis: How inductive use and measurement error lead to problematic inference. *Political Anal.* **2013**, *21*, 252–265. [CrossRef]
37. Marx, A.; Rihoux, B.; Ragin, C. The origins, development, and application of qualitative comparative analysis: The first 25 years. *Eur. Political Sci. Rev.* **2014**, *6*, 115–142. [CrossRef]
38. Schneider, C.Q.; Wagemann, C. Reducing complexity in qualitative comparative analysis (QCA): Remote and proximate factors and the consolidation of democracy. *Eur. J. Political Res.* **2006**, *45*, 751–786. [CrossRef]
39. Wu, J.; Zhang, Q. Multicriteria decision making method based on intuitionistic fuzzy weighted entropy. *Expert Syst. Appl.* **2011**, *38*, 916–922. [CrossRef]
40. Liu, Y.; Chen, S.; Zhao, R. Current Research and Future Prospect of Long-term Mechanism of Forestry Industry Based Poverty Alleviation. *World For. Res.* **2021**, *34*, 54–59. Available online: https://www.webofscience.com/wos/alldb/summary/20aeb043-0616-46fd-b2d1-8f116db91dfb-75857eba/relevance/1 (accessed on 18 March 2023).

41. Luo, Y.; Shi, Z.; Guo, D.; He, P. Toward universal health coverage: Regional inequalities and potential solutions for alleviating catastrophic health expenditure in the post-poverty elimination era of China. *Int. J. Health Policy Manag.* **2023**, *12*, 1–9. [CrossRef]
42. Ebell, M.H.; Shaughnessy, A. Information mastery: Integrating continuing medical education with the information needs of clinicians. *J. Contin. Educ. Health Prof.* **2003**, *23* (Suppl. 1), S53–S62. [CrossRef] [PubMed]
43. Dunn, A.; Julien, G.; Ernst, W.; Cook, A.; Doe, K.; Jackman, P. Evaluation of buffer zone effectiveness in mitigating the risks associated with agricultural runoff in Prince Edward Island. *Sci. Total Environ.* **2011**, *409*, 868–882. [CrossRef] [PubMed]
44. Tuytens, M.; Devos, G. Teachers' perception of the new teacher evaluation policy: A validity study of the policy characteristics scale. *Teach. Teach. Educ.* **2009**, *25*, 924–930. [CrossRef]
45. Festjens, A.; Janiszewski, C. The value of time. *J. Consum. Res.* **2015**, *42*, 178–195. [CrossRef]
46. Bertot, J.C.; Jaeger, P.T.; Grimes, J.M. Using icts to create a culture of transparency: E-government and social media as openness and anti-corruption tools for societies. *Gov. Inf. Q.* **2010**, *27*, 264–271. [CrossRef]
47. Bhattarai, S.; Regmi, B.R.; Pant, B.; Uprety, D.R.; Maraseni, T. Sustaining ecosystem based adaptation: The lessons from policy and practices in Nepal. *Land Use Policy* **2021**, *104*, 105391. [CrossRef]

sustainability

MDPI

Article

Current Impediments for New England DOTs to Transition to Sustainable Roadside Practices for Strengthening Pollinator Habitats and Health

John Campanelli [1,*], Yulia A. Kuzovkina [1,*] and Samuel Kocurek [2,3]

[1] Department of Plant Science and Landscape Architecture, University of Connecticut, 1376 Storrs Road Unit 4067, Storrs, CT 06269, USA
[2] Department of Mathematics, University of Connecticut, 341 Mansfield Rd., Storrs, CT 06269, USA
[3] Department of Environmental Science, University of Connecticut, 1376 Storrs Road Unit 4087, Storrs, CT 06269, USA
* Correspondence: john.campanelli@uconn.edu (J.C.); jkuzovkina@uconn.edu (Y.A.K.); Tel.: +1-860-486-3438 (Y.A.K.)

Abstract: Government agencies whose work impacts the environment have sought more sustainable policy changes that require the use of native plants for revegetation. However, New England DOTs have encountered hurdles in attempts to transition from using cool-season turfgrass to implementing the more complicated management practices required to establish native plant communities. Two of the most significant barriers have been the slow transition to mainly using native plants for revegetation and the lack of genetically appropriate ecotypic native plant material (ENPM) produced in the Northeast. Growers of ENPM have resisted increasing production because they struggle to gauge demand due to substantial communication gaps between local producers and end users. Therefore, we organized focus groups with New England state DOTs, Departments of Environmental Protection, and Fish and Wildlife Services to determine their demand for ENPM and to explore their relationships with producers. We determined that the subcontracting of DOTs hinders their ability to develop relationships with producers because it leaves seed source selection in the hands of third parties. Another impediment was found to be the resistance of some maintenance departments to adopt sustainable management practices, such as reduced mowing. In summary, we determined that DOTs would benefit from establishing more direct relationships with producers to communicate their ENPM needs.

Keywords: roadside vegetation; reduced mowing; pollinators; ecotypic seed; native plant material

Citation: Campanelli, J.; Kuzovkina, Y.A.; Kocurek, S. Current Impediments for New England DOTs to Transition to Sustainable Roadside Practices for Strengthening Pollinator Habitats and Health. *Sustainability* **2023**, *15*, 3639. https://doi.org/10.3390/su15043639

Academic Editor: Aaron K. Hoshide

Received: 16 January 2023
Revised: 10 February 2023
Accepted: 14 February 2023
Published: 16 February 2023

1. Introduction

In an era of shrinking government resources and a growing awareness of the environmental impacts of various practices, government, academic, and nonprofit organizations whose work impacts ecosystems and ecological conservation have sought alternative management practices that are more sustainable than those previously implemented. In particular, ecologists have realized that transportation and right-of-way corridors represent some of the most extensive networks of linear habitats on Earth, collectively constituting significant opportunities for connecting previously fragmented habitats and landscapes [1,2]. The U.S. National Highway System in the contiguous U.S. consists of more than 262,400 km (163,055 miles) of roads bordered by nearly 1.38 million hectares (3.41 million acres) of unpaved, vegetated land along rights-of-way [3]. Research estimates that roads ecologically impact between 15% and 20% of the US [4], considering the extent to which roadsides impact adjacent land, whether in the form of noise and chemical pollution, disruptions of hydrological cycles and water quality, or the introduction of noxious and invasive weeds. These extensive corridors provide important ecosystem services, such as runoff filtration, carbon sequestration, air quality improvement, and aesthetics [5–7], and they support a diversity of wildlife by providing shelter, food, and breeding opportunities for many species including presently threatened pollinators [8,9].

As a result, over the last few decades, policy changes on the federal and local levels have sought to require the use of native plants for revegetation in situations where introduced and invasive species were previously used. In 1999, President Clinton released Executive Order 13112—Invasive Species [10], which established the National Invasive Species Council and ordered the preparation of a National Invasive Species Management Plan that would review federal policies to prevent the introduction and spread of invasive species, such as along roadsides. In 2016, President Obama built upon this Executive Order with Executive Order 13751—Safeguarding the Nation from the Impacts of Invasive Species [11], which established the National Invasive Species Council. In addition, in 2014, President Obama released Presidential Memorandum—Creating a Federal Strategy to Promote the Health of Honey Bees and Other Pollinators [12], which established the Pollinator Health Task Force and National Pollinator Health Strategy to review governmental policies that impact pollinator health, including increasing and improving pollinator habitats along rights-of-ways.

More recently, state and local governments have passed legislation to increase the use of native plant material (NPM) to benefit pollinators and wildlife while also mitigating the effects of climate change. In 2016, Connecticut passed Public Act No. 16–17: An Act Concerning Pollinator Health [13], a portion of which encouraged the increased use of native plants to benefit pollinator habitats and health. On November 8, 2022, voters in New York state passed the $4.2 billion Environmental Bond Act—Proposal 1, which includes $300 million for open space land conservation. On the local level, municipalities are passing measures that further the usage of native plants. For example, in May 2021, Somerville, Massachusetts, passed a Native Planting Ordinance that requires the use of only native plants and trees when planting on city-owned parks, open spaces, and streets.

Native plants provide greater ecological benefits than introduced and exotic plant species due to several factors. Because native plants have evolved with local climate and social conditions, they generally require fewer resource inputs and less maintenance after establishment [14,15]. Compared with introduced cool-season turfgrass (ICST), which requires more frequent mowing regimens, native warm-season grasses (NWSGs)—the backbone of most biodiverse native seed mixes—can be mowed less frequently, reducing maintenance costs [16]. NWSGs have longer life spans and deeper, more extensive root systems than ICST [17], thus improving long-term slope stability and increasing regional biodiversity [18]. Native plant communities (NPCs), consisting of native grasses, forbs, and woody plants, provide long-term defense against invasive and noxious weeds. In addition, research has found that native forbs and flowering shrubs, which have co-evolved with native arthropods, provide superior nutrition for pollinators and other arthropods compared with non-native flowering plants [19,20].

Considering these ecological benefits and concern about adapting to possible droughts resulting from climate change, the New England Transportation Consortium (NETC), a research cooperative funded by all six New England state Departments of Transportation (DOTs), commissioned a study in 2013 to find the most affordable, reliable, and expeditious methods for establishing biodiverse NPCs along New England roadsides. The findings from this research were compiled in a manual, Effective Establishment of Native Grasses on Roadsides in New England (EENG) [21], which guided state DOTs' efforts to transform their roadside revegetation management practices. The NETC realized that transitioning to revegetation practices that use shorter-growing species of New England NWSGs would provide local governments with revegetation choices that would not only require less long-term maintenance than for ICST but also, because of their warm-season growth cycles, be better able to survive droughts. Since these NWSG species grow to between 0.3 and 0.9 meter (1 and 3 feet) tall at maturity and maintain strong vertical habits, they require less frequent mowing regimens—as few as once a year or even once every other year. Less frequent mowing schedules not only save money on labor, fuel, and machine wear but also subject workers to less exposure to dangerous work conditions, thus contributing to healthcare, disability, and insurance savings [22–24]. In addition, since these NWSGs are native to New England, they have adapted to the regional climate and soil profiles. Their deeper root systems enable more rapid recovery after frequent disturbances—an important quality considering the need for vehicles to occasionally pull off onto the side of the road [19,23].

Despite the ecological, environmental, and economic benefits NPCs can provide to New England roadsides, transitioning from established long-standing practices would likely encounter significant bureaucratic hurdles. Many past attempts by other state and federal agencies to integrate native species into roadside revegetation efforts have faltered or failed [25–27], resulting from multiple missteps, including the poor integration of roadside revegetation strategies in infrastructure planning, a lack of cooperation and information sharing among agencies involved in road construction, and the use of cheaper, faster-establishing exotic species misidentified as native species [28,29]. In addition to these bureaucratic hurdles, DOTs would likely encounter technical difficulties with the establishment of biodiverse NPCs. ICST establishes far more rapidly, taking as little as 4 to 6 weeks compared with NWSGs, which often take 3 to 5 seasons to form a dense, stable stand. Despite requiring less long-term maintenance, the initial installation of NPCs demands more laborious and time-intensive effort than monocultural ICST. Finally, any changes in maintenance practices require time for DOT bureaucracies to adapt to new regimes [24].

One of the other major concerns the authors of *EENG* found was the lack of ecotypic native seed supplies for the New England region. Ecotypic native plants are native plant species that have adapted to the climatic conditions of a region and share the genetic markers of local native plant species. To maintain their genetic composition, local ecotypes are grown within the specific seed transfer zones within which they originated [30]. There are two definitions of seed transfer zones. Empirical seed zones are areas within which plant materials can be transferred with minimal risk of poorly adapting to their new location, and they are developed by combining species specific information on local adaptation with environmental information. Provisional seed transfer zones are also believed to enable the transferability of native plant materials, but they are developed using climatic and other data, not species-specific information. [31]

Research has shown the importance of prioritizing the use of local ecotypes over non-local ones for restoration purposes [32–34]. First, native ecotypes have evolved to adapt to local environmental conditions. Therefore, they are more likely to successfully establish themselves and persist. For example, a red maple that has evolved in the deep South, where winters are milder and summers are more humid, would not fare as well in more northern climates, where winters are harsher and soil may be more acidic [30]. Second, native ecotypes co-evolve with local pollinators and wildlife populations, which depend on NPCs for food, nesting, and shelter. Research has shown that non-local ecotypes frequently have different phenological cycles than local ones. This difference can result in the misalignment of the presentation of foraging resources with the emergence of native pollinator populations [30,34]. Third, ecotypes are more resistant to disease and local herbivores [35]. Lastly, introduced genotypes could be established to such an extent that they become problematic. The interaction between introduced plants from remote provenances and local native populations could result in species interbreeding, which may result in problems such as founder effects, genetic swamping and outbreeding depression [36].

The chronic commercial shortage of locally harvested and grown ecotypic native bulk seed in New England presents a serious challenge. Therefore, several stakeholders whose work involves NPM in the Northeast have made coordinated efforts to rectify this lack of supply. In 2020, the Connecticut Northeast Organic Farmers Association (CT NOFA) initiated the Ecotype Project, which involved recruiting a collective of organic farmers to grow founders' plots to increase ecotypic seed harvested in ecoregion 59 (https://www.eco59.com/ Accessed 9 February 2023) by botanists using Seeds of Success protocols. This effort led to the creation of a commercial seed production company, eco59. While eco59 joins the Wild Seed Project in Maine (https://wildseedproject.net Accessed 9 February 2023)) as sources of ecotypic seed harvested and grown in two different seed transfer zones, both do not yet grow seed in volumes large enough to meet the needs of DOTs or other government agencies working with NPM.

Other stakeholders working with NPM have started efforts to strengthen the Northeast NPM supply chain network. In particular, Eve Allen, a master's in City Planning candidate at the Massachusetts Institute of Technology with financial support from the Ecological Health Network, conducted research on developing practicable solutions to increase native plant diversity in urban ecological restoration projects [37]. Drawing upon literature from the fields of supply chain management (SCM) and social network analysis (SNA), her research mainly focused on analyzing the current state of and best methods to strengthen the NPM supply chain network (SCN) in the Northeast. Allen approached her research with the belief that it is important to analyze and describe the network of actors who engage in NPM SCNs—government agencies, academic institutions, nonprofit organizations, private companies, and local citizens—to understand how to optimize processes that ensure adequate supplies of NPM to meet current and future demands. Therefore, she reached out to these stakeholders in an effort to gain feedback as to how best to foster more effective collaboration. It was through this outreach that our team eventually became engaged with NPM SCN stakeholders.

In 2021, the NETC released a request for proposal that sought solutions for the lack of ecotypic seed supply in New England and awarded the grant to the authors of this article for a proposal titled "Initiating Seed Production for Effective Establishment of Native Plants on Roadsides in New England." The research team defined two overall objectives: promote ecotypic seed production for the effective establishment of native plants on roadsides in New England and expand management practices that result in the proliferation of existing native plant communities along roadsides in New England.

2. Roundtable Discussions

Over the course of our team's research on native plants, we developed close working relationships with stakeholders working with native plant material (NPM), whom we decided to leverage. We collaborated with Ms. Allen to organize two virtual roundtables of 11 members each to share perspectives, facilitate connections, and explore and encourage opportunities for collaboration. Stakeholders included end users, producers, and intermediaries in the Northeast NPM supply chain network (SCN) representing academia, Departments of Transportation (DOTs), Departments of Environmental Protection (DEPs), nurseries, farmers, landscape designers, and conservation organizations. The organizers defined producers, end users, and intermediaries as follows:

PRODUCERS: Any actors who play a key role in any of the steps related to the collection, storage, propagation, or distribution of native plant materials, e.g., farmers, nurseries, and growers.

END USERS: Any actor who plays key role in the planning, design, implementation, or monitoring and maintenance of restorative activities or are the beneficiaries of those activities, e.g., landscape designers, conservation organizations, residential and commercial consumers, and governmental agencies such as DOTs and DEPs.

INTERMEDIARIES: Any actor who supports or could support the activities of two or more actors who are engaged in SCN processes or in the planning, design, implementation, or monitoring and maintenance of restorative activities. Sometimes, intermediary actors such as academics, botanists, and conservation organizations are not directly engaged in SCN processes or restorative activities but have unique expertise and resources that could be leveraged to advance restoration outcomes and strengthen the supply chain.

The roundtable discussions helped us to realize that a substantial communication gap exists between producers and end users, especially in the sector of the production of ecotypic NPM. Through efforts such as the Ecotype Project, an initiative started by the

Connecticut chapter of the Northeast Organic Farmers Association, farmers and nurseries in the region have begun raising awareness of the benefits and have encouraged production of the use of ecotypic NPM. However, small-scale ecotypic native seed companies such as eco59 in Connecticut and the Wild Seed Project in Maine had difficulty recruiting farmers to grow seed on plots larger than founders' fields because they were uncertain of the demand that exists in the market. Without this knowledge, growers were reluctant to make the investments needed to expand the number and amount of species grown. To bridge this gap, the team identified six substantial next steps to address regional needs. These included:

1. A Regional Strategy Document: This document would draw inspiration from the National Seed Strategy and provide regionally applicable strategies for strengthening seed and plant material supply chains in the Northeast.
2. Impact Map: This web-based map would show the georeferenced locations of supply chain end users, producers, and intermediaries.
3. Regional Seed and Plant Material Needs Directory: An online portal or marketplace where end users and other seed and plant material buyers could share their confirmed and projected seed and plant material needs with producers.
4. Regional Lobby Organization: A formalized lobbying group that brings together conservation organizations, government agencies, farmers, and other stakeholders so they can team up together and lobby regional state governments to direct funds toward these supply-chain-strengthening efforts and to coordinate policy around such issues as seed quality regulations.
5. Regional quasi-governmental organization or regional non-governmental organization: The formation of a governmental or non-governmental agency to step in and play a coordinating role, namely, the regional state Departments of Environmental Protection and organizations such as the DOT's New England Transportation Consortium (NETC), where they could pool their research dollars to find regional solutions related to our issues.
6. More roundtables, working groups, and meetings to bring together end users and producers: The generation of more opportunities to bring the largest end users such as DOTs, DEPs, and conservation organizations together with growers so they can better understand how each of their organizations work so they can better meet each other's needs.

From the roundtable discussions, the authors concluded that DOTs and other governmental agencies, such as Departments of Environmental Protection (DEPs) and Fish and Wildlife Services (FWSs), are among the most substantial end users of NPM, and determining the existing demand for these regional agencies would go far in closing the information gap between end users and producers. Therefore, the authors decided to organize focus groups with regional DOTs, DEPs, and FWSs to achieve three overriding goals:

1. Estimate the demand for NPM that exists for these agencies.
2. Explore the lines of communication that exist between these agencies and producers.
3. Determine the progress DOTs have made to transition to revegetating roadsides with NPCs.

3. Materials and Methods

We conducted virtual focus groups to interview participants and collect information and data from the state Department of Transportation (DOT), Department of Environmental Protection (DEP), and Fish and Wildlife Service (FWS) managers concerning demand for and use of NPM during the summer of 2022. Focus group data were used to inform recommendations provided by our team to state DOTs regarding the best methods for transitioning to roadside revegetation using native plant material (NPM).

3.1. Focus Groups

Our experience has shown that the most effective means to collect information about managers' experiences is to use an open phenomenological approach that focuses on each participant's individual experience navigating bureaucratic work structures [38]. We chose the semi-structured, life-world approach afforded by the focus group format [39] because we have found that this technique elicits the most comprehensive set of data and encourages an interplay of opinions among group participants [40]. Focus groups depend on the "negotiated order" method for studying organizations developed by Anselm Strauss [41]. This specific interactionist theory emphasizes that "any change arising within or imposed on the order will require renegotiation to occur" [42].

3.2. Ethical Considerations

Before conducting the focus group, we first asked participants' permission to record our sessions before we read an opening statement that described the goals and intent of our research and focus groups. Our statement made clear several points:

1. Participation was voluntary, and interviewees were not obligated to answer any questions with which they felt uncomfortable and could withdraw from participation at any time.
2. Participant contributions would remain confidential unless researchers later specifically requested permission to reveals sources by name.

Before conducting the focus groups, we received approval of the interview script by the University of Connecticut Institutional Review Board (IRB) to ensure that the inquiry of human subjects was conducted in accordance with legal requirements and ethical principles of Respect for Persons, Beneficence and Justice.

3.3. Setting and Data Collection

All focus groups were virtually conducted using Webex or Microsoft Teams videoconferencing programs. We allocated and asked permission to conduct the focus groups for 90 min.

We conducted the focus groups of DOT managers using a semi-structured survey instrument—a script—composed to explore the following topics related to demand for NPM and BMPs:

1. True Need for Native Plant Material.
2. History of Use of Native Plant Material.
3. Topic of Ecotypic Seed.
4. Procurement Process for Native Plant Material.
    a. Interaction with Growers.
    b. Prices for Native Plant Material.
5. Mowing Regimens and Strategies.
6. Invasive Species Removal.
7. Expectations of How Our Team Could Assist.

The script for the DOTs consisted of 47 questions (Appendix A). The script for the DEP and FWS managers had the same topics, except for topic 1, 4, and 5, and consisted of 29 questions (Appendix B). Questions included both open-ended and Likert-scale queries that explored the demand for and use of NPM, subject knowledge, beliefs, behavior, barriers to implementation, and existing resources and processes.

We recorded each focus group and used Descript, a transcription software, to transcribe each recording. Our team reviewed transcripts for accuracy and studied the recordings when the transcription indicated inaudibility.

3.4. Participants

Our team organized focus groups by corresponding with key individuals in state DOTs, DEPs, and FWSs. Key informants for DOTs were asked to select managers involved in landscape design and roadside maintenance. Key informants for DEPs and FWSs were asked to select managers involved in NPM procurement, who were usually wildlife biologists. For state DOTs, we were able to arrange focus groups for five of the six New England states—Connecticut (CT), Massachusetts (MA), New Hampshire (NH), Rhode Island (RI), and Vermont (VT)—that consisted of 11 participants in total. For state DEPs and FWSs, we were able to arrange focus groups for all six states. Each focus group only consisted of individuals from each agency.

3.5. Data Analysis

Our team analyzed the qualitative data of the focus groups using Kvale and Brinkmann's approach of meaning condensation (2009), an empirical phenomenological-based method consisting of five steps:

1. Thoroughly read transcripts to gain an overall sense of the whole.
2. Identify natural meaning units within the transcripts.
3. Decode the main themes articulated by participants.
4. Relate themes to the focus of the research.
5. Connect themes to construct descriptive statements.

4. Results and Discussion

4.1. Problems with Lines of Comunication between Producers and State Agencies

The most important priority for conducting these focus groups was to estimate the level of demand that exists for native plant material (NPM) among Departments of Transportation (DOTs), Departments of Environmental Protection (DEPs), and Fish and Wildlife Services (FWSs). We thought that the best way to make this determination would be to request that each agency review their records from the last five years. Realizing it would take each agency time to collect such data, we submitted the following requests to each agency for these records before conducting the focus groups: *"Please collect five years of data on previous use of native plants in roadside revegetation projects. Include species composition and estimates of the amount of seed and propagules used."* However, we discovered that fulfilling this request was very difficult, if not impossible, for the DOTs. While the work of DEP and FWS agencies centers around the use of plant material to conduct conservation and restoration, the revegetation of roadside landscapes is more ancillary to the DOT focus than the construction of hardscapes, such as roads and bridges. In addition, unlike DEPs and FWSs, DOTs usually subcontract roadside revegetation projects to private entities. To ensure that DOTs do not favor one NPM source over another, the selection of NPM sources is left to the discretion of the subcontractors as long as the NPM selection meets prescribed specifications. This subcontracting approach requires projects be put out to bid and is usually awarded to the lowest bidder. As a result, accessing accurate records for DOTs becomes challenging. In addition, because DEPs and FWSs prioritize the use of ecotypic NPM out of an abundance of caution to preserve the genetic integrity of specific ecosystems, they either work closely with growers or propagate their own NPM. On the other hand, DOT landscape designers, who keep records of projects, are several steps removed from interacting with producers.

Since we had determined from our roundtable discussions that the best way for producers to meet end user demand required improved lines of communication, we realized that the way DOT projects develop made communicating the NPM needs of DOTs to producers challenging. Unlike DEP and FWS projects, which are plant-centered, DOT projects are hardscape-centered. Designing roadside revegetation frequently occurs toward the end of the DOT infrastructure planning process. We concluded that, if possible, DOTs would benefit from moving roadside planning to earlier in the planning process so that DOTs could communicate their needs to producers early enough for the producers to meet their needs.

Since our team hypothesized that lines of communication would benefit from the implementation of two mechanisms (a Regional Seed and Plant Material Needs Directory and roundtables, working groups, and/or meetings to bring together end users and producers), we asked questions to explore the likelihood that DOTs would utilize such tools. We first explained to the focus group participants that producers had expressed reluctance to invest in producing large volumes of NPM unless they believed that they could predict that each species was in demand. We then asked: *"On a scale of 1 to 5, with 1 being not likely and 5 being very likely, how likely would your department be to opening lines of communication with native plant and seed growers about your native plant needs?"* While Massachusetts and Vermont both answered with 5s, Connecticut, New Hampshire, and Rhode Island expressed reluctance that such mechanisms would work. New Hampshire managers pointed to their subcontracting approach as an impediment since the choice of sources remains in the hands of the subcontractors. Since the Rhode Island DOT presently uses little to no NPM for revegetation and has a similar subcontracting structure to New Hampshire, it could not understand why it would need to open lines of communication with producers.

We then explained how a Needs Directory would work and asked: *"On a scale of 1 to 5, with 1 being not likely and 5 being very likely, how likely would your department be to using a Regional Seed and Plant Material Needs Directory for growers to assess what needs to be grown to meet your native plant needs?"* Once again, the New Hampshire and Rhode Island DOT managers cited their subcontracting systems and current lack of use of NPM for revegetation as reasons why they believe neither would use such a directory. While the other three states expressed a greater openness to using such a tool, they also cautioned that they leave roadside revegetation designing toward the end of infrastructure planning, thus possibly preventing them from giving producers enough time to meet their needs.

Since regional seed zones transect political boundaries of each state and, therefore, state DEPs and FWSs share ecotypic NPM, we thought that it would be helpful if state DEPs and FWSs shared their NPM needs with one another to better determine and communicate their collective needs with producers. In addition, since New England DEPs and FWSs do not have the political and governmental financial resources of their Western USA counterparts, which are subsidized by the Bureau of Land Management, we thought that it would benefit them to pool their resources and political voices. Therefore, we asked questions that explored the likelihood that they would participate in lobbying efforts and pooling research dollars: *"New England DOTs pool their research funds and use an organization called the New England Transportation Consortium to conduct research on various aspects of infrastructure. Do regional DEPs have a similar organization?"* A participant from the New Hampshire FWS mentioned the Northeast Regional Conservation Needs Program (NRCNP), which pools 4% of regional research and conservation funding to address critical landscape wildlife conservation needs. The participant gave the example of states pooling their resources to conserve the endangered New England cottontail rabbit. However, the fact that no other DEP or FWS participant cited the NRCNP suggests that such agencies would need prompting to leverage such an organization.

We then asked the following question: *"On a scale of 1 to 5, with 1 be not likely and 5 being very likely, how likely do you believe your department would be to participating in a political lobbying organization with other stakeholders to lobby for greater government funding to increase the use of native plants in government projects?"* Almost none of the DEP and FWS participants could cite such an organization. In fact, several mentioned that they could not lobby their legislatures by law. However, a participant from the Maine FWS mentioned that there are both Northeast and National Association of Fish and Wildlife Agency lobbying groups for just such purposes. Nevertheless, the fact that only one participant knew of these organizations indicated to us that these agencies have not felt the need to pool their voices to meet their NPM needs.

4.2. Lack of Urgency and Limited Knowledge

Although we believe that the lack of communication between DOTs and the producers regarding their NPM prevented the producers from meeting the needs of these agencies, our focus groups also revealed that a lack of urgency and limited knowledge regarding ecotypic NPM has contributed to the lack of progress in transitioning to the revegetation of roadsides using NPM.

Concerning the knowledge and use of native plants for roadside revegetation, we implemented a two-level approach. First, we explored the DOTs' true need for native plant material. Second, we explored their knowledge of and motivation to use ecotypic plant material. Our first question concerning the departments' true need was the following: *"On a scale of 1 to 5, with 1 being not important and 5 being of high priority, how would you rate the importance your state DOT places on transitioning to the use of native plant material, whether seed or plants, for the revegetation of roadsides?"* Representatives from two states, Connecticut and Massachusetts, gave ratings of 5. However, the other three interviewed states gave responses of 3 or lower, raising concerns about the state of such a transition. Two representatives from the Vermont DOT, one from landscape design and one from maintenance, gave ratings of 3. However, a third representative from landscape design answered: *"I would love it if we were 3 or even higher, but my perception is I was going to actually say 2 because you're asking about the mix of seed and other plant material . . . we try to do a really good job of making sure we're selecting native plants. But by and large, most of our revegetation when we have construction projects we're putting down our standard seed mixtures, which really are not natives. So, for that reason I'm skewing it to a 2."*

This response encapsulates what we found across departments: despite discussion about needing to transition to the use of NPM to revegetate roadsides, most departments are, in practice, still using quick-establishing introduced cool-season turfgrass (ICST) mixes following construction projects. A representative from Rhode Island, a landscape architect, emphasized that the two department landscape architects are also part of their environmental division and that they give the use of NPM a high priority. However, for those outside their small group, the emphasis is on funding hardscape work:

Interviewer: *"In terms of allocation of funds . . . for the people who do allocate funding . . . how much of a priority is it to them?"*

RI landscape architect: *"Probably close to 0 right now. Unfortunately, we're in a situation where the current group that makes a lot of those decisions about types of projects and funding are very hard material focused. Pavement and bridge work is their focus and environment and things like that are much lower down their list of priorities with the exception that something that has a permit, or we have to follow for legal reasons, federal reasons to do with our funding. And that's so they can't pick and choose whether or not they want to."*

This comment underlines two consistent themes we encountered throughout our interviews:

1. With scarce resources, DOTs will prioritize hardscape funding over roadside landscape funding
2. DOTs are more likely to change habits only when required to do so.

Concerning the second point, an environmental manager made the following point: *"Say this is a mitigation project, and we're doing wetland mitigation and we need to go through our department of environmental services required mitigation services, then we would need to look at native seed and we would need to develop a native seed mix and specify that in the contract for replanting in that area. But that would be a deviation from these normal specifications because of that particular need . . . Since I know we're talking about native ecotypes . . . I'm not aware that these specs specify that either."* As long as the drive to transition to revegetating roadsides using NPM remains merely aspirational rather than a mandatory requirement, such a transition appears elusive.

Another impediment we discovered to transitioning to the use of NPM for revegetation was a lack of knowledge concerning ecotypic NPM and the reasons for its use. One reason we explored was whether managers knew and understood how the term *ecotypic* relates to the objectives of the request for proposal (RFP) funding. The Project Background portion of the RFP cites a series of next steps suggested in a manual resulting from research conducted from 2013 to 2016 by the authors, one of which was the need to develop local seed sources, which the region currently lacks. Although native seed already exists from growers outside the New England region, such sources are typically ecotypes that have adapted to other regional climates. Considering the vast size of roadsides and the ecological impact roadside NPCs can have on surrounding NPC gene pools, DOTs would be wise to source seed from local populations to prevent any unintended consequences that may occur from a more aggressive or weaker strain of a particular native plant species mixing with local populations. Therefore, we asked DOT, DEP, and FWS managers the following question: *"Have you heard the term ecotypic in relation to native plant seed? If yes, what do you think the term ecotypic means?"* All the DEP and FWS managers had heard of the term and could explain what *ecotypic* means, which is understandable since the focus of their work centers around the restoration of local NPC-based ecosystems and habitats. However, more than a third of DOT managers had not heard of the term. More concerning were the responses to the following series of questions posed to DOT managers:

1. *"Is there internal pressure from DOT management to use ecotypic seed for revegetation projects? If yes, from whom? What are the justifications for using ecotypic seed?"*
2. "Is there any external pressure outside your DOT, such as community activism or legislation, to transition to using ecotypic seed?"
3. "On a scale of 1 to 5, with 1 being not a priority at all and 5 being a high priority, what level of priority does your department give to present or future use of ecotypic seed?"

In response to question 1, no respondent could recall any internalized pressure from DOT management to use ecotypic NPM. However, a manager from Massachusetts understood the benefits of using ecotypic NPM and intentionally tries to create seed mixes with ecotypes closest in regional proximity to Massachusetts. She mentioned that one of the reasons she is concerned with ecotypes relates to the previous use of a variety of switchgrass (*Panicum virgatum*) sourced from the Midwest that might have been engineered to grow more rapidly to better serve as biofuel. This variety rapidly spread in an aggressively invasive manner and formed roadside monocultures. Not only did their inclusion prevent other species, both grasses and forbs, from establishing a biodiverse NPC, but she also feared that this more aggressive variety would swamp the gene pool of neighboring switchgrass ecotype communities.

In response to question 2, no manager had encountered outside pressure. However, the same manager from Massachusetts did mention that several local groups with whom her department closely works, including Massachusetts pollinator working groups and networks, promote the use and benefits of ecotypic NPM.

In response to question 3, since no one could cite any internal or external pressures to transition to the use of ecotypic NPM, all answers were 2s or lower, except from the Massachusetts manager who gave a 4. Some mentioned that, even if there was pressure to use ecotypic NPM, the added expense and lack of bulk supply would probably prevent such a transition in the near future.

4.3. Resistance and Impediments to Transitioning to Best Management Practices

Research has shown that particular roadside maintenance best management practices (BMPs), such as reduced and strategically timed mowing, can stimulate roadside native seed bank germination, encourage the spread of native plant communities (NPCs), and benefit the health of pollinator and wildlife habitats. Therefore, we asked a series of questions to determine the current state of DOT roadside BMPs as they relate to the exploitation of existing roadside NPCs, starting with the following: *"Have there been changes in mowing strategies in the last few years? If yes, how have your mowing regimens changed and why?"* Only two DOTs, those of Connecticut (CT) and Vermont (VT), cited intentional efforts on the part of their maintenance departments to reduce mowing. Massachusetts (MA) has also reduced mowing. However, this trend was resource-based rather than ecology-based since their maintenance department was short-staffed. Rhode Island (RI) had had a similar reduction in mowing because of budget cuts, but once budgets increased, the maintenance department returned to regular mowing schedules with roadsides being mowed multiple times a year. RI DOT managers claimed that their maintenance department managers frown upon reduced mowing because they fear tall grasses attract roadside dumping and that the DOT receives complaints from the public that roadsides look unkempt. New Hampshire (NH) managers have also not seen reduced mowing, but they attributed this to the fact that NH does not own its own mowers but rather rents equipment or subcontracts its mowing services. This saves money but limits its ability to adjust the timing of the mowing sessions for optimal times.

Since the timing of mows influences growth cycles and the availability of nesting and foraging resources for pollinators and wildlife, we asked the following three questions:

1. What factors influence the timing of mowing?
2. Does your department strategize where and when to mow? If yes, how so?
3. Does your department consciously plan what sections of roadside are prioritized for mowing? If so, what sections are prioritized for reduced mowing?

For MA, the timing for mowing is most influenced by snow removal efforts and street cleanings since the same people who mow are involved in snow removal and street cleaning. Only CT and VT intentionally time their mowing regimens for ecological reasons. The Connecticut DOT has designated 116 conservation areas that it mows either once a year or every other year, except for the shoulders and perimeters. It more regularly mows the shoulders to allow vehicles to safely pull to the side of the road without causing fires and the perimeters to prevent the spread of woody plants from neighboring woodlands. The authors themselves, who reside in CT, have witnessed how DOT-reduced mowing regimes have resulted in the proliferation of previously rare and scarce species, including foxglove beardtongue (*Penstemon digitalis*) and various milkweeds (*Asclepias* spp.). In 2021, Vermont joined the Nationwide Monarch Candidate Conservation Agreement with Assurances (CCAA) for Energy and Transportation Lands, which requires those participating to adopt measures to create conservation benefits for the monarch butterfly and provides a model for other at-risk species. As a consequence, VT has reduced most roadside mowing to once every three years, with only 1/3 of roadsides getting mowed annually. Maintenance districts keep track of when they mow using a GIS-based maintenance activity tool.

To gauge the openness of DOT maintenance departments to changing their mowing strategies, we asked: *"If a manual was developed outlining insect friendly mowing practices, on a scale of 1 to 5, with 1 being not willing and 5 being definitely willing, how willing would your DOT be to adopt new mowing practices? Why or why not?"* CT and VT were the two states most likely to be influenced to change their mowing strategies. MA participants expressed reservations because they feared that reduced mowing could result in the encroachment of invasive species. NH saw the possibility of its department joining the CCAA, which would require its DOT to change their mowing strategies. Therefore, the managers could see the possibility that a manual could influence their mowing strategy. However, they would still encounter a problem resulting from their department not owning their own equipment. RI managers were the most definitive in believing that their maintenance departments would be less open to such changes because they characterized maintenance workers as "old school thinkers".

Invasive species can tend to outcompete NPCs. Therefore, to explore the degree to which DOTs prioritized invasive species removal, we asked: *"On a scale of 1 to 5, with 1 being not a priority at all and 5 being one of the top priorities, what level of priority does your department give to invasive species removal?"* With a rating of 3, CT gave the highest rating for invasive species removal. Both MA and NH asserted that their departments prioritized invasive species removal during construction but rarely following construction. RI managers said they prioritize the removal of Japanese knotweed, but they otherwise witness scant effort to remove other species. VT participants gave a rating of 1.

We hypothesized that, if maintenance departments were to confine mowing to times outside the growing season, they would have time available during the summer to devote to invasive species removal. Therefore, we asked: *"On a scale of 1 to 5, with one being not likely to five being very likely, how likely would your department be to adopt a strategy of invasive species control during the summer when mowing is less encouraged?"* Across the board, DOTs rated the likelihood as low, with lack of funding for the expansion of invasive species removal as the main reason for the resistance.

4.4. Other Impediments and Possible Solutions

We ended our focus groups with the following two questions that explored how our research could best help these agencies:

- *"Considering the subjects we have covered during this focus groups, how can our team most benefit your departments through work we are doing on this grant?"*
- "What steps would help your department to transition to greater use of native plant use for roadside revegetation?"

The most frequent recommendation concerned the need for seed mixes that consist of readily available, easy-to-establish workhorse species tailored to various roadside microclimates and regional seed zones. Respondents also requested an increased availability of ecotypic NPM and guidance as to where to find such seed. Two FWS managers believed that it would be helpful if we could make a more persuasive argument for the economic and sustainability advantages of using NPM for roadside revegetation, as DOTs would more likely transition to their use.

Several DOT, DEP, and FWS managers stressed the need for educational campaigns both within government agencies using NPM and for the public. For government agencies, several respondents suggested expanding the knowledge base regarding the benefits of using ecotypic NPM. They believed that a strong enough case for using ecotypic NPM could persuade agencies to pay the higher prices. Others recommended that we find examples of what worked in other states, whether the most effective establishment techniques or BMPs. For the general public, some argued for educational campaigns that explained the economic and ecological benefits of reduced mowing and the change in roadside aesthetics as a result.

Finally, the most often repeated recommendation was the need for mandates and the funding to meet such requirements. Several respondents believed that educational campaigns can be persuasive but cannot match the ability of mandates to counter bureaucratic inertia to change long-standing agency behavior.

5. Conclusions

- *Need for Improved Lines of Communication between DOT Managers and Producers*

We concluded from the roundtable discussions we conducted preceding our initiation of focus groups that a substantial communication gap exists between producers and end users. This conclusion was re-enforced by our focus groups. We found from interviews with Department of Transportation (DEP) and Fish and Wildlife Service (FWS) managers that these agencies have developed closer relationships with growers because their work is plant- and landscape-centric. Because DEPs and FWSs concern themselves with preserving the genetic integrity of native plant communities (NPCs), they need to work with growers to secure ecotypic native plant material (NPM) for their restoration and conservation work. On the other hand, DOT projects are hardscape-centric, leaving roadside design for later in project timelines. We concluded that Departments of Transportation (DOTs) would benefit from developing roadside landscape design earlier in the planning process because it would allow DOTs more time to communicate their NPM needs to growers, thus providing growers more lead time to grow appropriate NPM. In addition, we deduced that the subcontracting approach for revegetation projects made it difficult for DOT landscape designers to directly work and develop relationships with growers since the current subcontracting structure leaves the selection of NPM producers to the subcontractors. The tendency to choose the lowest bid for projects tends to eliminate the use of ecotypic plant material, which tends to be more expensive for the New England region. We believe the closer the NPM choices are to the landscape designers, the greater control designers will have over NPM selections.

- *Need to Prioritize Revegetation Using NPM*

Before DOTs can develop closer relationships with growers, however, we realized that several New England DOTs need to prioritize the use of NPM for revegetation. At least two of the five states we interviewed still mainly use turf grass and introduced-species seed mixes rather than native seed mixes for new construction projects involving revegetation. Managers consistently indicated that transitioning to the greater use of NPM would be more likely to happen if they were required to do so. In fact, the Vermont DOT transitioned to the greater use of NPM once it decided to join the Nationwide Candidate Conservation Agreement with Assurance for Monarch Butterfly (CCAA), which encourages landowners and land managers to adopt measures to create net conservation benefits for the monarch butterfly. Therefore, finding ways to mandate the use of NPM appears to be the most effective method for ensuring DOTs' transition to the use of NPM for revegetation. Only once DOTs further increase their use of NPM for revegetation could the use of ecotypic NPM be prioritized.

- *Need to Change BMPs to Exploit and Increase Existing Roadside NPCs*

Aside from transitioning to the use of NPM for roadside revegetation, the cheapest, most effective method for increasing the presence of native plant communities (NPCs) along roadsides is the adoption of mowing strategies that benefit existing native plant seed banks. The simplest best management practice (BMP) to exploit existing seed banks is reducing mowing during the growing season to allow NPCs to flower and go to seed. However, even this BMP met resistance from some DOTs. For example, the Rhode Island DOT participants expressed reluctance to reduce mowing because they fear the possible dumping of garbage in tall grasses. The New Hampshire DOT managers, on the other hand, have less control over the timing of their mowing schedules because they do not own their equipment. They either rent equipment and have their maintenance departments mow or they subcontract the mowing to outside organizations. All DOTs expressed concern that reduced mowing could lead to invasive species encroachment. Our team has tried to encourage maintenance departments to prioritize invasive species removal during the growing season since mowing would be reduced. However, since departments have relied on mowing as the first line of defense against invasive species and several states have restricted the use of herbicides, designating invasive species removal during the growing season has proven problematic.

We believe there are other specific actions our research team could undertake to encourage the greater use of NPM for roadside revegetation:

1. We are working with the Northeast Seed Network to develop seed transfer zones for New England and seed mixes for specific microclimates within each seed zone.
2. We intend to create educational campaigns both for DOT managers and the public. For managers, we want to educate them on the benefits of using ecotypic NPM and transitioning to BMPs that benefit the persistence and spread of roadside NPCs. In terms of educational campaigns for the public, DOTs have received complaints that roadsides that receive reduced mowing tend to look unkempt. We want to educate the public on how reduced mowing benefits pollinator and other wildlife health.
3. We are researching what has worked in terms of effective NPC establishment and educational campaigns. DOTs believe that they are more likely to adopt policies that other DOTs have proven to work.

Author Contributions: Conceptualization, methodology, and interviews, J.C., S.K. and Y.A.K.; data analysis, J.C. and S.K.; writing—original draft preparation, J.C.; writing—review and editing, J.C., S.K. and Y.A.K. All authors have read and agreed to the published version of the manuscript.

Funding: This research received funding through the New England Transportation Consortium/Maine Department of Transportation NETC 21-3.

Institutional Review Board Statement: The study was approved by the university Institutional Review Board and was performed in accordance with the ethical standards laid down in the 1964 Declaration of Helsinki and its later amendments or comparable ethical standards.

Informed Consent Statement: For DOTs: We intend to record this focus group interview. Do we have your permission to do so? This focus group will be conducted by Julia Kuzovkina, a professor of Plant Science at the University of Connecticut, and her doctoral student, John Campanelli. It will be witnessed by their undergraduate intern, Sam Kocurek. We are conducting this interview to obtain data on previous and future use of native plant material for roadside revegetation as well as changes in management practices related to this transition. We will be asking approximately 47 questions and we are allotting 90 min for the focus group. Any participation is voluntary, and each participant has the right to withdraw at any point during the study, for any reason, and without any prejudice. For DEPs and FWSs: We intend to record this focus group interview. Do we have your permission to do so? This focus group will be conducted by Julia Kuzovkina, a professor of Plant Science at the University of Connecticut, and her doctoral student, John Campanelli. It will be witnessed by their undergraduate intern, Sam Kocurek. We are conducting this interview to obtain data on previous and future use of native plant material for conservation and restoration projects conducting by DEP [or FWS]. We will be asking approximately 29 questions and we are allotting 90 min for the focus group. Any participation is voluntary, and each participant has the right to withdraw at any point during the study, for any reason, and without any prejudice.

Data Availability Statement: Not applicable.

Acknowledgments: The authors would like to thank Eve Allen for her collaboration in our preliminary roundtable discussions. We also want to thank all the managers who participated in focus groups from each New England state Department of Transportation, Department of Environmental Protection, and Fish and Wildlife Service. Finally, we want to thank the three anonymous reviewers for their useful advice and comments that helped to improve this text.

Conflicts of Interest: The author declares no conflict of interest.

Appendix A. Script for Department of Transportation (DOT) Focus Groups

I. True Need for Native Plant Material

We would like to determine if and why your department prioritizes the use of native plant material.

1. On a scale of 1 to 5, with 1 being not important and 5 being of high priority, how would you rate the importance your state DOT places on transitioning to the use of native plant material, whether seed or plants, for the revegetation of roadsides?
2. From what you know about native plants, what sort of tangible effects do you see them having on the environment?
3. From what you have heard and observed, what are the main reasons expressed by your state DOT management for using native plant material for revegetation?
    - Pollinator health
    - Wildlife health
    - Sustainability
    - Decreased maintenance costs
    - Erosion control
4. Has there been a significant increase in the use of native plant material for roadside revegetation over the last few years? If yes, for what kind of projects has your DOT prioritized the use of native plant material?
5. Who, if anyone, in your department is the biggest driver for an increased use of native plant material for roadside revegetation?
6. What is needed in your DOT to advocate the need to use more native plants?

II. History of Use of Native Plant Material

We would like to determine your previous, present, and future use of native plant material to determine the level of demand that exists in our region for native plant material.

7. Please collect five years of records on construction projects that included revegetation. Of these projects, what percentage involved native plants?
8. Concerning present and future new construction, what percentage of new construction projects involve revegetation? What percentage of these revegetation projects include native plant establishment in the specifications?
9. Please collect five years of data on previous use of native plants in roadside revegetation projects. Include species composition and estimates of the amount of seed used. How large were the construction projects in the last five years?
10. Unlike turf grass, which can be established at almost any time of the year, native plant establishment has more limited windows for optimal establishment. For example, May is the optimal month for native plant establishment. Do your native plant establishment specifications following new construction take into consideration the need to plant during particular times of the year, especially if a construction project is completed outside an optimal window?
11. Who determines the seed mixes your department uses?
12. What methods are used for establishing native plant seed?
13. Does your department have access to a Truax drill or do your contractors own or have access to a Truax drill?
14. What percentage of your revegetation projects use native plants rather than native seed?
15. On a scale of 1 to 5, with 1 be not willing and likely and 5 being very willing and likely, how willing and likely would your department be to increasing the use of native plant propagules rather than seed to establish roadside native plant communities?

III. Topic of Ecotypic Native Seed

16. Have you heard the term ecotypic in relation to native plant seed? If yes, what do you think the term ecotypic means? (Explain term if they do not fully comprehend its meaning. Explain its possible benefits)
17. Is there internal pressure from DOT management to use ecotypic seed for revegetation projects?

If yes, from whom?
What are the justifications for using ecotypic seed?

18. Is there any external pressure outside your DOT, such as community activism or legislation, to transition to using ecotypic seed?
19. On a scale of 1 to 5, with 1 being not a priority at all and 5 being a high priority, what level of priority does your department give to present or future use of ecotypic seed?
20. Having heard a definition of ecotypic seed and why its use should be prioritized, on a scale of 1 to 5, with 1 being not a priority at all and 5 being a high priority, what priority does your department would give to using of ecotypic seed in future projects?

If 3 or higher, why does your department prioritize ecotypic seed use?

IV. Procurement Process for Native Plant Material

Interactions with growers

21. Where does your department source most of your native seed?
22. What factors influenced your choice of these seed sources?
23. (Explain concept of workshops or roundtables to create predictability for growers) On a scale of 1 to 5, with 1 be not likely and 5 being very likely, how likely would your department be to opening lines of communication with native plant and seed growers about your native plant needs?
24. (Give Directory definition) On a scale of 1 to 5, with 1 be not likely and 5 being very likely, how likely would your department be to using a Regional Seed and Plant Material Needs Directory for growers to assess what needs to be grown to meet your native plant needs?

Prices

25. Do specification requirements limit the amount spent on plant material?
26. How willing would your department be to pay higher prices to use ecotypic seed? By how much would they be willing to pay more?
27. Do you know whether there has been Economic Impact Analysis of the level of funding that would be required to transition to a greater use of native plants during revegetation?
28. Have you witnessed any efforts or initiatives to secure greater funding to transition to the use of native plants for revegetation?
29. Do you have any knowledge as to whether your state has attempted to allocate fund from FAST Act block grants to fund roadside pollinator habitats?
30. Have there been discussions about using initiatives to raise funds to offset the increased costs of transitioning to the use of native plant revegetation, such as the license plate initiatives in NC and Iowa?

V. Mowing Regimens and Strategies

31. Have there been changes in mowing strategies in the last few years?

    If yes, how have your mowing regimens changed and why?

32. What are your and your colleagues' attitudes toward your department practicing reduced mowing?
33. What benefits, if any, have resulted from reduced mowing?
34. If your DOT has decreased mowing in the last few years, has your department experienced a public response to this decrease? If yes, please describe.
35. Has there been discussions about developing educational campaigns for public acceptance of reduced mowing to benefit pollinator health?
36. What factors influence the timing of mowing?
37. Does your department strategize where and when to mow?
38. Does your department consciously plan what sections of roadside are prioritized for mowing? What section are prioritized for reduced mowing?
39. (Explain possible plans for software-based mowing strategies) On a scale of 1 to 5, with 1 being not likely and 5 being very likely, how likely would your department be open and willing to adopting a GIS-based system for planning roadside mowing?
40. If a manual was developed outlining insect friendly mowing practices, on a scale of 1 to 5, with 1 being not willing and 5 being definitely willing, how willing would your DOT be to adopt new mowing practices? Why or why not?
41. Reduced mowing tends to result in greater biomass. Has your department had to transition to using equipment other than usual mowers, such as brush hogs, to mow?

VI. Invasive species removal

42. On a scale of 1 to 5, with 1 being not a priority at all and 5 being one of the top priorities, what level of priority does your department give to invasive species removal?
43. What methods are used to remove invasive species?
44. If your departments have decreased mowing over the last few years, has there been an increase in the removal of invasive species as a result of the freeing up of time from reduced mowing?
45. On a scale of 1 to 5, with one being not likely to five being very likely, how likely would your department be to adopt a strategy of invasive species control during the summer when mowing is less encouraged?

VII. Expectations of how we can assist

46. Considering the subjects we have covered during this focus groups, how can our team can most benefit your departments through work we are doing on this grant?
47. What steps would help your department to transition to greater use of native plant use for roadside revegetation?

Appendix B. Script for Department of Environmental Protection (DEP) and Fish and Wildlife Service (FWS) Focus Groups

I. True Need

We would like to determine your previous, present, and future use of native plant material to determine the level of demand that exists in our region for native plant material.

1. Please collect five years of records on construction projects that included revegetation. Of these projects, what percentage involved native plants?
2. Concerning present and future new construction, in your estimation, what percentage of new construction projects involve revegetation? What percentage of these revegetation projects include native plant establishment in the specifications?
3. Please collect five years of data on previous use of native plants in roadside revegetation projects. Include species composition and estimates of the amount of seed used.
4. Who determines the seed mixes your department uses?
5. What methods are used for establishing native plant seed?
6. What percentage of your revegetation projects use native plants rather than native seed?
7. On a scale of 1 to 5, with 1 be not likely and 5 being very likely, how likely do you believe your department would be to increasing the use of native plant propagules rather than seed to establish native plant communities?

II. Topic of Ecotypic Native Seed

8. Have you heard the term ecotypic in relation to native plant seed? If yes, what do you believe term ecotypic means?
9. Is there internal pressure from DEEP management to use ecotypic seed for revegetation projects? If yes, from whom? What are the justifications for using ecotypic seed?
10. Is there any external pressure outside your DEEP, such as community activism or legislation, to transition to using ecotypic seed?
11. On a scale of 1 to 5, with 1 being not a priority at all and 5 being a high priority, what level of priority does your department give to present or future use of ecotypic seed?
12. Having heard a definition of ecotypic seed and why its use should be prioritized, on a scale of 1 to 5, with 1 being not a priority at all and 5 being a high priority, what priority do you believe your department would give to prioritizing the use of ecotypic seed in future projects?

    If 3 or higher, why do you believe your department prioritizes ecotypic seed use?

13. Have you seen a significant increase in the use of ecotypic native plant material for roadside revegetation over the last few years?

If yes, for what kind of projects has DEEP prioritized the use of ecotypic native plant material?

14. Who, if anyone, in your department do you believe is the biggest driver for an increased use of ecotypic native plant material for roadside revegetation?
15. What do you think is needed in your DEEP to advocate the need to use more ecotypic native plants?

III. Procurement Process for Native Plant Material

Interactions with growers

16. Where does your department source most of your native seed?
17. What factors influenced your choice of these seed sources?
18. (Explain concept of workshops or roundtables to create predictability for growers) On a scale of 1 to 5, with 1 be not likely and 5 being very likely, how likely do you believe your department would be to opening lines of communication with native plant and seed growers about your native plant needs?
19. On a scale of 1 to 5, with 1 be not likely and 5 being very likely, how likely do you believe your department would be to using a Regional Seed and Plant Material

Needs Directory for growers to assess what needs to be grown to meet your native plant needs?

20. New England DOTs pool their research funds and use an organization called the New England Transportation Consortium to conduct research on various aspects of infrastructure. Do regional DEEPs have a similar organization?

If not, on a scale of 1 to 5, with 1 be not likely and 5 being very likely, how likely do you believe your department would be to organize a group like the NETC?

21. On a scale of 1 to 5, with 1 be not likely and 5 being very likely, how likely do you believe your department would be to participating in a political lobbying organization with other stakeholders (native plant users and producers, conservation, wildlife, and pollinator organizations, etc) to lobby for greater government funding to increase the use of native plants in government projects?

Prices

22. Do specification requirements limit the amount spent on plant material? (How do we go about asking about what prices would be tolerated?)
23. Do you know whether there has been Economic Impact Analysis of the level of funding that would be required to transition to a greater use of ecotypic native plants during revegetation?
24. Have you witnessed any efforts or initiatives to secure greater funding to transition to the use of native plants for revegetation?
25. Have there been discussions about using initiatives to raise funds to offset the increased costs of transitioning to the use of native plant revegetation, such as the license plate initiatives in NC and Iowa?

IV. Invasive species removal

26. On a scale of 1 to 5, with 1 being not a priority at all and 5 being one of the top priorities, what level of priority does your department give to invasive species removal?
27. What methods are used to remove invasive species?

V. Expectations of how we can assist

28. Considering the subjects we have covered during this focus groups, how do you think our team can best benefit your departments through work we are doing on this grant?
29. What steps do envision would help your department to transition to greater use of ecotypic native plant use for conservation projects?

References

1. Hopwood, J. The contribution of roadside grassland restorations to native bee conservation. *Biol. Conserv.* **2008**, *141*, 2632–2640. [CrossRef]
2. Wang, Y.; Jia, S.; Wang, Z.; Chen, Y.; Mo, S.; Sze, N.N. Planning considerations of green corridors for the improvement of biodiversity resilience in suburban areas. *J. Infrastruct. Preserv. Resil.* **2021**, *2*, 1–15. [CrossRef] [PubMed]
3. Ament, R.; Begley, J.; Powell, S.; Stoy, P. *Roadside Vegetation and Soils on Federal Lands: Evaluation of the Potential for Increasing Carbon Capture and Storage and Decreasing Carbon Emissions*; Federal Highway Administration. General Technical Report; Vancouver: Washington, DC, USA, 2014; Available online: https://www.montana.edu/spowell/FLMA%20Roadside%20Carbon%20Report%20Final_January2014.pdf (accessed on 10 February 2023).
4. Forman, R.T.; Alexander, L.E. Roads and their major ecological effects. *Annu. Rev. Ecol. Evol. Syst.* **1998**, *29*, 207–231. [CrossRef]
5. Forman, R.T.; Sperling, D.; Bissonette, J.A.; Clevenger, A.P.; Cutshall, C.; Dale, V.H.; Fahrig, L.; France, R.L.; Goldman, C.R.; Heanue, K.; et al. *Road Ecology*; Island Press: Washington, DC, USA, 2003.
6. Harrison, G.L. *Economic Impact of Ecosystem Services Provided by Ecologically Sustainable Roadside Right of Way Vegetation Management Practices*; General Technical Report; Florida Dept of Transportation: Tallahassee, FL, USA, 2014. Available online: http://rightofway.erc.uic.edu/wp-content/uploads/2018/05/2A4-FDOT-ecosystem-services-roadsides-report.pdf (accessed on 10 February 2023).
7. Dániel Ferreira, J. Linear Infrastructure Habitats for the Conservation of Plants and Pollinators-the Value of Road Verges and Power-Line Corridors for Landscape-Scale Diversity and Connectivity. Available online: https://pub.epsilon.slu.se/25947/1/daniel-ferreira_j_211025.pdf (accessed on 10 February 2023).

8. Hopwood, J.; Hoffman-Black, S.; Fleury, S. *Pollinators and Roadsides: Best Management Practices for Managers and Decision Makers*; Remley, D., Ed.; General Technical Report; Federal Highway Administration: Washington, DC, USA, 2016. Available online: https://rosap.ntl.bts.gov/view/dot/55914 (accessed on 10 February 2023).
9. Phillips, B.B.; Wallace, C.; Roberts, B.R.; Whitehouse, A.T.; Gaston, K.J.; Bullock, J.M.; Dicks, L.V.; Osborne, J.L. Enhancing road verges to aid pollinator conservation: A review. *Biol. Conserv.* **2020**, *250*, 108687. [CrossRef]
10. Clinton, W.J. Exec. Order No. 13112, 3 C.F.R. Volume 64: Number 25. Washington, DC, USA, 8 February 1999. Available online: https://www.invasivespeciesinfo.gov/executive-order-13112 (accessed on 10 February 2023).
11. Obama, B.H. *Presidential Memorandum: Safeguarding the Nation from the Impacts of Invasive Species*; The White House: Washington, DC, USA, 2016. Available online: https://obamawhitehouse.archives.gov/the-press-office/2016/12/05/executive-order-safeguarding-nation-impacts-invasive-species (accessed on 10 February 2023).
12. Obama, B.H. *Presidential Memorandum: Creating a Federal Strategy to Promote the Health of Honeybees and Other Pollinators*; The White House: Washington, DC, USA, 2014. Available online: https://obamawhitehouse.archives.gov/the-press-office/2014/06/20/presidential-memorandum-creating-federal-strategy-promote-health-honey-b (accessed on 10 February 2023).
13. CT Public Act 16-1; Section 12: "An Act Concerning Pollinators Health" CT Bill. 2016. Available online: https://www.cga.ct.gov/2016/act/pa/2016PA-00017-R00SB-00231-PA.htm (accessed on 10 February 2023).
14. Simmons, M.; Bertelso, M.; Windhager, S.; Zafian, H. The performance of native and non-native turfgrass monocultures and native grass polycultures: An ecological approach to sustainable lawns. *Ecol. Eng.* **2011**, *37*, 1095–1103. [CrossRef]
15. Alba, C.; Fahey, C.; Flory, S.L. Global change stressors alter resources and shift plant interactions from facilitation to competition over time. *Ecology* **2019**, *100*, e02859. [CrossRef] [PubMed]
16. Wagner, D.L.; Metzler, K.J.; Leicht-Young, S.A.; Motzkin, G. Vegetation composition along a New England transmission line corridor and its implications for other trophic levels. *For. Ecol. Manag.* **2014**, *327*, 231–239. [CrossRef]
17. Ott, J.P.; Hartnett, D.C. Contrasting bud bank dynamics of two cooccurring grasses in tallgrass prairie: Implications for grassland dynamics. *Plant Ecol.* **2012**, *213*, 1437–1448. [CrossRef]
18. Brown, R.N.; Percivalle, C.; Narkiewicz, C.; DeCuollo, S. Relative rooting depths of native grasses and amenity grasses with potential for use on roadsides in New England. *HortScience* **2010**, *45*, 393–400. [CrossRef]
19. Isaacs, R.; Tuell, J.K.; Fiedler, A.K.; Gardiner, M.; Landis, D.A. Maximizing arthropod-mediated ecosystem services in agricultural landscapes: The role of native plants. *Front. Ecol. Evol.* **2009**, *7*, 196–203. [CrossRef]
20. Shibel, Z.; Hendrix, S.D.; Heard, S.B. The co-evolution of herbivore impact: Field estimates of impact in older and newer gallmaker–Solidago interactions. *Evol. Ecol. Res.* **2018**, *19*, 43–59.
21. Kuzovkina, Y.A.; Campanelli, J.; Schulthess, C.; Ricard, R.; Dreyer, G. Effective Establishment of Native Grasses on Roadsides in New England. New England Transportation Consortium. 2016. Available online: https://digitalcommons.conncoll.edu/cgi/viewcontent.cgi?referer=&httpsredir=1&article=1001&context=arboretum_otherpubs (accessed on 10 February 2023).
22. National Research Council. *Assessing and Managing the Ecological Impacts of Paved Roads*; National Academies Press: Washington, DC, USA, 2005; Available online: https://nap.nationalacademies.org/read/11535/chapter/1#vi (accessed on 10 February 2023).
23. Berger, R.L. *Integrated Roadside Vegetation Management*; Transportation Research Board: Washington, DC, USA, 2005; Available online: https://books.google.com.nf/books?id=HIE8Kx_8-3UC&printsec=frontcover#v=onepage&q&f=false (accessed on 10 February 2023).
24. Storey, B.; Das, S.; McFalls, J.; Moran, R.A.; Dadashova, B. NCHRP 14-40 Report: Comparison of Cost, Safety, and Environmental Benefits of Routine Mowing and Managed Succession of Roadside Vegetation. Texas A&M Transportation Institute, The Texas A&M University System: College Station, TX, USA, 2020. Available online: https://onlinepubs.trb.org/Onlinepubs/nchrp/docs/NCHRP14-40FinalReport.pdf (accessed on 10 February 2023).
25. Steinfeld, D.E. *Roadside Revegetation: An Integrated Approach to Establishing Native Plants*; Federal Highway Administration, Western Federal Lands Highway Division, Technology Deployment Program: Vancouver, WA, USA, 2007. Available online: https://flh.fhwa.dot.gov/resources/design/library/roadside-revegetation-manual.pdf (accessed on 10 February 2023).
26. Poe, C.; Mast, T.; Colton, P. *Evaluation of Promoting Roadside Revegetation: An Integrated Approach to Establishing Native Plants (No. FHWA-HRT-17-008)*; United States. Federal Highway Administration; Office of Corporate Research, Technology, and Innovation: Washington, DC, USA, 2017. Available online: https://www.fhwa.dot.gov/publications/research/randt/evaluations/17008/17008.pdf (accessed on 10 February 2023).
27. Nemec, K.; Stephenson, A.; Gonzalez, E.A.; Losch, M. Local Decision-makers' Perspectives on Roadside Revegetation and Management in Iowa, USA. *J. Environ. Manag.* **2021**, *67*, 1060–1074. [CrossRef] [PubMed]
28. Keller, G.; Sherar, J. *Low-Volume Roads Engineering: Best Management Practices Field Guide*; General Technical Report, US AID and USDA Forest Service: Washington, DC, USA, 2003.
29. Nemec, K.; Stephenson, A.; Losch, M. How Engineers and Roadside Vegetation Managers Maintain Roadside Vegetation in Iowa, USA. *J. Environ. Manag.* **2022**, *70*, 593–604. [CrossRef] [PubMed]
30. Pike, C.; Potter, K.M.; Berrang, P.; Crane, B.; Baggs, J.; Leites, L.; Luther, T. New seed-collection zones for the eastern United States: The eastern seed zone forum. *J. For. Res.* **2020**, *118*, 444–451. [CrossRef]
31. National Academies of Sciences, Engineering, and Medicine. An Assessment of Native Seed Needs and the Capacity for Their Supply. 2023. Available online: https://nap.nationalacademies.org/catalog/26618/an-assessment-of-native-seed-needs-and-the-capacity-for-their-supply (accessed on 10 February 2023).

32. Gustafson, D.J.; Gibson, D.J.; Nickrent, D.L. Using local seeds in prairie restoration—Data support the paradigm. *Nativ. Plants J.* **2005**, *6*, 25–28. [CrossRef]
33. Germino, M.J.; Moser, A.M.; Sands, A.R. Adaptive variation, including local adaptation, requires decades to become evident in common gardens. *Ecol. Appl.* **2019**, *29*, e01842. [CrossRef] [PubMed]
34. Baughman, O.W.; Agneray, A.C.; Forister, M.L.; Kilkenny, F.F.; Espeland, E.K.; Fiegener, R.; Horning, M.E.; Johnson, R.C.; Kaye, T.N.; Ott, J.; et al. Strong patterns of intraspecific variation and local adaptation in Great Basin plants revealed through a review of 75 years of experiments. *Ecol. Evol.* **2019**, *9*, 6259–6275. [CrossRef] [PubMed]
35. Jones, A.T.; Hayes, M.J.; Hamilton, R.S. The effect of provenance on the performance of Crataegus monogyna in hedges. *J. Appl. Ecol.* **2001**, *38*, 952–962. [CrossRef]
36. Hufford, K.M.; Mazer, S.J. Plant ecotypes: Genetic differentiation in the age of ecological restoration. *Trends Ecol. Evol.* **2003**, *18*, 147–155. [CrossRef]
37. Allen, E.B. Cultivating Capacity in the Northeast's Native Seed and Plant Supply Chain. Master's Thesis, Massachusetts Institute of Technology, Cambridge, MA, USA, 2022.
38. Campanelli, J.; Kuzovkina, Y.A.; Ricard, R.M.; Schulthess, C.P. Attitudes toward and adoption of roadside revegetation using native plants by departments of transportation in New England states. *Nativ. Plants J.* **2019**, *20*, 31–47. [CrossRef]
39. Kvale, S.; Brinkmann, S. *Interviews: Learning the Craft of Qualitative Research Interviewing*; SAGE Publications: Thousand Oaks, CA, USA, 2009.
40. Barbour, R.S. Are focus groups an appropriate tool for studying organizational change? In *Developing Focus Group Research: Politics, Theory, and Practice*; Barbour, R.S., Kitzinger, J., Eds.; SAGE Publications: London, UK, 1999; pp. 113–126.
41. Clarke, A.E. Straussian negotiated order theory c. 1960–present. In *The Routledge International Handbook of Interactionism*; Routledge: New York, NY, USA, 2021; pp. 47–58.
42. Strong, P.M.; Dingwall, R. The interactional study of organizations: A critique and formulation. In *Context and Method in Qualitative Research*; Dingwall, R., Miller, G., Eds.; SAGE Publications: London, UK, 1997; pp. 139–154.

sustainability

MDPI

Article

Delving Deeper into Market Concentration of Poultry Feed and the Driving Factors for Brand Switching: Evidence from Commercial Egg Producers in Nigeria

Fatai Abiola Sowunmi [1], Abeeb Adetunji Bello [1], Adebayo Isaiah Ogunniyi [2] and Abiodun Olusola Omotayo [3,*]

[1] Department of Agricultural Economics, University of Ibadan, Ibadan 200005, Nigeria; fa.sowunmi@ui.edu.ng (F.A.S.); abeebbello2016@gmail.com (A.A.B.)
[2] United Nations International Fund for Agricultural Development (IFAD), Abuja 900211, Nigeria; a.ogunniyi@ifad.org
[3] Food Security and Safety Niche Area, Faculty of Natural and Agricultural Sciences, North West University, Mmabatho 2735, South Africa
* Correspondence: 25301284@nwu.ac.za; Tel.: +27-611649252

Abstract: An increase in the price of branded feed has been a major problem for the poultry subsector of Nigerian agriculture, and brand switching for a cheap feed at the expense of quality is a common strategy used by egg-producing farmers. Using descriptive statistics, the Markov chain model and the logistic regression model, our study shows that almost 96% of the egg producers use branded feed while 43.1% switched feed brands because of the increase in the price of their preferred brands. Most farmers used Chikun (39.3%) and Top feed (23.2%) six months before data collection and during data collection, respectively. Our study found that approximately 37% of the feed sellers sold at least 10–50 bags per day. We revealed great inequality regarding market concentration: 50% of the feed sellers accounted for 89.5% of the total bags of feed sold per day. Hybrid had the highest customer loyalty. The study shows that Chikun gained 23.7% and 7.1% from Hybrid and Top feed, respectively, while Hybrid gained 36.0% and 35.7% from Chikun and Top feed (change in loyalty by egg producers), respectively. Membership of an association, distance to feed sellers, flock size and the average price of feed per bag were factors that influenced brand switching of poultry feed among egg producers. The study recommends that the government assists in subsidizing the price of critical ingredients (maize and soya bean) in feed production to prevent the price of eggs (the cheap source of protein) from becoming out of reach for most Nigerians.

Keywords: feed brand loyalty; switching behavior of egg producers; egg-producing farmers; logistic regression

Citation: Sowunmi, F.A.; Bello, A.A.; Ogunniyi, A.I.; Omotayo, A.O. Delving Deeper into Market Concentration of Poultry Feed and the Driving Factors for Brand Switching: Evidence from Commercial Egg Producers in Nigeria. *Sustainability* **2022**, *14*, 8030. https://doi.org/10.3390/su14138030

Academic Editor: Aaron K. Hoshide

Received: 30 March 2022
Accepted: 23 June 2022
Published: 30 June 2022

1. Introduction

The importance of the poultry industry in the Nigerian economy cannot be overemphasized. The industry contributes about 25% of agricultural GDP to the Nigerian economy (CBN, 2019). Apart from employment generation, eggs, one of the products of poultry production, are more affordable for the populace sources of animal protein [1–3]. Interestingly, egg consumption has jumped from 366,000 tons in 2000 to 598,000 tons in 2015, and projected to reach 947,000 tons by 2030. The poultry meat subsector has grown from 158,000 tons in 2000 to 317,000 tons in 2015, and projected to reach 544,000 tons by 2030 [4]. However, despite the importance of the industry, various challenges of which cost of feed and others factors (high rate of disease and pest attack, lack of loan and credit procurement, lack of technical knowledge, high rate of mortality, supply of poor-quality chicks, inadequate poultry extension services, and inadequate access to and high cost of veterinary services) have prevented the industry from expected performance [5,6].

Feed alone accounts for 70% of the total costs in poultry production [7]. The unabated increase in the price of poultry feed in Nigeria led to the high cost of maize, which constitutes between 50 and 60% of feed ingredients [8]. According to Akinfenwa [9], the price of maize thatwas NGN 80 (USD 0.22)/kg in March 2020, crept to NGN 180 (USD 0.22)/kg and NGN 260 (USD 0.69)/kg in August 2020 (the official exchange rate of dollars to naira in March 2020: USD 1 = NGN 364.55 and August 2020: USD 1 = NGN 378.49) and December 2021, respectively. As a result of the sharp increase in the price of maize, the cost of feed per bag also increased from NGN 3000 (USD 7.87) in March (the average official exchange rate of dollars to naira from March–April 2021: USD 1 = NGN 381) to between NGN 4600 (USD 12.07) and NGN 5300 (USD 13.91) in April. In December 2021, the price of 25 kg poultry feed ranged from NGN 5300 (USD 12.60)–NGN 7950 (USD 18.90) (the average official exchange rate of dollars to naira from March–April 2021: USD 1 = NGN 381), depending on the type and the brand of feed. Several reasons have led to the increase in the price of maize. These include a ban on the importation of maize used to augment the shortfall in the local supply, the problem of armyworm infestation, insecurity and climate change, among others [10]. Financially viable poultry production in Nigeria is essential to keep the cost of protein (e.g., eggs) lower for consumers and is in line with the National Food and Nutrition Policy and Agricultural Food Security and Nutrition Strategy of the Federal Government [11].

The consequence of the high cost of maize has resulted in higher poultry production costs and alteration of the finished poultry feed quality (commercial feed millers often do not meet the requirements of the animals due to the high cost of conventional feedstuffs). This leads to inadequate animal protein intake as a result of poor performance of the host animals fed with the poor-quality diets by some feed miller [12]. According to Madubuike [13], the high cost of feed has remained the major constraint facing poultry production in Nigeria because of the high percentage it accounts for in the total cost of poultry production. The poultry feed industry in Nigeria operates under monopolistic competition where each producer claims product differentiation from others in terms of content (quality) and package. It is a type of imperfect competition such that many producers compete against each other, but sell products that are differentiated to Top feed, Animal care, Chikun, Breedwell, New Hope, Vital and Hybrid, among others. Since these brands are differentiated from each other in the market based on quality and other attributes, the market is not perfectly competitive but rather monopolistically competitive [14,15]. According to Chron Contributor [16], brand switching as an outcome of customer switching behavior describes customers/consumers abandoning a product or service in favor of a competitor. The high cost of feed encourages brand switching among poultry farmers.

With the ever-increasing cost of feed, the likelihood of the commercial egg producer switching to cheaper poultry feed that will make it possible to achieve the profit-maximization goal is becoming unrealistic. The consequence is that many egg farmer workers may opt out of egg production. This may lead to farmworkers losing their jobs. The inability of the remaining few commercial farmers to meet egg demand will increase the price. This would make eggs unaffordable for an average Nigerian. Prior research [17] revealed that a medium-sized egg/crate was sold for NGN 1200 (USD 3.16) in November 2020. The price increased to NGN 1500 (USD 3.94) in March 2021 and is currently sold for NGN 1900 (USD 5.00). In the long run, at the detriment of food security, the pressure on the natural resources (soil and water) for the production of feed ingredients (maize, soybean, sorghum and groundnut) would decline. Tropical deforestation through various human activities, such as intensive crop farming, would not only lead to biodiversity loss and soil degradation, but deforestation is also responsible for significant amounts of greenhouse gas (GHG) emissions [18,19]. The expansion of arable land has driven growth in crop production rather than adoption of improved technology to enhance productivity [20].

Many factors influencing brand switching have been identified in the literature. These include price, promotional activities, brand image, variety and packaging [21–24], involvement [25,26] and dissatisfaction [25]. The effects of an increase in the price of poultry feed include: a reduction in the number of poultry farmers, the emergence of different poultry feeds that may be marginally cheaper than others, increase in the price of poultry products regarded as a cheap source of protein (egg and chicken), which makes it out of reach for the high percentage of the populace, as well as losses of jobs and livelihoods [27–29]. Manipulation of feed ingredients by the feed millers, use of synthetic amino acids, control of feed wastage, enzyme supplements, use of flavors and sourcing for cheap poultry feed are some of the coping strategies being adopted by poultry farmers to cope with the increasing cost of feed [30].

Consumer brand switching behavior has been researched in several studies in the mobile telecommunications, cosmetics, toothpaste, soft drink and banking industries [30–32]. Past studies on poultry feed [33–38] have concentrated on profitability analysis, quality of poultry feed, alternative feedstuff and marketing. Our study's aim is to address the gap in the literature on the brand switching of poultry feed among egg producers in southwest Nigeria, which is the nation's poultry production hub. Here, brand switching is a coping strategy to adapt to higher poultry branded feed prices by commercial egg farmers.

The poultry industry is concentrated in southwest Nigeria, which is a geopolitical zone over six decades old with a poultry population that has steadily grown to 30 million or 60% of the national flock [9]. Our study is timely because the current increases (almost weekly) in the cost of feed has forced many farmers out of business and remaining farmers are searching for coping strategies that would keep them in business. There is a dearth of literature on feed brand switching by poultry farmers unlike other commodities, such as mobile phones, beer, gin, toilet soaps, etc. Therefore, this study seeks to fill these gaps and also provide answers to the following research questions: What are the socioeconomic characteristics of egg farmers/farms and poultry feed sellers in southwest Nigeria? What is the extent of market concentration in the sales of poultry feed in the study area? What is the pattern of brand switching of poultry feed among commercial egg-producing farmers? What are the factors influencing brand switching of poultry feed and the proportion of feed used in the long run by commercial egg-producing farmers? What are the implications of brand switching of poultry feed in the community? Evidence and recommendations from this study are important for policy trajectories and the development of the poultry industry in Nigeria, which enhances and ensures the possibility of household dietary diversity.

2. Theoretical Framework and Literature Review

Production and rational choice theories support our study. Production involves the combination of various material and immaterial inputs (plans, know-how) to produce something for consumption (the output). In the egg production business, feed, day-old chicks, drugs and depreciation on fixed items (the pen, cage, feeder, drinkers, wheelbarrow and shovel, among others) constitute production inputs, while the eggs are the output. Rational choice theory states that individuals use calculations to make rational choices and achieve outcomes in agreement with the objectives [39]. This theory is associated with maximizing an individual's self-interest. Using rational choice theory is expected to result in outcomes that provide people with the greatest benefit and satisfaction, given the limited option they have available. Many mainstream economic assumptions and theories are based on rational choice theory. Rational choice theory is associated with the concepts of rational actors, self-interest and the invisible hand [39,40]. As an entrepreneur, the motive of a commercial egg farmer is to have good returns on the amount invested through profit maximization. One of the ways to achieve this is to ensure low mortality and having access to quality feed feeds (Chikun, Top, Hybrids and Animal Care, among other brands) at a price that will not undermine profit. This is because feed costs are 60–70% of the cost of egg production. While egg farmers may not influence the cost of poultry feed per bag in the market, they can opt for cheaper poultry feed if their preferred brand of feed is costly [41].

In the literature, Herfindahl–Hirschman, Linda and Horwath indices approaches are methods used to measure the market concentration or of the extent of inequality in the market shares in a particular sector. However, HHI shows sensitivity to firm size, while the Linda index is based on the distribution of the largest firms, not the entire distribution within the sector. It is difficult to find the data required for the estimation of the Horwath index, especially the marginal cost [42]. The Lorenz curve and Gini coefficient were used in this study to measure the extent of market concentration in branded poultry feed. The Lorenz curve is an absolute measure of concentration in which firm size inequality is represented by the convexity of the curve [43]. The Gini coefficient was calculated from the Lorenz curve and this measures the magnitude of the area between the Lorenz curve and the absolute equation line. This area reflects the proportional effect of the firm's size and control share [42]. The Lorenz curve shows how the variable of interest is distributed among the population. It produces an alert for monopoly emergence.

Several methods (Autoregressive Integrated Moving Average (ARIMA); exponential smoothing and simultaneous equations) have been used in the literature for forecasting. The ARIMA model has been used to make future predictions and this model uses time-series data [44,45]. One of the limitations of the ARIMA model is that the parameters (p, d and q) need to be manually defined. Therefore, finding the most accurate fit can be a lengthy trial-and-error process. The exponential smoothing model has also been used, but it does not recognize seasonal patterns and cannot project trends [46,47]. Simultaneous equation models have also been adopted in many studies [48–51]. The limitation of this method is that the two-stage least squares (2SLS) estimator is just the ratio of two covariances and it has weak instrumental variables. The 2SLS or general instrumental variables estimator does not exist and is inconsistent. Generalized method of moments (GMM) parameter estimates are usually measured with more errors. GMM estimates have the limitation of small sample properties.

A Markov chain model was used in this study since two consecutive periods were considered (brand of feed in use during data collection and the brand of feed used six months before data collection). A Markov chain or Markov process is a stochastic model describing a sequence of possible events in which the probability of each event depends only on the state attained in the previous event [52]. The state of probabilities at a future instant given the present state of the process does not depend on the state it occupied in the past. Markov chain models have been extensively used in brand switching studies [53–57]. Markov models are generalized and the generated sequences look similar to a sample of the real usage as long as the model captures the operational behavior. The analytical theory of the Markov chain model presumes a formal stochastic process [58].

Analytical Frameworks of the Markov Chain Model

The theory of Markov process assumes the existence of a physical system S, which has a number of possible systems, $S_1, S_2, \ldots, S_n$, and which at each instant of time can be in one of these states. The time after each successive trials can be denoted by $t_0, t_1, t_2, \ldots, t_n$, with t_0 representing the starting point in time, t_1 as the time of conducting the first trialfor Markov chain; the probability of passing to some state S_1 at a given time depends on the state that the system was at the preceding time and does not change if you know what the states were at the earlier times. In the Markov chain, P_{ij} is used to denote the transition from one state to another (i to j). The probability transition matrix can be simplified as:

$$p = \begin{pmatrix} p_{11} & p_{12} & p_{13} \\ p_{21} & p_{22} & p_{23} \\ p_{31} & p_{32} & p_{33} \end{pmatrix}. \tag{1}$$

To forecast the proportion of the variable of interest at time k:

$$p(k) = p(0)p^k \tag{2}$$

where:

$p(k)$ represents the probability transition matrix at time k, and
$p(0)$ represents the probability transition matrix (PTM) at the initial or time zero (0).

At the equilibrium or steady state, the proportion of variable of interest is equal to the proportion multiplied by the PTM given as:

$$e = ep \tag{3}$$

where:

e can be 1×2 when PTM is 2×2, 1×3 when PTM is 3×3, 1×4 when PTM is 4×4, etc.
p istheprobability transition matrix.

For 3×3 PTM, $e = ep$ is given as:

$$\begin{pmatrix} e_1 & e_2 & e_3 \end{pmatrix} = \begin{pmatrix} e_1 & e_2 & e_3 \end{pmatrix} \begin{pmatrix} p_{11} & p_{12} & p_{13} \\ p_{21} & p_{22} & p_{23} \\ p_{31} & p_{32} & p_{33} \end{pmatrix} \tag{4}$$

This gives three equations:

$$p_{11}e_1 + p_{21}e_2 + p_{31}e_3 = e_1 \tag{5}$$

$$p_{12}e_1 + p_{22}e_2 + p_{32}e_3 = e_2 \tag{6}$$

$$p_{13}e_1 + p_{23}e_2 + p_{33}e_3 = e_3 \tag{7}$$

$$e_1 + e_2 + e_3 = 1 \tag{8}$$

The total equation from $e = ep$ is four (4). Solving the system of equations for the *es* produce the required equilibrium probability vector. However, it should be noted that the system of equations provides a set of n + 1 equations. That is, the equations are more than the unknowns—four equations with three unknowns. This shows that one of the equations (first three) is not linearly independent of the others. Therefore, one of the firstthree equations can be removed (assuming Equation (11) is removed) to presentthree equations with three unknowns as:

$$p_{12}e_1 + (p_{22} - 1)e_2 + p_{32}e_3 = 0 \tag{9}$$

$$p_{13}e_1 + p_{23}e_2 + (p_{33} - 1)e_3 = 0 \tag{10}$$

$$e_1 + e_2 + e_3 = 1 \tag{11}$$

The solution to the system of equations producesthe equilibrium probability vectors of e_1, e_2 and e_3.

3. Materials and Methods

3.1. Study Area

The study was conducted in South West Nigeria (SWN), which is one of the six geographical zones in Nigeria. Three states (Ogun, Lagos and the Oyo States) were considered out of the six states that make up the zone. SWN falls within latitude 60° to the North and latitude 40° to the South (Figure 1). SWN is bounded in the North by Kogi and Kwara States, in the East by Edo and Delta States, in the South by the Atlantic Ocean and in the West by the Republic of Benin. The zone is characterized by a tropical climate with a distinct dry season between November and March and a wet season between April and October. The mean annual rainfall is 1480 mm with a mean monthly temperature range of 18–24 °C during the rainy season and 30–35 °C during the dry season [59]. The zone has a land area of about 114,271 square kilometers. The total population of the SWN was 27,581,992 in 2006 [60]. Occupations in this region are predominantly crop farming. In addition to crops, this region has the highest concentration of poultry farms in Nigeria and contributed to most of the 646,667 tons of eggs produced in 2020 [9,61].

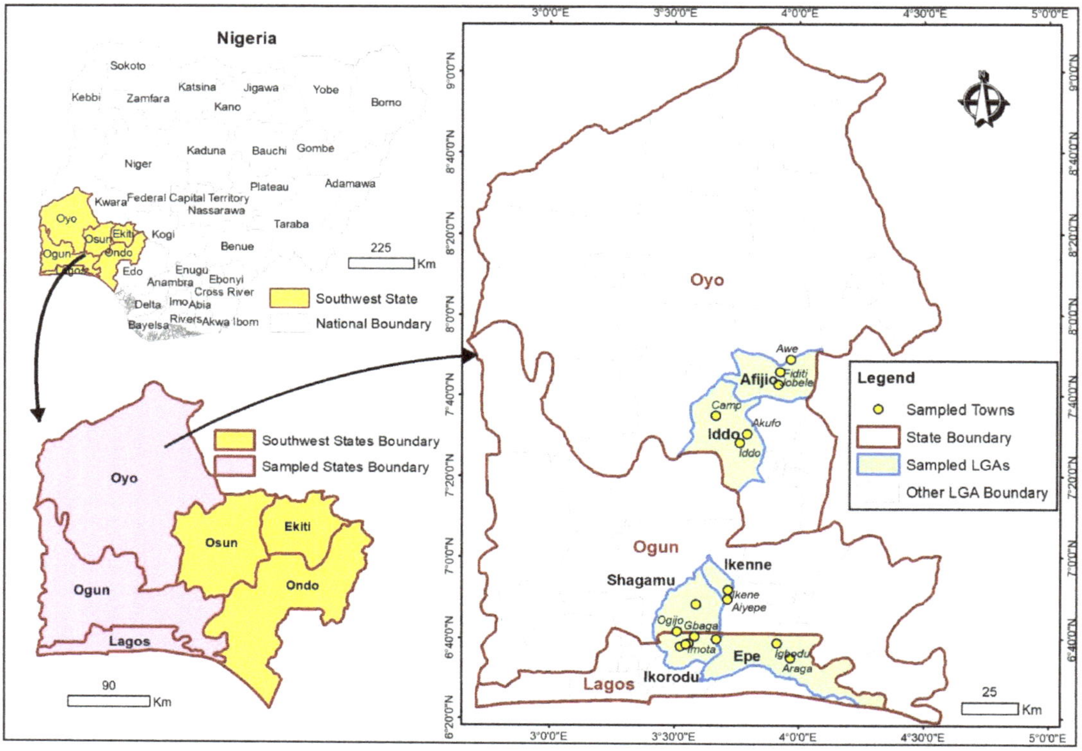

Figure 1. Geographical location of the selected states and communities in the South West, Nigeria.

3.2. Sampling Procedure and Sample Size

A four-stage sampling technique was employed. The first stage was the purposive selection of Ogun, Osun and Lagos States out of the six states that make up SWN, which is known for commercial egg production. In the second stage, two Local Government Areas (LGAs) with a high concentration of poultry farms (egg production) were purposively selected from each state (Ogun; Sagamu and Ikenne, Oyo: Afijio and Iddo, Lagos: Ikorodu and Epe). The third stage involved random sampling of towns/villages where poultry production is concentrated, based on proportionate size. The fourth stage was a random selection of commercial egg farmers proportionate to size based on the list obtained from the local chapter of the Poultry Association of Nigeria (Table 1). Eighty poultry feed sellers were randomly selected based on the list of the sellers in each location. The sample sizes for the egg producers (150) and feed sellers (80) were arrived at using the International Fund for Agricultural Development (IFAD) procedure [62].

The calculated sample sizes (egg producers: 138.2 ≅ 138; feed sellers: 59.0 ≅ 59) for the study were obtained using IFAD procedure based on the formula below. The final sample sizes (150 and 80 for egg producers and feed sellers, respectively) used included allowances for design effect and contingency. The allowance for design effect is expected to correct the difference in design, while the allowance for contingency accounts for contingencies, such as non-response or recording error.

The sample size was obtained using:

$$n = \frac{z^2 p(1-p)}{m^2} \tag{12}$$

where:

n = the sample size;
Z = the confidence level at 95% (1.96);
p = the estimated percentage of egg producers using branded feed in the study area (90%), estimated percentage of feed sellers with at least two different brands being sold (96%);
m = the margin of error (5% or 0.05).

Moreover, 150 and 80 copies of the questionnaires were administered to egg producers and feed sellers, respectively. One hundred and forty copies of completed questionnaires were collected. However, after cleaning the completed questionnaires, 145 copies of the questionnaire for egg producers were good enough for analysis. Additionally, 65 copies of the sellers' questionnaires were returned, and 62 properly completed questionnaires were used for the analysis (Table 1). Data were collected on the socio-economic characteristics of egg-producing farmers and feed sellers (age, sex, marital status, household size, educational status, sellers' membership of association and source of credit). Other data collected were farm characteristics that included flock size, bag(s) of feed consumed per day, duration of birds laying and distance to the feed sellers.

Table 1. Distribution of respondents based on sampling technique.

State	LGA	Town/Village	Number of Respondents of Egg Producers	Number of Respondents of Feed Sellers
Ogun	Sagamu	Ogijo	32	17
		Gbaga	9	5
		Shotumbo	4	2
	Ikenne	Ikenne	7	4
		Aiyepe	4	2
Lagos	Ikorodu	Erikorodo	29	15
		Gbaga	11	6
		Imota	3	2
		Laspotech	2	1
		Lucky Fibre	17	9
	Epe	Farm Settlement	3	2
		Araga	7	4
		Gbodu	3	2
Oyo	Iddo	Camp	3	2
		Akufo	4	2
		Iddo	4	2
	Afijio	Awe	2	1
		Fiditi	3	2
		Jobele	3	2
Total planned respondents			150	80

3.3. Analytical Techniques

3.3.1. Lorenz Curve and Gini Coefficient

The Lorenz curve was used to show the extent of market concentration in branded poultry feed graphically, while the Gini coefficient was used to complement the Lorenz curve by giving the empirical value of the market concentration or inequality in branded poultry feed markets. The Gini coefficient is expressed mathematically as:

$$G = 1 - \sum_{k=1}^{n} (X_k - X_{k-1})(Y_k - Y_{k-1}) \tag{13}$$

where:

X_k is the cumulated proportion of the poultry feed sellers.
Y_k is the cumulated proportion of the sales revenue of poultry feed seller per month.

3.3.2. Markov Chain Model

The probability transition matrix (PTM) was used to determine the pattern of brand switching of poultry feed among commercial egg producers. The study considered only three brands of poultry feed (Chikun, Hybrid and Top feed) that had consistent patronage before (6 months ago) and during the data collection (Table 2). The notation of PTM is given in Equation (1).

Table 2. The probability transition matrix.

Brand Switching of Poultry Feed		During Data Collection (t)		
		Chikun	Hybrid	Top Feed
Six months before data collection $(t-1)$	Chikun	P_{11}	P_{21}	P_{23}
	Hybrid	P_{12}	P_{22}	P_{23}
	Top feed	P_{13}	P_{23}	P_{33}

Where $P_{11}, P_{12}, P_{13}, \ldots, P_{33}$ represents the probability transition of the poultry egg farmers as theyswitch from one brand of poultry feed to the other, as well as where loyalty is maintained. The initial proportion of the branded poultry feed is given as:

$$p(0) = \begin{pmatrix} t_1 & t_2 & t_3 \end{pmatrix} \tag{14}$$

where t_1–t_3 represents the proportion of the feed sold at the initial stage.

3.4. Logistic Regression

The logistic regression was used to determine the factors influencing brand switching of feed among the egg producers in the study area. Following [63], the logistic regression gives each predictor a coefficient, which measures its independent contribution to variation in the dependent variable. The dependent variable Y takes the value 1 if the response is "Yes that brand switching of feed took place", and takes a value 0 if the response is "No that there was no brand switching of feed." The model form for predicted probabilities is expressed as a natural logarithm (ln) of the odds ratio:

$$\begin{aligned} \ln\left(\frac{P(Y)}{1-P(Y)}\right) = \; & \beta_0 + \beta_1 X_1 + \beta_2 X_2 + \beta_3 X_3 + \beta_4 X_4 \\ & + \beta_5 X_5 + \beta_6 X_6 + \beta_7 X_7 + + \beta_8 X_8 + \beta_9 X_9 + \beta_{10} X_{10} \end{aligned} \tag{15}$$

where:

β_0 represents the intercept;
β_1 to β_{10} represents the regression coefficient;
X_1 represents the age (year) of respondent;
X_2 represents the marital status of respondent (married = 1, others = 0);
X_3 represents the respondent's household size;
X_4 represents the respondent's engagement in other economic activities (Yes = 1, No = 0);
X_5 represents the respondent's membership of related organization (Yes = 1, No = 0);
X_6 represents the respondent's flock size (population of egg-laying birds);
X_7 represents the respondent's years of experience in egg production;
X_8 represents the respondent's duration (months) of egg laying by flock;
X_9 represents the average number of bags of feed (25 kg) fed to birds per day;
X_{10} represents the respondent's distance (km) to the nearest feed seller;
X_{11} represents the average price (NGN) of feed per bag (25 kg) used by the respondent.

4. Results and Discussion

4.1. Socioeconomic Characteristics of Egg Producers

Our study revealed that most commercial egg production was dominated by males (69.3%), while 29.9% of the respondents were within the age bracket of 38–47 years with an average age of 42.9 years. The average age of egg farmers obtained in the study agrees with the findings of [64] on access to credit by poultry farmers in SWN. Additionally, 2.1% of the poultry farmers had primary school education, while most egg producers (43.4%) had tertiary certificates. Moreover, 57.8% of the egg producers were members of associations. The breakdown of the association membership showed that 88.2% were members of the Poultry Association of Nigeria while 11.8% were members of the Poultry Egg Producers' Association. However, there was a marginal difference in the percentage of poultry farmers that belonged to associations in the findings of [65] on poultry farmers' willingness to participate in national agricultural insurance scheme in Oyo State. They recorded 60.3% poultry farmers. The benefit gained was prompt information on the current price of egg per crate through the social medium used by the members. Moreover, 46.2% of the egg producers had 6 to 10 years of experience. The average years of experience in egg production by respondents was 7.7 years. Most poultry farmers (85.2%) in the study area adopted the intensive method of egg production. The average laying period of birds per farmer was 9.3 months, and most egg producers' birds had been laying for 8 to 10 months. Our study revealed that 95.6% of the farmers used branded feed, which is higher than 55.6% recorded by [66]; most egg producers used 6 to 10 bags (25 kg/bag) to feed birds per day. The average quantity of feed used by farmers to feed their birds per day was 7.7 bags (Appendix A, Table A1). Moreover, 70.1% of the egg producers changed the feed brand used six months before data collection for the study, while 43.1% and 25.5% attributed the change in feed used to the increase in the price of their favorite poultry feed and the availability of the cheaper brand of feed, respectively (Appendix A, Table A1). Out of twelve brands of feed used by farmers in the study area, 39.3%, 18.0% and 14.8% of the egg producers used Chikun, Top and New Hope, respectively, to feed birds six months before data collection for the study. Ten brands of feed commonly used by poultry farmers, namely, Vital, Stellar, Animal care, Breedwell, Amo Byng, Top, Hybrid, Ultima, Chikun and New Hope, were identified by [67]. During the data collection, 23.2% and 21.6% of the farmers used Top and Chikun, respectively, to feed birds (Appendix A, Table A2). The average distance from the farms to the feed sellers was 8.9 km (Appendix A, Table A1).

4.2. Socio-Economic Characteristics of Feed Sellers

Our study showed that 56.5% of the sellers of branded poultry feed were male. This is similar to the finding of [68] conducted in Free State province, South Africa. The average age of the sellers was 32.5 years, and the majority (26.2%) of the feed sellers were within the age range of 28 to 37 years. This finding disagrees with [33,36] that found 45 and 37 years, respectively, in Delta and Imo States. Furthermore, 1.7% and 37.3% of the feed sellers we surveyed had no formal education and Higher National Diploma (HND)/Bachelor of Science (BSc) certificates, respectively. In [69], 1.4% of the feed sellers had no formal education. The average experience in feed sales was 5.3 years, and most respondents had experience in feed sales below the average (positive skewness). Similar studies on feed marketing in Delta and Imo States [33,36] recorded 15 and 12 years experience in feed marketing, respectively. The differences in the years of experience may be attributed to frequent increases in the price of feed over the past year, which has driven many sellers out of the business. Most sellers sold between 10 and 50 bags of feed per day, and the average bag (25 kg) of feed sold was 136.1 bags per day (Appendix A, Table A3). The average number of bags sold per day was higher than 3.3 bags.

The most expensive branded poultry feed was New Hope while Spring was the cheapest six months before the data collection. New Hope and Bonka brands were the most expensive and cheapest, respectively, during data collection (Appendix A, Table A4). Tables 3 and 4 shows there is asignificant difference in the average price of the different brands of poultry feed in the two periods ($p < 0.05$). According to [70], the average price of 25 kg bag of poultry feed was NGN 7500 (USD 19.68) (the exchange rate of dollars to naira as at the time of data collection (March 2021): USD 1 = NGN 381).

Table 3. Difference in average feed prices before and during data collection.

Parameter	Variable	Mean	Variance	Z-Cal	Z-Tab
Average price (NGN) per bag (25 kg) of different brands of feed	Six months before data werecollected (N = 32)	5131.07	482,004.22	−2.31 **	1.98
	During data collection (N = 10)	5476.95	73,034.30		

Note: ** means significant at 5%.

Table 4. Brand switching behavior of egg producer on feed.

Poultry Feed Brands		T		
		Chikun	Hybrid	Top Feed
T − 1	Chikun	0.360	0.360	0.280
	Hybrid	0.273	0.591	0.136
	Top feed	0.071	0.357	0.571
$P(K=1) = (0.233 \quad 0.451 \quad 0.316)$, $P(K=2) = (0.229 \quad 0.463 \quad 0.307)$				

4.3. Extent of Market Concentration (Inequality) in the Number of Bags of Feed Sold per Day

In the Lorenz curve, the farther away from the curve from the diagonal (line of equality), the more inequality will be encountered. Being away from the diagonal shows more monopolization tendency in the industry [42]. Figure 2 shows that 50% of the poultry feed sellers in the study area controlled about 89.5% of the total sales (bags of feed). The result was confirmed by a Gini coefficient of 0.5377. This showed that there was high inequality (concentration) in the bags of feed sold per day. This means that few sellers were controlling the feed sales in the study area. The high inequality in daily sales may lead to monopoly, thus forcing other sellers out of business. This may be attributed to reduced price compared to others, having assorted brands most of the time, location close to where many farms were sited and good customer relationships, among others. The market concentration of bagged of poultry feed sold was lower (0.302) in Osun State in 2018 [71].

4.4. Pattern of Change in Feed Brands and the Proportion of Change in the Brand in the Long Run

Table 4 shows that consumers of Hybrid feed had the highest rate of brand loyalty (59.1%), followed by Top (57.1%). This means that 59.1% of the egg farmers that bought Hybrid six months before the data for the study were collected used the same feed during data collection. Chikun had the least loyalty of customers (36.0%). Additionally, 57.1% of the egg farmers that used Top feed six months before the data for the study were collected used the same feed during data collection. Moreover, Table 4 reveals that 36% of the egg-producing farmers that purchased Chikun before maintained the brand loyalty status after (during data collection). Chikun gained 23.7% and 7.1% from Hybrid and Top feed, respectively. That is, 23.7% (7.1%) of the egg producers that used Hybrid (Top feed) before changed to Chikun six months after. Hybrid gained 36.0% and 35.7% from Chikun and Top feed (change in loyalty by egg producers), respectively. On the other hand, Top feed gained 28.0% and 13.6% from Chikun and Hybrid. Generally, the change in brand/loyalty appeared to be attributed to an increase in price, a drop in egg production, as well as the scarcity of preferred brands, and many farmers complained during data collection, which is consistent with a study conducted in Irepodun LGA of Kwara State in 2017 [72]. Additionally, [73] revealed that Top feed is prone to more switches than any other brand while Chikun users expressed the least switches. They posited that the Top feed brand has more latent competitors and may be more prone to switching experiences in case of failed customer experiences or when the brand is out of stock in the retail outlets.

The predictions for $k = 1$ and $k = 2$ showed that six months after the data collection, 23.3% of the egg producers would purchase Chikun, while 45.1% and 31.6% would use Hybrid and Top feed, respectively. Additionally, twelve months after the data were collected,

22.9%, 46.3% and 30.7% of the egg producers in the study area would purchase Chikun, Hybrid and Top feed, respectively. Moreover, our study affirmed that at equilibrium (long-run prediction), the proportion of egg farmers buying the three brands of feed would be 23.0% for Chikun, 46.8% for the hybrid and 30.2% for the Top feed.

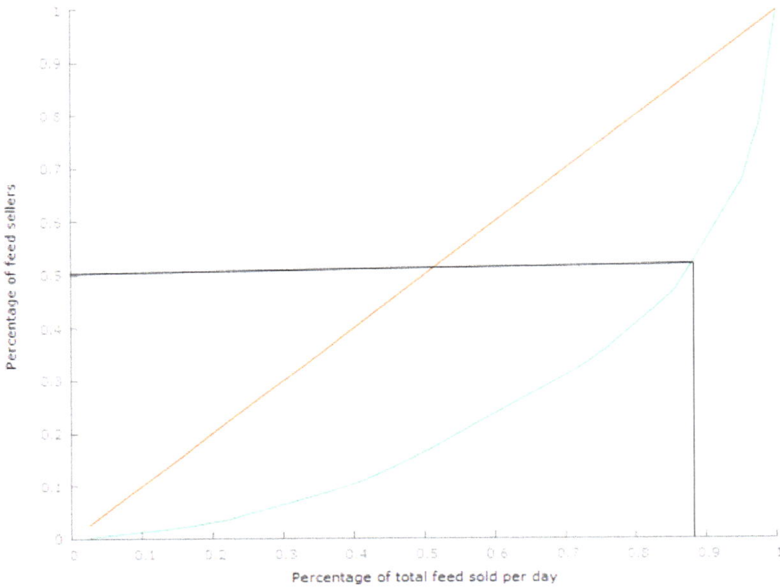

Figure 2. Extent of inequality in bags of feed sold per day.

4.5. Determinants of Brand Switching in Poultry Feed among Egg-Producing Farmers

The independent variables considered in the logistic regression model were age (years), marital status, household size, engagement in other activities, membership of association, flock size, experience in egg production, duration in bird laying, bags of feed given per day, distance to feed sellerandthe average price of poultry feed. The log-likelihood value of −52.820865, LR chi^2 (11) = 32.73, Prob > Chi^2 = 0.0006 and Pseudo R^2 = 0.237 affirmed that the explanatory variables in the model predicted the outcome of the model effectively. Out of the eleven independent variables captured in the model, the coefficients of five variables (membership of an association, flock size, bags of feed used per day, distance to feed sellers and average price of feed) significantly influenced consumer brand switching. Specifically, other variables included in the model were not significant (Table 5).

The membership of association variable had a significant ($p < 0.1$) and positive relationship with brand switching of poultry feed by the farmers in the study area. The result implies that farmers' memberships in the association would increase the probability of brand switching of feed by 19.5%. This may be attributed to the access to information and new technology by members of associations. The result shows that as the flock size increases, the probability of brand switching increases, especially when there is an increase in the price of feed or a drop in the birds' production. As the number of bags used per day increases in an inflationary economy, the likelihood of the farmer opting for cheaper feed increases. The coefficient of distance to feed sellers was significant at 10% and had a positive relationship with brand switching of poultry feed by the farmers in the study area. This shows that the further the feed seller is from the farm, the higher the probability that the farmer would engage in brand switching of feed. This may be the case to reduce the cost of production, most especially the cost of transporting feed from a distance far away from the farm. This finding agrees with a non-agricultural study [74] in Indonesia that an increase in distance to the preferred brand increases the likelihood to switch to another brand nearby. The result also revealed that the average price of feed had a negative relationship with brand switching of poultry feed by the farmers in the study area. This shows that as the price of the feed band increases, the probability that a farmer would switch the brand of feed used increases by 1.6%. The negative relationship with brand switching agrees with a study conducted in Manado by [75] that an increase in the price of a branded commodity increases the decision of the buyer to switch brands.

Table 5. Logistic regression output for factors associated with poultry feed brand purchases.

Variables	Coef.	Std. Err.	z	*p*-Value	dy/dx
Age	0.033	0.028	1.17	0.241	0.006
Marital status	−1.009	0.753	−1.34	0.180	−0.162
Household sizes	−0.144	0.106	−1.36	0.173	−0.027
Engagement in other activities	0.666	0.519	1.28	0.199	0.127
Membership of association	0.988 *	0.579	1.71	0.088	0.195
Flock size per farmer	0.001 **	0.000	2.36	0.018	0.000
Experience in egg production	0.103	0.081	1.28	0.201	0.020
Duration of egg laying by flock	0.098	0.087	1.12	0.264	0.186
Average bags of feed used per day	−0.105 **	0.051	−2.05	0.040	−0.020
Distance to the nearest feed seller	0.084 *	0.046	−1.81	0.070	0.016
Average price (NGN) of feed per bag	0.003 ***	0.001	2.99	0.003	−0.001
Con.	14.8200	5.148504	2.88	0.004	

Note: *** indicates significance at 1%, ** indicates significance at 5%, * indicates significance at 10%.

4.6. Community Sustainability Implications

The expected positive impact of the commercial egg farmers on the immediate environment hinges on sustainable production. A good return on the capital invested is important for economic sustainability. Since feed accounted for the major cost of production, the availability of poultry feed at a reasonable price is germane to achieve the profit maximization objective. However, brand switching may be able to keep poultry farmers in business with the increasing cost of poultry feed, as confirmed in Table 3. This may lead to farmers opting out of egg production, and by extension, the farm workers who serve as an important component of the rural economy would be jobless. As many as 350,000 poultry farmers were forced out of business in Ogun State alone, while others are reducing their flocks (reduction in labor) due to the high cost of feed making egg production unattractive [76]. Though, to the detriment of food security, the reduction in the number of poultry farmers may reduce the pressure of clearing vegetation all the time for the cultivation of maize, soybeans, sorghum and groundnut used in poultry feed production to meet increased demand. Intensive crop management practice is accompanied by the use of fertilizers, herbicides and heavy farm machines. These inputs escalate rates of land degradation as well as soil and water deterioration [77–80].

5. Conclusions and Recommendations

The importance of the poultry industry in the Nigerian economy cannot be overemphasized. This study estimated the extent of market concentration of the sales of poultry feeds, the pattern of brand switching of poultry feed among egg-producing farmers, and factors influencing brand switching of poultry feed and the proportion of feed used in the long run among the egg-producing farmers. The result shows that there is great inequality in the number of bags of branded feed sold in the study area, which suggestamonopolistic market in branded feed may force feed sellers out of the business. A substantial percentage of poultry farmers engaged in brand switching of poultry feed attributed to an increase in the price and the distance to the sellers of choice feed. However, the buyers of Hybrid and Top feeds were more loyal to the feed sellers compared to other brands. Further research on the impacts of brand switching in poultry feed on the profitability of commercial egg farmers is advocated.

Since most egg-producing farmers attributed the increase in the price of feed to brand switching behavior, it is recommended that the government assists in subsidizing the price of critical ingredients (maize and soybean) in feed production to reduce the frequent increase in the price of feed. The Poultry Association of Nigeria should mandate state branches to embark on the backward integration of the production of maize and soybean. The Feed Dealers Association should ensure that the emerging monopolists in the feed market are addressed to protect the small feed sellers. This may be accomplished by engaging the expertise of marketing specialists. However, as long as the cost of feed continues to increase, poultry egg farmers will keep on searching for cheaper brands of feed that they believe will help reduce their cost of production to sustain their business.

Author Contributions: Conceptualization, F.A.S. and A.A.B.; methodology, F.A.S. and A.I.O.; software, F.A.S.; validation, F.A.S. and A.O.O.; formal analysis, F.A.S. and A.I.O.; data curation, A.O.O. and F.A.S.; writing—original draft preparation, F.A.S. and A.O.O., writing—review and editing, F.A.S., A.I.O. and A.O.O.; supervision, F.A.S. All authors have read and agreed to the published version of the manuscript.

Funding: This research received no external funding.

Institutional Review Board Statement: Not applicable.

Informed Consent Statement: Informed consent was obtained from all participants in the study.

Data Availability Statement: Data are available upon request from the first author.

Acknowledgments: The authors appreciate their respective institutions for providing an environment to conduct this research study.

Conflicts of Interest: The authors declare no conflict of interest.

Appendix A

Table A1. Socioeconomic characteristics of egg producers.

Socioeconomic Characteristics	Frequency	Percentage (%)
Age (Year) Distribution		
18–27	14	10.2
28–37	31	22.6
38–47	41	29.9
48–57	40	29.2
58–67	9	6.6
68–77	2	1.5
Total	137	100
Experience (years) in egg production		
1–5	41	31.1
11–15	25	18.9
16–20	5	3.8
6–10	61	46.2
Total	132	100
Average experience	7.7	
Educational level		
Primary education	3	2.1
Secondary education	36	24.8
OND/NCE	28	19.3
BSC/HND	63	43.4
Post-graduate	15	10.3
Total	145	100
Sex of respondents		
Female	42	30.7
Male	95	69.3
Total	137	100

Table A1. *Cont.*

Socioeconomic Characteristics Age (Year) Distribution	Frequency	Percentage (%)
Average bag (25 kg) of feed birds consumed per day per farmer		
1–5	56	41.2
6–10	50	36.8
11–15	15	11.0
16–20	11	8.1
21 andabove	4	2.9
Total	136	100
Average bags consumed	7.7 bags	
Flock size distribution		
At most 100 birds	3	2.2
101–500	36	26.7
501–1000	42	31.1
1001–1500	20	14.8
1501 and above	34	25.2
Total	135	
Average	1309.3	
Membership of association		
No	38	42.2
Yes	52	57.8
Association		
Poultry Association of Nigeria PAN	75	50.0
Poultry Egg Producers' Association	10	6.7
Benefits from association		
New sales point	34	
New sales point and price preference	14	17.9
Price preference	30	38.5
Total	78	100.0
Management Practice		
Intensive	115	85.2
Semi-intensive	20	14.8
Total	135	100
Laying period (month)		
2–4	10	7.8
5–7	33	25.6
8–10	43	33.3
11–13	25	19.4
14–16	13	10.1
17–20	5	3.9
Total	129	
Branded feed usage		
No	6	4.4
Yes	128	95.6
Total	135	100.0
Reason for change in brand used		
Availability of cheap feed	26	25.5
Decline in birds productivity	19	18.6
Increase in feed price	44	43.1
Reduction in egg size	6	5.9
Scarcity of the preferred brand	7	6.9
Distance (km) to feed sellers		
1–5	43	31.2
6–10	55	39.9
11–15	20	14.5
16–20	8	5.8
20 and above	12	8.7
Total	138	100.0
Average distance	8.9	

Table A2. Distribution of branded feeds used by farmers before and during data collection.

Brand	Six Months Before Data Collection		Brand	During Data Collection	
	Freq	Percent (%)		Freq	Percent (%)
Amobyng	4	3.3	Amobyng	6	4.8
Animal care	2	1.6	Animal care	2	1.6
Breedwell	2	1.6	Biacom	1	0.8
Chikun	48	39.3	Bonka	1	0.8
Hi pro	3	2.5	Breedwell	1	0.8
Hybrid	13	10.7	Chikun	27	21.6
Livestock	5	4.1	Cornerstone	2	1.6
New hope	18	14.8	Hi pro	4	3.2
Spring	2	1.6	Hybrid	24	19.2
Top	22	18.0	Livestock	12	9.6
Victory	1	0.8	New hope	11	8.8
Vital	2	1.6	Spring feed	2	1.6
Total	**122**	**100**	Top	29	23.2
			Vita feed	3	2.4
			Total	**125**	**100**

Table A3. Socioeconomic characteristics of branded poultry feed sellers.

Socioeconomic Characteristics	Frequency	Percentage (%)
Sex of respondents		
Male	35	56.5
Female	27	43.5
Total	62	100.0
Age (year) range of feed sellers		
18–27	15	24.6
28–37	16	26.2
38–47	16	26.2
48–57	10	16.4
58–67	4	6.6
Total	**61**	**100.0**
Average age	32.5	
Experience (year) of feed sellers		
1–4	28	48.3
5–8	21	36.2
9–12	6	10.3
Above 12	3	5.2
Total	58	100
Average	5.3	
Educational level		
No formal education	1	1.7
Primary education	3	5.1
Secondary education	17	28.8
OND/NCE	12	20.3
BSC/HND	22	37.3
Post-graduate	4	6.8
Total	59	100
Bags of different brands sold per day		
10–50	15	36.6
60–100	13	31.7
105–160	7	17.1
270–305	4	9.8
585–1200	2	4.9
Total	41	100
Average	136.1	

Table A4. Distribution of price per bag of branded feed used by farmers before and during data collection.

Brand	Price (NGN) per 25 kg Six Months Before Data Collection	Brand	Price (NGN) per 25 kg during Data Collection
Spring	4300.0	Amo	5625.0
Animal Care	4650.0	Animal care	5562.5
Bonka	4600.0	Bonka	5150.0
Chikun	4635.0	Chikun	5625.6
Hybrid	4571.6	Hybrid	5250.0
New Hope	5400.0	Livestock	5325.0
Top	4787.5	New hope	5950.0
		Top	5781.4
		Hi pro	5200.0
		Victory	5300.0

N.B: Exchange rate of dollars to naira in March 2021 when the data were collected was USD 1 = NGN 381.

References

1. Heise, H.; Crisan, A.; Theuvsen, L. The poultry market in Nigeria: Market structures and potential for investment in the market. *Int. Food Agribus. Manag. Rev.* **2015**, *18*, 197–222.
2. Aboki, E.; Jongur, A.A.U.; Onu, J.I. Productivity and technical efficiency of family poultry production in Kurmi Local Government Area of Taraba State, Nigeria. *J. Agric. Sustain.* **2013**, *4*, 52–66.
3. Ojo, S.O. Productivity and Technical Efficiency of Poultry Egg Production in Nigeria. *Int. J. Poult. Sci.* **2003**, *2*, 459–464.
4. Chiekezie, N.R.; Nwankwo, E.C.; Ozor, M.U. Analysis of Small Scale Broiler Poultry Production in South East Nigeria, West Africa. *Int. J. Anim. Livest. Prod. Res.* **2022**, *6*, 1–16.
5. Anosike, F.; Rekwot, G.; Owoshagba, O.; Ahmed, S.; Atiku, J. Challenges of poultry production in Nigeria: Areview. *Niger. J. Anim. Prod.* **2018**, *45*, 252–258. [CrossRef]
6. Hafez, H.M.; Attia, Y.A. Challenges to the poultry industry: Current perspectives and strategic future after the COVID-19 outbreak. *Front. Vet. Sci.* **2020**, *7*, 516. [CrossRef]
7. Netherlands Enterprise Agency. Poultry Sector Study Nigeria. 2020. Available online: https://www.rvo.nl/sites/default/files/2020/10/Poultry-Sector-Study-Nigeria.pdf (accessed on 29 March 2022).
8. Poultry World. Nigeria: Shortage of Maize Is Crippling the Poultry Industry. 2020. Available online: https://www.poultryworld.net/poultry/nigeria-shortage-of-maize-is-crippling-the-poultry-industry/ (accessed on 29 March 2022).
9. Akinfenwa, G. High Cost of Feeds Threaten Southwest Poultry-Farms. 2021. Available online: https://guardian.ng/features/high-cost-of-feeds-threaten-southwest-poultry-farms/#:~{}:text=PHOTO%2FKHALED%20DESOUKI-,As%20predicted%20last%20December%20by%20the%20national%20body%20of%20the,of%20the%20industry%20anytime%20soon (accessed on 29 March 2022).
10. Šūmane, S.; Kunda, I.; Knickel, K.; Strauss, A.; Tisenkopfs, T.; des Ios Rios, I.; Rivera, M.; Chebach, T.; Ashkenazy, A. Local and farmers' knowledge matters! How integrating informal and formal knowledge enhances sustainable and resilient agriculture. *J. Rural Stud.* **2018**, *59*, 232–241. [CrossRef]
11. Youssoufane, A. Shaping national food and nutrition policy in Nigeria. *Nutr. Exch.* **2017**, *8*, 28.
12. Jiwuba, P.; Dauda, E.; Ezenwaka, L.; Eluagu, C. Replacement value of maize with sweet potato (*Ipomoea batata*) root meal on growth performance and haematogical characteristics of broiler starter birds. *Arch. Curr. Res. Int.* **2016**, *53*, 1–7. [CrossRef]
13. Madubuike, F.N. *Bridging the Animal Protein Supply and Demand Gap in Nigeria*; Inaugural lecture; Serial No.: 7; Imo State University: Owerri, Nigeria, 2012.
14. Krugman, P.R.; Obstfeld, M. *International Economics: Theory and Policy*; Addison-Wesley: Boston, MA, USA, 2008.
15. Poiesz, T.B.C. The Free Market Illusion Psychological Limitations of Consumer Choice. *Rev. Bus. Econ. Lit.* **2004**, *49*, 309–338.
16. Chron Contributor. Concept of Consumer-Switching Behavior. 2020. Available online: https://smallbusiness.chron.com/concept-consumerswitching-behavior-65092.html (accessed on 29 March 2022).
17. Abdulkareem, M.; Udegbunam, O. Special Report: Nigeria's Poultry, Egg Prices 'Highest Ever' amidst Inflation, Insecurity. 2021. Available online: https://www.premiumtimesng.com/news/headlines/450185-special-report-nigerias-poultry-egg-prices-highest-ever-amidst-inflation-insecurity.html (accessed on 29 March 2022).
18. Henders, S.; Persson, U.M.; Kastner, T. Trading forests: Land-use change and carbon emissions embodied in production and exports of forest-risk commodities. *Environ. Res. Lett.* **2015**, *10*, 125012. [CrossRef]
19. Busch, J.; Engelmann, J.; Cook-Patton, S.C.; Griscom, B.W.; Kroeger, T.; Possingham, H. Potential for low-cost carbon dioxide removal through tropical reforestation. *Nat. Clim. Chang.* **2019**, *9*, 463–466. [CrossRef]
20. Silva, G. Feeding the World in 2050 and Beyond—Part 1: Productivity Challenges. 2018. Available online: https://www.canr.msu.edu/news/feeding-the-world-in-2050-and-beyond-part-1 (accessed on 29 March 2022).

21. Manzoor, U.; Baig, S.A.; Usman, M.; Shahid, M.I. Factors affecting brand switching behavior in telecommunication: A Quantitative Investigation in Faisalabad region. *J. Mark. Inf. Syst.* **2020**, *3*, 63–82.
22. Veloutosou, C.; Gioulistanis, E.; Moutinho, L. Own label choice criteria and perceived characteristics in Greece and Scotland: Factors influencing willingness to buy. *J. Prod. Brand Manag.* **2004**, *13*, 228–241. [CrossRef]
23. Ness, N.; Gorton, M.; Kuznesof, S. The student food shopper. *Br. Food J.* **2002**, *104*, 506–525. [CrossRef]
24. Dick, A.; Jain, A.; Richardson, P. How consumers evaluate store brands. *J. Prod. Brand Manag.* **1996**, *5*, 19–28. [CrossRef]
25. Shukla, P. Effect of product usage, satisfaction and involvement on brand switching behaviour. *Asia Pac. J. Mark. Logist.* **2004**, *16*, 82–105. [CrossRef]
26. Baltas, G. Determinants of store brand choice: A behavioral analysis. *J. Prod. Brand Manag.* **1997**, *6*, 315–324. [CrossRef]
27. Awodola, B.; Agyeno, O. Terrorismin Northern Nigeria: A threat to food security in Maiduguri. *Mediterr. J. Soc. Sci.* **2015**, *6*, 11–17.
28. Onuoha, C.; Onyemaobi, O.O.; Anyakwo, C.N.; Onegbu, G.C. Effect of filler content and particle size on the mechanical properties of corn cob powder filled recycled polypropylene composites. *Am. J. Eng. Res.* **2017**, *6*, 72–79.
29. Falodi, F.; Adeoye, G.; Ajaja, T.; Alagbe, J. Chickens Forced to Skip Meals as Cost of Poultry Feed Rises by 100% n.d. Available online: https://punchng.com/chickens-forced-skip-meals-cost-poultry-feed-rises-100/ (accessed on 28 March 2022).
30. Hussain, K.; Rizwan, M. Customer loyalty and switching behavior of customer for Pepsi in Pakistan. *J. Public Adm. Gov.* **2014**, *4*, 129–145. [CrossRef]
31. Umoh, G.; Awa, H.O.; Ebitu, T. Markovian Application to Brand Switching Behaviour: A Survey of Toothpaste. *Eur. J. Bus. Manag.* **2013**, *5*, 110–124.
32. Nimako, S.G. Consumer switching behaviour: A theoretical review and research agenda. *Res. J. Soc. Sci. Manag.* **2012**, *2*, 74–82.
33. Gbigbi, T.M.; Chuks-Okonta, V. Performance analysis of poultry feed marketers in Delta State, Nigeria. *Int. J. Rural Dev. Environ. Health Res.* **2020**, *4*, 145–154. [CrossRef]
34. Wali, A.F.; Wright, L.T.; Uduma, I.A. Customer relationship management for brand commitment and brand loyalty. *Br. J. Mark. Stud.* **2015**, *3*, 45–58.
35. Dejene, M.; Bediye, S.; Alemu, D.; Kitaw, G.; Kehaliw, A.; Assefa, G.; Tadesse, G. Livestock feed marketing in Ethiopia: Challenges and opportunities for livestock development. *J. Agric. Sci. Technol. A* **2014**, *4*, 155–168.
36. Emenyonu, C.; Nkemka, C. Profitability analysis of poultry feed marketing in Imo State, Nigeria. *World Poult. Sci. J.* **2013**, *69*, 407–413. [CrossRef]
37. Idahor, K.O. Alternative feed stuffs utilization in Nigerian poultry industry: Potentials, problems and prospects. *World Poult. Sci. J.* **2013**, *69*, 666–675. [CrossRef]
38. Achoja, F.; Ofuoku, A.; Okoh, R. Linkages between socio-economic variables and the efficient marketing of poultry feeds in Delta State, Nigeria:Implication for extension services. *World Poult. Sci. J.* **2006**, *62*, 709–715.
39. Paternoster, R.; Jaynes, C.M.; Wilson, T. Rational choice theory and interest in the"fortune of others". *J. Res. Crime Delinq.* **2017**, *54*, 847–868. [CrossRef]
40. Levin, J.; Milgrom, P. Introduction to Choice Theory 2004. Available online: https://web.stanford.edu/~{}jdlevin/Econ%20202/Choice%20Theory.pdf (accessed on 26 April 2022).
41. Cabanias, J.O. Why Does Choice Matter? *Eur. J. Educ. Pedagog.* **2021**, *2*, 21–37. [CrossRef]
42. Ukav, I. Market Structures and Concentration Measuring Techniques. *Asian J. Agric. Ext. Econ. Sociol.* **2017**, *19*, 1–16. [CrossRef]
43. Clarke, R. *Industrial Economics*; Blackwell: Cambridge, MA, USA, 1993.
44. Mgaya, J. Application of ARIMA models in forecasting livestock products consumption in Tanzania. *Cogent Food Agric.* **2019**, *5*, 2–29. [CrossRef]
45. Kaltoch, R.; Sidhu, A. An Application of ARIMA Model to Forecast the Dynamics of COVID-19 Epidemic in India. *Glob. Bus. Rev.* **2021**, 1–14. [CrossRef]
46. Kumar, M.; Anand, M. Anapplication of time series ARIMA forecasting model for predicting sugarcane production in India. *Stud. Bus. Econ.* **2014**, *9*, 81–94.
47. Shastri, S.; Sharma, A.; Mansotra, V. A Model for Forecasting Tourists Arrival in J and K, India. *Int. J. Comput. Appl.* **2015**, *129*, 32–36.
48. Evans, R.W. Generalized Method of Moments (GMM) Estimation. 2018. Available online: https://notes.quantecon.org/submission/5b3b1856b9eab00015b89f90#:~{}:text=GMM%20allows%20for%20most%20flexible,enough%20to%20identify%20the%20parameters (accessed on 29 March 2022).
49. Roberts, M.J.; Schlenker, W. World supply and demand of food commodity calories. *Am. J. Agric. Econ.* **2009**, *91*, 1235–1242. [CrossRef]
50. Lin, C.Y. Estimating supply and demand in the world oil market. *J. Energy Dev.* **2008**, *34*, 1–32.
51. Epple, D.; McCallum, B.T. Simultaneous equation econometrics: The missing example. *Econ. Inq.* **2006**, *44*, 374–384. [CrossRef]
52. Gagniuc, P.A. *Markov Chains: From Theory to Implementation and Experimentation*; John Wiley & Sons: New Jersey, NJ, USA, 2017.
53. Adekoya, A.; Oyatoye, E. Modelling online retailing repurchase and switching behaviour of nigeria millennials using continuous time Markov chain (ctmc). *UNILAG J. Bus.* **2021**, *7*, 216–225.
54. Meher, N.; Afzal, H.; Zakir, H.; Bhuiyan, H.; Kabir, C. Markov analysis for assessing consumers' brand switching behavior: Evidence from telecommunication sector in Bangladesh. *Int. J. Educ. Soc. Sci.* **2021**, *8*, 55–67.

55. Neger, M. Consumers' Attitude towards Brand Switching: A Study on Telecommunication Sector in Bangladesh. Ph.D. Thesis, University of Dhaka, Dhaka, Bangladesh, 2018. Available online: https://repository.library.du.ac.bd:8080/handle/123456789/1392 (accessed on 25 March 2022).
56. Brezavšček, A.; Bach, M.P.; Baggia, A. Markov analysis of students' performance and academic progress in higher education. *Organizacija* **2017**, *50*, 83–95. [CrossRef]
57. Shujaat, S.; Ahmed, U.; Syed, N.A. Factors behind brand Switching in telecom sector of Pakistan. *IBT J. Bus. Stud.* **2015**, *11*, 29–40.
58. Wohlin, C.; Höst, M.; Wesslén, A.; Runeson, P. Software Reliability. *Encycl. Phys. Sci. Technol.* **2001**, *15*, 25–39.
59. Aderogba, K.A. Substantive causes and effects of floods in south western Nigeria and sustainable development of the cities and towns. *J. Emerg. Trends Educ. Res. Policy Stud.* **2012**, *3*, 551–560.
60. Makama, S.I.D. Report of Nigeria's National Population Commissionon the 2006 census. *Popul. Dev. Rev.* **2007**, *33*, 206–210.
61. Knoema. Nigeria—Production of Eggs Primary. 2021. Available online: https://knoema.com/atlas/Nigeria/topics/Agriculture/Live-Stock-Production-Production-Quantity/Production-of-eggs-primary (accessed on 29 March 2022).
62. International Fund for Agricultural Development (IFAD). Calculating the Sample Size. n.d. Available online: http://www.ifad.org/gender/tools/hfs/anthropometry/ant_3.htm (accessed on 29 March 2022).
63. Boateng, E.Y.; Abaye, D.A. A review of the logistic regression model with emphasis on medical research. *J. Data Anal. Inf. Processing* **2019**, *7*, 190–207. [CrossRef]
64. Akintunde, O.; Coster, A.; Nwigwe, C.; Agboola, T. An analysis of factors influencing access to credit by poultry egg farmers in Southwestern, Nigeria. *Niger. J. Anim. Sci.* **2020**, *22*, 204–212.
65. Adeyonu, A. Determinants of poultry farmers' willingness to participate in national agricultural insurance scheme in Oyo State, Nigeria. *Appl. Trop. Agric.* **2016**, *21*, 55–62.
66. Okoli, I.; Anyaegbunam, C.; Etuk, E.; Opara, M.; Udedibie, A. Entrepreneurial characteristics and constraints of poultry enterprises in Imo state, Nigeria. *J. Agric. Soc. Res.* **2005**, *5*, 25–32. [CrossRef]
67. nigerianfinder.com. 10 Best Broiler Feed Brands in Nigeria. 2022. Available online: https://nigerianfinder.com/best-broiler-feed-brands-in-nigeria/ (accessed on 29 March 2022).
68. Nkonki-Mandleni, B.; Ogunkoya, F.T.; Omotayo, A.O. Socioeconomic factors influencing livestock production among smallholder farmers in the free state province of south Africa. *Int. J. Entrep.* **2019**, *23*, 1–17.
69. Okonkwo, K. Analysis of Poultry Feeds Marketing in Ahiazu-Mbaise Local Government Area of Imo State. *J. Mark. Consum. Res.* **2013**, *1*, 76–82.
70. Adebowale, O.; Makanjuola, M.; Bankole, N.; Olasoju, M.; Alamu, A.; Kperegbeyi, E.; Oladejo, O.; Fasanmi, O.; Adeyemo, O.; Fasina, F.O. Multi-Drug Resistant Escherichia coli, Biosecurity and Anti-Microbial Use in Live Bird Markets, Abeokuta, Nigeria. *Antibiotics* **2022**, *11*, 253. [CrossRef]
71. Kassali, R.; Tanimonure, V.; Owabumuwa, J.; Isioye, A. Analysis of Poultry Feeds Marketing in Osun State, Nigeria. *IFE J. Agric.* **2019**, *31*, 92–101.
72. Isitor, S.; Ehien, A.E.; Makinta, U.; Olude, D.A. Socio-economic factors influencing egg production in Irepodun Local Government Area of Kwara state. *FUW Trends Sci. Technol. J.* **2019**, *4*, 196–201.
73. Jolaosho, S.O.; Ogunpaimo, O.R.; Dipeolu, A.O.; Mudasiru, Y.A. Switching Among Brands of Poultry Feeds: The Egg Farmers' Acuity in Ogun State, Nigeria. *Asian J. Soc. Sci. Manag. Technol.* **2020**, *2*, 68–78.
74. Arafah, W.; Mutis, T. The Research of Customer Switching Behavior at Some Leasing Companies. *Int. Rev. Manag. Mark.* **2018**, *8*, 115–118.
75. Pantow, E.F. The Effect of Price and Advertising on Brand Switching Scooter Motorcycle in Manado. *J. EMBA J. Ris. Ekon. Manaj. Bisnis Dan Akunt.* **2019**, *7*, 3428–3437.
76. Linden, J. High Feed Prices Harm Nigeria's Poultry Farmers. 2020. Available online: https://www.feedstrategy.com/africa/high-feed-prices-harm-nigerias-poultry-farmers/ (accessed on 29 March 2022).
77. Kanianska, R. Agriculture and Its Impacton Land-Use, Environment, and Ecosystem Services. In *Landscape Ecology—The Influences of Land Use and Anthropogenic Impacts of Landscape Creation [Internet]*; Almusaed, A., Ed.; Intech Open: London, UK, 2016; Available online: https://www.intechopen.com/chapters/51201 (accessed on 15 June 2022).
78. Omotayo, A.O. Economic synergy between rural off-farm income and households' poverty in Ekiti State, Nigeria. *J. Hum. Ecol.* **2016**, *56*, 99–106. [CrossRef]
79. Omotoso, A.B.; Daud, A.S.; Adebayo, R.A.; Omotayo, A.O. Socioeconomic determinants of rural households' food crop production in Ogun state, Nigeria. *Appl. Ecol. Environ. Res.* **2018**, *16*, 3627–3635. [CrossRef]
80. Daud, S.A.; Omotayo, A.O.; Aremu, A.O.; Omotoso, A.B. Rural infrastructure and profitability of food crop production in oyo state, Nigeria. *Appl. Ecol. Environ. Res.* **2018**, *16*, 4655–4665. [CrossRef]

sustainability

MDPI

Article

Training Sources and Preferences for Agricultural Producers and Professionals in Middle-North Mato Grosso, Brazil

Jordane Aparecida Vieira dos Reis [1], Aaron Kinyu Hoshide [1,2,*], John Robert Vreyens [3], André Soares de Oliveira [4], Vanessa Aparecida Moreira de Barros [1], Wininton Mendes da Silva [5], Luana Molossi [1], Jessica Lima Viana [1], Daniel Carneiro de Abreu [1,4] and Ronaldo Alves de Oliveira [1]

1 AgriSciences, Universidade Federal de Mato Grosso, Caixa Postal 729, Sinop 78550-970, MT, Brazil
2 College of Natural Sciences, Forestry and Agriculture, The University of Maine, Orono, ME 04469, USA
3 College of Food, Agricultural and Natural Resource Sciences, University of Minnesota Extension, Saint Paul, MN 55108, USA
4 Instituto de Ciências Agrárias e Ambientais (ICAA), Universidade Federal do Mato Grosso, Campus Universitário de Sinop, Avenida Alexandre Ferronato, 1200, Sinop 78550-728, MT, Brazil
5 Empresa Mato-Grossense de Pesquisa, Assistência e Extensão Rural (EMPAER-MT), Centro Político Administrativo, Cuiabá 78049-903, MT, Brazil
* Correspondence: aaron.hoshide@maine.edu; Tel.: +1-207-659-4808

Abstract: Brazil's midwest has rapidly expanded large-scale commodity crops such as soybeans and maize. We surveyed both agricultural producers and agricultural professionals in the middle-north region of Mato Grosso state, Brazil. Agricultural professionals provide technical support to agricultural producers and both are served by Assistência Técnica e Extensão Rural (ATER) with nationally and internationally recognized extension outreach. Our objectives were to define and contrast agricultural producer and professional characteristics, especially source(s) relied upon for agricultural training. There were 94 agricultural producers and 89 agricultural professionals that responded to our surveys, which were summarized and contrasted using statistical software. There was a predominance of male farmers, married with a broad age range. Agricultural professionals who advise producers had a high educational level. Producers and professionals were most reliant on private sector agricultural companies and business support organizations for agricultural training, versus public institutions such as universities and state/federal agencies. In the state of Mato Grosso, extension outreach can involve joint efforts by public and private sector entities. However, more targeted efforts are needed to ensure that public sector research is more equally used by agricultural producers and professionals in the region, especially during field days and face-to-face technical lectures during the off-season.

Keywords: agricultural education; agricultural producers; agricultural professionals; education and learning processes; questionnaires; rural extension

Citation: Reis, J.A.V.d.; Hoshide, A.K.; Vreyens, J.R.; Oliveira, A.S.d.; Barros, V.A.M.d.; Silva, W.M.d.; Molossi, L.; Viana, J.L.; Abreu, D.C.d.; Oliveira, R.A.d. Training Sources and Preferences for Agricultural Producers and Professionals in Middle-North Mato Grosso, Brazil. *Sustainability* **2023**, *15*, 4712. https://doi.org/10.3390/su15064712

Academic Editor: Emanuele Radicetti

Received: 24 January 2023
Revised: 26 February 2023
Accepted: 3 March 2023
Published: 7 March 2023

1. Introduction

The Cerrado–Amazon ecotone is a macro-region composed of sixteen municipalities with relief, soil, and climate conditions that are favorable to the cultivation of annual crops. According to the Instituto Mato-Grossense de Economia Agropecuária (IMEA) in 2017 [1], this is an important region for grain production and distribution in the state of Mato Grosso and for the Brazilian economy. Technological innovations in Mato Grosso have quadrupled agricultural production from 1990 to 2020, while land area used to grow grain has only increased by 68% in the same period [2]. Brazilian agribusiness development, especially in the state of Mato Grosso, has seen increased demand for qualified labor and professional skills and competencies. Brazil's federal government has stimulated agricultural development in middle-north Mato Grosso state since the 1970s and 1980s.

Migrants from southern Brazil (e.g., Parana state) were encouraged to occupy the Cerrado (i.e., savannah) and Amazon, expanding cattle ranching and crop production [3,4].

Mato Grosso is a state located in the midwest region of Brazil spanning 903,207 km^2. In 2021, it had 3,567,234 people with a population density of 3.94 inhabitants/km^2 [5]. The state of Mato Grosso has the largest cattle herd in the country, with 32,424,958 heads. In 2017, Mato Grosso state had 118,679 rural properties totaling 54,922,850 hectares intended for agricultural production [6]. Mato Grosso is the largest Brazilian soybean-producing state, having 10,909,400 hectares planted producing 39,961,100 metric tons (t), with a productivity of 3.663 t/ha in 2022. For maize in the year 2022, the state of Mato Grosso had a planted area of 6,547,400 hectares producing 41,620,100 t with a productivity of 6.357 t/ha [7].

To increase productivity gains in agricultural regions, independent extension projects such as those developed by universities are very important [8]. However, despite the potential for agricultural technological innovation in the region and access to the Universidade Federal de Mato Grosso (UFMT) and its partners, there has been low participation in extension programs (innovation programs, technical lectures, and workshops) offered by these institutions. AgriSciences at UFMT is a program responsible for applied research and rural extension programming (Table 1) focused on sustainable agriculture established in 2016 [9]. Through public/private partnerships, AgriSciences develops extension actions focusing on producers, technical professionals, and rural youth. From 2019 to 2022, more than 3600 people were directly reached through extension activities (Table 1).

There are numerous challenges to establishing successful agricultural training programs even with a lot of extension agents per capita [10]. It can be difficult for rural agricultural extension to customize education to different types of producers (e.g., row crop, cattle ranchers, subsistence farming) with different socio-economic backgrounds [11]. Previous studies in developing nations suggest more limited participation in agricultural training for older, less educated farmers living further away from extension services [12,13]. Agricultural producers also have limited time and resources to devote to agricultural training. A previous study [14] surveyed 355 farmers in northern Greece where 47.6% could attend less than 2 days of Agricultural Education Programs annually. On-farm demonstrations require extensive and multi-faceted planning [15]. More modern methods of rural extension can be novelties, both for agricultural producers and professionals and institutions in Brazil [16].

While production challenges, mainly in grain production, have encouraged new investments and increased technology diffusion in agriculture [17], addressing environmental and community sustainability can remain challenging. Brazil has had challenges successfully developing agro-ecological extension approaches for agriculture at a federal level [18]. Agricultural education has been shown to be more challenging for environmental management in New Zealand [19] and for rural development policies in Italy [20]. For Peruvian dairy farmers, private agricultural advisors tend to increase farmer reliance on the external inputs these advisors are selling [21]. Acknowledging farmers' rational barriers to adopting more environmentally friendly practices [22] and engaging them in developing individualized solutions can improve sustainability [23].

Due to the middle-north region of Mato Grosso's relevance for agriculture and importance in contributing to Brazil's national and international economy, it is important to better understand farmers' and agricultural professionals' sources and preferences for agricultural training. This can reveal potential drivers of farmers' low level of interest in participating in public innovation programs. Addressing these shortfalls in program interest can meet the goal of enhancing the engagement of agricultural producers and professionals in this region in Brazil.

Table 1. AgriSciences program components in middle-north Mato Grosso state, Brazil.

AgriSciences Program Components	Description	Methods	Participants (2019–2022)
Demonstrative Unit (DU)	Commercial rural property, where research and extension actions are already established. Offers producers and technicians local training programs for practical application of technologies and develops and evaluates research on advantages and disadvantages of such technologies.	• Field day • Technology showcase • Course • Technical visit • Lecture • Meeting	2261
Multiplier Unit	Rural property close to DU disseminating and transferring one or more technologies developed and/or validated at the DU. Promotes technology adoption and enables exchange of experiences between producers. Disseminates knowledge in environment close to their familiar daily life.	• Course • Technical visit • Lecture • Knowledge exchange • Meeting	33
Community Engagement	Community formed by academics, rural producers and professionals, young people, and rural children. Intended to bring together younger people and rural dwellers to disseminate academic knowledge, scientific production, and technological innovation using correct approaches.	• Workshop • Lecture • Course • Seminar	1301
Exchange of Leaders	International, practical experiences by academics and recent graduates. Allows exchange of rural youths, higher education students, and graduates in the United States. Focus on practical teaching of new practices for Mato Grosso.	• Course • Hands-on • Internship	54

Thus, there were three general research objectives of our study. The first objective was to develop baseline profiles of crop farmers and agricultural professionals in middle-north Mato Grosso. Our second objective was to identify the sources of and preferences for public and private agricultural knowledge and education by both agricultural producers and professionals in this region. Finally, our third research objective was to determine the extension method (e.g., field day, course, workshop, seminars, etc.) that our audience of farmers and professionals prefers in order to receive such knowledge and education. Customizing agricultural training to better meet the preferences and needs of producers and professionals can improve agricultural production practices and cropping system management in the long run.

2. Materials and Methods

2.1. Study Area and Methodological Approach

The area selected for study was the middle-north macro-region of Mato Grosso state (MT) because of its central location within this state in Brazil. The region is important for the production, marketing, and distribution of commodity crops (e.g., soybeans, maize, and cotton) in MT [1,2]. Our study used qualitative, quantitative, and descriptive data and analyses using the participatory rural appraisal (PRA) methodological approach. PRA is a research and data collection process intended to include the perspectives of all groups of interest of a community, causing a change in the traditional surveyor/surveyed roles so that multiple stakeholders participate in determining how data collection will be conducted [24]. The method also serves as a communication channel between those who share a common problem. We solicited feedback from agricultural producers and professionals to refine our surveys (Supplementary Materials) as well as collection of survey data.

2.2. Sampling and Survey Administration

Our research team surveyed 94 medium-to-large grain producers and 89 agricultural professionals during the spring and early summer of 2019. Producers were mainly from the cities of Nova Mutum (13.82681 S, 56.07165 W), Lucas do Rio Verde (13.073898 S, 55.91885 W), Sorriso (12.54209 S, 55.72081 W), and Sinop (11.85984 S, 55.50723 W), which lie on Route 163, which is among the larger cropping areas in South America. The middle-north region of Mato Grosso transitions from the Cerrado to Amazon biomes moving south to north from Nova Mutum to Sinop (Figure 1). Our survey design, implementation, and analyses are summarized in Figure 2.

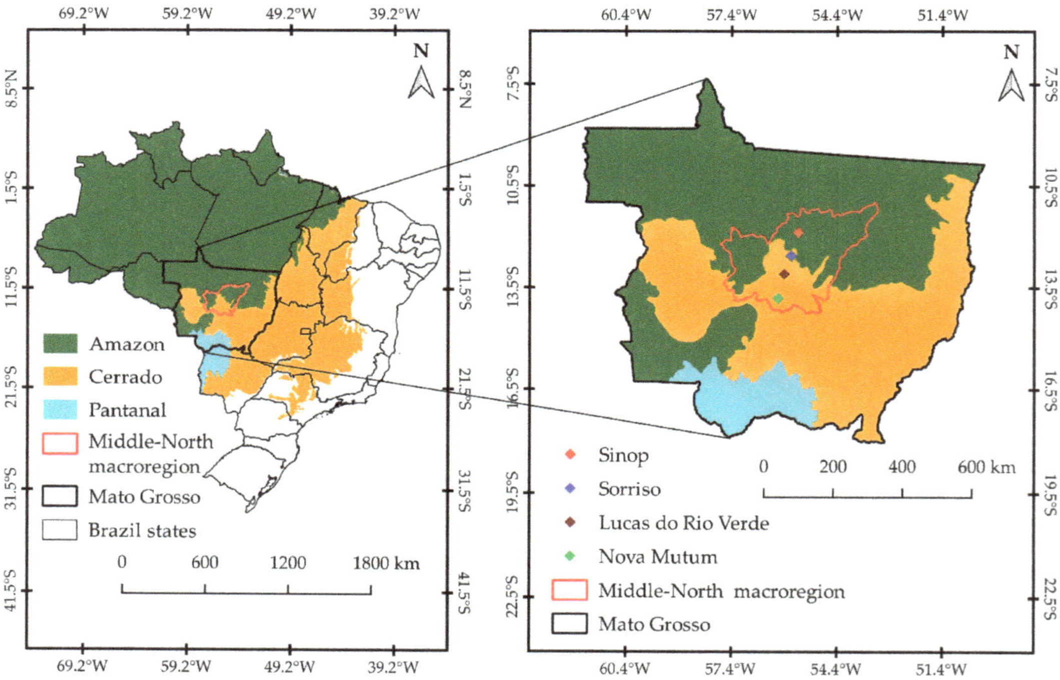

Figure 1. Survey city locations in the middle-north of the state of Mato Grosso, Brazil. Both maps show the Amazon, Cerrado, and Pantanal biomes in Brazil.

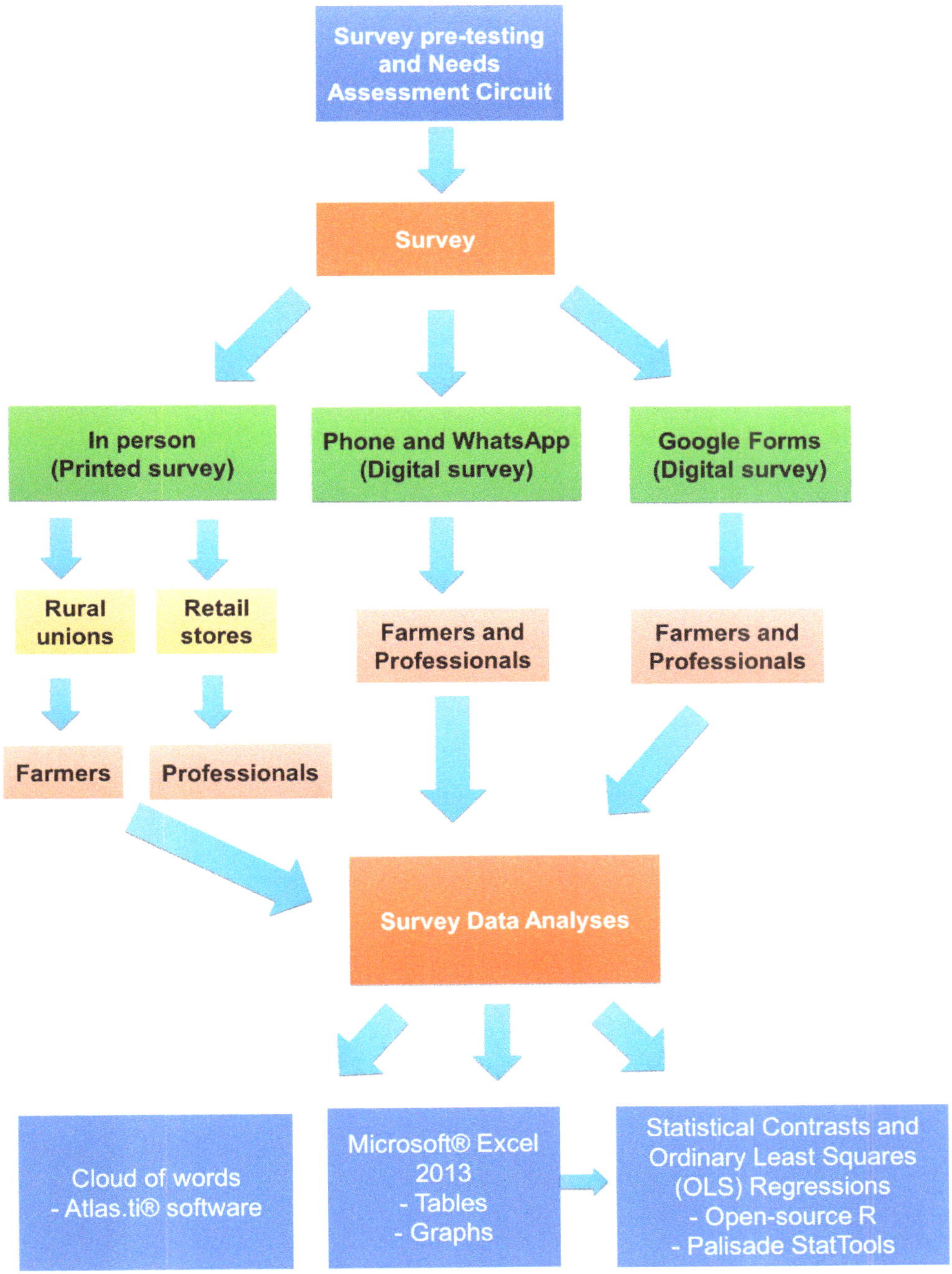

Figure 2. Survey flow chart for agricultural producers and agricultural professionals in middle-north Mato Grosso state, Brazil.

Surveys of agricultural producers and professionals were administered both in person and online, using the Google Forms platform. Survey questions were designed to take around half an hour to complete. A team of researchers and graduate students from the Universidade Federal Mato Grosso (UFMT), Empresa Brasileira de Pesquisa Agropecuária (Embrapa), and the University of Minnesota revised survey questions before final administration. We used the "Needs Assessment Circuit" (NAC) [25] which estimated time needed to answer survey questions and number of target responses for each town. NAC was selected based on its use in previous studies of agricultural stakeholder involvement [26]. Surveys of agricultural producers and professionals started during the spring of 2019. Due to low initial response to our online survey after a month, we directly surveyed both agricultural producers and professionals via unions and retailers over three months from June to August 2019.

Our project was presented to the trade unions of each town to facilitate communication to our target audience of agricultural producers. Producers answered questions on paper and sometimes by phone. Support of rural trade unions helped ensure at least two working days were available for direct contact with producers. Our team visited all four cities (Figure 1) and coordinated with agricultural companies to forward e-questionnaires to interested agricultural professionals and their teams. Presentations about our survey were given to at least ten agricultural companies in each city. This facilitated survey data collection in person using paper surveys and, in some situations, by using the online version of the survey and quick communication applications (WhatsApp®).

Our agricultural producer survey was divided into three sections. The first section included socio-demographic questions on age, gender, marital status, educational level, and residence. Additionally, respondents were asked about their length of experience or service in the agricultural sector. Agricultural producers were asked if they lived on-farm and, if not, how often they visited their farms, and who was responsible for the farm's technical decisions. The survey's second section asked questions on farm area, crop production and management, and participation in farm programs and agricultural training. Finally, questions in the third section asked producer preferences for agricultural extension topics, extension methods (e.g., field day, course, workshop, seminars, etc.), as well as general comments and suggestions.

The agricultural professional survey was also divided into three sections. The first section asked about the professionals' general background such as where they were located, education, occupation, number of farmers served, and crop area and types covered. The second survey section involved technology transfer to clients, probability of using information through current communication channels, frequency of reading specialized literature, and the amount of training carried out in a year. The final survey section asked about preferred extension methods, outreach duration and scheduling, and availability for in-person and online training. General suggestions and comments were also solicited.

2.3. *Data Organization and Analysis*

Survey data were entered into Microsoft® Excel 2013 to generate tables and graphs. We coded survey responses to categorical questions in order to generate summary statistics and to use for multi-variate regression analyses [27]. Both agricultural producers and professionals specified the crop systems with which they were involved. Responses to questions commonly asked to both agricultural producers and professionals were tested for statistically significant differences between these two respondent groups using the Z-test for significant differences between proportions in open-source R. These common questions included residence, education, and source(s) used for agricultural training. We arranged open-ended responses for comments and suggestions as a running text and then built a cloud of words using the Atlas.ti® software. This program identified frequently used words by both producers and professionals for visual comparison. We ran Ordinary Least Squares (OLS) regressions in both open-source R and Palisade Decision Tools Suite's StatTools to determine independent variables significantly impacting the proportion of public (versus private) training sources for both agricultural producers and professionals.

3. Results

3.1. Survey Respondent and Farm Area Characteristics

3.1.1. Agricultural Producers and Their Farms

Characteristics of agricultural producers are presented in Table 2. About one-quarter of producers (24.7%) were between 56 and 65 years old, while those aged 25 to 35 years comprised 23.6% of the sample. Most producers were male (91.4%) and were married (82.6%). Based on a Brazilian agricultural and livestock censuses conducted between 2006 and 2017, the cultivation of cotton, sugarcane, seeds and seedlings, planted forests, and aquaculture had minimal participation from women. These activities employ poorly paid wage labor, are highly mechanized, and are common on patriarchal farms [28].

Most surveyed producers (61.3%) reported having ≥20 years of farming experience (Table 2). Agricultural producers had a range of educational backgrounds, where 46.7% had up to a high school degree and 27.2% had an undergraduate college degree (Table 3). Of the 40 producers specifying other higher education or an undergraduate degree, 50% were agronomists (Table 2). Advanced degrees (6.5%) and technical education (3.3%) comprised the remaining balance of educational backgrounds (Table 3). Our results are consistent with a previous study reporting that rural producers in Brazil differed with regard to schooling and how technical information is obtained [29].

Most interviewed producers (68.8%) did not live on-farm (Table 2) and were more likely to manage farms located in areas outside of the four major cities compared to the farms serviced by agricultural professionals (p = 0.005, Table 3). Typically, producers maintain a residence in an urban area, which allows access to a better living standard for their families (e.g., education, healthcare, entertainment, commerce, and other services). Finally, management decisions were made individually by the producer (29.2%) or were made together with some member of the family (42.7%) or with an external individual (24.7%). Thus, 96.6% of producers' agricultural property and enterprises are managed by the farmer, his wife, and children (Table 2).

Most farms managed by agricultural producers fell in the 500-to-1499-hectare (ha) size class (34.4% of producers), with 77.4% of the farms greater than or equal to 500 ha (Table 1). Most farms managed in our study were larger than observed in other regions of Brazil where only 0.9% of the farms are over 1000 hectares [30]. Mato Grosso is Brazil's largest agricultural state, producing 21.1% of the country's agricultural value [31]. Commodity crops include maize (*Zea mays* L.) or cotton (*Gossypium* sp.) grown as a second crop following soybeans (*Glycine max* L.) in the same production year [32]. Commodity crop production has historically transitioned from frontier development to greater intensification in order to reduce deforestation through collaboration of the private and public sectors with non-governmental organizations (NGOs) [33].

Table 2. Characteristics for crop producers from middle-north region, Mato Grosso state, Brazil in 2019.

Producer Characteristic	*n*	%
Age (*n* = 93):		
<25	5	5.4
25–35	22	23.6
36–45	18	19.4
46–55	18	19.4
56–65	23	24.7
>65	7	7.5
Gender (*n* = 93):		
Male	85	91.4
Female	8	8.6
Civil Status (*n* = 92):		
Single	14	15.2
Married	76	82.6
Divorced	2	2.2
Undergraduate area (*n* = 40):		
Agronomist	20	50.0
Other professions	15	37.5
Administrator	2	5.0
Veterinarian	1	2.5
Economist/Accountant	2	5.0
Live on farm (*n* = 93):		
Yes	29	31.2
No	64	68.8
Administrative decisions made (*n* = 89):		
Individually	26	29.2
With husband (wife)	15	16.9
With husband (wife) and children	23	25.8
With someone outside the family	22	24.7
Outside the family	3	3.4
Years in agriculture (*n* = 93)		
0–10	16	17.2
10–19	20	21.5
20–29	23	24.8
30–39	20	21.5
40–50	11	11.8
>50	3	3.2
Farm area managed (hectares) (*n* = 93):		
<500	21	22.6
500–1499	32	34.4
1500–3499	20	21.5
3500–9999	19	20.4
>10,000	1	1.1
Uses farm credit (*n* = 93):		
Yes	76	81.7
No	17	18.3

Table 3. Characteristics and preferences of agricultural producers and professionals in middle-north region, MT, Brazil in 2019.

Respondent Characteristics and Preferences	Producers		Professionals		Z-Test
	n	% resp.	*n*	% resp.	*p*-Value
Location:	91		89		
Sinop	19	20.9	29	32.5	0.108
Sorriso	23	25.3	28	31.5	0.450
Nova Mutum	16	17.6	12	13.5	0.580
Lucas do Rio Verde	9	9.9	9	10.1	>0.999
Other	28	30.8	11	12.4	0.005
Educational level:	92		89		
Up to high school	15	16.3	2	2.3	0.003
Completed high school	28	30.4	1	1.1	<0.0001
Technical education	3	3.3	14	15.7	0.009
Other higher education	15	16.3	0	0	0.0002
College Degree (B.A., B.S.)	21	22.8	47	52.8	<0.0001
Graduate Degree (M.S., Ph.D.)	6	6.5	25	28.1	0.0003
Works with crop system type(s) [a]:	93		85		
Just soybeans (S)	6	6.5	1	1.2	0.155
S-Maize (M)	55	59.1	56	65.9	0.439
S-Cotton	7	7.5	28	19.7	<0.0001
S-M-Pasture	28	30.1	26	18.3	>0.999
Other crop(s)/enterprise	31	33.3	31	21.8	0.778
Sources for agricultural training [a]:	83		86		
Public training sources:					
Regional extension	39	47.0	22	25.6	0.006
Universities	10	12.0	16	18.6	0.333
State research	7	8.4	6	7.0	0.947
Federal research	21	25.3	18	20.9	0.623
Private training sources:					
Chemical companies	37	44.6	71	82.6	<0.0001
Fertilizer companies	17	20.5	55	64.0	<0.0001
Retail companies	40	48.2	44	51.2	0.816
Independent consultants	15	18.1	29	33.7	0.032
Other private entities	3	3.6	4	4.7	>0.999
Preferred topics for training [a]:	90		88		
Physical soil qualities	40	44.4	44	50.0	0.554
Soil conservation	47	52.2	40	45.5	0.451
Soil–plant–atmosphere relations	26	28.9	17	19.3	0.188
Soil preparation	38	42.2	48	54.5	0.135
GPS and precision agriculture	35	38.9	34	38.6	>0.999
Weed management	55	61.1	63	71.6	0.187
Plant mineral nutrition	42	46.7	61	69.3	0.004
Correcting soil acidity	39	43.3	53	60.2	0.035
Plant fertilizer management	50	55.6	64	72.7	0.026
Soil biology	49	54.4	37	42.0	0.132
Soil fertility assessment	49	54.4	62	70.5	0.040
Other topics	10	11.1	8	9.1	0.843
Optimistic on future of agriculture:	91		88		
Yes	68	74.7	83	94.3	0.0007

[a] Producers and professionals could specify multiple categories.

3.1.2. Agricultural Professionals and Advisory Area

Agricultural professionals were statistically more highly educated than producers, with 52.8% being college graduates, 28.1% having a post-graduate degree, and 15.7% completing technical education beyond high school. Only 3.4% had some high school education or just a high school diploma (Table 3). Other research has highlighted the need for educated workers since they have to perform different jobs in the urban or rural agribusiness sector [34]. Table 4 summarizes responses to questions specifically asked to agricultural professionals. Here, 43.8% of professionals provided assistance to 0 to 19 clients, 33.7% to 20 to 39 clients, and 18% to more than 60 clients. This was similar for farm area serviced with 59.1% of professionals being responsible for under 40,000 hectares (ha) and 25% for more than 60,000 ha. Large-scale soybean production is one of the most economically important crops in Brazil and crop extension plays a critical role in promoting scientific advances and new production technologies [35].

Table 4. Characteristics of agricultural professionals from middle-north region, Mato Grosso state, Brazil in 2019.

Professional Characteristics	*n*	%
Number of rural producers assisted (n = 89):		
0–19	39	43.8
20–39	30	33.7
40–60	4	4.5
>60	16	18.0
Total area covered by professionals' work (n = 88):		
0–19,000 hectares	24	27.3
20,000–39,999 hectares	28	31.8
40,000–59,999 hectares	14	15.9
>60,000 hectares	22	25.0

3.2. Agricultural Systems

Based on the responses of agricultural producers and professionals regarding the crop(s) that they work with, five annual cropping systems were defined: (1) soybeans (S) only, (2) S-maize (M), (3) S-cotton, (4) S-M-pasture, and (5) systems involving other crops/enterprises. The S-M double-cropping system had the highest involvement from both producers (59.1%) and professionals (65.9%). For producers, this was followed by other crops/enterprises, S-M-pasture, S-cotton, and just soybeans. Professionals were similar, with the exception of S-cotton being more commonly worked with than the S-M-pasture integrated system. Professional involvement with S-cotton was statistically greater ($p \leq 0.0001$) than producers (Table 3), which may be due to the management complexity of cotton [36] requiring more agricultural professional advisory services.

Most producers plant one, two, or three annual crops in a year, typically following each other from the wet season (October to February) to dry season (March to September). Only 6.5% of producers and 1.2% of professionals are specialized in just soybeans. More common are involvement in either (1) S-cotton due to longer maturity for cotton or (2) S-M where a third sequential crop could be added, such as S-M-pasture, where the pasture is under-seeded during maize planting in February. Most producers (89.2%) and professionals (84.2%) were involved with soybeans double-cropped with maize as a second crop (*safrinha*). Fewer producers (33.3%) and professionals (21.8%) dealt with second/third crops such as bean, sorghum, and *Crotalaria juncea* (rattle pods grown as a cover crop). Furthermore, 30.1% of farmers cultivated pasture, typically as the third crop, while 18.3% of professionals worked with this type of integrated cropping system (Table 3). Third crops grown during the dry season (June to September) with no or very limited rainfall require more technology (e.g., irrigation), typically used by larger producers [37].

3.3. Contrasting Use of Public versus Private Agricultural Training

Table 3 compares surveyed agricultural producers and professionals with respect to their sources for public or private agricultural training. Public sources of agricultural training include regional extension (e.g., Fundação MT, IMA, SENAR-MT), universities (e.g., UFMT), state research (e.g., EMPAER, INDEA), and federal research (e.g., Embrapa, MAPA) entities (Table 5). Private sector training sources were companies making agricultural chemicals (e.g., Bayer, Syngenta, Pioneer), fertilizers (e.g., Mosaic, Yara), retailing agricultural inputs (e.g., Agroamazônia, Agroinsumos), independent consultants, and other private entities (Table 4). Agricultural producers were more likely to rely on regional extension ($p = 0.006$) compared to professionals.

However, professionals were significantly more dependent on chemical and fertilizer companies ($p \leq 0.0001$) as well as independent consultants ($p = 0.032$) for agricultural training compared to producers (Table 3). Of the 83 producers and 86 professionals that answered our question on source(s) of agricultural training, less than half of each group relied on public sources (7% to 47%). Less than half of producers relied on private sources as well (3.6% to 48.2%, Table 3). Over half of surveyed professionals relied on companies producing or retailing agricultural inputs (51.2% to 82.6%, Table 3).

These private entities had greater budgets on average compared to regional/state-level public entities. Private entities had budgets ranging from USD 15,655,000 to 1,630,600,000, while regional/state-level public sources only had budgets of USD 25,756,833 to 180,261,746 (Table 5). For the Universidade Federal de Mato Grosso (UFMT), this is the entire university system budget, where most resources are not directed to farmer outreach unlike the Serviço Nacional de Aprendizagem Rural de Mato Grosso (SENAR-MT), Empresa Mato-Grosssense de Pesquisa, Assistência e Extensão Rural (EMPAER-MT), and the Instituto de Defesa Agropecuária do Estado de Mato Grosso (INDEA). For example, the AgriSciences program at UFMT, which conducts direct outreach to area farmers, has an average annual budget of USD 169,523, of which only about USD 11,808 is directly from the UFMT university system.

Table 5. Public and private agricultural training entities and 2021 budgets in Brazil.

Type	Class	Abbreviation	Entity Name	2021 Budget (USD/Year)			
				International	Brazil	Mato Grosso	Source
Public	Regional extension	Fundação MT	Fundação de Apoio à Pesquisa Agropecuária de Mato Grosso	-	-	n/a	-
		IMA	Instituto Matogrossense de Algodão	-	-	n/a	
		SENAR-MT	Serviço Nacional de Aprendizagem Rural de Mato Grosso	-	-	25,756,833	[38]
	Universities	UFMT	Universidade Federal de Mato Grosso	-	-	180,261,746	[39]
			AgriSciences	-	-	169,523	
	State	EMPAER-MT	Empresa Mato-Grosssense de Pesquisa, Assistência e Extensão Rural	-	-	30,468,664	[40]
		INDEA	Instituto de Defesa Agropecuária do Estado de Mato Grosso	-	-	43,059,919	[40]
	Federal	Embrapa	Empresa Brasileira de Pesquisa Agropecuária	-	657,227,049	n/a	[41]

Table 5. *Cont.*

Type	Class	Abbreviation	Entity Name	2021 Budget (USD/Year)			
				International	Brazil	Mato Grosso	Source
		MAPA	Ministério da Agricultura, Pecuária e Abastecimento	-	787,328,902,333	n/a	[42]
Private	Chemicals	-	Bayer (includes Monsanto)	-	2,041,688	n/a	[43]
		-	Corteva (includes Pioneer)	15,655,000	n/a	n/a	[44]
	Fertilizers	-	Mosaic	1,630,600,000	n/a	n/a	[45]
		-	Yara	384,000,000	n/a	n/a	[46]
	Retailers	-	Agroamazônia	-	n/a	n/a	-
		-	Agroinsumos	-	n/a	n/a	-
	Independent	-	Independent Consultants	-	n/a	n/a	-
	Other	-	Other private entities	-	n/a	n/a	-

Our results are comparable with Carbonera et al. 2020 [47], who found that agricultural producers in Santa Rosa, Rio Grande do Sul state in southern Brazil relied upon both private and public sources for agricultural training. Here, farmers were technically assisted by agricultural input suppliers, finance and other professionals, and private cooperatives. Public entities included the Association of Technical Support and Rural Extension Enterprises in Rio Grande do Sul (Associação Riograndense de Empreendimentos de Assistência Técnica e Extensão Rural—EMATER/RS) and City Administration. Producers also used non-governmental organizations (NGOs) for training. Increased producer demand for more technical assistance was observed in some of the relationships examined.

Agricultural producers had a lower number of responses (481) compared to professionals (594) regarding preferred topics for agricultural training. Word clouds of open-response questions answered by both agricultural producers and professionals are shown in Figure 3. Agricultural producers emphasized "soil" versus "management" for professionals. The three words most often mentioned by agricultural producers were soil, field, and research (Figure 3a). Agricultural professionals most often used the words management, soil, field, and producers (Figure 3b). Greater emphasis on "management" was consistent with professionals having statistically greater preferences for preferred topics for training having to do with more technical aspects of crop management, such as plant mineral nutrition ($p = 0.004$), fertilizer management ($p = 0.026$), soil fertility assessment ($p = 0.040$), and correcting soil acidity ($p = 0.035$, Table 3).

Ordinary Least Squares (OLS) regression results for both agricultural producers and professionals explaining the proportion of public training used (dependent variable) are shown in Table 6. Years of education, years in agriculture, number of crops grown, and those specializing in only soybeans were not significant in explaining use of public (versus private) training sources. For producers, years spent in agriculture and being from the city of Sorriso were also not significant. Agricultural professionals who received training from private companies selling agricultural inputs (resale) did not have any significantly different proportion of public training used compared to other professionals. Like agricultural producers, years of education was also not significant. Agricultural professionals that helped producers specializing in soybeans and that received training from resale companies (i.e., distributors) were not more or less likely to use more public training sources (Table 6). All other variables surveyed from both agricultural producers (Tables 2 and 3) and agricultural professionals (Tables 3 and 4) were not significant in influencing use of public training sources.

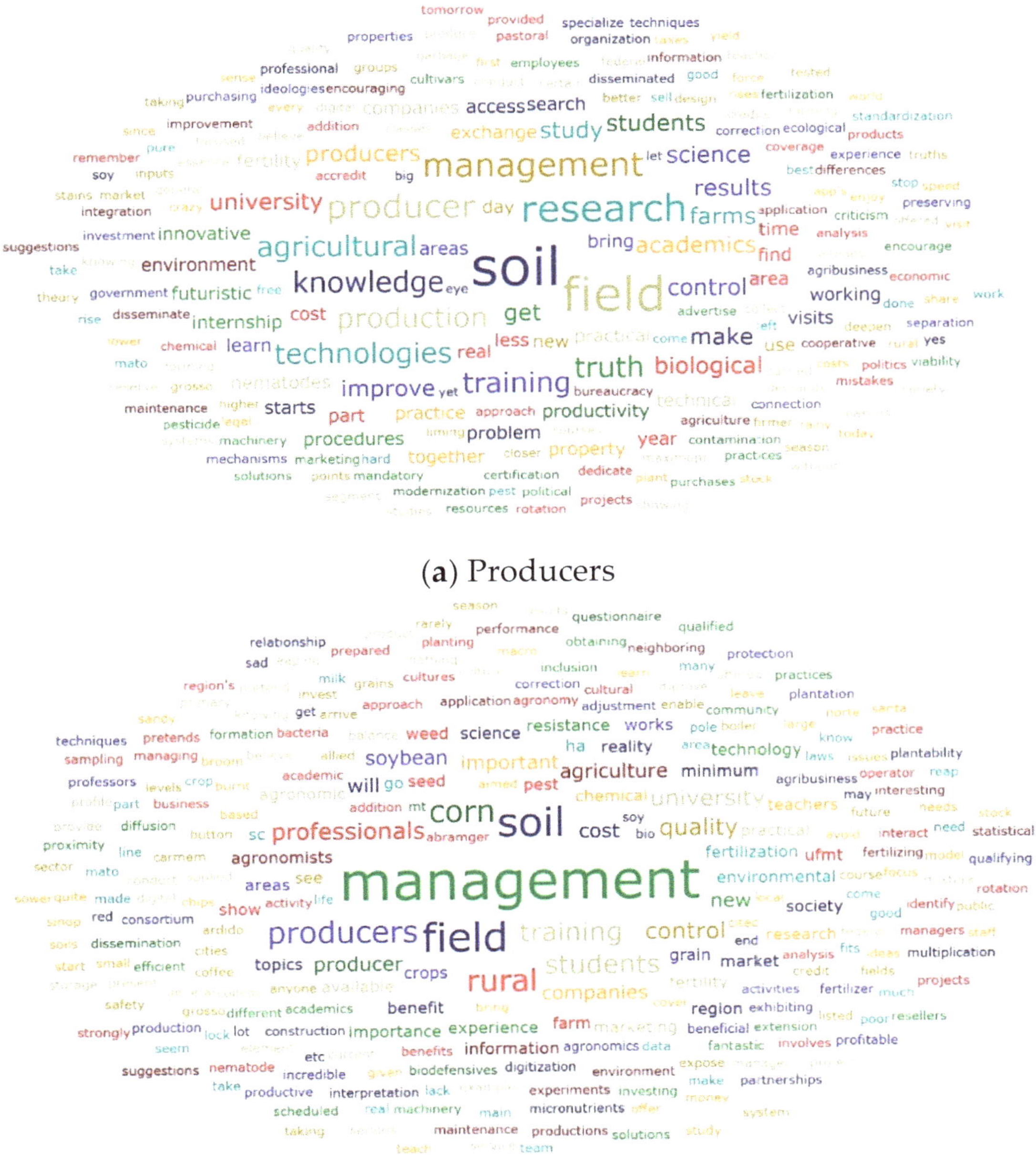

(**a**) Producers

(**b**) Professionals

Figure 3. Word cloud using Atlas.ti for (**a**) agricultural producers and (**b**) professionals in the middle-north region of Mato Grosso state, Brazil surveyed during the spring of 2019.

Table 6. Ordinary Least Squares (OLS) regression for proportion public training for both agricultural producers and professionals from middle-north region, Mato Grosso state, Brazil in 2019.

Proportion Public Training	**Producers**			**Professionals**		
OLS Model: Independent Variables	**Coefficient**	**Standard Error**	***p*-Value [a]**	**Coefficient**	**Standard Error**	***p*-Value [a]**
Constant	0.65826	0.25337	0.0115 **	0.56507	0.20079	0.0063 ***
Private training source:						
Chemical companies	−0.20100	0.07393	0.0083 ***	−0.35957	0.07259	<0.0001 ***
Resale companies	−0.27186	0.07207	0.0003 ***	−0.05143	0.05403	0.3444
Residence:						
Sinop	−0.29254	0.10318	0.0060 ***	−0.09202	0.06498	0.1554
Sorriso	0.03919	0.08784	0.6569	−0.16216	0.06462	0.0144 **
Socio-demographics:						
Female	0.32599	0.13429	0.0179 **	_ [b]	_ [b]	_ [b]
Single	0.22590	0.09779	0.0239 **	_ [b]	_ [b]	_ [b]
Years of education	−0.00565	0.01277	0.6595	−0.00854	0.01233	0.4907
Agricultural background [c]:						
Managed area (ranges)	−0.00002	0.00001	0.0725 *	−0.000002	0.000001	0.0801 *
Soybeans only	−0.30665	0.19923	0.1284	−0.34075	0.24407	0.1670
Number of crops	0.01103	0.02657	0.6795	0.06672	0.02151	0.0028 ***
Years in agriculture (range)	0.00192	0.00296	0.5178	_ [a]	_ [a]	_ [a]
Number articles read/year	_ [b]	_ [b]	_ [b]	0.00032	0.00018	0.0812 *
Model summary:						
Sample size (*n*)	94			91		
Degrees of freedom (df)	11			9		
R-squared (R^2)	0.4159			0.4631		

[a] Denotes significance at confidence level (α) of 0.10 (*), 0.05 (**), and 0.01 (***). [b] Independent variable question not asked. [c] For producers, this is managed by them, while for professionals this is related to the producers that they consult.

Agricultural producers that had significantly higher proportions of public training sources were less likely to receive training from chemical and resale companies, but they were more apt to be from the city of Sinop, female, and single. Producers that managed less area were more likely to use more public training sources, through this was marginally significant (α = 0.10). Similarly, agricultural professionals that used more public training sources were less likely to receive training from private chemical companies and be from the city of Sorriso. Agricultural professionals that used public training sources significantly read more technical articles per year, consulted on a larger number of crops, and managed less agricultural area (Table 6).

3.4. Financial and Technical Assistance for Producers

As summarized in Table 7, most surveyed producers use rural credit (81.7%) and technical assistance (83.5%) to support their farms. Banco do Brazil and SICREDI combined to make up 80.7% of responses for sources of rural credit, with bartering (11.6%), other types (6.2%), and Agricultura de Baixo Carbano (ABC) or the Low-Carbon Agriculture program (1.6%) making up the difference (Table 7). ABC is used by farmers in Brazil who are implementing sustainable agriculture, such as low-carbon practices [48].

Table 7. Producer use and preferences for financial and technical services in middle-north region, Mato Grosso state, Brazil in 2019.

Producer Characteristics and Preferences	*n*	%
Rural credit:		
Rural credit used (*n* = 93):		
Yes	76	81.7
No	17	18.3
Type [a] (responses = 129):		
Banco do Brasil (FCO, etc.)	54	41.9
SICREDI	50	38.8
ABC	2	1.6
Other type(s)	8	6.2
Barter	15	11.6
Technical assistance		
Receives technical assistance (*n* = 91):		
Yes	76	83.5
No	15	16.5
Source(s) [a] (responses = 176):		
Regional extension (SENAR)	21	11.9
Universities	5	2.8
State research (EMPAER)	2	1.1
Federal research (Embrapa)	14	7.9
Private companies	58	33.0
Producer associations	45	25.6
Rural unions	29	16.5
Non-Government Organizations (NGOs)	1	0.6
Other sources	1	0.6
Preferences [a] (responses = 99):		
Understanding cost–benefit analysis	37	37.4
New technology and science	28	28.3
Basic technical knowledge	21	21.2
Communication skills	8	8.1
Other	5	5.0
Preferred extension methods [a] (responses = 245):		
Field days	74	30.2
Fairs and exhibitions	45	18.4
Presentations	71	29.0
Congresses and seminars	48	19.6
Other types	5	2.0
None	2	0.8

[a] Producers could specify multiple categories.

Responses to sources of technical assistance were led by private companies (33%), followed by other private entities such as producer associations (25.6%) and rural unions (16.5%). Public sources of technical assistance (SENAR, Universities, EMPAER, and Embrapa) only totaled 23.7% of responses. Preferred technical assistance topics were cost/benefit (37.4%), new technologies/science (28.3%), and basic technical knowledge (21.2%) (Table 7).

Producers indicated a preference for presentations with social gatherings, exchange of experiences, and emphasis on practical dynamics in a familiar environment. Preferred agricultural extension formats as a percentage of responses were field days (30.2%), presentations (29%), congresses or seminars (19.6%), and fairs/exhibitions (18.4%) (Table 7). From October 2016 to May 2022, AgriSciences at the University Federal de Mato Grosso has organized numerous presentations, four field days (2019–2022), and one international congress (2016). There were 538 people that presented and/or attended the VIII SIMBRAS (Simpósio Brazileiro de Agropecuária Sustentável) from 5 to 8 October 2016 [49]. From February 2019 to December 2021, AgriSciences held 56 training events involving a total of 3649 people.

A total of 263 + 786 + 275 = 1324 people attended the first, third, and fourth Agricultural Field Days (2019, 2021–2022). Even though the second live event was canceled in 2020 due to COVID-19, more than 800 people downloaded the inaugural book for this [50], with book publication occurring the year after the field day (e.g., the third live event in 2021 had a book published in 2022 [51]). The total number of AgriSciences program participants have been 3649 + 2124 + 538 = 6311 since 2016. The most participants attended presentations (57.8%), followed by field days (33.7%) and congresses (8.5%).

3.5. Agricultural Professionals' Training

Agricultural professionals' responses to questions on their baseline training as well as time availability for training are summarized in Table 8. The number of trainings that agricultural professionals attend per year was bi-modal. There were 40.9% of respondents that attend 0 to 3 trainings annually, while 40.9% attend 5 or more. Most professionals also read technical articles and books. Only 1.1% of respondents do not read technical articles, while 98.9% read such articles at least occasionally. Almost half (47.2%) of professionals that responded read 1 to 2 books per year (Table 8).

The duration (number of hours) that professionals considered ideal for technical and online courses was also bi-modal. While over one-third of professionals (36%) considered one to four hours as the ideal time, 23.6% favored 8 h and 29.2% of respondents preferred trainings more than 8 h long. A minority (11.2%) did not know. Most professionals (51.7%) reported having only up to one hour per day available for online trainings. Others reported having one to three hours available (29.2%) or a half day (6.7%). The remaining respondents (12.4%) reported having no interest in online trainings (Table 8).

Agricultural professionals preferred trainings to be scheduled during times of the day, day(s), and month(s) that they were less busy, presumably when they were not consulting with producers. Saturday mornings were the most preferred, followed by weekday evenings compared to mornings and afternoons (Figure 4a). July and August are the most preferred months (Figure 4b) after second-crop maize (*safrinha*) has already been harvested or prior to cotton harvest. These two months correspond to the commodity cropping off-season in Mato Grosso during the middle of the dry season.

Table 8. Professionals' participation and availability for training in middle-north region, Mato Grosso state, Brazil in 2019.

Professional Participation and Preferences	*n*	%
Participation in training and reading		
Trainings attended annually (*n* = 88):		
None	3	3.4
1–3 training(s) per year	33	37.5
4–5 trainings per year	16	18.2
>5 trainings per year	36	40.9
Technical articles read (*n* = 89):		
Daily	23	25.9
Weekly	31	34.8
Monthly	10	11.2
Occasionally	24	27.0
Do not read technical articles	1	1.1
Books read annually (*n* = 89):		
None	29	32.6
1–2 books	42	47.2
3–5 books	11	12.3
>5 books	7	7.9

Table 8. *Cont.*

Professional Participation and Preferences	n	%
Preferences for training:		
Ideal length for agronomic course (n = 89):		
1–4 h	32	36.0
8 h	21	23.6
>8 h	26	29.2
Do not know	10	11.2
Hours available for online training (n = 89):		
Up to 1 h	46	51.7
1–3 h	26	29.2
A half day	6	6.7
Not interested in participating in online program	11	12.4
	0	0.2

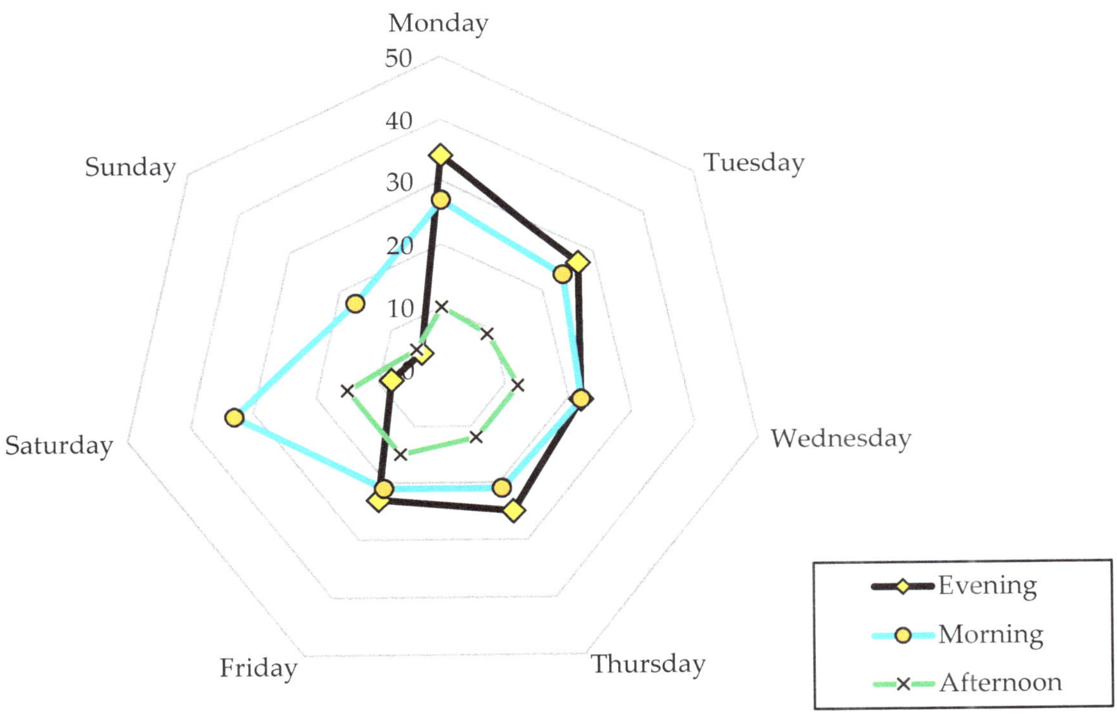

(**a**) Days

Figure 4. *Cont.*

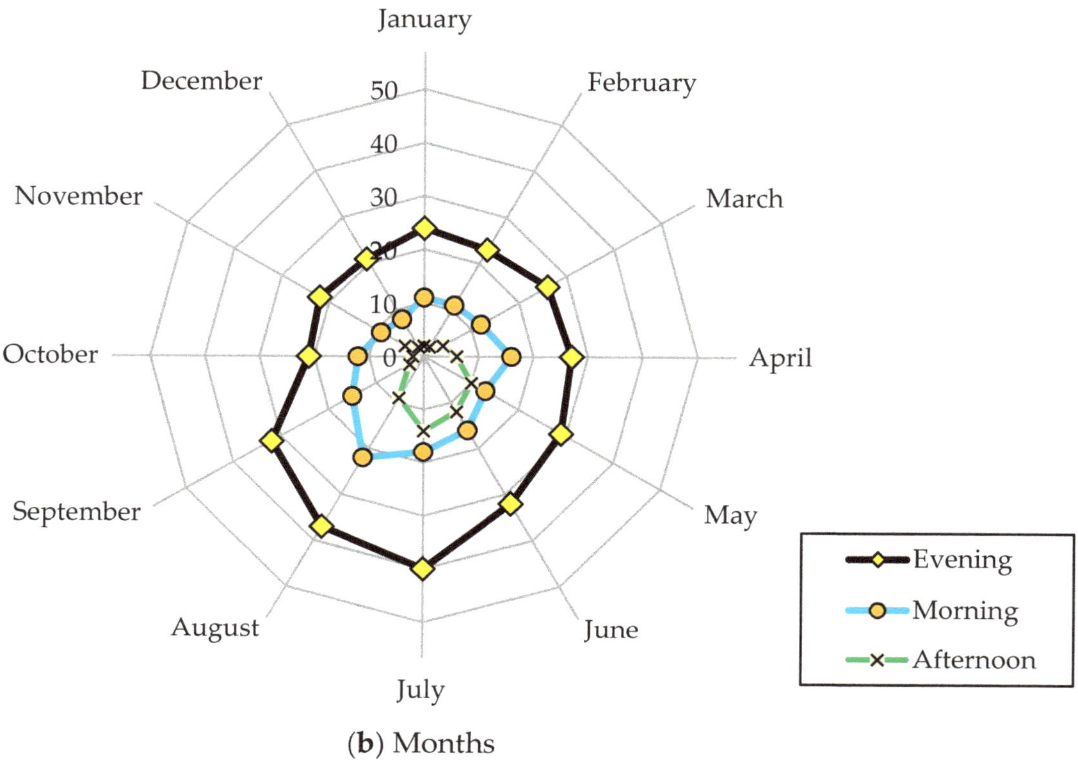

Figure 4. Agricultural professionals' preferred (**a**) days and (**b**) months for training.

4. Discussion

4.1. Improved Participation for Sustainable Agricultural Systems

According to prior research, the Brazilian challenge is to maintain the capacity to produce and use soybeans profitably and sustainably [52]. Greater sustainability here involves crop rotation diversification and/or integrating commodity crops like soybeans and maize with perennial pasture for livestock grazing [53,54] or using soybeans during the pasture finishing of beef cattle in the region [55]. Integrated crop–livestock (ICL) systems have had low participation due to the management complexity of adding another enterprise or due to the coordination needed between specialized crop and livestock producers [56,57]. In addition to encouraging greater adoption of ICL, addressing social factors can also more broadly impact farm persistence and growth [58].

Access to rural credit in the Brazilian state of São Paulo increased the probability of adopting the ICL-Forest System (Sistema Integrado Lavoura-Pecuária-Floresta—ILPF) [59]. In our study, 81.7% of producers depend on financing for rural credit as a source for investment (Table 1), with 30.1% of surveyed producers using ICL systems (Table 3). Our ICL percentage was less than a prior assessment of integrated systems in Mato Grosso state where 89% of integrated systems involved just crops and livestock (rather than agro-forestry) and such ICL systems involved roughly 40.5% of 134 total producers (61 integrated + 73 non-integrated) surveyed [60]. Integrated systems are beneficial, diversifying rural revenues, recovering degraded pastures, improving physical, chemical, and biological attributes of soil, and reducing greenhouse gas emissions [61].

Increased participation of private companies in the development of agricultural markets in the 2010s was a major driver of production cost increases, which interfered with the direction of technologies and producers' decision-making power [62]. Our results suggest that while most rural producers use strategies that may have more limited sustainability, many have adopted sustainable practices such as cover cropping. Sixty-nine percent of producers we surveyed used no-till which relies on glyphosate (i.e., Roundup®) [63]. Glyphosate use in Brazil for soybeans has been shown to reduce erosion but increase herbicide resistance in weeds [64]. The prevalence of producers using *Crotalaria juncea* as a cover crop (46.3%) was similar to those using biological nitrogen fixation (47.6%). More complex and expensive strategies such as integrated crop–livestock systems and recovery of degraded pasture lands were used by only 17.9% and 7.1% of producers, respectively.

4.2. Increasing Participation in Public Sources of Agricultural Training

Our study identified factors influencing the use of public (versus private) sources of agricultural training by agricultural producers and professionals in the middle-north region of Mato Grosso state, Brazil. The entities that most contributed advisory support and capability toward developing programs were private agricultural companies and business support organizations (e.g., syndicates, associations). Public entities were less utilized, such as universities like Universidade Federal de Mato Grosso (UFMT), as well as state (e.g., EMPAER-MT) and federal (e.g., Embrapa) institutions. Brazilian agricultural commodity production lends itself to more dependence on the support industries for Brazilian agribusiness [65].

Our results suggest that UFMT and EMPAER-MT should continue to prioritize professional development activities for producers in rural areas (Table 3), such as on-farm field days [66]. Producers are more likely to be from smaller towns (30.8%) outside of the four major cities in the region compared to professionals (12.4%) and this difference is significant ($p = 0.005$, Table 3). In addition, AgriSciences also needs to hold on-farm events closer to Sinop where area producers are more likely not to use public training sources ($p = 0.006$, Table 6). Similarly, public training outreach to professionals from Sorriso needs to improve, since this group has lower participation with such extension ($p = 0.0144$, Table 6).

Well-structured technical assistance and rural extension can enhance field activities and improve rural development [67]. Publicly and privately funded agencies working in technology diffusion tend to become mediating or border organizations, assuming hybrid configurations such as multidisciplinary participatory platforms rich in partnerships, building innovations [68]. When training rural extension professionals, it is important to have continuity in conceptual discussion, systematization of innovative educational practices, and research on education, training, and results [69].

Hybrid learning platforms are also important, such as online videos and audio lessons, although such media are not widely deployed due to lack of internet connectivity in rural areas in Brazil [70]. Improved development of social media to enhance agricultural innovation and rural development in Canada has been limited by mobile telephone quality, having compatible technological equipment, and specialized technical assistance [71]. Keeping up with increasingly fast technological advances can ensure that rural producers and professionals are able to access updated information.

Both agricultural producers and professionals in the middle-north macro-region of Mato Grosso are more receptive to practical, in-service training while still being receptive to online training such as webinars, videos, etc. Professionals indicated the off-season period as the best time to improve their education and competencies. Therefore, the characterization provided in this study can be an excellent support to institutions in Mato Grosso state, Brazil and beyond that provide technical training, development, and assistance. Our results presented here can provide insight on how similar institutions can improve their training programs to both agricultural producers and professionals.

5. Conclusions

Identifying how to best meet the agricultural training needs of both agricultural producers and agricultural professionals from public extension sources is critical to improving the economic and environmental sustainability of commodity cropping systems. We surveyed 94 such producers and 89 agricultural professionals in the middle-north region of Mato Grosso state in Brazil's midwest, which specializes in the production of soybean, maize, and cotton. Most producers were male, married, and varied in age, with 59.1% using a soybean–maize annual rotation. Relative to producers, agricultural professionals were more educated, with 52.8% having a college degree compared to only 22.8% for producers. Agricultural producers were more likely to use public versus private sources of training if they received less training from chemical companies and retailers, were from Sinop, female, unmarried, and if they managed less farm area. Agricultural professionals were more likely to use public agricultural training sources if they received less training from private chemical companies, were not from Sorriso, read more technical articles, and if they consulted producers on managing more crops on less farm area.

Our research is a start for better understanding the agricultural training needs of farmers and farm consultants in our region. However, our analysis could be repeated in the future to allow for pre–post testing of improved extension outreach over time, since we have contact information for all survey participants. Other regions in Mato Grosso (MT) state as well as Brazil can conduct similar surveys of agricultural producers and professionals in the future to compare to our results. Future public research and extension activities (e.g., field days on Saturday mornings and presentations on weekday evenings) in the middle-north region of MT need to continue to support producers and professionals involved in more diversified cropping systems. However, more targeted outreach to more specialized producers is needed, ideally on-farm in more rural areas. Future work can improve our understanding of how to better balance private sector training with agricultural extension outreach from the public sector (e.g., universities, state, and federal agencies) to encourage greater sustainability in commodity cropping systems in Brazil and elsewhere.

Supplementary Materials: The following supporting information can be downloaded at: https://www.mdpi.com/article/10.3390/su15064712/s1, Farmer's survey (pp. 1–4); Professional's survey (pp. 5–7).

Author Contributions: Conceptualization, D.C.d.A. and J.A.V.d.R.; methodology, D.C.d.A. and J.R.V.; software, D.C.d.A. and J.R.V.; validation, D.C.d.A., J.A.V.d.R., J.L.V. and V.A.M.d.B.; formal analysis, A.K.H., J.A.V.d.R. and L.M.; investigation, D.C.d.A. and W.M.d.S.; resources, A.K.H., D.C.d.A. and W.M.d.S.; data curation, J.A.V.d.R. and J.R.V.; writing—original draft preparation, A.K.H. and J.A.V.d.R.; writing—review and editing, D.C.d.A. and A.K.H.; visualization, R.A.d.O.; supervision, A.S.d.O. and D.C.d.A.; project administration, D.C.d.A.; funding acquisition, D.C.d.A. and W.M.d.S. All authors have read and agreed to the published version of the manuscript.

Funding: This research was funded by Serviço Nacional de Aprendizagem Rural de Mato Grosso (SENAR-MT), Programa Global REDD Early Movers de Mato Grosso (REM-MT), Projeto Rural Sustentável—Cerrado (PRS—Cerrado), Coordenação de Aperfeiçoamento de Pessoal de Nível Superior (CAPES) for granting a masters scholarship, Grupo Bragança Agronegócios (Mato Grosso) and Grupo Osvaldo Sobrinho.

Institutional Review Board Statement: Not applicable.

Informed Consent Statement: Not applicable.

Data Availability Statement: Study data can be obtained by request to the corresponding author or the second author, via e-mail. They are not available on the website, as the research project is still under development.

Acknowledgments: We thank all cooperating producers and professionals in addition to the Universidade Federal de Mato Grosso (UFMT) in Sinop, MT, Brazil, Empresa Mato-grossense de Pesquisa, Assistência e Extensão Rural de Mato Grosso (EMPAER-MT), and the University of Minnesota for institutional support. Special thanks the Serviço Nacional de Aprendizagem Rural de Mato Grosso (SENAR-MT), Programa Global REDD Early Movers de Mato Grosso (REM-MT), Projeto Rural

Sustentável-Cerrado (PRS-Cerrado), Coordenação de Aperfeiçoamento de Pessoal de Nível Superior (CAPES), Fazenda Taguá Agropecuária (Grupo Osvaldo Sobrinho) and Fazenda Bragança (Grupo Bragança Agronegócios) for financial support. We also thank four anonymous reviewers whose comments and edits improved this work.

Conflicts of Interest: The authors declare no conflict of interest. Supporting entities had no role in the design of the study; in the collection, analyses, or interpretation of data; in the writing of the manuscript; or in the decision to publish the results.

References

1. Instituto Mato-Grossense de Economia Agropecuária (IMEA). Metodologia—Justificativa da Divisão do Mapa de Regiões. 2017. Available online: https://www.imea.com.br/imea-site/view/uploads/metodologia/justificativamapa.pdf (accessed on 22 September 2020).
2. Instituto Brasileiro de Geografia e Estatística (IBGE). Brazilian Institute of Geography and Statistics. 2021. Available online: https://www.ibge.gov.br/ (accessed on 19 September 2020).
3. Muchagata, M.; Brown, K. Cows, colonists, and trees: Rethinking cattle and environmental degradation in Brazilian Amazonia. *Agric. Syst.* **2003**, *76*, 797–816. [CrossRef]
4. Wesz Junior, V.J. O Mercado da Soja no Sudeste de Mato grosso (brasil): Uma Análise das Relações entre Produtores Rurais e Empresas a partir da Sociologia Econômica. *Dados Rev. Ciên. Soc.* **2019**, *62*, 1–36. [CrossRef]
5. Instituto Brasileiro de Geografia e Estatística (IBGE). Cidades e Estados. Mato Grosso. 2021. Available online: https://www.ibge.gov.br/cidades-e-estados/mt.html/ (accessed on 23 February 2023).
6. Instituto Brasileiro de Geografia e Estatística (IBGE). Censo Agropecuário de 2017. 2017. Available online: https://cidades.ibge.gov.br/brasil/mt/pesquisa/10100/95260 (accessed on 23 February 2023).
7. Companhia Nacional de Abastecimento (CONAB). Acompanhamento da Safra Brasileira. 2022. Available online: https://www.conab.gov.br/info-agro/safras?view=default (accessed on 23 February 2023).
8. Fumagali Junior, G.N.; da Costa Silva, M.M. Efeitos dos cursos de Ciências Agrárias na Produtividade da soja no Mato Grosso: Uma Análise Especial. 2020. Available online: https://brsa.org.br/enaber-2020/#artigos (accessed on 15 October 2021).
9. AgriSciences. 2022. Available online: https://www.agrisciences.org/ (accessed on 20 June 2022).
10. Ragasa, C.; Ulimwengu, J.; Randriamamonjy, J.; Badibanga, T. Factors Affecting Performance of Agricultural Extension: Evidence from Democratic Republic of Congo. *J. Agric. Educ. Ext.* **2016**, *22*, 113–143. [CrossRef]
11. Gido, E.O.; Sibiko, K.W.; Ayuya, O.I.; Mwangi, J.K. Demand for Agricultural Extension Services Among Small-Scale Maize Farmers: Micro-Level Evidence from Kenya. *J. Agric. Educ. Ext.* **2015**, *21*, 177–192. [CrossRef]
12. Mittal, S.; Mehar, M. Socio-economic Factors Affecting Adoption of Modern Information and Communication Technology by Farmers in India: Analysis Using Multivariate Probit Model. *J. Agric. Educ. Ext.* **2016**, *22*, 199–212. [CrossRef]
13. Suvedi, M.; Ghimire, R.; Kaplowitz, M. Farmers' participation in extension programs and technology adoption in rural Nepal: A logistic regression analysis. *J. Agric. Educ. Ext.* **2017**, *23*, 351–371. [CrossRef]
14. Charatsari, C.; Papadaki-Klavdianou, A.; Michailidis, A. Farmers as Consumers of Agricultural Education Services: Willingness to Pay and Spend Time. *J. Agric. Educ. Ext.* **2011**, *17*, 253–266. [CrossRef]
15. Adamsone-Fiskovica, A.; Grivins, M.; Burton, R.J.F.; Elzen, B.; Flanigan, S.; Frick, R.; Hardy, C. Disentangling critical success factors and principles of on-farm agricultural demonstration events. *J. Agric. Educ. Ext.* **2021**, *27*, 639–656. [CrossRef]
16. Estevão, P.; de Sousa, D.N. A Web como ferramenta de capacitação para a extensão rural. *Cad. Tecnol. Ciênc.* **2021**, *38*, 26656. [CrossRef]
17. Mazzali, L. O Processo Recente de Reorganização Agroindustrial: Do Complexo à Organização em Rede. PhD Thesis in Business Economics, Fundação Getúlio Vargas, São Paulo, Brazil, 1995. Available online: http://hdl.handle.net/10438/4638 (accessed on 16 December 2021).
18. Diesel, V.; Dias, M.M. The Brazilian experience with agroecological extension: A critical analysis of reform in a pluralistic extension system. *J. Agric. Educ. Ext.* **2016**, *22*, 415–433. [CrossRef]
19. Botha, N.; Coutts, J.; Roth, H. The role of agricultural consultants in the New Zealand Research, Development and Extension system. In Proceedings of the New Zealand Agricultural and Resource Economics Society Conference, Nelson, New Zealand, 25–27 August 2006; 11p. Available online: https://ageconsearch.umn.edu/record/31971/?ln=en (accessed on 6 March 2023).
20. De Rosa, M.; Bartoli, L. Do farm advisory services improve adoption of rural development policies? An empirical analysis in GI areas. *J. Agric. Educ. Ext.* **2017**, *23*, 461–474. [CrossRef]
21. Faure, G.; Huamanyauri, M.K.; Salazar, I.; Gómez, C.; de Nys, E.; Dulcire, M. Privatisation of agricultural advisory services and consequences for the dairy farmers in the Mantaro Valley, Peru. *J. Agric. Educ. Ext.* **2017**, *23*, 197–211. [CrossRef]
22. Vanclay, F.; Lawrence, G. Farmer rationality and the adoption of environmentally sound practices; A critique of the assumptions of traditional agricultural Extension. *Eur. J. Agric. Educ. Ext.* **1994**, *1*, 59–90. [CrossRef]
23. Hunt, W.; Coutts, J. Extension in Tough Times—Addressing Failures in Public and Private Extension, Lessons from the Tasmanian Wool Industry, Australia. *J. Agric. Educ. Ext.* **2009**, *15*, 39–55. [CrossRef]

24. De Freitas, A.F.; de Freitas, A.F.; Dias, M.M. O uso do diagnóstico rápido participativo (DRP) como metodologia de projetos de extensão universitária. *Rev. Ext.* **2012**, *11*, 69–81. [CrossRef]
25. Benge, M.; Warner, L. *Conducting the Needs Assessment #2: Using Needs Assessments in Extension Programming*; Publication #AEC684, 5 December 2019; The Institute of Food and Agricultural Sciences (IFAS), University of Florida: Gainesville, FL, USA, 2019; Available online: https://edis.ifas.ufl.edu/publication/WC347 (accessed on 22 December 2022).
26. Diori, H.I. A Critical Insight into Needs Assessment Technique and the Way Social Needs are Actually Assessed. *Adv. J. Soc. Sci.* **2021**, *8*, 3–9. [CrossRef]
27. Fávero, L.P.; Belfiore, P. *Manual de Análise de Dados: Estatística e Modelagem Multivariada com Excel®*; SPSS®e Stata®; Elsevier: Rio de Janeiro, Brasil, 2017; pp. 1–1216.
28. Balsadi, O.V.; da Mota, D.M. Diversidade de vínculos de trabalho de mulheres no censo agropecuário brasileiro de 2017. *InterEspaço Rev. Geogr. Interdiscip.* **2017**, *7*, 202113. [CrossRef]
29. Da Conceição, J.C.P.R. Capital Humano e Obtenção de Informações Técnicas na Agricultura: Perfil e Diferenças Regionais a Partir dos Dados do Censo Agropecuário de 2017. Instituto de Pesquisa Econômica Aplicada. 2020. Available online: http://repositorio.ipea.gov.br/handle/11058/10474 (accessed on 20 March 2022).
30. Global Forum for Rural Advisory Services (GFRAS). Activities—Global Good Practice Initiative. 2021. Available online: https://www.g-fras.org/en/ggp-home.html (accessed on 16 January 2022).
31. Statista, Inc. Leading States for Agricultural Production in Brazil in 2021, Based on Share of Production Value. 3 World Trade Center, 175 Greenwich Street, 36th floor, New York, NY 1007, USA. Available online: https://www.statista.com/statistics/1072317/agricultural-production-value-brazil-state/ (accessed on 23 December 2022).
32. Picoli, M.C.A.; Maciel, A.; Simões, R.; Santos, L.A.; Sanches, I. Agricultural production gains in Brazilian commodity hotspot: Case study state of Mato Grosso. In Proceedings of the XIX Brazilian Symposium on Remote Sensing, Santos, Brazil, 14–17 April 2019; Volume 19, ISBN 978-85-17-00097-3.
33. Arvor, D.; Daugeard, M.; Tritsch, I.; De Mello-Thery, N.A.; Thery, H.; Dubreuil, V. Combining socioeconomic development with environmental governance in the Brazilian Amazon: The Mato Grosso agricultural frontier at a tipping point. *Environ. Dev. Sustain.* **2018**, *20*, 1–22. [CrossRef]
34. Junqueira, V.H.; Bezerra, M.C.d.S. The new requirements of reproduction of workforce qualification for agribusiness. *Trabalho Educação* **2015**, *24*, 221–238. Available online: https://periodicos.ufmg.br/index.php/trabedu/article/view/9462 (accessed on 2 February 2022).
35. Ventura, M.V.A.; Batista, H.R.F.; Bessa, M.M.; Pereira, L.S.; Costa, E.M.; de Oliveira, M.H.R. Comparison of conventional and transgenic soybean production costs in different regions in Brazil. *Res. Soc. Dev.* **2020**, *9*, e154973977. [CrossRef]
36. Barros, M.A.L.; da Silva, C.R.C.; de Lima, L.M.; Farias, F.J.C.; Ramos, G.A.; dos Santos, R.C. A Review on Evolution of Cotton in Brazil: GM, White, and Colored Cultivars. *J. Nat. Fibers* **2020**, *19*, 209–221. [CrossRef]
37. Richetti, A.; Ito, M.A. Viabilidade Econômica da Cultura do Feijão-Comum, Safra da seca de 2016, em Mato Grosso do Sul. Embrapa. 2015. Available online: http://www.infoteca.cnptia.embrapa.br/infoteca/handle/doc/1001428 (accessed on 12 March 2022).
38. Serviço Nacional de Aprendizagem Rural de Mato Grosso (Senar). Portal da Transparência—Transparência Senar. Available online: http://app3.cna.org.br/transparencia/?gestaoOrcamento-MT-2021-863 (accessed on 28 June 2022).
39. Universidade Federal de Mato Grosso (UFMT). Portal da Transparência. Controladoria-geral da União. Available online: https://www.portaltransparencia.gov.br/orgaos/26276?ano=2021 (accessed on 28 June 2022).
40. Secretaria de Estado de Fazenda (Sefaz). Governo do Estado de Mato Grosso. Orçamento Cidadão 2021. Available online: http://www5.sefaz.mt.gov.br/documents/6071037/11388742/Or%C3%A7amento+Cidad%C3%A3o+2021.pdf/7172b6c5-90e3-97f9-cbab-d88831597240 (accessed on 28 June 2022).
41. Empresa Brasileira de Pesquisa Agropecuária (Embrapa). Portal da Transparência. Controladoria-Geral da União. Available online: https://transparenciapublica.gov.br/orgaos/22202?ano=2021 (accessed on 28 June 2022).
42. Ministério da Agricultura, Pecuária e Abastecimento (MAPA). Portal da Transparência. Controladoria-Geral da União. 2022. Available online: https://www.portaltransparencia.gov.br/orgaos-superiores/22000?ano=2021 (accessed on 28 June 2022).
43. Bayer, S.A. e Controladas. Relatório da Administração. 2021. Available online: https://www.bayer.com.br/pt/balancos-financeiros-do-grupo-bayer-brasil (accessed on 28 June 2022).
44. Corteva. Corteva Annual Report 2021. Available online: https://investors.corteva.com/static-files/fb19f308-4766-4ca6-a3d6-bd0a8035f09a (accessed on 28 October 2022).
45. Mosaic. Performance Highlights. Annual Report 2021. Available online: https://s1.q4cdn.com/823038994/files/doc_financials/2021/ar/2021AnnualReport_FINAL.pdf (accessed on 28 October 2022).
46. Yara. Integrated Report 2021: Growing a Nature Positive Food Future. Available online: https://www.yara.com/siteassets/investors/057-reports-and-presentations/annual-reports/2021/yara-integrated-report-2021.pdf/ (accessed on 28 October 2022).
47. Carbonera, R.; Basso, N.; Buratti, J.B.L.; Kovalski, C.H.; Scheer, M.R.; Oliveski, F.E. Níveis de reprodução social e estratégias para a agricultura familiar. *Redes* **2020**, *25*, 2035–2059. [CrossRef]
48. De Magalhães, M.M.; Braga Júnior, S.S. Evolução recente e potencial da agricultura de baixo carbono. *Periód. Eletrônico Fórum Ambient. Alta Paul.* **2013**, *9*, 100–118. [CrossRef]

49. De Abreu, D.C.; Morales, M.M.; dos Anjos, A.F.T.; Felipe, R.T.A.; de Lima Dias, M.P.; de Paula Lana, R. (Eds.) *VIII SIMBRAS, Proceedings of the 8th Simpósio Brasileiro de Agropecuária Sustentável and 5th International Conference on Sustainable Agriculture, Mato Grosso, Brazil, 6–8 October 2016*; Universidade Federal de Mato Grosso: Sinop, Brazil, 2016; pp. 1–214. ISBN 978-85-921803-0-0.
50. Sobrinho, O.R.; de Abreu, D.C.; de Lima Dias, M.P.; da Silva, W.M.; de Souza Santos, D.M.; Molossi, L.; Somavilla, A.; Baldan, A. (Eds.) *2a Vitrine Technológica Agrícola: Atualidades na Pecuária de Corte para Baixada Cuiabana*, 1st ed.; Fundação UNISELVA: Cuiabá, Brazil, 2021; pp. 1–272. ISBN 978-65-86743-42-5.
51. De Abreu, D.C.; de Lima Dias, M.P.; Boscoli, D.Z.; da Silva, W.M.; de Paula Alberto, F.; Martins, A.R.R.; Pinheiro, D.T. (Eds.) *3a Vitrine Technológica Agrícola: Atualidades na Cultura do Milho em Sistema Soja e Milho-Safrinha*, 1st ed.; Fundação UNISELVA: Cuiabá, Brazil, 2022; pp. 1–230. ISBN 978-65-86743-50-0.
52. Cattelan, A.J.; Dall'agnol, A. The rapid soybean growth in Brazil. Embrapa Soja-Artigo em periódico indexado (ALICE). *Oilseeds Fats Crops Lipids* **2018**, *25*, D102. [CrossRef]
53. Pereira, C.H.; Patino, H.O.; Hoshide, A.K.; de Abreu, D.C.; Rotz, C.A.; Nabinger, C. Grazing supplementation and crop diversification benefits for southern Brazil beef: A case study. *Agric. Syst.* **2018**, *162*, 1–9. [CrossRef]
54. Pedrosa, L.M.; Hoshide, A.K.; Abreu, D.C.; Molossi, L.; Couto, E.G. Financial transition and costs of sustainable agricultural intensification practices on a beef cattle and crop farm in Brazil's Amazon. *Renew. Agric. Food Syst.* **2019**, *36*, 26–37. [CrossRef]
55. Molossi, L.; Hoshide, A.K.; Pedrosa, L.M.; de Oliveira, A.S.; Abreu, D.C. Improve pasture or feed grain?: Greenhouse gas emissions, profitability, and resource use for Nelore beef cattle in Brazil's Cerrado and Amazon biomes. *Animals* **2020**, *10*, 1386. [CrossRef]
56. Hoshide, A.K.; Dalton, T.J.; Smith, S.N. Profitability of coupled potato and dairy farms in Maine. *Renew. Agric. Food Syst.* **2006**, *21*, 261–272. [CrossRef]
57. Asai, M.; Moraine, M.; Ryschawy, J.; de Wit, J.; Hoshide, A.K.; Martin, G. Critical factors to crop-livestock integration beyond the farm level: A cross-analysis of worldwide case studies. *Land Use Policy* **2018**, *72*, 184–194. [CrossRef]
58. Inwood, S.M.; Sharp, J.S. Farm persistence and adaptation at the rural–urban interface: Succession and farm adjustment. *J. Rural Stud.* **2012**, *28*, 107–117. [CrossRef]
59. Carrer, M.J.; Maia, A.G.; Vinholis, M.d.M.B.; de Souza Filho, H.M. Assessing the effectiveness of rural credit policy on the adoption of integrated crop-livestock systems in Brazil. *Land Use Policy* **2020**, *92*, 104468. [CrossRef]
60. Gil, J.; Siebold, M.; Berger, T. Adoption and development of integrated crop-livestock-forestry systems in Mato Grosso, Brazil. *Agric. Ecosyst. Environ.* **2015**, *199*, 394–406. [CrossRef]
61. Garrett, R.D.; Niles, M.; Gil, J.; Dy, P.; Reis, J.; Valentim, J. Policies for reintegrating crop and livestock systems: A comparative analysis. *Sustainability* **2017**, *9*, 473. [CrossRef]
62. Da Silva, R.F.; Batistella, M.; Moran, E.; de Melo Celidonio, O.L.; Millington, J.D.A. The soybean trap: Challenges and risks for Brazilian producers. *Front. Sustain. Food Syst.* **2020**, *4*, 12. [CrossRef]
63. Possamai, E.J.; Conceição, P.C.; Amadori, C.; Bartz, M.L.C.; Ralisch, R.; Vicensi, M.; Marx, E.F. Adoption of the no-tillage system in Paraná State: A (re)view. *Rev. Bras. Ciênc. Solo* **2022**, *46*, e0210104. [CrossRef]
64. Cerdeira, A.L.; Gazziero, D.L.P.; Duke, S.O.; Matallo, M.B.; Spadotto, C.A. Review of potential environmental impacts of transgenic glyphosate-resistant soybean in Brazil. *J. Environ. Sci. Health B* **2007**, *42*, 539–549. [CrossRef]
65. Ministério da Agricultura Pecuária e Abastecimento (MAPA). Acesso à Informação. 2020. Available online: https://www.gov.br/agricultura/pt-br (accessed on 12 October 2022).
66. Pinheiro, D.T.; Santos, D.M.S.; Martins, A.R.R.; da Silva, W.M.; de Araújo, C.V.; Hoshide, A.K.; Abreu, D.C. Produtividade e qualidade nutricional dos principais híbridos de milho no cerrado mato-grossense. In *Chapter 3 in 3.a Vitrine Tecnológica Agrícola*; AgriSciences and Fundação UNISELVA: Cuiabá, Brazil, 2022; pp. 43–64.
67. Gonçalves, A.C.; Junior, L.R.; Fonseca, M.I.; Nadruz, B.V.; Bürger, G.R. Technical assistance and rural extension: A case study that demonstrates its importance for the improvement of milk production. *Rev. Bras. Hig. Sanid. Anim.* **2014**, *8*, 47–61. [CrossRef]
68. Cristóvão, A.; Koutsouris, A.; Kügler, M. Extension systems and change facilitation for agricultural and rural development. In *Farming Systems Research into the 21st Century: The New Dynamic*; Darnhofer, I., Gibbon, G., Dedieu, B., Eds.; Springer: Dordrecht, The Netherlands, 2012; pp. 201–227. [CrossRef]
69. Landini, F.; Brites, W.; Rebolé, M.I.M.Y. Towards a new paradigm for rural extensionists' in-service training. *J. Rural Stud.* **2017**, *51*, 158–167. [CrossRef]
70. Chowdhury, A.; Odame, H.H. Social media for enhancing innovation in agri-food and rural development: Current dynamics in Ontario, Canada. *J. Rural Community Dev.* **2013**, *8*, 97–119.
71. Kolling, C.E.; Rampim, L. Agricultura de precisão e digital: Perspectivas e desafios dos produtores rurais do estado do paraná. *Rev. Uningá* **2021**, *36*, eURJ3981. Available online: https://revista.uninga.br/uningareviews/article/view/3981 (accessed on 6 March 2023). [CrossRef]

MDPI
St. Alban-Anlage 66
4052 Basel
Switzerland
Tel. +41 61 683 77 34
Fax +41 61 302 89 18
www.mdpi.com

Sustainability Editorial Office
E-mail: sustainability@mdpi.com
www.mdpi.com/journal/sustainability

www.ingramcontent.com/pod-product-compliance
Lightning Source LLC
Chambersburg PA
CBHW061459220926
51850CB00004B/124
9783036581781